People as Living Things

Chuck —
appreciate your support

Dag

Living Control Systems Publishing
2740 Gamble Court
Hayward, CA 94542–2402 USA
www.livingcontrolsystems.com

People as Living Things

The Psychology of Perceptual Control

Philip J. Runkel

Living Control Systems Publishing
2740 Gamble Court, Hayward, CA 94542–2402 USA

Copyright © 2003 by Philip J. Runkel

All rights reserved. No portion of this book may be reproduced, by any process or technique, without the express written consent of the publisher.

First published in 2003.
Printed in the United States of America.

Library of Congress Control Number: 2003106389

Runkel, Philip Julian, 1917–
    People as Living Things; The Psychology of Perceptual Control /by Philip J. Runkel

ISBN 0-9740155-0-4

♾ The paper used in this book meets the requirements of ANSI/NSIO Z39.48-1992—Permanence of Paper. Paper has 30 percent post-consumer recycled content.

Acknowledgements
of permission to quote

Some credits to publishers appear in the text next to the figures or tables they permitted me to reproduce. Some credits are more conveniently placed in the list below. I am grateful to the publishers mentioned below (and to the other publishers noted at appropriate places in the text) for their permission to quote from the publications listed. When not given here, the full bibliographic information for a book can be found in the "References" at the back of the book.

Andrew Thomson Publishing (Australia) for excerpts from *RTP Intervention Processes* by Carey and Carey (2001).

Science News for excerpts from "Incriminating Developments" by Bruce Bower. Reprinted with permission from SCIENCE NEWS, the weekly newsmagazine of science, copyright © 1998 by Science Service Inc.

John Isaacs of the Council for a Livable World for excerpts from an e-mail message.

Phi Delta Kappa International for excerpts from the article by Wayne Jennings and Joe Nathan in the *Phi Delta Kappan* (1977).

Random House for A HISTORY OF WARFARE by John Keegan, copyright © 1993 by John Keegan. Used by permission of Alfred A. Knopf, a division of Random House, Inc.

W. W. Norton & Company for *Accelerating the Shift to Sustainability* by Gary Gardner, from STATE OF THE WORDL 2001: A WORLD WATCH INSTITUTE REPORT PROGRESS TOWARDS A SUSTAINABLE SOCIETY, edited by Lester R. Brown, et al. Copyright © 2001 by Worldwatch Institute. And *The battle of behaviorism: An exposition and an exposure* by Watson, John B. and William McDougall (1929).

Quotations attributed to Kirk and Kutchins in Chapter 31 are reprinted with permission from Stuart A, Kirk and Herb Kutchins. THE SELLING OF DSM: THE RHETORIC OF SCIENCE IN PSYCHIATRY. (New York: Aldine de Gruyter.) Copyright © 1992 Walter de Gruyter, Inc., New York.

Excerpts reprinted by permission of Waveland Press, Inc. from Clark McPhail, Collective action and perceptual control theory. Pages 461–465 in David L. Miller (Ed.), *Introduction to Collective Behavior and Collective Action* (2nd ed.). Prospect Heights, IL: Waveland Press, 2000. All rights reserved.

Clark McPhail, William T. Powers, and Charles W. Tucker. Simulating individual and collective action in temporary gatherings. *Social Science Computer Review*, 10(1), pp. 1–28, copyright © 1992 by Duke University Press. Figures 4, 5, 7, and 10 reproduced by permission of Sage Publications, Inc.

Quotations attributed to Mirowsky and Ross in Chapter 29 are reprinted with permission from John Mirowsky and Catherine E. Ross. SOCIAL CAUSES OF PSYCHOLOGICAL DISTRESS (New York: Aldine de Gruyter.) Copyright © 1989 by Aldine de Gruyter.

Quotations attributed to Shipper and Manz in Chapter 34 are reprinted from *Organizational Dynamics,* 20(3), An alternative road to empowerment, pages 48–61, copyright © 1992, with permission from Elsevier Science.

William T. Powers for excerpts from his various writings.

Excerpts from Marvin R. Weisbord, *Productive Workplaces.* Copyright © 1987 by Jossey-Bass, San Francisco. This material is used by permission of John Wiley & Sons, Inc.

For completeness, I repeat here my thanks to persons who have allowed me to quote or paraphrase their messages to the CSGnet: W. Thomas Bourbon, Timothy A. Carey, Bruce Gregory, Richard Kennaway, Mark Lazare, Richard S. Marken, Hugh Petrie, Frans X. Plooij, Mary Powers, William T. Powers, and Martin M. Taylor.

This book is dedicated to every person who,
aware that it was happening,
has struggled through a large reorganization
to find out what was on the other side.

Contents

Acknowledgements v
Preface xii

PART I
CONTROL OF PERCEPTION 1

1. The springs of action 3
2. Living things 13
3. Inside and outside 19
4. The loop 31
5. Beware how I write 51

PART II
RESEARCH 59

6. Do it yourself 61
7. Some foundations 67
8. Some models of control 81
9. Some social interaction 103

PART III
SCIENCE 121

10. Don't fool yourself 123
11. Falsification and confirmation 131
12. Models and their worlds (Bourbon & Powers) 137
13. Quantitative measurement of volition (Powers) 155
14. Is behavior probabilistic? 167
15. Where's the reality? 171
16. Explaining other theories 177
17. Beware how anybody writes 183

PART IV
HIERARCHIES OF PURPOSE 191

18. The neural hierarchy 193
19. Memory and imagination 215
20. Reorganization 227
21. Emotion 243

PART V
THE HIGHER ORDERS 251

22. More about the control loop 253
23. Internal conflict 257
24. Up at the top 263
25. Logic and rationality 275
26. Personality 287

PART VI
DYADS AND GROUPS 303

27. Degrees of freedom I 305
28. The social environment 313
29. Helping 333
30. The method of levels 345
31. Psychotherapy 353
32. Language and communication 371
33. Influence 385

PART VII
THE SOCIAL ORDER 391

34. Degrees of freedom II 393
35. Coordination 411
36. Planning 427
37. Schooling 437
38. Mental testing 457
39. Ford's Responsible Thinking Process 467
40. Society 477

References 489
Name index 509
Subject index 517

x

Again the fixed end, the varying means!

—William James, 1890, page 7

*When we perform a voluntary action
what we select voluntarily is a specific purpose,
not a specific movement.*

—Rosenblueth, Wiener, and
Bigelow, 1943, page 19

Preface
or
What you can expect of this book

This book offers a theory of human functioning. The theory does not claim to predict the acts humans will produce, or be induced to produce, or be prevented from producing—though that topic will come up. Rather, the theory will explain how humans function regardless of the acts they choose—how acts serve the functioning. The book will also tell how we can stop demanding impossible behavior from humans, ourselves and others, and thereby free ourselves of the costs of many sorts of conflict.

Unlike authors of many popular books claiming to offer psychological knowledge, I will not tell you how to win friends and influence people. In fact, I will advise you to avoid trying to do that. And I will tell you that you can sometimes be influential without trying. I will certainly not tell you how to outwit others or bend them (or break them) to your will.

Unlike authors of most texts in psychology, I will not drag you through the traditional topics that label courses and resound in the lecture halls of universities—though I will say some things about some of those topics as I go along. And I will not try to display for you in any systematic way the multifarious shapes and guises of human behavior, either in the popular manner or the academic; I leave that task to historians, cultural anthropologists, novelists, and other chroniclers.

I will not indoctrinate you with the conceptions, theories, and passwords that will get you into graduate school. I spent three years in graduate school and then about 30 years unlearning most of what I had learned there. The three years were difficult, and the 30 years were even more difficult. I would not want you to go through all that.

Though this book is not academic in the usual sense of repeating what most academic psychologists have believed during the past several decades, I do claim it to be scientific in the sense that a good many of the claims I make about human functioning can be put to experimental test—can be tried out in tangible, physically demonstrable ways that can be reproduced or extended by anyone who takes the trouble. The theory I offer here is Perceptual Control Theory, or PCT for short. Its core postulates have indeed been tested, the results of the tests have been published in the scientific literature, and the core assumptions are being extended in the designs of further experimental tests. Furthermore, the experimental tests have been far more demanding than the experimentation in the mainstream psychology books, as you will see. I am not saying that *everything* I say here has been tested empirically, but I do make that claim about the fundamental postulates and about a good many derivations from them.

I will disagree in serious ways with most of the widely accepted psychological theories you encounter in popular literature, in textbooks (of whatever discipline), and in the halls of academe. I will agree with the other theories at some points, but the underlying assumptions of the theory here (Perceptual Control Theory) are not those you will find either printed or implied on many of the pages printed about psychology. In that sense, this book is disputatious. I do not, by the way, claim that those other authors and lecturers are immoral or mentally deficient. I claim only that they are wrong.

This book is about what life is like for humans—how we function, what we can and cannot do with our brains and bodies, when we are happy and unhappy, and the like. It is not only about what human life is now and has been like, but also about what it *can* be like—about what I *want* it to be like.

As I have said, the book is not bound by the customary topics and rituals of academic psychology. It has, however, other limitations. Despite my efforts to keep my assertions close to the logic of the theory, some of what I say will inevitably be tainted by my various insularities. My education, for example, has been mostly modern. I am unable to quote extempore from Lao-tzu (6th century B.C.), Confucius (551–479 B.C.), Socrates (470?–399 B.C.), or Mencius (4th century B.C.), not to speak of Democritus (460?–370? B.C.) or Aristotle (384–322 B.C.). My professional training and work have been largely in the field of social psychology, although I did not enter graduate school until the age of 34 and therefore have the benefit of some other occupations in my earlier years. Further, though I have been poor and in one period suffered some debilitating pangs of hunger, most of my life has been, economically, one of white, middle-class comfort. I am thoroughly North American, and speak only English. I lived in Central America for some years, but I have visited Asia (Japan) only for three weeks, and I have never traveled to Europe. I have no children of my own flesh, but I have known the deep satisfactions and glories of mutual love with two wives and the terrors and grief during the long dying of one.

Some conditions of my life were given me at birth. I chose the profession of psychologist, however, after I had some experience of adult life. I learned some things about social psychology from Calmer Batalden, a colleague and friend in the Panama Canal Zone, and in the summer of 1948 I attended summer school at the University of Nebraska, where I was fortunate that the book by Krech and Crutchfield (1948) came into my hands. I went back to the high school in the Canal Zone where I was a department head, opened the book to the how-to-do-it section, and did what the book said to do. The results were remarkable. It was that experience, and similar experiences in later years, that gave me a commitment to a psychological view of work and daily life. After my graduate studies at the University of Michigan, however, I found that I had to shake off the academic attitude toward the study of psychology if I was to make use of the useful parts of what I had learned. I am grateful to my colleague and friend Richard A. Schmuck for showing me some of the ways I could do that. I found, too, as I read further works in psychology, that the books that deviated from the established academic patterns were the books that gave me the most help. Indeed, the 1973 book by William T. Powers enabled me to make sense of all my previous discontent about psychological study. The Perceptual Control Theory (PCT) originated by Powers serves as the backbone for this book. Well, as more than that, actually. You'll see.

THANKS

Quite aside from my greedy use of their ideas about PCT, I am very grateful to the following for their helpful criticisms of various parts or all of the manuscript: W. Thomas Bourbon, Timothy Carey, Chris Cherpas, Kurt Danziger, Hank Folson, Dag Forssell, Bruce Gregory, Richard Kennaway, Len Lansky, Richard S. Marken, Bruce Nevin, Mary Powers, William T. Powers, Richard Robertson, and Mary Claire Runkel.

I am grateful to Dag Forssell for redrawing and improving most of my figures and for a great deal of generous help in further ways to make the book readable.

PART I

CONTROL OF PERCEPTION

Back in 1733, the English poet Alexander Pope wrote, "The proper study of mankind is man." Pope was describing the benefits of understanding ourselves and our fellow humans. A great many of us are still engaged in that proper study. This book is one more of the thousands that have followed Pope's advice.

Those who write about people go about the task in various ways. One way to study people would be to circle the globe in a spaceship and observe the evidences of human activity: lakes growing behind dams, clusterings of the lights of cities at night, black clouds rising from burning oil wells, and so on, much as one might study ants by watching the heaps of soil-particles rising around the holes of their burrows. Information about such large-scale events is welcomed by economists, demographers, and geographers, not to speak of public-health professionals, farmers, and astronomers. Another way would be to sit on the side of a mountain for a few centuries, look out across the plain, and watch farmers growing their crops and selling them in markets, road builders opening routes from one horizon to another, houses clustering and cities growing, herds of bison or gnu dwindling, and armies slaughtering each other—the "pageant" of history. Information about events at that scale is welcomed not only by those I mentioned just above, but also by historians, political scientists, and sociologists, not to mention politicians, people in businesses of all sorts, and the military.

Or one could listen at political gatherings, attend public lectures, eavesdrop in hotel lobbies and bus depots, and read newspapers. One could attend rock concerts, conventions, football games, board meetings, classroom meetings, and conferences. One could listen to strollers in the park, conversations by the drinking fountain or in the parking lot, families at dinner, and so on. Information from those settings has been useful to historians, politicians, anthropologists, linguists, psychologists, and sociologists. One could watch a single person for a month or a year, observing the kinds of dealings the person had with others, emotional attachments made and broken, deceits practiced or given up, physical exercise undertaken, and visits to physicians. Information at that scale is welcomed by anthropologists, psychologists, physicians, and novelists, among others. One could also read records such as those kept by physicians and get information interesting to neurologists, physiologists, and some psychologists.

Information about human life at those various scales—from movements of masses of people to the small doings of individuals and the smaller doings of their internal organs—is useful in many ways. But every kind of information is more useful for some purposes and less useful for others. When we watch the actions of other people, we learn the sorts of actions of which they are capable and the circumstances in which they are more capable and less capable. We learn the frequencies with which, this week, they take various sorts of actions and the circumstances in which the various frequencies appear. We do not learn, however, *how* the people can be capable of those actions. We do not learn anything about the internal functioning that *enables* people to do all those things. How is it, for example, that a person can stand upright? How is it that we *can* manage, as a wind pushes on us, as we move to wave at someone, as we stand on the deck of a wallowing ship, and as our muscles tire, to remain upright instead of toppling over, as you would naturally expect a mere assembly of loosely jointed bones and yielding flesh to do? And when we get distracted from our purposes, as we do repeatedly every day, how is it that we can repeatedly return to those previous purposes instead of staying with the new direction into which the distraction (so some would think) has sent us?

Thousands of books have been written about human behavior that do no more than describe what can be observed of the movements of humans. Nevertheless, if a book inquires into the springs of human acts or the human uses of them, the author must inevitably make some assumptions about the functioning of individuals. And of course every author does. Everyone, in fact, author or not, has some belief (some theory) about "what makes people tick." There are two common theories the world over. One is that people do what they do because of the kinds of persons they are. Psychologists call that the theory of personality. The other is that people do what they do because of the stimulation they get—because they are pushed on by something. Psychologists call that the theory of behaviorism. (I am simplifying here, but not much.) Both theories ignore important and obvious features of behavior.

The most obvious thing, it seems to me, is that living and nonliving things obey different laws of behavior. The flesh and blood of living things is as subject to the laws of physics and chemistry as all other materials, but the behavior of the whole living creature arises from causes lying both without and within. Those causes, without and within, act jointly and simultaneously. Accordingly, I spurn theories that rely only on forces from outside the person; the behavior of the whole creature does *not* obey the laws of physics. If you push on a rock, it will roll over and lie there uncomplainingly. If you push on a person, the person is very likely to push back, remain standing, and utter a complaint something like, "Who d'you think you're shoving?" Living things push back. Why do I mention such an obvious thing? If you have never read a book on psychology, you might naturally suppose that every psychology book would begin with the fact that humans and other living creatures typically act to oppose disturbances from the environment, to maintain conditions favorable to them. Actually, few books do, at the beginning or anywhere else.

When we look for the "stimulus" that will "cause" someone to "react" in the way we desire, we are using a conception substantially the same as that of pushing an object to where we want it to go—a conception that works well with rocks, footballs, and dead bodies, but not with living creatures. That conception leads to the belief, when our relations with others are unsatisfactory, that things can be set right by pushing other people into their presumed proper places—into the behavior we think suitable. At the extreme, that conception leads to the murderous use of force. But that conception gets us into trouble long before it gets murderous, because the people we push on are going to push back, in one way or another, at the first smallest hint that we are disturbing the perceptions they want to maintain. They cannot help doing so. All living creatures are built that way.

You can see now why I titled this book "People as Living Things." Living things have purposes, goals, criteria, standards. They want to perceive certain conditions and not others. They are always ready, 24 hours a day and 365 days a year and another on leap year, to react against disturbances to what they want to perceive. *How they can do that* is what this book is about.

The theory I use here to explain how people can maintain their perceptions of what they want to perceive is called Perceptual Control Theory (PCT).

FOOTNOTES AND REFERENCES

Skip this explanation
if you are an old hand with footnotes
and references to literature

Often in these pages, you will encounter mentions of other authors, like this: "Estervern (1832)." Sometimes you will find the mention written out, something like this: "In 1832, Estervern wrote a book in which. . . ." I will always give some clue to the reason I am mentioning Estervern so that you can judge whether you care to read what Estervern has to say. You can find the full bibliographic specifications for Estervern's book, article, or other writing in the list of "references" at the end of the book.

But I will not always be able to tell you in only a few words the reason I have mentioned Estervern. Once in a while the explanation, if I were to put it in the text, would become too long and be a nuisance to you. In those cases, I will use a footnote indicator in the form of a superscripted numeral[1]. The "footnotes" will not appear at the foot of the page, but instead will appear in a list at the end of the chapter as "endnotes." An endnote will always refer to literature, and it will often contain a comment. I will never write an endnote without a reference to other literature.

If you don't care what other writings came to my mind as I wrote, or if you feel no urge to read something further on the topic, you can just let your eye glide past the superscript or Estervern's name.

---

[1] Like the "1" you just saw after the word "numeral."

Chapter 1

The springs of action

People have always been fascinated by the actions of others. We like to hear about the doings of neighbors, friends, fellow citizens, even strangers in far-off places. We exchange news orally; we read newspapers and magazines; we watch television. We have, most of us, such eagerness for tales about others that we devour not only news about actual people, but also tales about wholly fictional people in books, in theaters, and on television. Many of us listen avidly to storytellers.

Some of our desire for news about others comes from the sheer practical importance of the information. Has George come home yet with the groceries? Does the teacher approve what I wrote in my essay? Are the people at that company offering employment of a kind I want? Have the people at the bank credited the check to my account? Is this person welcoming my attentions? A great deal of the time, however, we seem to seek information about people, real or imaginary, for the sheer fascination of it. Much of the time, we listen to the current gossip, go to the theater, read a novel, or watch television not to be instructed, but merely to enjoy watching people choose actions that get them into or out of interesting situations. Some of the pleasure is esthetic, and some is simply that of "Gee whiz! Imagine that!"—like the pleasure of discovering the unending variety of stones and shells on the seashore. Much of the pleasure of the sciences is of that last sort—a pleasure any collector knows. Here is a fine specimen—and it goes into a case with others of its kind or perhaps into a case by itself.

Often, merely watching the passing parade is not enough. Often, we want to fit our new experience into our earlier experience to make categories and sequences, connections and patterns, so that our memory becomes more than a jumble of items. We want to explain, interpret, understand, or "find meaning in" the behavior of our fellow humans. "Is this the kind of behavior I should expect from Alfred in the future?" "Will Maisie always be angry in situations of this sort?" "What caused Veronica to wait so long?" "Why did Joe choose to do *that*?" Sometimes we think a work of fiction fails to match our own experience: "Would any real person do *that*?" Questions of that sort and also answers to them come into out minds repeatedly, every day. I write in this book about understanding the actions of others and of ourselves.

Some of us spend the greater part of every day dealing with actions of others and ourselves—anticipating actions, estimating the consequences of the actions that occur, adapting our own actions to those actions, and reflecting on what happened. A salesperson does that, and so does everyone who deals much with other persons—clerks at check-out counters, taxi drivers, counselors, clergy, librarians, hairdressers, nurses, politicians, managers, social workers, police, waitpersons, and so on. When we are with our families, we deal with their actions and our own almost constantly.

Some of us spend long hours alone, out of reach of others. Some people spend hour after hour in front of a computer-screen writing programs. Some spend hours in a lonely laboratory working with chemicals or bacteria. Some people go off into the wilderness to be "away from it all." Charles Proteus Steinmetz (1865–1923), the famous electrical engineer and inventor, liked to get in a canoe and paddle into the middle of a lake to think about electrical machinery. I do not know how much Steinmetz, afloat, thought about his relations with other people. But some people go off to a quiet place expressly for the purpose of reflecting on relations among people more deeply than is possible while interacting with others. Henry David Thoreau (1817–1862) went off to Walden Pond and wrote a good many essays about his relations with other people in the world and their relations to one another. (Actually, he was not as isolated as the popular tale has it. One biographer writes, "... hardly a day went by that Thoreau did not visit the village [of Concord] or was not visited at the pond.") People who think even a little about the complexities of social life often feel the urge to get away from the constant demands of others upon their attention—to get to a place where they can sort through their own thoughts about all those demands. Sometimes they think about the subject while riding on the subway, sitting in a waiting room, or lying in bed. Sometimes they go into a study, close the door, and write books.

We do not, any of us, think about the actions of ourselves and others at every opportunity. I might go up to a clerk in a store and ask, "Do you have any sport shirts cut straight across the bottom?" If the clerk says "Yes, right over there," or "No, sorry," I go on to the next step without pausing to wonder about how the clerk came to say "yes" or "no." If, however, the clerk says, "Who cares?" and walks away, I am likely to wonder how that reply could have come about. Most of us, I think, reflect upon human action now and then, fitfully, and unsystematically.

Some of us, sometimes, take pains to think about human action systematically—to search for features of human action that always stay the same, to examine carefully the reliability of the information we get, to examine the logical connection between the information and our beliefs about the constant features of action, and to look for instances of behavior that could *contradict* the beliefs we are forming. When we take all that care with information, logic, evidential connections, and disproof—when we search for statements about human action that will hold up against all conceivable observations of action in the actual world—then we are thinking about human action in the way *scientists* are presumed to do. I will try in this book to maintain the scientific point of view, but I will try also to give as much respect to information from everyday experience as to information from the laboratory. You saw me do that in the introduction to this part of the book when I wrote that it is obvious that living and nonliving things obey different laws of behavior. By "obvious," I meant that nobody has to do a systematic experiment to ascertain that fact.

When I want to refer to the scientific study of human action, the available terminology is awkward. "Psychology," "social science," and "life science" all have their advantages and disadvantages for use here. I will be writing mostly in the domain where those subjects overlap, but I will also stray now and then to one side or another. Instead of trying to be precise, I think it will be best merely to ask you to understand that I will be using those labels and some others, too, all somewhat sloppily.

TWO SPRINGS OF ACTION

In thinking about human action, we must pay attention to two sources of the need to act. One source is the person, or more precisely, the mind—the many patterns of electrochemical activity in the nervous system (or neural net) of the individual. The other source is the environment—the many events out there producing energies that impinge upon our sense organs and the many objects and materials we can use in our daily pursuits. Action or inaction depends on the interaction of those two sources. Neither source alone can produce action in a living creature. I will give several examples here of the way acts result from the linking of the two sources. I'll begin with the intensity of light striking the retina of the eye.

Example: Amount of light

Our nervous system is connected to the light-receptors in the eye in such a way as to regulate the amount of light falling on the retina. When the light is brighter, the nerve bundles from the retina send electrical pulses toward the brain at a more rapid rate than when the light is dimmer. The nervous system maintains a memory of the range of light intensity within which the retina will function properly. When the light is too bright, the iris contracts at the center so that the pupil becomes smaller, admitting less light. If the light is so bright that shrinking the pupil cannot keep out enough light, we can close our eyelids, hold a hand in front of our eyes, turn the head away from the source of the light, put on dark glasses, pull down a shade, walk to the shady side of the building, and so on. We act to maintain or achieve a desired perception—to maintain the desired level of some incoming energy that can be sensed. But if the light intensity is just right to match the internal standard for brightness, we do none of those things. The "act" we choose then is to leave things as they are; an observer would think we were paying no attention to the intensity of the light. When the incoming light is within the "right" range of brightness, we usually pay the fact no conscious attention. We become aware of "too much" or "too little" light when the action of the iris and the eyelid no longer suffice to bring us the intensity we want.

Note the necessary sources of the act. First, there must be an internal standard for a level or range of incoming energy that we want to maintain. In this example, we want the incoming light to be neither too bright nor too dim. That standard is the (1) *internal source* of the act. Second, there must be a disturbance of that level or range so that the perception of the incoming energy no longer matches the standard. In this case, I gave the example of the environment supplying light that is too bright. That disturbance of the desired level or range of perception is the (2) *external source* of the act. Neither of those necessary sources of the act, however, specifies a particular act by which the individual will bring the level of light back to the level wanted.

We choose acts (a) that we conceive to be likely to alter the magnitude of the disturbed perceptual variable so that it matches the internal standard and (b) that make use of objects that we perceive actually to be present in the environment. For example, I might think of reducing the general level of light in the room by covering a window. I might notice that the window is flanked by drapes pulled to each side. But if I believe the drapes to be purely decorative, and do not conceive that they might be brought together to cover the window, I will not act to do that. That is an example of the first case (a) conceiving a way to use a chunk of the environment. For an example of the second case (b), I might think of putting on dark glasses, but I could not do so if I had brought none with me and saw none nearby.

There is a third necessity if a particular act is to occur. Even though we judge some feature of the present environment to be suitable for use in restoring a perception to the level we desire, we will not choose that line of action if doing so will threaten some other variable we are controlling. For example, a friend might be sitting beside a window reading a book. In that case, I might not pull down the shade to reduce the general level of light in the room, because doing so would disturb what I perceive to be a comfortable environment for my friend. I would choose some other act instead.

Here I want you to think back (or even look back) at what I have been expecting you to find interesting about light falling on your retina. I have not been asking you to imagine yourself an experimenter watching someone else's actions when the experimenter shows the person certain kinds of things. Nor have I been writing about kinds of actions or conditions that an experimenter might find going along frequently with some other kinds of action on the part of some people the experimenter was watching. Instead, I have been asking you to imagine light falling upon your eye, to imagine what you might care about when that is happening, and to imagine what you might do to keep the light at an intensity you prefer. You will find that emphasis on perception and on *your* point of view (not an experimenter's) throughout this book. That emphasis is characteristic of Perceptual Control Theory (PCT). That is not to say that Perceptual Control theorists disdain experimentation. Quite the contrary. In their experiments, however, they are not seeking to learn how the world is experienced by experimenters, but how it is experienced in everyday life by anyone.

In this first example, I have set forth three necessary features of the two springs of action, though the matter gets complicated in a situation containing many persons, each controlling many perceptions simultaneously, those persons sharing an environment rich in opportunities to take action to restore levels of disturbed perceptions they want to control, cherishing differing understandings of the possibilities in the environment for restoring their disturbed perceptions, and sometimes acting in ways that interfere with the actions of others. Later parts of this book will be devoted to the ways we deal well and poorly with such a complicated situation. The early chapters, however, will explain how certain characteristics of individuals give them their *capacity* for dealing well or poorly with disturbances of their controlled perceptions.

Requisites for a Particular Act

Before going on to further examples, I will review now what I have said so far about the springs of action, but in a more formal way, and add some comments. For a *particular* act to occur, it is first necessary that the person be motivated to take *some* act—that is, it is necessary that the person experience (not necessarily consciously) a mismatch between (a) an internal standard for what the person wants to perceive and (b) the actual perception. That mismatch or discrepancy motivates action. The first two Requisites for action, therefore, are

1a that the person be controlling a perceptual variable (such as intensity of light).

1b that some environmental event disturb the controlled variable; more exactly, that the environmental event have an effect on the controlled variable such that, if the variable were *not* controlled, the variable would underreach or overreach the internal standard.

In stating those two Requisites, I have become a little more technical in writing *variable* instead of *perception*. When you perceive something, you perceive some aspect of it to some degree. The degree can *vary*. Light can vary in the aspect of intensity and the aspect of blueness. The sound of a trumpet can vary in brashness. I will sometimes write *perception*, sometimes *perceptual variable*, and sometimes just *variable*, as convenient.

If only one or neither of those first two Requisites exists, the person will not act. If both those conditions exist, the person will act. The direction of the action will be such as to bring the controlled variable back to the level or limits specified by the internal standard. The first two Requisites determine only whether *any* act, *some* act, will occur; they do not determine the particular act.

What do I mean by a particular act? I have not been able to think up a neat, concise definition. Let's just leave it to be an undefined term.

All of this can be conscious—or none of it. Most of the time while you are reading these words, you are unconscious of having an internal standard for light intensity and unconscious, too, of the actual intensity you are experiencing. If the intensity diminishes, you may turn the book more toward the light without being conscious that you are doing so—if you are sufficiently interested in what you are reading. On the other hand, when you are settling down to begin reading, you will often be conscious of all those perceptions.

I have made here a sharp distinction between what is internal (the perceptual standard) and what is external (the event that disturbs the controlled perception). Actually, disturbances can arise within the nervous system, too, as when we are examining our own thoughts. But I'll say more about that sort of thing later.

You may have noticed that I have been trying not to refer to causation. I have been trying not to say that an internal standard causes something or that an environmental disturbance causes something.

By itself, an internal standard causes nothing, and by itself, a disturbance causes nothing. Something happens only when a disturbance affects a variable that the person seeks to hold to a standard. Even then, however, a particular act is not caused. All we can say, when a controlled variable is disturbed, is that the person begins acting toward the goal of returning the variable to the standard. The coupling of the first two Requisites does not cause a particular act; rather, it sets off a search for a way of matching the perception to the standard.

Next it is necessary that the person find an object or event or feature of the environment that can be used in carrying out an act that will affect the controlled variable. This necessity has three and sometimes four aspects. It is necessary

2a that some means (object, event, or feature) suitable for affecting the controlled variable be available in the environment,

2b that the person come upon or believe it possible to come upon a suitable means,

2c that the person be capable of carrying out an act with an object or other means that will affect the controlled variable (this includes being capable of conceiving or imagining the act, when that is a necessary step),

2d and sometimes (if the act begins in the conscious state) that the person estimate the likelihood to be sufficiently high that a feasible act will aid in controlling the perceived variable.

Here are two examples of Requisite 2a: If a person you are about to ask does not in fact know anything about the binomial theorem, it doesn't matter whether you think she does; you will not find out about it from her. If you want to be warmer by putting on a coat, you will not do that if you have no coat.

Under Requisite 2b, I mean, for example, that a suitable object may not at the moment be visible, but the person may correctly believe it possible to find a suitable object by opening a drawer or looking in a catalog. If you do not know there is a pair of pink socks in your drawer, you will not choose to wear pink socks this morning. If indeed there actually is no pair of pink socks anyplace about, it doesn't matter whether you think there is; you will not wear pink socks this morning.

Here are examples for Requisite 2c. I might want to move a piano up the stairs and into my living room. I might imagine doing it by myself, but if I tried to do so, I would find the piano far too heavy. I would not attain my goal by using only my own body, no matter how vivid my imagination. Some uses of the environment are unavailable because of lack of knowledge or envisioning. If I wanted to polish brass, I would not try to use hot vinegar and salt if I did not know that those ingredients would do it. If I wanted to go to China from Europe but conceived the world to be flat, not round, I would not choose to go west to China. If I wanted to protect myself from illness, but had no conception of germs, it would not occur to me to boil water before drinking it. Depending on the emphasis you have in mind, you could also classify these last examples under 2b.

When I speak of the requisites for a particular act to occur, I mean a purposive act, not an accidental act. Bumping a lever or putting on someone else's hat can happen accidentally, without having to be conceived beforehand. And when I speak of conceiving the act, I do not mean that the conception must be conscious. An image in memory can guide an act even though the image does not rise to consciousness. I am using here mostly conscious acts as examples, but the Requisites for a particular acts apply just as well to unconscious acts if you ignore the conscious aspects.

Here are examples for Requisite 2d. If you think the likelihood is very small that any female knows anything about the binomial theorem, you will not think to ask any female to explain it to you (and you will never find out that many females *do* know about the binomial theorem). If you think a rope bridge is possibly too rotten to bear your weight, you will find some other way to cross the river.

Finally, the third Requisite for an act to occur is:

3   That the chosen act not disturb some other controlled variable.

I do not claim that my categories of restrictions on one's choice of an act are the best. What you see here is my third revision, and the scheme still has faults. Feel free to revise the categorization to fit it better into your own way of thinking.

Example: Amount of Sound

As a second example of the coupling of the two springs of action, suppose you are listening to music on your hi-fi. You prefer the music to be coming to you at a certain loudness. But that perceived variable (loudness) is disturbed by the fact that your hi-fi is set to less loudness than you prefer. Obviously, you can act to bring your actual perception into match with your preference by adjusting the volume control. Just as you reach out to do so, however, the thought comes to you that you would like to maintain, too, your friendly relations with your neighbors. You might reach your goal of loudness by putting on earphones. This example illustrates the third requirement.

Example: Conflicts

In the previous example, you solved the conflict between two internal standards by finding another path through the environment (the earphones) by which to maintain your control of perceived loudness. It is also possible to avoid the conflict by altering the domain or meaning of an internal standard. I might act, for example, to maintain a perception of myself as an honest person. I might give back the money the clerk in the grocery store gave me in excess of the proper change, or I might tell the automobile salesperson the defects of my automobile when I trade it in on a new one, or I might tell my wife that it was I, not the cat, who dropped the new vase on the floor. But I might also tell myself that letting the cat take the blame causes no harm to anybody; I might maintain my belief that I am an honest person and at the same time maintain my wife's tranquility (and mine) by telling myself that I lie only when it hurts no one and telling my wife that the cat knocked the vase onto the floor. Those are examples of requirements 1a and 3.

Sometimes you hear someone (perhaps yourself) saying something like, "He could do it if he would just get up off his butt." Or, "She had to do it because they gave her no choice." Both are wrong. He can do it if he gets up off his butt *and* if the right environmental opportunities are there. And choices are not made for us by other people. Other people can put into our environment conditions that make choices more difficult for us, but we make our own choices. Sometimes, because of their internal standards, people even choose death in preference to what other people believe to be reasonable.

Simple though those ideas may be, they are very important in social life, for when we are busily pursuing our own purposes, we often forget them. We forget that others may not find their way to the same particular acts we might choose, or that they may not reject the same acts we might reject. We forget (1a) that others may not have the same internal standards we ourselves cherish. Others may have standards of honesty that are stricter than ours or standards of neatness that are looser than ours or a standard of maximum light energy to the retina that is lower than ours.

We forget (1b) that the environments of others may or may not produce events that disturb the controlled variables of the people there. Others may feel urged to act when we ourselves do not, and they may not feel urged to act when we do. We forget (2a) that the environments of other people may or may not offer the opportunities and limitations that our own offers to us, and (2b) that other people may or may not believe certain acts to be possible, and (2c) that they may not conceive the acts that we conceive, and (2d) that other persons may not agree with our judgment that a particular act will be likely to improve control of a controlled variable. And finally (3) we forget that others may or may not judge a path of action likely to threaten more of their controlled variables than just the one on which we have our attention. The less we are aware of those differences in the ways particular acts can come about, the more often we will be baffled by the behavior of others. Furthermore, if we are inept at communicating about these matters (and most of us, I think, are inept), mistakes and conflicts will multiply. Finally, we will be more prone to mistake the unintended side-effects of acts for their intended consequences.

In trying to predict someone's action, we are often wrong. Where is Al? I think he might be at Isabel's Bar. I call Isabel's Bar, and sure enough Al is there. I have predicted correctly. But I try the same thing tomorrow, and Al is *not* there. It is not foolish to estimate likelihoods. Al is always more likely to be at Isabel's Bar than at the Tuesday Music Club. But it is beyond doubt wrong to suppose that you can know enough about a person and her environment to predict her action correctly every time. Even when people say they will make an action predictable (make a promise), they often fail to do so.

Traditional texts in psychology commonly say that the goal of scientific psychology is to predict and control the behavior of other living creatures, including humans. Subscribers to perceptual control theory will not adopt that goal. Indeed, perceptual control theory tells us that except in the most severely restricted circumstance (such as a small cage bare of everything but a lever and a hole in the wall), it is *impossible* to reliably predict or control the behavior of other living creatures—not just undesirable or difficult, but *impossible*. You can influence the purposes and even the behavior of others to some extent, sometimes briefly and sometimes for a long time depending on their purposes and the environmental opportunities, but even then it is very chancy to try to predict their particular acts. Now and then you can guess right, but there is no reliable way to know whether you are going to guess right the next time (except in the severely restricted circumstances).

When I say it is impossible to predict particular acts (or to cause them), I am speaking of a particular means of reaching a goal taken by a particular individual within a reasonable limited time period. In predicting the route my wife will take to the grocery store next time she goes, I will often be right, maybe more often than not, but I will often be wrong, too. I will say more about this later.

Example: Opportunities

I remember an experiment[1] I read about many years ago that I admired very much. It occurred in full "real life," it was done at almost no expense to anybody, and it ran a very low risk of doing even small harm to anyone. At a building on the campus of a university, the experimenters put up a sign a few feet in front of the main door at the center of the building. The sign read, "This door closed." Then, of the people who came up the steps, they counted the people who, seeing the sign, turned and went away; they also counted the people who went past the sign and into the building. Then they repeated all that on the same day of the next week, during the same hours of the day. This time, however, the sign read, "This door closed. Please use door at end," with an arrow pointing of to the side. As you might suppose, a much smaller percentage of people, during the second trial, violated the sign.

I do not remember the explanation the authors gave. The simple explanation, I think, is that the second sign made it easier for most people coming up the steps to find an alternate route to their goal than did the first sign. This experiment illustrates the second Requisite.

Example: Inside and Outside

Now I will give an example of how, even when several people all seem to seek very similar goals in the same environment and even when interacting with the same person to reach those goals, they can nevertheless choose very different acts. You may think I am going to too much trouble to make the point that there are many ways to skin a cat. I do so to counterbalance the experiments paraded in psychology books—experiments in which only one or a few internal standards are presumed to be working and in which only one or a few kinds of variations in environmental events are allowed, and in which only one or a few kinds of action are noted. For reasons I have explained elsewhere (Runkel 1990, Chapters 4 through 7), I think it fruitless in a complex and unpredictably changing world to try to build an explanation of behavior (a psychology) from such scraps.

My fictitious example will display what seems to me an ordinary range of diversity in the actions people might choose by which to maintain similar desired perceptions in a work environment. I think the diversity I have written into this example is quite ordinary, but it is nevertheless much greater than we typically find among the variables in experimentation or field studies in the social sciences, even in extended programs of studies.

Let us imagine visiting a machine shop and observing four machinists. Of the four, Angie is the quiet one. She seems taciturn, sometimes even dour. The foreman says to her, "The day you say something more than 'Yes, sir,' we'll all take an hour off to celebrate."

Paul is the most methodical and the neatest. He wipes off his bench and sweeps the floor around it several times a day. He spoils the fewest pieces of work. Today he spoiled one. The foreman says to him, "Well, what do you know! Old Perfect Paul slipped up today!"

Darrell is the most loquacious. His garrulity seems to soar when the foreman comes to talk to him. The foreman can hardly get a word in edgewise. The foreman says to him, "All right, all right, hold still while I say something."

Catherine does good work, but she is often absent. She calls in sick often enough to keep her remaining sick leave close to zero, though she always seems to be in good health when she is at work. She uses up compensatory time as soon as possible. She was absent yesterday. Today the foreman says to her, "Well, I see you're honoring the company with your gracious presence today."

After observing a few more hours, we find that the remarks of the foreman are typical. He has a sharp tongue. Only Darrell ever initiates conversation with him. Sometimes the machinists make derogatory remarks about him to one another.

In my fictitious example, I am supposing that all four machinists find the behavior of the foreman disturbing to one or more of their internal standards, and all want to reduce their experience of his biting remarks to zero. They all take actions containing some feature which, in interaction with some feature of the foreman's behavior, will enable them, they hope, to perceive a reduction in the foreman's biting remarks. The four choose different kinds of actions.

Angie tries to attract the foreman's attention as little as possible. If he talks to her, she uses replies that she hopes will end the conversation quickly. She reduces her own talking to a minimum in the hope of discouraging the foreman from talking. Paul tries to give the foreman no occasion for talk. He tries to do his work so well that the foreman will have no reason to talk to him. Darrell tries to give the foreman no opening in his own flow of words. When the foreman comes around, he prattles on, hoping the foreman will give up trying to get a word in and go away. Catherine simply stays away from the foreman by staying away from work as much as possible. The four are all trying to maintain the "same" perception—a low frequency of the foreman's biting remarks—but they use the environmental resources in different ways.

If you were trying to see a simple causal connection of the machinists, those four machinists would certainly discourage you. Even if you had been lucky enough to hit upon the abrasive features of the foreman's behavior as your "independent variable" (the variable by which to predict something), the ensuing actions would seem to scatter in very different directions. One person chooses taciturnity, another talkativeness, another proficiency, and another absence. The same stimulus seems to produce very different responses. No lawfulness is apparent.

You might say, if you were thinking in terms of customary experimental design, well, we can measure and calculate the effects of all the "moderating" and "intervening" variables. Or we can conduct a long series of experiments with all those moderating and intervening variables balanced out. To use either of those two strategies, you could rather easily think of circumstances and personal qualities (variables) that could make it easier for one machinist to choose one kind of action and another to choose another. You might speculate that Angie chose reticence because, unlike Paul, she did not have exceptional skill as a machinist and because, unlike Darrell, she was not facile in speaking (maybe English was her second language), and because she was not as healthy as Catherine and needed to save up her sick leave for actual illnesses. You might speculate, too, that Angie had a desperate need for her job and couldn't risk initiating a grievance procedure. That for the same reason, she didn't complain to the foreman's boss. That she couldn't very often hear what the foreman was saying to the others, thought the foreman was picking mostly on her, was ashamed of it, and didn't want to attract the attention of the others by more noisy or visible actions. That she didn't try to persuade the foreman to be more polite, because she thought that would only give him more opportunity to make nasty remarks. That she didn't try to hit the foreman with a baseball bat, because there was no baseball bat on the premises. That she didn't hit him with a wrench, because she thought she might lose the fight. That she didn't ask to be transferred to another foreman, because those four machinists, with their foreman, comprised the only machining department in the plant. That she didn't offer the foreman sexual favors or put arsenic in his lunch, because she had internal standards restraining her from those actions. And so on.

You can probably think of twenty more kinds of action Angie might have taken, along with corresponding conditions that could have discouraged her from taking them, to reduce her suffering the foreman's biting remarks. And you can probably think of twenty more kinds of action that Paul, Darrell, and Catherine might have taken. To measure and calculate all those moderating and intervening conditions, or to design experiments to compensate for them, you would have to know a lot about Angie, about the people around her, about the organization of the plant and its norms, about the physical layout, and so on.

And I have illustrated some of the possibilities only of "responses." Considering "stimuli," we would have to cope with a similarly large number of possible features of the foreman's behavior that Angie and the others might perceive in addition to his abrasiveness.

Turning to other internal standards that might conflict with the desire to reduce the foreman's remarks differently in Angie, Paul, Darrell, and Catherine, we would again find multitudinous possibilities. Internal standards for being helpful to others, for taking interpersonal risks, for enduring harassment, for maintaining employment opportunities, for maintaining calm, for suffering persecution on earth as the path to peace in heaven, for proper respect for superiors—those standards and many others could compete with the foreman's abrasiveness for attention and action; they could compete for the use of environmental paths through which to reduce the abrasiveness.

Multiplying the possible relevant aspects that four or four hundred humans might perceive in an environmental event by the events an environment might produce, then by the possible internal standards that might be compared with the perception, then by the possible degrees of discrepancy between perception and standard, then by the available resources in the environment through which to select suitable action, and then by the effectiveness of the various actions, we get a number of cross-combinations of conditions so large as to be not merely formidable, but dumbfounding and preposterous to contemplate. Even so, my examples of variables were all static ones. I said nothing about dynamic interactions among variables. But though the number of cross-classifications dumbfounds the experimentalist trying to work only with the concepts of stimulus and response, it also brings us awe and humility in contemplating the adaptability and ingenuity of humankind in coping with disturbances to controlled variables. Despite the varieties of internal standards we find in other people and the acts they choose for controlling their perceived variables, we do nevertheless manage a great deal of the time, even in stressful situations much worse than the situation I have pictured for Angie, Paul, Darrell, and Catherine, to carry out joint work effectively.

I will make no attempt in this book to try to explain the behavior of humans (or of rats or of Escherichia coli) by explaining the behavior in each of all those thousands of cross-classifications of variables that I described above. My strategy will be quite different. I will describe the internal functions with which humans seem to be naturally endowed and to describe the ways humans use that endowment to keep events in the environment from disrupting the integrity of their bodies and minds—that is, how they control their perceptions. The early chapters will lay out the reasoning and some evidence. Later chapters will show how the control of perception can be seen in complex social interaction.

SUMMARY

We often find ourselves wanting to "make sense" of the actions we are witnessing or being told about. Some of us, sometimes, take pains to think about human action systematically—to search for features of human action that always stay the same, to examine carefully the reliability of the information we get, to examine the logical connection between the information and our beliefs about the constant features of action, and to look for instances of behavior that could *contradict* the beliefs we are forming.

One source of human action is the person, or more precisely, the mind—the many patterns of electrochemical activity in the nervous system (or neural net) of the individual. The other source is the environment—the many events out there producing energies that impinge upon our sense organs. Action or inaction depends on the interaction of those two sources. The Requisites for action, therefore, are:

1. that some environmental event disturb a perceived variable the person is controlling.
2. that the person find an object or event or feature of the environment with which to affect the controlled variable.
3. that the chosen act not disturb some other controlled variable.

There are many ways to skin a cat.

ENDNOTE

[1] The experiment was reported by Freed, Chandler, Mouton, and Blake (1955). Blake and several colleagues engaged in a series of clever experiments in Austin, Texas about that time. See, for example, Blake and Mouton (1957), Lefkowitz, Blake, and Mouton (1955), and Rosenbaum and Blake (1955).

Chapter 2
Living things

If the rest of this book is to be of any service to you, the first axiom that you must accept is this: living things do not act like nonliving things. If you deny that, this book can only annoy you. That axiom may seem so obvious to you as to be not worth mentioning. Yet most of us very often act as if we expect other people to behave like rocks. And when we act toward other people as if they were rocks or blankets or typewriters or even potted plants, we make unending trouble for ourselves. It is true that people do have some features in common with rocks and typewriters. There are, however, important differences between living and nonliving things that most of us overlook time and time again, and to our sorrow. One of my chief purposes here is to make it easier for you to call to mind the important differences between people and rocks.

I do not say that we often mistake people for rocks. We do tell them apart, but how do we, indeed, do so? How do we distinguish living from nonliving things? What features are most telling? Some familiar features are not wholly reliable indicators. A slug can look very much like a little rock. A slug moves, but so does a river. A human talks, but so does a radio.

PURPOSES

The crucial difference between living and nonliving things is purpose. Physical laws are sufficient to describe the behavior of a nonliving thing. Physical laws are necessary to describe the behavior of a living creature, but they are not sufficient; you must also take account of the intentions or *purposes* of the creature. You must know what the creature *wants* to do.

Think of throwing a ball to Cora. Your hand pushes the ball in Cora's direction. Your hand "tells" the ball to go to Cora. When the ball leaves your hand, it must move in a ballistic arc described by Newton's laws of motion; it can do nothing else. The ball will not stop until something physical stops it—Cora's hands or the ground or something else. The ball has no character enabling it to stop or turn in midair. If you have thrown the ball so high and fast that it will go over Cora's head, it will do you no good to shout at it. The ball will continue over her head.

Now think of sending Charles with a message for Cora. You tell him, "Take this to Cora," and you give him a shove in the right direction. If Charles accepts your purpose as his own, he will go off to Cora with your message. But if he has no care for your purpose, he may turn aside to watch a croquet game or to go fishing. Our ball contains within itself no purpose of its own. It moves as your hand "tells" it and as its own mass and the earth's gravitation tell it—as a physical thing must. But Charles contains within himself his own purpose, either one he generates (such as going fishing) or one he borrows or accepts from you (carrying the message to Cora).

Even if Charles accepts your charge to him and carries the message faithfully, his trajectory will not be analogous to that of the ball. Charles will turn left or right at his own initiative to avoid obstacles. He will look farther afield if he does not find Cora where he looks first. If he does not at first find Cora, Charles does not (as the ball would) fall down on the ground and lie there.

Or think of walking up to a statue—a nonliving statue of stone or wood or papier-mâché. Push on it. If you push hard enough, the statue will tip and then fall over. As the statue begins to tip, it will not increase its backward push against your hand; it will do nothing that would help it remain upright. And after it topples, it will not get up again. It will just lie there until someone moves it. If, however, you walk up to a person and push, the person, intending to remain upright, will immediately push back. If you push harder, the person will push back hard or jump away to avoid your pushing. If you push suddenly and hard so that the person falls, he or she (the person who wants to be upright) will immediately leap upright.

The nonliving statue is passive; it obeys Newton's laws of motion. It continues in its state of rest or uniform motion until disturbed by an outside force. Then it moves in whatever direction the force pushes, and remains in its new state. The material structures of human bodies also obey Newton's laws, but people go beyond reactions to external forces, initiating their own forces to guide their actions toward their own purposes. People at all times act to maintain their preferred states. If a person wants to remain vertical, the person exerts an immediate counterforce against any disturbance to that preferred state.

In Chapter 1, I said that we choose actions because of internal urges or standards and because of what is possible in the environment, the two sources of causes acting simultaneously. What I said there may have sounded as if the two sources are equal in their effects upon our actions—or as if each realm, the living and the nonliving, puts shape on events in the other. That is not the case. Living creatures bring about events in the environment that nonliving things cannot produce. Nonliving things always go downhill unless thrown uphill by an external force. Living things go uphill whenever they wish. If they have the strength, they go directly uphill; if they do not, they find some way to circumvent that difficulty. The environment does not put shape on living behavior; the environment offers opportunities and restrictions, but it does not determine the particular actions of living things. I will put more detail on this asymmetry in Chapter 3.

Rocks, balls, and statues do not push back; people do. Rocks, balls, and statues do not start, stop, or alter their paths to carry out inner purposes; people do. Living creatures counteract disturbances to preferred or intended states; nonliving things do not.

TALKING ABOUT PEOPLE

All that, I hope, seems simple and straightforward. Yet our ways of talking about other people often sound as if we think people will behave like statues. Here are some things we say:

Put some pressure on him.
We'll push to get it done.
She can pull strings.
She made me do it.
They resisted.
She carries more weight than he does.
I've got to get the upper hand.

You can imagine saying those things about nonliving objects: *Push* the rock out of the way. *Take hold* of that thing. *Pull* the strings of the marionette. The heavy rock *resisted* our efforts. But we speak about people that way, too, as if our dealings with them were as simple as that. Some of us even speak of "knocking some sense into" people as if bringing about understanding, or at least compliance, could be brought about in a manner analogous to kicking a vending machine that isn't giving out an expected candy bar. We speak of "getting the upper hand" or of "handling" someone carefully. The word *manage* comes to us from the ancient Latin word for "hand." We speak of a person who assists in the training of a prizefighter as his "handler."

Managers and administrators often speak with metaphors implying that "hard" things or methods are more effective or reliable or true than "soft" ones. "Now, don't go soft; you're going to have to be hard with him." "These are the hard facts." To many people, hard, strong, and tough sound good, while soft, weak, and tender sound bad —or at least incompetent. As Kenneth Boulding (1990, pp. 77–78) has pointed out, however, actions called strong and hard are usually those using threats, and threats often have less lasting effect than negotiating or purchasing, even though talking and paying for benefits are sometimes considered weak behavior. Taking a "hard line," Boulding says, usually signals unwillingness to learn. To be "tough," he says, is usually to defend oneself against change, whereas "it is the softies who are adaptable," and often "the softies who survive and have the greatest power" (p. 78).

Among nonliving things, it is sometimes true that the harder things are the more effective. One uses harder metals to shape softer metals. But even with nonliving things, the opposite is also sometimes true. Water wears away rock. A dinner plate is easily broken with a fist. An edge of paper will cut tough skin.

My point is simple: Our language provides us with many easy ways to speak misleadingly about living things, including ourselves. If we go by the surface of the words, if we do not beware the metaphors, we can find ourselves behaving stupidly toward others.

TREATING PEOPLE LIKE THINGS

Not only do we often talk about people as if they were nonliving objects, but we often act toward them that way. Here are a few ways we do that.

Suspended Animation

We treat people like inanimate objects when we act as if people, like rocks, do nothing between the times they are in our presence—as if nothing goes on with other people except those matters in which we have a hand. Few teachers (at least those in colleges where I have studied or taught) try to find out what students already know before starting their lectures.

We often think, too, that if other people do not seem to be doing anything, then indeed they must be doing nothing, or at least nothing that could be important to us. We often think that children who are quiet are not doing anything important to themselves, either. I remember a day at school when I was in the fifth grade. It came time for music appreciation, and I discovered, to my joy, that the teacher was about to play Rossini's overture to "William Tell" on the phonograph. It was one of my favorite pieces. I put my arms on my desk, put my head on my arms, and closed my eyes, ready to hear every treasured note with full concentration. I heard the teacher's shoes clacking along the aisle. "Philip!" she said. "Sit up and listen to the music!"

Another example: I was once giving a long lecture to a large crowd and noticed some signs of inattention. Even a dear friend, sitting near the front, was staring unseeingly out a window. I wanted to do something to recapture everyone's attention (I am a glutton for attention). I thought my friend, knowing my ways and having often indulged my quirky behavior, would forgive me for a remark that would embarrass a stranger. "Mary Ann!" I said, "Sit up and pay attention!" She turned to me with dreamy eyes, smiled a small smile, and said, "I was thinking about something you said earlier." I was doubly embarrassed. I had acted as if, since she had been sitting unmoving as a rock, she had been giving me no more attention than a rock would give. But on the contrary, she had gone away in her mind to tuck something I had said alongside her other prized ideas.

Employers often ask questions of a prospective employee about what he or she can do, but rarely ask questions like that after the person is hired. Twenty years and more can go by without a query. Some forward-looking organizations, it is true, do keep track of the changing capabilities of their employees; some actually encourage their employees to expand their capabilities, and many of those latter organizations actually make use of the employees' new capabilities. That appreciation of human resourcefulness, however, is not common. It is so rare that organizations that show it are written about admiringly in magazines and professional journals.

Employers treat people like objects, too, when they act as if nothing relevant happens to employees between work shifts. Sometimes employers say, "Don't tell me your troubles," or "Work and home don't mix." The idea seems to be that an employee who admits to having human capabilities and purposes is a defective part in the machine. Some employers are learning that all of us bring all of our characteristics with us no matter where we are, at home, at work, or elsewhere, and we can only pretend not to do so. Most employers and managers I have met, however, seem not yet to have learned that.

Job Descriptions

Job descriptions are typically used in ways that treat people like objects. Many organizations, perhaps especially governmental bureaucracies but thousands of others too, write out descriptions of the duties required in jobs placed at the lower and middle levels of the hierarchy. When personnel officers hire people for those jobs, they seek people who seem likely to act according to those descriptions. Their conception seems to be that an organization is built of all-but-unalterable jobs operated by people, and you need operators who will slide smoothly into the jobs. Writing a job description is like telling a supplier how a part must fit into a machine. You tell the supplier as exactly as you can just what the machine can and must do at the point where you need a new part. You don't think of the part or the machine as changeable, lively, or resourceful; you don't think that together, the new part and the old machine might produce a new and improved way of functioning. On the contrary, you want a part that will operate exactly as the last part did, and that will certainly not require any alterations in the functioning of the machine as a whole.

Despite a lot of talk about working conditions, "human relations," "empowerment," and so on, in many ways we go on treating employees like cogs in the machine. Even when managers come to a realization that they want to stop treating employees like objects, usually the managers sit down only with one another to plan the new policy. It is rare that managers ask even one lower-level employee to join the planning, and even rarer that they ask a group of them to do so.

You may be wondering, about now, to what extent my claims about the ways most organizations are managed are backed by systematic data. The assertions here are so commonly supported in the research literature that I won't even bother to give you specific references. Just look on any shelf of scholarly literature about organizational management. I am also supposing that you yourself have had experience with the kinds of management I describe here, but if you have not yet been an employee in an organization of some size, just think back to your experience with schools.

The employees at the lower levels of organizational hierarchy are those who more often get treated like nonliving things. Most of us think very differently about the executives at the top of the pyramid. At that level, the job description is very loose and frequently not even put on paper. Often, indeed, a new executive is expected by board members and top bosses *not* to fit into the organization's present way of functioning, but rather is expected to alter purposely both the job and the organization. Most of us somehow think of the executives at the top as creatures of a species unlike those near the bottom. (Maybe the executives near the top are themselves especially likely to think like that.) We seem to think we should not try to hold top executives to predictable behavior. On the contrary, we seem to think it entirely reasonable not only that top executives should act differently from previous ones, but that they should feel free to call upon those around and below them to change *their* previous ways of doing things—*including all those who were hired to fit particular job descriptions*. We often seem to think that sort of "shaking up" will somehow be good for the organization.

Managers seem to intend detailed job descriptions to act like fences, to limit the domain of behavior. Even though those fences are made of words, not of stone or steel, even though we don't expect fences of that sort to restrain the behavior of top executives, nevertheless most managers seem to believe that it is reasonable to command employees to stay inside such a fence in just about the same way they expect beans or bottles to stay in a bin. Some managers get very angry when an employee acts like a living creature and steps over the fence. In a grocery store one day, I saw a sign that read, "49 cents, two for 99 cents." At the check-out counter, I described the sign to the checker and suggested that she might want to tell the manager about it. She said, "Not me! I told him about something like that once, and he said, 'Listen, when you get to be manager of your own store, you can run it the way you want to!'"

Stereotypes

We treat persons as objects, too, when we treat them as stereotypes. When we think, "She is a black" (or female, or Hungarian, or Baptist, or any other classification) "and therefore she will act thus-and-so," we are treating the person as an object. That kind of thinking works pretty well (not perfectly) for nonliving objects. To think that this thing under our hand is a rock and therefore will be hard not only today but also tomorrow and tomorrow is also pretty good thinking. Pretty good, though not always reliable. I once brought home a beautiful, wave-rounded chunk of sandstone from the seashore and put it in a place of honor on my patio. A couple of weeks later, I looked at it only to discover that it had fallen into two pieces. And as those chunks dried out, they fell apart too, until I had only sand where I once had a rock.

Living creatures are far more changeable than nonliving things. Not only do they change their behavior in reaction to changing circumstances (different temperatures, different teachers, different jobs, and so on), but they change themselves at their own initiative. We often expect simple classifications of people to be useful: this person is a plumber, that one a preacher, this one a male, that one a female, this one a reporter, that one a Republican. But if this plumber or this female is ignorant of algebra, you can make a bad mistake by talking or acting as if the next plumber or female you meet will be ignorant of algebra. And plumbers or females who are ignorant of algebra may not remain so. If this Republican or this black person is poor, you can make a bad mistake by talking or acting as if the next Republican or black person you meet will be poor. And so on.

When we think in stereotypes, we fill in information out of our own heads where we have no information from the outside world. Our sense organs and our brains are built to operate that way (as well as in other ways). In many situations, filling in information out of our heads as if we are seeing it with our eyes is necessary and reliable; in others, it is wrong and dangerous. Stereotypic thinking is pervasive in social science, for example, and there it is wrong and dangerous; I will say more about that in Chapter 5 (under "Reification"), Chapter 17 (under "Stereotypy"), and 18 (under "Seventh Order: Categories").

PREDICTING BEHAVIOR

Most of us believe we can predict the "behavior" of nonliving things very well. We put a bowl on a table and expect it to stay there. If we return to the table and find the bowl gone, we do not think it might have moved itself. Barring hurricane or earthquake, we do not ask *what* moved the bowl; we ask *who* moved it. Our prediction about the behavior of the bowl itself is very reliable, and so is our knowledge about the conditions under which the bowl will stay put or move.

Things stay put or continue in uniform motion until acted on by an external force. That is what Isaac Newton (1642–1727) told us long ago. With sizes and speeds appropriate to lemons or locomotives, uninstructed common sense or a groping understanding of Newtonian physics is good enough, as in the case of the bowl on the table. We can make a meal, put the food on the table, capture it with forks and spoons, and put it into our mouths. We can do all the ordinary doings of daily life and rarely be surprised at the behavior of nonliving things.

We become very confident in our expectations of many kinds of events in the physical world. Billiard players expect the ball to go where they send it *every time*. When the ball misses, they almost always blame the miss on their own skill, not on unreliabilities in the caroming of the balls. Players might on a rare occasion suspect a sloping table or an unbalanced ball, but they *never* claim that the laws of physics have failed. Golfers who lose sight of the ball while it is rising into the sky go looking for it where they think the trajectory would bring it to the ground. Even when they have trouble finding it, they *never* think the ball might still be up in the air.

We expect to encounter *no* exceptions to the lawful workings of physical events.

You may think, as I write here about the unpredictability of particular acts, that I am exaggerating. You might think you could not get capably through the day if social behavior were not highly predictable. Didn't you predict that someone would be at the check-out counter when you went to the grocery store to buy food? Didn't you predict successfully that your spouse would serve you lunch at noon? Didn't you predict that all those people would be driving on their own side of the street so that you could drive safely on yours? Are not those dozens, even hundreds, of successful predictions? (Enough, maybe, to make a professional research psychologist green with envy?)

You do not succeed in getting food at the grocery by predicting that Sylvia Sanderson would be at that check-out counter when you would get there. You needed only for *someone* to be there. Sylvia could be at another counter, off on a rest break, home sick, or departed for Norway. You might or might not think about whether you had expected (predicted) Sylvia to be there. You might not remember that instance when, reading this page, you think how well you cope with your daily tasks. Yes, you do cope well with them. But you do so by coping with whoever is at the check-out counter, not by predicting that Sylvia will be there at that time.

Yes, today you predicted successfully that your spouse would serve you lunch at noon. Yesterday, however, you got home only to find a note from your spouse saying that you should get your own lunch, because he had to pick up a visitor at the airport. You don't remember that failed prediction, because you didn't think about going home for lunch as a prediction. Anyway, it was "only natural" for your spouse to pick up a visitor at the airport, so you wouldn't think of that as anything gone wrong, but just as an ordinary event that required a little alteration in routine. Certainly. That is the way we cope with our daily tasks—by little alterations in our routines, not usually by predicting when those alterations will become necessary.

It is true that I have never driven out onto the street and there found other drivers all wandering to one side of the street or the other as the urge struck them. But neither have I ever expected drivers to stay on their own side with the same confidence that I expect billiard balls and golf balls to go where they are struck. It is not rare, in my experience, to find a car coming toward me on the wrong side of the street. I have ridden in several taxicabs whose drivers took me unexpectedly from one side of the street to the other. I was never able to predict which drivers would do that or when they would do it. Furthermore, I make my way along the street not by predicting that George Gatling and Gertrude Johnson and Graham Garrison will be going past me on their proper side of this street at this time, but by assuming that whoever is going by will have the goal of getting safely to wherever she wants to go.

I make this point because some psychologists think it reasonable to want to predict and control human behavior. Those psychologists, however, never undertake to predict when George Gatling will be driving down the street toward me or even toward them. Nor when a particular taxi driver will swerve across the center line. Nor when I will encounter Sylvia Sanderson at the grocery. Instead, they predict that I will answer "yes" to a particular questionnaire item. In fact, they rarely look at *who* answered "yes," but only at *how many* people answered "yes." And if I leave that item unanswered or write beside it, "This is silly," they do not even count me in their tally, but act as if I was never in their experiment.

Rats, too, cope very well with their daily tasks. But psychologists have usually put their rats into small boxes or cages or mazes where the acts that will bring the rats food are severely limited. The psychologists then count the pushes on a lever, for example, within a particular interval of time following the rat's last meal. In brief, psychologists find behavior in the natural environment to be so various and unpredictable that they resort to the severely restricted environment of the laboratory to improve their chances of making a correct prediction. That is a reasonable strategy. Even then, however, psychologists feel proud when their predictions are correct more often than you would expect by sheer chance.

When we know a person's goal (what he wants to do, such as get home safely), we can usually predict well the *consequence* of his actions (he will get home safely). But we still cannot predict the *particular acts* he will use to get himself to the goal (when he will press on the accelerator or brake, which streets he will traverse, or whether or when he will swerve to the left side of the street, for example).

In general and in principle, the particular acts of living creatures are not predictable. When I speak of predicting a particular act, I mean doing so with the same confidence in the outcome you have when you predict the particular course a billiard ball will take after being struck at a specified spot. That is the confidence we require when we build bridges, airplanes, and television sets. But no general has that kind of confidence when he orders his troops into battle, not even any restaurateur when she orders her waiters into the dining room.

There is no hope, ever, no matter how much we learn about the behavior of living things, that we shall be able to predict the particular acts of a living creature. There *is* hope that we can learn how to predict the *consequences* of a person's acts *if* we can first learn the person's purposes. If we know that a person has the purpose just now of getting some food into himself, we may not know just what actions he will take to do that, but we can predict, *if* the hunger persists, *if* his muscles remain strong enough, and *if* food remains available, that his actions will take him closer to food, will move some food closer to his mouth (or his mouth closer to some food), and will eventually result in his swallowing some food.

SUMMARY

Physical laws are sufficient to describe the behavior of a nonliving thing. Physical laws are necessary to describe the behavior of a living creature, but they are not sufficient; you must also know the intentions or purposes of the creature. Living creatures counteract disturbances to preferred or intended states; nonliving things do not. In general and in principle, the particular acts of living creatures are unpredictable.

Our language provides us with many easy ways to speak misleadingly about living things—including ourselves. Not only do we often talk about people as if they were nonliving objects, but we often act toward them that way. All of us use stereotypic thinking every day in almost every realm of our thinking. Stereotyping living creatures is dangerous.

Chapter 3

Inside and outside

So far, I have described how acting is initiated by the person (not by the environment) and how it is enabled and restricted by the available environment. And I have said that the basic assumptions made by almost all social scientists are those of physics—the assumptions of a science of nonliving things. For example, one of Newton's laws of motion, stated in language one might find in a high-school text, is: A body will continue at rest or in uniform motion unless acted upon by an external force. In Chapter 2, I gave several examples of how easy (and wrong) it is to think as if human behavior obeys Newton's law. I will say a little more here about linear causation (the physical kind), and then I will go on to describe circular causation (the kind that enables living creatures to live).

Almost everyone, psychologist or not, thinks of action as starting *here* and ending *there*. Our language seems to make it easy to talk about delimited events. We say that we "did something." We started doing it, did it, and stopped doing it. When we want to draw attention to a particular portion of what was happening, we usually put it into the subject of the sentence and put it within a frame of beginning and ending. When we want to report that we made a trip to Ashland last week, we almost always say, "I drove to Ashland last week," not, "I was driving to Ashland." If we were to say, "I was driving to Ashland," our listener would probably say, "Yeah? What happened?"—thinking we were giving the background against which we would now describe the event (with a beginning an ending) we had primarily in mind. In an effort to take attention away from beginning and ending, we might say something like, "During my behaving last week, I spent some time driving to Ashland." Or, "During my comings and goings last week, Ashland was one of the places I passed through."

You can see that I find it difficult to construct a sentence that shies away from putting a beginning and ending on an experience while still sounding like ordinary conversation. That is my point. We usually do talk as if what we do has a *thing* character, and when we try to describe our doings as part of a continuous flow, the result has a strange, weak, and uncomfortable sound. But behavior, actually, does flow. Our muscles never completely relax while we are alive; the constant tension is called "muscle tone" and is necessary to proper functioning. Our brains, awake or asleep, never stop their electrochemical moiling. Our *purposes* in acting come and go as we achieve their intent, but our acting goes on seamlessly. One's life is not a succession of actions, but a ceaseless maintenance of one direction and another, a ceaseless pursuit of goals. Sometimes the goals help each other; sometimes they conflict. In brief, actions are not really separated by inaction; action varies between violent and quiescent, between this purpose and that, but action never ceases. The importance of this point will swell as we go along.

Almost everyone, psychologist or not, thinks of the causes of behavior as acting linearly—in a line that starts here and ends there. Almost everyone, psychologist or not, believes that when some sort of energy impinges on a person, it causes an action. The stimulus goes in that side, and the response comes out this side, and that's the end of the matter—the end of this linkage of cause-and-effect. The physician says, "Say ah," and you open your mouth and say, "Ah!" Your parent says, "Stop that!" and you take your finger away from the cake frosting. You see a picture of a pretty woman smoking a Virginia Slim and you run out to buy a package of Virginia Slims. The bell rings, and you run out of the classroom. We all, including the psychologists, know that those sequences do not always play themselves out, but we continue, most of us, including the psychologists, to think as if they do. We continue to hunt for the stimulus that will set off the reaction we want. Sometimes we think a *stronger* stimulus will do it. Buy the reluctant woman a bigger box of candy. Beat the disobedient child more severely.

Figure 3–1 symbolizes this linear kind of causation.

Figure 3–1.

Something in the environment impinges on the person, something happens inside the person, and the person acts on the environment. In this kind of thinking, those three steps complete the story. The story says nothing about what happens next. Presumably the person must wait for something further to happen in the environment.

We do not always think in such simple one-two-three sequences. Sometimes we say, "Oh, he won't respond to that; he's not the type." The speaker is dividing people into two or more types and saying that despite the indifference of this person, there are persons of another type who *will* respond. As I said in Chapter 2, this is the strategy of the moderator variable (or intervening variable, or contingency). The reasoning is that if you divide a collection of people by this criterion, and again by this variable, and yet again by this contingency, and so on, you will eventually find a subgroup in which all the people do react in the same way when you offer them a certain incentive or threaten them with a certain punishment. In an earlier book (Runkel 1990, Chapter 7), I called this tactic "fine slicing" and explained why it is hopeless for use in research on human functioning. I explained there, too, that this sort of reasoning does have some practical use. When you don't care whether everyone will act as you hope, or who will do so, but only whether a sufficient number will do so in the near future in a specified population—as in advertising—this reasoning is useful. But my topic here is individual human functioning, not shotgun advertising.

When we look for the type of person who will react to certain conditions, circumstances, sentences, incentives, or some other "input" from the environment, we often think about the "personality" of the person. Some people use that word to mean attractiveness—"Oh, she really has personality." Psychologists, however, use the word merely to mean the make-up or clusters of dispositions of the person, leaving the particulars to be specified—attractive, repulsive, lively, quiet, or whatever. Psychologists have invented hundreds of personality types. Their hope has been that if you can isolate people who are of this or that type, you will be successful in predicting what they will do in certain situations. Many still hope.

I hope the paragraphs above will be sufficient to give the flavor of the assumption of linear causation, the assumption almost universally found in whatever you read or hear about human motivation. Linear causation, however, is not the way living creatures function. Almost everyone believes in linear causation, but in respect to the actions of living creatures, that is a mistake.

In the conception of circular causation, the person and the environment are in constant, unceasing interaction. Purposes start and stop, but actions flow. Imagine two figure skaters whirling together on the ice. Each skater is part of the environment of the other. When the two are holding each other and whirling about a point between them, each is both enabled and limited by the other. Neither could whirl in that pattern, leaning backward with an arm up like that, without the other. Notice, too, that movements do not have any obvious starting and ending points. One glide slides into another. One gesture of an arm wafts smoothly into another, so smoothly that you realize that "another" is happening before you realize that the "one" is no longer happening. At all moments, what one person is doing opens some possibilities and closes others for what that person, and the other person, too, can be doing at later moments.

The relationship of person and environment is always like the relationship between the one person and the other on the ice. The environment pulls or pushes on us, and we pull or push back to keep what we care about in balance. We can choose, too, to pull or push with a force greater that necessary to maintain the balance—to change the direction of the whirl, to alter position in relation to environment or to alter the environment. But all the time, we can move only *because* we can push on the environment, and we do move only because we choose to alter the way the environment pushes on us (or sends energies to others of our senses).

PERSON AND ENVIRONMENT

The environment acts on us, and we act on the environment. Winds buffet us all, the rain falls on the just and the unjust, some of us are mashed by automobiles or falling trees, and some are swept way by floods.

> There once was a singer named Hannah
> Who was caught in a flood in Montana.
>   She floated away,
>   And her sister, they say,
> Accompanied her on the piana.

At every moment, too, we are affected by forces less harsh than floods but part and parcel of our living. Light interacts with the cells of the eye's retina in such a way that electrical pulses run along the fibers of nerve cells, and as hundreds and thousands of those fibers act in parallel, the effect is one of electrical currents reaching thousands and millions of more complicated circuits deep in the brain. Pressure waves in the air impinge on our ear drums, moving bones that move liquid in the cochleae of our ears, where the throbbing liquid shakes fine hairs against nerve cells, which in turn send currents to deeper (or "higher" if you prefer) neural circuits. Other sensors tell us of many other energies coming from outside and inside our bodies that we interpret as taste, odor, balance, pressure, muscular fatigue, pain, heat, cold, and so on.

Our sensors enable us to perceive some of the effects of our actions on the environment. If those perceptions are not yet the perceptions we want, we take further action in the hope of bringing our perceptions into match with our internal standards. Usually, we succeed. Living requires the continuous control of perceptions through actions on our environment, including actions within our bodies.

We live only because we interact with an environment that is in part nonliving. The nonliving environment provides us a planet to which we cling. We feel the pressure of our feet upon the ground and the swell of our lungs as we breathe. We step into the sunlight to feel its warmth. We live, too, because of the living part of our environment. We feed upon other living creatures—beginning with our mothers. We sit at table and savor a taste of melon or shrimp. We join others to accomplish our daily work. We look up to enjoy the smile of a lover. We give attention to a supervisor, hoping for words of encouragement. In a million ways, we look for signs that things are going the way we want—or are not.

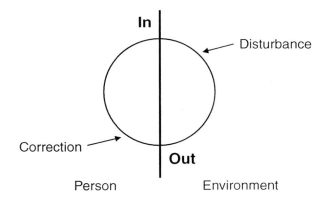

Figure 3–2.

Figure 3–2 symbolizes the circular causation through which life is possible. The circle has two parts, one in the person and one in the environment. The half of the circle in the environment represents the person's actions. The beginning of that semicircle at the bottom of the figure (at "Out") represents muscular or glandular action that will affect some aspect of the environment perceived by the person. At the upper part of the semicircle, the arrowhead (beside "In") symbolizes the sensing of an environmental energy by a sense organ—the perception of a variable quantity in the environment. At the same time that the person is acting to bring a perception of a variable close to an internal standard, the environment is continuing to have an effect on that variable; it is *disturbing* the energy. For example, when you are driving a car along a highway, you want to perceive your car to be near the middle of your lane. So you move the steering wheel a little to maintain that perception close to that in-the-middle standard. But at the same time that you are adjusting the steering wheel, the wind is blowing your car away from the position you want to maintain. You must act to steer where you want to go, but you must also steer to *counteract* the disturbing effect of the wind. The joint result of your muscular action (at "Out") and the disturbance of the wind produces the position you perceive (at "In"). Another example: When I chew food, my hearing aids amplify the sound of my chewing well beyond the level at which people with normal hearing hear their chewing. As a result, when I converse with my wife at mealtime, I must stop chewing while she is speaking if I want to hear her words (which I do). The joint result of the sound of my wife's words and my muscular effort (at "Out") to hold my jaws still to counteract the noise that would otherwise disturb my attention produces my desired success in hearing (at "In").

At the same time that things are happening in the environment, things are also happening in the person's nervous system. At "In," a sense organ transduces an external energy such as light into a current of neural pulses. In the language of Perceptual Control Theory (PCT), that current of neural pulses is called a perception. Powers (1973, p. 286) defines a perception as "a perceptual signal (inside a system) that is a continuous analog of a state of affairs outside the system," a perceptual signal as "the signal emitted by the input function of a system; an internal analog of some aspect of the environment," input function (on p. 284) as "the portion of a system that receives signals or stimuli from outside the system and generates a perceptual signal that is some function of the received signals or stimuli," and function (p. 284) as, "a rule making the state of one variable dependent on the states of one or more other variables." At "Out," neural currents set off muscular contractions or glandular secretions, or both, to bring about action in the environment.

Figure 3–2 displays a *feedback loop*. Cause and effect do not simply go in one side and out the other. Cause and effect chase each other around the loop. Internally, between "In" and "Out," a multitude of neural events can happen. Sometimes, in a low-level reflex such as a knee jerk or an eye blink, the neural part of the loop is short and uses few paths. At other times, especially when conscious thinking is part of the loop, we make use of most of the cerebral cortex and take many actions on the environment before we feel that our experience is coming to match our internal standards. Sometimes, in pursuing a long-term goal such as saving enough money for a trip around the world (this is long-term for most of us, at any rate) or undergoing the training required for entry into a profession, a great many loops of perception and action are required over a number of years. When the perception matches the internal standard, we take no action to better the match. When the match is not good enough, we do something to narrow the gap. At some point in the loop, therefore, a function must exist to send out a neural current to keep things as they are or to change things by glandular or muscular action. That corrective function is symbolized in Figure 3–2 at "Correction." (In Figure 3–4, I will change the name of that function to "Comparator.")

The relation, then, of the organisms to the environment is one in which the organism makes use of the opportunities in the environment to control the perceptions it prizes. The perceptions most obviously valuable to any organism are those that signal the acquisition of nourishment (food, water, oxygen) and the maintenance of the optimum conditions for processing the nourishment (heat flow in and out, ratio of water content in the body, and many others). To assure itself of the proper quantities of those necessities—to stay alive—the organism must carry within itself the internal standards for the perceptual analogs of them. For the continuation of the species, still further perceptions (and internal standards) are vital: configurations and other signs of the opposite sex, signals of readiness for copulation, signs from infants of the need for care, and so on. A minimal degree of stability in family and group life, for example, seems vital to bringing up healthy children and inculcating the social norms that ensure another cycle of doing that in the next generation, but no one knows in what ways—or whether—those social patterns are influenced by internal standards existing at birth.

In addition, however, to activities that seem obviously vital to continuation of the species, many animals spend a good fraction of their time in pursuits that seem in no obvious way necessary to preserve the life of the individual or the continuation of the species. Otters, dogs, and humans are notable for their playfulness. I suppose a devoted Freudian would insist that every activity, whether playing baseball, playing chess, painting voluptuous females, or composing abstract music, has something to do with sexuality. In the perception of some people, I am sure that is the case. But whether you take the Freudian view, or Richard Dawkins's (1982) view that behavior has the purpose of maintaining the reproduction not of whole animals but of the "selfish gene," or the view of many religious that our behavior serves to enable God more easily to choose those of us who will be sentenced to spend eternity singing his praises, or some other all-encompassing explanation of our daily life, surely most of us will agree that humans show unending ingenuity in finding ways of using the resources in the environment to serve their purposes, satisfy their longings, and reach their goals.

Here is an example of operating on the environment to keep oneself comfortable. On the CSGnet (an e-mail discussion group on the Internet), I read a contribution from Bruce Abbott on 16 December 1997 in which he described a study he had seen portrayed in a TV program called "Scientific American

Frontiers." As I understand it, the experimenters, working in France, undertook to find out the size of cage in which hens would feel comfortable. Perhaps the experimenter's further purpose was to discover whether the hens would slow their egg-laying in too-small cages. One wall of the cage moved slowly but constantly inward. But the cage of each hen contained a button that the hen could easily peck; pecking the button caused the wall to move outward at a rate that overcame the rate of inward motion. If the hen wanted the cage to be at least as large as some minimum size, it could maintain that size by repeatedly pecking the button. That is what every hen did. The exact perceived variable that was disturbed by the wall moving inward is unknown—whether it was the distance from the moving wall to the wall opposite, or the number of cubic inches within the cage, or the auditory reverberation, or something else. But clearly, there existed some perceived quantity for which the hens had an internal standard and which could be controlled by pecking the button that moved the wall outward. The study illustrates nicely the continuous, simultaneous, circular causation that enables us to control our perceptions. It illustrates, too, how well chickens can keep themselves comfortable while understanding nothing of the detailed cause-and-effect chains through which their actions affect their perceptions—and presumably how we do so also.

In some ways, the loop is symmetrical. The effects of the loop come in and go out of the inside half and also of the outside half. A crucial input to the loop occurs outside at "Disturbance" and another inside at "Correction." Causation in the loop is circular. The organism's output at "Out," *together* with the effect of "Disturbance," affects some aspect of the environment that becomes input at "In." The incoming neural current at "In," *together* with the effect at "Correction," becomes an outgoing neural current at "Out." Furthermore, all parts of the loop act simultaneously. The inside part of the loop is busy at the same time that the outside part is busy.

The action of the loop is not cyclic; if you are lifting a cup to your lip, your eye does not wait to see how accurately your hand has moved upward before sending another perceptual signal forward for correction. Signals are continuous, one sort telling the position of the cup in relation to the lip at all moments, another the rate of movement toward the lip, others the muscular accelerations that keep the arm in positions to provide good leverage on the cup, and whatever others are needed for the total operation. The arm does not stop to wait for another output signal; the eye does not wait for another movement from the arm; the correction does not wait for another signal from the eye. All currents inside the nervous system are continuous. All feedback action outside the person is continuous, even when the needed action consists of "no action"—such as continuing to sit still (which actually requires a good deal of muscular action to stay in position). The functions in the loop are consecutive, with causal effects traveling only in one direction from one to the next, but all functions are active simultaneously, with the functions inside the person all sending neural currents to the next function at the same time.

Here I had better interpose a couple of subtleties. For one thing, you might have been wondering how I could be talking about a continuous neural current in the face of the fact that a neuron does not put out a continuous current, but "fires" in discrete pulses; it sends out an electrical burst and then must wait through a "recovery period" before it can fire again. The functions, however, in a neural feedback loop are not connected by single neurons; they are connected by bundles of parallel fibers of neurons. A connection to a muscle, for example, may be effected through hundreds of neural endings at the muscle. The effective unit of action in the loop is not the cell's firing, but the neural current, which Powers (1973, p. 22) defines as "the number of impulses passing through a cross section of all parallel redundant fibers in a given bundle per unit time."

For another thing, you might be wondering how the loop gets completed through the environment when the person cannot think of anything to do in the environment, just now, to move closer to a goal. In the case of the two whirling ice-skaters, the perceptions are easy to imagine. Each person wants to see the partner at the right distance and feel the right tugs on arms and legs, the right pressure of foot against ice (against the sole of the skate-shoe, actually), the right perception of movement over ice, and so on. It is easy to imagine the continuity of all those signals. But what about the goal of winning the first prize? The dance is over; the partners go to the sidelines to await the judgment of the judges. What is the loop doing while they are sitting there waiting? The loop is doing just what the skaters are doing—waiting. To reach the goal of winning the prize, the skaters must wait to hear the announcement of the judges. They have a wide choice of things they might do. They might go back to the ice and do a few more whirls, hoping to get extra credit. They might hold up a sign to the judges reading, "One hundred dollars for a vote for us." You can think of a dozen other things they might do. Of all the possibilities, the skaters in this example chose to use their muscles to put themselves down on the bench at the sidelines and stay there until the judges announced their ratings. Doing "nothing" is doing something, too.

And a caution. Figure 3–2 is not a diagram of organs in the body, and it is not a specification of correlations; it is rather a loose diagram of the ordering of events. I'll tighten the specifics later.

The fact of an intimate connection between person and environment has surely been noted for thousands of years. But many kinds of connection can be conceived, and the crucial necessity is to specify a kind of connection that can enable a properly functioning model to be built. Unfortunately, typical statements about this connection that one finds in today's textbooks of psychology are like this one by D. G. Myers (1986, p. 409): "Our behavior is influenced by our inner dispositions, and is also responsive to the situation."

Note Myers's language: "is influenced by" and "is responsive to." With Perceptual Control Theory (PCT), we would say that behavior is the way we go about bringing our perceptions into a match with our inner dispositions; we would not say that inner dispositions merely somehow "influence" our actions. With PCT, we would say that actions chosen with which to control a perception must depend on what is available in the environment to act upon that will alter the perception; we would not say that we "respond" to the situation. Furthermore, note that syntactically, Myers's sentence is equivalent to these two disconnected sentences:

Our behavior is influenced by our inner dispositions.
Our behavior is responsive to the situation.

In this chapter, in contrast, I have tried never to separate the left and right halves of the circle in Figure 3–2. I have tried, in fact, to make it clear that either half by itself is meaningless and, more important, inoperable. But Myers reaches for a very different connection between his implied sentences. He goes on to say, "... one of the most important and long-standing questions in all psychology—is which is *more* important."

To anyone looking at person and environment as inextricably interdependent in the functioning of the living creature, that sentence of Myers's is indeed disheartening. The question of which is more important is like asking whether a horse's front legs or hind legs are more important to its running. Or whether a bird's left wing or right wing is more important to its flying. Psychological investigation is indeed forlorn when such a misconceived question is "one of the most important and long-standing questions in all psychology." The interaction between the person and environment is not one of adding one ingredient to another as in making granola or cough medicine. The interaction is of the kind shown in Figure 3–2; it is folly to call one part of that feedback loop more "important" than another. A thing cannot be a living thing without the ability to control its perceptions, and it cannot continue to be a living thing without an environment by means of which to affect its perceptions. To ask whether person or environment is more important is like asking which is the most important point of the compass or the most important tooth of a gear. The nature of the interdependence will become more clear, I think, in the next section.

But I should be clear that I did not pick out Myers's sentence as being egregious; you can find similar misconceived assertions about motivation in almost every current text. Gary Cziko (2000, pp. 169–172) finds it so, too. He gives a clear explanation of the interaction between the genetic and the environmental in his *The Things We Do*.

ASYMMETRY

I said that the two halves of the loop, inside and outside, have some similarities. But there is one way in which the two halves are critically different. The difference arises from the fact that living creatures can greatly magnify the effects of energies that impinge upon them, but the nonliving environment does not typically do that. If you slap a rock, it moves an amount that soaks up the energy you delivered to it, and that's that. But if you slap me, I might pick up a rock and crack your skull with it. Or I might use the explosive power of gunpowder to fire a lead pellet into you. About the year 1630, after gazing with grief upon his dead wife, a maharajah set thousands of people to work building the Taj Mahal.

Sixty-some years ago, some people looked upon the Golden Gate and thought about that expanse of water as an obstacle. They then expended a great deal of energy—energy that no solely physical effect of the light reflected from those waters and into the eyes of the viewers could have brought about—to bring together the efforts of other people, thousands of them, along with a great many chemically powered machines, and the result was the astonishing bridge, still doing its duty at this writing. Energy expended on the physical environment has effects in accord with Newton's laws; energy detected by the senses of a living creature can set in motion results that expend a thousand or a million times that detected energy. In brief, there is asymmetry in the effects diagrammed in Figure 3–2; the physical effect of the energy going from the person to the environment at "Out" is typically much greater than the energy entering the person at "In." The living creature is an amplifier of energy; the physical environment is not. The physical environment obeys entropy; the person does not. This asymmetry is another way of expressing the difference between living and nonliving things.

Figure 3–3 is a more precise and specific version of Figure 3–2. Figure 3–3 shows one kind of thing (we will call it **K**) that happens inside the person and another kind of thing (we will call it **E**) that happens in the environment. It also shows how those happenings (or functions) are connected—how they affect each other. I borrow Figure 3–3 and my explanation of it from Powers (1989, pp. 251–252). I could simply tell you in words about the functions and connections, but I hope the figure will serve as a summary to which you can refer without having to hunt through a paragraph or two for the connection you might want to recall. Another virtue of the diagram is that it carries far fewer of the superfluous connotations to which words are susceptible.

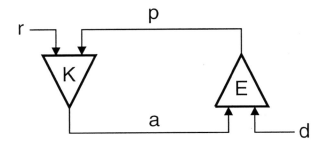

Figure 3–3.

Figure 3–3 is not an arrangement of correlations such as the "path diagrams" that appear in psychology books. It is not a diagram of bosses telling underlings what to do, as in an organization chart. It is very like a wiring diagram, a tracing of electrical circuits. Like Figure 3–2, this figure shows something (**K**) going on in the person and something (**E**) going on in the environment, but this diagram is more detailed; it shows certain physical quantities and functions. We can put numbers on the quantities connecting the functions and therefore show the relations among the functions mathematically. It is not yet as detailed as similar figures you will see in later chapters; it contains just enough functions and connections to make clear the asymmetric relation between person and environment. I will now walk slowly around the diagram with you, even at the risk of seeming tedious to some readers.

In Figure 3–3, **K** and **E** stand for agencies, or functions, or places in the circuit where a change can be made. The symbol **E** stands for the function in the environment through which the opposed forces of the action **a** and the disturbance **d** result in an energy that can be sensed by a sensory organ and be converted into the perceptual signal **p**. (The sensory organ is only implicit in this diagram.) The symbol **K** stands for the function in the person through which the perceptual signal **p** is compared with the reference signal **r** to produce the right quantity of signal **a** that sets off the muscle action against **E**. It is in the action of muscles, of course, that the person makes use of the energy provided by taking in food and oxygen. The diagram is meant to correspond one-to-one, to be homologous, with actual neural functions and currents—though it is not meant to be a picture of them. Please read all diagrams in this book having lines with arrows on them in this same way—as connected functions.

In Figure 3–3, think of the lines as neural channels or fibers that carry pulses between places where things happen. **K** and **E** are the places where things happen; they are the *functions* in the loop. Because pulses flow at a certain rate, and because forces and energies can be measured in the environment, we can apply numbers to the "signals" or energies that flow along the paths symbolized by the lines labeled **r**, **p**, **a**, and **d**. The relation between the person and the environment, remember, is maintained by what the person does with the energies taken in. Food and oxygen enable the person to move muscles; other energies, sensed, enable the person to stabilize (control) his or her relation to the environment. The uses the person makes of the environment are continuously disturbed by the environment, and the person acts continuously against the disturbances to maintain perceptions that conditions are what the person wants them to be. By putting that idea into a quantifiable model that will actually run (in the sense that a machine runs) organized as in Figure 3–3, we can test whether we have understood how it is possible for a person to maintain a stable relation to the environment—that is, to survive disturbances and thereby stay alive.

The signal **a** activates muscles to affect the environment E. There are also processes or events going on in the environment that affect the results of the action **a**. (When the connection from K to E runs along neural paths, we can call the connection a *signal*. When it eventuates into an action on the environment, we can call that same connection an *action*.) For example, the action **a** points the automobile ahead along the road, but a crosswind **d** would change that pointing were it not for the fact that the effect of the wind on the car is sensed (as represented by **p**), and the action **a** is modified to keep the automobile pointed the way the driver wants it.

The perception **p** is compared with the reference signal **r** by the person K. The reference signal **r**, for example, could be the perceived distance wanted between the front corner of the automobile and the edge of the pavement. The function **k** converts the comparison to a quantity **a** (by subtraction, for example) that will act on E, despite **d**, to alter **p** so that **p** approaches **r** in magnitude. So goes the loop. Now I turn to the quantitative relations.

The output **a** from the person is equal to some constant K (representing some quantitative conversion of input and output at K) times the sum of the inputs to K (namely **r** and **p**):

$$a = K(r + p)$$

And the output **p** from the environment is equal to some constant E times the sum of the inputs to E, namely **d** (the disturbance affecting the physical variable being attended to by the person) and **a**:

$$p = E(a + d).$$

In the diagram and in those equations, you now see the "negative feedback loop," the organization of neural currents indispensable to perceptual control. Such a loop is negative because E and K have opposite signs; when E or K increases a signal, the other decreases it, and vice versa, thus keeping the whole operation within bounds. K, however, is capable of amplification, which is not obvious in the diagram. The asymmetry of the loop results from that amplification, and the asymmetry is expressed in the equations. Let us begin by solving the two "system equations" simultaneously, first putting E(a + d) in the place of **p** in the equation for **a**, giving

$$a = K[r + E(a + d)]$$

which, after simplification and rearranging terms, gives

$$a = \frac{KE}{1 - KE}(d + r/E).$$

Similarly, putting K(r + p) in the place of **a** in the equation for **p** and rearranging terms, we have

$$p = \frac{KE}{1 - KE}(r + d/K).$$

Those rearrangements of terms will make it easier for us to see the asymmetry of the loop. To do so, let us pretend, for the moment, that both K and E are very large numbers—that is, if both amplified their inputs—then the fraction KE/(1 − KE) would be very close to one (a negative one, as we would expect in a negative feedback loop), and so the value of the action **a** would depend, as closely as makes no difference, on **d** + r/E. But since we are pretending for the moment that E is a very large number, r/E would be very small, and as closely as makes no difference, **a** would equal **d**. Similarly, **p** would equal **r**. That is, the environment E would make the output **a** from the person K equal to the environment's input **d**, and the person K would make the output **p** from the environment to the person equal to the person's internal input **r** to itself. The relationship would be symmetrical, each magnifying the input from the other.

But E is not a large number. Environments do not ordinarily amplify the inputs that affect them. Because of entropy, in fact, there is usually some loss of effect; E is usually less than one. On the other hand, as long as the organism lives, K is a large number. Organisms are highly sensitive to the environment and act with much more energy than their sense organs receive. If K is large and E is close to one, then as close as makes no practical difference, we have **p** = **r**. That is, the person succeeds in *bringing the perception signal p into match with the reference value r*, the perception remaining unaffected by the disturbance **d**. But since E is close to one, *a does not come to be equal to d*; instead, **a = d + r**. The action **a** is affected not only by the disturbance **d**, but also by the person's reference value **r**. The organism's reference signal **r** affects the environment, but the disturbance **d** in the environment does not have the corresponding effect on the organism. Most psychologists know that the environment affects the person and the person affects the environment. But the effects are not equal. The person controls the perceived relation to the environment and takes purposive action to do so. The environment does not reciprocate. Any working model must contain this asymmetry.

A LITTLE FLESH AND BLOOD

So far, in describing the feedback loop, I have been writing abstractly, using diagrams that look nothing like actual flesh and blood. Perhaps it is time to remind you that I am actually writing about real flesh and real neurons. I will not get down to the detail you would see in a book on physiology or neurology, both because I don't know enough about neurology to risk writing much about it and because the fine detail is not necessary to a good understanding of the functioning of the negative feedback loop. Look now at Figure 3–4, which I have copied from Powers (1973, p. 83); with some help from Dag Forssell; it shows a single feedback loop within the body.

In Figure 3–4, the neural current runs clockwise. At the bottom of the diagram, we see the Golgi tendon receptor, which converts a degree of stretching in the tendon into neural pulses that run both to the comparator and, in branching fibers, "up" to "higher" control loops. The comparator performs the function that I called the "correction" in Figure 3–2. The comparator receives not only the perceptual signal from the Golgi receptor, but also a reference signal (internal standard) from "higher" control loops, and subtracts the one from the other. The result is the error signal, which runs down the motor nerve to the end plates that serve as the output function. When the error pulses reach the end plates, the muscle is caused to contract (but remember that this is continuous

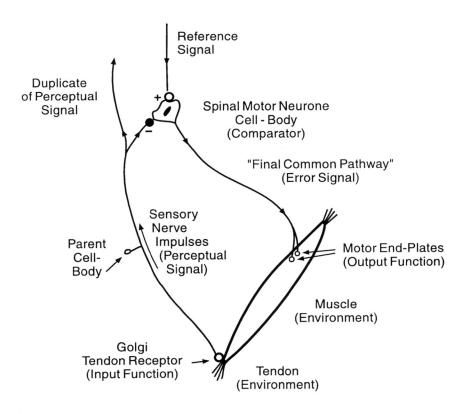

Figure 3–4.

action, not an on-and-off event). That action stretches the tendon, and the pulses from the Golgi receptor change in frequency. The muscle is the environment on which the neural system acts to control the perceptual signal (which is an analog of the amount of contraction in the muscle)—to bring the perceptual signal into a match with the reference signal.

The feedback loop in Figure 3–4, as in previous figures, runs partly inside and partly outside the nervous system. (In Figure 3–4, the muscle and tendon are outside the nervous system.) There are millions of such loops in the human body. These loops at the boundary of the nervous system make contact with the environment through output functions (such as the motor end plates on a muscle) and input functions (sensory nerve endings). Beyond these loops or control systems at the boundary are millions of further control systems lying "higher" or "deeper" in the central nervous system (brain and spinal cord). The perceptual signals passed up to the higher loops are the outputs from the lower systems such as the perceptual signal labeled "duplicate" in Figure 3–4. The outputs from the higher loops become reference signals to the lower loops such as the reference signal so labeled in Figure 3–4. The reference signals sent down from the higher loops serve to coordinate the actions of the lower loops by resetting the values of their reference signals. Loops or control systems are arranged in immense, highly interconnected hierarchies, but hierarchies nonetheless, with each loop or control system (except those at the "bottom" in contact with the environment) taking incoming (perceptual) signals upward from many lower loops and sending output (error) signals downward to lower loops where they act as reference signals for those lower loops. Inputs (perceptual signals) always come upward (brainward) from lower levels or laterally from loops at the same level; none turn to go downward. Perceptual input signals go "upward" not only to the next level, but, through branching, directly to still higher levels. Outputs (error signals) always go downward to become reference signals for any number of lower loops or go laterally to become inputs to loops at the same level; none turn to go upward to higher levels. The loops at the "higher" levels take inputs from the loops interfacing with the environment and from loops between; they send their outputs to those lower loops; none of the higher-level loops sends outputs directly to the environment.

A higher loop receives information via perceptual signals from many lower loops; it weights and combines those signals into one which, acting as a resultant perceptual signal, will be compared to its reference signal. That is the way several lower loops become coordinated; the higher loop says, so to speak, "I will pay attention to 80 percent of what loop A sends me, to 40 percent from loop B, and to 10 percent from loop C." A higher loop also sends information via error signals to many lower loops to contribute to the reference signals of those lower loops; each lower loop weights and combines those descending signals into a single reference signal.

In Chapter 18, I will describe the hierarchical arrangement of the layers of loops (control systems) more fully. At this point, I want to say only that the neural loops I am writing about are not hazy abstractions; I am writing about actual electrochemical circuits that exist in the body and brain. Though not all the necessary sorts of connections have yet been traced, you can see magnified photographs of some of the layers and loopings in texts on physiology and neurology. You should keep in mind, of course, that the actual "wiring" of the nervous system is far from as simple as these diagrams with which I introduce the idea of the feedback loop. There are millions upon millions of these busy little loops keeping your functions functioning and enabling you to fasten upon a goal and actually get there—that enable you, in short, to live. The great bulk of our brain consists of loops that deal with inputs branching up from lower loops. The loops at the lowest level receive signals only from the sensory organs, and the loops at all other levels receive signals only from other loops. All the observing of the world around us and all the thinking, talking, and book-writing we do about our experiences—all of that begins, miraculous though it may seem, with the simple pulses that go into our nervous systems from our few sorts of limited sensory organs. It is no surprise that humankind has been discovering, during the last three or four centuries, an external physical world that our senses had not led us to suspect—one requiring, to understand it, concepts quite beyond those our unaided senses had shown us.

I have been describing a figure that looks biological, but it is not pictorial; it is schematic. The diagrammatic neurons in Figure 3–4 do look somewhat like what you would see in a microscope; you can see in the diagram a blob that represents a cell-body, and you can see some long lines representing neural fibers ("processes"). But neurons branch far more profusely than the figure suggests. Furthermore, and most important, the actual loop controlling the signals from the Golgi receptors contains hundreds or thousands of neurons, not merely the paltry two or three represented in Figure 3–4. The figure is meant to show functions; it is meant to show how things work, not to show actual shapes and structures. There is, however, no doubt that there are functions in the body and the nervous system like those named in the figure and connected neurally in the order shown. I suppose every book on neural anatomy reports the existence of feedback loops—though the few I have actually read failed to tell me how the loops work.

SUMMARY

Living creatures act upon their environments with the *purpose of controlling perceptual inputs*. An organism affects the environment, and environmental energies affect the organism. Organisms amplify incoming energies; nonliving things typically do not. Organisms control perceived variables by means of negative feedback loops in the neural net. Feedback loops are arranged hierarchically, with the "higher" loops setting standards in "lower" loops.

In Chapter 4, I will begin to describe the negative feedback loop.

Chapter 4
The loop

Psychologists have talked a lot about the "black box," that being a metaphor for any internal part of an organism that you cannot see into or that seems forbiddingly complex to look into. Or that stops living whenever you cut it open to see what makes it live. Some psychologists, perhaps chiefly the sensory and the physiological psychologists, have wanted to look into the tiny parts of the human body that might carry the internal events that ultimately cause a muscle to contract and act upon the outside world. Other psychologists have disdainfully called that strategy "reductionism" and have claimed that action cannot be understood by looking at its tiny components. They say that we can understand action only by observing whole actions, and we should not only leave the black box unopened, but we should not even speculate about what goes on in there.

Indeed, early in the twentieth century, the famous psychologist John B. Watson (1878–1958) exhorted his colleagues to pay no attention to any concepts that implied internal processes—concepts such as sensation, desire, purpose, and so on. Watson called himself a Behaviorist—and wrote the word with a capital B. In 1929, Watson and a famous British psychologist of the time, William McDougall, published a little book containing the text of a debate between them. The title of the book was *The Battle of Behaviorism: An Exposition and an Exposure*. In it, Watson said:

> In 1912 the Behaviorists reached the conclusion that they could no longer be content to work with the *intangibles*.... The Behaviorist began his own formulation of the problem of psychology by sweeping aside all medieval conceptions. He dropped from his scientific vocabulary all subjective terms such as sensation, perception, image, desire, purpose, and even thinking and emotion as they were originally defined (pp. 16–17).

> Now what can we observe? Well, we can observe *behavior—what the organism does or says*.... we can keep [the animal] without food, we can put it in a place where the temperature is low ... or high, where food is scarce, where sex stimulation is absent, and the like, and we *can* observe its behavior in these situations.... We soon get to the point where we can say it is doing so and so because of so and so. The rule ... which the Behaviorist puts in front of him always is: "Can I describe this bit of behavior I see in terms of stimulus and response"? (pp. 18–19).

Watson had very great influence on American academic psychology. I do not think he had great influence on popular thought; what he took as scientific justification for a "theory" was a supposition already widely cherished by people everywhere. Throughout history, most people seem to have believed that a great deal of human action does result from what happens *to* the person. Most people still seem to believe that you can cause a person to do a certain thing by hitting her or by threatening to hit her. Or hit him first to show that you mean business, and then threaten him with a worse blow—a favorite method of people throughout history who have had soldiers at their command. As another example, many people will explain why someone does what he does at the age of 30 by the fact that in his childhood, his father was frequently out of town.

Did Watson try to influence non-psychologists? He certainly did. He wrote direct advice for teachers and parents. I am sure that teachers, parents, and others picked up a good deal of his vocabulary. I do not, however, think he had much convincing to do. I think most people who seized upon his vocabulary were happy to have a scientific endorsement for their already existing beliefs. Be that as it may, the history of academic psychology since Watson makes it seem reasonable to say that he had immense influence on the work of academic psychologists. With them, too, it is possible that he had picked up a flag that most of them were eager to follow anyway.

In the same book, McDougall had his turn:

> I place my hand upon the table, and Dr. Watson sticks a pin into the tip of one finger. My hand is promptly withdrawn; that is a behavioristic fact. I say that I felt a sharp pain when the pin was stuck in; Dr. Watson is not interested in my report of that fact. His principles will not allow him to take account of the fact, nor to inquire whether my statement is true or false. He repeats this experiment on a thousand hands, hands of babies, men and monkeys; and, finding that in every case the hand is promptly withdrawn, he makes the empirical generalization that sticking a pin into an extended hand causes it to be promptly withdrawn—and that is as far as his methods and principles will allow him to go in the study of this interesting phenomenon. He maintains with some plausibility that my introspectively observed fact of painful feeling is quite irrelevant and useless to him as a student of the human organism. But now I ask Dr. Watson to repeat the experiment on myself. He sticks in the pin once more; and this time the hand is not withdrawn, but remains at rest; and I continue to smile calmly upon him. What will Dr. Watson do with this new fact, a fact so upsetting to his empirical generalization which appeared to be on the point of becoming a "law of nature"? He can do nothing with it (pp. 55–56).
>
> The narrower formulation runs: . . . . Every human activity and process . . . is strictly determined by antecedent processes and therefore, in principle, can be predicted with complete accuracy. . . . In the sphere of human nature and conduct, this mechanistic assumption has never shown itself to have any value or usefulness as a working hypothesis. Rather, it has in very many cases blinded those who have held it dogmatically to a multitude of facts, and has led to various extravagant and absurd views of human nature, of which views Watsonian Behaviorism is one (pp. 66–67).
>
> The most fundamental fact about human life is that from moment to moment each one of us is constantly engaged in striving to bring about, to realize, to make actual, that which he conceives as possible and desires to achieve, whether it is only the securing of his next meal, the control of his temper, or the realization of a great ideal. Man is fundamentally a purposive, striving creature. He . . . longs for what is not (p. 72).

You can see that I would have sided with McDougall. (That is easy for me to say from my present PCT viewpoint. But how can I know with whom I would have sided had I been listening to the debate in 1924?) Most American academic psychologists of the time sided with Watson. McDougall tells us so in a postscript he wrote in 1927, three years after the first two parts of the book were presented in debate. During those three years, McDougall saw Watson's views welcomed by more and more psychologists. He was not happy about it:

> . . . in America Behaviorism pursues its devastating course, and Dr. Watson continues, as a prophet of much honor in his own country, to issue his pronouncements. . . . Dr. Watson, consistently pursuing his wise policy of abstaining from all attempt to reply to criticisms, has issued a new book [*Behaviorism*], a restatement of his views as bald as the palm of my hand, and more bare of any indications of regard for reason and good sense. . . . the book goes far to justify Dr. Watson's contention that his thinking processes are nothing more than the mechanical interplay of his speech-organs (p. 87).
>
> Meanwhile in America the tide of Behaviorism seems to flow increasingly. The press acclaims Dr. Watson's recent volume in the most flattering terms. One leading daily says: "Perhaps this is the most important book ever written. . ." (p. 94).

The trouble with Watson's strategy (and the continuing trouble with most research into human functioning today) is that the functions that shape what humans can do are the functions that go on inside the skin and especially in the circuitry of the nervous system. Would you be satisfied with a physician who refused to take a blood sample or an X-ray and who had never read a book on anatomy? Would you go to a radio repairman who refused to look inside the cabinet or to an automobile mechanic who refused to raise the hood?

Some people say, in talking about the functioning of a system, that the whole is not equal to the mere sum of the parts. I agree with that; everyone should. I like the example that Roger G. Barker used (recounted in P. Schoggen, 1989). Imagine, he said, a video camera focused on just one player in a game of baseball, let us say on one of the basemen. The baseman stands there a while, maybe dusts off his hands, then fastens his attention firmly on something off-screen, and suddenly puts his foot on the base and catches a ball that zips into the picture. Without pausing, he throws the ball out of the picture. Another player runs into and out of the picture. The baseman stands around some more. Imagine making a video, a separate video, of the playing of every player on the diamond, but without ever showing two or more players at the same time. Now imagine that you know nothing about baseball, and you want to understand how that game is played. And imagine that someone shows you those videos of the players in a game. You would never get a glimmering of the game until you saw what was going on with other players at the same time as *this* player. I think that is a very good way to illustrate how the purposes of the individual players are achieved in the patterns of interaction of everyone and how the game is more than an arbitrary assembly of "parts."

Certainly we can say that understanding the functions of the separate components is not sufficient to understand the whole. But the converse is true, too. Knowing the external pattern is not sufficient to understand the required internal functions. Here we need a different example, one with a black box. Think again of the example of the TV set. Suppose you obey Watson's exhortation and experiment with the TV only from the outside. You fiddle with a switch or two and a picture appears on the screen. You fiddle with a knob or two and the picture changes. Aha! you think, I am causing things to happen. All I have to do is make a careful list and I'll know what stimulus causes this "animal" to give any certain response. But whoops!—sometimes this very same knob shows us galloping horses, and sometimes pirouetting dancers. Sometimes the printed names of people come on after half an hour, sometimes after an hour or even two hours, no matter how many knobs you turn. You pull out the wires from the wall and everything stops. Does the picture come in through that wire, or is the picture already inside the box, with the wire serving only to stimulate the picture into visibility? You could turn knobs and throw switches for the rest of your life, and you would never learn about the functions of tuners, amplifiers, antennas, and so on. You would learn a little about what is twistable and about a few effects you could produce, at least sometimes. You would learn, for example, how to kill the device. But you would learn nothing about the functions necessary and sufficient *to build an apparatus that can behave that way.*

I'll say once more that what you can learn from outside a device or a living creature can be very useful. From outside a TV set, you can learn how to get some interesting pictures. You can learn how to make the sound loud and soft. After you have learned those things, you can have a good time watching the thing. To learn that much, you don't have to know a thing about what goes on inside. You don't even have to know that the shapes and colors of the pictures come into the set via electromagnetic waves from elsewhere. But if you want to *make* something that acts like that thing, then you have to know how to produce the functions that can change those electromagnetic waves into a picture that your eye can translate for you. Or in the case of living creatures, if you want to make a model that can test your understanding of the functions necessary inside a living nervous system, you must be able to build parts which, when connected in the right order into a functioning whole, will do some things characteristic of living things. A model specifies the functions the researcher believes must be components of the living nervous system, and it specifies the order of connections among the components. Running the model tests whether those components connected in that way can actually "behave."

Notice, please: I am *not* saying that if you want to find the invariant processes that function in any kind of action by the living creature, you must, after explaining outward actions by functions within the nervous system, explain functions there by functions in the neurons, and then explain those functions by chemical changes, and those by interchanges of electrons, and so on without end. Let's go back to the TV set for an analogy. For the purpose of the analogy, let's assume that the set is in good working order (as we assume about a living "subject," too) and let's assume that the power from the wall outlet won't fail (just as we assume that a "subject" is adequately nourished during a study). And to make the analogy still simpler, let's assume that no program guide exists. Playing with the set, one of the invariant effects we soon discover is that a little light goes on when you push the on-off switch (and usually you get a picture, too, though not always). Another invariant is that the picture changes when you push the channel-change buttons. But you cannot find any way of doing something that will assure your seeing a picture—Elizabeth Taylor on the slopes of Kilimanjaro, let's say, or Donald Duck in the White House. You do find, if you poke around long enough, that a particular title-screen comes on at eight o'clock: "The Joneses." But you don't know what the picture will look like when that title goes away. You know by then that certain characters are likely to appear, but you don't know when they will appear or just what they will do. Furthermore, one day "The Joneses" disappears and never comes back. You push buttons like crazy, you push them harder, you scream at the set, you kick it, but "The Joneses" won't come back.

Now suppose you learn something about the components of the set. You learn that the programs come from outside the set (just as the opportunities for actions come from outside the person). You learn that the buttons will give you channels and volume of sound, but they won't give you "The Joneses" if the network is not sending it out. That is not much new knowledge, but even that much is very profitable; right away, you can *stop trying to get a particular scene on your screen*. You will understand that it is beyond the capability of the set to give you any particular picture that you might want. You can save yourself hours of hunting. You know that you can never find a way to stimulate the TV set to respond with a picture of Donald Duck in the White House.

I pursue this analogy with the TV to show that knowing something about the components of a system enables you to draw inferences and take actions of which you would be incapable without that knowledge, and getting and using that knowledge does not require you to understand the components of the components and so on down into atoms and particles. There is knowledge at every level that is useful without having to have the knowledge at the other levels.

Maybe I should be more careful at this point about my language. I wrote: ". . . knowing something about the components of a system." But when you are faced with a black box, as is so often the case in studying living things, you often cannot "know" something about components in the sense of watching them do what they do. Often, you must be satisfied to make a guess (hypothesis, if you want to be formal) about the function you believe must be there, without knowing anything about the structure, location, or anything else about the actual, tangible component. Even with only guesses, however, you can test whether your hypothesized components could act together like an actual living creature by *building a model* containing those components. If your model behaves properly, your confidence skyrockets that you have learned something about how the living creature functions. Furthermore, you feel encouraged to look for actual living tissues that might perform the functions of your hypothesized components.

Since each level of functioning yields its own kind of knowledge and since each kind of knowledge helps us to understand the functioning at the next "higher" level, we should not fear a cry of "Reductionism!" when we turn to internal functions to learn how a creature operates. Knowing the functions inside the person enables you to know what the person can and cannot do. You can then, for example, *stop trying to predict particular acts of individuals* and *stop trying to "make" people do things*. Inside living creatures, there are no functions that connect particular "stimuli" to particular "responses."

Furthermore, to discover the internal functions, you do not need to be able to build an actual flesh-and-blood nervous system. You need only build something that will do what you think is a function inside the black box, hook it up to the others in the way you think the functions inside the black box are connected, and try out the whole thing to see whether it acts like a living thing. (That is not a simple thing to do, but it will suffice until we learn how to build flesh-and-blood neurons.) Then you can judge whether you are headed in a propitious direction. If nothing else, you know a lot about what *not* to waste time speculating about. In sum, the internal side of the feedback loop is a black box of which we do want to make a model—a tangible model that works the way a living creature works.

INTERNAL FUNCTIONS

Here I resume describing the internal functions of the loop. Figure 4–1 will serve as the map for this section. Figure 4–1 here is another form of the same loop you saw in Figures 3–2, 3–3, and 3–4 of Chapter 3. You can also find an excellent diagram, equivalent to Figure 4–1 here but laid out differently, in the engaging book by Cziko (2000, p. 77).

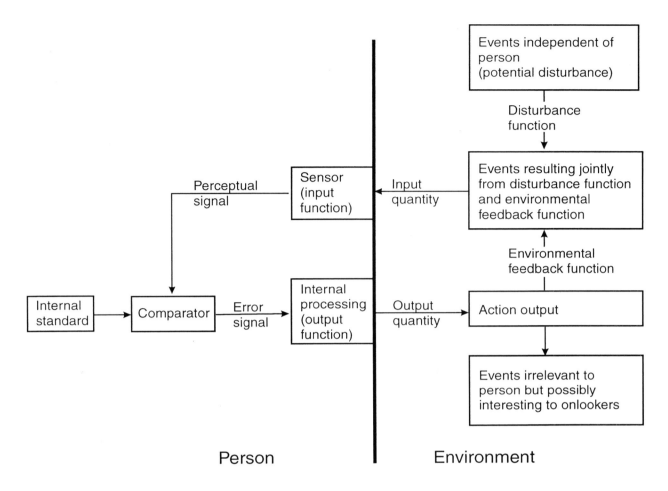

Figure 4–1.

I remember a cartoon showing two scientists standing before a chalkboard upon which one of them had written a long mathematical equation. Near the middle of the long string of characters, he had written, "Here a miracle occurs." The other scientist was saying something like, "I think you're going to have to be more explicit here in the middle." The first thing I want to say about Figure 4–1 is that I do not want you to read it as having a miracle someplace in the middle or as saying that something has some sort of vague effect on something. I want you to read it as you did Figure 3–4 to refer to specific functions that can eventually be found in human flesh. Figure 4–1 specifies some black boxes, but no miracles.

The boxes in Figure 4–1 do not stand for variables; they stand for places where functions occur; that is, where an incoming signal is transformed in some way to produce an outgoing signal. In the "person" part of the diagram, the lines having arrowheads stand for the paths of neural signals; those paths would be where measurable neural signals would be found. The sensor converts ("transduces") an external physical energy into electrochemical pulses, into the neural current we call the perceptual signal. The comparator subtracts the perceptual signal from the reference signal and sends on the difference. The output function converts the neural current into muscular contraction or glandular secretion. In the "Environment" part of the diagram, the lines having arrowheads indicate the transmission of actual physical energies. The diagram describes a model in the sense of a working analog. Many such models have actually been built inside computers; they mimic human action (in limited domains) very closely.

INPUT QUANTITY

That said, look at Figure 4–1 again, and find the arrow labeled "Input quantity." That arrow indicates energy affecting a sensor. It is called a quantity, because it can vary in amount. The effect of that energy is a direct physical phenomenon. I am not talking here about a mere correlation that leaves you wondering how an effect gets from one variable to the other. Nor about some mysterious "psychic energy," some unspecified "vibe," or other form of magic. Input quantities are simple, straightforward stuff: ordinary light, for example, that is sensed by the eye, or simple pressure on the pressure sensors in the flesh of your hand when you shake hands with someone, or pressure-and-rarefaction waves in the air that vibrate the eardrum, or the airborne chemicals in your nasal passages that excite the olfactory sensors there, and so on. Each sort of sensor transduces a particular kind of outcome from the events in the environment into pulses in neural fibers. Eyes transduce the light that is reflected from the events. Ears transduce the tickling of receptors by the hairs in the cochlea, and those hairs are set in motion only by sound waves in the air—waves that are set in motion by events in the environment such as horns blowing, people talking, or a coffee cup crashing to the floor. Olfactory sensors do not respond to pressure or light or sound waves, but only to chemicals; the chemicals may drift into your nose from a flower, or from someone's perfume, or from someone's whiskeyed breath.

SENSOR

Sensors change (transduce) the incoming energy into electrical pulses in neurons. Remember, too, that every pulse in any neuron is like every other pulse in that neuron and in any other neuron, though neurologists do speculate about "codings" (see, for example, Uttal, 1978, Chapters 6 and 7). The eye does not send a picture to the brain; it does not send "sight." It sends simple electrical pulses to "higher" loops. Some neurons will be sending more frequent pulses than others because they are receiving light of a greater intensity. The pulse in every neuron, however, and the flow of pulses in a bundle of neural fibers, are simply electrical. They convey no quality except rate of pulsing (firing). Before we can become conscious of seeing a configuration or an event, a lot of activity must go on in loops higher than those to which the eye sends perceptual signals. A lot of perceptual signals get compared with a lot of internal standards for red, for purple, for brilliance of hue, for edge-transitions between areas of color, for relationships (proportions) of dark and light—and I am not even mentioning a myriad of standards for controlling perceptions of motions—and a tremendous but ordered tangle of weightings and comparisons must be run and revised and balanced and revised again and still again before we feel that we "know" what we are seeing. And all that goes on not in an iterative manner, but continuously and seamlessly. Ordinarily, all that stupendous adjusting (control) of the perceptual signals goes on so rapidly and seamlessly that we are not aware of anything happening at all. Now and then we might squint and say, "What's that down there in the lower left corner?" Or, "I can't tell whether I'm seeing a big one at the back or a little one at the front." At such times, the standards of the lower loops fail to bring full control of the visual imagery, and higher loops come into play—loops enabling comparisons with a great many stored images and sometimes loops enabling language to be used to find more successful comparisons.

The nature of a sensor is to pass on currents of neural pulses from thousands of nerve-endings and to leave to the higher loops the task of organizing (so to speak) all those present currents along with many remembered images to produce a picture or image.

THE PERCEPTION

A line in Figure 4–1 is labeled "Perceptual signal." Through that pathway, the contact with the environment gets into the part of the loop where it can do some good, so to speak. The sensor sends electrochemical signals to the comparator, where a comparison with the internal standard can call for or not call for a rectification of the perceptual signal. If you are looking in from outside, you can think of the signal as the perception. But from inside looking out, we are almost never aware of a perception as such. We never think, "Ah, that's what the *perception* of it looks like." Instead, we think, "Ah, that's what it looks like," where by "it" we mean what we think is the real thing out there.

"Look at your hand," says Powers (1998, page 19). "There it is, with fingers and skin and wrinkles. You can wiggle the fingers, turn the hand palm up and palm down, make a fist. As you do these things, you are, of course, perceiving that they are happening. So you can see your hand and what it is doing—but where is the *perception* of your hand?" You know that the perception of your hand must be in your brain. You know there can be no perceptions without neural signals—no clairvoyance, no telepathy, no extrasensory perception. You know, too, that the only experience you can have of your hand is the experience your perceptions bring you. Consequently, says Powers (p. 21), "You don't need to look inside your head to find perceptions: *when you look at your hand, you're already looking at them.* You're directly experiencing the signals in your brain that represent the world outside you." Your experience of your hand through your eyes (or through the smell of it, or through a touch to it from your other hand) is not wispy or gossamer; it is as rich and convincing as any "humanistic" psychologist could wish a representation to be. "[W]hen we *control* something," Powers (pp. 23–24) continues, "what we control, necessarily, is one or more of the perceptions that make up this world of experience. Our only view of the real world is our view of the neural signals that represent it, inside our own brains."

Here is a word about the word "perception." The standard texts say that the sense organs, together with the brain, provide us with two kinds of experiences—sensations and perceptions. Those texts make only occasional and tenuous connections between sensations and perceptions, on the one hand, and experiences such as reasoning, planning, and acting on principle, on the other. Powers uses the term "perception" in a different way. He applies the term to any incoming neural current. In later chapters, especially Chapter 18, I will explain how Figure 4–1 can be multiplied to construct a model having many levels of control. Loops at one level control the incoming currents sent "upward" from "lower" levels. At the higher levels, perception are controlled such as configurations, transitions, events, relationships, and more. A few examples of what can be called perceptions in PCT are the taste of strawberry ice cream, the spatial relationships among soldiers on parade, a series of instructions (a program) for operating a VCR, a principle of neighborliness, a self-concept, and an explanation of how an airplane can fly.

THE CONTACT WITH REALITY

We do not "know" or "experience" external events any more intimately than through the transmission and transduction of energies such as those described above. Our experience, however, does not seem to us attenuated or diaphanous, because the energies that we do detect are those that fill our attention, and they do so in the only way we have ever known and ever can know. We have no way of comparing our experience of the energies we do detect with what our experience could be of those we cannot. We cannot know how events would appear to us if we had the eyes of a bee or the ears of a dog or the pressure-sensing organs of a fish or the navigational equipment of a pigeon. We can get a few hints. I once saw a video showing the patterns invisible to us that insects can see in a flower because their eyes respond to wavelengths ours do not. But the video, of course, sent light to my eyes that I could see, not light that only insects could see. The light patterns at the shorter wavelengths had been translated into the longer wavelengths my eye could see. I saw the patterns, but not the colors (wavelengths). Seeing the colors to which I was accustomed but knowing that the insect would see something beyond those colors was, conceivably, like seeing a black-and-white photograph of an orchid instead of seeing an orchid. Or like trying to understand the life of a fish after swimming under water for part of a minute. Or soaring in a glider and trying to feel the attitude of an eagle looking for prey.

We know by our senses only a small portion of any reality. We know the narrow band of light that comes to us from an orange, we know the feeling of weight and inertia as we handle the orange, we know the feel of its rind and of its inner flesh to the sensors of our hands as we peel it and break apart the sections, we know the texture as we chew it and the taste of the juices, we know the feel of swallowing the pieces, and we know the swelling of the stomach. Once in a while we might hear some very small sound as we peel it. We know something of the effects on taste of combining it with other foods. We know by our senses those sorts of sensory experiences, and that's about all. Except for the small range of visible light, we sense nothing directly of the vast ranges of electromagnetic waves reflected from the orange. We know nothing of the sounds from it too faint for our ears, nothing of the effects of the cosmic rays passing through it, nothing of the tensions in its membranes when we squeeze it, nothing of the bacteria living within it. I can list those experiences we do not have (but which some insects, perhaps, might have) only because of the knowledge *about* things that have come to me from people who have studied the spectrum of light with instruments that extended the powers of their eyes, from people who have looked through microscopes at bacteria, and so on. Neither I nor those people can ever know those things with our unaided senses. But even with those instruments to aid our senses, our knowledge remains fragmentary. The rest is words. Or diagrams of an artist's imagination. I "know" there are bacteria in the orange not because I have ever seen any there, not because of my faint memory of having seen some squiggly blobs through a microscope in a high-school laboratory, and not because I have tasted a faintly acid taste when biting into a soft spot on an orange. I "know" about the bacteria only because several people have told me (with words and diagrams) that I should interpret my limited experiences with oranges and with that long-ago microscope as if there were invisible creatures making their living in the orange.

We are not born with a brainful of internal standards with which to make sense of all the sensory experiences we may later encounter. The infant must learn to fit together the signals from the various kinds of neurons in the retina in ways that result in variable neural signals that can be controlled by changing the size of the iris, by moving the eyeball, eyelid, neck, and so forth. It is not surprising that it takes some time for the infant to be able to recognize a cluster of visual, auditory, tactile, and olfactory patterns as a repeating cluster, and a lot longer for the infant to put "mama" together with that cluster.

Even the experiences that feel very "sensory" to us, such as the taste of an orange, come to us from loops some distance "up" from the loops containing the sensors. The taste of an orange does not come to us from a sensor devoted to detecting oranges. It comes from higher loops that get perceptual input from lower loops—from the loops containing the sensors (as in Figure 4–1) in the taste buds that detect various chemicals. The sensation we call "sweet" comes to us from the perception of chemicals such as sucrose, sodium saccharin, and others that excite certain of the neurons in the taste buds and not other neurons. Similarly, other neurons send upward signals from other chemicals. The few books on sensation that I have looked into give me only indirect evidence somewhat mixed with speculation, but they seem to be claiming that the higher loops make complex tastes out of only four kinds signaled by the lower loops: sweet, sour, bitter, or salt. We taste an orange, presumably, when the higher loops are "reporting" a particular combination of those four basic tastes. A friend has told me of reading about some very recent research claiming to find two more tastes just as basic: astringent and umami; maybe next year's texts will include those.

Many substances seem to us "tasteless." Like the substances we do taste, they too are composed of chemicals, but the neurons at our taste buds are not sensitive to those chemicals. What would an orange taste like if our taste buds told us about some of those other chemicals? What would our foods taste like if we could taste more of the things that are poison to us? The world of taste, like the world of sight, is a tiny portion of what is there to be tasted. Our experience of it is small and biased.

Our knowledge of the world rests at bottom on a very small foundation of sensory experience—small in comparison with what is surely there to be sensed by more commodious sense organs. We can be confident of the existence of that wider world because of the reasoning we do *about* the experiences we do and do not have, such as the simple reasoning in the previous paragraph, but we rarely reason about the matter at all. We are confident that there is something real out there to be perceived because events so often turn up where we expect them to turn up. We see a glass of water, and when we reach for it, the sensation of touch in our fingers confirms what we saw. I call a friend and arrange to meet her at 3 P.M. under the clock at the Vanderbilt. The two of us read our watches sufficiently in the same way, and we have sufficiently the same map of the city in our minds and of the place of the Vanderbilt on the map that we find each other in the agreed hotel lobby at the agreed time.

Not only do the perceptions of two or more persons very often confirm the shape of the external world, but we also find evidence of the external reality when the evidence of our senses contradicts our wishes. When we go downstairs in the dark and imagine there is one more step where there is not, the reality jolts us. It doesn't matter that we would prefer not to be jolted—that we would rather that last step had been there. When someone gets run down by an automobile, it doesn't matter whether the person saw the automobile coming; the smash is just as mangling. We can resist and reshape the physical environment in many ways, but there are limits. We cannot reshape the environment at all unless we conform to the physics and chemistry of doing so.

INTERNAL STANDARD AND COMPARATOR

What makes people act? That question is an example of the trickery of language. It implies that something pushes upon an otherwise passive creature. A better question (not the only better question, but one of the better ones) is: Toward what do people act? People act when they find a discrepancy between the condition they perceive themselves to be in and the condition they want to be in. They turn the faucet when the water is too cold or too hot. They act "toward" their goals. In PCT, motivation lies between the two arrows you see in Figure 4–1 pointing into the box labeled "Comparator."

Every loop contains the comparator function. The comparator subtracts the perceptual signal from the reference signal and sends the resulting error signal to the output function. If, as a result of ensuing action in the environment, the incoming perceptual signal comes closer to the standard but does not yet reach it, the error signal continues at a value greater than zero, calling for continued action—though a lowered error signal may call for slowed action. If the perceptual signal becomes farther from the standard, then, depending on the consequence desired, higher loops may alter the internal standard so as to call for more intense action from this loop, or they may call for no action at all. Higher loops are not shown in Figure 4–1; I will describe the hierarchy of loops in Chapter 18. Here I will ask you to be content with mere hints about the hierarchy of control systems (loops) so that I can continue without delay this first journey around the archetypal loop diagrammed in Figure 4–1.

The topic we have come to in these paragraphs is motivation, a topic that has long been important to psychological theorists, who have argued and argued. In PCT, motivation is still central, but it is simpler conceptually, and it is no longer a separate topic: behavior is always action to control perception. Whenever we wish to control a perception, we must act. (Sometimes the "act" is hard to see; I'll return to that matter in later chapters.) In PCT, motivation lies in the negative feedback loop. You might say that motivation is a name for the way the loop is constructed. Or you could say that motivation is a name for the "rules of motion" among living things.

In the past, motivation has been made into a puzzle by the assumption that an outside force must move us. Think of anything you have felt forced to do, or that someone else has seemed forced to do, in your own acquaintance or in your knowledge of history, and you will be able to think of some other thing that you or the other person could have done in that instance. It is true that swords and guns are very persuasive. But there are times, not rare, when people do put other values before pain and even death. Barbara Frietchie, according to John Greenleaf Whittier, stared into the muskets and cried, "Shoot, if you must, this old gray head, but spare your country's flag!" Every day, somewhere, a parent runs in front of an automobile to rescue a child.

The comparator is the heart, the core, of PCT. We act when what we perceive to be happening to us does not measure up to our standards. We add pepper when the food is not spicy enough; we push the food away when it is too spicy. We ring a friend's doorbell when we have not had enough of his company. We ask the student to let us see a revision of his essay when what he has written does not yet match our criterion for an understanding of the topic. We say, "I am not yet satisfied," to explain why we are still practicing Beethoven's *Für Elise*. We say, "Yes, I'd like a little more, please—about half a portion" to describe our desire for still more rice pudding, but not too much more. We say we would like to be home on the range, where we seldom hear a discouraging word. Such statements, clearly, indicate an internal standard and a perceived mismatch.

(Instead of "internal standard," Powers prefers the term "reference signal"; that is the term used by electrical engineers when they talk about control circuits. That term is fine with me, and you will find it repeatedly in this book. I also like "internal standards." You can invent your own label; pick some word with the flavor of criterion, goal, intention, preference, purpose.)

This is what we mean by pursuing a *purpose*: moving a perception from where it comes upon us to where we want it to be, or keeping it where we want it once it is there. Purpose resides in the comparison between perception and standard. PCT does not claim to be unique in having the concept of purpose as a cornerstone or in defining purpose as a discrepancy. PCT is unique, however, in conceiving purpose as a component of a negative feedback loop in such a way that a quantitative, mathematical model can be tested and the degree of success quantified for a single individual. I do not want you to conclude, when you see the word *purpose* here, that PCT has some significant similarity to older theories just because the word *purpose* or some synonym for it appears there also. Why should I care if you connect the term *purpose* (or perhaps *comparison* or *dissonance*) with other theories? If you have read little about psychology and do not intend to read more widely in the literature, then I do not think I need worry. But if you have already read a good deal of psychological writing, or if you intend to do so, then I feel compelled to issue a warning about what you will find in that wider literature. I do not say that you will be somehow wrong to read it or that doing so will endanger your soul; I say only that the literature may seem to say more than it does. By the way, isn't that a strange way to talk? How can it seem to say more than it does? I suppose I mean that I hope it will not seem to you to say more than it seems to me to say.

OTHER APPEARANCES OF PURPOSE

To compare a theory fairly with PCT, you should look first to see whether the theory has been tested by actual *physical modeling of individuals*. If the author claims that has been done, it will then be worth your while to read on to see whether the theory specifies a negative feedback loop along with circular, simultaneous causation and whether it contains the functions that appear in Figure 4–1. If the theory lacks any of those requirements, it is not a competitor of PCT. To make clear the ways other theories typically fall short of those requirements, I will describe some appearances of the idea of *purpose* in other literature and explain what is missing in those other conceptions.

The idea of an internal standard which, when compared to a perception of what is actually happening, will motivate action to bring "reality" or a perception of it into congruence with the standard —that idea is to be found not only in commonplace, everyday thought, but, not surprisingly, also in the literature of psychology and other social sciences. You saw at the beginning of this chapter that the British psychologist William McDougall thought that way in 1929. And I could have mentioned that the American William James thought that way in 1890. But in case you dismiss those two as old-fashioned and therefore possibly feeble-minded, I'll describe briefly some of the places the idea has appeared more recently in the literature of psychology.

To turn to more modern times, I rely on the thorough review of the literature done by Mortimer H. Appley, who wrote a long chapter in 1991 about the widespread appearance in the psychological literature of the "equilibration theory of motivation." On page 8, Appley wrote:

> The theme of maintenance or restoration of equilibrium, balance, stability, or consistency after disequilibrium pervades the literature of motivation, development, personality, social psychology, and psychology generally. It describes "self-regulatory processes functioning through negative feedback mechanisms to reduce the differences between some preferred internal state and the organism's current state"....

On page 3, Appley named "four quite different homeostatic/equilibratory models": (1) G. L. Freeman's *neuromuscular homeostasis*, (2) Harry Helson's *adaptation level*, (3) Wiener and Ashby's *cybernetics and feedback models*, and (4) Lewin's *psychological field theory*. Appley classified Powers's (1973, 1973a) PCT under cybernetics and feedback models. On other pages, Appley described or mentioned more than a couple of hundred articles and books dealing with "equilibration theory." He also described a 1964 book of his (with Cofer) on the same topic containing almost a thousand pages and a great many more references. His review reached back to Claude Bernard in 1859, though publications on the topic accelerated after about 1940. Appley demonstrated thoroughly and incontestably the fact that the idea of equilibration or comparing input with internal standard has been widely available in the psychological literature for several decades.

Unfortunately, Appley wrote in the usual manner of the academic reviewer of psychological writings—as if every theory is an exhibit in a sort of zoo, every one alive and worthy of the attention of the passerby. Appley missed the actual modeling against which PCT has been tested—that is, the *quantitative* comparison of the behavior of a human individual with the "behavior" of a model in a computer. This sort of tangible, quantitative model-building to match human behavior occurs in some fields of engineering, but in no field of psychology (not even in neurology, as far as I can tell from my reading of a couple of texts in that field) except PCT. Appley missed, too, another crucial difference between PCT and almost every other field of psychology: the use of the appropriate mathematics to yield a model testable with a single subject; that is, simultaneous equations (to represent effects from both organism and environment) and the integral calculus (to calculate quantitatively the expected motions). Almost all other experimentation in psychology uses only the mathematics of statistics for testing outcomes, and statistics cannot describe a dynamic model of the behavior of a single individual; it can only count instances. This matter of tangible modeling is a difference that goes beyond the personal preferences of the strollers in the zoo. Letting the evidence of tangible, measurable facts outweigh the preferences and desires of the researcher is the principle at the foundation of science. Researchers in psychology should be pledged, I think, to pay more respect to a theory that can model accurately every individual than to a theory that can make no tangible model and can predict the behavior only of a nonchance proportion of actors during a small moment of social history. Finally, Appley missed the crucial distinction between linear, successive causation and circular, simultaneous causation.

If it seems strange that Appley missed those crucial distinguishing features of Powers's work, a possible explanation may be discernible in the fact that Appley made no comment of his own about PCT; the work of Powers was mentioned in Appley's article only in quotations from other authors. The authors quoted by Appley mentioned Powers's fundamental book of 1973 and his article in *Science* in that same year. By the time those authors were writing (a book in 1981 and an article in 1989), the important 1978 article by Powers had appeared in the *Psychological Review*. By the time of publication of Appley's review (1991), at least a dozen more papers by Powers were in print, and Marken had published nine reports of experimentation on PCT beginning in 1980. One pair of authors quoted by Appley published their book in 1981, too soon to have mentioned the later experimentation. The other pair of authors wrote in 1989, but may not have mentioned Marken or the later writings by Powers.

Also published in 1991 was the book *Feedback Thought in Social Science and Systems Theory* by George P. Richardson. While Appley's article stayed almost entirely within the mainstream psychological literature, Richardson's book inspected many sectors of biological and social science and paused for a moment now and then in engineering. Richardson gave considerable space to Powers's 1973 book and a few of his earlier writings. In my reading of Richardson's book, I was surprised to find no recognition of the fact that Powers and his followers have been the only persons to have built actual operating models that closely mimic the functioning of living creatures—with the possible exception of Ashby (1952). Nevertheless, Richardson's book is fascinating for its own sake, and it serves as an excellent intellectual history of PCT.

Reactance

One line of research mentioned by Appley, one that attracted for a while considerable attention from researchers, is the research on *reactance* initiated by Sharon S. Brehm and Jack W. Brehm (1981). The Brehms were impressed (as I am, too) with the care we all seem to take to maintain a margin of free movement or free choice. We say we want to keep our options open, to maintain elbow room, to have room to maneuver, to have a margin of safety, to be free or untrammeled or emancipated. The opposite of that free condition, of course, is restraint, coercion, and confinement. I have used phrasing here that may connote to you primarily physical or bodily freedom or restraint, but the idea applies to mental matters, too, such as freedom of speech or other expression. The Brehms (1981, p. 4) say:

> ... a threat to or loss of a freedom motivates the individual to restore that freedom. Thus, the direct manifestation of reactance is behavior directed toward restoring the freedom in question.

I take some extra space here for the Brehms, because they were explicit about action to oppose a disturbance. They were not explicit about maintaining *perceptions*, but they were explicit about maintaining desired conditions. The Brehms did not actually write explicitly about internal standards, either by that name or any other, but you can see in the quotation above a clear implication of an internal standard guiding the action: "behavior directed toward restoring the freedom." The Brehms did not discuss, either, the possibility that people can form an internal standard concerning any perception whatever; they wanted to concentrate on what seemed to them the internal standard of wanting to maintain some minimal room to choose acts or thoughts.

In their book, the Brehms devoted a lot of space to various kinds of situations in which they might expect stronger or weaker reactance. They did not offer any detail (such as Figure 4–1, for example) about how reactance might be connected to actions. They recounted many studies of situations in which they expected people to exhibit reactance. Typically, a study compared the behavior of persons who, in the judgment of the experimenters, had their options reduced or threatened with the behavior of those who did not. For example, the Brehms (1981, p. 351) recounted a study by Edney, Walker, and Jordan (1976), who went to a beach where people were sitting on the sand. They interviewed some of them (the Brehms do not tell us how many) about their feelings of "control over the situation" and about how far the area of sand extended that they felt "to be their own." The experimenters also measured (the Brehms do not say how) the actual distance to the nearest neighbors on the beach. But I won't take space to tell more about the studies reported.

In no study reported by the Brehms was there a search for a controlled variable. The experimenters chose or constructed a situation in which they thought most people would feel a loss of freedom of choice and then counted the people who acted to maintain or restore that freedom. No model (in the PCT sense of the word) was tested with any individual; the averages or proportions of groups were the numerical data, as is still customary in almost all psychological studies.

The Brehms did not discuss control of perception. They seemed, as far as I can tell, to be discussing control of the presumed objective situation, and took for granted the perception of freedom in it. They wrote, in effect, as if the subjects in the study would act to maintain or restore freedom of choice if the *experimenters* perceived in the situation a threat to the subjects' freedom—an idea quite foreign to PCT. The Brehms' theory differed from PCT, too, in their insistence that ". . . the person must *know* that he or she *can* do X" (1981, p. 12) if the person is to show reactance. In PCT, awareness can affect the particular action the person takes to counter a disturbance, but awareness is not a necessary feature of control.

Dissonance

Dissonance theory guided the research of many researchers in social psychology for some years. The original book under that label was Festinger's (1957). Other important early work was that of Heider (1946, 1958) and Newcomb (1953, 1959). From the flood of writings that ensued, I'll mention only my own piece in 1956a and the serviceable collection by Abelson and others in 1968. Cooper and Fazio (1984) reviewed later research. Most of the writing was devoted to the simplest form of the theory; an insightful but largely ignored proposal for a more sophisticated theory was that of Newcomb in 1959. Runkel and Peiser (1968), using symbolic logic, showed the simplest form of the theory to be in fact trivial.

Briefly, dissonance theory holds that if a person finds that two perceptions seem to contradict each other, the person will do something to remove the contradiction (dissonance) and bring the perception of the situation into consonance. The idea of an internal standard was not usually explicit in this research, but you can see that one perception could serve as a standard for the other, or the standard of the logic of classes could be at work.

Those researches and many later ones were of the nose-counting sort (using the method of relative frequencies), and do not demonstrate any invariant about human functioning. They demonstrate only that some people in some situations can behave in ways that those words seem to some of us to fit. But the researches do hint that with proper use of PCT method, people might be found to be controlling a perceived variable similar to what all those researchers have called cognitive dissonance.

In 1989, Berkowitz and Devine wrote, "One of the high points of the 1987 American Psychological Association meeting was the symposium commemorating the 30th anniversary of dissonance theory" (p. 493). Berkowitz and Devine went on to say that dissonance theory was losing its popularity among researchers. They wrote on page 502, ". . . we tabulated the number of articles explicitly dealing with this theory in the 1967, 1977, and 1987 volumes of the *Journal of Personality and Social Psychology*. . . . there were seven such articles in 1967, two in 1977, and none in 1987." (Some high point!) Berkowitz and Devine said that they thought the "cognitive perspective" was supplanting dissonance theory.

As far as I know, some psychological researchers may still be occupying themselves with cognitive dissonance, though it was back in 1974 that William Dember announced it as "the cognitive revolution," and it may be outstaying its welcome by now. In his article, Dember paid special attention to the powerful effects of political and religious ideology, plainly matters of internal standards.

Living Systems

Oddly, Appley omitted mention of J. G. Miller's monumental *Living Systems*, which appeared in 1978. Miller included negative feedback loops as one of the "adjustment processes among subsystems or components, used in maintaining variables in steady states" (p. xxviii). He described (on pages 462–464) numerous feedbacks in the human body, both neural and endocrine. He wrote of the *echelon*, a concept similar to Powers's *hierarchy*, though Miller seemed to think of echelons, as can be seen on his pages 423–424, more as patterns of connections among clusters of neurons than as controls for types of perceptions. On page 29, Miller wrote:

> In living systems with echelons, the components of the decider . . ., an information processing subsystem, are hierarchically arranged. Certain types of decisions are made by one component of that subsystem and others by another, each component being at a different *echelon*.

Furthermore, Miller described some concrete, tangible models; see, for example, his pages 493–499. And his description of motivation certainly leaned toward the cybernetic:

> Motivation . . . is a message or messages from a lower echelon or subsystem that a higher echelon or system at a higher level should carry out an action to restore some steady state or maintain one that is threatened. . . (p. 429).

At first glance, that looked to me upside down. In PCT, actions are carried out by the lowest level, not by higher levels! But I think my trouble in understanding lay with what Miller meant by "carry out." On page 425, Miller wrote, ". . . in human beings higher brain echelons . . . can on occasion take over control of processes usually regulated by lower echelons. . .". That sentence is right side up, though it shows again how Miller missed the idea of controlling *perceptions*.

Miller noted feedbacks meticulously in cells, organs, and whole creatures (and even in groups and organizations) and saw motivation cybernetically as purposeful. Yet motivation was a very small topic for Miller; in his book of more than a thousand pages, he gave motivation about one page in his chapter on the whole organism and barely mentioned it in a few other places. I think Miller conceived the organism to consist of 19 subsystems, each performing its particular function (boundary, distributor, transduce, decider, and so on), and the whole somehow managed by the decider. But Miller's prodigious tome left me wondering what I should do with all that information. In his concluding chapter, Miller seemed most concerned about whether his conceptions of subsystems were fitting descriptions of all of his seven "levels": cell, organ, organism, group, organization, society, and supranational system. As to what we should expect from living systems, what we should be ready to observe, or how we should try to interact with them, he says there on page 1025 only that his theory "is eclectic. It ties together past discoveries from many disciplines and provides an outline into which new findings can be fitted." Well, I could see how Miller could reasonably classify this or that study's finding under one or another of his subsystems, and for a year or two after I read the book, I sometimes thought about those subsystems, but Miller's scheme never pointed me in one direction instead of another in my research or in my organizational consulting. PCT tells me immediately what I should *not* try to do in research or in practical life; Miller's Living Systems theory did not do even that.

Though Miller gave serious attention to feedback loops at all his seven levels and frequently wrote of goal-directed action, I had the feeling always that I was getting incidental or peripheral information. He never told me that the negative feedback loop with its internal standard was the key to understanding the nature of the beast. He would tell me about some feedback loops and then, on the same page, he would tell me of some findings from traditional (correlational) stimulus-response research as if the two were unquestionably compatible; that is, he seemed to me to be giving equal billing to circular and linear causation. In his final summary of the characteristics of living systems on page 1027, Miller again wrote almost as one might expect a follower of PCT to write, but then somehow did not see that all that feedback he described enables the organism to *control perception*, did not note that the tangible models he described were unable to operate in an unpredictably changing environment, did not see the uniqueness of the internal standards of individuals, and did not see the vital difference between the stance of the PCT investigator trying to see how behavior is chosen from inside and the traditional stimulus-response investigator trying to find the average behavior of *other* people that conforms to his own view of how behavior is shaped from outside.

Other Mentions of Purpose

So far, I have been exhibiting writings that implied "equilibrating" processes. I will turn now to the less specific literature which nevertheless specifies or implies purpose.

Psychologists still seem divided as to whether the concept of purpose is worth attention. Those who follow J. B. Watson's lead shudder at the very thought. Others omit purpose as a requisite feature of motivation, but may, for example, interpret experimental results by speculating about what subjects might have been "trying to do". Many other psychologists deal explicitly with purposes (intentions, goals, aspirations, needs, preferences). A tradition in psychological theory reaching back to the 1890s and giving explicit attention to purpose is *functionalism*. Mark Snyder (1993, p. 254) wrote:

> . . . functionalism was the purposive psychology of [what an act] *is for*. . . . Functionalist themes pervade psychological perspectives as diverse as psychoanalysis, behaviorism, psychobiology, and evolutionary biology—each of which emphasizes, in its own way, the adaptive and purposive pursuit of ends and goals. . . .

I suppose many psychologists would disagree with parts of those sentences, but the quotation shows again how the idea of purpose pops up, sometimes unexpectedly, in many branches of psychological theory.

Finally, to be sure I was reasonably up to date with my remarks in this chapter, I spent half a day in my university's library looking at books published in 1990 or later—introductory texts and books on general psychological theory. In that time, I got about halfway through the alphabet of authors and found myself with eleven books to examine[1]. In each, I looked at what the author said about control, feedback, circular causation, and perception. Five books mentioned W. B. Cannon's (1932) idea of homeostasis; two of those each devoted a sentence or two to the topic, the other three about a page. Three books mentioned control, but only in the sense of control of the environment and other people, not control of perception. Four books mentioned feedback, but used the word to mean merely one person giving information to another. One book printed a diagram showing a feedback loop through a muscle, but said nothing about it in the text. Two books contained a few pages on biofeedback. One book had several pages on causality, but no mention of circular causality. In the book by Gray (1991), Figure 6.9 looked like this:

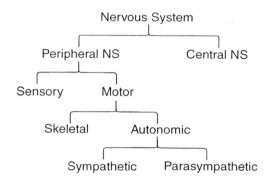

"Sensory" nerve fibers are those leading inward from sense organs; "motor" fibers are those going outward to muscles and glands. The diagram implies (to me, anyway) that Gray thought the Central Nervous System (the brain) has no fibers reserved for input or output. And the diagram also implies (to me) that the skeletal and autonomic parts have no fibers devoted to input signals. I don't think such a nervous system could keep us going. I have no idea what the meaning could be of the lines connecting the words in that diagram.

Books that apply psychology to practical affairs have long dealt with purpose without apology. An example is the psychology of sport; another is the field of organizational management. Athletes are always assumed to have purposes, and so are managers in business and industry. In the latter field, even the workers are sometimes assumed to have purposes.

I think I have now devoted enough space to the point that the psychological literature of recent decades has contained a good sprinkling of publications containing the idea of action being taken to bring a perception into congruence with an internal standard. And I hope I have made it clear that a good idea or two is not sufficient to compose a theory that can be tested quantitatively with a single individual living creature, human or otherwise. It is a good idea to note the widespread occurrence of "equilibration" of various sorts in human behavior, but noting that much is not sufficient for selecting the proper mathematics and building a working model to demonstrate equilibration. It is a good idea to note that people often react against incursions upon their freedoms, but that is not enough for building a model, either. Now that Powers and other researchers have shown us how to build models with PCT, we can state the minimal requirements of theory that can produce a working model; namely,

1 a negative feedback loop
2 circular and simultaneous causation
3 functions arranged so that the perception is controlled to match a reference signal (internal standard).

A theory with those features is the only sort that so far has enabled a working model to be built that will control a perception (hold to a purpose) despite unpredictable disturbances to the perception from the environment. Let us return now to our journey around the loop.

INTERNAL PROCESSING

Look again at Figure 4–1. The comparator delivers an error signal. If the incoming perceptual signal matches the reference signal (the internal standard), then the value of the outgoing error signal (in rate of pulses) will be zero; otherwise, the value of the error signal will be greater than zero. The greater the error signal, the greater will be the effect on the output function, the output quantity, and then the action output.

Now time out for a technical note. I put this here not because it is vital to your understanding of the connections among the functions in the feedback loop, but because I want to remind you that technicalities do arise in building a model, as this small example will show. I wrote in the preceding paragraph about the perceptual signal's "matching" the reference signal. What do I mean by "matching"? At the comparator, the value of the perceptual signal is subtracted from the value of the reference signal, and the difference goes out as the error signal. But by "signal," we mean a neural current measurable in pulses per second. If we have a reference signal of 1000 pulses per second and an incoming perceptual signal of 600, then the error signal is 1000 minus 600 or 400 pulses per second. But if the incoming signal is 1200, we cannot say that the error current will be 1000 minus 1200 or a negative 200. A negative neural current is a meaningless phrase. All neural currents flow away from the cell bodies, no matter what sets them off, so the idea of positive and negative directions cannot apply. The comparator has no choice, so to speak, but to interpret any difference that would yield a negative number (I'm not saying that the comparator works with numbers) as zero. To be properly precise, therefore, the sentence about matching would have to read like this:

> If the incoming perceptual signal is equal to or greater than the reference signal (the internal standard), then the outgoing error signal will be zero. If the perceptual signal is less than the reference signal, then the error signal will be greater than zero.

And now I'll add a small complication to this small technicality. The description just above is fine for many kinds of variables; for example: Are my muscles able to counteract the force with which gravity pulls this object downward? As long as I can pull upward with that force *or more*, the answer is yes. Or suppose you call out, "Quiet! Quiet! I think I hear something!" Presumably the other people cannot then be *too* quiet to suit you. But other kinds of variables have reference values that can lie between too little and too much. I like my coffee, for example, not too cool and not too hot. A model could cope with that by having one system send out a non-zero error signal if the sensation from the coffee is too cool, and another system if the coffee is too hot. You can think of further ways that a model could cope with such a nonmonotonic reference value.

Let us go back now to the loop.

The error signal actuates the output function, which is the transducer between the neural net and the world outside. Output functions (or transducers) activate muscle contractions (as indicated in Figure 3–4); they also regulate the production of secretions by some glands. Glandular secretions act chiefly on the body, though they sometimes figure rather directly in events far beyond the body, as, for example, when a person feels some controlled variable to be disturbed by the sight of another person's tears or when semen from one person joins with ova from another to start growing a new creature. Muscular contractions bring about by far the greater part of our actions on the environment.

In connection with Figure 3–4, I described the layers or levels of control loops that set reference values (internal standards) for loops lower down. I said that above the "lowest" loops, whose outputs actuate the output function and enable us to act on the environment, are thousands upon thousands of loops that enable us to coordinate the outputs of those lowest loops. The higher loops make it possible for all those muscles in the leg to move with the right order and force so that we actually walk, and not just lurch this way and that. The higher loops make it possible for us to tell a rose from a skunk, recognize our friends, read a book, and wonder whether a tree falling in the forest makes a sound if no one is there to hear it. The higher loops make it possible for us to move a fork accurately from plate to mouth and also to plan a 500-mile trip so that we can eventually enjoy seeing grandmother's smile.

The higher loops make it possible for us to choose a particular act through which to control a perceived variable. We can control the flow of heat from our bodies by choosing to put on a sweater or to turn up the thermostat. "Internal processing" is the term I used in Figure 4–1 to cover all those controls of the lower loops. Figure 4–1 makes it look as if the internal processing for the loop is somehow contained within that loop itself—within that little box on the diagram. But the diagram is schematic. It shows not how some particular loop would be "wired up," but rather the kinds of functions and connections that would have to be "wired up" in any loop, regardless of the level in the hierarchy of the loop or the number of connections it might have with other loops. I will describe the hierarchy in Chapter 18.

OUTPUT QUANTITY

When the output function converts a neural signal into a muscle contraction (for example), the muscle exerts a certain amount of force on something. The force is an example of *output quantity*. Some output quantities are amounts of secretion, but most of the output quantities with which we affect our environment are muscle forces. Outputs from the higher loops are neural signals to the comparators of lower loops.

Many of our actions occur within the body. The muscles of the heart pump blood within the body. Various glands put their chemicals into the blood stream. Through glands and muscles, our nervous systems maintain their bodily environments, including the environments containing the ions that move the electrons to the neural fibers, enabling them to fire repeatedly and to send their pulses in a thousand directions and still maintain their electrical capabilities. Keeping the body functioning well enough to supply the physical needs of the brain and its neural extensions is itself a large topic; it comprises the entire science and practice of medicine. That is not my topic.

With only a few exceptions, everything we do to affect the outer environment we do by pushing or pulling on things. We can push a branch or a person out of our way. We can pull on things by hooking our hands or arms or legs around them; pulling is only pushing from behind. We can pull down a branch to pluck a fruit from it. We can push from two sides at the same time; that is, we can pinch, grasp, twist, bite, and hug. The most conspicuous things we do are done with our muscles: eating, waving, walking, hollering, talking, writing. We can make sound waves in the air with our vocal cords and by clapping our hands. We can blow on hot coffee and into trumpets. We can suck soda through a straw.

ACTION OUTPUT

We act with the purpose of controlling some perceived variable. The violinist tightens a string or loosens it with the purpose of hearing the pitch that matches the pitch given out by the oboe or piano. We usually name our acts with the purpose we have in mind. I see my wife putting on her coat. "What are you doing?" I ask. If she wanted to be literal or wanted to tease, she could say, "I'm putting on my coat." If she wanted to be even more literal than that, she could say, "I'm holding my left arm steady while I contract the biceps in my right arm, and I'm doing just the right things with a lot of other muscles to keep from falling over." But instead (I'm glad to say) she tells me her larger goal: "I'm going out to buy some food."

Claire (my wife) wants to perceive herself bringing home some food. Putting on her coat is an early part of a complex program through which she can eventually perceive herself bringing home some food. In other parts of the program, Claire must perceive herself walking to our automobile, she must perceive herself moving closer and closer to the automobile, she must perceive the motor making the right noises when she turns the ignition key, she must perceive the car moving close to (but not touching) the edges of the garage door, and so on. She must operate the car in such a manner as to see herself arriving at the market. On the way, she must turn the steering wheel to counteract unexpected bumps in the road and gusts of wind that would otherwise throw her off course. She must slow or stop when other cars come into view. And so on. All those small actions within the larger action of bringing home some food are actions to control some perception that is instrumental in eventually perceiving the food at our house. This journey is an example of the control of perceptions by loops at a higher level—a level that Powers has labeled "program." Through complicated programs such as this and through even more encompassing levels of control, we are able to complete control loops that require long periods of time and unpredicted uses of opportunities in the environment.

The action on the environment produces a feedback function. The action affects some energy in the environment that is delivering to a sense organ an input quantity that the person wants to control. Claire puts on her coat for the purpose of keeping warm; that is, she wants to sense a low rate of heat flow away from her body, and she will sense that low rate when she puts on her coat. If the rate becomes too low (if she comes to feel too warm), she will unbutton the coat or take it off. Claire wants to see herself bringing home the food. At the end of her journey, her actions enable her to see herself walking into her kitchen carrying sacks of groceries. Her purpose has been achieved.

So the feedback loop is completed.

Notice that taking action to control a perception requires energy. Organisms get energy from the food they ingest and the oxygen they inhale. Energy is needed to maintain the tissues and substances of the body—to replace damaged cells and replenish enzymes, for example. But those are internal maintenance functions. By themselves, the maintenance functions do not enable the organism to go on living. Those functions maintain the systems (such as muscles and sensory organs) that the organism uses in controlling perceptions. Without sensory organs, the person could not tell whether he or she was getting the right kinds and amounts of food. Without muscles, he or she could not get any kind of food. Even the *Escherichia coli* has to be able to wiggle its cilia. I will say more about calling up energy in Chapter 21 on emotion.

ENVIRONMENTAL FEEDBACK

Notice what we mean by *feedback*. Many people use the word *feedback* to mean information given by one person to another: "I gave him some feedback about how he was coming across," or, "I was glad to get her feedback." As a technical term in PCT, *environmental feedback* designates the results of output actions that affect the controlled input quantity and thence the controlled perception. Putting on a coat affects the movement of warmed air near the body and thence the perception of heat flow from the body. Feedback is the part of the loop that feeds back energy to the input quantity and thence energy to the controlled perception. All those various actions Claire took to get home with the groceries constituted the environmental feedback function of the program loop. The loop did not require that anyone give her any information, helpful or not. In PCT, strictly speaking, no one else can give you feedback; you must do it yourself. In Figure 4–1, environmental feedback occurs where you see the label "Feedback function." If, for example, you want to perceive that your necktie is straight, one way you can make use of the environment to achieve that perception is to look in a mirror. Another way is to ask your friend, "Is my necktie straight?" And you might be satisfied if she answers, "Yes." So you might succeed in imagining (which is a kind of perception) the proper straightness of your necktie by using her report. Or if she says, "No," you could make use of that report, too, by adjusting the necktie and asking her again for a report.

Now suppose you do not ask your friend for her opinion, but she proffers the remark, "Your necktie is crooked." You might say, "Thanks," and straighten it. You might not actually care about the necktie; you might straighten it because you care about what *she* wants to see. Or you might reject her comment; you might reply, "Who asked you?" Or you might just ignore her remark. In these examples, you are not using what your friend says to control your own satisfaction with your necktie. If you straighten your tie only because you want it to look the way *she* likes it, you are not controlling your perception of the necktie; you are controlling your perception of *her* satisfaction. What *you* do to control your perception of your necktie is what you do to affect the input quantity; that is where the feedback must have its effect. But a remark initiated by another person may or may not be useful to you as a feedback function.

The other person may intend it to be useful to you, and you may indeed find it so, but the other person cannot be sure the remark will become a part of your feedback function. You are the only one who can know whether it does so.

You can, of course, make use of what other people do, including what they tell you, to control your perceptions. As another example, if Claire could not find her coat, she could ask, "Where's my coat?" and I could tell her it is in the other closet. My telling her the whereabouts of her coat is not properly called feedback in the technical PCT sense. Claire's question to me is part of her feedback function, and so is her use of my answer. So in a minor sense, you could say that my answer to her is a part of her feedback function, but it is not my talking that is the feedback. My talking is not feedback from me to her; it is feedback from her to herself. She must find the path or get the information she needs to reach her goal.

DISTURBANCE

The feedback affects the input quantity, but the feedback is not the only force affecting the input quantity. The feedback is necessary because energies in the environment are usually acting to disturb some input quantity. As we walk, we are always in effect falling forward, and we must always be putting a foot forward to catch ourselves. As we walk, the forward falling must be constantly counteracted. The input quantity, in this case the feeling of rate of falling, is always a resultant between the feedback force, pushing upward with the legs, and the disturbing force, gravity. As we drive a car, the position we perceive the car to be in on the road is a resultant of the force we apply with the steering wheel and (for example) the force of a crosswind that would otherwise blow us off course.

We do not always act because of an immediate disturbance to a controlled variable. Sometimes disturbances are long lasting or counteractions are long delayed. We have continuing occupations, ambitions, aspirations. But we still act to change the present state of affairs into one we believe we will find more pleasing. Sometimes we have goals that are continuing, never to be culminated, such as having a preference for keeping busy. We could conceive of "keeping busy" as a controlled variable for which the reference signal is "yes" or, equivalently, we could conceive of "idleness" as a controlled variable for which the reference signal is zero, and the person who keeps busy is succeeding in keeping that variable at the reference level.

IRRELEVANT EFFECTS

When we act to control a perceived variable, we affect not only that variable, but also other parts or aspects of the environment. We may notice some of those other effects, but never all of them. I walk up to the check-out counter with my groceries, and the person there greets me, but I am too occupied thinking about some difficulty to respond with more than a mumble. As it happens, two or three other people among the next four or five respond in the same way. Discouraged, the check-out person mumbles her greeting to the next customer, who doesn't bother to answer. Check-out person and customers alike feel glum and disaffected for the next fifteen minutes until an ebullient customer shows up to break the gloom. A tree in our yard dies, and we have it cut down and hauled away. We may not notice that a habitat for woodpeckers is also gone. We drive to Ashland, adding to the air our portions of carbon monoxide and dioxide, discouraging the passenger service of Amtrak, breaking up a certain amount of the pavement, and annoying some drivers who think we are going too slowly. Some industrial managers capture a large share of the market for their product and note that they are making a large profit for themselves and their shareholders and are employing 10,000 people. They may not notice that they are polluting air, river, and trash dump and damaging the health of a large fraction of those 10,000 people. When you conceive a child, you put a claim on the space and resources of the planet that the child will eventually use.

I will say more about irrelevant effects in Chapter 22.

SUMMARY

Seeking to understand the functions in the control loop does not obligate you to use the quantum theory of atoms and molecules to explain your desire to play the bassoon. Have no fear of people who cry, "Reductionism!"

To test the usefulness of a model, words or even mathematics are not enough. A model should be capable of being built with tangible materials and set into motion. The functions and connections in Figure 4–1 have provided the assumptions for numerous models that have "behaved" in the same way as actual human individuals. I will describe some models in Chapters 7, 8, and 9. To make a psychological theory that can be tested with a working model, it is not enough to have a good idea such as purpose or acting against disturbances. It is necessary to specify a negative feedback loop with circular and simultaneous causation having functions arranged so that perception can be controlled to match a reference signal (internal standard).

What we do affects more than the perceptual variables we want to control. What we do also has effects to which we pay no heed. Some of those "irrelevant" effects come back to haunt us.

ENDNOTES

[1]The books were M. W. Eysenck (1994), M. I. Friedman and G. H. Lackey, Jr. (1991), A. Furnham (1996), Peter Gray (1991), C. R. Hollin (1995), K. Huffman and others, 3rd ed. (1994), J. Kagan and J. Segal, 8th ed. (1995), G. A. Kimble (1996), J. V. McConnell and R. P. Philipchalk, 7th ed. (1992), R. Ornstein and L. Carstensen (1991), and J. Rodin and others (1990).

Chapter 5

Beware how I write

I think the key ideas in PCT are fairly simple. The interlacings of those simple ideas, however, can become complex. Furthermore, some of the ideas sharply contradict ideas widely accepted in our culture; one contradiction with the culture, for example, is PCT's core idea that whatever we do, we do by controlling perceptions. Accordingly, I will sometimes fail to say something in a way that enables you to grasp my meaning, because you and I differ in how we have learned to combine ideas and in the aspects of our culture we have come to cherish. Worse than that, you may come upon something that I wrote when I was mixed up in my thinking and did not catch before publication. I do take great care not to let my manuscript be wounded by erroneous statements, and several of my friends will have read through the manuscript looking for faults of whatever kind, but every book I have written got past the author, editorial readers, and editors with some embarrassing blemish.

If it were possible, scientists would prefer to communicate entirely through experimentation. They would invite one another to come and witness their experiments, letting the lessons learned bloom behind the eyes of their guests without any verbal interference from the host. You can see how impractical that would be. Scientists would have to use words to issue the invitations, and the invited scientists would then ask, "What's it about? How long will it take? Who is going to pay my expenses?" Scientists therefore write articles and books not only to tell one another about their experiments, but also to persuade one another that the others ought to *want* to know about the experiments and even join in the experimentation. (This is one of those books.)

The writing becomes part of the scientific enterprise. It is the way that scientists try to join hands in building a science which, in turn, can be offered to the members of the embedding culture as a way of understanding and coping with the world. The writing, therefore, should be done with the same care with which the experimentation is done. But joining hands and understanding the world do not stop with the writing. The joining and the understanding cannot begin until after the reading. The reading too, therefore, should be done with care.

Details and nuances can be important.

For example, my very first sentence in Chapter 1 was, "People have always been fascinated by the actions of others." That sentence is unsound. How many people have been fascinated? Everybody? If not, which people? Have they been fascinated by *all* the actions of others? If not, by what fraction of actions or by what sort? I like to be precise, but I let that wobbly sentence stand.

If I had let my desire to be precise rule my writing, I would have written a sentence without those ambiguities. But there I was at the first sentence of the first chapter, wanting to invite your attention to the things I thought you would be in the mood to find fascinating if you had picked up this book; namely, other people. I wanted a short sentence followed by some other short sentences, none of them steering your attention to logical niceties and syntactical precision, but instead keeping your focus on people and their doings. More than anything else, I wanted to make it easy for you to imagine some of the many ways in which we observe the doings of other people. So I gave up precision in favor of a quick and easy focus on human doings, leaving more precise statements to come later.

I hope you will at least occasionally scrutinize my sentences to see whether you find them believable. When you do not find me believable, ask yourself whether I might have let myself be vague so as to get on with the story. But if you don't think that is the case, then write something in the margin like,

"Some! Not everybody, you idiot!" When you came in the previous paragraph to my phrase ". . . steering your attention. . .," you might have thought, "He can't steer my attention. I steer my own attention. Runkel can offer something for my attention, but I myself choose whether to give it." If you did say that to yourself, I can only agree with you. And you will every now and then, I fear, find sentences like that in this book.

Look out even for single words. You may discover (though I hope not) that I am using some word in a nonstandard manner. As the years go by, I still discover now and then that I am using a word for a meaning of which my dictionary is ignorant. Usually I convert to the dictionary's belief, but sometimes I backslide. So look out for the possibility that I have picked the wrong word.

Other authors do it, too. In an article in a psychological journal, the author said that a great many studies on a particular topic had been done, and "Sawyer reviewed a plethora of those studies." I think the author meant only that Sawyer reviewed a large portion of the studies; but according to the dictionaries, the sentence meant that Sawyer reviewed more articles than he really cared to. Dictionaries say that *plethora* is used to mean not merely a lot, but an excess or superabundance.

Positive and *negative* can be ambiguous. Most people most of the time nowadays, it seems to me, use those words to mean simply good and bad. But they can be used in other meanings. Dictionaries use half a column to list them all. *Positive*, for example, is often used to mean being confident of one's opinion, being in no doubt. In reviewing a book, the reviewer may write, "My opinion of this book is a positive one." Does she mean that she likes the book? Or does she mean that she is in no doubt about her opinion, though she is refraining from saying whether she likes it?

Some words that bear upon theorizing recur in this book. While I am on the topic of word usage, let me tell you the meanings I have in mind for some in this family of words: assumption, axiom, conjecture, guess, hypothesis, postulate, premise, presumption, presupposition, principle, theorem, theory, thesis. A guess, hypothesis, or theorem is (as I interpret my dictionaries) a statement formed for the purpose of putting it to the test. Looking for evidence for the statement clarifies not only that statement but also a connected, larger body of assertions. An assumption, axiom, postulate, or presupposition is an assertion taken to be correct or true without question while investigating other (even though connected) matters. An author may want you to accept certain assumptions as axiomatic just while you read his book, or he may expect you to accept the assumptions that are widespread in a discipline or school of thought. Some authors, especially mathematicians, will try to set forth their axioms explicitly for you at the outset. Without help from an author, assumptions often lie implied, unseen. Finally, a theory contains both assumptions and hypotheses. I have tried to stick close to what my dictionaries tell me about these words, though I lean somewhat toward the usages of writers on scientific theory and mathematics. You will no doubt find other authors using some of these words (especially "assumption") in other ways.

Now and then writers hurry too fast past their sentences. The following appeared in *Science News* in 1998: ". . . women who had gained 22 pounds or more since age 18 ran an increased risk of dying." Every woman will die, no matter how few pounds she gains. I suppose the author meant that the women who gained 22 pounds or more would die *sooner*, on the average, than those who gained fewer. Here is another hurried sentence: "Some of the meteorites have been in the ice for more than a million years, possibly longer."

But usages of words and hurried sentences will be much smaller dangers than the dangers of implied assumptions.

ASSUMPTIONS

Writing is always shaped to a considerable extent by the author's beliefs about how the world works, about how things function—by what Powers calls our "system concepts." If we believe that unexpected events can come about by chance, we write to a friend, "If I'm lucky, I'll get there on Tuesday." If we believe that although we can be surprised by events, nothing happens except at the will of God, we write, "I will arrive, God willing, on Tuesday." If you believe that persons who break laws have sinned, and that when they are put in prison with little to do, they will reflect on their sins and become penitent and therefore resolve not to break a law again, then you will chisel the word "Penitentiary" over the door of the prison. If you believe that subjecting lawbreakers to a restrictive and coercive discipline of obedience will cause them to maintain that mood of obedience after they leave the prison, and if you believe that teaching them a manual skill will enable them to make a legal living after getting out of prison, the two "treatments" together reforming or correcting their behavior, you will chisel "Reformatory" or "Department of Corrections" over the door.

If you believe that the acts of a person are caused largely by events in the person's environment, you will write, "The new incentives instituted by management resulted in a 15 percent increase in production," and you might even believe that not just some, but all the workers on that production line were influenced in that direction by those incentives. You will write, "The prick of the pin caused her to jerk her hand away," a sentence William McDougall would not have written, as you can tell from the quotation from him that I put at the beginning of Chapter 4. You will write, "Classical conditioning reinforcement strengthens a response," even though that explanation seems to fit only nonhuman animals deprived of food and imprisoned in an environment (such as a Skinner box) offering severely restricted opportunities for controlling vital perceptions such as hunger and nourishment. If you are B. F. Skinner, you will write (as he did on page 35 of his 1953 book):

> The external variables of which behavior is a function provide for what may be called a causal or functional analysis. We undertake to predict and control the behavior of the individual organism. This is our "dependent variable"—the effect for which we are to find the cause. Our "independent variables"—the causes of behavior—are the external conditions of which behavior is a function. Relations between the two—the "cause-and-effect relationships" in behavior—are the laws of a science. A synthesis of these laws expressed in quantitative terms yields a comprehensive picture of the organism as a behaving system.

If you believe that the acts of a person are caused largely by the kind of person he or she is, you will write, "We need a strong leader at the head of this company," and you will urge the new leader to recruit a better class of worker. To recruit those workers, you may advocate using screening tests of manual dexterity, obedience, honesty, intelligence, or some other desirable quality. You will write, "Vote for Jones—a man of probity and experience." You might write an article in a magazine explaining that a particular politician's leadership was demonstrated by the fact that the public debt was reduced during that politician's incumbency, and you might even believe that the public debt would *not* have decreased if some other politician had been in office. To improve the social order, you will urge measures to change the inner qualities of people—perhaps their morality, practical knowledge, patriotism, or team spirit. You may believe those inner changes can be brought about by the shining examples of morality and citizenship to be encountered at church or school. Or you may believe that those inner qualities are given at birth and are unchangeable or are too slowly changeable. In that case, you may advocate improving society by killing off the undesirable people, a procedure put into practice in our own time by Hitler, Stalin, Pol Pot, and others.

Most people appeal to both those sources of action, even if alternately. People who want a strong president often believe he or she will have strong influence on leaders of Congress, executive departments, the military branches, and industry. In other words, they believe the president will act from his or her inner qualities, but that the members of Congress and the others will act because of being skillfully or forcefully prodded by the president. Persons with that combination of belief rarely, it seems to me, wonder who prods the president (or king or other top boss). Once in a while I have heard the speculation that a president is influenced by some "power behind the throne" such as a wife, and in the days when actual thrones were numerous, the ruler usually claimed to be guided by God; rulers still occasionally make that claim today.

I do not want to leave the impression with you that assumptions are bad. They cannot be avoided. To learn something, you must assume that you already know something. To learn how far it is from Chicago to Omaha, you must assume that there are such places as Chicago and Omaha. You must act as if those places exist while you are hunting for the information about distance. If you find no trace of those places under those or other names, you may then relinquish your assumption. But you will never find out the distance between them unless you assume that they do exist.

The trick is to find assumptions that match fact. We make trouble for ourselves when we assume that the world is flat, that a fever is always bad, that tobacco is good for us, or that nothing moves until pushed by something else.

TRAITS, NEEDS, AND MOTIVES

Psychologists and non-psychologists who believe action to be caused by the person almost always hypothesize the existence of traits, needs, or motives within the person, and psychologists almost always describe a trait, need, or motive as being common to all people (or to most people or to a large specifiable fraction of people) though existing in varying degrees from one person to another. Examples of traits that have been proposed in psychology books are introversion-extroversion, surgency, emotional stability, conscientiousness, intelligence, submissiveness-dominance, and psychasthenia. Examples of motives and needs are power, affiliation, and novelty in experience. In our everyday speech, you and I and Uncle George propose dozens more: courage, generosity, insolence, laziness, persistence, pigheadedness, sensitivity, stubbornness, and on and on. In seeking evidence for the "existence" of those internal qualities, researchers almost always do so by claiming that persons high on trait X will perform certain acts more often than persons low on trait X. The researchers then put a lot of persons in situations where those acts are possible and tally the proportions of the people who perform those acts. There are great difficulties in interpreting an experiment of this sort; I have described the chief of them in Chapters 4 and 5 of my 1990 book. In this book, I will describe difficulties in assessing intelligence and personality traits in Chapter 26, psychological disorders in Chapter 31, and academic aptitude in Chapter 38.

I hope you will be suspicious of claims about what can be told about the likely behavior of people from assessments of their traits and motives. The person's internal standards are only part of the story, and ascertaining internal standards by the traditional methods of assessment are very uncertain when not absurd. So as not to drift too far afield at this point, I will omit to mention the many arguments and counterarguments that can arise concerning traits and other similarly conceived internal standards. I will ask you only to remember the Requisites for the occurrence of a particular act (I gave these in more detail in Chapter 1 of this book under the heading "Requisites for a Particular Act"). The Requisites lie *both* in the person and in the environment:

1. That some environmental event disturb a controlled variable.
2. That the person find some means in the environment with which to affect the controlled variable.
3. That the chosen act not disturb some other controlled variable.

CONTROLLING OTHERS

Watch out, too, for the assumption that it is possible for one person, psychologist or not, to "cause" another person to do some particular thing—pick up his socks, get married, or grow up to be a minister of the Gospel. Most of us, I think, would like to have power not only over nonliving things such as chairs and automobiles, but also over animals and people. Fairy stories, myths, so-called science fiction, biographies, self-help books, religious books, histories, and psychology books are sprinkled with putative ways, magical or otherwise, to compel others to our will. I do not claim that we are all genetically endowed with a lust for power. But there can be no reasonable doubt that we all *do* affect the environment, living and nonliving, and no reasonable doubt, either, that almost all of us consciously hope and try now and again to influence other people.

Hardly a day goes by without hearing someone (or oneself) yearning to influence the acts of other people in one way or another.

"I hope I can persuade him to do it."
"I'm going to offer her a better price than he did."
"He'll do as I say if he knows what's good for him."
"When she realizes how much we care, she'll want to do the right thing."

Most people do seem to act as if they think they can, if they are clever enough or forceful enough, cause other persons to perform particular acts. Bosses issue specific orders every day. Politicians promise to coax legislators to pass certain kinds of laws. Parents tell children to pick up their socks. When some people doubt their own ability to control others, they often think someone else has the ability. One parent may say to another, "I can't make him do it; *you* make him do it." I am not saying, by the way, that it is never useful to do the kind of thing of which I have given examples. I will go into the topic of interpersonal influence in Part VI.

Psychologists, in their professional work, are not exempt from the desire to control other people. Earlier in this chapter, I quoted Skinner's 1953 book, in which he said, "We undertake to predict and control the behavior of the individual organism." Myers (1986, p. 18) wrote, ". . . psychologists attempt to . . . predict . . . and perhaps control behavior and mental processes." L. D. Smith (1992, p. 216) wrote, "Behaviorists have long held that the aim of science is the prediction and control of phenomena. . .". Myles I. Friedman and George H. Lackey, Jr., whose book I mentioned in an endnote in Chapter 4, seem to see the desire for prediction and control in everyone, and wrote their 1991 book, following up two earlier volumes, to give evidence for their belief. On page xiv of their 1991 book, they say,

> We now contend that people want to control the world around them, and a large preponderance of their behavior is directed to that end. The mental ability that is largely responsible for that control is predictive ability—the ability to make accurate predictions about the future. . . .

Presumably the term "people" there includes psychologists. Maybe it includes sociologists, too. Jack P. Gibbs wrote a book called *Control: Sociology's Central Notion* (1989). The phrase "behavior modification" almost always means an intervention by an experimenter or therapist intended to cause altered behavior on the part of a subject or patient; a 1997 book about behavior modification, for example, is entitled "Change, Intervention and Consequence." Books on techniques of persuasion and books on advertising are also full of presumed methods of controlling other people—methods, that is, of causing particular changes in their attitudes or behavior.

Many books written for organizational managers are about organizational change. Some of those books tell how organizations are changing in relation to the surrounding society, but many purport to tell managers how they themselves can cause their organizations to take on a new way of functioning.

Ellen J. Langer wrote a book (1983) on the consequences of a person's perception of having or not having control over events in the environment. Among other findings, she writes that "perceiving control apparently is crucial not only to one's psychological well-being but to one's physical health as well" (p. 13). That word "perceiving" might let you wonder whether Langer was getting close to PCT, but she was not.

In sum, psychological writings (like the writings of non-psychologists) assume much more often than not that particular acts of other people can be produced (controlled), and they assume much more often than not that prediction of particular acts is the first and essential step in learning how to produce them. From the viewpoint of PCT, these two assumptions are pernicious, and I urge you to beware of them.

Despite the record of paltry success during the last hundred years, psychologists, like non-psychologists, persist in trying to predict the particular actions of other people. Most psychologists write proudly of their achievements in doing so, even though in almost all their published studies, they can claim only with less than full statistical confidence that the fraction of their subjects whose actions they predicted correctly was greater than they could have expected to get by pure chance. It seems to me reasonable to call that a paltry record. (In contrast, investigators of PCT publish their findings only when the experimental results are far too strong to require calculations of statistical confidence.)

As to predicting behavior, I have no quarrel with psychologists who begin an experiment with a predictive hypothesis. My quarrel concerns what sort of behavior is predicted. To make clear what I mean, here is the form that a hypothesis in PCT might take:

> I have built a model such that, when I ask an actual person to perform the same task I have given the model, the person will behave very much as if his or her internal functions are connected in the same way I connected them in my model.

Notice that there is no mention of particular acts in that hypothesis. Notice, too, that the prediction is made for a single individual, not for a proportion or an average among many. The chief criterion for arranging functions in a model built with PCT is that the connected functions succeed in controlling a perceived quantity—a perceptual variable. A more precise statement of a typical hypothesis, therefore, goes like this:

> I have built a model such that, when I ask an actual person to control his or her perception of a particular variable (or more than one), the person will do so in very much the same way, measurable quantitatively, that my model does it.

When such a hypothesis is confirmed, we do not conclude that one behavioral variable is correlated with another or that the occurrence of a particular act is correlated with some environmental event or stimulus. Instead, we conclude that the *model* behaves closely enough to the way humans function so that it is worth further exploration—worth further investment of thought and money. If the model sits still while the person waves an arm, or says, "Stop! Stop!" while the person says, "Go! Go!" then we should, after checking all the connections in the model, give up the theory and look for a better one.

In the usual psychological experiment, the hypothesis predicts a correlation (a statistical association) between two variables perceivable by the *experimenter*. For example, B. G. Fricke (1956) reported that the best way to find high-school graduates who will do well in college is to take those with high academic rank in high school. Koslowsky and Locke (1986) reported a way of classifying credit-card holders so as to find persons among which a proportion much greater than among all credit-card holders were likely to buy insurance by mail. Mohandessi and Runkel discovered in Illinois in 1958 that secondary schools with higher mean scores on academic aptitude lay farther from coal mines, on the average, than those with lower. Such studies deliver information that can be very useful for many social purposes, but they do not enable us to test models of human functioning. It is useful to college admissions officers, for example, to know that academic rank in high school is the best predictor (on average) of grades in college, but the information tells us nothing about how students can or do go about "getting" grades, about how they cope with conflicting demands from parents and teachers, or about any other sort of internal functioning.

I urge you to be critical when I mention correlations between variables—correlations, that is, between particular kinds of actions or between environmental conditions and particular actions. When I mention such a correlation, do I seem to think I am writing as a social statistician or as a theoretical psychologist? If the first, that is what I intend. If the second, I will have slipped up, and you would be right to rebuke me.

As I read books and articles on psychology, I find most authors saying in so many words or clearly implying that if they can predict behavior, they can then control it. That is, they imply that the "independent" or "predictor" variable will *cause* the action predicted as the "dependent" variable. If you are dealing with nonliving things, that faith in the connection between prediction and control is usually justified; if you are dealing with living things, it is a delusion.

REIFICATION

Josiah Royce (1913, p. 27) wrote:

> The creator of the English speech is the English people. Hence the English people is itself some sort of mental unit with a mind of its own.

Josiah Royce (1855–1916) was a "noted metaphysician," a professor of philosophy at Harvard, an author of many learned tomes—in short, a person from whom I would expect careful thinking. Notice my stereotypy. I just admitted that because Royce belonged to the class of metaphysicians, the class of professors, and the class of authors, I expected him to exemplify my stereotype of such classes and turn out to be, as an individual, a careful thinker. And now when I discover that my stereotype has led me astray, I feel the urge to complain—as if Mr. Royce has betrayed me. But Mr. Royce demanded no opinion from me. My opinion is my own doing, and it turned out to be wrong because of my own unsubstantiated preconceptions. I will say more about the pervasiveness of stereotypy in a later chapter. Here I want only to remind you that I am as susceptible to it as you, and I am sure I will fall prey to it again before the end of this book. But let me return to the English people.

How is it that English is being spoken? Each of us learns to speak English by copying our parents and others nearby in our infancy. We do not "create" the language. The English we speak today has come about through continuous modification over thousands of years and through dozens of distinguishable languages. Royce's argument seems to have this form: If we make use of something that has come to us from earlier people, that is evidence that we who use it are "some sort of mental unit with a mind of its own." By that reasoning, all the tenants of the Empire State Building are some sort of unit with a mind of its own. The people who drive on the nation's highways are possessed of a group mind of some sort. Everybody who attends the University of Oregon. All the people who buy hamburgers at McDonald's. And even if we limit ourselves to people who "create" something, the argument is as silly. The people who built the Empire State Building? The people who strung the telegraph wires?

Royce seems to have convinced himself that because he could conceive "the English people," there existed "some sort of unit." Confusing a conception in the mind with a tangible thing in the environment is called reification.

During the 1700s in Europe, personification ran riot. Poets wrote odes to gods and goddesses who personified love, war, courage, history, art, cooking, or anything you can name. Kings erected statues of themselves personifying victory. And so on. Even today, you can still see statues of justice on some of our public buildings, and in New York harbor you can see (on a clear day) the Statue of Liberty. Personification is perhaps the ultimate reification. Reification and stereotypy must be as old as language. Look out for them, no matter who perpetrates them.

PATHETIC FALLACY

In Chapter 2, I protested against speaking of living things in the same way we speak of nonliving things. Now I will protest against the converse. We say that this peg "doesn't want" to fit into this hole. But it is the person who perceives the lack of fit, not the peg. In a novel, I read, "the branches of a massive oak tree flailed helplessly against the elements." Trees can't lash their branches, either helplessly or otherwise; the wind does it. The writer John Ruskin, in 1856, called that kind of talk the "pathetic fallacy," (referring to "pathos," the arousal of emotion). That kind of writing is fine for novelists and poets, but for scientists, it can be dangerous.

The novelist does not mislead us. Few of us expect that a tree, on a calm day, will suddenly flail helplessly at us. In science, however, the same kind of thinking can indeed mislead us. Psychologists often say that food has "reinforced" a response such as turning left in a maze. But food is not a purposive creature, and it has in itself no capability of influencing the rat. Turning left after having found food down that alley (when the rat is once again hungry) can be described better as the rat choosing to go again to a place where it had found food. The food doesn't make the choice; the rat makes it. The food does not have the purpose of getting rat and food together; the rat does.

It is easy to fall into the pathetic fallacy, especially if we believe that things in the environment, by themselves, can cause us to do something. I can say, "That book enraged me." But the book didn't do anything. I was the one, not the book, who opened the cover. Similarly, I, not the carrot cake at the GoodEats Restaurant, have the purpose of putting myself at a table on which the people at the GoodEats Restaurant will put a piece of their carrot cake. I, not the carrot cake, will draw me to it.

Scientific writers fall into the fallacy easily. One often sees, "This experiment confirms. . ." instead of "I interpret this experiment to confirm. . .". Maybe the author wants to convey to the reader that he welcomes no interpretation by any other human mind.

I suppose I will fall into writing, now and then, as if experiments, assumptions, theories, or attitudes can themselves *do* something. If I do, I hope you will not follow my example.

SUMMARY

By the time this book is published, some of the books I call "recent" may seem not very recent. Don't let that worry you. Change is slow in psychology. You can see in these pages how sentences in books in the early 1950s sound very much like sentences in books from the middle and late 1990s. You can see how certain underlying assumptions have stayed the same ever since Wilhelm Wundt established the first psychological research laboratory at Leipzig, Germany in 1879—the assumptions that causation is linear (as it is with nonliving objects), that it is possible in principle to predict and produce, arbitrarily, particular acts and thus to control the behavior of others, and that there exists one or more internal standards common in some degree to all humans (or to some specifiable, large fraction of humans) that can be discovered by predicting how persons possessing a high degree of the standard will act in certain situations. Several assumptions about method have also held sway since early in the 1900s, despite the fact that they have not produced reliable knowledge about how living creatures function. I will not complicate this chapter by describing those assumptions here; I described them in my 1990 book, and I will touch upon them again later in this book. I will, however, take space here to repeat once more the assumptions underlying PCT. Perhaps you will wish to COVER THE REST OF THIS PAGE with your hand and see to what extent you can recall the assumptions before you read them here.

1. Causation in the human neural net is circular and simultaneous.
2. Action has the purpose of controlling perception. Controlling perception produces repeatable consequences by variable action.
3. A controlled perception is controlled so as to match an internal standard (reference signal). Every internal standard is unique to the indi-

vidual, though two individuals can have very similar standards.
4. Particular acts are not, in general, predictable.

Actually, I should not call the fourth statement another assumption, since it is derivable logically from the first three. But I assert No. 4 explicitly because it is such a good, quick test of whether a theorist believes any of the first three statements. When you encounter someone (psychologist or not) claiming that an event in the environment (perhaps an act by another person) has caused or will cause a person to do a particular thing, you know that the psychologist (or other sort) does not believe the first three statements.

In Chapter 4 and in this chapter, I have given some illustration of how various psychologists have seized upon one or another feature of theory that is also part of PCT. None, however, except W.T. Powers and his followers, has adopted theory or experiment based on the assumptions just above, consciously or unconsciously. So far, only those investigators employing PCT have succeeded in building working models of the living creature. By the time you read this book, PCT will not yet have displaced the older assumptions and theories in the minds of a large fraction of psychologists. You will still profit from watching for the older assumptions in what you read here and elsewhere.

A NOTE ON ENGINEERING

In the paragraph above, I said that only investigators employing PCT have built working models of the living creature. There is a sense in which that is not true. Mechanical and electrical engineers have built mechanisms and electrical circuits that produce forces and motions in machinery that control sensed quantities in very much the same way a person does it. Sometime in the late 1700s, James Watt invented the mechanical "governor" to control the speed of steam engines. The principles of organization in the governor were analyzed mathematically by James Clark Maxwell in 1868. In 1934, R.L. Hazen published his *Theory of Servomechanisms*. In the same year, H.S. Black published his paper on "stabilized feedback amplifiers," setting forth the basic principles of negative feedback systems and inspiring the systematic development of control-system engineering. Nowadays, we have complex electronic feedback controls in every automobile, in the rockets that "lock on" to a planet and guide the rocket steadily toward it, and in a thousand other applications. All those mechanical and electrical control systems have been behaving like living creatures in the sense that they have been controlling inputs (perceptions) by the use of negative feedback loops. The designers were not, however, trying to understand living creatures. The connection between control-systems engineering and living creatures was first made in print, as far as I know, in 1948 in Norbert Wiener's *Cybernetics: Control and Communication in the Animal and the Machine*, but Wiener's description of the connection persuaded few readers.

Unfortunately, as Powers explained in his article in 1978 in the *Psychological Review*, Wiener and others failed to get the functional components hooked up in the way they are hooked up in living things. The oversight was easy, because it lay in the conception of motivation. In machines, the electrical engineer supplies the goal; that is, the reference signal. Living creatures supply their own. Powers illustrated how upside-down some people managed to see things with a quotation from the president of the Society of Engineering Psychologists:

> The servo-model, for example, about which there was so much written only a decade ago, now appears to be headed toward its proper position as a greatly oversimplified inadequate description of certain restricted aspects of man's behavior.... Whenever anyone uses the word *model*, I replace it with the word *analogy* (Chapanis, 1961, p. 126).

Despite disdain from many quarters, some of the ideas of control theory continued to show up in the psychological literature. In Chapter 4, I spent the section headed "Other Appearances of Purpose" commenting on a literature review by M.H. Appley, in which he showed how widely those ideas had appeared, especially after about 1940.

When, therefore, I said that only those investigators employing PCT have succeeded in building working models of the living creature, I meant investigators who were trying to do such a thing. Electrical engineers have built devices that use negative feedback loops to control inputs, but they were not trying to build models of living creatures. They did not know they were pointing their machinery in the direction of a new science of living things.

Part II

Research

So far, I have been offering you nothing but words. The next chapter will urge you to *do* something. Here I offer a brief review of some things I have said so far.

The Requisites for action by a person are:

1. That the person be controlling some perceived variable(s).
2. That the person find a feature of the environment suitable for controlling the variable.
3. That the chosen act not disturb some other controlled variable.

Physical laws are not sufficient to describe a person's behavior. The behavior hinges on the person's purposes. Living creatures counteract disturbances to intended states; nonliving things do not. But we often act toward other people as if they were nonliving objects.

Living creatures act upon their environments with the purpose of controlling perceptual inputs; they do so by means of negative feedback loops in the neural net. Organisms amplify incoming energies; nonliving things typically do not. Living things act against entropy. Causation in the loop is circular and simultaneous. To make a psychological theory that can be tested with a working model, it is not enough to have a good idea such as purpose or acting against disturbances. It is necessary that the theory specify a negative feedback loop with circular and simultaneous causation having functions arranged so that perception can be controlled to match a reference signal (internal standard).

Chapter 6 will describe some demonstrations you can do with the help of a friend or two. Do them. You will see perceptual control at work.

In Chapter 7, I will turn to actual modeling and some further matters of theory.

Chapter 6

Do it yourself

I have put before you a lot of pages of theory and argument. It is time to give you relief from stretching your imagination and let you stretch something with your hands. I will describe a few games you can play with a friend. I urge you to do them. The games will give you some experience with (a) consciously observing yourself controlling, and (b) observing another person controlling. You will get an understanding of the basic principles that words alone cannot convey. My description here follows very closely Powers's Chapter 5 in his 1998 book, even to using many of his sentences (for which thanks).

THE RUBBER BANDS

Get two rubber bands just alike, three or four inches long. Knot them as shown in Figure 6–1 by passing one through the other and pulling them tight. You will also want a table where you can sit across from your friend or side by side. And you will need a mark on the table between the two of you. You could put a mark on a piece of paper and lay the paper between you. Or use a dent or mark already on the table. (You can do this exercise without a table, but a table is comfortable.) Each person now hooks a finger through an end of the rubber bands, stretching them horizontally an inch or so above the paper. If you sit side by side, use your outside hands to avoid bumping into each other.

Designate one person as Experimenter and the other as Controller. (Change roles from time to time so that both people can see what's going on from both viewpoints.)

Figure 6–1. The Rubber Bands

The task of the Controller (C) is simply to keep the knot that joins the rubber bands exactly over the mark. The *internal standard* that C must adopt to perform this task is the relation between the knot and the dot—namely, the knot holding directly over the dot.

The Experimenter (E) uses E's end of the rubber bands to disturb the position of the knot. E can do that by moving the finger forward or back, left or right—in any horizontal direction (not up toward the sky or down toward the earth). E should understand that the object of this experiment is *not* to prevent C from controlling the position of the knot. You cannot keep the knot stationary (exercise control) if the other player moves faster than your natural reaction time can compensate. Move smoothly, not too fast. The lessons to be learned will be much more obvious to both of you if C is able to keep the knot always close to the mark. Of course, after the basic observations are made, E can try all sorts of things to see what control looks like under difficult conditions. But especially at first, we want to keep the conditions easy by letting C learn to get good control of the knot. E moves the disturbing end of the rubber bands around in any kind of slow pattern, while C concentrates on keeping the knot accurately over the dot. A few minutes' practice should be enough.

You will notice very soon that every motion of E's finger is reflected exactly by a motion of C's finger. When E pulls back, C pulls back. When E moves inward, C moves inward. When E circles left, C circles left. C must do that, of course, to keep the knot stationary. Discounting small control errors, at every moment C's hand is exactly as far from the dot as E's hand (if the rubber bands are identical). The action illustrates very plainly the phenomenon of control—that we act in opposition to a disturbance.

If a third observer happened on this scene, what would the first impression of these actions be? It would be that C is mirroring the movements of E symmetrically around the dot. It would not be obvious which person is putting in the disturbances and which one is counteracting them. Even if E confessed to being the disturber, it would still not be obvious that control is happening. Much more likely, the third observer would see E doing things and C reacting to them: stimulus and response. The third observer would say that what E does causes the acts of C. The third observer might not notice that the knot stays over the dot.

This interpretation, based on a quick judgment, would be reasonable. The third observer might well lose interest at this point, and leave with the impression that control theory is just the same old stimulus-and-response idea that's been around since great-grandfather's day. But a quick glance is not enough to grasp that control is going on.

Remember the basic organization proposed by PCT: perception, comparison of the perception with an internal standard, detection of error, and conversion of error into an action that affects the perception. C is perceiving the present position of the knot relative to the dot. The perceived relationship is compared with an internal standard—knot over dot. The difference (the perceived horizontal distance of the knot from the dot) is converted into an action (a motion of C's end of the rubber bands) that will bring the perception of the knot-to-dot distance to the distance required by the internal standard—zero.

How could we test whether the PCT model is right, or whether the stimulus-response interpretation is just as good? According to PCT, what is being controlled is a perception of the knot and dot. The stimulus-response interpretation (in one form) says that C is responding to movements of E's hand. So the two theories are actually claiming that C is responding to different perceptions of the situation, and we ought to be able to decide which claim is right.

An easy test would be to get a piece of cardboard and use it to keep C from seeing first E's hand, and then the position of the knot. If C has been responding to movements of E's hand, then blocking the view of E's hand while still allowing the knot to be seen should greatly modify C's behavior. On the other hand, if C is perceiving the relationship of knot to dot, blocking the view of E's hand should have no effect on C's actions, while blocking the view of the knot and dot should make control much worse, if not destroy it. If you want to be sure what would happen, you can get a piece of cardboard and actually do those two things, though it would be easier simply to ask C, "Are you watching E's hand or the knot?" C will deny paying attention to E's hand.

Doing this test more formally, using instrumentation and computers, shows that control of the knot-to-dot distance depends critically on the controller's being able to see the knot and the dot, and not at all on the ability to see the cause of disturbances of the knot. I will show several examples of this fact, demonstrated by the use of computers, in the next chapter. Recognition of this fact is one of the crucial differences between PCT and other psychological theories. Other theories try to explain how it comes about that people perform particular acts—such as moving the end of a rubber band in a particular direction. PCT tries to explain how it can come about that people maintain a particular *perception*—such as the relation between a knot and a dot. Recognizing the fact makes a huge difference in the success of the explanation.

As well as using a piece of cardboard to hide the knot, there is another way to test for control. The idea here is simply to find out whether the knot is doing what it would be doing under solely physical effects. Let C, for a moment, hold C's end of the rubber bands stationary. Let E start with the rubber bands almost slack, and then pull directly away from the dot by about six inches. Watch the knot. The knot will move half as far as E's end of the rubber bands moves. This shows us the effect on the knot that E's disturbance has when C does nothing. E could figure this out without any help from C at all. E wouldn't need C's finger to hold one end of the rubber bands in place. C could go to lunch, and E could use a dowel in the table to hold C's end in one position, and E could watch the knot move half as far as B's finger moved.

But with C's finger hooked into a rubber band and with C acting to control the position of the knot, E can now apply exactly the same disturbance as before and observe what the knot does. Now, of course, pulling back by a calibrated amount will have essentially no effect on the position of the knot. The knot will move only a tiny fraction of the amount that it moved when there was no control system attached to the other end. This failure of the disturbance to have the physically predicted effect is a strong clue that there is a control system acting. It is not infallible as a proof that control exists, because you still have to rule out simpler explanations for the lack of effect, but it is infallible in the other direction. If the amount of movement of the knot is exactly what you would predict under the assumption that there is no control system, then you have ruled out the existence of a control system. This test can eliminate wrong guesses very quickly, which is almost as helpful as being told what the right guess would be. Indeed, these two tests—cutting off C's sight of the knot and cutting off C's control of the knot—are essential parts of the procedure known in PCT lore as The Test for the Controlled Quantity, which is the core of experimental method in PCT. You can see that this method is eminently suitable to examining control on the part of an individual. I will say more about The Test in Chapter 7.

I have mentioned in earlier chapters that PCT includes multiple levels of feedback loops, though I have not yet explained much about that. We can, however, illustrate two levels of control with the rubber-band game. To do so, let C make the knot move very slowly and uniformly around the dot in a circle, with a radius of about one inch. The knot should take at least ten seconds to go once around the circle. E, of course, continues to move the other end of the rubber bands in big, smooth, slow, random patterns. If E sees that C is having trouble, E should slow down the disturbances. We want to see the controller succeeding, not failing.

Obviously, the internal standard is no longer "knot on dot." Perhaps, as many theoreticians in this field have done, you unconsciously assumed that the dot was specifying the internal standard—that the knot was the controlled perception, and it was brought to the standard set by the dot. Now, however, we can see that the controlled variable was really the *relationship* between the knot and the dot. Now the knot is being maintained in an ever-changing relationship to the dot. And if you still think the dot is not simply part of the controlled perception, we can let E choose to move the piece of paper as well as the rubber band—the two simultaneously. C is controlling a relationship between two perceptions, one of the dot and the other of the knot, and keeping this relationship in a match with an internal standard that now involves continuous motion.

If you are only reading this description, this won't be obvious, but if you are actually doing the experiment, you will realize that the experimenter, all this time, has been moving the disturbing end of the rubber bands around in big continuous patterns. You may have been thinking that to make the knot move in a circle, C has to make the hand holding the rubber band move around in a circle—bigger than the knot's circle, but a circle. Actually, if C were to hold a marking pen through the loop in the rubber band so as to leave a record of hand movements on the paper (this is worth trying), the trace would show not circular movements but a random mess.

In the movements of the knot relative to the dot, we are seeing the internal standard that C has chosen. The internal standard determines what the controlled perception will do. But in the movements of C's hand, we see a composite of the effect of the internal standard and the even larger effect of the disturbances. The hand movements correspond neither to the internal standard nor to the disturbance; they represent what has to be done to maintain control as the disturbance changes.

Let C now stop the motion of the knot at a point one inch to the left of the dot while E continues to apply disturbances. Now we are back to the original case where C's hand movements are symmetrical with those of E—but C is now maintaining the knot in a different and now stationary relationship to the dot. The control process is just like the first one, but with a different internal standard. We can call this one level of control.

The second level of control is the one that perceives continuous change. When the internal standard for this kind of change is the perceptual equivalent of "one revolution every 10 seconds," the knot moves in a circle because the *internal standard for knot position is being changed* so as to maintain that perceived circular movement. The first level of control, which is concerned with maintaining a particular, relative position of the knot and dot, is being used as the output of the second level of control, which is being used to maintain a perception of circular movement. The position control system is being used as part of a motion or trajectory control system. C could use a different trajectory control system, and make the knot write C's name. Many different higher-level control processes could be carried out using this same position-control system (although not at the same time).

Many more variations are possible, involving various internal standards, simultaneous control of more than one perception, more than two people, and multiple rubber bands. They are fun to explore. You can also do these experiments against paper on an easel, so that an audience can watch. Here, I will describe two further uses of the rubber bands.

Two Controllers

This is a demonstration of conflict. On the piece of paper, add a second dot about ¼ inch to one side of the dot that's already there. Now E disappears and becomes another controller, C_2; we have C_1 and C_2 controlling the same knot. The experiment is simple. C_1 aims to hold the knot exactly over the old dot, and C_2 aims to hold it exactly over the new dot. Their internal standards differ by 1/4 inch. If both controllers insist on keeping the knot over the "right" dot, there's only one possible outcome. A rubber band will break.

This seemingly innocent situation exemplifies the most serious problem that can arise between control systems, whether they are in different people or inside one person—conflict. PCT explains how conflict works and how it can cause immense difficulties; I will return to this topic in later chapters, especially 9, 23, 28, 29 and 33.

Four Controllers

This game can demonstrate cooperation, too. Neither paper nor pencil is needed. It is convenient to do it standing. Get eight rubber bands. Connect four of them in a circle, and attach the other four to the four knots. Find four obliging people. Ask each to take hold of one of the four rubber bands attached at the knots. Tell them, "Make a square" (of the first four bands). They will quickly do so, without needing to talk about it. Think for a moment about all the ways that the other three people can disturb the corner that is held by any one person. Despite the fact that any motion by one person will to some extent disturb the positions of all the other corners, the four people, without consultation, will somehow move into positions that result in a reasonably accurate square!

You can try various experiments with this layout. You can have someone give instructions about how to go about making the square. Will that square be made faster or better? You might have one group of four do it as described in the previous paragraph and another group (who have not watched the first group) do it after discussing the task and agreeing on how to do it. How would the performances differ? You might hook two rubber bands at each of the joints and use eight people. What would the additional people do? You can think of more variations.

Try to imagine for a moment all the sorts of little motions the four people around the rubber-band circle might make while bringing the bands into a square. Many people who design artificial intelligence for robots believe that a robot (or a person) cannot act without having inside itself, before it acts, a detailed "map" of the environment in which it is going to act. Can you imagine each of the people with the rubber bands trying to anticipate what each of the other three might do next? Each small motion by any person changes the environment for the other three, and all do that simultaneously and continually. Any map would become out of date the moment that anyone made any motion whatever. Yet people do this task of squaring the rubber bands with very little difficulty. I will say more about the idea of making an internal map in Chapter 24 under the heading "Model-Based Control." And I will say more about trying to anticipate specific future acts in Chapter 36 on planning.

You can find other descriptions of the game in Powers (1973, pp. 235–236 and 241–244), in Robertson and Powers (1990, Chapter 4), in Runkel (1990, pp. 105–108), and in Cziko (2000, pp. 87–89).

THE COIN GAME

Get four coins, a flat surface (a table-top or a patch of sand at the beach), and a friend. The four playing-pieces need not, actually, be coins. They could be checkers, or chess pieces, or little shells. They can be alike or different, as you choose. Charles Tucker, who teaches PCT at the University of South Carolina, prefers to use paper disks or poker chips, all alike. But here I'll suppose you will be using coins. As before, let one person be the Experimenter and the other the Controller. Let C arrange the coins on the table in any pattern C chooses. C might choose to have all the coins in a straight line. (That would be pretty easy for E to discern.) C might choose to have three of the coins in a cluster while leaving one isolated. Or C might choose to have the imaginary line joining one pair of coins always crossing through the imaginary line joining the other pair. You can think of a dozen other patterns, some obvious, some subtle.

The task of E is to discover, *without any discussion about it*, the pattern (internal standard) that C is exemplifying in the way C has laid out the coins. C should write down a description or definition of the pattern the coins are exemplifying. Now E can begin probing to discover C's pattern. E pushes a coin (or more than one) to a new position. If the result changes the pattern away from C's internal standard for the pattern, C must *correct the error*—that is, push the coin back to its previous position or to some position that corrects the error. If E's push of the coin does not take the pattern away from C's internal standard, C can merely wait or can say, "No error." ("No error" means "You have not caused me to feel that the pattern is now in error.") This process continues until E becomes certain of being able to make three moves that will bring corrective moves from C and three moves that will bring only a "no error" response. If C corroborates E's certainty, E and C compare their definitions.

Typically, E will begin the game feeling reasonably confident of eliciting a correction from C, and will be surprised when C says, "No error." Playing this game, it becomes very obvious how easy it is to think up explanations of "what C is doing" and how easy it is to be wrong about it. The game demonstrates, too, the relation between doing and talking. The three correction-eliciting moves and the three "no error" moves demonstrate that E can now *do* what C was doing, but E's oral description of the pattern may not sound very much like what C wrote down at the beginning of the game. C might have written down "Large to small," and E might have called it "a string of drops of water." An observer might say, "You are doing the same thing; I don't care what you call it." Or, after the three correcting moves and the three "no error" moves, E might say, "You were making a Z." And C might say, "No, it was an N." And an observer might say, "I thought it was a zig-zag."

Playing this game as E, you come not rarely to the point where you are sure of the pattern the other person has been controlling only to discover that the pattern was something very different. You might have settled on a geometric pattern when C was actually keeping the coins in order by date, or by size, or alphabetically by name: dime, nickel, penny, quarter.

This procedure, which is a variant of The Test for the Controlled Quantity, can make it easy for you to understand what it means to say, "You cannot tell what people are doing just by watching what they are doing." But I will phrase that more transparently: You cannot guess very accurately what people's purposes are just by watching their actions. That sounds reasonable, but most of us most of the time, I think, are too ready to believe we can descry the purposes of others. The coin game will help you to look at your own belief.

In psychological experimenting, as in other domains of social life, the pitfalls of language leave us very uncertain whether we have arrived at the condition we sought or have gathered the facts we envisioned. I devoted Chapter 6 of my 1990 book to the weakness of language, and in Chapter 11 there I told about some researches that were carried through with a minimum of language. The Test for the Controlled Quantity can often be carried out with no talking (or writing) at all; the coin game, after you have agreed with the other person on the procedure, can be played that way—silently. Saying "no error" speeds the game, but it is not necessary; you can just say nothing and let E conclude that your perception of the pattern is not disturbed, because you have not pushed a coin. You can see that The Test is not limited to humans; it can be used with any sort of creature.

When you play the coin game, remember that you are using it to see how control on the part of another person can be discerned. If you are playing the part of C, you want to see how E can discover the pattern you have in mind. Sometimes, maybe out of habit with games, a player seems to want to "win" the game by choosing a pattern that will be impossible for E to guess. If you do that, you will lose your chance to learn about control.

Do actually try these games. They yield insights you will never get by trying to imagine what the words here mean. The games will help you to discern control and non-control in everyday life. It is fun, too, to make up your own variations of these games. If rubber bands or coins seem beneath your dignity, remember that Galileo Galilei (1564–1642) discovered the shape of gravity by rolling little balls down a slanted piece of wood.

There is still another thing you can do without having a laboratory or a budget. Richard Marken has constructed demonstrations of several features of PCT. If you can get on the World Wide Web, you can interact with his models, and read his write-ups of them, at

http://www.mindreadings.com/demos.htm.

You can also find demonstrations and models by Powers at

http://home.earthlink.net/~powers_w/

and at

http://faculty.ed.uiuc.edu/g-cziko/twd/demos.html

and at

http://www.PCTresources.com

And you can get a script booklet and its accompanying video entitled "Rubber-Band Demonstration Introducing Perceptual Control Theory" dated 1993 and updated 2000. The script is by Dag Forssell, based on an outline by William T. Powers. For information, go to www.PCTresources.com.

Chapter 7

Some foundations

In this chapter and the two following, I will describe some research that has been done to test whether behavior as described by PCT—that is, the control of perception—can actually bring about the consequences that humans do bring about. In this chapter, I will describe some studies that demonstrate phenomena important to the theory. I begin by demonstrating that one cannot be sure of discerning the goal a person is pursuing in the acts you see the person performing. In Chapter 6, I invited you to demonstrate this to yourself and a friend by using the coin game. Here you will see another kind of demonstration.

WHAT IS THE PERSON DOING?

According to PCT, every act is a step toward a further goal; the act itself can never be a final purpose. The final purpose is always a perception. I put my foot forward not because the stepping brings a consummate satisfaction, but because the step enables me, now, to perceive myself shaking the hand of my friend, or to perceive myself conforming to the command of my lieutenant, or to perceive myself closer to the dinner table. Or perhaps I am recovering from a broken leg, and I am testing whether I can yet take a step with my healing leg. Still it is not the movement of the leg itself that is the goal, but the sensation that the movement brings, such as the absence of pain. When I smell the rose, it is not the rush of air into my lungs that I seek, but the sensation of the rose's perfume as the air passes through my nose. People initiate acts and guide the progress of their actions, but only to control the *perceptual consequences* that result.

It is not always easy for an observer to divine the further purpose an act is satisfying. It is true that the immediate purpose of a person—what the person is trying to do just now—is often obvious. Examples are a person swinging a club at a golf ball, putting a coin in a turnstile, or putting an air hose to the valve of a tire. But the further purposes remain uncertain. Why does the person want to play golf? Why does she want to take the subway instead of a taxi? Where does he want to go in his automobile? We learn early not to jump to conclusions about the further consequences people are intending as they act—though of course we differ from one another in our alertness to possible purposes. PCT makes it obvious that all of us, no matter how observant or intelligent, must *necessarily* find difficulty every day or even every hour in guessing what people are trying to do—what perceptions they are trying to hold steady by their actions. The possibilities simply become too numerous. I explained the sources of possible actions in the section headed "Requisites for a Particular Act" in Chapter 1. (You may remember, by the way, that J. B. Watson urged us to ignore most of these possibilities.)

William Powers contrived the following demonstration of how easily we can be fooled by what someone seems at first glance to be doing. All experiments designed according to control theory exhibit the fact of control, but this demonstration is especially dramatic in showing that people control the *perceptual consequences* of their acts. Powers presented this demonstration at the meeting of the American Society for Cybernetics in St. Gallen, Switzerland, in March of 1987. An updated version of this program, including commentary, is available at http://home.earthlink.net/~powers_w/. Look for *Squaring the circle*.

Procedure

The participant sits before a computer screen with a hand on a joystick. A joystick is a small lever poised vertically and moored at the bottom by a ball joint.

The top end can move in any direction. When the joystick is moved this way and that in this demonstration, a dot on the right of the screen mirrors the motion, since the computer is programmed so that the connection between the top end of the joystick and the position of the dot on the screen is direct and proportional. One unit of motion of the joystick produces one unit of movement of the dot on the screen and in a corresponding direction.

The screen also shows a dot at the left side. When the top end of the joystick describes a circle, the left dot follows the motion directly, just as the right dot does. But when the joystick moves radially, either inward toward the circle's center or outward away from the center, something else happens. Actually, a "home" circle is specified by the program in the computer. The speed of radial movement of the left dot changes according to where the joystick (and the right dot, too) lies in relation to the home circle. For radial motion, the link between the motion of the joystick and the left dot is not proportional (linear), but accelerated. When the joystick (or the right dot) lies on the home circle, the left dot does not move radially. But when the joystick lies away from the circle (either inside or outside), the distance of the joystick or right dot from the home circle specifies the *rate* of radial motion of the left dot. The farther the joystick lies from the home circle, the *faster* the left dot will move radially.

The right dot provides a direct record of the movements of the joystick. But between the joystick and the left dot, the relation itself shifts as the position of the joystick changes either toward or away from the home circle.

The participant is asked to draw a figure with the *left* dot. The experimenter does not tell the participant the nature of the connections between the joystick and the dots. The experimenter tells the participant only to draw whatever figure the subject wishes with the left-hand dot. The participant must then move the joystick in whatever manner necessary to produce the figure the subject has chosen.

Now, in the next few sections, I am going to talk about this demonstration as if it is an investigation of what people can do and how they can do it. I will be talking about what we can learn from this demonstration—as if it is an experiment. Accordingly, I will here and there compare the PCT point of view with the traditional point of view you find in most books and articles about psychological research.

Results

Figure 7–1 shows what one participant did. On the left, we see that the participant chose to draw two squares and a triangle. On the right, we see the direct record of the movements the participant made with the joystick to produce the figures at the left.

The experiment demonstrates dramatically the fact that the participant was controlling his perceptions of the movements of the left dot. He moved the joystick in whatever way necessary to enable himself to perceive that the dot was describing the figures he wanted to see it describe. If you were watching only the participant's hand or only the right side of the screen, you wouldn't have a clue to the geometry the participant had in mind.

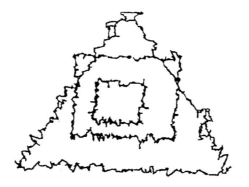 

Figure 7–1.

In our ordinary way of thinking about purposive action, we would expect the right-hand trace to give us some clue to what the subject was "doing." If, for example, a supervisor had told a worker sitting at this screen to draw a square, we would not be surprised if, watching only the right-hand trace, the supervisor were soon to say, "Hey, I thought I told you to draw a square!" Here, however, as in a good many ordinary situations, we see no clue to the person's intention in the detailed acts of his hand. In every case, no matter what figure the participant chooses to draw, we see the participant's hand moving, with some seeming inaccuracies, around an apparent circle. Except in those cases where the participant actually does choose to draw a circle, we would always guess wrong about what the participant is doing.

And we would be wrong, too, about the "inaccuracies." The small deviations from the circle at the right side of the screen look like errors, but they are not. They are in fact the small, necessary, purposeful movements by which the subject moves the left dot in the pattern he wants to see. The right side of the screen shows the person's *moves*, but the left side shows the person's intended results.

Assumptions

We take for granted, of course, the fact of control of perception. If the person could not see (perceive) the left side of the screen, the person could have no way of knowing how much or in what direction the dot was moving; he could draw a figure only in imagination. If the participant were blindfolded, given a pencil, and asked to draw a square on paper, he could do a fair job of it because of the direct connection between the sensed hand-motion and the line on the paper. But with the complicated connection Powers built between the hand of the participant and the cursor movement, no vividness of imagination would be sufficient for success.

The experiment assumes that *all* humans act this way. Can you imagine, once the person accepts the task, that any physically normal person would perform differently? You don't need to count proportions of people who behave as predicted or test for statistical significance. You don't need to run a "control group." Would it help us to understand this demonstration if we were to ask a participant to do the task blindfolded? No, it would not. We did not enter the experiment, as a traditional methodologist might have done, with the hypothesis that, on the average, participants with uncovered eyes would do better at drawing a figure than blindfolded participants. Our hypothesis was that *every* participant with uncovered eyes would succeed at drawing the figure despite the unrevealing pattern described by the participant's hand.

The experimenter does not need to wonder whether the participant understood the instructions. Once the participant produces a figure with the left dot, it doesn't matter whether the participant actually heard the experimenter talking or just happened to feel an urge at that moment to draw a figure.

An important assumption is that we can learn how behavior is managed only if we track it on the same time scale that it actually occurs. Suppose someone had given the instruction, had then walked out of the room, had kept no record of the participant's hand movements, and had come back later to find a square showing on the screen. That observer would naturally suppose that the participant's hand had moved in a square to produce the square on the screen, and it would be easy to conclude that the instruction had set in motion a square-drawing routine for the hand to carry out. But the moment-to-moment record made by the computer tells a different story.

What's Remarkable?

I think this demonstration shows with remarkable clarity the fact that people control their perceptions. Furthermore, the experiment shows the hierarchy in the neural net. The internal standard for the intended figure is necessary to set the standards for directions and amounts of hand movement, but the reverse is not true. Our ability to move our hands in various directions and amounts does not tell us what figures to draw. In other words, at one level there is control of the consequences of muscle contractions (speed, direction, and duration of movements of the hand) and at a higher level there is control of the consequences of those movements (production of the intended figure). I will say more about the hierarchy in Chapter 18.

The demonstration shows plainly how appearances can deceive us: (1) how we can go wrong by focusing on acts (the right side of the screen) instead of purposes (the left side) and (2) how we can go wrong if we take the line through the middle of the dots (in this case the home circle amid the dots at the right) as the real thing and call the deviations from it "error."

CHEMOTAXIS

I said in Chapter 3 that the negative feedback loop is always on—that the organism continuously monitors the status of the controlled variable, though action may be intermittent, taken as the opportunity arises. I have also said that a hierarchy of millions of those loops can enable an organism to do marvelous things. Here I want to show how that continuous monitoring and that kind of loop can do marvelous things even in a tiny hierarchy in a microscopic creature.

R.S. Marken read a book by D.E. Koshland (1980) about the behavior of the bacterium *Escherichia coli*, which lives in the intestines of various mammals and swims this way and that to find denser concentrations of nourishment. Marken was struck by the difficulty this behavior posed for reinforcement theory, but here I will set that question aside; I want only to use Marken's experiment to show how powerful the negative feedback loop can be in enabling even the simplest creatures capable only of the simplest of acts to maintain a purposeful progress. The *E. coli* moves through chemical gradients by wiggling its cilia—the hair-like appendages extending outward from the cell wall. It moves in one direction by coordinated movements of the cilia—by rowing movements, if you like. When it wants to change direction, however, it has no way to coordinate its ciliary movements to produce a particular new direction. It can only flail randomly with its cilia and produce thereby a random change in direction of movement. Marken (1985) and later Marken and Powers (1989) carried out experiments in which a human participant was limited to the same capabilities as *E. coli*.

Procedure and Results

In Marken's 1985 experiment, he showed participants a computer screen on which were to be seen three little squares and a small dot serving as a cursor. The squares stayed put, but the cursor moved constantly, without any signal from the participant to do so. The participant could alter the *direction* of the movement of the cursor by pushing the space bar on the keyboard. The new direction of the cursor was wholly unpredictable; it was selected randomly by a program in the computer. Participants were asked to choose one of the targets and keep the cursor near it by pushing the space bar at moments of their own choosing. All participants were able to do that. Participants succeeded at the task by letting the cursor continue as it was when it was moving toward the chosen square and by pushing the space bar when they saw the cursor moving away from the square.

That is just the way the bacterium called *E. coli* proceeds. When the nourishment in the surrounding fluid is sufficient or increasing, *E. coli* swims straight ahead. When the nourishment decreases, *E. coli* reverses some of its cilia, causing itself to tumble randomly for a moment, and then proceeds as before, but now on a new and random direction. This method of navigating is surprisingly effective; *E. coli* spends much less of its time tumbling and swimming the wrong way than the right way. Marken's humans operating the space bar performed with similar effectiveness.

Marken and Powers (1989) built a model to behave like *E. coli* and reported the experiments they carried out to test the model. I will not repeat those reports here, but I will show a couple of their figures to give you the flavor of the performances of their participants.

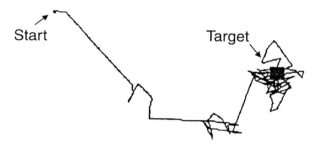

Figure 7–2.
Typical behavior of spot produced by a person

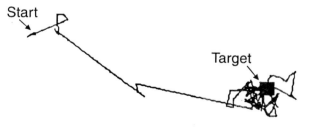

Figure 7–3.
Typical behavior of spot produced by a control-system model

Note:
For a demo you run in your browser, see http://www.mindreadings.com/ControlDemo/Select.html. For a PC version, download *E. Coli reorganization* from http://home.earthlink.net/~powers_w/

Figure 7–2 shows the way one of the human participants succeeded in controlling the perceived proximity of the cursor to the target, and Figure 7–3 shows the success of the model in doing so. You can see in those figures how rapidly the method of random-change-of-direction brings the cursor to the neighborhood of the target and how well it then keeps the cursor close to the target and repeatedly on it. Marken and Powers (1989, p. 93–94) say:

> By adjusting parameters, we have been able to make this process as much as 70% as efficient as a straight-line motion to the target, in terms of average velocity in the right direction. . . . Nothing seems to faze it. . . . This mode of action presumes little about the properties of the world surrounding it. Where a systematically behaving organism depends on the world's maintaining its properties reasonably constant, this randomly acting system can work even under radical changes of conditions. . . . The method [of *E. coli*] is the only feasible way for an organism to maintain control over important effects on itself when its environment is totally beyond its comprehension.

Assumptions

This demonstration requires some of the assumptions made in the previous demonstration, and I won't repeat them. Another assumption clear in this demonstration, however, is that planning is not necessary for the successful pursuit of a goal. Many psychologists and researchers in artificial intelligence believe that a living creature can act only by first making an internal map or some sort of representation of the external world, then figuring out a successful path through that map, and then act by moving *as if it were moving through that map*. If Marken or Powers had believed that, it would never have occurred to them to design this demonstration. Obviously, the bacterium makes no map. It simply senses the gradient of nourishment—the change of concentration—and compares this sensed signal with an internal standard such as "not lessening." If the gradient is indeed not lessening, *E. coli* proceeds as before; if it is lessening, *E. coli* tumbles randomly. (Marken and Powers explain this in more technical detail in their 1989 article.) The model built by Marken and Powers works in the same way, without a plan, and the human participants perform exactly like the model.

What's Remarkable?

Escherichia coli does have some organization of sensory and motor functions, primitive though it may be. It has some sort of memory, brief though it may be; it can say to itself, so to speak, "A moment ago the food concentration was that; now it is this." It compares those sensings to an internal standard, and it tumbles or not according to its error signal. What a marvel. The poor thing has no neurons; all those vital, delicate functions must operate chemically.

The main lesson I hope you will see in this experiment is that a living creature can be very effective even with very simple, primitive internal functions and the simplest of on-off external actions when they are built upon the negative feedback loop operating continuously. Critics of PCT sometimes complain that the negative feedback loop is "too simple" to explain the complicated behavior of living creatures, especially humans. In saying that, they neglect three ideas: (1) the complications possible when millions of loops are put together, (2) the subtle and multifarious patterning that is possible when the loops are organized hierarchically, as Powers postulates, and (3) the astonishing effectiveness (shown in this demonstration) of even the most simple and primitive loop organization. I will write again about this process of seeking randomly for the right direction (so to speak) in Chapter 20, where I describe the restorative process that Powers calls "reorganization."

INTENTION

You will remember from the first pages of Chapter 4 that J. B. Watson urged us in 1912 and again in 1929 to sweep aside all "medieval conceptions" such as purpose and note only "what the organism does or says." And if the person observed were to say something about purpose, Watson would urge us to ignore those remarks, because we could never know whether such a thing as purpose, such a medieval conception, could actually exist. Many other psychologists, while wanting to grant purpose to living creatures (perhaps especially to humans), have claimed that we can have no inkling of anyone's purpose or intention from direct observation, but must fall back on what the person tells us.

If, however, there is a central postulate of PCT, it is that we act to *counteract* events in the environment that would cause something to be the way we do not want it to be—events that would, if we did not act, push some quantity away from the internal standard we have set for it. When you see something in the environment that is not in a place where inanimate forces would have left it, you know that purposeful acts have occurred. When you see a steep mound of earth with nothing growing on it, with holes in it here and there where ants go in and out, and having a kind of shape you would never expect from geological forces, you conclude that those ants have been acting with purpose. When you see a field of wheat with very few other kinds of plants growing in its midst, the "unnatural" pattern tells you that a farmer has been purposefully at work.

When you see letters of an alphabet on paper not scattered randomly, but arranged in ordered rows, you know that a purposeful writer has been busy. When you watch the *Escherichia coli* making its random changes of direction, but note that it moves through its environment *not* randomly, you know that the tiny creature is moving with purpose. Purpose, whatever it may seek in detail, always works to maintain a perceived quantity against disturbances from the environment. This logic, to look for a feature of the environment that does *not* change when you would ordinarily expect events in the environment to change it, is the core logic in the Test for the Controlled Quantity, which I will explain in more detail later in this chapter. The Test provides the core logic for experimentation with PCT, as you will see as we go along.

(By the way, you will remember that in Chapter 5 I warned you to beware how I write. In the paragraph above, I wrote, "Purpose, whatever it may seek in detail, always works . . . ." I hope you raised an eyebrow at that sentence. Purpose is only an idea in our heads; it is not a live thing that can *do* something. There does not exist a thing called "purpose" that can "work" in some way. Our brains can work. Purpose can't seek; *people* seek. That sentence I wrote is not scientific. But I left it in because I don't want to be scientific *all* the time; I think my prose needs attention to esthetics now and then.)

Though we cannot look inside a person's nervous system and announce, "This fellow is looking for some banana-cream pie," we can come close, sometimes very close, to describing the variable the person is controlling by using The Test. It is true that sometimes The Test requires a good many trials to eliminate variables we thought might have been controlled but turn out not to be so, just as the Coin Game (in Chapter 6) requires a series of guesses). By using a simplified environment in a computer, however, we can rather easily demonstrate how to use The Test to divine the controlled quantity. Richard S. Marken (1982) did that, indeed, in the demonstration I will now describe.

Procedure

The participant sat at a video monitor on which two vertical lines were displayed as in Figure 7–4. The participant was asked to choose one of the lines, the upper or the lower, and move it back and forth across the screen by pressing the left and right arrow-keys. The lines did not move, however, in direct linear relation to key presses. The computer was programmed to insert slow random disturbances between the key and the lines. Furthermore, pressing either key caused *both* lines to move, though at different rates. Pressing the left-arrow key, for example, caused *both* lines to move to the left, but at randomly different rates.

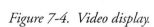

Figure 7-4. Video display.

It was impossible to tell the line the participant had chosen by watching the screen. The participant was not told to maintain any regular pattern, but was free to change direction and speed at will. Nevertheless, if the participant was following directions, the movement of one line was by intent, and the movement of the other was an irrelevant side-effect.

The crux of the experiment is the fact that the participant must, indeed, act to control a perception and therefore to act *against* disturbances. If the participant were to do nothing—to touch neither arrow-key—then of course the chosen line would follow exactly the random movements programmed for it. The correlation over moments in time between the programmed disturbance and the position of the line on the screen would then be 1.0, since there would be no other effect on the line. But as soon as the participant acts to move one of the lines, then that line will move according to the sum of the effects of the randomization and the key presses. And to make the line go where the participant wants it to go, the participant must of course *counteract* the effects of the random disturbances. Since a randomly moving point must deviate randomly from almost any regular pattern of points, the participant's counteractions to produce a regular pattern of movement of the line will produce positions of the line on the screen that have a correlation close to zero with the programmed random positions.

(If you are unacquainted with the idea of correlation, it will be good enough at this point if you just take it to mean connected or going along with, where 1.0 means as tightly connected as a connection can get, zero means no connection at all, and –1.0 means fully connected in the other direction. In Chapter 26, under "Correlations," I will say more about the technicalities of correlations.)

Marken's prediction was that the participant would succeed very well in controlling the movement of the chosen line despite the random disturbances, with the result that the correlation between the positions of the line and the random disturbance would be very much less than 1.0. But the key pressing by the participant would have much less effect on the other line, since the participant would let it go wherever it went without trying to prevent it. Marken predicted that the correlation between the random disturbances and the other line would always be higher than that between the disturbances and the chosen line.

Results

Marken ran two participants, each in ten "trials" of one minute each. The correlations between line positions and disturbances are shown in Table 7–1.

Before each trial, participants wrote down the line they intended to move. The lines are indicated in the table by "U" for the upper line and "L" for the lower. You can tell the line the participant intended to move by comparing the correlations. As Marken predicted, the average of the differences in correlation was large, though two of those for participant LH were small—a difference of only .08 (that is, .14–.06) in trial 4 and of .03 (.28–.25) in trial 7. The largest differences were those of .50 for RM at trials 3 and 9 and .73 for LH at trial 3.

	Participant RM		Participant LH	
Trial	Intended line	Other line	Intended line	Other line
1	.06 L	.40	.12 U	.22
2	–.04 U	.31	–.03 L	.43
3	–.13 L	.37	–.16 L	.57
4	–.02 U	.28	.06 L	.14
5	–.04 U	.40	.09 U	.26
6	–.09 U	.31	.25 L	.40
7	–.07 L	.34	.25 U	.28
8	–.02 L	.32	.03 L	.38
9	–.04 U	.46	.23 U	.36
10	–.11 L	.41	.06 U	.34
Means	–.05	.36	.09	.34
Range	–.13 to +.06	+.28 to +.46	–.16 to +.25	+.14 to +.57

R 65, T 8

Table 7–1. Correlations between disturbances and positions of upper and lower lines. "U" and "L" tell whether the participant reported intending to move the upper or lower line.
Adapted from Marken (1982, table 1, p. 649).

Why was the correlation with the other line *always* higher than that with the intended line? There was nothing in the set-up that would have brought the correlation between the positions of one of the lines and the random disturbances close to zero except the control—the opposition to the disturbances—being exerted by the participant. The two correlations would have been higher and more alike in value, for example, if the participant had merely pressed the keys lackadaisically, from occasional urges to relieve the boredom of sitting in one place.

The more unremitting the participant's insistence on opposing the random disturbance to the chosen line, the closer to zero the correlation would go. To me, the impressive feature of the outcome is not the fact that one or another correlation was always lower—that was inevitable if only by the nature of arithmetic. The impressive feature is that one of the correlations was so often very close to zero. Then, when we get the information that the lesser correlation was *always* attached to the line the participant intended to control, the outcome is still more impressive.

Marken published a report of an experiment also showing intention but with a different task in 1983, reprinted in 2002. In 1989, Marken published a report of another ingenious experiment on divining intention in which he gave the onlooker a way of seeing the intention of the participant simply by looking at the computer screen. He also showed there another way to put numbers on the results. I will let you look up these two reports for yourself.

Assumptions

As always with PCT, Marken's experiment assumes circular causation in a feedback loop; the participant acts at every moment to maintain the desired movement of the line despite disturbances from the computer that would otherwise disrupt the motion. The experiment also makes the assumption of species; that is, that every normal member of the species functions by the same principles. In this case, the assumption is that every normal member of the species can exert control in a way that brings the correlation between the line position and the random disturbances close to zero.

What's Remarkable?

The experimenter instructed the participant to choose a line and move it back and forth across the screen. The participant did that. What is remarkable about that? The point of this experiment, of course, was not that the experimenter succeeded in getting the participant to follow instructions. The points were (1) to show that humans behave as if they have purposes or intentions, (2) to show that you can discover an intention even in a situation where the naked eye cannot discern the part of the environment the person is acting upon, and (3) to show how PCT enables you to make that discovery.

The experiment seems simple. It will seem less so if you read the original report. It will seem even less so if you try to design a similar experiment yourself. And it will seem still less so if you think for a moment about all the argument in the psychological literature about purposes and intentions and other inner states.

SELF-CONCEPT

Organisms act to oppose disturbances of perceived variables that they want to control. It is fairly straightforward to demonstrate that principle with movements of dots and lines on the screen of a computer. Special difficulties arise, however, when demonstrating it with a variable conveyed with oral language between humans. Yet Robertson, Goldstein, Mermel, and Musgrave defied those difficulties in carrying out a series of experiments on the self-concept. I will recount here one group of four of their experiments. My account follows their paper of 1987; a slightly revised version was published in 1999.

Procedure

In these experiments, the researchers gave students in college psychology classes 80 cards bearing adjectives describing personality characteristics. The researchers asked participants to pick 16 adjectives from the 80 that they could confidently judge to be like them or not like them. They asked the participants to sort the 16 adjectives into piles, putting just one into the pile labeled "most like me," a certain number into the next pile, and so on.

Once that was done, the students met in pairs. One student in each pair was labeled the "experimenter," the other the "participant." Without the participant's knowledge, the "experimenter" in each pair had previous instructions. Following those instructions, the "experimenter" looked over the participant's sorting, then read aloud the most-like-me adjective, and said, "No, you're not _____," pronouncing the adjective as the last word in that sentence. The "experimenter" then wrote down exactly what the participant said immediately after that.

Results

The researchers had postulated that all of us carry about self-images that act as internal standards in higher-level control systems. (Here again, I am referring to the neural hierarchy.) Like all higher-level standards, the researchers argued, self-images "tell" lower-level systems the kinds of standards they should "require" of incoming perceptions. Since every participant had picked, from a large variety, his or her own most-like-me adjective, the researchers predicted that every participant would act to oppose the statement by the "experimenter." The statement "No, you're not _____" would threaten to disturb the self-image, and the participant would counteract the disturbance.

Robertson and his colleagues coded all the utterances of the participants that the "experimenters" had written down. The researchers reported four experiments conducted in this manner, having a total of 35 participants. They found only one utterance that did not seem to oppose the presumed disturbance. They found two utterances they were unable to code as opposing or unopposing:

Opposing utterances:	32
Unopposing utterances:	1
Uncodable utterances:	2
Total participants:	35

Assumptions

I think several features of the design worked out by Robertson and his colleagues helped the experiment to work well. First, the researchers did not (as others might have done) pick out a particular dimension (intelligence, for example) and assume that all participants would care about it during the experiment. Instead, from a highly multidimensional collection of 80 adjectives, they asked the participants to pick 16 they were able to say with some firmness of opinion were like them or not like them. That is, they allowed every participant to pick his or her *own* salient dimension.

Second, Robertson and colleagues did not pick a particular kind of action to indicate opposition to the disturbance. They gave no instruction whatever at the point when the participant's self-image was presumably threatened. They knew, of course, that the handiest use of the environment for almost all the participants would be some sort of oral act with words. Every participant, nevertheless, was free to choose his or her own use of the environment in counteracting the disturbance—whether words, hostile stares, expectorations, or punches in the nose. Actually, in writing about their coding, Robertson and colleagues mention only verbal utterances.

Third, Robertson and colleagues kept self-image salient by allowing only a few moments between the sorting and "No, you're not _____" and by enabling the counteraction to occur only a split second later. The short times reduced the chance that some other high-level standard would come into play.

Fourth, Robertson and colleagues reduced to a minimum the use of language and therefore assumptions about the efficacy of communication. The "items" they used required no agreement about meaning between the participants and the researchers or among participants. Neither the researchers nor participants needed to understand anything about how any participant sorted the adjectives. At the point of sorting, the only common understanding necessary was the understanding between researchers and participants of the words "like you" and "not like you."

Robertson and colleagues admitted that so far they had not achieved perfect results. They implied that they would continue to seek improvements in their methods, to which they referred as "primitive."

What's Remarkable?

In a demonstration such as mimicking the behavior of the *Escherichia coli*, it is easy to tell whether the participant is controlling a perception of a relational standard such as nearness to a target. When working with a high-level standard like self-image, however, and with words, it is not easy to be sure the high-level standard you are testing is always the one in control. Robertson and his colleagues did not track the maintenance of self-image over a number of minutes, but only at the one instant of the reply to the "experimenter." It would be very difficult to design an experiment that would track a particular high-level standard over a period of time, even a short period.

According to control theory, the person acts on the environment only when the maintenance of a standard is threatened and when the person can find an action that restores the desired perception. We can, therefore, see a particular higher-level standard acting over a period of time to control perception only when the person can find counteractions to take during that period and when no standard at a still higher level takes charge during that period. In ordinary life, those conditions do not hold for very long periods of time except in situations the person experiences as stressful or as a period of severely focused concentration. To use strong stress in the laboratory to hold a high-level standard in place would be unethical, and to find fascinating activities that can dependably draw severely focused concentration uninterrupted for some minutes is very difficult.

Considering those features of the high-level control of perception, I think the achievement of Robertson and colleagues—the score, so to speak, of 32 out of 35—is remarkable. It seems to me that several kinds of other high-level standards could have come into control in one or another of the participants at the crucial moment when the "experimenter" said, "No, you're not _____."

One kind of standard other than self-image could have been something like, "I want to say something now that will please Dr. Robertson." A second kind could have been, "I want to be nice to my fellow student." A third could have been, "I want to protect myself against the possibility that this student and the professor are in cahoots to deceive me about something." A fourth could have been, "Part of my picture of myself is my understanding that other people do not always see me as I see myself. This person is entitled to his view of me. No comment is necessary." I cannot, however, imagine how these other standards could be predicted to produce such a high percentage of statements in opposition to "No, you're not _____."

Experiments designed according to the linear assumption of input-output are always weakened by the influences of standards like those I just listed—standards experimenters ordinarily do *not* want to be in control. I think Robertson and colleagues showed great ingenuity in working out an experimental design that reduced to only 3 out of 35 the chances for those unwanted standards to come into control.

Finally, I think it is remarkable how much Robertson and colleagues were able to reduce the degree to which the outcome relied upon agreement between researcher and participant about the meaning of words. We need more experiments designed to reduce reliance on semantic agreements. Of course, Robertson and colleagues must still rely on words to convey a picture of their experiments to *you* and *me*. I don't see any way out of that.

Bryan Thalhammer (2000) carried out a study on the effect of threats to the self-image in an educational setting. "Participants," he wrote, "reframed their perceptions of the interaction around [a] computer task to regain justification in their self-image as good learners and subjects."

THE TEST

Marken's study of intention and the study by Robertson and colleagues of the defense of the self-concept both demonstrate uses of the Test for the Controlled Quantity, which is the basis for method in all PCT research. Marken's purpose was to show how purpose or lack of purpose can be discerned in the relation between environmental events and the perception the participant brought about of the movements of a line on the computer screen. In Table 7–1, we can look for the connection (the correlation) between the position of a line and the disturbances given it by the environment. In every trial by either participant, when we look for the line having the *lower* correlation with the environmental disturbance, that line turns out to be the one whose position the participant intended to control. The other line follows the disturbances to a much greater degree. The correlation of the other line with the disturbance given the chosen line was not high—did not approach 1.0—because the position of the other line was also connected to the arrow keys. The correlations between the intended line and the disturbance were small, averaging –.05 for one participant and .09 for the other. That pattern exemplifies The Test, which tells us to look for a variable (one perceivable by the person) that is acting as if a purposeful influence is acting on it. It is often clearer, actually, to say this in the negative: we look for a perceived variable that is *not* behaving as it would in a nonliving environment. We look for a perceived variable that is *not* behaving as it would if there were *not* a purposeful influence acting on it.

Robertson and colleagues wanted to test the opposition of action to disturbance in maintaining a perception very high in the neural hierarchy—the self-concept. They did that by having the "experimenter" present the participant with an idea that they thought would contradict or disturb the participant's self-concept. That is the core idea of The Test: disturb the presumed perceived quantity and see whether the person opposes the disturbance.

Notice how different The Test is from traditional psychological research. Traditionally, psychologists have looked for strong correlations between input (the "independent variable") and output (the "dependent variable"). In Marken's experiment, that would be the correlation between the disturbance (input) and the position of a line the participant can act upon (output). In Table 7–1, the correlations with the line-position the participant does act upon are those in the columns headed "Intended line"; contrary to the traditional assumption, they are *not* strong, but are very small. Under PCT, they are of course predicted to be small; the intended line is the line moved purposely by the participant and not abandoned to the influences of the environmental disturbances. The correlations with the line-position the participant ignores, shown under "Other line," are much larger, also as predicted.

I turn now to a more formal and detailed description of The Test for the Controlled Quantity. The procedure is contained in the following nine steps. I have rephrased them from Powers's 1973 book, pages 232–246, and his 1979a (vol. 4 no. 8, September) article, pages 110, 112.

1. Select a variable that you think the person might be maintaining at some level. In other words, guess at an input variable. Examples: light intensity, sensation of skin temperature, admiration in another person's voice. (Powers often speaks of the input *quantity*, because one usually looks for an amount or degree of some variable—such as temperature—the perception of which is controlled.)
2. Predict what would happen if the person is *not* maintaining the variable at a preferred level.
3. Apply various amounts and directions of disturbance directly to the variable.
4. Measure the actual effects of the disturbances.
5. If the effects are what you predicted under the assumption that the person is *not* acting to control the variable, stop here. The person is indeed not acting to control it; you guessed wrong about the variable.
6. On the other hand, if the effect is markedly smaller than the predicted effect, look for what the person might be doing to oppose the disturbance. Look for a cause of the opposition to the disturbance which, by its own varying, can counterbalance variations in the input quantity (such as pulling as necessary on the handle of your umbrella to keep the wind from carrying it off). That cause may be caused by the person's output. You may have found the feedback function.
7. Look for the way by which the person can sense the variable. If you can find no way by which the person can sense the variable (the input quantity), stop. People cannot control what they cannot sense.
8. If you find a means of sensing, block it so that the person cannot now sense the variable. If the disturbance continues to be opposed, you have not found the right sensor. If you cannot find a sensor, stop. Make another guess at an input quantity.
9. If all the preceding steps are passed, you have found the input quantity, the variable the person is controlling.

When you find the controlled variable, you can then usually make a very good guess about the nature of the internal standard controlling it. But describing the internal standard in precise words is not of first importance. The important thing, both for further experimentation and in practical affairs, is to have found how to disturb the controlled variable and how to avoid doing so.

Sometimes, both in research and in everyday life, you can ask people to adopt temporarily an internal standard that you describe to them. If they have themselves freely chosen to comply with your request and if you can describe the internal standard clearly and objectively (as in the examples of research I have given so far), you are off and running. At other times, you may not be able to persuade the person to adopt the internal standard you have in mind. Even if the person is willing, the person may not understand your request sufficiently well. In that case, you must start from scratch and use all the steps of The Test to discover what variable the person is indeed controlling.

Guessing Wrong

To use The Test, you must make a guess about an internal standard and then change something in the environment that the person senses. If you succeed in changing the thing—that is, if the person does *not* act to maintain it the way it was—then you have guessed wrong. If the person *does* act against the change you try to make, then you have guessed right, or at least you are on the right track. You know something about the person you did not know before. But you may have guessed wrong about the *aspect* of the change, the input quantity, that you think the person is wanting to maintain. You will discover that fact if later steps in The Test go wrong. Then you have to guess again. You will, however, be ahead of the game, because you know that the input has something to do with the change you tried to make in the environment.

It is easy for an onlooker, watching someone ward off a threat to a controlled variable, to make a wrong guess about the variable the person is trying to maintain. Cries of "No! No!" or "I won't do it!" or "You think you're pretty clever, don't you?" or a stony silence—those are all good indicators that some variable is being disturbed, but poor indicators of what the variable might be like. For example, what perception might a person be defending who said to you, "Don't talk to me that way!"? Here are some possibilities:

> I've been trying to be helpful to you, and now you tell me I've been actually been doing you harm. That's exactly opposite my intent, and it hurts me to hear that; I don't want to hear that.
>
> I don't want people to hear you speaking disrespectfully to me.
>
> That's a frivolous way to talk, and I want the people here to believe you are taking this seriously.
>
> You sound desperately discouraged; please don't give up hope. I want to hear optimism.
>
> You talk as if I have done something bad! I am not going to think of myself as a bad person! I don't want to hear you telling me I am a bad person!

Once you have made a guess, you can then hunt for something you can do that might disturb that variable. Then you have to be careful about interpreting the person's reaction. The person might want you to stop talking out of fear that you will disturb the variable you have hit upon, or simply because you are distracting the person's attention from a task the person wants to resume.

Even a simple physical action can be perplexing. You move through a crowded hotel lobby. You step aside to avoid someone and find yourself pushing against a third person. The third person makes a quick contrary shove that opposes your push. Is the person simply trying not to fall over, is the person maintaining his manliness, or is the person wanting to communicate an antipathy toward physical contact with strangers?

It is rarely possible, in the natural setting, to hit upon a good guess at the first try. Narrowing the possibilities requires several tries, sometimes a good many. People do, of course, learn a good deal about the internal standards of others after making wrong guesses for several months or years. Still, people can live together for decades, giving careful attention every day to evidences of disturbance, and still be surprised at the reactions of family members. People who claim, "I know what you are thinking!" after brief acquaintance are being fatuous; so are those who say, "Well, you ought to have known what I was thinking!"

Sometimes we are not sure whether a person is intending to control a variable. Sometimes, after we have "defined" a variable in such a way that we can recognize changes in it (for example, the brightness of light on a page or the number of people talking at once in a conversational group) and have tried to alter it, we find that the person pushes back, but not skillfully. That is, the person seems to show poor control. The person may be trying to control that variable, or the person's effect on it may be a side-effect of the person's intent to control a different variable. In a message to the CSGnet on 16 October 2000, Rick Marken said this:

> [If] The Test tells you that a variable is *not* being controlled very well, then there are at least three possible reasons for this finding:
>
> 1 The variable, as defined, is not a controlled variable. This is the default hypothesis when a variable fails The Test. The next step is to try a different definition of the possible controlled variable and Test again.
> 2 The variable, as defined, *is* a would-be controlled variable; the behaving system [for example, a human] is trying to control this variable but [has not yet found an effective way to do it]. This might be our hypothesis if we have reason to believe that most control systems of this type *do* control this variable.
> 3 The variable, as defined, *is* a controlled variable; the behaving system .. is not controlling it very well, because there is a conflict. This might be another hypothesis is we have reason to believe that most control systems of this type *do* . . . control this variable.

Ethics

Sometimes people new to The Test worry that they might do damage to the people they want to Test. It is essential, in carrying out The Test, to make sure, when you take an act you think will disturb a controlled variable, that you do *not* move (or speak) so strongly that the person will be unable to counteract what you do (or say). If you move or speak too strongly, you will not discover what you want to discover. Knocking the person over with a bulldozer does not tell you anything useful about the ability of the person to stand upright. The Experimenter with the rubber bands always wants to keep the amplitude of the disturbances small enough so that the Controller can easily maintain control. The piano teacher always wants to keep the bad news about the pupil's fingering small enough so that the pupil can quickly rectify the faults.

I know a couple of people who were one day talking about perceptual control while they were whizzing along a freeway at about 60 miles per hour. The passenger offered to demonstrate to the driver how people cope with disturbances. He took hold of the side of the steering wheel and pulled down, very gently at first, then more strongly, while the driver, of course, resisted that disturbance and kept the car going along in its lane. Then the passenger gradually, slowly released the wheel. You can see that the passenger in that Test certainly had no wish to exceed the driver's ability to keep the car in its lane. That is a good example to keep in mind when you are thinking of the degree of disturbance you want to apply. Just imagine that you might be killed if you pull too hard on the steering wheel.

In the next chapter, I will tell about some research that has been done to test whether behavior as described by PCT—that is, the control of perception—can actually bring about the consequences that humans do bring about.

Chapter 8

Some models of control

I turn now to a few examples of model-building. The models I will describe will be neither flesh-and-blood nor Tinkertoy; they will be models built in computers. The computers used are not supercomputers; they are the PC sort owned by millions of people nowadays. You can do this sort of thing yourself. It takes some study, of course. But if you can read well, if you are willing to learn some not-very-advanced mathematics, and if you can be patient with computing manuals, you can do the sort of thing you will read about here. I say again that these are tangible, operating models, not verbal arguments about how things might work. Furthermore, these models are built to work in unpredictable environments, just as do living creatures. I begin with a study showing the dependability of this kind of research.

ACCURACY AND RELIABILITY

Bourbon, Copeland, Dyer, Harman, and Mosley (1990) undertook to demonstrate the accuracy and reliability of predictions of tracking made with PCT. I recount here some parts of their project.

Bourbon and eight of his students performed a task of tracking a target on a computer screen. Every person performed the task more than once; all told, the nine persons replicated the task 104 times. And here I must pause to make sure you do not pass lightly over that statement—that the task was replicated 104 times. In psychological literature, replications are hard to find. Most experiments in the literature are never replicated; there are dozens of reasons, and I will not go into them here. Some experiments have been replicated once or a few times, almost never with results that could be shown quantitatively to be close to previous trials. (And in the methods of traditional psychological research, experiments called "replications" typically differ so much from the original that they might better be called "somewhat similar" experiments.) A few experiments, chiefly psychophysical experiments, have been replicated (so to speak) some dozens of times. I had never heard of an experiment, until I came upon PCT, that had been replicated 100 times. So please reach for your yellow highlighting pen and highlight "104 times."

Procedure

Figure 8–1 (from Bourbon and others, 1990) shows the experimental arrangement. The larger oval on the right side represents the computer. The oval at the top represents the computer screen; you see there three short lines labeled T, C, and T. The two lines labeled T were, together, the target; they were moved up and down in unison by the program in the computer. The task of the participant was to move a Handle (symbolized by the letter H underneath the ovals) forward and backward so as to keep the short line labeled C (the Cursor) as accurately between the Target lines as possible. Each replication (or "run") of the task contained two parts. In the first part, the Handle was the sole cause of the position of the Cursor between the Target lines; when the Handle was moved, the Cursor moved a proportionate distance. In the second part, a random disturbance was added by the computer program to the effect of the Handle. The effect of that, when the Handle was moved, was to cause the Cursor to move at an unpredictable speed, and even sometimes in an unexpected direction.

The duration of each part was one minute. The first part was used to determine the idiosyncratic performance characteristics (but not the particular acts) of the participant. I will explain what I mean by an individual's performance characteristics below under "The Model," and I will clarify some features

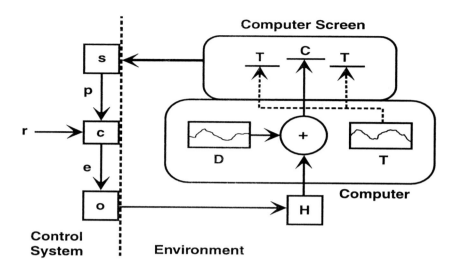

Figure 8–1.
Relations among variables in the environment and in the model.
In the environment, T=Target, C=Cursor, H=Handle, and D=Disturbance. In the model, the functions are s=sensor, c=comparator, and o=output; the signals are p=perceptual, r=reference (internal standard), and e=error.

of procedure there, too. The second part was the test of the theory; in it, the numbers characterizing the participant were inserted into the computer's model, the performance of which would be compared to the human participant's performance.

The random disturbance of the effect of the Handle is a feature of all tests of PCT with a computer simulation; the random disturbance simulates the unpredictable occurrences that interfere with the effects of our acts in non-laboratory environments. An example is the effect of unexpected gusts of wind when we point an automobile down the highway. Another is the distraction of an interruption from a third person when we are carrying on a conversation.

There were two forms of task (each run in each form containing the two parts so far described). Four students and Bourbon (five persons in all) carried out the first form. In this form, the Target moved up and down at an unchanging rate. Each person ran through both parts of each run ten times, with a few minutes between the first and second parts while the idiosyncratic constants were inserted into the computer model. The disturbance to the effect of the Handle added to the second part was different for every replication and every participant. When I say that the disturbances followed different patterns, I am not thereby saying that the replications were different in a substantive way. The unpredictability of the precise effects of the Handle is necessary to test the theory. If you want to know whether a person (or a mechanical substitute) is capable of driving an automobile from here to there, you will want to see the journey succeed more than once despite changing winds that affect the course of the automobile. Using this same task, Bourbon, acting only as a participant, ran four more replications of the first part, but then waited a year before running the second part.

Four more students and Bourbon participated in the second form of the task. This form was the same as the first except that the Target did not move at a regular rate; now, the Target moved according to a table of random numbers. This was presumably a more difficult task than the first form. Again, as in the first form, the Target paths in the first and second parts were different in every replication. Here again, each participant ran through ten replications.

Table 8–1.

Mean correlations between indicated pairs of variables calculated over fifty replications of the task with both target and handle randomly disturbed. Data from five participants, each giving ten replications with 1800 data-pairs calculated within each replication.

	By the participant	In the model
Between Cursor and Target	.984	.993
Between Handle and Cursor	.701	.707
Between Handle and Target	.710	.708
Between Handle and disturbance	–.682	–.696
Between Cursor and disturbance	.032	–.001
Between Handles of participant and model	.996	
Between Cursors of participant and model	.992	

Figures 8-1, 8-2 and Table 8-1 reproduced with permission of authors and publisher from: Bourbon, W.T., Copeland, K.E., Dyer, V.R., Harman, W.K., and Mosley, B.L. On the accuracy and reliability of predictions by control-system theory. Perceptual and Motor Skills, *1990, 71, 1331–1338. Copyright Perceptual and Motor Skills,*

Results

Table 8–1 shows some mean correlations among positions of Cursor, Target, Handle, and disturbance produced by the five participants in the later form of the task in which both the Target and the Handle were randomly disturbed.

To keep this narration simple, I am omitting the data from the simpler task performed earlier; those data (which you can see in the original article) tell much the same story as these later data. Figure 8–2 shows graphically the performance of one of the participants who contributed data to Table 8–1.

In the left graph, labeled "Person," you can see how very closely the participant was able to cause the Cursor C to track the Target T. You can see that same accuracy when you look at the first line of Table 8–1 under "By the participant." The correlation you see there between Cursor and Target is .984, only .016 away from a perfect score of 1.000. The participants achieved correlations like that despite the unpredictable disturbances given the Target and Handle. Since the mean is only .016 away from the maximum possible, it is obvious that all five participants gave very accurate performances. Those performances should not, however, surprise us. A lion chasing a gazelle must do that well or go hungry. The sailor of a sloop does that well to keep the sail full of wind. The driver of an automobile does that well or runs off the road. If a psychological theory is to be tested quantitatively, the theory should be able to mimic action that is as accurate as the action we see here. Indeed, in Table 8–1 under "In the model," we see the number .993, which is the average of the correlations between Cursor and Target produced by the model in the computer. The model's average correlation of .993 is very close to the mean of .984 among the live persons. Looking at the upper part of the graph labeled "Model" in Figure 8–2, we see that the records for C and T are so close together that it is difficult to tell that there are two records there. It is easy to see, too, that the records at the right, produced by the model, are similar indeed to the records at the left, produced by the human.

You can easily infer from Figure 8–2 that the correlation is somewhat positive between the Handle and either the Cursor or the Target, because some of the larger ups and downs go somewhat together. But you cannot pick up the record of the Handle and fit it perfectly to the record of Cursor or Target. The correlations between the Handle and the other records were not zero or negative; as you see in Table 8–1 in the second and third lines, those correlations were

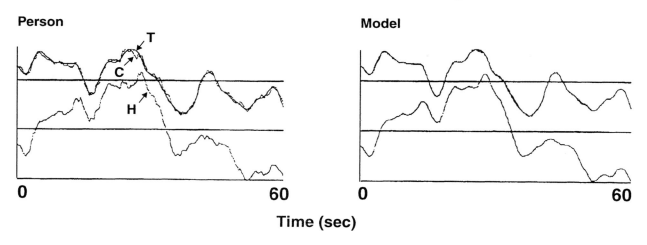

Figure 8–2.
Results of pursuit tracking by a person (at left) and by the corresponding model (at right). T = Target, C = Cursor, H = Handle.

somewhat positive, averaging .701 and .710 for the human participant and .707 and .708 in the model. Since the Handle was moved with the purpose of counteracting the disturbances put to it, the negative correlations of –.682 and –.696 between Handle and disturbance therefore account for almost all of the movement of the Handle.

Because of the action of the participant with the Handle, almost all of the movement of the Cursor followed the movement of the Target, and the relation of the Cursor to the disturbance given the Handle becomes irrelevant; the irrelevance is shown by the correlations lying very close to zero; namely, .032 and –.001. Finally, as we would expect, the correlations we see in the two columns of the table are very similar. The behavior of the model follows closely that of the person; .993 is very close to .984, and so on. Similarly, comparing the simulated Handle with the Handle operated by the participant, we see in Table 8–1 a correlation of .996. And the correlation between the simulate Cursor and the participant's Cursor was .992. Those results show the accuracy of predictions made by PCT for a pursuit task.

Reliability

By reliability, we mean being able to count on a phenomenon to repeat. In the trials (replications) by Bourbon and others shown here, the phenomenon repeated 104 times (including replications both within and between persons), and did so with the strong quantitative similarity shown in Table 8–1 and illustrated by the records in Figure 8–2. (You can see records produced by three other people and their models in the original article.) The reliability within one person was further demonstrated here by the four replications in which Bourbon was the participant and in which he waited for a year to pass before running the second part (the test) of the task. At the close of that year, the mean correlation between the modeled positions of the Handle and the actual (Bourbon's) positions was .996.

In addition to that, in 1988 Bourbon ran the first parts of eight runs, each against a different pattern of disturbance, and, as before, calculated the constants (reflecting the characteristics of the person) for the model. Inserting those constants into the equations for the model constituted a prediction of how the participant, Bourbon, would move the Handle in the future. In 1993, Bourbon performed the task against two of those eight disturbance patterns and compared the movements he gave the Handle with the predictions made five years earlier; he reported the results in the *Psychological Record* in 1996. The correlations between predicted and actual Handle positions in the two runs were .998 and .997. In January of 2001, nearly 13 years after Bourbon recorded the constants characterizing his capability in this tracking task, Bourbon performed two runs again as a demonstration for a group of educators in Phoenix, Arizona. A couple of months later, he sent the numbers to me by e-mail. The correlations between prediction (the model) and actual (Bourbon's behavior) for the cursor and for the handle were these:

	Run 1	Run 2
Cursor	.981	.964
Handle	.998	.999

The Model

As well as showing you the reliability of a model built from principles of PCT, I want to introduce you to the manner in which a PCT model is constructed. To do so, I will continue to use the 1990 report by Bourbon and others. The mathematics of the model follows the connections shown in the left part of Figure 8–1. The counterclockwise loop in Figure 8–1 shows the connections and functions that you saw in Figure 4–1 of Chapter 4. In Figure 8–1 here, you see on the representation of the computer screen the same kind of event that was labeled in Figure 4–1 "Events resulting jointly from disturbance function and feedback function," and you also see in Figure 8–1 the arrows leading from the disturbance and the Handle to be added together to affect events C on the screen. You can see here the arrow (input quantity) going to the left to the sensor **s**. And the arrow carrying the perceptual signal **p**. And the comparator **c** with its internal standard **r**. And the error signal **e**, the output function **o**, and the arrow (output quantity) affecting the action output **H**.

Now, to describe the features of the model, I am going to copy almost word for word from Bourbon and others (1990) on their page 1333. There are two constants in the model. One is the reference signal (internal standard) **r**—the distance the person desires to see between the Cursor and the Target. In each replication, Bourbon and colleagues estimated **r** to be the mean difference between Cursor and Target in the first part of the task. The other constant is the "integration factor" **k**, which is the "gain" applied by the function **o**. The symbol **k** does not appear in Figure 8–1, because it is not a function or a signal, but an amplifying factor applied by the output function **o**. This gain is the "amplification" I mentioned in Chapter 3 in connection with Figure 3–3 there. Here, **k** represents the velocity with which the person moves the Handle when there is error—when the person perceives the Cursor to be departing "too far" from the Target. The output function **o** multiplies **e** by **k**. To estimate **k**, Bourbon and colleagues substituted values for it into the equations for the model until the positions of the Cursor reached the highest possible similarity with those produced by the person. (Powers gives a more thorough treatment of **k** and other constant multipliers in the appendix to his 1973 book.)

In a run, all calculations were repeated 1800 times, once every $1/30^{th}$ of a second. For each iteration, the model subtracted the Cursor position **C** from the Target position **T** to estimate the perceptual signal **p**. The reference value (internal standard) **r** was subtracted from **p** to yield an error signal **e**, then **e** was multiplied by **k**, and that product was subtracted from the previous Handle position to yield the modeled position of the Handle in the next time interval. In the first part of the iteration, the modeled position of the Cursor was identical to that of the Handle; in the second part, it was the sum of the modeled Handle and the disturbance **D**. The following steps calculated the values in the model. For the position of the Cursor on the screen,

$$C = H + D.$$

For the assumed perceptual signal in the person,

$$p = C - T.$$

And for the person's actions on the Handle,

$$H(new) = H(old) - k(e), \text{ where } e = p - r.$$

The first equation above says succinctly what you read in words in several places in earlier chapters: that the intended consequence **C** is compounded of what the person does (with the Handle) and what the environment **D** does. The second equation says that the person is giving attention (**p**) to the distance between Cursor and Target. The third equation shows the two constants characterizing the person, **k** and **r**, especially if we substitute (**p** – **r**) for **e**:

$$H(new) = H(old) - k(p - r).$$

That says that the person moves the Handle to close up the gap between **p** and **r** by moving the Handle at rate **k**. (And here I must add a technical note. The 1990 article reads "r − p" on page 1333, but Bourbon tells me that was an error that would give the wrong algebraic sign to the loop, and the expression should have been "p − r," which is the way I have put it here.)

The three equations put PCT into a nutshell. They do not by any means elucidate all the features of the theory or its implications even as it now stands in print; the most obvious omission from these equations is the internal hierarchy of control at which I have hinted now and again. But the three equations here contain the core terms and relations necessary to build a simple, minimal model that will actually function and whose functioning can be assessed quantitatively. The necessary terms and relations appear in the equations as follows.

The first equation, connecting **C**, **H**, and **D**, can be called the environmental equation. All the quantities in it can be measured outside the person; any experimenter or onlooker can ascertain them directly, needing no help from the person. If this were a more complicated situation, perhaps with more persons seeking to control more perceived variables and perhaps more sources of independent disturbances, we would begin with more terms and more equations. But with this simplest of situations, those three terms are enough to represent the relevant environment.

The second equation can be called the internal equation or the equation of the living control system. It portrays what we theorize to be the representation inside the person of the relevant features of the task. The difference between Cursor and Target is the internally perceived quantity that the person cares about (the caring being what is represented by **r**). At this point, an important assumption is necessary—but one which Bourbon and his friends found to be justified when they examined the data from the experiment. The assumption is that the quantities **C**, **T**, and (**C − T**) seen and measured by the experimenter have quantitative analogs in the perceptual signals (**p**) inside the person. Another way to express the assumption is to say that the person whose performance is being modeled perceives (**C − T**) very much the same way as the experimenters and onlookers do.

That kind of assumption may very well not hold when the controlled variable lies at a higher level in the neural hierarchy—for example, when the person is trying to play the cello part of a trio by Brahms so as to blend well with the other two players. But as Bourbon put the matter in a letter to me, "For tracking tasks, that bit of modelers' legerdemain seems to work." By "work," he meant that a computer program built upon these equations succeeded in mimicking the actual person's tracking behavior to the very close approximation you saw in Table 8–1 and Figure 8–2.

The third equation connects the person with the environment. It contains the environmental terms H(old) and H(new) and the internal terms **p**, **r**, and **k**. It tells how the perception (**p**), the internal standard (**r**), and the amplification factor (**k**) enable the action (or consequence) H(new) to come about. You might want to compare that third equation with Figure 3–3 in Chapter 3, where I also talked about amplification of energy. The correspondences are as follows. The terms **p** and **r** here are the same as in Figure 3–3. The term **D** here is the same as **d** there. The term **k** here is the same as **K** there. H(old) and H(new) here are successive values of **a** there. And E there is assumed here to have a value close to one and is therefore not explicit. The mathematics there and here were written for somewhat different purposes, but they are thoroughly compatible.

Assumptions

Bourbon and colleagues assumed, among other things, that all humans are alike in the way they function to perform this task. This is a good place for me to illustrate what I mean by "alike." Briefly, I mean that the three equations above apply to everyone, but that the values of **p**, **r**, and **k** will differ from person to person. The three equations are the *invariants*; **p**, **r**, and **k** are the individual's idiosyncrasies. Everyone who agrees to do the task will control the cursor continuously (without discernibly separate acts), but some will be more accurate than others. In driving an automobile, everyone will continuously move the steering wheel (though in some moments the rotation of the steering wheel is zero) so that the car continues in the lane the person has chosen, though some drivers will prefer to drive closer to the edge than others, and some will wander within the lane somewhat more than others. To take conversation as another example, every speaker in a conversation will be sensitive to interruptions, though some will allow more interruptions before acting to squelch the interrupter.

Notice that this assumption of universality means that the theory will collapse if the researcher discovers that *not everyone* will behave according to the three equations. Finally, circular causation is of course assumed. Anyone using PCT and writing simultaneous equations such as those above is unavoidably assuming circular causation in a feedback loop. I have described opposing assumptions in more detail, with illustrations, in Chapters 4, 5, 7, and 11 of my 1990 book.

I mentioned earlier the assumption that the person whose performance is being modeled perceives the difference C – T very much the same way that the experimenters and onlookers do. This is an assumption that is meant to hold only within this particular sort of tracking task. It is not a part of PCT generally.

What's Remarkable?

Why do I ask you to give attention to the series of experiments by Bourbon and colleagues (1990)? My reason is the same reason (insofar as two minds can have the "same" reason) that led Bourbon to initiate the series; namely, the wish to show that PCT can indeed be used to make a statement about the functioning of the human creature that is borne out repeatedly in what we can observe in the effect of the person on the environment. This is not to say that we are predicting "behavior" in the usual sense of particular acts as seen by an onlooker such as an experimenter. Rather, we look for the controlled variable, such as the closeness of the cursor to the target. We do not predict particular motions of the handle. Indeed, we find that the person's handle motions have relatively low correlations (here .701 on the average) with the motions of the cursor. To test our prediction, we try to *disturb* the movement of the cursor; we try to push the cursor away from the target by applying unpredictable disturbances to the movement of the target and to the effects of the handle. But to no avail. The person counteracts the disturbances we apply and keeps the cursor very close to the target, producing an average correlation of .984. That is what we predict to test PCT: the successful counteraction of disturbances to the variable the person wants to control. This series of experiments demonstrates with every individual tested (not merely with an average of many) that a model constructed with PCT can make those quantitatively close predictions with any arbitrary person who wanders in off the street—well, in this case, with arbitrary persons volunteering from the halls of academe. The experiments demonstrate, too, that the prediction can be made with success repeatedly with the same person. Finally, they demonstrate that the test can be successful with delays of one year, five years, and 13 years following the prediction.

Psychologists remind us of the prodigious diversity of human behavior, and so do all sorts of other commentators on the human scene: novelists, dramatists, historians, anthropologists, sociologists, and so on. Some writers in every generation offer us simplifying concepts, perhaps a few instincts or drives or motives that seem to the writer to explain in some satisfying way a great range of human acts. None of those explanations, however, has so far become axiomatic in the scientific world, not to speak of the world of everyday discourse. As you keep in the back of your mind that age-old and frustrated yearning for understanding, I hope this series of experiments by Bourbon and colleagues will inspire you with admiration and wonder. The experimenters offer us three equations of astonishing simplicity that enable us to mimic some behavior of an actual person and measure numerically our degree of success. Though the domain of behavior here is the small world of pursuit tracking, its success is nevertheless the sort for which scientific psychologists have been yearning for well over a century.

CONFIRMATION

I have just said that the study by Bourbon and colleagues provides impressive confirmation of perceptual control theory. The studies you read about in Chapter 7 provide more confirmation, and you will read about still more in later chapters. Here I will note, too briefly, an ingenious and massive study by Martin M. Taylor (1995). Here is what Taylor wrote to the CSGnet (an e-mail discussion group on the Internet) on 6 December 1995:

> Using programs supplied by Bill [Powers], we ran last year some 15,000 tracking runs with people who were working for 64 hours without sleep, helped by [various] drugs. For many of these tracks, the model-[to-]real correlations were above 0.99. But for some, the correlations were as low as 0.3 or even less.
>
> Does this repeal PCT? Of course not. For some of these poorly predicted runs, you could see the subjects sitting staring at the screen for longish

periods, not moving the mouse, or perhaps were waving the mouse wildly.... the observer on the scene would agree that someone lolling with his head on his shoulder, eyes closed, is probably not trying very hard to track a moving cursor on the screen.

The poor performance by some of those drugged and sleepless people is a demonstration of The Test itself. Step 7 of The Test (for which see Chapter 7 under the heading "The Test") tells us to look for the means through which the person can sense the variable we think might be under control, because a variable that cannot be sensed cannot be controlled. Those people Taylor described who had their eyes closed, I am sure, were the ones who produced the poorest tracking performance.

Remember, too, the demonstrations that R.S. Marken offers at

http://www.mindreadings.com/demos.htm

and that Powers offers at

http://home.earthlink.net/~powers_w/

and that Cziko offers at

http://faculty.ed.uiuc.edu/g-cziko/twd/demos.html

and Forssell offers at

www.pctresources.com.

We can suppose that PCT is confirmed once again every time anyone runs any one of those demonstrations; no one has yet told us of a failure of one demonstration.

HIERARCHY

At various places, I have mentioned the neural hierarchy postulated by PCT. While postponing to Chapter 18 an overall description of the hierarchy, I will tell you here about an experiment in which Marken (1986) demonstrated the way a portion of the hierarchy could work.

Procedure

The participant sat before a computer screen displaying three vertical lines arranged as in Figure 8–3. Also in front of the participant were two paddle-handles. The handles could be rotated. The participant was asked to turn the handles as necessary to keep the horizontal distance from the left line to the middle line as equal as possible to the horizontal distance from the middle line to the right line, and to try to keep both those distances as close to two centimeters as feasible. That was the only instruction given.

Figure 8–3. Video display

The participant held a handle in each hand. The computer was programmed so that the left handle affected the left line, and the right handle affected the middle line. Neither handle affected the right line. In addition, the positions of the lines were varied by three slowly varying random disturbances, all different, one applied to each line. The three lines, therefore, moved continuously and unpredictably left and right on the screen, and the participant's task, in effect, was to add movement to the left and middle lines, or subtract movement, in such a way as to keep the distances from the middle line to the outer lines equal, no matter what the right line did. At all times, the positions of the three lines were all that were visible on the screen. Each participant was tested in several two-minute sessions. The random disturbances were changed from session to session. Six adults served as participants.

Prediction and Results

Marken's reasoning was that the participant would have to control the positions of the lines *as the means of* controlling the distances among them. Therefore, he predicted that the participant would of necessity *allow more variation in the positions of the lines than in the magnitudes of the distances.*

The results for one participant for the last 90 seconds of one run are shown in Figure 8–4.

Part (a) of the figure shows the records of the three lines. Q1 labels the record of the left line, Q2 the middle line, and Q3 the right line. You can see that the two outer records keep very much the same distance from the middle one. Part (b) shows in another way the participant's success in maintaining the two distances equal. There, L1 (L for length) labels the curve representing the distance from the left line to the middle one, and L2 labels the distance from the middle line to the right one.

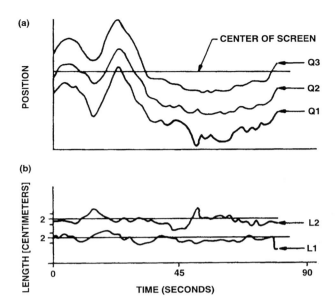

Figure 8–4.
Position and distance traces for one participant Adapted from Marken (1986, Figure 3, p. 272).

The perception the participant undertook to maintain might seem a simple one—to keep three little lines equally spaced. It is no more complicated, surely, than keeping an automobile in its lane while driving along a curving road. But the experiment illustrates beautifully how much delicately coordinated bodily movement a living creature brings even to a simple act.

Part (a) of Figure 8–5 shows the records of the three random disturbances (D1, D2, D3) against which the participant (the same participant as in Figure 8–4) had to act. Part (b) of the figure shows how the participant's two hands (H1, H2) did it. To maintain the equal distances, it is clear from Figure 8–5 that the two hands had to act independently. Indeed, the correlation between the two handle-records averaged over participants was only .25.

Note that the participant could control only the *positions* of the left and middle lines. That is, one hand controlled one line and the other another, independently. But giving the two hands control of two lines did not guarantee equal distances. The two hands could produce an infinity of positions of the two lines that would not give equal distances. Marken reasoned that not only control of the positions of the two lines, but also another feature of control had to exist that would select only those positions that would

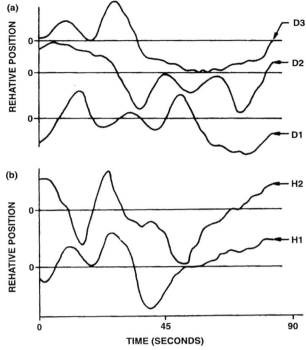

Figure 8–5.
Disturbance traces (D1, D2, and D3) and handle-position, traces (H1 and H2) for one participant Adapted from Marken (1986, Figure 3, p. 272).

give equal distances. The lowest level of control would receive perceptions of the positions of the three lines. Then there had to be a higher level of control that would compare those perceptions and check whether the two distances were equal. That higher level of control would alter the internal standards of the lower level (alter, that is, the participant's notion of where a line "ought" to be) so as to maintain the two distances equal and near to two centimeters.

One way Marken demonstrated the existence of hierarchical control was through the use of a measure he calls the "stability factor." Let V_e be the expected variance of a controlled variable (such as the position of a line or the distance between two of them) and let V_o be the observed variance. (If you are unacquainted with the concept of variance, you will be losing only a very small part of this discussion. Variance is a measure of the amount of variation in the values of a variable during an experiment.) For example, the expected variance of the distance from the left line to the middle one is equal to the sum of the variances of (1) the random disturbance applied to the left line,

(2) the random disturbance applied to the middle line, (3) the positions of the handle affecting the left line, and (4) the positions of the handle affecting the middle line. The observed variance of that distance is simply the variance of the distance between the two lines on the screen over the course of an experimental run. Variances of the other distances can be arrived at similarly. The formula for the stability index is:

$$S = 1 - (V_e/V_o)^{1/2}$$

If the observed variance is fully as great as the expected variance, then S is zero, which means that the participant is effecting no control. If S is less than one (that is, negative), then the participant's behavior is reducing the observed variance and counteracting disturbances.

Table 8–2 below shows the stability indexes for two experimental runs of each of two individual participants and also the averages for all six participants. In the table, L1 stands for the distance from the left line to the middle one, L2 the distance from the middle line to the right one, Q1 the position of the left line, and Q2 the position of the middle line.

The *distances* L1 and L2 are controlled variables and so are the *positions* Q1 and Q2. We see in the table that the average stability indexes for L1 and L2 were –11.0 and –10.7. The numbers in the body of Table 8–2 are expressed in statistical units measuring the probability of an event of the underlying magnitude occurring by chance. In conventional psychological research, a number of this sort as high as 3 is rare, and one as high as 4 is almost unheard of, even suspect. A number greater than 10, such as the mean indexes here for L1 and L2, is as rare in social science as hen's teeth. That rarity, however, is of small moment here. The usefulness of the stability index here is in *comparing* the amounts of control at the two levels of the hierarchy. The average stability indexes for Q1 and Q2 in Table 8–2 are both about –1.0. That figure indicates a good deal of control (the ratio of V_e to V_o is four to one), but a much smaller amount than that at the higher level. This result fits the prediction; the higher reference signal, distance, sets the lower reference signal, position. Therefore the reference values for position were varied somewhat over time by the higher reference signal, and the observed variance of the positions Q1 and Q2 was therefore greater than that of the distances L1 and L2.

One thing we have seen so far is that the participants did, every one, succeed in keeping the two distances very closely the same. That seems hardly strange, because we are accustomed to witnessing such dexterous capabilities of living creatures. But we have also seen that the control over the perception of *distances* (which was the task the participants accepted) was greater for every participant than the control over the perception of *positions*—just as the theory predicted.

The Model

Marken built a model of this behavior in the form of a computer program. Let us turn to the equations that specify what the model must do. As before, L1 and L2 stand for the two distances. Similarly, Q1, Q2, and Q3 stand for the positions of the three lines on the screen. Further, D1, D2, and D3 will stand for

Table 8–2. Stability indexes.

	Variable			
	L1	L2	Q1	Q2
Participant 1				
Run 1	–10.2	–9.2	–1.1	–1.3
Run 2	–12.2	–13.8	–1.1	–1.2
Participant 2				
Run 1	–7.6	–8.9	–0.88	–0.72
Run 2	–9.8	–10.3	–0.99	–0.87
Averages for six participants				
Means	–11.0	–10.7	–1.0	–1.1

Excerpted from Marken (1986, table 1, p. 273).

the three random disturbances of line position, and H1 and H2 will stand for the effects on line position of the two handles operated by the participant.

Like the participants, the model is required to keep the two distances the same, or equal to a constant value c, which in this case is two centimeters:

$$L1 = L2 = c \qquad (1)$$

Since the positions of the first two lines are determined by the sum of the random disturbance and the handle position, while the position of the third line is determined only by the random disturbance applied to it, we have the following three specifications for dependencies within the model. Here (t) indicates temporal variations.

$$\begin{aligned} Q1(t) &= D1(t) + H1(t) \\ Q2(t) &= D2(t) + H2(t) \qquad (2) \\ Q3(t) &= D3(t) \end{aligned}$$

The distances, of course, must be connected to the line positions as follows:

$$\begin{aligned} L1(t) &= Q2(t) - Q1(t) \\ L2(t) &= Q3(t) - Q2(t) \end{aligned} \qquad (3)$$

Substituting appropriately from equations (1) and (2) into (3), we get these two equations for the ways the two handle positions will counteract the effects of the random disturbances in the model:

$$\begin{aligned} H1(t) &= D3(t) - D1(t) - 2c \\ H2(t) &= D3(t) - D2(t) - c \end{aligned} \qquad (4)$$

In effect, the task of the human participants was to solve the equations (4) simultaneously. That ability must now be added to the model. To show how Marken did that, I turn to Figure 8–6.

Figure 8–6 is a diagram of the model. The lines and arrows stand for actual electrical signals, and the boxes, ovals, and circles from and to which the arrows point stand for actual functions. Nobody claims that those functions are carried out in the computer by components that are built like living tissue, but only, if the model is to mirror human behavior, that the same *functions* must be carried out both in the computer and in the human, just as something must turn axles both in a real railroad engine and in its model. Running across the figure near the bottom is a horizontal line with the word "system" (meaning the organism) above it and the word "environment" below it. Below the line you see the Handles H1 and H2, the positions Q1, Q2, and Q3 of the lines on the screen, the disturbances D1, D2, and D3, and the distances L1 and L2. L1 is the distance from Q1 to Q2, and L2 from Q2 to Q3.

Above the horizontal line are diagrammed five control systems (loops) connected in such a way as to operate on two levels. One system appears at the lower left. It comprises Q1, H1, and everything labeled (1,1). System (1,2) lies to the right; it consists of Q2, H2, and everything labeled (1,2). Both systems (1,1) and (1,2) correspond, component for component, to Figure 4–1 in Chapter 4. Beyond system (1,2) to the extreme right lies system (1,3); actually, it is only a part of a system. In this model, system (1,3) is needed only to perceive Q3, and since Q3 is never moved in this experiment, system Q3 needs no comparator and no output function.

Above system (1,1) lies system (2,1), comprised of everything labeled (2,1). Inputs come up to system (2,1) from all three systems at level 1, and system (2,1) sends outputs down to both systems (1,1) and (1,2). Similarly, system (2,2) lies to the right of system (2,1) and above system (1,2). The systems at level 2 are not complete in themselves; that is, a complete circuit cannot be made in this model within level 2. A complete circuit through level 2 is made by going down through level 1, into the environment, back into level 1, and up again to level 2. The brain is, of course, far more complex than this. In Part IV, I will describe various complexities, including ways that loops can be complete within one level.

To examine the connections in Figure 8–6, let us begin at the lower left corner. In the environment, we see Q1, a physical quantity providing input to the neural net—or in the case of the model, a quantity that will be treated as input in the subsequent calculations the computer is programmed to perform. Q1 is determined partly by the random disturbance D1, which is put into the programming by the model-builder to simulate unpredictable events in the environment. Q1 is also partly determined by the handle position H1, which will be simulated by the output signal from this level–1 control system.

The little box labeled S(1,1) stands for the sensor at level 1 sending the signal on in system (1,1). The signal stays the same at this point in the model; think of it as the incoming signal being multiplied by one. The oval containing the summation sign, sigma, is there only for completeness; it will make more sense when I talk about it at level 2. P(1,1) stands for the

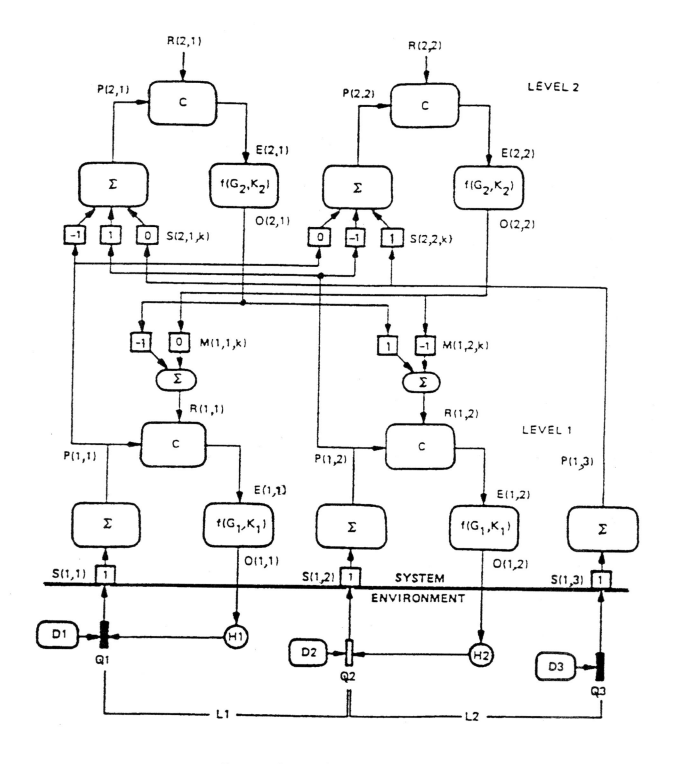

Figure 8–6. A model of two-level control

perceptual signal coming in system (1,1). Note that it goes not only to the comparator in system (1,1) but also on to the summation function at level 2. Comparator C compares perception P(1,1) with reference signal (internal standard) R(1,1) and sends out error signal E(1,1). Effector or output function $f(G_1, K_1)$ converts the error signal into instructions for action. In the human, the effector function would send signals to the muscles at H1. In the model, since there are no muscles, it sends the compensatory signal directly to Q1. O(1,1) stands for the output signal to H1.

Now let us go to the top of the diagram, at level 2. The reference signals (internal standards) for distances, R(2,1) and R(2,2) are equal constants (two centimeters) put into the program by the model-builder. This is the specification given by equation (1). The input "perception" comes into system (2,1) from the three systems at level 1. The signals from below are weighted at the boxes labeled S(2,1,k) and summed by the summation function indicated by the sigma. The input from system (1,1) is weighted at −1, and that from system (1,2) at +1. Their summation, then, subtracts the position of Q1 from the position of Q2, yielding the distance between them. The perceptual signal from system (1,3) is weighted at zero, and has no effect on the upper-level perceptual signal P(2,1).

The output O(2,1) from the upper-level system (2,1) goes downward to modify the reference signals of the level–1 systems. At the lower-level system (1,1), the output signal from upper-level system (2,1) is weighted at −1, and the output from system (2,2) is weighted at zero. The output from the upper-level system controlling the distance from Q1 to Q2 sets the reference signal R(1,1) to tell the lower-level comparator how much Q1 should be moved to stay at the right distance from Q2. The output from the upper-level system (2,2) has no effect (weighted zero), because Q1 will be most accurately placed if its position depends only on Q2.

You can trace out the other connections similarly. There are some technicalities connected with the effector functions that are needed to make the computer simulate a continuously acting feedback loop, but I omit them from this description. The information put into the programming of the model by the model-builder to simulate the conditions under which the human participant worked enables the model to solve equations (4) from moment to moment. The information includes the reference values R(i,j), the weights S(i,j,k) and M(i,j,k), and the effector functions $f(G_i, K_i)$.

In a more complex "creature," other systems at level 1 would be complete loops, not truncated like system (1,3) in the diagram, and the perceptual inputs at level 2 would not stop there, but would have branches going on to higher-level systems, just as the perceptions from level 1 in the diagram branch off to go to level 2.

The Fit of the Model

How well did Marken's computer model succeed in behaving like his human participants? To test the fit, Marken calculated the correlations between the positions of line Q1 produced by the model and the positions produced by the participant when both model and participant were working against the same three disturbances. In the same way, he calculated correlations for Q2 and the handle-positions H1 and H2. For mathematical reasons for which I will not take space here, Marken could not calculate useful correlations for L1 and L2. Table 8–3 shows the correlations for Q1, Q2, H1, and H2, each calculated from 400 data-pairs. The table shows data for two experimental runs of each of two individual participants and also the average correlations for all six participants. You can scarcely ask for a better fit than the table shows.

Assumptions

The assumptions here, those needed and not needed, are very much the same as those I have mentioned earlier. I'll repeat here only two points.

The use of words is reduced to the simple instructions given to the participant; no words whatever were needed from the participant. You can check whether the participant, every participant, is following the instructions by looking at the data. Indeed, when you see a participant behaving as in Figure 8–4, you do not need to assume that the participant understood the instructions or even heard them. You do not care.

The success of the experiment is not measured by the correlation between environmental input and action output. Instead, success is measured by the *lack* of correlation between environmental input (the random disturbances) and the controlled input (the perception of equal distances).

Table 8–3. Correlations between human behavior and model's behavior.

	Variable			
	Q1	Q2	H1	H2
Participant 1				
Run 1	.992	.971	.996	.995
Run 2	.983	.972	.989	.992
Participant 2				
Run 1	.968	.982	.986	.992
Run 2	.983	.972	.989	.992
Averages for six participants				
Means	.979	.976	.990	.991

Excerpted from Marken (1986, table 1, p. 273).
See text for explanation of symbols.

What's Remarkable?

What's remarkable, I think, is the almost perfect fit between the behavior of the human participant (every one of them) and the behavior of the model in the computer. That fit was achieved even though every participant had to cope with three random disturbances acting simultaneously, none of them ever repeated in a subsequent experimental run, and had to do so by using two hands acting independently. It was achieved, too, even though a model had to be built that would act at two hierarchical levels and to which it was impossible to supply a formula for achieving the equal distances, because the model (like the participants) had to cope with unpredictable disturbances.

TWO PEOPLE

All the examples of research I have shown so far in Chapter 7 and in this one have been studies of single persons. Now we turn to a study by Bourbon (1990) in which two persons interfere with each other while they pursue their own goals successfully.

Procedure

The laboratory set-up for this experiment was similar to the set-up diagrammed in Figure 8–1 in the earlier section called "Accuracy and Reliability". The most important difference was that in this experiment, there were *two* control systems (that is, participants), one at the left and one at the right. Correspondingly, there were on the screen two cursors, and there were also two handles, one to be moved by the person at the left, the other by the person at the right. Each handle caused *both* cursors to go up and down, but not equally. Each handle had a direct effect on the person's own cursor, but only half that much effect on the other person's cursor. At any one moment, the vertical position of the left cursor was calculated as the sum of (a) the position of the left handle, (b) one-half the position of the right handle, and (c) the value at that moment of a random disturbance generated at the start of the experiment. The position of the right cursor was the sum of (a) the position of the right handle, (b) one-half the position of the left handle, and (c) the value of another random disturbance.

As you can see, this arrangement gives the actions of both persons side effects on a variable (position of the cursor) the other person is controlling. Bourbon (1990, p. 96) says that a person's actions

> while controlling one variable nearly always affect other variables, including many that the person does not know exist.... Unintended consequences are frequent sources of annoyance with oneself ("That wasn't what I meant to do!") and of frustrating exchanges ("Why do you always do that?" "Do what?" "You know what I mean!").... But these ... confusing circumstances are the material from which people fashion all their interactions, from conflict to cooperation or, as Shakespeare knew, from blood feud to young love.

Results

Each person adopted the goal of keeping his or her own cursor at a position about half an inch below the center of the screen. As usual, the participants had no difficulty in achieving precise control. Figure 8–7a shows the records. The wiggly lines just below the zero-line (the center of the screen) are the records of the two cursors. The lines that deviate far beyond the center are the two records of the handle positions. There was no way the participants could sense separately the two disturbances to the effects of their handles: (1) the disturbance applied from the random numbers generated by the computer and (2) the disturbance applied by the movements of the other person's handle. The participant perceived only that the movements of his or her handle had varying effects—unpredictably varying effects—on the cursor, and that was that. To keep the cursor in the desired position, each person moved the handle a little or a lot as necessary to counteract the net disturbances. The mean correlation between handle position and net disturbance reported by Bourbon in his 1990 article (p. 101) was –.982. The correlation between handle position and cursor position (because of the disturbances) was of course very small: –.009.

Living creatures take varying actions (handle) to bring about unvarying consequences (holding the cursor in place). Bourbon (1990, p. 96) says that control brings about a transfer of variability from one part of the world to another:

> The fact that variability disappears from the part of the world that is controlled, only to appear in the behavior of the person who controls it and in variables accidentally affected by that person, causes unending confusion in the behavioral sciences.

THE MODEL

Bourbon constructed a model of each of the participants. (I have changed the notation here somewhat from Bourbon's so as to make it more parallel with the modeling I showed earlier in this chapter in the section headed "Accuracy and Reliability.") Let **CL** stand for the vertical position of the left cursor, **HL** the position of the left handle, **HR** the position of the right handle, **TL** the position of the target, **dL** the value of the random disturbance to the left cursor, **k** the output or gain factor, **p** the perceived position

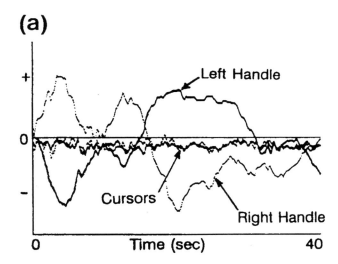

Figure 8–7(a).
Actual control of two cursors by two persons.

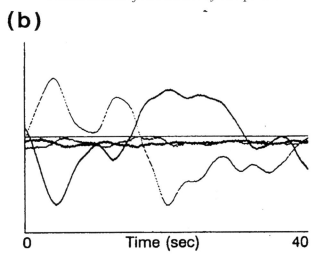

Figure 8–7(b).
Simulated control of two cursors by two persons. Adapted from Bourbon (1990, Figure 3).

of the cursor, and **r** the reference value or intended position of the cursor. Then the modeling equations for the left person were:

$$CL = HL + .5HR + dL,$$

$$p = CL - TL, \text{ and}$$

$$HL(\text{new}) = HL(\text{old}) - k(p - r).$$

The equations for modeling the right-hand person were similar. Further technicalities can be inspected in Bourbon's (1990) article.

As in the experimentation I reported under "Accuracy and Reliability," values of **k** and **r** were estimated from early runs of the living participants. The runs of the model, using those values, used the same number (1200) of predicted positions, one for each $1/30^{th}$ of a second of each 40-second run. Figure 8–7b shows the record of the modeling of the two people whose record appears in Figure 8–7a. Did person and model operate their handles in the same way to cope with the same disturbances? For the person and model on the left, the correlation between positions of the actual and simulated handles was .996; for those on the right, .995.

Assumptions

I will not repeat here the assumptions that apply to all experimentation of this sort. There is, however, one detail I want to mention that follows from the basic theory. That detail is the modeling of influence from one person to another. In common language, we speak of influence "on" another person's action or even the person's body. "Here," we say, "hold the knife like this," and with our own hands, we wrap the other person's fingers around the knife. (I am not saying that is always a bad thing to do; I am saying only that it is not what it looks like on the surface.) But Bourbon did not model the influence by one person on the other by connecting the output of one person (handle movement) to the *output* of the other person—that is, to the line running out from **o** in Figure 8–1.

He added it instead to the *disturbance*, the effect of which came about through the motion the person perceived on the screen. Note, too, that the effect of one person on the other here was *not* on what the other wanted to perceive (the position of that person's cursor), but only on that person's handle motion—a thing that person was *not* trying to control. Each person influenced the *acts* of the other, but not the purpose—what the other was "trying to do" or achieve.

What's Remarkable?

This study shows quantitatively how one person can maintain a desired condition—can control a variable—even though someone else actively "interferes" with the control of the variable. As long as the interference leaves the first person enough latitude to act on the variable without coming too close to his or her limits of skill or physical ability, the first person may not even notice what to an onlooker or experimenter seems an additional disturbance. In the natural world, control of a variable is usually being disturbed by more than one source of disturbance. The feeling of being in balance when we are walking is disturbed by how we choose to apportion speed, length of step, and angle of leaning, by changes of course, by gusts of wind, by distractions of something interesting to look at, by changes in footing—all those and more. If our attention is focused on our progress toward the goal, or if we are focused on our thoughts about some puzzling matter, we may notice consciously none of those sources of disturbance, not to speak of recognizing the resultant of them as a resultant. If our bodies are fit, walking does not feel to us like a struggle against multiple disturbances. On the contrary, many of us go walking for the pleasurable feelings it gives us. So it is with most action when we do not encounter physical pain or internal conflict.

The instructions to the participants could have been different. One cursor could have been put on the screen. One participant could have been asked to keep the cursor near the top of the screen, and the other to keep it near the bottom. That would have made control by either very difficult, since neither could achieve a satisfying degree of control if the program permitted them to have equal power over the cursor. That condition is one of competition. I will show a simulation of one sort of competition in Chapter 9 under "Collective Control of Perception," and I will discuss competition and conflict more generally in later parts of the book.

Most of us make plans for dealing with other people. When plans do not satisfactorily anticipate our actual interaction with the other people, we often scold ourselves for having planned poorly. Notice, in this experiment, that the control being exerted by an individual was being disturbed by another person *without the individual's knowledge*. When I say "disturbed," I don't mean that the disturbance actually overwhelmed the individual's ability to control the variable (the position of the cursor). I mean that the individual had to take wider action to counterbalance the disturbance and maintain the cursor in place. That sort of unpredictable disturbance always occurs in social situations, and, as here, its source may not be recognized when it does happen. Sometimes we succeed easily in coping with unpredicted disturbances. We then have the feeling, especially when we are unaware of the sources of those easily-opposed disturbances, that we did a good job of planning. Sometimes, however, the unanticipated disturbances are so difficult to cope with as to overwhelm our capabilities. Sometimes we blame fate, sometimes the interference of other people, sometimes our own inadequacies. Often, whatever we may choose to blame, the blaming does us no good.

Bourbon (1990, p. 104) says:

> With its elegant simplicity and effectiveness, the control theory model stands in sharp contrast to the complexity and nonspecificity of most theories of coordination. Control theory offers the possibility of using the same few principles to explain coordination at every level, from movements of parts of our own bodies to interactions like those in simple tracking tasks, in infant-parent dyads, in social gatherings, in marriages, and on the job.

Bourbon (1990) investigated a couple of questions in addition to the one I have told about here. If you are becoming captivated by PCT, I urge you to read the original.

MORE MODELING

Much more modeling has been done than I have described here. To learn more about how modeling is done, read the originals of what I have reported here, and look up also Bourbon (1989), Bourbon and Powers (1993), Richard Kennaway (1999), Kent McClelland (1994), Marken (1990a), Marken and Powers (1989, 1989a), Pavloski, Barron, & Hogue (1990), Powers (1973, pp. 273–282; 1978; 1979a; 1983; 1989b; 1994; 1999, 1999a), and Runkel (1990, pp. 93–99), Young and Illingworth (1999), and Wolfgang Zocher (1999).

MODELS AND THEORIES

I am sure you can tell that I believe a working model to be the ruling, essential criterion for justifying, validating, defending a theory. If person M has a theory that enables the construction of a model that behaves like a living creature, and person W has a theory that has not produced a model, the choice between the two is easy. The model gives tangible, nonverbal evidence that the theory has seized upon some of the ways of functioning necessary to the living creature. The theory is validated when we see the model actually functioning in the same way as the living creature. A theory embodied only in words, no matter how convincing the words may sound, no matter how logical, must remain a speculation. If theories inspire models, we can then watch the models and compare their performances quantitatively—not how they might conceivably perform, but how they *do* perform. But until a theory can produce a model, the theory remains only one among many, with the adherents of each hoping to convince the others, through argumentation, of the superiority of their own.

Almost all introductory texts in psychology—and many advanced texts, too, for that matter—are written as compilations. The contents go something like this:

> Jacob's theory of motion perception.
> Jansen's theory of motion perception.
> Jacquard's theory of motivation.
> Jeremiah's theory of motivation.
> Jimenez's theory of motivation.
> Joachim's theory of personality.
> Johnson's theory of personality.
> Josephson's theory of personality.

. . . and so on.

As you read along, you find that the theories of one kind of phenomenon (such as motion perception) seem to have little or no connection to the theories of another kind (such as motivation). The author gives you no way to fit together the various theories of the various phenomena to form a theory of the functioning of the whole, integrated, unitary organism. Furthermore, the author presents competing theories of each phenomenon as if they are all equally worthy of your attention. You are left with the impression that you should refer to one theory of motivation when you are thinking about eating, another when you are thinking about getting married, and still another when you want to reduce costs in your business. And even within one sub-domain (getting married, let's say), you are often left with several theories and no criterion for choosing among them. It is, in fact, nearly impossible to compare the excellence of one theory against the excellence of another in the standard textbooks, because it is nearly impossible to judge the excellence of any single theory.

Most experiments and other types of studies in psychology are assessed by comparing the number of people who behave as predicted with a calculation of the number you would expect by chance if the influence the experimenter cares about were not acting. If the chance of the result coming about without the influence acting is very small, then the experimenter concludes that the influence did cause the behavior. But this method and the reasoning that goes with it are very unsatisfactory, because chance events *do* affect different individuals in the study in different ways, despite the efforts of the experimenter to rule out those effects. Therefore, when another experiment similar to the first experiment is performed, the results are always somewhat different, because the chance influences are different. Often, when the result in a first experiment is better than chance, the result in the next similar experiment is no better than chance or even worse.

Physical Science

Compared to psychology, it is impressive how quickly a new concept or theory has sometimes supplanted an older one in the physical sciences. In the eighteenth century, almost all the scientists of Europe believed that fire was produced by a substance called phlogiston that was present in every burnable thing. That theory may have been as widely accepted at that time as today the theory is accepted that stimuli cause responses. The theory of phlogiston was taught in every school that taught anything about burning. In 1772, however, Lavoisier published reports of his experiments with sulfur and phosphorus, and Priestley his on nitrous air. Contrary to the theory of phlogiston, which predicted that substances should be lighter after burning than before (because of burning away the phlogiston), Lavoisier and Priestley reported that their burned substances were *heavier*. Three years later, in 1775, Lavoisier published his famous Easter memoir with further evidence that combustion was an additive process, not a subtractive one, and Priestley discovered oxygen, though he didn't say it just that way. Three years after that, in 1778, Lavoisier revised his Easter memoir. By 1783, the composition of water (hydrogen and oxygen) was established. By then, the phlogiston theory was widely discredited; few persons were trying to defend it.

In 1789, Lavoisier published his *Elementary Treatise on Chemistry*, which set forth the new chemistry clearly and systematically. From the first experimental challenge to the phlogiston theory in 1772 to its effective demise in 1783 was only 11 years. If you want to count the publication of Lavoisier's text as the year of demise, the elapsed period was 17 years. J. B. Conant (1956) speaks of "the revolution of 1775–1789." The number of experiments sufficing to bring about that revolution in chemistry was astonishingly small. I get the impression from Conant's book that the crucial beginning experiments were surely fewer than a dozen, and a few dozen more, including those Lavoisier reported in his *Treatise*, sufficed to relegate phlogiston to museums. Incidentally, those scientific giants were amateurs; none of them had a degree in chemistry; none had a doctorate in any field.

Einstein's theories of relativity, too, both the special theory of 1905 and the general theory of 1916, were accepted with lightning speed by most physicists the world over. When I say "lightning speed," I mean within two or three decades after publication.

The examples of Lavoisier and Einstein stand in contrast, it is true, to work that lay ignored for a century or more—the work, for example, of Copernicus and Mendel. I am not saying that the history of physics and chemistry has been one of unceasing lightning strokes. But I am saying that one need not hunt there long to find examples of the acceptance within a part of one human lifetime of theories that upset previous basic assumptions.

What of the science of psychology? Wilhelm Wundt established the first psychological laboratory at the University of Leipzig in Germany in 1879. I am writing these words about 120 years later, after a period in which thousands upon thousands of reports of psychological experiments and theoretical papers have been published. During this period, too, there have been *thousands of times* more psychologists at work than there were chemists at work during the fall of the phlogiston theory, and the amount of money spent on research has been several orders of magnitude greater. Furthermore, psychologists have had available the histories of the successes in physics and chemistry, and psychologists often claim to have learned something from those sciences. Surely, one would expect to see several strokes of lightning during those 120 years.

Let me tell you about just two explanatory concepts that were highly admired a hundred years ago in my own field of social psychology—instinct and group mind. I want to tell you about them not because I think all ideas ought to be given up after 20 years or so, or even 120 years. I do it because I think a science that is going about its business in the right way ought to have done some winnowing during a hundred years or more, and if a concept was receiving wide attention at one time, it should usually, after a few decades, have been shown *either* to be worthy of a greater share of the investigative pie or to be a small matter much overshadowed by other more powerful concepts, if not indeed a dead end. I can think of a couple of ancient ideas that did fade away into oblivion among psychologists if not among the public at large: astrology and the humors of Galen (2nd century A.D.). I thought of mentioning too the phrenology of F. J. Gall (1758–1828), but some psychologists and neurologists are still claiming that certain kinds of actions are "controlled" in certain regions of the brain, though others (for example, W.R. Uttal, 1978, pp. 253, 345) deny it.

Aside from those two or possibly three ancient ideas, I cannot think of an idea about psychology that was once widely admired and has since become widely ignored. So let me turn to the two ideas that were Big Ideas a hundred years and more ago, then went out of fashion, but somehow are still in use.

Instinct

The concept of instinct was once regarded as capable of accounting for almost all of behavior, or at least of great swathes of it. William McDougall in his 1908 *Introduction to Social Psychology* and in his 1923 *Outline of Psychology* treated instincts as dispositions residing within the person to perceive or behave in certain ways. That is very much what psychologists mean nowadays by the term *trait*. I have no idea how old the concept of trait may be. It is probably as old as there have been adjectives applying to animals in any language. Modern psychologists have given up the four humors of Galen not because they have given up the idea of traits, but because they don't like Galen's four traits or the idea that they arise from fluids in the body. Following the usage of their time, S. Freud used the term *instinct* in his *Instincts and Their Vicissitudes* (1915), and W. Trotter used it in his *Instincts of the Herd in Peace and War* (1917). At about that time, the term *instinct* began to go out of style, and various psychologists took the trouble as the years went by to tell readers that *instinct* was no longer in favor.

Hilgard, Atkinson, and Atkinson, for example, wrote in 1975 (p. 303) that *instinct* had been replaced by *drives* during the 1920s. In 1991, however, D. M. Senchuk apparently believed that the question of instinct had not been firmly buried in the 1920s or even by 1975, but still had some interest for some readers; he published a book entitled *Against Instinct*. In the meantime, a good many books were published that continued to deal with instinct in a serious and scholarly manner just as if it had not been declared dead. I discovered this fact easily by leafing through my university library's catalog. Books containing the term *instinct* in their titles appeared in every decade following the 1920s. The more recent ones were dated 1962, 1967 (two), 1970, 1982, 1994 (two), and 1997. All but three of those were classified by the Library of Congress under psychology.

Though many or even most psychologists may believe that *instinct* is no longer with us, they have no doubt about the presence of *traits*, which have been proposed by the dozens as the years have gone by. Instinct, trait, motive, drive—those concepts seem to have, in the minds of the authors I have read, the character of what I call internal standards. They seem to denote internal urges that pull everyone to a greater or lesser degree in one particular direction. Almost all researchers using those concepts seem to conceive those urges as having an existence somehow apart from the individual person. That is, traditional researchers put a name on a particular instinct, trait, motive, or drive and look for amounts of that named thing in more than one person—everybody, usually. In PCT, we believe that some similarities can certainly be noticed among the evidences of reference signals (internal standards) in various individuals, but we also believe that every reference signal is unique not just in quantity, but in quality, too, by that person's heredity and experience. I will elaborate on the uniqueness in later chapters.

Psychologists always measure traits on scales that sum up points or numbers from multiple items (in, say, a questionnaire) to yield an overall number. Some people obtain higher numbers than others. So it is easy to talk about one person as having a "stronger" trait or having "more of" a quality than someone else. The concept of trait thus becomes very much like that of phlogiston. If a thing contains more phlogiston, it burns more fiercely. If you have more Conscientiousness inside you, you will more often keep your promises. This is not at all the way internal standards in PCT are postulated to function.

Group Mind

The third concept popular a century ago and still with us is that of the group mind—the idea that a group of persons is somehow "more" than a collection of individuals, that some quality can "emerge" that is beyond the qualities of the individuals. I have already mentioned in Chapter 5, under the heading "Reification," the belief of Josiah Royce in the mental unity of the English people. I should confess that I myself once believed in the separate reality of the group. I can remember clearly the feeling of certainty that I had. I remember showing social psychologist T. M. Newcomb how I had (so I thought) proved it mathematically.

W. S. Sahakian (1982, p. 4) says that Hegel wrote in 1807 about a world spirit or mind directing the evolution of civilization, that Espinas wrote in 1877 about the group mind or collective consciousness, and Schaffle, writing about 1875, attributed purpose and consciousness to the group mind. Emile Durkheim, in his *Suicide* (1897), wrote:

> Collective tendencies have an existence all their own; they are forces as real as cosmic forces, though of another sort... (p. 309 of the 1951 edition).

Gustave LeBon is famous for his book *The Crowd* (1895). In it, he wrote:

> ... the fact that [individuals] have been transformed into a crowd puts them in possession of a sort of collective mind which makes them feel, think, and act in a manner quite different from that in which each individual would... (p. 30 of the 1896 edition).

Sociologists have been coping with the idea of "a sort of collective mind" for a long time. The ancient Greeks and Romans (not to speak of Egyptians, Mesopotamians, and Chinese) must have had something to say about crowds, but I have not bothered to hunt up a quotation. In the prologue to his 1991 book *The Myth of the Madding Crowd*, Clark McPhail points out that we have the phrase "madding crowd" from Thomas Gray's "Elegy Written in a Country Churchyard," published in 1750[1]. McPhail mentions, too, a book of 1852 by Charles Mackay, entitled *Extraordinary Popular Delusions and the Madness of Crowds*. Though the most-quoted promoter of the idea was LeBon, McPhail says that the idea was easy to find in later decades; he mentions Festinger, Pepito, and Newcomb (1952), Zimbardo (1969), Diener (1980), and Moscovicci (1985). But by about 1970, McPhail says in his Chapter 4, some influential psychologists were "explicitly recharacterizing individual crowd members as purposive and rational actors rather than as individuals transformed by the crowd..." (p. 109). Among other writings in that vein, he mentions Couch (1968), Berk (1974), Tilly (1978), and Lofland (1985). McPhail's own book is a thorough examination of the idea of collective mind and a careful rejection of it. In the next chapter I will tell you about some of McPhail's research that made use of PCT. I will also tell you about some modeling of collective *control* done by Kent McClelland.

Sahakian says (p. 5) that *The Group Mind* (1920) by William McDougall signaled the death of the concept. You can see that McPhail does not agree with that, and neither do I. It is true that you can find many books in addition to Sahakian's that tell you that nobody believes any more in a group mind. Yet a very strong restatement of the idea appeared in 1978. J. G. Miller says on page 515 of his *Living Systems* (1978):

> The view that a group is a concrete reality has weakened in recent years.... Certain social psychologists consider a group to be no more than a collection of individual organisms. I do not agree, but hold that groups are concrete entities....

In that paragraph, Miller seems to imply that most sociologists agreed with him in his belief that a group has a concrete reality of its own—that only "certain social psychologists" would disagree. Though I know of no actual tally of that opinion, and despite Sahakian's opinion, I believe that most social psychologists still share Miller's belief. Not only do my friends among social psychologists talk that way, but almost all of the organizational consultants I have known have talked that way.

It seems to me that trait is occasionally a useful term for everyday conversation, but I think the idea of a group or organization having some kind of life greater than the lives of its members is just wrong.

There you have two examples of concepts about which, a hundred years ago, scholars were writing weighty tomes, and which later were declared by scholars just as weighty to be scientifically unjustifiable, but to which, as you see, still more tomes are being devoted. When no criterion is used for winnowing other than statistical preponderance and argumentation, that sort of cycling can go on for still more centuries.

A concept will not be rejected by many of one's colleagues unless they are offered another idea that lies on the better side of a criterion that can capture the allegiance of most of one's colleagues. In science, that criterion has always been the empirical test. Unfortunately, the empirical test in psychology has long been shaped by a research method that yields only a probability. When the experimenter's calculation says that the result of an experiment is "significant," it means only that there is but a small chance that a result as good as the experiment gave could have been obtained by flipping a coin instead. The experimenter can say only, "I am probably right." And since other experimenters with other theories also get "significant" results, they too can say, "I am probably right." And experimenters can go on saying "I am probably right" for a hundred years, and no concept gets firmly discarded.

Now, however, it has become possible to build working models, and therefore it has become possible to adopt the working model as the criterion for a successful theory. If the model works, the theory on which it is built deserves attention. If it does not work, one has no obligation to listen to arguments in its favor. Whether a model "works" needs no calculation of probability. If the behavior of a model does not match the behavior of the person within an error of a fraction of one percent, then the model should be improved or the theory should be given up. The match, in other words, is required to be so close that calculating a probability of a chance result would be absurd. PCT has at last enabled that kind of model to be built. (The existence of electronic computers has been necessary, too.)

I might also mention that social psychologists sometimes sneer at books like Trotter's *Instincts of the Herd in Peace and War* and LeBon's *The Crowd*, in which the author proposes one concept with which to explain large domains of behavior. Efforts like those have been scorned as "simple and sovereign" theories. But there is nothing of which to be ashamed in trying to explain a lot with a little. That is what Watson and Skinner tried to do, too. A theory that explains a lot with a little is superior to one that explains less or uses more concepts. A theory is especially desirable if its simple concepts can be used to build a working model.

Concerning PCT, Powers (1973, p. 78) says:

> To the extent that the model has been carried to completion, covering all aspects of behavior, subjective experience, and brain function, every attempt to apply the model will test it and, where it fails, point to what needs modification.
>
> Only a complete model that is supposed to apply all of the time and in all circumstances can really be tested by experiment. If one limits the scope of a model, failures of prediction or explanation can always be attributed to effects of what has been omitted.

NOTE ON ANOTHER KIND OF THEORY

When I write of theory in this book, I almost always mean a theory about how something in the tangible, "real" world functions. Examples are a theory of fluid flow (as in pipes or over airplane wings), of stress and strain (as in the deformation of beams under load), of the circulation of the blood in animals, and of the control of perception in animals. But there are also theories about intangible things—about systems of thought that exist only in the mind. One branch of mathematics, for example, is called the theory of numbers. The theory consists of definitions and axioms from which one can derive logically sets of numbers having certain characteristics. Another example is the theory of chess—an examination of the kinds of configurations of pieces on the board that give strategic advantage. A board and pieces are not necessary. Chess players often play "blindfold" chess, in which they keep all the moves in their memories and announce to each other the "moves" they are making. The word theory is as properly applied to those conceptual systems as to tangible systems, but when I write here about what theory ought to be like, I am writing about tangible systems that require confirmation in the observable "outside" world, not about systems that exist only in the mind.

ENDNOTE

[1] In the list of references at the end of this book, I have not included, from the rest of this paragraph, the writings mentioned by McPhail.

Chapter 9

Some social interaction

In Chapter 7, I told about some laboratory studies demonstrating some of the key features of PCT; namely, (1) You cannot be sure what a person is "doing" by watching the person act. Actions are steps along a path to a goal. Even when you successfully divine one of another person's goals (or your own), there are always further goals lurking unseen. (2) Living creatures can make their way toward a goal effectively even when using very simple and primitive ways of finding a path. *Escherichia coli* can "steer" only by random flip-flops, but nevertheless finds food enough. (3) When a variable changes the way you would expect the external physical world to cause it to change, you can conclude that no living thing is keeping it from changing. That is the essence of The Test for the Controlled Quantity—which, in turn, is the basis for experimental method in PCT. (4) Control is effected at many "levels." The Test applies to maintaining such a complex "variable" as self-concept as well as to such a simple, "sensory" quantity as distance between marks on a computer screen. In Chapter 7, I described The Test in some detail.

In Chapter 8, I described three examples of model-making. In this chapter, I will describe, first, a use of PCT in studying stages of development in infancy. This study illustrates further the use of The Test. It also gives some corroboration of Powers's proposed levels in the neural hierarchy. Then I will describe a simulation of movements in a crowd, and after that a simulation of the collective control of perceptions. These simulations are not models of unique single individuals, but rest on models of hypothetical individuals.

DEVELOPMENTAL STAGES IN INFANCY

It is characteristic of almost all living creatures that they change as they age. The common term for this change is "development." Some of the changes occur in all the individuals of a species. Examples in humans are growth in size, maturation of the sexual organs, babbling, walking, and speaking words. (When I say "all individuals," I mean all genetically normal individuals who encounter the necessary environmental opportunities.) Specific changes occur within certain time periods. When individuals do not find the necessary environmental matrix for the changes at the right ages, abnormal and pathological developments occur. Some changes that occur as we grow older do not expand our capabilities, but instead reduce them. Instead of calling those changes development, we often call them degeneration.

Other changes occur in some individuals but not in others. I speak here of changes in the sense of types of capabilities that the individual could not have exhibited earlier. In this sense, changes can be called development even though they occur only in some individuals. We speak of "developing" a skill such as using a carpenter's saw. Not all humans develop the skill of sawing. It is obvious, nevertheless, that genetic endowment has something to do with sawing: at a very early age, the human cannot even recognize the saw as a "thing." At a later age, the human can perceive the "thingness," but is too small and weak to pick up the saw. At a still later age, the human can pick up the saw and even place it against the wood, but the idea of purposefully moving it back and forth may be beyond his imagination. And so on. The creature's genetic endowment makes sawing eventually possible, but does not cause it to happen, to be tried, or even to be imagined. We grow into a bodily structure that enables us to climb trees, but if we grow up where there are no trees, we do not engage in tree-climbing motions.

A great many of the opportunities for action that we encounter are provided by our culture. Our culture (in the U.S.A.) makes it easier for us to get food by buying it at a grocery store than by throwing a spear at a wildebeeste. Our culture makes it easier (for most of us) to satisfy sexual desires by becoming acquainted with possible sexual partners in customary social activities, using language to test possible intentions, and so on, instead of—but what can I write here? My imagination fails when I try to imagine how sexual behavior could go if there were no cultural paths provided for it. What I want to say in this paragraph is that some of our behavior is strongly specified by genetic requirements (urinating every day, for example) and some of it is not specified at all by genetic requirements (where we urinate, for example). But I think it is a bootless question whether a given action partakes more of heredity or environment. I wrote about that futility in Chapter 3 under "Person and Environment."

There is the old joke about the English couple who adopted a baby born to French parents; the couple enrolled in a course in conversational French so that they would be able to understand the child when it began to speak. Our genes do not require us to speak French, but they do seem to require us to speak. Do our genes require us to live in groups? In families? To produce children? Once a child has reached the stage of babbling, is language then entirely learned, or do our genes specify some of the structure of language? The arguments about nature versus nurture seem to go on forever. But not here.

The Neural Hierarchy

Investigators of the functioning of humans and other animals often want to distinguish capabilities that are necessarily developed by the maturing organism from those that are merely enabled or permitted, but not required, by maturation. Since PCT conceives behavior as the varying means of controlling perceived quantities, the developing capabilities of the organism are conceived in PCT as developing capabilities of *perception*. PCT conceives the levels of the neural hierarchy, accordingly, as successive levels of perceptual capability or of the inclusiveness of control of perception. One can describe the hierarchy as one of successively more encompassing control systems—control systems that control a collection of lower control systems. Earlier, I gave a few examples; one was the experiment by Marken (1986) in Chapter 8 under the heading "Hierarchy."

The key to thinking about the levels of the hierarchy is that upper levels control the control (the internal standards) at lower levels, but not vice versa. If you want to walk to the library, your purpose of getting to the library controls the way you make use of your ability to walk, but your ability to control your walking muscles will not necessarily steer you to the library. As long as you maintain your purpose of getting to the library, your walking will take you in that direction, but your walking control is not built to take you only to the library. That is, your internal standard for seeing yourself at the library is at a "higher" level than your internal standard for perceiving yourself walking. I will give here a very quick sketch of the "lower" part of Powers's postulated neural hierarchy. You will find much more detail and a view of the entire structure in Part IV.

The "lowest" kind of system in the hierarchy is the kind that is in direct contact with external energies—the system such as the one diagrammed in Figure 4–1 of Chapter 4, showing a sensor in the organism and action in the environment. The perception at this lowest level is the electrical excitation in the afferent neural bundle going from the sensor to the comparator and is a perception only of *intensity*. It is not a perception of light, or warmth, or sound, or pressure, but of mere intensity—of rate of neural pulsing. At this level, there is no distinction among sensory organs. Plooij and van de Rijt-Plooij (1990) report that newborn chimpanzees react with the same sort of "staccato" and "uh-grunt" vocalizations to every kind of change of intensity:

> These vocalizations are produced in relation to any disturbance, any sudden change in intensity regardless of the . . . physical variable in which the change in intensity occurs, such as a sudden change in light by a. . . shadow moving across the baby or a sudden sound, such as the creaking of a tree branch, sudden thunder, or "breaking wind" by the mother (p. 70).

The second level is that of *sensation*. Systems at this and all higher levels send their outputs not to actions on the external world, but only to the comparators in lower systems. The output of a second-order system always contributes to a reference signal (internal standard) for a first-order system (or several of them). Perceptions at the second level enable us to distinguish from one another (and therefore to control separately) sensations such as light, sound, pressure, warmth, muscular effort, balance, and so on. We can also recognize combinations of sensory signals; Powers (1973, p. 108) gives this example of a sensation that can be recognized by cooks and diners:

> The "taste" of a steak is recognized as maximum when a whole array of intensity signals is present, including tastes, smells, temperatures, efforts of biting, and even (sizzling) sounds all in just the right proportions.

The third level is that of *configuration*, which is a static arrangement of sensations. We see shapes not so much by seeing a boundary of a thing as by seeing a difference between one visual sensation and another. To say it another way, we know that a thing has a boundary only if we can experience a difference between sensations. If we are in the middle of a large field of wheat, we can be wholly unaware of passing from the U.S. into Canada. If we look at a cloud in the sky, there is not a line drawn around it to tell us where it is; we see a cloud because we can see a patch of white against a field of blue. The edge of a cloud is often tantalizingly vague.

The fourth level is *transition*, which is what we see when a configuration changes. Transition gives us a perception of motion. We get a feeling of motion not only from seeing the scenery go by, but also from the run of a melody, from a sequence of fingertips walking across the skin, and from the changing tensions in our muscles as we swing a foot.

Fifth is *event*—a perception having beginning, middle, and end, but a rather short one having the character of unity, not having parts. Powers (1998, p. 144) gives examples:

> ... the bounce of a ball, the explosion of a firecracker, the opening of a door, the serve in a tennis game, a fragment of a song, a spoken word.

Powers's examples are instructive, but we also think of longer events such as a baseball game. It is even possible to conceive events longer than a human lifetime; historians have put dates on the rise and fall of the Roman empire.

Sixth is *relationship*. Our language contains a large number of words that label relationships: Near, behind, bigger, approaching, sweeter, louder, inside, after, away from, beloved, bossed, infiltrated, and so on and on. The sociologists' conception of *role* names a class of reciprocal social relationship: father-child, physician-patient, teacher-student, employer-employee, and so on.

In Chapter 18, I will tell you about more of the neural hierarchy: the levels of categories, sequences, programs, principles, and system-concepts, and I will give more detail than I have given here. I am telling you here only enough to help myself tell you about the studies of chimpanzee and human infants by Frans X. Plooij and Hedwig van de Rijt-Plooij.

Evidences of Control

How can we know when we are seeing, in everyday life, actions that a person is using to control a particular perception? Plooij and van de Rijt-Plooij (1994, p. 3 ff.) describe three ways. One way to look for control is The Test (for which see Chapter 7). We can look for—guess at—conditions in the environment that the person is causing to stay the same despite events in the environment that would otherwise cause them to change. If, despite the pulls of gravity and centrifugal force as the mother chimpanzee moves this way and that, and despite the displacement of the baby chimpanzee as the mother changes her hold on the baby, if the baby repeatedly moves across the mother's bosom to the nipple, we can have some confidence in the guess that the baby wants to maintain a position at the nipple. How can the neonate find its way to the nipple? It is possible that the neonate moves toward more warmth, since the nipple is the warmest spot on the mother's chest.

A second way to look for control and its level in the hierarchy is to look at the speed of an action. Control systems lower in the hierarchy act faster than those higher, since the changes in sensing by the higher systems must wait for the changes sensed by the lower systems. Powers (1973, p. 74) quotes a passage from a report of measurements of human tracking motions; certain muscular movements in the tracking sequence required .07 and .10 of a second. Plooij and van de Rijt-Plooij (1990, p. 69) say that first-order control systems "are very fast—about 0.1 second or less." They then go on to say,

> Furthermore, since control systems oscillate when they become unstable, with higher-order systems oscillating more slowly than lower-order ones, the frequency of oscillation provides information about the order of control involved. For example, "clonus" oscillations result from unstable first-order systems when muscles exert excessive effort. They oscillate at about 10 [cycles per second]. Several types of "tremors," such as in Parkinsonianism, oscillate at approximately three

[cycles per second], evidencing second-order instability. Finally, overcorrection, such as over- and undershooting the target while reaching out for something, results from third-order instability.

A third way to look for control is to compare the variability at adjacent levels. Control results in keeping a variable unchanged—even if the variable is rate of change. A variable being controlled by a system at a certain level in the hierarchy will remain controlled from that level as long as a control system at a higher level does not send a changed reference signal to the lower system. That is, the system maintaining a variable unchanged is the system at the highest level doing any controlling of lower systems. In Chapter 8 under "Hierarchy", I told how Marken used a "stability index" to compare the variability of control between two levels of perception.

I sit here at my keyboard, and my desire to watch the words that are appearing on the page sets the reference values for the lower systems that control my place of sitting, seating posture, the inclination of my head, my rate of breathing, and other matters that maintain my concentration on the page. But why do I watch the page so intently? I do so because I want to see a certain meaning taking form and finally standing out from the page, revealing itself in scintillating clarity and significance. As I see the words getting closer to that moment when I will thrill myself with the flowering of my thought, my posture and my breathing keep me firmly oriented to the page.

But now, as my ability to seize upon the right words slackens, as the right shade of meaning seems suddenly to elude me, and as I wonder whether I should doubt even the fragrance of the meaning itself, the blank white paper repels my eye, the keyboard becomes a fractious goblin, and my lungs pull in a deep draft with which to strengthen my spirit amid my mind's disorder. I abandon now my seated place. I pace the floor. I ask myself in a dozen ways how I can find once again that clear path to that clear idea I want to convey. Control has left the level of the keyboard and has gone up to a level of hunting for paths to meaning.

Chimpanzees

Plooij and van de Rijt-Plooij studied the onset of ability to control perceptions at various levels in chimpanzee infants. I draw what I say here from Plooij (1990), Plooij and van de Rijt-Plooij (1989b, 1990, 1994), and from van de Rijt-Plooij and Plooij (1987, 1988, 1993). Plooij and van de Rijt-Plooij began this series of studies in 1971 when Plooij went to the Gombe National Park in Tanzania, East Africa, to observe free-living chimpanzee infants. Plooij found that Powers's hierarchy of control was very helpful in organizing his observations of the chimpanzee infants. I quoted Plooij and van de Rijt-Plooij a couple of pages back concerning the uniform and undiscriminating vocalizations of the chimpanzee neonate at the occurrence of any sudden change in physical energy affecting its sense organs. Plooij concluded that control systems higher than the second order were not active in the neonate. The behavior of the neonates were, however, very much ruled by *sensations*—the second order of control:

> For instance, thermoregulation plays a part in the "comfort-contact search" and rooting toward a nipple. That higher than second-order systems are *not* yet achieved can be concluded by employing two of the three strategies [for ascertaining level of control]. Lack of variability in the "comfort-contact search" is the first indication that no more than two orders of control are functional. As long as the temperature is deviant from the optimal state, the "comfort-contact search" proceeds; as soon as the optimal temperature is obtained, the search stops. This implies that there is one fixed target value or reference value for this system.
>
> The speed with which the neonate's head oscillated from one side to the other during rooting provides the second indication that no more than two orders of control are operating. This occurs with a frequency of 2 to 3 [cycles per second]. . . . If this frequency is compared with the frequency of a clonus or a tremor, rooting may be considered to result from an unstable second-order control system (Plooij and van de Rijt-Plooij, 1990, p. 70).

The third order—configurations—comes at about two months of age:

> The speed with which the head oscillates during rooting has changed. Rooting has been replaced by the head-turning response. One turn of the head from one side to the other lasts 2 seconds instead of 1/3 to 1/2 second. The oscillation has become slower; thus a higher order must have become operative. The chimpanzee baby 2 months old and older does not whimper anymore when ventro-ventral contact is broken. Thus variability has appeared in the "comfort-contact search." This variability in second-order control indicated that

a higher-order control system must have become functional, allowing the second-order reference signal to vary (p. 70).

Plooij and van de Rijt-Plooij go on to give several other evidences for the control of configuration. Then they turn to the fourth level—*transitions*. This comes at about three or four months, and all movements of the infant become much smoother; tremors and jerky movements were no longer to be seen. At five months, the fifth level appeared, and *events* were observed to be under control—examples are walking, climbing, and picking fruit. Here, too, Plooij observed the first serious conflicts between mother and infant. The mother for the first time restricted the infant's access to the nipple and sometimes even used force to pull the baby off. The baby then had to take the initiative to maintain ventro-ventral (front-to-front) contact. Next, at seven to nine months, a sixth level arrived—the control of *relationships*. The infants began to place an "object *on top* of head, object *into* neck pocket, object *against* belly, and so on" (p. 72). Finally, toward the end of the first year, the infants began to show the ability to control *programs*. Examples were "fishing" for termites with a stem of grass, building a nest, and gathering food. At this same time, the infant began to spend more time off the body of the mother and even to make occasional forays beyond the mother's reach. Plooij (1990), Plooij and van de Rijt-Plooij (1990, 1994), and van de Rijt-Plooij and Plooij (1996) give considerably more detail on this than I have taken space for here. I urge you to read the original writings.

Van de Rijt-Plooij and Plooij (1987, 1988) conducted studies, too, on the progress of chimpanzee infants in achieving independence from the mother. The Plooijs discovered that independence increased in jumps, each jump preceded by a conflict between mother and infant. In recounting their findings, Plooij and van de Rijt-Plooij (1994, p. 368) say,

> . . . this sequence of regression-conflict-jump started at exactly the same ages at which the new control systems emerged.
>
> For instance, the ability to perceive and control so-called "configurations" emerged at two months of age. . . . The . . . infant was able [then] to cling to the mother, and the mother abruptly stopped . . . supporting and carrying the infant.

Plooij and Rijt-Plooij go on to describe several other stages at which the infant's ability to recognize more encompassing kinds of perceptions was matched by a greater independence from the mother in bodily position. They say then:

> We think that the maturation of the new types of control systems only create new potentials or, in other words, learning instincts. Actual reorganization of the infant's overt behavior and the actual learning of new skills depend on the interaction with the situational-social context.
>
> So, the infant may never develop certain skills if it does not get the opportunity or if it is not forced to do so. For instance, . . . chimpanzee babies in captivity [can] show a delay of many months in taking their first step, because they were not forced to do so.
>
> . . . [the] mother contributed by demanding that the infant reorganize its behavior according to its new potentials. Furthermore, during the conflict periods, the mothers never rejected the infant as a whole, but only certain aspects of its behavior. Thus, mother-infant conflict and even maternal aggression has its positive effects in that it enhances development (p. 368).

Van de Rijt-Plooij and Plooij (1988) reported also that conflict was followed by the infant's illness when a mother asked too much of the infant.

Humans

The Plooijs also studied the development of human infants. Van de Rijt-Plooij and Plooij (1993), p. 230) say,

> Similar processes seem to be at work in human infants. They are found to be ill more often at certain ages . . ., and also human infants appear to experience "regressive periods" regularly from week 5 onwards. . . . It was found that normal infants go through 10 regression periods at surprisingly similar ages.

The Plooijs took data weekly by questionnaire and interview from 15 mothers with babies under 20 months old. They discovered "distinct periods of conflict between interests of mother and infant [peaking] around the ages of 10, 13, 20, 27, 40, 48, 55, 64, and 78 weeks" (p. 233).

The studies carried out by the Plooijs show, even without constructing working models, how PCT can guide an investigation. PCT can suggest kinds of perceptions for which to look and methods of detecting them. In showing the connection between illnesses and "deficiencies and excesses of caretaking" (Plooij

1990, p. 133), PCT can be very specific about ages and the nature of "deficiencies" and "excesses." In a summary of their study, Plooij and van de Rijt-Plooij (1994) say,

> The big surprise was the fact that we found no less than ten regression periods! [in 85 weeks]. At least three more than the developmental transitions . . . reported so far in the literature. . . . This is understandable if one realizes that most studies did not sample frequently enough (p. 371).
>
> Apparently something fundamental is going on ten times in the first twenty months. We believe this something fundamental is the emergence of new types of control systems with new types of perception and new types of learning instincts. Each time this happens, the infant is off-balance and has to adapt, to reorganize. . . . when the mother starts being annoyed and mother-infant conflict follows, the infant is forced to start using the new learning instincts. It enters the zone of next development. In doing so, it shapes the new potential into new skills. . . . It is during these developmental transitions, brought about by the learning instincts, that the roots of culture are formed (p. 372).

What's Remarkable?

Notice how the studies by the Plooijs differ from the usual studies of development. Instead of listing or cataloging actions, the Plooijs look for expanding capabilities of perception—that is, for successively more complex *purposes*. The Plooijs must look for actions, of course, such as moving toward the nipple, moving away from the mother's body, showing distress, and so on, but the Plooijs do not simply catalog those and other actions according to superficial similarities. They look for actions they can use in one of the three ways they described earlier of *hunting for a controlled variable*. As the chimpanzee or child comes to use some actions as a means through which higher-level perceptions can be controlled, and then as still higher-levels begin to take over control, the sequential development of more and more encompassing capabilities can be traced. At the same time, the behavior of the mother can be understood as a part of the environment that provides prodding (or fail to provide it) at the time when the child has developed the capability for a new level of learning, and then as a part of the environment provides support (or fails to do so) as the learning proceeds. Finally, the hierarchy of perception provided by PCT enables the Plooijs to draw their insightful parallels between chimpanzees and humans in their developmental behavior. One of the crucial keys to profitable research is knowing what to look for.

In a communication to the CSGnet of 9 August 1999 on "Extending a thought on learning," Powers sums up:

> From what the Plooijs seem to have found, the [chimpanzee's] hierarchy has reached the sequence level by the age of 18 months. . . . What the Plooijs reported was that just after a new level comes into view, its operation looks stereotyped, as if the reference signals were not being varied by a higher level of control system. But as the next level begins to form, *random* variations appear, so that for a time behavior seems to become *less* organized, even regressing to earlier stages. The likelihood that a child will become sick increases greatly during this period. Then, as the new level starts to get organized, signs of the new control capabilities are seen, and order reappears while the new kind of control is practiced.
>
> I don't doubt that reorganizations continue through life, at all levels. However, as each level acquires a broader range of controlled variables, and as all the levels acquire increasing skill, the occasions for reorganization become fewer—that, after all, is the point of acquiring the systems in the hierarchy, to learn systematic ways of controlling one's experiences so that the large errors that drive reorganization don't happen any more.

The Plooij's (writing as Vanderijt and Plooij, 2003) have written a book whose title is just as fascinating as the rest of the book: *The Wonder Weeks: How to Turn Your Baby's 8 Great Fussy Phases into Magical Leaps Forward*. It is a treasure trove for anyone contemplating parenthood; their own or other's.

THE CROWD

The animal mind looks for associations, relations, patterns. It is no surprise that evolution produced animals with that capacity. If animals did not have it, they would have to hunt anew for water whenever they got thirsty, because they could not perceive patterns they could remember as signs of the path to water. They would not be able to find their own offspring. Like all our capacities, however, this one sometimes leads us astray. We look so persistently for connections that we sometimes see connections where there are none. We flip on a light switch and the doorbell sounds; it takes us a moment to realize that a coincidence has happened, not a causal sequence. Earlier today, I was carrying a large book in front of myself as I walked toward my wife. I suddenly thought of something I wanted to say to her, and I stopped to face her, holding the book between us. She reached out and took the book from me.

"What do you want with that?" I asked.

"Nothing," she said, "I thought you were handing it to me."

Seeing connections where there are none happens to all of us several times a day—and you have my permission to change "several times a day" to any number you prefer. It happens to scientists in their work as scientists, too. One of the long-standing topics in sociology is The Crowd. It is easy to see patterns in the behavior of people in crowds. Toward the end of Chapter 8, I told a little about the idea that crowds have a life of their own quite beyond the lives of their members. That idea is very much the same as the idea that groups are living systems in the same ways that individuals are—an idea of long standing among social psychologists.

It is easy to see and hear patterns. A person moving through a crowd may be trailed by a dozen others. A cheerleader waves a pompom and hundreds of people shout in unison, "Rah! Rah!" A hundred people move down an aisle of a church and spread out into the seats on both sides in very orderly fashion, with or without ushers. The priest intones, "Let us pray," and the members of the congregation recite the prayer in unison. To what extent must "the group" seize hold of the individual, somehow, for those patterns to appear? Clark McPhail and some colleagues have explored this question by examining some modeling done with PCT of individuals moving among other individuals, each individual having internal standards about its relation to other individuals, but with no internal standard specified for the group as a whole—with no postulation, that is, of the group as a whole having any effect on the individuals. The question McPhail and colleagues posed was whether a model could exhibit some typical patterns of "crowd behavior" without the model containing any influence from the crowd as a whole on the individuals.

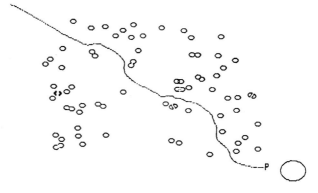

Figure 9–1.

Figure 9–1 is taken from a report by McPhail, Powers, and Tucker (1992, Figure 4, fourth panel) of simulation of various kinds of movements within a crowd. Figure 9–1 shows a single individual (P) moving from upper left to the large oval at the lower right, while being careful (so to speak) not to collide with the persons (symbolized by the small ovals) standing about. The program is written to give P the goal of reaching the large oval. The program also gives P an internal standard for P's proximity to the other "people"—not too close! (The program represented by Figure 9–1 actually contains more than those two reference signals; it contains reference values for direction of movement and speed of movement, for example. The original article by McPhail, Powers, and Tucker gives detail on such matters and a technical appendix as well. Similar research is reported by Tucker, Schweingruber, and McPhail, 1999.) Figure 9–1 shows nothing about a pattern within a crowd. I include the figure here to show you what the movement of a single individual will look like in further figures. Figure 9–1 shows the kind of element or building-block from which further simulations were assembled.

Figure 9–2 (from McPhail and others, Figure 5) shows the movement of two persons, each of which is defined (programmed) in exactly the same way as the person in Figure 9–1. The two Ps avoid bumping into each other just as they avoid the stationary persons among whom they make their ways. They begin near each other at the upper left, and they move along fairly close to each other. An observer might easily conclude that the two were moving together purposely. They were not, however, programmed to do so; they were programmed to act independently. You will have noticed that the placement of the

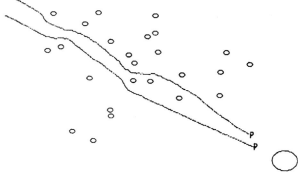

Figure 9–2.

stationary "people" in Figure 9–2 differs from their placement in Figure 9–1. They will take still further positions in later figures; they were placed in new random positions at every run of every simulation. As you have seen in earlier chapters, trying a model in an environment containing randomly influential events is a standard feature of research in PCT.

Figure 9–3 (from McPhail and others, Figure 7) again shows two people moving through a crowd of stationary people. This time, however, each moving person was programmed not only to move toward the target at the lower right, but also to move toward the other moving person. These two persons seem to keep themselves closer to each other than did the two in Figure 9–2; indeed, they now and then cross each other's paths. The association is tighter than in Figure 9–2, though I am not sure to what degree it would seem so if we were looking at the moving Ps on the computer screen unaccompanied by the trails that show here in the figures. Be that as it may, the pattern in Figure 9–3 is still one given by the controls within the individuals. Each individual P "wants" to be near the other as it moves along. The program for the simulation did not include a criterion for closeness or maximum distance apart as a characteristic of the moving *pair*. Each moving P was separately programmed to move toward the other moving P. Nothing hangs in the air between them or in the heavens above to keep them chummy. The chumminess results from the individual actions of each.

Figure 9–4 (from McPhail and others, Figure 10) shows four moments in an episode of following a leader. The leader, D, is programmed to move from the lower left to the oval target at upper right. Fourteen other persons begin at upper left. They are programmed only (a) to pursue D and (b) to avoid contact with one another. The clustering of the followers and the final ring around D look very much like what we see when followers move in an open space to stay close to, say, a politician or a preacher. Or one can imagine the fans of an entertainer before a hotel or a theater. Again, this very orderly process results wholly from the internal standards (purposes) of the individuals.

McPhail, Powers, and Tucker (1992, p. 7) say:

> ... there are at least three ways in which two or more purposive actors ... can generate similar reference signals that result in ... collective action. . . . First, two or more individuals may generate independently the same or similar reference signals. . . . Second, two or more individuals may generate [them] interdependently. . . . Third, two or more individuals may adopt voluntarily or obediently the reference signal(s) offered by a third party [italics omitted].

Figure 9–2 illustrates the first way: independent action. Figure 9–3 illustrates the second: interdependence. We can think of Figure 9–4 as illustrating independent action, and we can also take it to illustrate voluntary adoption of a reference signal, since each P takes its direction of movement from D. I will write more about this view of coordinated action in Chapter 35 under "McPhail." The patterns in the figures here are only a few of the patterns you can see in the article by McPhail, Powers, and Tucker. In all the multi-P patterns, we can see movement that looks very much like the kind of thing called "system" by people who write about general systems theory; for example:

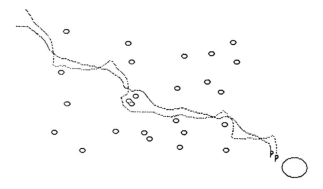

Figure 9–3.

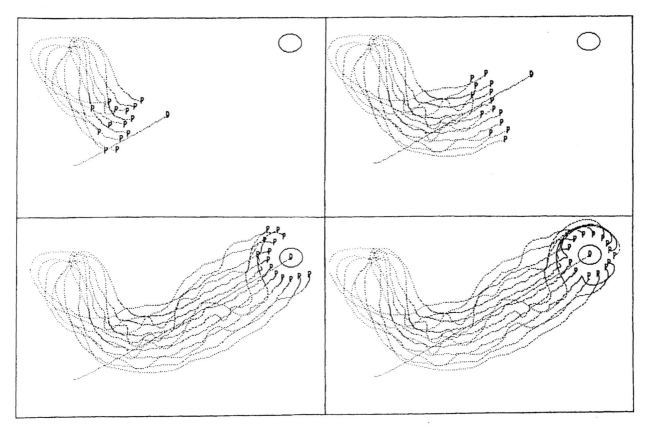

Figure 9–4.
Download the program *Crowd Simulation* at http://home.earthlink.net/~powers_w/

A whole which functions as a whole by virtue of the interdependence of its parts is called a *system*... (Rapoport, 1968, p. xvii).

A system is a set of objects together with relationships between the objects and between their attributes. Our definition does imply of course that a systems has properties, functions, or purposes distinct from its constituent objects, relationships, and attributes (Hall and Fagen, 1956, p. 81).

[In living systems,] their subsystems are integrated together to form actively self-regulating, developing, unitary systems with purposes and goals (J. G. Miller, 1978, p. 1027).

That talk of interdependence, relationships, and purposes, it seems to me, would describe the impression almost anyone would get from the motions depicted by Figure 9–4. The impression would be illusory, tempting though it might be. W. Thomas Bourbon (1995) says of the study by McPhail, Powers, and Tucker:

The social phenomena illustrated here were [earlier] reported in studies by the sociologist Clark McPhail and several colleagues of people in gatherings of many kinds—including situations in which it is popular to say that people are "out of control." ... simulated stationary people occupied randomly different locations, so the paths followed by D and P were necessarily different on each run, illustrating one of the defining features of control: unvarying ends created by variable means in a variable world. Control was achieved with no central, hierarchical commands and with no formal decision rule.... The various trajectories and the arc that formed around D at the destination are characteristic of many instances of social interaction and organization: they occur with no plan or *advance* intention, and there is no need for the actors to be aware their actions produce those externally observable consequences (p. 162).

... these results suggest an answer to the question of whether animals in social groups such as flocks, schools, or swarms must know they are part of such a structure or whether their actions might simply become coordinated with those of their immediate neighbors when each of them controls its own perceptions (p. 165).

What's Remarkable?

Earlier, I asked to what extent "the group" must somehow seize the individual if such patterns are to appear. Judging by the simulation carried out by McPhail and colleagues, typical patterns can appear without any influence whatever from "the group" as an entity existing beyond its individuals. The complexity of the path of a P in this simulation does not come from environmental stimuli; it comes from P's possession of several simultaneous goals: to avoid collisions, to remain at a specific minimal distance from another person, and to increase the proximity to a goal-position. These internal requirements combine with the positions of external obstacles to produce the path taken, the consequence. The program contained no influence from "stimuli"—no active environmental forces of any kind. Further research remains to be done, of course, including research into interaction that uses words and other symbols.

Finally, the pattern in Figure 9–4 gives me the excuse to mention a complaint that critics of PCT research sometimes make: that so much of it consists of studies of tracking. My answer to that is, "What is life?" (There I go again, falling into a foolish way to ask a question.) More exactly, "How else do we manage to stay alive?" Tracking is what we do at every moment. Think of the reference value as the target—the reference value of whatever controlled variable you wish to think about. And think of the value you are perceiving the variable to have now as the marker you want to bring into match with the reference value. And think of the action you take to bring the discrepancy to zero as what you do with the game-handle or the mouse when you are tracking on the computer screen. The negative feedback loop does not know whether you are tracking with a cursor on a computer screen or tracking the expression on the face of your beloved or tracking the meaning of the words you are typing.

Look again at Figure 9–4. If you think of leadership as something that one person does to other people (perhaps through the magic of charisma), or if you think of leadership as energizing other people in much the same way as you might wind up a toy and send it across the floor, then you will not see tracking in Figure 9–4. You will see one person pulling others, as on strings. But if instead you believe that individuals act because of the discrepancy between their perception of (a) where they are and (b) where they would like to be, then tracking is precisely what you will see in Figure 9–4. The interpersonal pattern we call leadership comes about by tracking, though charisma, inspiration, instruction, directives, scoldings, goal-setting exercises, quality-control discussions, planning retreats, and all the rest can serve as targets to the potential followers (trackers) that they may choose to accept and track, if their internal standards permit. To be followed a leader must find or provide opportunities for others to track the goals the leader is offering, if they wish to do so. I will say more about leadership in Part VI.

COLLECTIVE CONTROL OF PERCEPTIONS

In Chapter 8, you saw how tracking a target on a computer screen could show the functioning of human perception and action and could display the theorized connection between the two with an accuracy never achieved before the advent of PCT and the computer. In this section, I will show still another sort of insight that can be reaped from tracking a target, this time when two persons (or more) undertake to control the same perceived quantity while their capabilities are constrained in some way. The results illuminate some matters about which sociologists have speculated for a long time. Here I am drawing upon a series of simulations carried out by Kent McClelland (1996). The key finding is that "the collective control of perceptions can stabilize variables in a shared environment, even when interactants conflict" (p. 1).

Procedures and Results

McClelland began by recording the tracking done by an actual person.

Figure 9–5 (from McClelland, 1996, Figure 2, p. 8) shows the record. The vertical axis shows deviation above and below the target position (marked as zero). The horizontal axis indicates time in sixteenths of a second. The program applied a random disturbance to the cursor, a disturbance that could

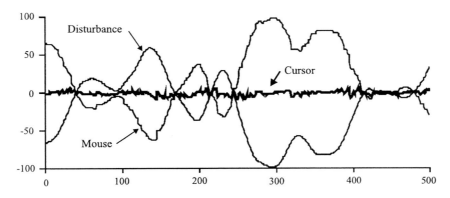

Figure 9–5. Data from a tracking experiment

be countered by moving the computer's "mouse." The pattern of the disturbance is shown by the dotted line in the figure. As you can see, the person was able to counter the effects of the disturbance by moving the mouse almost exactly in opposition to the disturbance (the mouse movement being shown by the thin solid line), with the result that the cursor was held very closely to the zero target position (the thick solid line). Recalling Chapter 8, you will not be surprised by Figure 9–5.

McClelland then built a model in the computer to simulate the person's behavior. He did that in the same way the researchers of Chapter 8 did. When McClelland ran the model against the same curve of disturbance shown in Figure 9–5, the result (as you might expect) was that the model produced movements of the simulated mouse and the simulated cursor that were very close to the movements the person had produced. Indeed, the correlation between the mouse positions produced by the person and the model was 0.998.

The next step was to build a simulation of cooperation. McClelland did so by putting *two* simulated "persons" into the computer. They were identical except for the rapidity with which they acted to correct their perceived errors. In the simulation of a single person, McClelland used a "gain" of 500. The "gain" factor in the equations of the model determines the speed of recovery from perceived error; the units of gain are arbitrary. In the simulation of two cooperating persons, McClelland gave one "person" a gain of 200 and the other a gain of 300. But now I will stop using the word "person" in quotation marks. I am now referring to computer circuits that are negative feedback systems and are given gains and a few other constants that enable them to behave like the actual person whose traces we saw in Figure 9–5. Accordingly, in speaking of the circuits in the computer that *simulate* the actions of persons, I will hereafter call them *systems*.

The two simulated persons (systems) were given the same controlled variable: the vertical distance between the cursor and the target at the middle of the screen. And they were given the same reference value: a distance of zero. These instructions in the models would result in both systems taking counteractions against the disturbance, and the counteractions would have the effect of reducing any distance between the cursor and the zero position. McClelland then ran the two systems simultaneously against the same disturbance curve as before.

The result is shown in Figure 9–6 (from McClelland, 1996, Figure 4, p. 10). You can see immediately that the two systems held the cursor just as close to zero as the single person had done. You see, too, that the degree of counteraction of neither system is equal to the degree of the disturbance, but the *sum* of the two counteractions is just what is necessary; the reason, of course, is that 200 + 300 = 500. This result, so easily seen in the figure, matches our everyday recognition of the fact that cooperating people can produce results of which a lone individual would be incapable. In Figure 9–5, you can see that at one place the person had to exert an effort (on the vertical scale) of almost 100 units, but in Figure 9–6, you can see that neither person had to be capable of exerting more than about 60 units. (The ratio of 60 to 100 is the same as the ratio of 300 to 500.) There is certainly a lot more to be said about cooperation; I'll say more in later parts of the book.

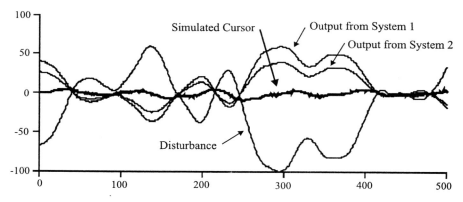

Figure 9–6. Simulation of Cooperative Control

Now suppose the two persons, although both want to control the same variable, want to hold it at different values. McClelland's next simulation used the same two systems (model "persons"), but now changed the reference values away from zero. He gave one system the reference value for the cursor of +1.0 and the other a reference value of –1.5. Again, the same disturbance was applied as before.

The outcome is shown in Figure 9–7 (from McClelland, 1996, Figure 5, p. 13). Two features leap to the eye. First, the two systems held the cursor just as close to zero as they had done when they were cooperating. They stabilized the perceived variable even though they were in conflict over it in the sense of having different positions at which they "wanted" to hold it. Second, that equilibrium was maintained only by efforts at counteraction that became greater and greater as time went on. The efforts of both systems were escalating without limit. Both systems were working not only against the random disturbance, but also against the disturbance put to the cursor by the other system.

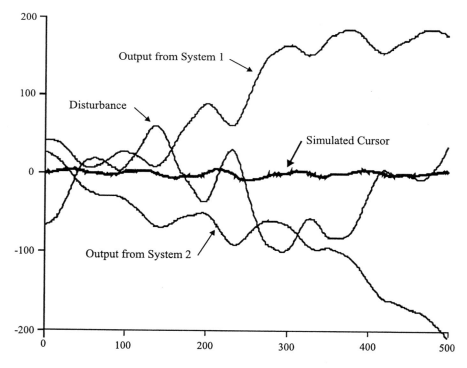

Figure 9–7. Simulation of Conflictive Control

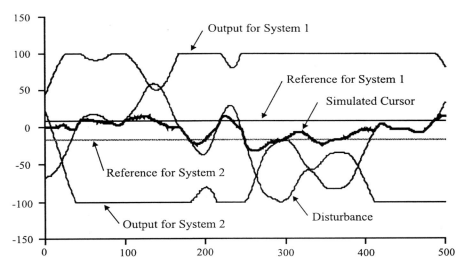

Figure 9–8. Simulation of Conflict with Limits on Output

The models behind Figure 9–7 were built to produce conflict—and did. Nevertheless, the result *on the cursor* of their conflictive efforts looked exactly like the cooperative effort we saw in Figure 9–6! Here we see both conflict (evidenced in the escalating efforts of the systems) and a kind of coordination that produces the same stabilizing result as cooperation. McClelland (1996, p. 13) says:

> People often engage in conflict and cooperation simultaneously. Any number of groups such as political parties, "dysfunctional" families, or academic departments, are fraught with internal conflicts, yet carry on their activities from year to year with little change. Such high-tension arrangements may satisfy none of the participants, but still can provide enough stability for everyone to carry on.

Real persons, however, cannot interminably increase their efforts against competitors. McClelland next altered the models so as to put limits on the degree of counteraction that could be brought to bear. He made two changes in the models. First, he put a maximum of 100 units on the output of each system. Second, he set the reference values farther apart. He gave a reference value of +10 to the system having the gain of 300 and a reference value of –15 to the system having a gain of 200. The gap between reference values was then wide enough to be discernible on the plot, as you can see in Figure 9–8 (from McClelland, 1996, Figure 6, p. 14).

The figure shows very well how the fortunes of competitors fluctuate, sometimes overcoming the onslaught of environmental change (the disturbance) and sometimes being overwhelmed by it. It shows, too, how a variable can stay within narrow limits despite the strong tugs from competitors. McClelland (1996, pp. 14–15) explains:

> As long as the two outputs are equally balanced against each other, the only force leading to any change in the environmental variable is the disturbance, and the environmental variable begins dutifully following the disturbance, until the disturbance pulls the variable outside the disputed region between the reference lines. Whenever that happens, the system whose reference line has been crossed can relax enough to move [its output] away from its output limit and thus begin again to control. So the variable stays near the reference line [of] the system in control. [Look at the traces, for example, between about 40 and 80 time-units and again between about 180 and 210.] That control lasts, however, only until the disturbance begins pulling the variable back toward the other system's reference line, at which point the first system once again runs into its output limit and loses control.
>
> Looking carefully at the figure, one can see that the system pulling in the upward direction, aided by the disturbance which is pulling the same way, keeps the cursor near its own reference line from about 50 to 175 on the horizontal (time)

scale of the graph. From then until about 250, the two systems trade control momentarily while the disturbance fluctuates near zero, until the disturbance (by veering sharply in the negative direction) hands over control of the variable to the system pulling downward. Finally, from about 400 until almost the end of the run, the disturbance moves back inside the "dead zone" (Powers, 1973, p. 255) between the reference lines, and the variable is no longer controlled by either system but simply follows the path of the disturbance.

Comparing Figures 9–6 and 9–8, two differences are plain. First, control of the perceived variable is much more stable when both persons have the same goal—the same reference value. Second, the amount of effort expended when the two persons are striving toward separate goals (as in Figure 9–8) is much greater than in the cooperative case. Still, although the controlled variable goes up and down somewhat under conflict, it often remains, in ordinary life, sufficiently between limits so that organized social life can cope with the variation without too many people being pushed past their output limits. On the other hand, although controlled variables often remain sufficiently controlled, the values at which they are controlled sometimes draw more effort from individuals than they can tolerate, and they reorganize their control hierarchies in ways that many of us deplore. That, at any rate, is one way I interpret the rising populations of prisons and of the homeless in the United States.

Finally, McClelland examined one of the ways of resolving conflict.

Figure 9–9 (from McClelland, 1996, Figure 7, p. 16) shows the result of agreeing to adopt the same reference value, so to speak. This simulation began with the same conditions used for Figure 9–8, but after the first 50 time units had passed, the program changed the reference values of the two systems so that both were now zero. This is analogous to two persons, after struggling against each other for some time, abruptly agreeing to a compromise position.

For the first 50 units of time, you can see that Figure 9–9 is exactly like Figure 9–8. But after that, the two courses of events differ. In Figure 9–8, the output curves repeatedly go to their limits of +100 and –100; in Figure 9–9, the two output curves gradually approach each other. At the start in Figure 9–9, the output curves go to their extremes in about 40 time units, but after the two systems adopt their compromise, the output curves do *not* return to something like Figure 9–6 (the cooperative case) in 40 time units or even in 500. This slow approach to cooperation after conflict is common in ordinary life—in families, business organizations, legislatures, international relations, even "friendly" games with their "sore losers."

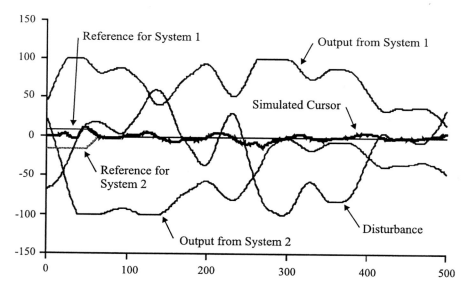

Figure 9–9. Simulation of Limited Conflict and Conflict Resolution

Parallels and Extensions

The slow approach has also been found in studies of actual people in reasonably natural events; one of the most notable is the one by Sherif and others (1961), in which adolescent boys in two summer camps were observed while they dealt with some arranged difficulties. The experimenters brought together boys from two summer camps for a series of athletic contests; competition and hostility developed quickly. After that, the experimenters arranged conditions in which the boys could get what they all badly wanted only by bringing to bear all the hands from both groups. For example, Sherif and Sherif (1969, p. 256) summarize one event as follows:

> One day the two groups went on an outing at a lake some distance away. A large truck was to go for food. But when everyone was hungry and ready to eat, it developed that the truck would not start (the staff had taken care of that). The boys got a rope—the same rope they had used in [an] acrimonious tug of war—and all pulled together to start the truck.

That event was one of several events with "superordinate goals" (the Sherifs' term) that enabled the boys to revise their internal standards about cooperating with members of the other group. At the end of the camping period, the numbers of friendship choices made to the other group had risen a great deal in comparison to the earlier competitive period (from six percent to 36 percent in one group and from eight percent to 23 percent in the other; see pp. 119 and 187 of Sherif and others, 1961). Nevertheless, as you can see from those percentages, far more than half the choices remained in the chooser's own group, even after the lengthy and compelling experiences in which cooperation was unavoidable. (I could not find any information about the numbers of days that passed in the various stages of the experiment.) In my opinion, by the way, this experiment, widely known as the Robbers Cave experiment, is one of the jewels of social-psychological research.

Oddly enough, though everyone has experienced this slow return to trust and cooperation after conflict, stress, and distrust, almost everyone seems ready to forget that necessarily slow return to trust when mounting a program of social change. After a period of stress—for example, after a strike by employees against managers—it is common for someone to stand up and say something fatuous like, "Let's put the recent unpleasantness behind us and all get together now and put our shoulders to the wheel!" Another sort of example occurred in the 1970s when the people at the National Institute of Education became discouraged at the failure of many of the projects they had funded in the hope of bringing about improvements in the public schools. Many projects they had funded for three, four, or even five years left behind them very few schools that were making use of what the research had presumably discovered. In the hope, therefore, of increasing the ratio of benefit to cost, the Institute changed to a policy of funding most research for only a year, so that if no benefit was shown by then, they could easily stop that funding and put the money elsewhere. That was a strategy by which they hoped to make important changes in educational and social practices that had been in place for decades and even centuries, and in which thousands, even millions of persons had large investments in money, careers, and self-regard.

I am not saying that it is always impossible for social change to occur quickly; all of us can cite examples of social change that did occur quickly, such as the patterns of urban living made possible by the advent of millions of automobiles—an example, by the way, of just the kind of coordinated action resulting from individual motives that McClelland has simulated. I am saying, however, that it is folly to have much confidence in advance that a large-scale change can be brought about at a rate much faster than the rate at which most changes in society do take place, especially in cases where a number of interdependent people risk losses in the change.

It is easy to see in Figure 9–9 why the return to the cooperative pattern of Figure 9–6 is slow. When, after about 60 time units, the internal standards of both participants have moved to zero, the two systems succeed (from about 60 to about 120) in keeping the cursor very close to zero. At about 120 on the time axis, you can see that the output of System 2 reaches its limit of 100 units of effort and stays there until about time 140. During that short span, the disturbance is rising sharply, and the output of System 1 is falling sharply in counteraction. Indeed, the output of System 1 is falling more sharply than it would if the output of System 2 had not reached its limit. The result is that the outputs of the two systems draw together slightly. Their approach is difficult to see (for me, anyway) without actually measuring the vertical distance between the two curves. But if you do actually measure the distance between the output curves at about 120 and at, say, 200, you will see that they have come toward each other by a good bit. And that happens again between about 270 and 310 on the time axis, when the output of System 1 has reached its limit. If you look closely at the output curve for System 2 at the point where the curve of System 1 reaches its limit (at about time 270), you will see right there a sudden upward increase in the slope of System 2's curve—an increase sufficient to counteract the difference between the disturbance and the counteraction that System 1 is able to supply. And the result there, too, is to bring the two curves closer together, as you can see if you measure the distances at about 310 and anywhere later. These occasional approaches of effort can occur, because neither system wants to pull the cursor away from where the other wants it. There are some further interesting technicalities about Figure 9–9 that I won't go into here; you can read about them in McClelland's (1996) fascinating article.

Suppose you have been working at a job for some years, and the management announces that now you are going to become a member of a "quality circle" in which you are going to talk about how work is carried out in this company and ways in which the work or the working conditions could be improved, and the chairperson of the quality circle will report those ideas to the managers. (Or suppose any other kind of sudden and unusual change in any kind of organization you wish.) Some few people will be delighted at the prospect and will plunge eagerly into such discussions. Typically, however, employees will wonder what kind of action will be safe to take. They will have become accustomed to the levels of variables they want to control at work, and they will be accustomed to the methods they use to do so—or to try to do so. They will have learned that certain kinds of action—perhaps making suggestions to managers about how to do things—control the variables no better and sometimes worse. They will, typically, wait warily before they try new sorts of actions. What can happen, however, is what you see in Figure 9–9.

They (these typically cautious people I am talking about) will continue to pull in their usual ways against disturbances to variables they want to control, but a time arrives when they see a variable about to get out of control even though they are pulling against it as hard as possible—and then others come to the rescue; they put out extra effort to keep the variable where

all want it to stay! Then the person can see a way of controlling the variable that he or she did not previously think was available—namely, get a little help from your friends! It takes time for that kind of thing to happen. There are, of course, many other kinds of events that can happen that can enable persons to discover new ways to control the variables they care about. My point is that those events do have to happen before the typically cautious person can discover that the other methods are safe.

The clarity that Figure 9–9 shows concerning the slow recovery of cooperation is due to the strong and simple connection between the two simulated systems. First, the two systems "care" about only one variable (position of the cursor). They do not care what color the cursor is or whether the other system is a friend. Neither "knows" in any way that the other exists—which is often the case in ordinary social life. Neither cares about the rate of recovering cooperation—again, they have no way of "knowing" that such an idea (or variable) as cooperation exists. Second, both systems are set to control the *same* variable. So the "world" of these two systems is as simple as it can be, and what happens anyplace in it immediately affects the perceptions of both systems. In contrast, coordination of large numbers of people can be much looser in both the nature of the variables controlled and in the values of those variables acting as reference values. But even so, many perceived variables are controlled in society *despite* great differences in the variables and their reference values, as we will soon see. To repeat, a variable held close to a certain value by the combined outputs of a large number of people can now and then change rather quickly because the ties among a large number of people are much looser than the simple tie illustrated by Figure 9–9. Often a group of people can make changes quickly in a social routine simply because other people do not care (have an internal standard) about what the group wants. On the other hand, the very features of the models illustrated in Figure 9–9 are sufficient to hold a social value in place for along time and allow it to change only very slowly.

McClelland (1996, pp. 17–18) comments:

> ... conclusions drawn from simple control-system simulations about social conflict and cooperation can apply to interactions involving not just two or three people, but thousands or millions, all seeking to control their imperfectly shared perceptions of, for instance, the outcome of a political contro-

versy. The most important finding from these simulations has been that interacting control systems need not hold the same reference standards in order to achieve cooperative outcomes. . . . As long as systems with conflicting reference values can stay inside their maximum limits of output, their interaction results in joint control of the contested environmental variable. As Powers (1973, p. 255) puts it, conflicting control systems will stabilize the environmental variable around the "virtual reference level" of the group as a whole, even when all the individuals within the group are experiencing uncorrected errors.

Consider what happens when large numbers of people all become involved in controlling a single perception, say the outcome of a national political decision. The participants in such a widespread cooperative-conflictive interaction can represent all shades of opinion, or in other words might hold as many different reference values as there are participants. People will then pull in many different directions to adjust their perceived reality to fit their preferences. Because everyone is experiencing perceptual errors, except those few whose preferences approximately match the virtual reference value of the group as a whole, nearly everyone may end more or less frustrated with the situation. As PCT points out [see Chapter 21], prolonged errors generate negative emotions. Some participants may throw maximum effort into the interaction, pulling as hard as they can, while others act as free riders, observing but adding nothing to the collective output, and still others become totally apathetic, turning their attention elsewhere. Others may cope with perceptual error simply by adjusting their reference values to match the status quo as they perceive it. The action depends on participants' reference values, their energy and endurance, other disturbances affecting them, and other perceptions they seek to control. If some participants with extreme views "max out" in their output, it will have little effect on the collective outcome as long as those with more moderate views can maintain their control by keeping their own output within reasonable bounds. . . .

A virtual reference value will emerge from the interaction and decide the outcome. To an individual participant, it might well seem as if some invisible superhuman control system is imposing stability, since no action by that single individual will have noticeable impact. The individual can pull, give up, or start pulling just as hard in the opposite direction without perceptibly budging the virtual reference point. But even if no particular individual's contribution counts for much, the collective outcome is wholly determined by the joint efforts of the participants. . . (p. 17).

To an individual it may feel as if some external force has determined the outcome, and it may even seem as if the control which emerges is the responsibility of some virtual collective actor, personified, perhaps, as "the group," "the family," "the corporation," "society," the "great leader," or even "God." Although illusory, this virtual collective actor may appear to have human and even superhuman qualities, in that the outcome of the collective control is indistinguishable from the stability that could have been produced by a single, powerful control system. No wonder, then, that sociologists in their explanations of social structure have invoked hidden collective entities! Sociologists, too, have experienced the same illusion. . . .

The stability produced by any widespread collective control is also likely to last longer than the results of any single individual's efforts. Thus, the virtual actor will seem not only to be stronger than the individual, but usually slower as well. If only a few individuals are involved, an episode of collective control may be over in minutes, but collective control involving large numbers of people generally takes longer to get organized and longer to die out. Individuals participating in the effort of collective control may come and go, but the attention span of the virtual actor will appear to last from the time the first two participants get together till the time the last two quit. Some efforts of collective control—for example, the maintenance of a building or public monument—may even continue for centuries, as generations of caretakers succeed each other. . . . virtual social actors can appear to change their minds, though typically more slowly than individuals do. Consider the slow drift in the currently fashionable vocabulary of a language. . .(p. 18).

McClelland (1996) also points out that once in a while a few persons can start a sequence of events that makes it possible for the "virtual reference level" to change appreciably in a surprisingly short time.

For example:

> If enough participants in an effort of collective control can coordinate their changes in reference values, thus generating the social power of a new alignment (see McClelland 1994), the resulting lurch in the collectively controlled variable can be revolutionary. A dramatic example occurred across Eastern Europe in 1989–92, as Communist governments toppled almost overnight, not because of any armed conflict, but because millions of inhabitants of the countries agreed with new leaders that a change had taken place. Not everyone needed to come to that agreement, but when enough did, the outcome was no longer in doubt (p. 19).

Now it is easier to explain why some large-scale changes can occur quickly and some require decades to get started and decades to die out. The toppling Communist governments are a good example. Virtual actors come into being to hold in place a value of a variable that people *care* about—that is, that they have reference values about. And virtual reference levels can change if the levels of enough individuals change; this circumstance is what McClelland mentioned in the quotation just above. There were many ways in which many people were dissatisfied with the existing governmental operations, and many people, including many political leaders, were thinking that Western forms of government might bring better control of the variables they cared about. Furthermore, those opposed were too few and not persuasive.

But change can also happen quickly in a society when a change that begins to happen (for whatever reason) fails to affect variables that many people care about. The advent of the automobile, for example, eased the satisfaction of many purposes while failing to obstruct immediately many purposes that many people cared about. Very few people could envision the changes that later came upon their neighborhoods, their streets, the air they breathed, the nation's arable land, and so forth. No virtual actor was protecting us from the adverse effects of millions of automobiles. In brief, large-scale change can occur comparatively quickly (1) if few people care about the changes that begin to occur or (2) if many people, including many influential people, begin to conceive an attractive direction of change while, at the same time, the people opposing that direction of change cannot gather effective support. In this second case, where large numbers of people do care about the direction of change, they must, if change is to occur, have some degree of agreement about the direction. If everybody wants intensely to change, but various factions rush off to all points of the compass, only turmoil will result.

I will take up the point of organizational and social change again in later chapters. Here I have wanted chiefly to let McClelland show you how PCT can illuminate (and model) a puzzle of social life that has bedeviled (and begodded) people since time immemorial—including sociologists when they came into existence a century or two ago (or three?—it is hard to pinpoint the emergence of their profession).

McClelland (1996) goes on in his article to talk about further fascinating matters. Using the viewpoint of PCT, he puts some interesting meanings on cooperation, conflict, obstruction, and accommodation. I will postpone my remarks about those matters to later chapters. The big point here is McClelland's point that "most episodes of collective control of perceptions will involve a mixture of cooperation and conflict" (p. 19). A great deal of social life, in other words, can look like Figure 9–7, in which the participants are spending a good deal of effort (sometimes a great deal) to pull the value of the variable toward their reference values against the opposite pull of people having other reference values; all (except those who don't care) experiencing a good deal of frustration in the process, but continuing to spend the effort because to give up would allow the value of the variable to depart even more from their reference value and increase their sense of wrongness (error).

What's Remarkable?

Finally, let me point out once more that the patterns in all the figures are there only because of the model of *individuals*. It was not necessary to make any assumptions, any theorizing, about something lying between the individuals—only about what lay within each. The assumptions behind the PCT model of the negative feedback loop were sufficient to show these patterns of collective behavior that we recognize in our society.

Part III

Science

In Chapters 6 through 9, I exhibited some ways that researchers have tested, through actual observations, some assertions made by PCT. I said, too, that you can carry out more precise demonstrations on your own computer by going through the World Wide Web to

http://www.mindreadings.com/demos.htm
and
http://home.earthlink.net/~powers_w/
and
http://faculty.ed.uiuc.edu/g-cziko/twd/demos.html
and
www.pctresources.com.

As I went along, I made some remarks about research method and scientific assumptions. Those are important matters for this book, because PCT makes some assumptions about science and research that are very different from those common in traditional academic psychology, and I want to be explicit about them. That is the purpose of the chapters in Part III.

Traditional psychologists test their understanding of human behavior by predicting acts. Examples read something like this: More people among those who answer "yes" to certain questionnaire items will also prefer certain kinds of recreation than among those who answer "no." Or this correlational form: People who score high on a test of Phebephobia will also score high on a test of Pontiphilia, and those low on the one, low on the other. It is no surprise to anyone that only *some* of the people observed turn out to conform to the predictions. Traditional psychologists feel vindicated when the portion who do conform is larger than one would expect from pure chance. That is the reason I refer to that sort of research as nose-counting. In my 1990 book I gave it a more formal label: the Method of Relative Frequencies.

Adherents of PCT do not try to predict particular acts such as scoring high on something. Neither do they count noses. PCTers insist that the correctness of an assertion derived from PCT must be found justified in *every* individual tested. The demonstrations and experiments described in Chapters 6 through 9, for example, were published even though they were performed with few participants, because the authors (and other PCTers) are ready to discard the theory or revise it radically if *one* person shows up reliably behaving contrary to the PCT prediction.

PCT is tested by modeling (in the manner I described in Chapter 8 under the heading "Models and Theories") and by using The Test for the Controlled Quantity that I described at the end of Chapter 7 under "The Test." The Test is used to examine every sort of question about PCT. It is used to investigate how nerves work together (physiological psychology), how people can see transitions (sensory psychology), how two or more people can interfere with one another's purposes (social psychology and sociology), and so on.

PCT does not claim that all animals have the same number of layers of control (humans probably have the most), and it does not claim that the nervous system of an octopus has the same gross morphology as that of a human, but it does claim that the negative-feedback control loop (Figure 4–1) reigns supreme. Furthermore, the claims of PCT about behavior are pertinent to all sciences (and to all lore, too) that deal with living creatures, because all those sciences make assumptions about the functioning of individuals—ethology, sociology, political science, economics, medicine, and all the rest. It is with that attitude that the chapters in Part III are written.

Chapters 10, 11, and 14 through 17 discuss some assumptions and procedures in regard to which PCT differs from conventional psychological science.

Chapters 12 and 13 are reprints of two articles that I think reveal with special clarity the view of science embedded in PCT. I print them here unedited, despite their technicalities, because I want you to see these scientific reports in their pristine beauty.

REVIEW

Here is a nutshell review of what has gone before. If you forget everything else you have read so far, please remember the following as you read the chapters in Part III.

Action springs from the circular causation between internal standards and environmental disturbances to controlled variables.

The relation between nonliving things and the environment is very different from the relation between living things and the environment. Living things initiate action, and they expend much greater energy than the energy received by the sense organs.

The distinctive characteristics of living things are (1) they act with purpose, to control perception, (2) they operate through negative feedback loops, and (3) causation in the loop is circular and simultaneous.

Chapter 10

Don't fool yourself

How do you know what you know? I have asked that question (or words to that effect) of a good many persons. Asking it about some piece of presumed knowledge a person has offered me, I have got various answers:

I just *know*, that's all!

(Somebody) told me so (or I heard it on the radio).

I read it (someplace).

I saw it in a movie (or on the TV).

Well, it stands to reason.

I saw it myself (or it happened to me once).

I read a report of a study (or experiment).

I did a study (or an experiment) on it.

I don't claim that those examples have sharp boundaries among them; I mean merely to say that the answers ranged from a vague faith (even if heartfelt) in a verbal statement to a careful personal inspection of palpable events. There are many ways to come to know something.

KNOWING SOMETHING

And what is it to "know" something? Everybody (I think) is aware of the distinction between knowing something and knowing *about* something—more precisely, the distinction between having the direct experience of something and having ideas about it or being able to say things about it. Someone might ask, "Do you know the Fiji Islands?" and you might reply, "Well, I've read about them, but I've never been there." With that reply, you are implying that while you have memories of what you have read *about* Fiji, there are experiences *of* which one can acquire memories only by having been there. Or someone might ask, "Do you know how to ride a bicycle?" and you might reply, "Well, I've seen a good number of people riding bicycles, so I have a pretty good idea how to go about it." But no matter how confident you may be of your knowledge, you might not succeed in wobbling down the street on your first try or your second or even your third. Do you know the fragrance of the frangipani flower? Well, you can be told about it, read about it, or smell it yourself.

The kind of knowing to which I give the most attention in this book is the kind that enables you to control a perception of some variable that is affected by the "thing" the knowledge is about—and to control that perception by acting on the world outside your own neural net. If you know the location of the Fiji Islands, you can control your perception of your distance from them. The "thing" your knowledge is about is your distance from Fiji. A couple of perceivable variables (among many possible) that would be perceivable aspects of that distance are (a) the distance you read or calculate from an atlas and (b) an announcement by a flight attendant of the name of the next airport you will be landing at and your translation of that information in your mind into approximate miles yet to go to Fiji. If you know the fragrance of frangipani, you can buy some of that kind, instead of lilac, if what you smell in the bottle matches your olfactory memory.

How can you know where Fiji is? If you have not been there, you can go by what someone tells you or by what you read in a book or see on a map. Those words or maps constitute instructions for getting there. Suppose you live in Chicago. The words or maps tell you, in effect, that one way you can get to Fiji is to buy a ticket that will take you first to San Francisco, then Hawaii, and then Fiji. But how do you know that the words or maps can be trusted?

On current city maps, I have found streets that do not exist and blanks where streets that do exist should have been drawn. In the end, the only way you can be sure that you can get to Fiji by going through San Francisco and Hawaii is by trying it yourself. That kind of knowledge (seeing it yourself) is what is customarily called "scientific." Also "empirical."

Science goes further than speculating about where Fiji might be or what it might be like or how you might get there. Science is about getting there. Science offers criteria for knowing when you have arrived at Fiji. It also offers guides for telling other people how to get there. To abide by those criteria, it is not sufficient to answer, "I just *know*, that's all!" And though it may be interesting for many purposes, it is not scientifically sufficient to say, "My travel agent told me how."

Sometimes people complain about the scientist's insistence on wanting to see for himself or herself. "Nobody can always see for himself," they say. "Most of the time, you have to take somebody's word for it." That's true. You can't get your daily work done if you are always off to Fiji or Bulgaria or the moon, checking on whether they actually are where people say they are. But if I want knowledge that *can* be verified, then I want instructions on how to verify it. If someone tells me that the moon revolves around the earth, I want the person to tell me how I might check up on that myself, even if I do not intend to do so. If the person cannot tell me how to check for myself, then I must take her assertion as merely one more speculation among others. If someone tells me that the earth is four billion years old, more or less, I want the person to tell me about the procedures through which I can reach such a figure for myself. And if someone tells me the earth is about four thousand years old, I want to know that person's procedures, too. What the person tells me will be maximally useful if the information is in the form of the functions and organization in a model.

Insisting on verifiable assertions is the first necessity in the procedure we call science, but of course carrying out a verification can be very complex. A lot of this book is about the complexities. My point here is simply that science deals with the external, verifiable world, and therefore a scientific inquiry must begin with an empirically verifiable assertion. Still, scientific procedures are often difficult and subtle, and scientists sometimes honestly think themselves to be on the road to Fiji when they are actually heading elsewhere. Once in a long while, too, a scientist fabricates data. That is sad and dangerous—though I think the proportion of scientists who do that is very much smaller than the fraction of manufacturers who pollute the water supply, and usually, I think, the perfidious scientists do less harm to public health and welfare.

The scientist's point is that if there is no way to compare an assertion about the external world directly with that external world, then there is no way to resolve competing claims. You can appoint a referee, but that only postpones the difficulty. You can have someone tell you the Revealed Truth, but that too only postpones the difficulty; you may find yourself having to admit, after people have thrust contrary evidence upon you for some 350 years, that you should not have insisted that the sun revolves around the earth.

TAKING A VOTE

You can take a vote. You may burst out laughing at that suggestion, but it has been made seriously many times. Petr Beckmann (1971) tells us that in 1897, a bill was introduced in the state legislature of Indiana entitled, "A Bill Introducing aNew Mathematical Truth." The bill declared the value of pi (the ratio of the circumference of a circle to its diameter) to be 9.2376. . ., which, Beckmann wrote, "probably represents the biggest overestimate of pi in the history of mathematics" (p. 174). The bill was actually passed by the Indiana House of Representatives, and was about to be voted on by the Senate when the fact came by sheer chance to the attention of a professor of mathematics at Purdue University; he "coached the senators," Beckmann says, and the Senate voted to postpone further consideration of the bill. It may seem strange that those persons to whose hands the welfare of the state of Indiana was entrusted should believe a geometrical or physical fact to be susceptible to legislation, but I should mention that in recent years, articles have appeared in psychological journals and in journals devoted to the philosophy of science in which physical facts such as the acceleration of gravity have been claimed, if I understand the authors correctly, to be no more than conventions or matters of "social reality"—that is, an agreement among a large number of people that objects approach each other in that way.

Some time in the 1940s, the Illinois legislature passed a bill to establish a statewide testing program for high schools. The bill specified that 70 should be the passing score for the test! In contrast to the Indiana case, that specification by the Illinois legislature could actually be carried out, although it was completely meaningless, since the bill made no specification about the nature of the test items, how many items the test should contain, how the items should be weighted in the scoring, or anything else that could affect the meaning of "70" or its effect on the lives of students and teachers. I cannot imagine, either, what the legislators could have meant by "passing." As far as I know, and I was associate director of that program for seven years, no one ever used the test as a gate through which students were to "pass" from one condition into another; educators used it chiefly for academic counseling.

When I was teaching an introductory course in social psychology (a good many years ago), I formed the students into groups of four to six persons and asked them to think of something they would like to know about the social world on campus but would not likely find in books. I wanted them to learn how one could go about getting observable information directly from the observable world. One group told me they would like to find out whether belonging to a fraternity or sorority caused students to get lower grades, on the average, than students who did not belong to those organizations. I told them that sounded feasible for research, and I asked them to come back in a few days with a plan for finding the answer to that question. They returned in a few days and told me that their plan was to go to some fraternities and sororities and ask the members whether they (the members there) thought they were getting lower grades than students who did not belong to fraternities or sororities. I don't remember any more of the conversation, but if I had asked them what they proposed to do about the differences in opinion they would inevitably gather, I suppose they would have said they would count the responses in the manner of a vote and declare the winner. I think it is sad, by the way, that people can get to be sophomores in college and still have no other conception of getting knowledge from the observable world than asking somebody else for the answer (or reading some author's answer). It is possible, of course, that when a professor asks college students to get some information, almost all students immediately think of asking someone for the information (or asking a book), because that is the way almost all professors and other teachers have almost always told students to get information.

I had an instructive experience when I was one of a faculty of a high school. At the opening of the school year, we learned that the superintendent wanted us and the faculty of the other high school in the district to discuss curriculum revisions once a month and make recommendations for change at the end of the school year. As the meetings came and went, it became clear that one member of the other faculty and I were together in disagreeing with all the other teachers on a fundamental point or two. At the end of each monthly meeting, a committee would put before us, for a vote, a proposed recommendation that seemed to sum up what the majority found pleasing. That other fellow and I would often vote against the proposal. At the last meeting of the year, when all the recommendations were bundled together to be forwarded to the superintendent, we two said we would submit a minority report. At that, one person stood up huffing and puffing in outrage. After we had been outvoted at every vote, he asked rhetorically, how could we possibly still hold to our opinion? As far as I could tell, he did honestly believe that any normal person, seeing that he or she was in opposition to a firm majority, would be convinced that his or her opinion was simply wrong. I am not sure whether that fellow thought we were physically or morally defective. Maybe both.

What happens in traditional psychological research seems to me something like that. A majority of participants, or enough to be beyond mere chance, act as the experimenter predicted they would, and the experimenter then reports, typically, that "the subjects" acted that way. Or the experimenter says that the participants were "tending" to act as predicted. That way of talking (and subsequent acting) seems to me very much as if each participant's act is taken as a vote for or against the experimenter's hypothesis. Some of my colleagues, upon hearing my dissent from that method of coming to a conclusion from data, react with very much the same outrage as the faculty member at the high school.

SHIFTING PROPORTIONS

The fact that psychological experiments continue, decade after decade, to turn up behavior that goes contrary to prediction has of course been bewailed by many researchers. Here I will quote only Cronbach's (1975, p. 123) lament. He remarked about the fact that the conditions under which observations of behavior are made keep changing. He offered this analogy:

> The trouble, as I see it, is that we cannot store up generalizations and constructs for ultimate assembly into a network. It is as if we needed a gross of dry cells to power an engine and could make only one a month. The energy would leak out of the first cells before we had half the battery completed.

A good example of the kind of change Cronbach had in mind is exhibited in an ingenious study by Urie Bronfenbrenner (1958), who reviewed the studies that had been made of child rearing practices in the lower and middle classes between 1932 and 1957. The earlier studies had found that lower-class parents were more permissive with their children, in several ways, than middle-class parents. Bronfenbrenner said that researchers in the earlier years typically characterized "the working class . . . as impulsive and uninhibited, the middle class as more rational, controlled, and guided by a broader perspective in time" (1958, p. 422). Later studies, however, found the differences between the two classes to be less than the earlier studies had found, and by the middle 1940s, the differences had vanished! Was this finding, so confidently proclaimed during the 1930s, merely one more social-science mirage? Was it perhaps merely the product of sloppy research? No. Bronfenbrenner showed that the direction of change was a reliable one, and as the years went by, studies increasingly showed that the middle class had become more permissive than the lower! The change, however, was not one of exchanging positions. Parents in *both* classes had become more permissive, but parents in the middle class had changed the more rapidly. Here are excerpts from Bronfenbrenner's summary:

> Over the past quarter of a century [1932–1957], American mothers at all social-class levels have become more flexible with respect to infant feeding and weaning.

> Class differences in feeding, weaning, and toilet training show a clear and consistent trend. From about 1930 till the end of World War II, working-class mothers were uniformly more permissive than those of the middle class. . . . After World War II, however, there has been a definite reversal [of the difference].

> Shifts in the pattern of infant care—especially on the part of middle-class mothers—show a striking correspondence to the changes in practices advocated in successive editions of U.S. Children's Bureau bulletins and similar sources of expert opinion.

> . . . socialization practices are most likely to be altered in those segments of society which have most ready access to the agencies or agents of change (e.g., books, pamphlets, physicians, and counselors).

In brief, what Bronfenbrenner's study showed was that at one period, lower-class parents were more permissive in certain of their child-rearing practices than middle-class parents, at another period there was no difference, and at another period the reverse was true. The research did not show, as most researchers in the 1930s and early 1940s mistakenly thought, that being in a certain social class caused parents to adopt certain child-rearing practices. It did not show the reverse, either—that being predisposed to certain child-rearing practices caused persons to move, by the time they had children, into a certain social class. The research showed that parents in both classes were capable of choosing their child-rearing practices, and they did so partly with the aid of what they read and heard from presumably knowledgeable people. Research of this head-counting sort is useful for discovering the current balance of opinion (which is what Bronfenbrenner did with impressive skill), but (as Cronbach properly pointed out) it tells us nothing that we did not already know about the nature of humans.

The mistaken conclusion that most psychologists (and some sociologists, too) adopted in the 1930s about child-rearing practices illustrates another way we often fool ourselves. We ascertain the present practice or state of affairs and then conclude that what we observe to be the case now is what *must* be so at every time and place—or at least in many times and places that we think are similar in some way to the present case. In the example I am using, the wrong conclusion was that certain child-rearing practices were *characteristics* of social class. The attitude at the time had the flavor of: Look at those people; that's the way those people *are*.

That kind of reasoning, combined with the assumption that the cases going contrary to the researcher's prediction somehow do not count, produce a strange conclusion that I have often found in the journals on business management. For example, a researcher might classify thirty companies according to their style of management and then look at their profit record during a relevant period. The researcher predicts that those using management style M will show higher profits than those using style Q. Let us say that the researcher then finds nineteen of the thirty companies conforming to that prediction; nine companies using management style M have above-average profits, whereas ten companies using style Q have below-average profits. The researcher then recommends to managers that they *not* use management style Q. But we also see in the data that six companies (let's say) using style Q also have above-average profits! (This is the sort of data-pattern and conclusion I have found every now and then in journals such as *Administrative Science Quarterly, Group and Organization Management, Academy of Management Journal, Organizational Dynamics,* and the like.) In such an array of results, the plain fact is that some of the companies (nine) are highly profitable while using management style M, and some of them (six) are highly profitable while using style Q. I see no reason to tell managers to stay away from style Q. If six companies can profit from it, maybe others can profit also. Maybe style Q fits your company better than style M. It might be better to judge by what you know about the capabilities of your company than by the "vote" of nine to six reported by the researcher.

SIMPLE SCIENCE

The word *science* is used in many ways. Sometimes people use it to label any body of knowledge, as *social science* or *library science*. Sometimes people use it to label any repeatable, systematic endeavor ("She has it down to a science"). One meaning my 1982 American Heritage Dictionary gives is "the observation, identification, description, experimental investigation, and theoretical explanation of natural phenomena." That, I suppose, is the meaning preferred by most people who call themselves scientists.

Some scientists say that no description is scientific that is not stated in mathematics. I certainly do not want to argue about what science "really" is. I will be satisfied to claim that the kind of endeavor most of us call science is shaped by the urge many people feel *not to fool themselves about what they think they know.* In a communication to the CSGnet of 13 February 1995, Wm. Powers wrote the following:

> For me, science is simply trying to know about things in a way that is influenced as little as possible by what I want to be true, hope is true, or believe is true. Scientific methods are mainly tricks and techniques that help to keep us from fooling ourselves, which even the most famous scientists have done quite frequently. People who don't take precautions against fooling themselves, of course, do it even more frequently.
>
> The real pay dirt in science comes when you try to *disprove* a theory, particularly your own theory. You say "If this theory is true, then by its own logic if I do X then Y HAS TO HAPPEN." So you immediately arrange to do X, and you look very critically to see if Y happens. If it doesn't, you're finished: you've at least put the theory into deep trouble, and at best have destroyed the theory. I say "at best" because if a theory can be so easily disposed of we should do so immediately to avoid wasting any more time on it.
>
> The problem is that doing this doesn't come naturally to human beings. . . . Once we start to BELIEVE a theory, it becomes very difficult to get up the motivation to try to disprove it.
>
> One thing you can do is to keep it as simple as possible. If you can think up a simple theory like PCT in which you can do tests involving only a few variables, and make predictions in a way that clearly shows failures if they occur, and if no test you can think of (within the rules of the theory) is failed, then you're more or less forced to accept the theory, for the time being, because you just don't see any way out of it.

HISTORICAL AND AHISTORICAL METHODS

One claim found in many books on psychological research method is that many causes of present action lie in the past—psychoanalysts are especially wont to say that. If you bought yourself a hat yesterday, or a month ago, does that fact cause you to wear a hat today? Well, if it is a cool day today, and you want to go outside for a while, you are much more likely to wear a hat than you would be if you had no hat. Statistically, people who buy hats are somewhat more likely to wear hats at a later date than people who do not buy hats. But, at that later date, the hat owner is under no causal necessity to wear a hat. The hat owner, on that fateful day, is free to choose whether to wear it or not.

On the average, gun owners are more likely to fire guns than people who do not own guns. But the gun owner is not physically caused, pushed, fated to pull a trigger.

Some sciences devote a large amount of attention to the history of the materials with which they deal. Geology, for example, discerns what is possible and what is impossible by finding evidences of changes in the earth's crust over billions of years. The Himalayas exist because an antarctic continental plate moved northward during millions of years and crashed (so to speak) against the Eurasian continental plate. (When I said "because" in that sentence, I did not mean anything about causes. I meant only to mention the sequence of events that ended with the Himalayas where they are. What causes went on during that plate movement, I do not know.) The geologic history of plate movements tells geologists, by extrapolation, the kind of large-scale movements that are likely and unlikely now. But that history cannot tell us where or when in the Himalayas to expect a landslide this year. The geologists can predict landslides better by examining the rocks, soils, interfaces of strata, ground water, and rainfall in a particular locality and judging from those *present* conditions the threat of landslide.

In January of 2000, for example, a landslide occurred on the coast a few miles north of Florence, Oregon, that blocked the coastal highway, U.S. 101. The Department of Transportation immediately began clearing the highway, but when the workers got the highway cleared, the engineers did not permit traffic to resume. From what they knew of the stability in wet weather of strata of that local sort, they judged that further slides were likely before long. They were correct; further slides did occur. The first traffic was not allowed through that stretch until about five weeks after the slide in January. I doubt very much that the engineers, before they made their judgment, looked up the history of the northwest coast a million years ago.

To predict the functioning of a person—that is, to model the functioning—PCT does not require us to know anything about the person's history. The ever-ready research method for PCT is, of course, The Test for the Controlled Quantity (for which see Chapter 7). The person's history may give us a hint or two about the nature of an internal standard the person may have formed in the interim, but it can never tell us unequivocally what the internal standard is like or whether there is a disturbance acting on the controlled variable at this moment. To ascertain the standard with any precision, we must use The Test, and we can use The Test effectively without any knowledge of the person's history. As for predicting *action* on the part of the person, actions always depend on the Requisites for a Particular Act that I set forth in Chapter 1. To simplify, the act that will be taken depends both on the variable being controlled by an internal standard and on the opportunities available in the environment for controlling it. Neither of those conditions can be ascertained by inspection of the person's history. What a person can do right now depends wholly upon the person's present state: on the perceptions being controlled right now and upon the environmental opportunities present right now for controlling them.

Suppose you have come to believe that Woodrow has a strong internal standard for neatness among his physical surroundings. (Maybe you have consciously used The Test, or maybe you have observed him informally for a long time.) If you move something on his desk, he soon moves it back. The clothing in his closet is stored in meticulous categories. The food in his refrigerator is arranged in rows and columns. You are confident that you can predict pretty well the kinds of situations in which he will be happy and unhappy. For example, you know that he likes to be courteous to friends and colleagues. Therefore, if Woodrow visits a friend whose parlor or office is messy, he will simultaneously want and not want to begin straightening things up. (Notice that we are not predicting particular actions here; we are predicting what Woodrow will *want* to perceive.)

Now let ten years go by. Here you are with Woodrow again. Are you going to use your knowledge from ten years ago to predict Woodrow's behavior today? Yes and no. Knowing that Woodrow controls the neatness of things around himself, you know that he will take action to bring things closer to his standard for neatness *when the environment and his other internal standards permit him to do so.* One answer to the question, then, is yes, you do know something useful for predicting Woodrow's behavior. But what you know is the kind of perception he *wants* to obtain. You do not know what particular acts he will use to bring about the perception nor when he will have the opportunity to use those acts. (Again, I refer to the Requisites for a Particular Act listed in Chapter 1.) Furthermore, one should always be cautious about the stability of internal standards; it is possible that Woodrow's standard for neatness has changed its character in ten years. Finally, my chief point in talking about Woodrow is that you didn't know how his standard for neatness came into being or when; all you needed to know was whether it was there, and you found out by using The Test. Ten years later, you got no help from knowing that it was there ten years earlier; still all you needed to know was whether it was there now.

CODA

Perceptual control theory claims that behavior controls perception—at every time, in every place, in every living thing. The theory postulates that control operates through a negative feedback loop—neurally, chemically, and both. The theory postulates the growth of layers of control both in the evolution of the species and in the development of individuals of the "higher" animals. Those are the crucial postulations of invariance in PCT. They are asserted to have been true for the single cells floating hither and thither a billion years ago, which might have had only two layers of control, and they are asserted to be true for you and me with our many layers. They are asserted for all races, nations, sexes, and indeed all categories of humans—and indeed all categories of creatures. Furthermore, if *one* creature is found reliably to violate any one of those postulations (and yet go on living), the theory will immediately be revised.

Do you know of another theory of such sweep anywhere in the sciences of living creatures?

Chapter 11

Falsification and confirmation

A sentence found in almost all books on methods of research in social science goes something like this: Empirical research (the study, that is, of observable, tangible things) can never *prove* a proposition, a claim; the research can only *disprove* it. In other words, by ascertaining *wrong* ideas—the ideas to shy away from—we keep ourselves headed, even if erratically, in the *right* direction.

Consider again, for example, the theory of phlogiston that I described in Chapter 8. When Lavoisier came to doubt the theory and Priestley to defend it, it occurred to them to weigh the mercury before and after it had oxidized ("calcined"). The theory of phlogiston predicted that the phlogiston in the mercury would be drawn off by oxidation or burning, leaving *less* weight behind, while the postulation of what was later called oxygen led to the prediction that the weight after burning would be *greater* than before. By itself, the comparative positions of the pointer on the scale after the two weighings *disproved* the phlogiston theory: if phlogiston was drawn off by the oxidation, then the weight after oxidation *could not* be greater than before. But the comparative positions of the pointer told Priestley and Lavoisier nothing about what was going on during the oxidation except that something was adding weight to the mercury. There was no mark on the scale labeled "Hey! It's oxygen!" Lavoisier and Priestley had to invent some way (some model, some sequence of events) to explain how the mercury could have become heavier than it had been. (Priestley could not invent an explanation better than Lavoisier's, but nevertheless he held to phlogiston to the end of his life.) Those experiments falsified phlogiston; they showed that was *not* the way to go. But they did not show the better way to go; that direction had to be found in the brain of Lavoisier.

Following upon the careful measurements made in those experiments, Lavoisier was later able to invent the "chemical equation" that has since remained the encompassing model for chemical investigations. The chemical theory we have now is a structure of inventions, no part of which has been shown (or can ever be shown) to be the *best* that could be produced. Every part holds the allegiance of chemists only until it is disproved or bettered.

This is not to say that physical and chemical theories get revised every week (even if the popular reports on cosmological hypotheses sometimes make it seem so). The physics given us by Newton is still serving with great exactitude after 300 years. It is the theory used in the calculations for the needed strength of automobile axles, the tensions in the Golden Gate bridge, and the navigation of rockets to the moon, Mars, Jupiter, and beyond. Revisions made by Einstein and others have revised ideas about the very small and the very large and have improved accuracy where velocities approach that of light, but they have not displaced Newton's laws. Similarly, the chemistry of Lavoisier has been vastly expanded, but not displaced.

FALSIFICATION

Falsification—the ever-present possibility of disproving a hypothesis—is of course important to understand and remember; it is the basis, the first cut, the springboard of scientific progress. But the literature on social science often gives the impression that it is the *only* technique for making one's way toward theories and models that can entice some continuing allegiance from others. That is not the case; when predictions are quantitative, it is possible to compare the accuracy of theories and conclude that one is better confirmed than another. Quantitative confirmation is a far faster way than falsification to find a theory that will withstand further testing.

Many hypotheses in social science predict only that an average measure of behavior in one group of people will be *different* from that in another group, sometimes even without predicting in which group the measure will be greater. Analogously, a chemist in 1775 might have hypothesized that after heating, the mercury would be either lighter or heavier than it had been before heating. That experiment could not have decided the question whether phlogiston or oxygen was at work. But the phlogiston theory predicted that something would go *out* of the mercury, and the competing theory predicted that something would go *into* the mercury.

When an experimental outcome rules out Theory A, it leaves Theory B as *only one* of many that might be correct—as one of an unknown number of theories that fit (to an acceptable degree) the data produced by the experiment. Later experiments may produce data that will be fitted better by some later theory. Theories often change as they become more quantitative. The experiments of Lavoisier showed that the mercury was heavier after heating. But was it heavier by the predicted *amount*? Lavoisier and Priestley and others discovered that elements changed from one compound to another in predictable ratios of weight and laid the underpinnings for later atomic theory. They enabled the quantitatively correct proportions to be found for H_2O, CO_2, $NaCl$, H_2SO_4, and the thousands of other known compounds.

(Here I have been using what I suppose are the modern terms—though I learned them in high school and college in the 1930s, and they may be modern no longer. Lavoisier and Priestley, of course, had to grope as best they could with words that took their meaning from the old theories, not from the new theories still unformed even in their own minds. It is fascinating to read accounts of the ways the 18th-century chemists and physicists posed their problems. Try, for example, J. B. Conant 1956 and T. S. Kuhn 1969.)

CONFIRMATION

When a theory is quantitative, the support of data can be more precisely tested. When not only a difference in weight is predicted, but the direction (heavier or lighter) also, the confirmation of the theory can be more precise. When not only the mere fact of decrease in weight can be predicted and ascertained, but also the ratio of the weight before to the weight afterward, then the confirmation of the theory can be as precise as the accuracy of the measurements possible with the instruments at hand. That, of course, is the degree of accuracy that physicists and chemists require in testing their theories: an accuracy that exploits the reliability of their measuring instruments.

Let me take gravitation as another example. Let s stand for distance (space) and t for time. When a theory of falling bodies says only that as time goes by, the distance traversed by a freely falling body will continue to increase, the theory cannot help us choose among these quantitative models:

(1) $s = f(t)$
(2) $s = (1/2)kt^2$
(3) $s = (1/2)gt^2$

That vague, verbal "theory" would be satisfied by a movement specified by any of those three formulas. The first formula says only that over a longer period of time, you will see the body falling a greater distance—that the distance traversed has some *unspecified* relation "f" to time. Galileo, however, was able to devise a way of timing the effect of gravity. He slowed the falling by rolling little balls down a groove in a slanting board and timed the distances by the amount of water flowing out of a thin tube while a ball rolled from one mark to another on the board. Galileo (in Stillman Drake's translation of 1974, p. 170) said that in repeated measurements of the elapsed times, "we never found a difference of even the tenth part of a pulse-beat." Drake adds (p. 170, fn. 25) that in some other experiments with inclined planes, Galileo "obtained results within one percent of modern theoretical values." Galileo's model was that of the second formula above, in which k is a constant.

You can see that Galileo's theory, because it could put numbers on the observed quantities, was far more useful than the first vague formula—and far more reliably testable. Indeed, Galileo speaks of "heavy" objects, by which I believe he meant to rule out falling leaves and handkerchieves. The theory enabled him to specify effects from which an experiment should be protected—wind and other obstructions; but if an "obstruction" such as an inclined plane was found useful, it should offer as uniform an effect as possible during the time of the experiment—the grooves should be very smooth, for example. Given those precautions, Galileo's theory could be tested by any careful worker, and a deviation of a fraction of a pulse beat would be enough to discredit the theory. Therefore, if two theories were to predict the accelerations with which balls would roll down inclined grooves, and one predicted the arrival of the balls at certain marks beside the groove with an accuracy of one second, but the other with an accuracy of a tenth of a second, the first theory would be rejected not because it was plainly wrong, but because the second theory did better. The data would have confirmed the second theory better than the first. Building upon Galileo's work, Newton (1642–1727) gave in his *Principia Mathematica* the third formula, in which the constant for the acceleration of gravity at the surface of the earth is quantified: g = 32 feet per second per second, approximately.

Engineers are not concerned with falsifiability. An engineer designing a bridge cannot be satisfied with a theory that is good enough only to have so far avoided falsification. If you hand the engineer a handbook on bridge design and say that the theory that guided its compilation has not been clearly disproved—well, you can imagine the engineer's reaction. The engineer wants a theory that will yield an unambiguous, precise prediction of the way the steel posts and girders will compress, stretch, twist, and bend under various kinds of loads.

In sum, if you predict that a certain variable quantity will increase as a result of your experiment, but it turns out to decrease instead, you have *falsified* your theory, and you should discard it or revise it radically. If, on the other hand, you predict that the quantity will be twice as large after the experiment, and it turns out to be as close to twice as large as your measuring instruments are capable of measuring, then your theory is *confirmed*. And if both Theory A and Theory B predict that the quantity will be larger, and the quantity that actually comes about differs from the value predicted by Theory A by one percent of the quantity predicted, but differs from the value predicted by Theory B by only a tenth of one percent, then the superiority of Theory B is confirmed, and you will reject or radically revise Theory A. In the physical sciences, that is the logic that has been in use since the time of Galileo (1564–1642).

In the introduction to Part III and in Chapter 10, I described some differences between traditional research method and the research method of PCT.

A test using the traditional method of nose-counting cannot produce a quantitative outcome that can be compared with another quantitative test to determine which approaches more closely to the theoretical prediction. In contrast, you saw in Part II several experiments in which outcomes produced by a human were compared quantitatively with outcomes produced by a model. The traditional method is limited to falsification; PCT can use confirmation.

THE HAZARDS FOR PCT

No method of research is foolproof. All have pitfalls. I have described some of the difficulties encountered with traditional method in this book; I described more in my 1990 book. Method in PCT, too, has its hazards. To show another aspect of the intertwining of theory and research method, I will describe here. briefly, two important awkwardnesses in designing research on PCT.

Reorganization

The prime method for PCT research is The Test for the Controlled Quantity. Most of the examples of PCT research you will find in this book, however, do *not* undertake to discover an unknown controlled variable. They are designed to test the validity of models; the question they ask of the data is: Does this model organize functions in a way that enables it to control a perceived variable in the same way a person does? In these experiments, the experimenters wanted to begin with a known controlled variable, and they asked the participant to maintain control of a specified perceived variable—such as the distance between two marks on a computer screen. The do-it-yourself experiments described in Chapter 6, on the other hand, were direct uses of The Test to ascertain the variable being controlled. In either kind of query, reorganization can flood the experiment with uncertainty. I will devote Chapter 20 to reorganization, but I will say a little here.

When an inner conflict cannot be reduced—when two controlled variables are in error, but reducing one error makes the other larger, and conversely—Powers postulates that the nervous system begins casting about for a reorganization that will reduce the total error. The casting about, according to PCT, is wholly unsystematic—for all practical purposes, random.

Accordingly, no one can predict the outcome of reorganization. We can in no way anticipate the assortment of variables that will have come under control when the reorganization slows and ceases. If we ask a participant to maintain a distance of one inch between marks on a computer screen, the participant may agree to do so with all sincerity, but then a stray word or something seen out the window may bring to the person a realization of a conflict that reduces the inch on the screen to a tremulous lack of priority. This threat to validity becomes more severe, of course, as we try to document control in less disciplined situations or over long periods of time.

Environmental Disturbances

Perceived variables are controlled by opposing the disturbances to them. We must usually observe control by observing events in the environment that would otherwise proceed uncontrolled. In the laboratory, we can reduce the possible uses of the environment to very few kinds of acts—pressing a key on a keyboard, for example. But in general, particular acts are not predictable. Robertson and others, in the experiment I told about in Chapter 7 under "Self-concept," hoped that all the participants would respond to "No, you're not" with an utterance that could be understood as either a clear defense of their perception or a clear portrayal of indifference. But a couple said something that the experimenters were reluctant to call one or the other. Even in that restricted situation, those two participants found a way to use the environment that the experimenters hoped they had made very unlikely. The predictability of particular uses of the environment becomes especially low when people are interacting freely in normal situations. To enable the behavior of individuals to produce measures of relevant variables, the studies of social interaction by Bourbon (in Chapter 8 under the heading "Two People") used very restricted social situations indeed.

Unpredictability in the use of the environment is exacerbated by the fact that the stability of physical states (not to speak of social states) in the environment is always low. A rock trips the foot. A rainstorm blows up. A landslide blocks the road. A sewer plugs up. A seam in the trousers rips. At every turn, the likely choices for acts with which to control variables become reordered. These alterations do not ordinarily destroy the competence of individual action or the integrity of social organization; our capabilities are organized to cope with just such a world. But the alterations sometimes make life difficult for the experimenter who wants to test PCT.

NATURAL REGULARITIES

Given those two sources of variability—in intention inside and in means outside—it is easy to see the reason PCT investigators have made use of computer screens and experimental tasks that can be performed in a minute or two. Those techniques reduce the hazards a great deal.

It is possible, however, to apply The Test to life outside the laboratory. It is often possible to make a Test, even it not strict, of a guessed-at controlled variable. The first principle, you will remember, is to try, gently, to disturb the variable you guess is being controlled. If it yields to your disturbance, then you have guessed wrong.

Suppose a man goes into a clothing store and says he wants a pair of trousers. He tries on a pair having a waist of 34. He tells the clerk that waist is too small. The clerk hands the man a pair having a waist of 36. The clerk thinks the man wants to perceive a feeling at his waist that is not one of constriction, but not one of looseness, either. And he wants to see in the mirror that same sort of fit. But imagine the clerk's astonishment when the man says, "Oh, no, this size 36 is far too small." After some confusing conversation, it turns out that the man is a clown in a circus and wants trousers into which he can fit a hoop so they stand out about six inches from his stomach. The man rejects size 50 as too large and settles on size 44 as just right. That would be a case where the clerk made a correct guess about the variable the clown was controlling, but not the level or quantity of the variable. Size of waist was the variable, but the clerk guessed 36, while the clown wanted 44. The clerk found out what the man wanted by finding out what the man would act *against*.

Maybe you guess that a young friend of yours has come to prefer to eat with a fork. You fail to put a fork beside his plate, and he says, "May I have a fork, please?" When you give him some French-fried potatoes, you say, "You may eat those with your fingers, if you like," but he eats them with his fork. And so on. In sum, when you try (gently, please) to separate him from his fork, he resists. And when you see him hold down a piece of bread with his fork while he butters it with his knife, you can be pretty sure that he is putting a high value on eating with his fork.

Let me relay to you a story told by Marvin Weisbord in a 1985 issue of *Organizational Dynamics*. In the 1960s, Weisbord was vice president of a mail-order printing company. It was organized in the standard manner—that is, in the manner handed down 60 years earlier by Frederick W. Taylor. The earmarks, as Weisbord said, were "time clocks; narrow work rules; jobs so subdivided that even an idiot would be bored; grown people treated like children, never let in on decisions, having no consequential information about the business or even their own work. . .". From 200 to 300 orders arrived every day at the order -processing department. With the work organized into sequential small functions, however, the absence of one or two persons would cause a bottleneck. That naturally frustrated not only those trying to perform the understaffed function, but those on either side of it, too, whose work was also slowed. Rancorous conversation was common.

The supervisors asked that a wall be built across the large room housing the order-processing department. Their reasoning was that the wall would reduce the communication among the workers and the hostility, too. The supervisors did not want to see and hear the hostility and see its effect in slowing the work. They reasoned that they could reduce the frequency of hearing hostile communication if they reduced the frequency of communication. But paper still had to be passed from one desk to another, and plenty of occasions for the expression of hostility remained. The wall did not improve the relations among the workers.

About this time, Weisbord learned about semi-autonomous work teams. He had read a book by McGregor, and he had a friend who was acquainted with new methods of management. Weisbord organized the order-processing department into teams of four or five people, each team having its own customers, telephones, and other equipment. Each team could set its own goals and priorities. Weisbord told them to teach each other their jobs, so that every member of a team could perform every needed job. An absence could increase the work load somewhat, but it could no longer cause a bottleneck. No one person would be frustrating the rest.

Not surprisingly, many difficulties arose during the changeover. Weekly meetings were held to work out ways of dealing with them, and the meetings themselves were difficult, taxing, even frightening. By the fifth meeting, Weisbord was ready to give up. At that meeting, however, no one said anything. When Weisbord asked for that week's problems, one person said, "We don't have any this week." Another said that they had dealt with the week's difficulties by the methods that previous meetings had generated. Weisbord wrote, "I understood [then] that the essence of effective organization was *learning* . . . trial, error, give, take, and experimentation."

Employees, if they are to be mutually helpful in an organization, must accept certain minimum goals from the managers—perhaps to produce certain products at a certain average rate, perhaps to furnish certain kinds of knowledge to customers, to transport people or objects to certain places at certain times, and so on. But within the overall purposes, employees all have idiosyncratic purposes of their own. One is usually to make some money, but that is itself always a means to further goals, and beyond that every person must perforce act to maintain his or her own controlled variables at their reference values. Every person's goals are myriad, of course, ranging from putting a point on a pencil to the feeling that there is a point to one's life (or choose your own examples). In the literature of industrial management, the "higher" purposes are often expressed as assumptions about what "most people" want; Weisbord, harking to McGregor, put it "that most people will take responsibility, care about their jobs, wish to grow and achieve and, if given a chance, do excellent work." In the semi-autonomous work teams, most of Weisbord's employees found that they could help one another carry out those higher purposes. The indicators were that productivity soared by 40 percent within six months, and absenteeism and turnover dropped almost to zero. One employee of 15 years said, "I used to hate coming to work in the morning. Now I can't wait. I love it."

Weisbord knew nothing about The Test, and as far as I know still knows nothing about it. But he understood a lot about idiosyncratic purposes and grew in that wisdom as the years went by. His book *Productive Workplaces* (1987) is the best book I know on organizational consulting. In helping his employees, back in the 1960s, to find their way into cooperative work in teams, he helped them at the same time to learn to help one another maintain their individual purposes. In the cooperative groups, they could find out what actions or procedures their coworkers would act *against* the actions or procedures that were disturbing the variables their coworkers were controlling. In such a setting, though it helps if internal standards can be accurately described, doing so is not necessary to doing the work in a manner that satisfies both the boss and the worker. The necessary thing is for the workers (and their boss) to continue to maintain, every day, *ways of working that do not exceed the ability of individuals to maintain their own controlled variables.* To do that requires workers to be alert to signs of stress from others and to take them seriously—that is, to *care* about the welfare of others and to have the freedom to act for their welfare.

Every person acts *continuously* in controlling variables. For the nervous system, life is not episodic. As I will explain in Chapter 18, there are levels of control at which we perceive things as events and categories, but the control loops enabling those perceptions throb continuously to do so. There is an analogy there (though not a direct connection) with the control of the variables individuals care about in the social group (or work group). It is commonly said that problems in the work group are never solved; you have to solve them again every day. You can never cease doing something about your need for air, and you can never cease doing something to maintain a fruitful cooperation with your coworker. Nevertheless, if you and your coworkers are caring and alert and your boss does not get in the way, you can maintain cooperation not only with facility, but with deep gratification. But I am anticipating later parts of this book.

To make life easy for one another, coworkers approximate The Test. They take note of what the other person resists, rejects, or feels threatened by. They form hypotheses about the variable the other person is controlling when resisting or rejecting an action or proposal. They help one another control variables in ways that will not interfere with working cooperatively. They find that one person wants a lot of help with details, while another resents a lot of help. They find that one person wants to form friendships that extend beyond quitting time, while another finds friendly relations at work to be sufficient.

To build a science, it is necessary to conduct The Test in as logically rigorous a situation as possible, but The Test can also be used with benefit in approximate ways in ordinary life.

EXAMPLES

Chapter 12 consists of an article by Bourbon and Powers that illustrates the strategy of confirmation. In the article, Bourbon and Powers compare how well models built from three theories predict data taken from an actual human individual. The three theories are the stimulus-response model, the "cognitive" model, and the perceptual control model. The article also elaborates on some of the points I have made in this chapter.

Chapter 13 is an article by Powers; it illustrates how an internal standard can be deduced from observations of behavior. It also describes some technicalities in designing a computer simulation (a model).

Chapter 12

Models and their worlds

Except for this introduction, this chapter is a reprint of an article by Bourbon and Powers (1993). I include it because it is a paragon of testing a hypothesis straightforwardly, rigorously, quantitatively, and conclusively. It shows the clarity with which a hypothesis in PCT can be confirmed or rejected. If it seems surprisingly simple in a place or two, remember that scientific method must be explicit about every step of procedure, no matter how simple it may seem to some.

If you do not wish, at this point, to devlve into the kind of detail contained in Chapters 12 and 13, feel free to skip on. You can return when you feel the urge.

As Richard Marken says, the tracking task is simple in the same way as were the little balls and inclined tracks used by Galileo in his seminal studies of the acceleration of gravity. We do not intend the tracking task to show what particular acts people take when they are driving a car or drinking water or building a house or painting a picture. We do intend to say that the tracking is controlled by the *same sort of neural organization* that is used in those other pursuits and all others, too, that act upon the environment. No matter how simple they are, experiments in PCT are remarkable because (1) both person and model produce quantified results that can confirm the match quantitatively, (2) the model is a material, functioning device that *can* produce quantified results, and (3) the model is tested not against an average over many people but against a single person.

In a posting to the CSGnet on 14 September 1995, here is what Powers had to say about the article below:

> In the physical sciences, the common way to test a theory is to examine it as a logical or quantitative structure, and see where you could vary conditions in a way that the theory would have to predict has some new kind of effect, something that hasn't been observed before.
>
> You'll see this strategy exemplified in the paper "Models and their worlds".... The control-system model is matched to behavior under the condition where a target moves in a regular way and the person makes a cursor track the target. Once the model's parameters are set for this condition, we then change the conditions. First, we vary the regular movements of the target so they become irregular. The same control model, with the same parameters, predicts that the behavior will change in a specific way that maintains the tracking, and in fact the real person does change the behavior in just the same way as the model, quantitatively. Then we introduce a smoothed random disturbance added to the cursor position, so now the position of the cursor depends both on the handle position and on an independent arbitrary variable. The control model predicts that tracking will continue, and that the handle movements will now differ from the cursor movements in a specific quantitative way. When the real person does the same task, the predictions are upheld with good accuracy. So now the control-system model has been challenged twice; it could have failed in either of the latter two experiments. All that would have been necessary to make the model fail would be for the person to have moved the handle in some way other than the predicted way. Since there were no constraints on how the person could move the handle, the success of the prediction was highly significant. It was significant because the model's behavior could have failed to match the real person's behavior....
>
> Sooner or later, we would think of a way to change the conditions that results in the model's doing something radically different from the real

person. Rick Marken and I [Marken and Powers, 1989a] did that when we did an experiment in which the sign of the connection between handle and cursor was reversed in a way that gave no sensory indication of the reversal (i.e., no bumps or joggles at the moment of reversal). The model and the person both showed a very similar exponential runaway after the reversals—for the first 0.4 seconds or so. Then the person did something to regain control, BUT THE MODEL DID NOT. So by thinking up the right change of conditions, we succeeded in making the model fail.

Of course that failure was simply a signal that we had to modify the model, which we did. We added a second level of control that could reverse the sign of the first-level control action when a runaway condition was sensed. That naturally restored the model to working order, and it once again was able to predict behavior correctly. So by finding a way to make the model fail, we learned how we could improve the model so it would no longer fail under that set of conditions, and of course continued to work properly under all the other changes in conditions we had already tried.

The article that follows appeared originally in the now-defunct journal *Closed Loop*, 1993, 3(1), 47–72. Another version of it appeared in the *International Journal of Human-Computer Studies*, 1999, 50, 445–461. *Closed Loop*, 1993, 3(1), along with several other issues has been restored and is available as a PDF-file at www.PCTresources.com

Models and Their Worlds

W. Thomas Bourbon
(Department of Neurosurgery, The University of Texas Medical School-Houston, 6431 Fannin, Suite 148, Houston, TX 77030)

William T. Powers
(73 Ridge Place, CR 510, Durango, CO 81301)

ABSTRACT

Many seemingly plausible models of behavior demand implausible models of the physical world in which behavior occurs. We used quantitative simulations of a person's performance on a simple task to compare the models of causality and of how the world works in three theories of behavior: stimulus-response, cognitive, and control-theoretic. Our results demonstrate that if organisms in fact functioned like the first two models, they could survive only in implausibly stable worlds; if like the third, they could survive in a changeable world. Organisms inhabit a changeable world that does not satisfy the demands of popular behavioral theories. For the sciences of behavior, the implications are clear: either cling to theories that do not mesh with knowledge of how the world works, or abandon many cherished notions about how and why behavior happens in favor of models that deal adequately with change.

MODELS AND THEIR WORLDS

The question usually addressed by behavioral theorists is "Why do organisms behave the way they do?" One group answers "Because the world outside them is the way it is"; another group answers "Because the minds or brains inside them are the way they are." In either case, behavior is at the end of a linear sequence of cause and effect, a consequence of antecedent stimuli from the environment or antecedent commands from the mind or brain. As an alternative, one can propose that organisms behave to control what happens to them. In the process, their actions affect the world outside of them. 'Why is the world the way it is? Partly because organisms behave the way they do."

"The world" is the part of the surroundings on which an organism can act, and which, in turn, affects the organism. Every statement about the antecedents or consequences of behavior either includes or implies notions about how the world operates. Every theory of behavior is, in part, a theory about the world in which behavior occurs.

In this paper, we reduce three models of behavior to elemental form to identify and test their ideas about causality. Two models represent core assumptions in most popular theories; the third is the model from perceptual control theory (PCT). We require each model to simulate and predict the same behavioral events that occur when a person performs a simple task, but we go a step further. For each model, we determine whether its implications about how the world and behavior affect one another are reasonable and true to what is known about the physical world.

Three Models

For convenience, we call the two popular models the "stimulus-response" (S-R) model and the "cognitive" model. Our simple versions of these models are not intended to represent, in detail, any specific variations on those two themes, but we believe they faithfully represent core assumptions about causality embraced in those themes. Our method of testing requires that each model predict moment-by-moment values of several continuous environmental variables, a challenge to which behavioristic and cognitive models are rarely subjected; hence, simple computational versions of those models are not readily available, and we constructed our own. Anyone who rejects our versions of those theories should identify acceptable versions and then require their models to duplicate the quantitative results we report here.

The stimulus-response model. Our S-R model represents all theories that say external influences determine behavior. Such models sometimes (but by no means always) recognize that motor actions produce environmental consequences, but all insist that action is a dependent variable. A behavioral episode begins with an independent antecedent (stimulus, context, event, occasion, relationship, or treatment), followed (in some theories) by an effect on the organism, then (in all theories) a behavior as a dependent variable, and finally the consequences of that behavior. Environmental consequences of action simply follow from what the environment did to the organism; if any consequences of action modify subsequent influences on the organism, that is merely another change in the independent variable, followed in a lineal causal chain by another action and another consequence.

We expect most behaviorists to say that our S-R model is "reflexological"—a version of behavioristic theory many behaviorists disavowed years ago—and to echo the comment: "There may not be a reflexologist alive" (Shimp, 1989, p. 163). Protests aside, at the core of every behavioristic theory is a claim that the environment controls behavior. From the beginning, behaviorists have asserted, like Donahoe and Palmer, "Although the organism is the locus of environmental action, it is the environment, and not the organism, that is the initiator and shaper of behavior" (1989, p. 410). When Hayes and Brownstein (1986) discussed prediction and control as criteria for evaluating behavioristic analyses of behavior, they said, "One could ask, for example, how do we know that *this is* the relevant stimulus for *this* behavior? The answer is of the general form that when we change *this* stimulus (and not *that* stimulus), we get a change in *this* behavior (and not *that* behavior)" (p. 178, emphases in the original). And Skinner claimed, "The ways in which behavior is brought under control of stimuli can be analyzed without too much trouble..." (1989, p. 14).

Here, we merely test results that would ensue were it in fact true that independent environmental stimuli specify instantaneous details of behavior and its consequences.

The "cognitive" model. Our cognitive model stands for all theories that say actions originate not from current external events, but from internal causes, inner traits, tendencies, propensities, sets, plans, attitudes, aspirations, symbol-generating processes, programs, computations, coordinative structures, or some kind of systematic endogenous brain activity. No major theory of this sort proposes that behavior is entirely spontaneous; in one way or another they say the internal causes of present behavior formed and changed slowly, during past experience with the outside world—the recent past in some theories, the geologically distant past in genetic theories of behavior. In cognitive theories, the link between present behavior and influences in the present external world ranges from weak to almost nonexistent. In many texts on cognitive theory, there is no mention of overt action, much less an attempt to explain such actions. When there are explanations, the causal chain runs from input to cognition to command to action to consequence.

Kihlstrom (1987) succinctly identified the linear causal model in cognitive theory: "Cognitive psychology comes in various forms, but all share an abiding interest in describing the mental structures and processes that link environmental stimuli to organismic responses..." (p. 1445). Each step of the assumed chain from stimulus (input) to response (output) is described in detail by various cognitive theorists. For example, Real (1991) describes how inputs from a variable world would be transformed, in three sequential stages, into cognitive "representations":

> ... three stages may be viewed... as three components of a single dynamical system mechanistically tied to the organism's nervous system. The encoding of information would... correspond to initial inputs, computational rules correspond to transient dynamics, and representations would correspond to the equilibrium configurations resulting from the transient dynamics. The animal reaches a representation of the environment through the operation of specific computational rules applied to a particular pattern of incoming sensory information (p. 980).

In a discussion of computations which they assume cause movement, Bizzi, Mussa-Ivaldi, and Giszter (1991) complete the chain between representations and actions: "... the central nervous system must transform the neural representation of the direction, amplitude, and velocity of the limb, represented by the activity of cortical and subcortical neurons, into signals that activate the muscles that move the limb" (p. 287).

Some theories combine cognitive and S-R models. In their simplest forms, hybrid models say that the mind-brain receives "inputs," then produces direct transformations of coordinates from "perceptual spare" to "action space" that are required to initiate commands to move the body or part of the body to a point specified in the input (as examples, see P.M. Churchland, 1986; P.S. Churchland, 1986). Such models reduce cognition and neurology to a simple table-look-up.

A more complex hybrid S-R/cognitive model was endorsed by the cognitive theorist Allen Newell (1990) in the 1987 William James Lectures. Newell spoke of how "It is possible to step back and treat the mind as one big monster response function from the total environment over the total past of the organism to future actions. . ." (p. 44). On a more immediate scale, he said, "The world is divided up into microepics which are sufficiently distinct and independent so that the control system (that is, the mind) produces different response functions, one after the other" (p. 44). For strategic purposes, Newell places his theory in the category of cognitive theories that he says do not effectively explain how perception and motor behavior are linked to central cognitive processes. Then he says that such theories ". . . will never cover the complete arc from stimulus to response, which is to say, never to tell the full story about any particular behavior" (p. 160). In his allusion to the reflex arc, Newell remarkably implies the equivalence of the causal models in his cognitive theory and in reflexological theory.

In either their simple or complex forms, hybrid S-R/cognitive models produce results identical to those of S-R models, so we will not discuss them further.

The perceptual control theory model. The PCT model, which we discuss later at some length, is the least familiar of the three models. In brief, it proposes that there is a simultaneous two-way interaction between organism and environment (see Hershberger, 1989; Marken, 1990; and Powers, 1973, 1989, 1992). In PCT, the basic unit of behavior is not the linear input-output chain, but the negative-feedback loop, which has properties different from the units of the other two models and implies interesting consequences about the way an organism's actions alter the outside world.

"Models"

We use the term "model" in the very narrow sense in which an engineer would use it: a precise quantitative proposal about the way some system operates in relation to its environment. Most behavioral scientists use *descriptive* models, which merely rephrase (usually in words; sometimes in mathematical form) previously observed relationships between organism and environment. There are unlimited ways to restate behavioral data. If each of them passes as a *model* of behavior, then the list of seemingly plausible models is also limitless. The availability of many equally plausible descriptive models is behind the mistaken assumption, common in behavioral science, that models are poor substitutes for real understanding—that if one understood the phenomenon at hand, one would state the facts, not a "mere" theory or model.

But "model" also means, in the present context, a *generative* model, in which the proposed organization is stated in a way that can be used to calculate behavior as a function of moment-by-moment variations in the independent variable. By that usage, a model does not substitute for knowledge. To the contrary, simulation of a well-posed model rigorously tests one's presumed knowledge of the causal principles at work in behavior.

S-R theory as a model. Calculations of the correlation between a dependent and independent variable produce a correlation coefficient, a regression coefficient, and an intercept. In most behavioral research, little attention is paid to the regression coefficient and intercept, one reason being that the typical scatter of the data is large enough to make a linear regression line almost useless for predicting behavior. But, by the logic of the S-R approach, the regression equation constitutes both a generative model and a description. It is a first approximation to a proposed law of behavior: at every moment, the behavioral measure is proportional to the magnitude of the independent variable. If that law is true, one can vary the independent variable and calculate (predict) the dependent one strictly from the previously determined regression equation.

It can be argued that this strict interpretation of a regression equation is inconsistent with the state of the art in behavioral science—all we can hope for now, in most cases, is to establish the presence or absence of a statistically significant relationship.

Our reply gives the benefit of the doubt to the theory underlying the S-R concept. If, given as many years as necessary, methodologies improve, sources of variance are eliminated, and better data are obtained, then regression equations will become meaningful. When they do, there will be an obvious test for whether a proposed regression coefficient is a law of behavior. In the regression equation, one can impose a new pattern of the independent variable and calculate the resulting pattern of behavior, the dependent variable. The modeled result can be compared against what happens when the organism encounters the altered independent variable. In more elaborate form, this process of testing a model against actual events is the basic methodology of the physical sciences. Used in this way, the regression equation is a generative model.

We use an alternative to waiting for years for data to improve: we apply this method in an experiment so simple that the regression line is highly meaningful, and random variation is a minor factor. We subject the S-R model to a test under conditions that should make it work as well as it ever will.

Cognitive theory as a model. We give the cognitive model a similar treatment. Cognitive models are more difficult to test and defend than S-R models; there is no simple way to determine whether a given cognitive model is correct, as well as plausible. No matter how well a model proposing a specific organization of the mind-brain predicts behavior, one cannot test the model objectively by, for example, deriving a regression line based entirely on observable variables. There is no way to know whether some other cognitive model would not work as well or better. There is only one regression line that best fits the behavioral data, but there are many seemingly plausible cognitive models.

Kugler and Turvey (1987) aptly described the problem of non-unique computational models for behavioral output:

> Whereas physical events are said to follow uniquely from their causes, internally consistent, logical descriptions of the causal process are multiple.... How does one get from the existence of multiple (logical) descriptions to a unique (causal) description? Dressing up logical formulae in instantiable programs does not resolve the uniqueness problem. Many programs can give rise to the same sequence of machine outputs (p. 28).

To avoid problems of this sort, we give cognitive models the same benefit of the doubt that we give S-R models. Given proper knowledge of the history and properties of the environment, and the correct internal computations, the ideal cognitive model should calculate exactly the motor outputs required to produce a preselected result. Of course, even a perfect cognitive model would require experience with an environment to build up knowledge of its properties: if the environment changed, the model would need new interactions with the altered form before it could again compute the correct action.

We test the cognitive model by assuming that it is perfect: it makes optimal use of information and computes the same required action on successive trials, and the motor systems perfectly obey its commands.

The reasoning behind our approach to the models is simple: in a well-defined experiment, if quantitative predictions by both the S-R and cognitive models, given the benefit of every doubt, are incorrect, and the PCT model predicts correctly in the same experiment, there will be excellent reason to say that the control-theoretic model is right and the other two are wrong, for that experiment. How far one generalizes the result depends on how clear are the parallels with other experiments and the simple one we use: we leave such judgments to the reader.

Perceptual Control Theory as a Model

Perceptual control theory always considers two simultaneous relationships: (a) the observed dependence of stimulus inputs on behavioral outputs and independent events, and (b) a conjectured dependence of behavioral outputs on stimulus inputs.

The environment equation. The first relationship the PCT model describes is how the input to an organism depends on the organism's actions and on disturbances arising simultaneously with behavior but independently of it in the external world. To simplify this part of the model, we restrict all variables in the experiment to change in a single dimension, described later. Consequently, the variable at the organism's input is simply the sum of a physical effect from the organism's output and another physical effect from an independent disturbance. The apparatus (a computer system) records exactly what these relationships are and exactly what disturbance is acting at any moment. This part of the model is completely determined by the experimental setup; it is a statement of fact, not a conjecture, and it is illustrated in detail by Bourbon, Copeland, Dyer, Harman, and Mosely (1990).

The organism equation. Perceptual control theorists assume an organism can be modeled as a system that senses some aspect of the environment that is then represented internally as a one-dimensional perceptual variable. The magnitude of this variable is compared continuously against a reference signal (or reference magnitude) inside the organism or the model of the organism. Any difference between the reference signal and the perception is a non-zero "error signal" which drives action, again in a single dimension of variation.

This part of the model can be treated exactly as a regression equation. The slope of the regression line represents the incremental ratio of output to input, and the intercept represents the setting of the internal reference signal. The slope reflects measured output as a function of measured input; the intercept is the magnitude of input for which the output does not change. Control theorists assume that the value of the input for which the organism produces no change in output is the input that the organism specified in advance.

The system equations. The organism and environment equations form a system of equations; for examples, see Pavloski, Barron, and Hogue (1990, pp. 33–37); Powers (1973, pp. 273–282; 1978, pp. 422–428); and Runkel (1990, pp. 93–99). There are two system variables (the input and output variables) and two equations. The input and output variables appear in both equations, and each must have only one value at a time. Consequently, the system can be solved for each variable as a joint function of any system constants and the values of the two independent variables (the external disturbance and the internal reference signal).

Our experiments use random disturbances that cannot be represented by any reasonable analytic equation. Consequently, in the PCT model, we calculate numerical solutions of the system equations. Numerical solution of system equations, with time as a parameter, is called simulation.

Simulation. Simulation recreates, through computation, a continuous relationship among system variables and independent variables. The experimenter causes a pattern of changes in the independent variables, while the equations for the model continuously compute the states of dependent behavioral variables at the input and output. For a good model, the results of a simulation look very much like a recording of an organism's actions in an experiment where the independent variables change in exactly the same way as during the simulation; for a bad model, the results of the simulation do not resemble those produced by the organism.

Simulation involves at least two stages. The first matches simulated behavior to real behavior, after the fact, by adjusting the parameters in the model. The second stage uses a new pattern of variation in the independent variable, with the model's parameters set as previously determined, and records the behavior of the model. Then the new pattern of variation is applied to the person, whose behavior is recorded and compared with the model's behavior. In the sciences and in engineering, models are often tested in a third stage (as we do here), with both a new pattern of variation for the independent variable and a new kind of environmental disturbance, not used in the original parameter determinations. In this third stage, the model predicts, in simulation, relationships not previously observed.

Reduced to its essentials, the logic of simulation resembles more familiar ways of studying relationships and testing to see if they generalize. It is, however, much more exacting: it compares modeled and actual behaviors instant-by-instant, rather than in terms of static data sets. For the present experiments, the models predict thousands of values for several variables, all of which are compared with the values produced by a participant. The success or failure of a prediction is immediately obvious.

Some people argue that models which work properly in very simple situations might not work when complexities occur. The converse of that hypothesis, also sometimes offered, is that failure of a behavioral theory in a very simple experiment doesn't necessarily mean that it will fail in more realistically complex studies. But engineers, who deal with both simple and complex systems, would not agree. Certainly, a model that works in a simple situation might need considerable revision to work in a more complex situation. But if a model fails to work in the simplest possible circumstances, there is no chance that it will successfully predict more complex phenomena. Complexity can be an excuse for failures of a model in a complex situation, but not in a simple one. If the core assumptions of a model fail in simple experiments like ours, there is no chance the model will work in more complex circumstances.

THE EXPERIMENT

The Task

Participants in this three-phase experiment move a control handle in one dimension, forward and backward. On a computer screen in front of them is a short horizontal bar, the "cursor," distinct from the background, which moves up as the handle moves forward and down when it moves back. Flanking the path of the cursor are two more bars, the "target," that remain even with one another and move slowly up and down the screen, following a path generated by the computer. The person's task in all phases of the experiment is to keep the cursor exactly between the target lines. (There is nothing special about that relationship between cursor and target; the person could easily select any other.) This task is known as "tracking." When the target is stationary, it is called compensatory tracking; when the target moves, as it does here, it is called pursuit tracking.

We can easily modify the experiment to include perceptual variables other than spatial position. For example, the handle can be set to alter the size or shape of a geometric figure, change the magnitude of a number displayed on the screen, or alter the pitch of a sound. And tracking can occur across stimulus attributes and sensory modalities, as when a person uses the handle to make the pitch of a sound match the magnitude of a number or the vertical position of a target. All relationships observed during a simple tracking experiment are found in these other tasks; any of them can be used to make the points we make here.

The Conditions: Three Phases

Phase 1. In Phase 1, the target moves up at constant speed to a preset limit, then down at a constant speed to another preset limit, and so on, in a triangular wave. Each excursion up or down takes 2.8 seconds. The person practices as long as necessary to keep the cursor between the targets with an error of no more than three per cent of the total movement averaged over one minute. Data from the final minute of practice when this criterion is reached are saved as the data for the experimental run.

The relevant parameters are estimated for each model, and then the models reproduce the person's behavior. In the next two phases, we use the parameters thus determined to create a simulated run before the person runs a single one-minute trial. No model is altered, in any way whatsoever, from this point on.

Phase 2. Conditions in Phase 2 are the same as in Phase 1, except that there is a probability of 2/3 that the target speed will differ from the last speed on any given up or down excursion. The speed of each excursion is selected randomly from 1.4, 2.8, or 5.6 seconds per excursion, with a mean of 2.8 seconds over the one-minute experimental run (the same mean excursion time as in Phase 1). The person must still move the handle to keep the cursor between the target marks. A few minutes prior to the person's run, each model is run with the same randomly generated pattern of variations in target speed that the person will experience. The person gets no practice: the first run under these new conditions is the only run for Phase 2.

Phase 3. Conditions are the same as in Phase 2, except that now a smoothed random disturbance also acts on the cursor. The disturbance is created at the start of the entire experiment by smoothing the output of a random-number computer algorithm and storing the resulting waveform. The same disturbance is used in runs by the models and the person. Cursor position is determined by the sum of handle displacement from center and the momentary magnitude of the disturbance. Again, the person does a single one-minute run with no practice. A few minutes before the person's run, each model predicts the results, with a new pattern of target excursions and with the disturbance acting on the cursor.

The experimental variables. During each 60-second experiment, each variable is sampled every 1/30 second, for a total of 1800 values per variable. In the figures illustrating the results, every third value is plotted. There are three measured variables: the positions of the target (T), handle (H), and cursor (C).

Phase 1

The person's data. The person kept the cursor even with the target, as shown in Fig. 1A. The perfectly regular triangular wave in the upper part of the figure is the vertical target position across time. The slightly less-regular wave that closely follows it is the cursor position created by the person. In the lower part is the handle-position record, identical to the cursor-position record because handle position directly determined cursor position. (The handle-position plot is scaled to be the same amplitude as the cursor-position plot; we use this scaling in all figures).

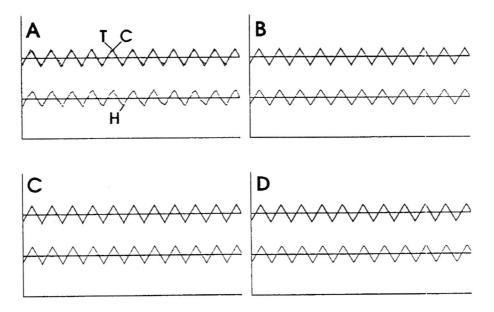

Figure 1. Results of pursuit tracking, Phase 1: data from the person (A); reconstructions of the data by the stimulus-response model (B); by the cognitive model (C); and by the control-system model (D). In A, H = handle, T = target, and C = cursor. For target and cursor, "up" in the figure is toward the top of the computer monitor; for handle, "up" is away from the person. The duration of each experiment is 60 seconds.

The mean vertical distance between the cursor and target was −0.8 units of screen resolution (S.D. −1.8; total vertical range on the screen = 200 units). The following Pearson correlation coefficients describe the relationships among variables in Fig. 1A: between positions of the cursor and target, .977; handle and target, .977; and handle and cursor, 1.0. In the regression of handle on target, the slope was 0.89 (the person moved the handle the equivalent of 0.89 screen units for every movement of one unit by the target), and the intercept was −0.8, identical to the average difference between positions of the cursor and target.

Testing the models: The rationale. In simulations of the models, computations begin with all variables set to the same initial values from the first moment of the run by the person and are repeated 1799 times, once for every 1/30 second in the run by the person. Each model produces handle positions in its unique way, but a common procedure determines cursor positions.

Establishing the S-R model. We remind readers that we do not compare the relative merits of the many varieties of behavioristic theory, nor do we examine or challenge behaviorists' descriptions of conditions in which learning occurs. We merely examine consequences that would ensue were behavior controlled by an independent antecedent variable—were behavior literally "under environmental stimulus control."

Our simple S-R model is rigorously true to the requirements laid down for laws of behavior by B. F. Skinner (1953):

> The external variables of which behavior is a function provide for what may be called a causal or functional analysis. We undertake to predict and control the behavior of the individual organism. This is our "dependent variable"—the effect for which we are to find the cause. Our "independent variables"—the causes of behavior-are the external conditions of which behavior is a function. Relations between the two—the "cause- and-effect relationships" in behavior—are the laws of a science (p. 35).

In our simple experiment, the only independent variable is the position of the target, determined solely by the computer program. The position of the handle depends on the actions of the person, so it is a pure dependent variable, which we model as a response to target position. In Phase 1, the handle determines the position of the cursor, which is a remote (from the person) consequence of behavior, not a cause.

Cursor movement is also a "stimulus," by any traditional definition, but it is not independent of behavior; it lies at the *conclusion* of the assumed causal chain. At best, it might be a "reinforcing" stimulus. Behavioral theorists claim that reinforcement produces long-term changes in the probability of a general class of actions (an "operant"). For example, some might say that, at an earlier time, cursor movement reinforced handle movement, which explains why the person uses the handle now. But reinforcement theory does not explain or predict how a person produces moment-by-moment changes in behavior and in its consequences.

We use a regression equation as our S-R model. For the handle and target positions in the person's data, shown in Fig. 1A, the slope (m) of the regression of handle on target is 0.89, and the offset (intercept, b) is –0.8. We represent target position as t, handle position as h, and cursor position as c. Therefore, the S-R model for handle position is of the form

$$h = mt + b,$$

and the position of the cursor is modeled as

$$c = h.$$

Results of running the S-R model. To "run" the S-R model, we start with all variables at their values during the first instant of the run by the person, then we multiply the remaining 1799 target-position values, in sequence, by the slope m and add the intercept b, and obtain the successive predicted positions of the handle and cursor, shown in Fig. 1B.

The positions of handle and cursor created by the model resemble those from the person: the correlation between modeled and actual handle positions is .977; between modeled and actual cursor positions, also .977. Our simple reflexological model accounts for 96 per cent of the variance (r-squared) in the behavioral data from Fig. 1A; the regression equation is highly meaningful.

Establishing the cognitive model. Our goal with the cognitive model is not to compare the many diverse computational algorithms studied by cognitive and brain scientists. We merely examine the consequences that would ensue, were it possible for a system to reliably compute the same output, no matter how it does the computation. Our cognitive model assumes that, during the practice period, some central process learns and models the amplitude and frequency of target movements and computes commands that cause the muscles to move the handle, and thus the cursor, in a pattern as close as possible to that of the target.

A detailed version of this model would use a program loop simulating a "higher cognitive process" to compute handle positions independently of target movements. It would generate commands for the amplitude, frequency, and shape of the movements. But severe phase errors (mismatches in timing between the positions of the target and the model's handle) would develop unless we gave the model exact information about the frequency of the target and started it at exactly the right moment with exactly the right initial conditions. To assure that there were no errors, we would tell the model exactly how to move the handle to re-create the results of Phase 1. To achieve the same result, without the complex computations, we simply assume that, however the cognitive model works, it works perfectly: it computes handle movements to match the average pattern of previous target movements. For the last minute of practice, it uses information accumulated earlier to command movements that reproduce the movements of the target (of course the model we use here does not actually need any practice).

This makes the cognitive model exceedingly simple: it is of the form

$$h = t.$$

Handle movements perfectly reproduce movements of the target that occurred during the practice run, and the resulting cursor movements also perfectly reproduce the movements of the target.

Results of running the cognitive model. A run of the cognitive model is extremely simple: since h = t and c = h, we simply plot the successive target position values as c and as h. The upper trace in Fig. 1C shows target and cursor positions perfectly superimposed; the lower trace of handle position is identical to the upper traces. The positions of handle and cursor created by the model are like those from the person: the correlation between modeled and actual handle positions is .977; between modeled and actual cursor positions, also .977.

Establishing the control-theory model. The environment part of the PCT model is just a description of the external situation: cursor position depends on handle position plus the magnitude of any possible disturbance. The environment equation is

c = h + d.

In Phase 1, the disturbance magnitude is zero.

The fact that the cursor is also a dependent variable wholly or partly determined by handle position is not a problem, because both the organism equation and the environment equation form a single system of equations. We symbolize the perceived separation of cursor and target, c - t, as p, which we take as the real input variable. This variable p is compared against a reference level p*, which specifies the state of p at which there will be no change in output; it is the value of p that the person intends to experience. Any difference between p and p* is called "error." The output, which is the handle position h, is the time-integral of error and takes the form

h = k[int(p* – p)].

The constant k is the "integration factor." It represents how rapidly the person moved the handle for a given difference between the perceived separation p and the reference separation p*; k is expressed in units of screen resolution the cursor would move per second for a given amount of perceived error.

To fit the model to the subject's behavior, we estimate p* and k, the only adjustable parameters of the model. We set p* equal to the average value of cursor-minus-target during the person's run in Phase 1. (By estimating p* from the data, we avoid claiming that we know the person is trying to keep the separation of target and cursor at zero. The person can maintain any reasonable separation-there is nothing special about p* = 0.) To estimate k, we insert the estimated value of p* into the model, then we insert an arbitrary value of k and "run" the model, a procedure we explain below. During each of several successive runs of the model, we insert a new arbitrary value of k and calculate the root-mean-square (RMS) difference between all of the cursor positions from both the model and the person. The best estimate of k is the one from the run with the smallest RMS difference.

To "run" the model, we start the handle position at the subject's initial handle position during Phase 1, and then do the following computer program steps over and over, changing the value of t on each step to re-create the target movements:

1: c: = h + d
2: p: = c – t
3: error: = p* – p
4: h: = h + k • error • dt

where dt is the physical duration represented by one iteration of the program steps. In all of the experiments reported here, each iteration represents 1 / 30 second, so dt = 1 / 30 sec. For the various terms in the program steps, k and p* are the system constants: k is the tentative value of the integration factor and p* is the estimated reference signal; t is the momentary target position, c is the cursor position, h is the handle position, and d is the disturbance magnitude—here, 0.

The fourth program step is a crude form of numerical integration; the notation means that the new value of h is computed by adding an amount (k • error • dt) to the old value of h. These are program steps, not algebra: do not cancel the h's! The "colon-equal" sign is the replacement operation, which replaces the previous value of the variable on the left with the new computed value of the argument on the right.

Results of running the PCT model. In the person's run during Phase 1, p* was estimated as –1 unit on the screen (–0.8 rounded), which means that, on average, the person kept the cursor slightly below the target. Following the procedure described above, the estimated best value of the integration constant k was 8.64 in units of resolution per second.

The results of a run of the model with those estimated values of p'" and k are shown in Fig. 1D. The positions of handle and cursor created by the model resemble those from the person: the correlation between modeled and actual handle positions is .989; modeled and actual cursor positions, also .989.

Summary of Phase 1. The person performed the tracking task reasonably well, and simulations of all three models produced results like those from the person. After this round of simulations, all three models remain defensible as explanations of the person's performance.

Phase 2

Next, we use the three models to predict behavior when one condition changes, then the person does a run under exactly the same conditions as those encountered by the models. The changed condition is that the target now moves up and down at randomly varying speeds. The mean speed is still 2.8 seconds per excursion, but on every successive excursion, there is a 2/3 probability of a change of speed that lasts until the end of the excursion, and then the next speed is selected randomly. The random changes are generated beforehand and recorded, so the same changes are presented to all three models and to the person. We have already established the three models, so our descriptions of the results are brief.

The person's data. Fig. 2A shows data from the person's run, after the models made their predictions. The person made the cursor follow the target about as well as in Phase 1. The mean vertical distance between cursor and target was −1.4 units of vertical screen resolution (S.D. = 2.2). The following Pearson correlation coefficients describe relationships among variables in Fig. 2A: between positions of the cursor and target, .966; handle and target, .966; and handle and cursor, 1.0.

Prediction of the S-R model. The linear regression equation developed after Phase 1 accurately predicts the positions of the cursor and handle despite the changes in target speed, as is shown in Fig. 2B. This is possible because, just as in Phase 1, the required handle movement is simply proportional to target movement at every instant. The positions of handle and cursor created by the model are like those from the person: the correlation between modeled and actual handle positions is .989; between modeled and actual cursor positions, also .989.

Prediction of the cognitive model. The results for the cognitive model, shown in Fig. 2C, reveal the first obvious failure of a model. The positions of handle and cursor created by the model are not like those from the person: the correlation between modeled and actual handle positions is .230; between modeled and actual cursor positions, also .230.

The reason for this failure is obvious. The cognitive model assesses properties of the environment and computes an action that will have a required result. But now the environment, in the form of target movements, is subject to unpredictable variation. The cognitive model gets no information about the next target speed before it is experienced. Thus, the best that a cognitive "central-process" model can do is command its output to match the best estimate of average target speed; in the present case, that average is the speed that occurred throughout Phase 1, when the motor plan was established. The cognitive model

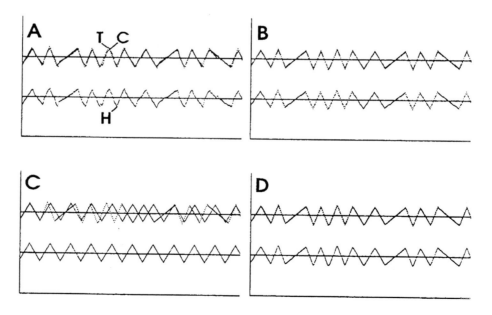

Figure 2. Results of pursuit tracking in Phase 2: data from the person (A); predictions of the data by the stimulus-response model (B); by the cognitive model (C); and by the control-system model (D. In A, H = handle, T = target, and C = cursor. For target and cursor, "up" in the figure is toward the top of the computer monitor; for handle, "up" is away from the person. The duration of each experiment is 60 seconds.

continued to produce a triangular wave of handle and cursor movement that conformed to the average waveform of target movement—a form not like the waveform of the target in Phase 2.

One might think of modifying the cognitive model so that the central processor re-assesses the environment's properties on an instant-by-instant basis. That would solve the problem, but only at the expense of converting the cognitive model into a control-system model intent on making its output match its input: the new model would be a control-system model acting like a stimulus-response model. The core concept of a cognitive motor plan would be abandoned.

Prediction of the control-system model. Fig. 2D shows the results for the control-system model. The program steps from Phase 1, using the same values for the parameters k and p*, successfully predict the person's handle and cursor positions. The correlation between modeled and actual handle positions is .981; between modeled and actual cursor positions, also .981.

Summary of Phase 2. The person performed the tracking task with reasonable accuracy, and simulations of the S-R and PCT models produced results like those for the person. However, the cognitive model continued to make its output follow the path 'learned" during Phase 1; consequently, its cursor did not follow the now-erratic waveform of the target. After this round of simulations, only the S-R and PCT models remain reasonable as explanations of the person's performance.

Phase 3

Now the three models predict behavior under a radical change of conditions. The target still moves up and down at randomly varying speeds, as in Phase 2, but for every time-interval, a new value of a random disturbance is added to the position of the cursor. Now, with the handle held still, the cursor wanders randomly up and down. When the handle moves, the net movements of the cursor are determined by the sum of handle movements and disturbance changes.

In both previous phases, the "d" in the cursor equation, c = h + d, was zero. Now it varies unpredictably, although not rapidly (the bandwidth of variations is about 0.2 Hz). This new disturbance enters after the motor outputs of the person and the accompanying handle movements, "downstream" in the causal chain. The cause of the disturbance is hidden; the only evidence the person has about the disturbance is the deviation of cursor position from the momentary equivalent of the handle position. At any moment, there is no practical way for the person to know the degree to which either the position of the handle or the value of the disturbance affects the position of the cursor.

The person's data. As we show in Fig. 3A, the person still made the cursor track the target (mean distance between cursor and target = −1.0 screen units, S.D. = 3.0), despite the unpredictable variations in target speed and the unpredictable interference of a disturbance. Had the person not moved the handle, the correlation between positions of the cursor and momentary values of the disturbance would have been + 1.0; that between positions of cursor and target, near 0.0. Instead, the correlation between the disturbance and cursor was only .101, while that between cursor and target was .940.

In Phases 1 and 2, the handle alone determined the position of the cursor: the correlation between handle and cursor was + 1.0. But in Phase 3, the person moved the handle any way necessary to cancel the effects of the random disturbance on the cursor: the correlation between positions of handle and cursor is only .294, that between positions of the handle and the disturbance that moved the cursor away from the target is −.992.

Prediction of the S-R model. As we show in Fig. 3B, the S-R model failed: the correlation between modeled and actual handle positions is .296; between modeled and actual cursor positions, .385.

Successful simulation can no longer be attained by moving the handle in synchrony with target movements. That is why the person moved the handle in a pattern that deviated radically from the pattern of target movements; the deviations were exactly the ones needed to counteract the effects of the new disturbance. But the S-R model responded to the target stimulus just as before, and moved the handle proportionately to any movement of the target. The simulated cursor, now subject to an independent disturbance, did not follow the target.

To salvage the S-R model, one might propose that the cursor, too, be included in the definition of the stimulus. However, the person's data in Fig. 3A show that the cursor moved in nearly the same pattern as the target, but neither pattern resembled what the handle did. To include the cursor in the definition of

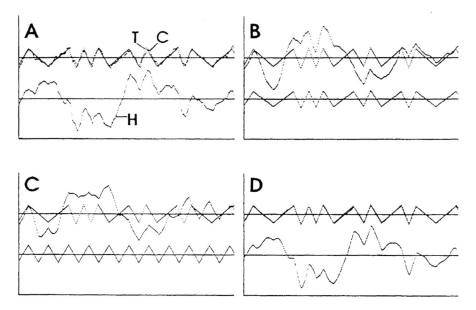

Figure 3. Results of pursuit tracking in Phase 3: data from the person (A); predictions of the data by the stimulus-response model (B); by the cognitive model (C); and by the control-system model (D). In A, H = handle, T = target, and C = cursor. For target and cursor, "up" in the figure is toward the top of the computer monitor; for the handle, "up" is away from the person. The duration of each experiment is 60 seconds.

the stimulus, we might conclude that the difference between the target and cursor positions is the stimulus. On further examination, we would find that this difference does not match the handle movements, either, but its time-integral does: perhaps the time-integral is the stimulus. That change is acceptable, but if we adopt it, we are left with the fact that cursor position depends, simultaneously, on handle position and the independent random disturbance: now there is no true independent variable in the causal chain, and the core premise of any model of stimulus control over behavior is abandoned. Neither the cursor nor any relationship between the cursor and any other variable can be described as a pure independent variable, because it is also, at every moment, a dependent variable.

Prediction of the cognitive model. Fig. 3C shows that the prediction by the cognitive model failed. The model followed its plan learned in Phase 1 and moved the handle to conform to the average behavior of the target. It should have moved the handle in the erratic pattern produced by the person, shown in Fig. 3A. The correlation between predicted and actual handle positions is .119; between predicted and actual cursor positions, .151.

Even if we gave the cognitive model more practice in the new situation (and the ability to learn), it would revert to essentially the same actions. The average deviation of cursor speed from 2.8 seconds per excursion is zero. The average amount of disturbance applied to the cursor closely approximates zero. Neither the next speed of the target nor the next variation in the disturbance is predictable. No matter how smart one wants to make the central processor when it comes to predictions, we can always make the disturbances still more random. Any cognitive model must compute output that is calculated to have a desired effect. It can base its computations only on experience with properties of the external world. When those properties contain significant instant-by-instant irregularities, as they do in our simple experiment, the core concept of the cognitive model cannot work. Unless, of course, it is modified to compare its plan of the world against its momentary perceptions of the world and to act so as to eliminate any discrepancy, but those modifications would make the model a control-system model.

Prediction of the control-system model. As we show in Fig. 3D, the control-system model produced precisely the outputs required to maintain a pre-selected target-cursor separation, despite two kinds of random variation that called for pronounced changes in the output pattern. The PCT model faithfully predicted the person's behavior. The correlation between actual and predicted handle positions is .996; between actual and predicted cursor positions, .969. Correlations as high as those here, between tracking behavior and predictions by PCT, are commonplace, even when the interval between predictions and behavior is as long as one year as is reported by Bourbon, Copeland, Dyer, Harman, and Mosley (1990).

To avoid drawing this paper out any longer, we omit analyses of other variations that the person and the PCT model can handle, with no change in the model's parameters. Both the person and the control-theory model continue to track accurately if we alter the scaling factor that converts handle movement into cursor movement; if we add a third or a fourth or a fifth independent source of disturbance to target speed or cursor position; if we put nonlinearity into the connection between handle and cursor (the person and the model still move the handle in an inverse nonlinear relationship to target and disturbance); or if we make the ratio of handle movement to cursor movement time-dependent (at a reasonable speed). None of these variations can be handled by the core concepts of the S-R or cognitive models. Yet all of these variations, as well as those shown in the three phases of our experiment, are commonplace in the real environments where real behavior must work.

DISCUSSION

We attempted to determine if core assumptions about the immediate causes of behavior in three different models of behavior are consistent with what is known about the world in which behavior occurs. We compared specific predictions made during simulations of the three models with the performance of a person for three phases of a simple task. We concluded that the causal assumptions in a control-theoretic model are consistent with what is known about the world, while those in any pure stimulus-response (stimulus-control) model, or any pure cognitive-control (neurological-control) model, are not. The control theory model assumes that, when organisms act, they produce correspondences between their immediate perceptions of selected variables in the world and internal (to the organisms) reference states (reference signals) for those perceptions.

We did not ask whether reference signals exist in any particular physical form, or, if they do, whether they are "gained" through interaction with the world, whether animate, inanimate, or social, or are inherited as part of a "genetic code." Robinson (1976) wrote of this issue in a discussion of Aristotle's concept of "final cause," which refers in part to a person's goals or intentions: "The issue is not how a given goal or intention was established. Rather, the issue or proposition is that outcomes are never completely understood until the final cause is apprehended, no matter what 'caused' the final cause" (p. 91, emphasis in the original). In our simulations, by hypothesizing and estimating the magnitudes of "reference signals," whatever their origins, that function in the manner of "final causes" within a control-system model of a person, we can understand and predict the outcomes when the person controls selected perceptions of parts of the unpredictably variable environment.

Modeling as a proper test of theory. The success or failure of our simulations immediately revealed the robustness, or lack of robustness, of alternative models of behavior. Other behavioral scientists recognize the importance of comparing the simulated behavior of models against the actual behavior of organisms. In a critique of conventional statistical methods in psychology, Meehl (1978) remarked:

> In my modern physics text, I am unable to find a single test of statistical significance. What happens instead is that the physicist has a sufficiently powerful invisible hand theory that enables him to generate an expected curve for his experimental results. He plots the observed points, looks at the agreement, and comments that "the results are in reasonably good agreement with the theory." Moral: It is always more valuable to show approximate agreement of observations with a theoretically predicted numerical point value, rank order, or function form, than it is to compute a "precise probability" that something merely differs from something else" (p. 825, emphasis in the original).

Similarly, Dar (1987) wrote:

> In physics... theories are tighter and lead to precise predictions. As a consequence, (a) if the numerical result is as predicted (that is, close enough to the predicted point value or curve), it will be very difficult, in contrast to the situation in psychology, to offer a reasonable alternative theory for that. This

is because it is difficult to imagine alternative states of nature that will lead to the exact same curve or numerical result. (b) If the experimental result is not as predicted, some serious revision of the theory would be required. This is because a tight theory simply does not allow for significant (I do not mean "statistically significant") discrepancies from predicted outcome (p. 148).

And in his review of a book on cognitive theory, the behaviorist Shimp (1989) declared:

A theory that behaves, that produces a stream of behavior, would seem in an intriguing way to fit better with Skinner's chief criterion for a good theory than do many more common sorts of behavioral theory. Skinner has argued that a good behavioral theory is a theory on the same level as the behavior itself. What is closer to the level of a behavior stream of an organism than a behavior stream of a theory? (p. 170).

We could not say it better. On any given experimental run, our simulations produced multiple simultaneous streams of behavior, altogether comprising thousands of predicted data points. The levels of agreement between the simulations and the behavior of a person allowed us to immediately assess the adequacy of the three models of behavior and of their implied models of the world.

The worlds implied by the models. For all three models, the results reported here would be general. Within its physical limits, any S-R system could make its movements match any target input, no matter how unpredictable. But, as happened with the cursor in Phase 3, if the consequences of those movements were disturbed, they would always deviate from the target by an amount equal to the variations in the disturbance.

Upon its first encounter with a new pattern of input, no cognitive system could compute commands to immediately make its behavior match the input. After some time, of course, an appropriately endowed cognitive system could search for a new pattern of commands. But if the input followed an unpredictable path or were presented only once or too few times for the system to "compute" an appropriate plan, learning would be impossible. Furthermore, if the consequences of its actions were continuously and randomly disturbed, no command-driven cognitive system could compute behavior to keep the consequences in any consistent relationship with the input. To do that, the behavior must deviate from its original pattern by precisely the amount needed to cancel the effect of the disturbance, but the source of the disturbance cannot be sensed in advance to allow anticipatory compensations in the commands for behavior.

The only ways to salvage the traditional models, short of turning them into control systems, rely on whimsical assumptions about the world. For example, the S-R model might still work if it were only necessary that changes in stimulation result in corresponding changes in behavior, with no regard for the consequences of behavior; and the cognitive model might still work, if it were only necessary that movements repeat, while their consequences were allowed to change at random. But those assumptions contradict any reasonable understanding of behavior and its role in survival: behavior is functional, and its consequences matter. An alternative defense is to assume that the antecedents of behavior never change, or that they conveniently change across a small enough set of discrete options so that we can always recognize which one is present and perfectly match it with computed outputs-either that, or we must anticipate the changes by "precognition." And nothing must ever disturb the consequences of behavior. The world demanded by those assumptions is not the one we know.

In contrast, within broad limits, any perceptual control system would vary its behavior to keep its perceptions of a controlled variable at the value specified by a reference signal, even if both the target event and the consequences of the system's actions were subject to unpredictable variations.

We live in a changeable world, in which organisms with behavior determined solely by environmental stimuli or solely by internal commands could not survive; but theories of behavior that postulate control by stimuli or by commands have survived for centuries largely because they are not systematically exposed to the test of modeling. To modify cognitive or S-R models so that, like living systems, they might thrive amidst change, we must abandon the core concept that behavior is at the end of a causal chain, wherever the chain allegedly begins. We must give each model an internal standard and a process for comparing present perceptions against that standard. But then the models would all be control systems, each controlling its input.

Conclusions. The sciences of life reflect a three-century commitment to linear models of cause and effect, with behavior as the final step in a causal sequence. If we are to advance our understanding of life, we must question those venerable models, however plausible they seem. We can no longer embrace them, knowing that they presuppose nonexistent worlds. To question our traditional models raises the specter of difficult change; but if we retain them, with their fanciful worlds, we risk the trivializing and decline of our science.

The search for alternative models of behavior can begin with a simple change in the question we ask, from "Why is behavior the way it is?" to "Why is the world the way it is?" The answer to the new question includes a long-elusive answer to the old one: the behavior of organisms controls many variables in the world.

ACKNOWLEDGMENTS

We thank Andrew C. Papanicolaou, Philip J. Runkel, Gregory Williams, and William D. Williams for their critical reviews of several early versions of the manuscript, and Glenn Millard for valuable assistance in preparing the figures.

REFERENCES

Bizzi, E., Mussa-Ivaldi, F. A., & Giszter, S. (1991). Computations underlying the execution of movement: A biological perspective. *Science,* 253, 287–291.

Bourbon, W.T., Copeland, K. C., Dyer, V.R., Harman, W.K., & Mosley, B. L. (1990). On the accuracy and reliability of predictions by control-system theory. *Perceptual and Motor Skills,* 71, 1331–1338.

Churchland, P.M. (1986). Some reductive strategies in cognitive neurobiology. *Mind,* 95, 279–309.

Churchland, P.S. (1986). *Neurophilosophy: Toward a unified science of the mind-brain.* Cambridge, MA: MIT Press.

Dar, R. (1987). Another look at Meehl, Lakatos, and the scientific research practices of psychologists. *American Psychologist,* 42, 145–151.

Donahoe, J.W., & Palmer, D. C. (1989). The interpretation of complex human behavior: Some reactions to *Parallel distributed processing,* edited by J. L. McClelland, D. E. Rumelhart, and the PDP research group. [Book review]. *Journal of the Experimental Analysis of Behavior,* 51, 399–416.

Hayes, S.C., & Brownstein, A. J. (1986). Mentalism, behavior-behavior relations, and the behavior-analytic view of the purposes of science. *The Behavior Analyst,* 9, 175–190.

Hershberger, W.A. (Ed.). (1989). *Volitional action: Conation and control.* Amsterdam: North-Holland.

Kihlstrom, J.F. (1987). The cognitive unconscious. *Science,* 237, 1445–1452.

Kugler, P.N., & Turvey, M.T. (1987). *Information, natural law, and the self-assembly of rhythmic movement.* Hillsdale, NJ: Erlbaum.

Marken, R.S. (Ed.). (1990). Purposeful behavior: The control theory approach. [Special issue]. *American Behavioral Scientist, 34(1).*

Meehl, P.E. (1978). Theoretical risks and tabular asterisks: Sir Karl, Sir Ronald, and the slow progress of soft psychology. *Journal of Consulting and Clinical Psychology,* 46, 806–834.

Newell, A. (1990). *Unified theories of cognition.* Cambridge, MA: Harvard University Press.

Pavloski, R. P., Barron, G. T., & Hogue, M. A. (1990). Reorganization: Learning and attention in a hierarchy of control systems. *American Behavioral Scientist, 34,* 32–54.

Powers, W.T. (1973). *Behavior: The control of perception.* Chicago: Aldine.

Powers, W.T. (1978). Quantitative analysis of purposive behavior: Some spadework at the foundations of experimental psychology. *Psychological Review, 85,* 417–438.

Powers, W.T. (1989). *Living control systems: Selected papers of William T. Powers.* Gravel Switch, KY: Control Systems Group.

Powers, W.T. (1992). *Living control systems II Selected papers of William T. Powers.* Gravel Switch, KY: Control Systems Group.

Real, L.A. (1991). Animal choice behavior and the evolution of cognitive architecture. *Science, 253,* 980–986.

Robinson, D. N. (1976). *An intellectual history of psychology.* New York: Macmillan.

Runkel, P. (1990). *Casting nets and testing specimens: Two grand methods of psychology.* New York: Praeger.

Skinner, B. F. (1953). *Science and human behavior.* New York: Free Press.

Skinner, B. F. (1989). The origins of cognitive thought. *American Psychologist, 44,* 13–18.

Shimp, C. P. (1989). Contemporary behaviorism versus the old behavioral straw man in Gardner's *The mind's new science: A history of the cognitive revolution.* [Book review] *Journal of the Experimental Analysis of Behavior,* 51, 163–171.

Chapter 13

Quantitative measurement of volition

The article that follows appeared originally as Chapter 13 in W.A. Hershberger (Ed.) (1989). *Volitional action: Conation and control*, Amsterdam: North-Holland.

QUANTITATIVE MEASUREMENT OF VOLITION: A PILOT STUDY

William T. Powers

In cybernetic control theory, overt intentional behavior is operationally defined as a controlled input or perceptual variable being maintained in a publicly-observable reference condition. In a control-system model the observable reference condition corresponds to a reference signal inside the behaving organism. The reference signal is the physical embodiment of the intention that is directing the volitional action. The volitional actions of others are not always obvious. Their discovery requires finding a variable that the person's actions are maintaining in some identifiable state despite disturbances that act directly on the variable. From the behavior of the controlled variable it is possible to infer the behavior of the internal reference signal and thus get a picture of the directing intention (Marken 1982, 1983). This inference is model-dependent, but as we will see in this study, it can be made with more internal consistency than might seem reasonable.

EXPERIMENTAL PROCEDURE

This analysis will be done in the context of a "compensatory tracking" task modified to include an interval of spontaneous behavior. The basic compensatory tracking task requires the participant to use a control handle to keep a vertically-movable cursor stationary on a display screen, centered between two fixed target marks. The cursor is continuously disturbed from inside the computer that runs the experiment, the disturbance varying randomly but smoothly in amplitude. About one third of the way through each run, a tone sounds to indicate the start of a period of spontaneous voluntary behavior, and two thirds of the way through, sounds again to end it. Runs last for 60 s, with a 2 s run-in period to allow control to be established before data recording begins. The screen is updated and a sample of handle position is taken 30 times per second, for a total of 1800 data points. The disturbance is generated and handle positions are measured with a precision of one part in 2000 relative to the maximum deviation from center, but cursor position is scaled down to fit on a screen with 200 lines of resolution.

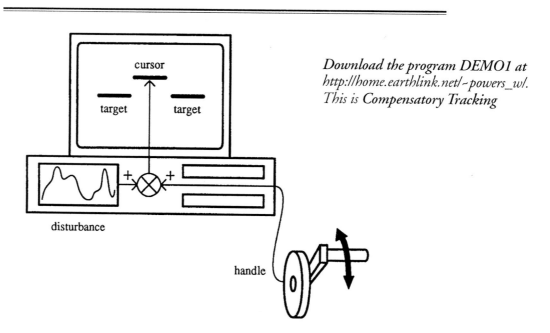

*Download the program DEMO1 at http://home.earthlink.net/~powers_w/. This is **Compensatory Tracking***

Figure 1. Experimental setup. Handle movements are added to a disturbance generated inside the computer to position the cursor. Two stationary target bars are placed in the center of the screen. The cursor can move up and down between them.

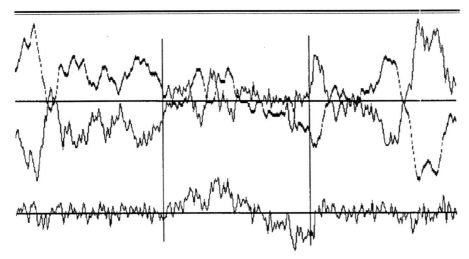

Figure 2. Results of experimental run. In the upper part of the figure, the solid line represents handle position, the broken line the magnitude of disturbance. Note the mirror symmetry in the first and third parts. In the lower part of the figure, the cursor position is shown. Deviations due to the disturbance appear throughout. In the center portion, the cursor rises slowly, then falls: the spontaneous part of the run, where the person is trying to make the cursor move in a slow regular way. The spontaneous part is delimited by vertical lines showing where tones occurred to signal start and end of spontaneous action.

The participant is instructed to keep the cursor aligned with the target for the first part of each run. When the tone first sounds, the participant is to start making the cursor move in a smooth and regular pattern of up-down movements. These spontaneous voluntary movements of the cursor are to continue until the tone sounds again (about 2/3 of the way through the run), at which time the cursor is again to be maintained level with the target marks. Choice of the pattern of spontaneous cursor movements is left to the participant. Note that the spontaneous voluntary behavior is defined in terms of cursor motion (a perceptual variable) and not in terms of a regular handle movement (an action).

The drawing of Figure 1 shows the experimental setup with the effect of the disturbance also indicated schematically. Figure 2, about which more will be said later, shows a plot of the results. Every third data point is shown. The handle behavior is the solid line in the upper part of the figure; the disturbance amplitude is shown by the intermittent line and the center of the screen is represented by the straight line. In the lower part of the figure, the cursor behavior is shown, again with a straight line indicating the center of the screen. The disturbance was made just difficult enough to result in appreciable movements of the cursor. The two vertical lines indicate the times when the tone sounded. Because of the disturbance, the changes in cursor position in the middle part do not resemble the handle movements that created them. For this run, the participant (the author) was trying to make the cursor move in a stairstep pattern, first upward, then downward.

THE MODEL OF THE ACTOR

The participant's organization (relative to this task) is represented as a system containing an input function, a comparison function or comparator, and an output function.

The input function converts the cursor position c into a perceptual signal p representing it inside the behaving system. The perceptual signal's magnitude varies as the cursor position varies, but with a slight exponential time-lag. Thus a step-change in the cursor position would result in a change in perceptual signal of the same numerical magnitude, but the perceptual signal would approach its final magnitude exponentially.

The form of the input function is given by a computer program step (in the Pascal language) that computes the perceptual signal **p** from the cursor position, **c**:

p := **p** + (**c** − **p**)/S

The programming symbol ":=" (colon-equal) means assignment or replacement, not equality. To introduce a lag we subtract the old value of perceptual signal from the computed new value, which is just **c**. This difference, **c** − **p**, is divided by a slowing factor S, and that fraction of the computed change is added to the old value of perceptual signal. The result replaces the old value of **p**. Thus **p** is allowed to change only a fixed fraction 1/S of the way from the old value to the new value on each repetition of this step. With no lag (S = 1), the perceptual signal would be numerically equal to the cursor position at all times.

This computation approximates an exponential lag with a time constant of tc = $(1/30 \text{ s})/\log_e(S/S-1)$.

The slowing factor S is one of the two adjustable constants in the model. The best value of S for the illustrated data proves to be about 5.50 (to the nearest 0.25), implying a perceptual time constant of 0.14 s. The method for evaluating parameters will be explained shortly.

If **p** is initially zero, and **c** is constant at 100, then with a slowing factor S = 5.50, the successive values of **p** (obtained by executing the above program step over and over) are 0, 18.2, 33.1, 45.3, 55.2, 63.3, 70.0, . . . 100.0. If the value of **c** changes during these computations, the value toward which the series is converging will be changing, so **p** will lag behind **c** by an amount that depends on how fast **c** is changing. This kind of lag is not a pure time-delay (or "transport lag") but simply a slowing or smoothing of the response of the input function.

The comparison function subtracts the value of the perceptual signal from the value of a reference signal **r**. It is the varying value of **r** during the spontaneous voluntary phase that we are attempting to estimate by the procedures outlined below. The outcome of the subtraction is an error signal **e**, the magnitude and sign of which continuously indicate the mismatch between the perceptual and reference signals. The comparator is represented by the program step

e. = **r** − **p**.

This sense of the subtraction was chosen to let all other constants be positive.

The output function receives the error signal and converts it into a value of handle position, **h**; that is, **h** = f(**e**). The particular function chosen makes handle velocity depend on the magnitude of error. If handle velocity is a constant K times the error signal's magnitude, then the handle position is calculated by a program step that does a crude numerical integration (over the 1/30 s interval):

h := **h** + K * **e**.

This step is not an equation; it means that K times the error signal magnitude is added to the current value of handle position to obtain the next value (on the left of the ":=" sign). The constant K is the second adjustable parameter in the model. The asterisk is the program-language version of a multiplication sign, which is always explicit in a written program. The single constant K absorbs all other possible constants of proportionality in the model of the participant.

These three steps result in a model whose dynamic properties approximate those represented by the "transfer functions" obtained in similar experiments done by engineering psychologists (See Osafa-Charles, Agarwal, O'Neill, & Gottlieb, 1980, for the conventional forms).

THE MODEL OF THE ENVIRONMENT

The handle position is sensed by an analogue-to-digital converter in the computer and is represented by a number that can range from −2000 to 2000. A second number is taken from a precalculated table of disturbances. The table is constructed by successive smoothings of a series of random numbers generated by an algorithm, and is scaled to a peak-to-peak amplitude of 2400 units. Adjusting the amount of smoothing changes the rapidity of variations in the disturbance amplitude, and so adjusts the difficulty of the task. The cursor position is determined every 1/30 s by sampling the handle position and adding to the result the next sequential entry from the table of disturbances. Thus the cursor position represents neither handle position nor disturbance alone, but only their sum. Using **d** for disturbance magnitude, we have the model of the environmental relationships in this experiment in the form of the final program step:

c := **h** + **d**.

This "model" of the environment correctly represents the actual environment, because the same program step is used to position the cursor when a human participant is moving the handle. When the model is running, the computer model of the participant is given the value of c directly where the real participant sees the cursor on the screen; the model gives back the value of h to the environment as a number where the participant moves a physical handle to generate that number. The same table of disturbances is used for both participant and model. This table can be changed easily to prevent memorization (by the person) of any patterns. All data given below are for a run in which a new disturbance pattern was experienced for the first time by the participant (although practice with other patterns preceded the live run).

It should be emphasized here that the object of this exercise is to demonstrate a principle and a method, not to show research results with many human participants. Such applications will be developed, but for now only a single participant is needed—the author. The reader may, however, assume with confidence that these results will be typical of any well-practiced participant. Control-system experiments are highly reproducible after learning is finished.

RUNNING THE MODEL

We now have a model of both the participant and an environment, consisting of five program steps arranged below in the sequence appropriate for computation (the fifth step repeats the calculations). These steps are executed 1799 times, with an index i (in brackets) advancing by one on each step. The index is used to access successive values of the disturbance, and also to point to locations in a table where the computed values of handle position are stored after being computed. Only handle position needs to be saved, as cursor position can be reconstructed exactly from $c := h + d$.

We set the model's reference signal r to zero at first, indicating that the model is attempting to keep the cursor at the zero position (corresponding to the position of the target marks on the screen).

Initialization:
$r := 0$
$i := 0$
$p := 0$
$h[i] := 0$

Step 1. $c\,.\, = h[i] + d[i]$
Step 2. $p := p + (c - p)/S$
Step 3. $e := r - p$
Step 4. $h[i + 1] := h[i] + K * e$
Step 5. Increment i; if $i < 1799$ go to step 1.

When a variable reference signal is used, a table $r[i]$ is substituted for the fixed value r. When the human being is doing a run, steps 2, 3, and 4 are replaced by a step that displays the cursor position on the screen and samples the physical handle position.

For a model run, two variables, h and p, must be set to initial values. The initial values are 0, a safe value because the same 2 s run-in period is used for the model and for the participant, and allows plenty of time for any starting transient to disappear. The run-in is not shown; it is accomplished simply by starting i at 60 and running it downward to 0 before starting to advance it upward again; the stored values of h are overwritten. In this way there are just 1800 values of h in the final table with no extras to discard.

Readers who are thinking of trying other models like this should be warned of a hidden difficulty. This model works primarily because of the time-integration in the output function. A digital computer model of a continuous closed-loop system, if constructed without any time integrations or other slowing factors, will not work properly because physical time is not properly handled. In the real system being modeled, the various functions all operate at the same time, not in sequence. Only when some suitable way of handling time is introduced can a model computed as a series of sequential steps give the right answers.

EVALUATING THE PARAMETERS K AND S

The parameters K and S are evaluated using a very simple, yet satisfactory, heuristic procedure.

Initially, the slowing factor S is set to 1 (no lag), the integration factor K is set to 0.1, and the model is given a trial run for comparison against the author's performance previously recorded, using the same series of disturbances. This occurs at high speed, taking about .1 s. The model's cursor behavior is then compared by subtraction with the author's cursor behavior and the sum of squares of the differences is computed. Only the first and last thirds of the data (before and after the tones, with a 1 s delay after the second tone) are used, because the model, at this point, can't generate a different pattern of behavior in the middle part.

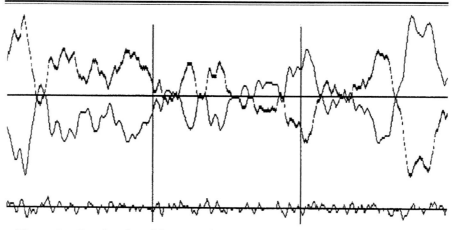

Figure 3. Results of model run with optimum perceptual lag (S = 5.50), optimum integration factor (K = 0.220), and reference signal set to zero. These values give the best fit (smallest least-squares difference) of model and participant cursor behavior in the first and third parts of the run. The model reproduces many features of the real data in Figure 2, but is generally smoother in its action.

Then the output integration factor is stepped upward from 0.1 in increments of 0.005 as the procedure is repeated over and over.

Each time the sum of squares reaches a new low, the values of K, S, and the summed squared error are saved. When the new squared error exceeds the minimum squared error by 3 per cent, the best parameter values are saved as the best values for that series. Then S is incremented by 0.25 and the entire sequence is repeated with K beginning 0.02 units below the previous best value. This procedure ends when the minimum squared error is 3 per cent greater than the value that went with the "best of the best" values of the parameters. The 3 per cent criterion for ending all runs was found by trial and error. While this method is not elegant, it is simple and takes less than one minute on a 10 Mhz IBM-AT-compatible microcomputer. More elegant statistical approaches do not give as good results because the statistical distribution of errors is not close enough to the usually assumed Gaussian distribution.

Figure 3 shows a run of the model with the reference signal constant at 0, the optimum value of S, 5.50, and the optimum value of K, 0.220. This model behaves reasonably well in the first and third parts of the run, although the behavior differs from the real run in the central part because of the constant zero reference signal.

DEDUCING THE REFERENCE SIGNAL

We now have a model with parameters that make it reproduce the participant's behavior in the first and third parts of the experimental run. This model has a reference signal of zero, meaning that the model is maintaining the cursor near the center of the screen where the target marks are, or would do so if the cursor were displayed. We have adjusted the parameters to make the model show nearly the same variations in cursor position and handle position that the subject produces in the first and last thirds of the run, where the target position is zero. In the middle part, the spontaneous action of the participant makes the two cursor traces very different.

To deduce the reference signal in the model, we apply the model's functions to the data taken from the participant. For each data point, we infer the error signal from the observed handle position, and the perceptual signal from the observed cursor position. Because the comparison process is defined as $e = r - p$, we can calculate that r must be $p + e$. The hypothetical perceptual signal can be obtained from the observed cursor position and the calculation representing the model's input function, by the program step $p := p + (c - p)/S$. The hypothetical error signal can be obtained from the observed handle position, the integral of the error signal: the error signal inside the participant would be dh/dt divided

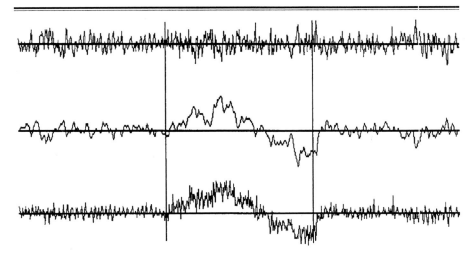

Figure 4. Deducing the reference signal. The upper trace is the person's error signal, the first time derivative of the actual handle position divided by the integration factor K. The middle trace is the person's perceptual signal deduced from the actual cursor position, assuming the same lag as in the model (see text). The bottom trace is the sum of the top two and represents the deduced reference signal. Notice the appearance of high-frequency variations in the reference signal.

by the integration factor K. As a program step, we have $e[t] := (h[t] - h[t-1])/K$. Having found e and p, we then calculate the value of $r = p + e$. The need to take a derivative is the main reason for recording handle position with such high resolution.

Figure 4 shows, from top to bottom, (dh/dt)/K, p, and r as deduced from the participant's data and the model's parameters and functions. The reference signal contains a variation higher in frequency than any variations seen in the handle or cursor traces of Figure 2. We could remove the high-frequency variations in r by a smoothing method that discriminates strongly against high frequencies, but we will accept them as real and see what the consequence is.

COMPLETING THE MODEL

As a check to see if the derived reference signal does in fact result in the right model behavior, we can use the pattern just obtained in the bottom trace of Figure 4 as the reference signal for a model run. The same program steps outlined above are used, but instead of initializing r to zero, we now use the result from Figure 4 as a table of reference-signal amplitudes and do the computations with r[i]. The index i picks out successive values of reference signal just as it picks out successive values of disturbance. The result of a model run is shown in Figure 5.

Comparison of this result with that of Figure 2 shows that the real run is duplicated. The correlation between handle positions, model and real, is .99841, and between cursor positions is .99991 (n = 1800). This does not indicate that we have made an extraordinarily accurate prediction, but only that the method of deriving the reference signal does generate just the signal needed to account for the observed behavior–in other words, that there has been no computational error and no cumulative rounding effect of consequence.

In Figure 6 we have, from top to bottom, the participant's cursor trace, the final model's cursor trace, the deduced reference signal used for the model run, and a version of the deduced reference signal smoothed with a four-pole low-pass filter.

The bottom trace in Figure 6, the smoothed version of the reference signal, shows a stairstep pattern more clearly than the cursor traces do, either for the model or for the participant. This is, presumably, a record of the intended positions of the cursor. The smoothed version of the reference signal shows a best estimate of the reference signal with the rapid

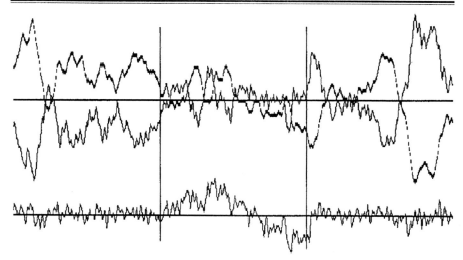

Figure 5. Model run using deduced reference signal from Figure 4. Handle-to-handle (model vs. real) correlation is .99841 (n = 1800). See Figure 6 for cursor-to-cursor comparison.

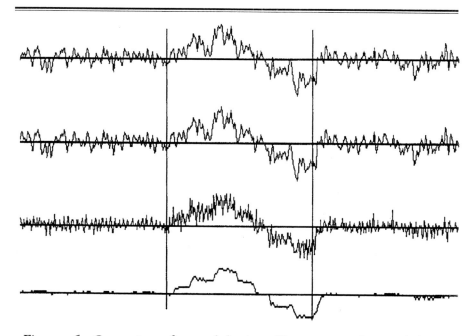

Figure. 6. Comparison of cursor behaviors. Top trace is real cursor behavior; center trace is model cursor behavior; next trace is the deduced reference signal used to run the model; lowest trace is the smoothed version of the deduced reference signal. Cursor-to-cursor correlation is .99991 (n = 1800). This correlation shows that the reference signal was deduced and applied consistently.

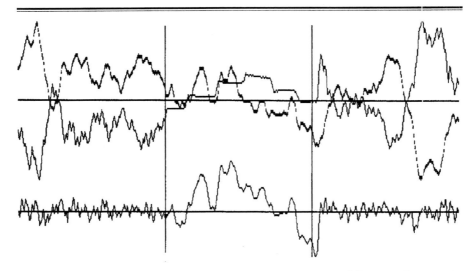

Figure 7. Experimental run when eyes are closed during middle part of run, and handle is moved by feel in a series of steps upward, then downward again.

variations removed. The same amount of smoothing applied to the actual cursor position makes almost no noticeable difference in its shape, because the smoothing cuts out only the highest-frequency variations.

Thus the reference signal could not be obtained simply by smoothing the cursor trace.

REVIEWING THE RATIONALE

Let us review the strategy. We first matched a model with real behavior for portions of the run in which the participant is presumed to have a reference signal constant at zero. In doing this we evaluated two constants, a perceptual lag constant and an output integration factor. This produced a model that could match the participant's behavior with normal accuracy outside the region of spontaneous voluntary behavior.

Then we used those constants under the assumption that the participant is organized as the model is. In the model, the simulated handle position is the time-integral of the internal error signal; hence the participant's assumed error signal is proportional to the first derivative of observed handle position. The model's perceptual signal is the lagged model cursor position; hence we assume that the participant contains a perceptual signal that is a similarly lagged actual cursor position. Still applying the model in a straightforward way, we then add the error signal to the perceptual signal to deduce the participant's reference signal. Finally, we run the model using that deduced reference signal (unmodified) to see if the resulting behavior matches that of the participant, to check that the derived reference signal does lead to reproducing the actual behavior—that is, to see if the derivation was correctly done.

The model allows us to see the behavior of a variable, the reference signal inside the person, that is not directly visible from outside.

AN INTERESTING VARIATION

The model, at least as it stands, cannot distinguish apparent from real intentions: it will always compute a reference signal. Because of the way the derivation is performed, this reference signal will always make the model reproduce the observed variables correctly. As a preliminary way of investigating this problem further, I performed the experiment with slightly different instructions.

Instead of the person at the controls moving the cursor in some regular pattern during the middle part of a run, the person now closes his or her eyes at the first tone and opens them at the second tone, continuing to move the handle (blindly) in a pattern something like the pattern in which the cursor was supposed to move (the same disturbance is used). The result of an experimental run is shown in Figure 7.

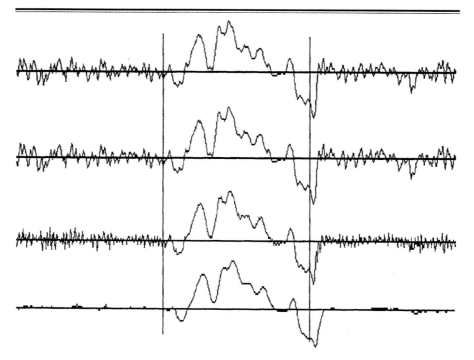

Figure 8. Comparison (like that of Figure 6). The reference signal is now spurious in the middle part. Cursor shows no regular pattern, high-frequency variations are missing from center part of unsmoothed reference signal trace. See Conclusions and Discussion.

Figure 8 shows the eyes-closed result that corresponds to Figure 6: the actual cursor behavior at the top, the model's cursor behavior next to the top, the "deduced" reference signal next to the bottom, and the smoothed version of the reference signal at the bottom.

The result of carrying out all our manipulations of data is a model that exactly reproduces the behavior throughout the run. But we know that the model can't apply during the middle interval–there is no perceptual signal representing cursor position. What we now have is a model that reproduces the cursor behavior throughout the run on the assumption that the cursor behavior was intentional. In other words, if the person had intended the cursor to move as it actually moved, this model would show the reference signal representing that intention.

In Figure 8, the high-frequency noise in the deduced reference signal disappears during the middle part of the run, although it is present before and after when the eyes are open, and tracking is actually occurring. We are still using the first derivative of handle position in computing the reference signal, so we know now that the noise does not originate in the output function or in the measurement. There is no similar noise in the perceptual signal at any time, so we have evidence that when tracking is really occurring the noise is probably associated with the reference signal.

Examining Figure 7, we see that the only stair-step regularity that appears is in the handle trace; the cursor trace shows no obviously regular pattern. Without physically disturbing the control handle, we cannot prove that the regularity in handle movements is intentional (in the terms of this theory), but clearly if such disturbances were used, we could apply this same analysis just as we have done here and deduce a reference signal for handle position. Thus we could build downward toward lower levels of a hierarchical model. Similarly, by making the tracking skill part of a task involving control of more general variables, we could build upward toward higher levels of a hierarchical model.

DISCUSSION AND CONCLUSIONS

About the present results we can say at least this: when a person deliberately makes the cursor move in some clearly-conceived way, the model will allow us to deduce a reference signal behavior that the person will agree represents the intended movement of the cursor more closely than the actual cursor movement represents it (when we smooth out the highest-frequency variations in reference signal). We still have to rely on the person to tell us that the cursor movements really were intentional, and that the deduced pattern is close to the intended pattern. While that is legitimate information, it would be better to obtain it some other way. The only way to do so is to expand the model to include more kinds of behavior and more levels of behavior—to find other ways of observing what we assume is the same phenomenon.

The ideal way to test this model might be through recording neural signals in appropriate parts of the central nervous system of the participant. In the present state of technology, however, doing this by non-intrusive and safe means is beyond us.

We are in much the same position as astronomers were before space travel became possible. When a telescope is pointed toward the tiny dot of a planet in the sky, we can see or photograph an image that shows a disk with markings on it. By referring to optical theory, and by analogy with observing objects on Earth that we can inspect by other means, we can infer that there really is something out there corresponding to the image. This inference, however, is unverifiable, because the same image could be generated in many ways other than by a planetary body located millions of miles away and illuminated by the Sun. All we can be reasonably sure of is that a collection of wavefronts of light enters the telescope and is subjected to consistent transformations caused by the optical elements: any phenomenon capable of creating the same wave-fronts at the open end of the telescope would produce the same appearance in the eyepiece or on film. A computer-generated hologram, for example, could reproduce the image exactly.

We have now sent spacecraft to Mars, for example, and their cameras confirm, generally, the fuzzy outlines we see from Earth. Or do they? Are we not in the same position as before when we look at the images generated by the cameras? All we can really say is that the assumption of a real body, given the laws of optics and extrapolation from phenomena on Earth, is consistent both with the spacecraft pictures and the Earth-based telescopic pictures. It would seem that we will not get final confirmation until a human being orbits Mars or lands on its surface.

Even then, the problem will not be solved, the inference will not become a fact beyond all doubting. All we could say is that the wavefronts of light reaching the pupils of the astronaut's eyes, transformed by the optical properties of the lens and interpreted by the computations in a human retina, create a perceived result consistent with the idea that a real body exists, and also consistent with the spacecraft pictures and the Earth-based pictures. With each step we take toward certainty, certainty itself recedes.

In short, we are faced with the same problem that greets all sciences that rely on models of reality for their understanding of nature—physics, chemistry, astronomy, geology, neurology, psychology, and so on. We assume models that seem to serve as instruments for observing formerly invisible objects. Then we try to find alternate ways of observing the same thing, which always turn out to be alternate models or alternate ways of applying the same model. When inconsistencies arise, we modify the models to remove them, or even invent new models when the old ones can't be made to work any more without changing their fundamental nature. The nearest we get to certainty that the models are true pictures of reality is a subjective conviction that what we see makes sense, looks simple, repeats itself, changes as it ought to when circumstances change.

We seem to be seeing reference signals here through an instrument called control theory. There are, no doubt, alternate explanations for what we see here. There always will be. The best we can do is expand our experiments and look for alternate views of the same phenomenon, either to increase our conviction that we are seeing something real, or to force us to change the model.

REFERENCES

Osafa-Charles, F., Agarwal, G. C., O'Neill, W.D., & Gottlieb, G. L. (1980). Applications of Time-Series Modeling to Human Operator Dynamics. *IEEE Transactions on Systems, Man, and Cybernetics, SMC–10,* 849–860.

Marken, R.(1982). Intentional and Accidental Behavior: A control theory analysis. 1982. *Psychological Reports 50,* 647–650.

Marken, R. (1983). "Mind reading:" A look at changing intentions. *Psychological Reports 53,* 267–270.

Chapter 14

Is behavior probabilistic?

Many psychologists say that predicting actions with certainty cannot be done, because behavior is probabilistic. That is, given any set of conditions in which we find a person, we will be able to say only that there is a probability less than one that the person will do a particular thing. Therefore, the argument goes, any experiment that shows more people doing the predicted thing than you would expect from the base rate—enough more than you would expect from chance variation—shows that you have learned something about the causes of behavior.

But is behavior randomly variable?

Well, consider walking. For the purpose of argument, let's suppose your chance of taking another successful step is 999 in 1000 or .999. To say it another way, let's suppose you would expect to fall once in every 1000 steps. Suppose, to make the arithmetic simple, you walk fairly briskly along the street at 100 steps per minute. You would fall, on the average, once in 10 x 100 steps or once every 10 minutes. Other people walking on the street at that same rate would also be falling down every ten minutes. If ten people were walking near you, going in your direction, one of them would fall, on the average, every minute. But 100 steps per minute is a little fast for most people out on most errands. Fifty steps per minute, on the other hand, is a mere saunter. Let's suppose that 75 steps per minute is fairly close to the average on a city street where people are not just hanging out, but actually going someplace. You would fall, on the average, once in 13.3 x 75 = 1000 steps or once every 13.3 minutes. Among 13.3 other people, one of them would fall, on the average, every minute. Is this an accurate description of your experience when you walk on a city street? Among the people around you, do you see one falling, or even stumbling, every minute or so?

Well, you might say, taking a step could still be probabilistic, but the probability could be very, very high. All right, suppose the probability of a successful step to be 999,999 in a million. You would expect to fall only once in a million steps. At 75 steps per minute, a million steps would take 13,333 minutes or 222 hours. Suppose you are pretty sedentary and walk only one hour per day (counting *all* walking) on the average. Then you should expect to fall once in 222 days. Is that your experience? Do you fall—not from interferences such as an unseen object underfoot, but from inexplicable misfunctioning—once or twice a year? Do you know anyone—any physically normal adult—who does? I don't fall that often even from stumbling over something—and even at my age. Indeed, if my leg were to fail to carry through the intent of my step, I would not think, "Oh, well, that's just the normal unpredictability of behavior." On the contrary, I would hasten to my physician.

From my own experience (and yours, I believe) it is obvious to me that we walk with much better odds than a million to one that we will take the next step successfully. When the probability is that great, it seems to me to be stretching things a great deal to claim that behavior is intrinsically shot through with only probabilistic regularities. It seems to me much simpler and more reasonable to suppose that walking is under very precise control and is not probabilistic at all. Environmental events that can interfere with walking, it is true, *are* probabilistic; we cannot know when a dog will dart between our feet or when we will step into an unseen hole in a meadow. But the management of walking, one foot after another, is highly controlled and successful almost without exception.

Tom Bourbon made some similar calculations some years ago to illustrate the tenacity and success with which humans can control the perceptions needed to negotiating the highways safely. On the CSGnet on 17 October 1994, in answer to a communication from Martin Taylor, Bourbon wrote:

> To summarize . . . reference perceptions might change from moment to moment. [Sometimes] forgotten, is the opposite possibility: reference perceptions *might not* change from moment to moment. Some reference perceptions might be stable over long periods of time and over many instances of control. . . . Here is an example of what I mean.
>
> At a CSG meeting a few years ago, I described some data on driving in the USA. The following are more up-to-date versions of those data. Even if some of the estimated values are off by factors of 2 or 3 or 10, the results will not change very much.

Year	Miles driven in USA in millions	Deaths	Avg. miles per trip	Trips in millions
1980	1,111,596	23,000	8.69 (est.)	127,916.7
1991	1,548,589	18,500	8.69 (est.)	178,203.6

> That is quite a bit of driving. PCTers often use driving as an example of perceptual control. At least we can't be accused of picking a trivial and unrepresentative example!
>
> How well did American drivers control their respective reference perceptions while they drove all of those miles? The answer will give us an idea of how well we are doing when we model control behavior. As indices of performance, I will use the probability that, during a trip, a person in the automobile became a traffic fatality, and the inverse—the probability that there was no fatality during a trip.

Year	p(fatality/trip)	p(no fatality/trip)
1980	.00000018	.99999982
1991	.000000104	.999999896

> Not bad. Those are the levels of precision people achieve while controlling the many perceptions that must be controlled during a trip in an automobile.
>
> Bill Powers [wrote:] "Incidentally, the '99.99% fit' is becoming a rather annoying exaggeration of the measures we actually get and the criteria we propose. A correlation of 0.996 (about the highest we ever see) corresponds to a mean error of prediction of about 0.03 or so, for a predictive accuracy of about 97%. The range of correlations we propose as being acceptable go with prediction errors from about 10% to about 3%."
>
> Your 99.99% . . . *overstates* the precision we achieve in our modeling, Martin, but it *understates* the precision people achieve in their "real world" control behavior, at least people in the USA when they drive their automobiles. It is obvious to me that we have a long way to go before our modeling comes close to the precision people achieve. It is also obvious that reference perceptions can be much more stable than you have implied and that sometimes our chances of identifying them might be easier than you have implied.

Bodily processes such as the psychophysics of perception, the control of muscular action, and so on, are indeed highly predictable in psychological laboratories the world over. Some psychologists, I suppose, would say that this more physiological kind of behavior (including walking and driving) is inherently more predictable and less probabilistic than the more mental kind of behavior. I would, however, think that explanation to be a modern version of the ancient claim that mind and body function by different laws. Books on psychology do not put that belief into words any more, but I wonder about it when I discover that the research methods in psychological subdisciplines differ in ways easy to see (in the insistence, or not, on random sampling, for example) and when I notice that specialists in one subfield know little about the work in another and show little desire to know more—when they seem to believe that knowledge of a colleague's specialty will not help them in their own. The implication is strong that most psychologists believe that the laws applying to other subdisciplines can safely be ignored—that those laws have very little application to their own subdisciplines.

To say that nature has arranged for one kind of functioning of the human animal to proceed with a random or probabilistic admixture and another kind to proceed in a controlled manner—to say that, it seems to me, violates not only the principle of parsimony. It also violates the principle of evolution that later forms grow from earlier forms, that nothing grows out of the void, that the new builds on the old. If the neocortex is a late evolutionary development, it should use the principles that enabled the older brain to work. To claim that a wholly new principle of functioning can appear in the brain at some point, a principle not built on an old principle, is to claim to construct a brain by supernatural means. It seems to me that we should demand of all varieties of psychology that if a law describes one kind of human behavior, the law should have its counterpart in other kinds of behavior. We should not have separate psychologies for walking, eating, naming colors, gambling, obeying, persuading, teaching, listening to Beethoven, and so on.

It is time for psychologists to give up facile excuses such as "behavior is intrinsically probabilistic" and "my specialty requires its own laws." It is time to return to the fundamental faith of science—to the principle that a function or effect possible or necessary at one place and time must be possible or necessary at all places and all times.

Note that in Bourbon's example and mine, we were not predicting particular actions. We were not saying that we or the walkers or the drivers were demonstrating an ability to predict when anybody was going to take a step or turn the steering wheel or decide to go via Seventh Avenue. We were making estimates of ordinary success in purposeful behavior so as to compare two assumptions: (1) that behavior is typically shot through with unpredictable variability and (2) that purposeful behavior is continuously controlled and, when environmental opportunities for control are sufficient, control is very precise and successful indeed.

Walking and driving seem to be examples of muscular movement toward a goal in which a misstep or a mistwist occurs once in millions or billions of instances. Toward some kinds of goals, steps are more frequently unsuccessful—making a million dollars or becoming a famous movie star. But even then, the variability is not random, and the missteps are due to lack of environmental opportunities as often as to lack of skill or poor choices of action.

Chapter 15

Where's the reality?

In my youth, I read a lot of stuff about what philosophers of science said about psychology and a lot of stuff about what psychologists said about the philosophy of science. I even repeated a lot of that stuff to my students, more's the pity. After some years, I came to realize that all those people, psychologists and philosophers alike, were accepting without question the causal assumptions that physicists used in studying nonliving things—accepting them as suitable assumptions about causation in studying *living* things. And since they accepted them without question, they did not even state them explicitly as assumptions. When I read the 1978 article by W.T. Powers, I discovered what those unuttered assumptions had been and what assumptions were required for living things. No doubt you can guess whose words on assumptions about reality I am going to be quoting in this chapter.

Here is William Powers beginning at the beginning in a contribution he sent to the CSGnet on 22 February 1996. He was writing about how theories lie between imagining and observing:

> All theories begin the same way, as a story about nature that we make up out of our imaginations—a story that *might* be true if nature is arranged as we imagined it to be arranged. At this stage, all theories are created equal: they are equally worthless.

Powers went on to say that Aristotle had a theory (so to speak) that men have more teeth than women. Perhaps there was a lot of argument about that in the agora. Whether anyone bothered to look in a good number of men's and women's mouths and count, no one can know. Anyway, I am sure the people standing about in the agora had more fun arguing the matter than they would have had going around asking to look in people's mouths. You can think up a lot of good arguments, beginning, for example, with the idea that men are superior to women and therefore deserve more teeth.

Powers was saying that a theory is not worth arguing about unless it can be tested against observation. A proponent of an untested or untestable theory might persist, "But it *could* be that way, couldn't it?" The answer, very often, is yes, it *could*—and so could a dozen or a hundred other theories, all equally worthless. You might think this has been said often enough. Often enough, certainly, in book after book, you see this demand for testing theory against the evidence of the senses. But I let Powers repeat it here, once again, for the same reason that Powers wrote it. He was writing to some participants on the CSGnet—professional psychologists and other social scientists—who wanted to argue for their ideas for revising PCT but who did *not* choose to undertake an observable, experimental test of their ideas, or even to design a test for someone else to carry out. They wanted only to argue, "But it *could* be that way, couldn't it?" Designing and carrying out a test of a model is a chore. But until you are prepared to do so, or at least sketch how it might be done, it is courteous to refrain from using up the precious hours of experimenters in badgering them with your idea of how it "*could* be that way."

This is not to say that one should refrain from theorizing. No one can refrain from explaining things to herself. And it is difficult to resist the urge to tell others one's ingenious explanation. It is to that urge, indeed, that we owe literature, art, and science. Powers's point (and mine) is that if your explanation makes a claim about an observable reality, the efficient (and peaceful) way to settle a disagreement about it is to observe the reality. But if your explanation is of something unobservable, then arguments are in principle unresolvable, because all the criteria lie inside the arguers, and there can be no guarantee that the arguers will hold to similar criteria. You can, it is true, sometimes convince another person of your belief by sheer argumentation, but if there is no external criterion holding from day to day, that person can be convinced of an opposite belief tomorrow by someone else.

WHAT CAN WE KNOW?

We cannot know all the ways that the external reality does or does not correspond to our experience. Because of observational tests, we know that there is *some* correspondence, some covariation, between the activity in the brain and the energies that impinge on our sense organs, and some correspondence between neural signals and our experience, but we know only a tiny fraction of what there is to know, and no one yet conceives a strategy for discovering the correspondences between neural activity and subjective experience. Whatever the best strategy, it must be one that uses a model that includes both brain and environment in its terms and is testable by experiment.

Now I will quote from a chapter by Powers (1992a) entitled, "The Epistemology of the Control System Model of Human Behavior"; this excerpt is from pages 226–228:

> In the Drug Decades, with which we have not finished, there grew up an epistemological position that became very popular. It can be stated simply: my reality is as good as your reality. With a little chemical aid, epistemology became laughably easy. What is reality? It is simply whatever pops into your head. The principal effect of the various chemical substances ... seems to have been instant destruction of the capacity to be skeptical about the propositions that constantly occur to any creative person.
>
> I do not think that anyone's reality is as good as anyone else's reality. There are competent and incompetent realities, workable and unworkable realities, realities we can experience and realities we can only imagine. Notice that I do not say there is only one reality. If there is any positive result to be found from the Drug Decades, perhaps it is the widespread realization that people construct many different worlds for themselves out of the raw material with which [they] all begin. But this diversity of worlds implies more than freedom; it also implies a serious problem, a scary problem when you think about it.
>
> Constructing a workable, competent, and above all communicable reality is a task that seems more impossible the more closely one examines what we start with. All the findings of physical and biological science conspire to tell us that we are each alone, isolated inside a nervous system. To live in the world, and more impossibly, to live with other similar (?) people in a world, the brain must solve an unsolvable problem. It, with the rest of the central nervous system, is presented with a world consisting of millions of tiny signals, each telling the state of a single sensory receptor quite independently of the state of any other sensory receptor. This world contains no objects, no relationships, no sequences, no transitions, no logic, no persons, no science, no nothing—nothing but intensity, the report of how much of something is present. From that, the brain must create shoes and ships and sealing wax, not to mention capitalism and communism. Impossible. Science tells us that science is impossible.

Here Powers refers to the hierarchy of control when he mentions intensity, transition, relationship, and sequence. Chapter 18 will be devoted to that topic. Powers is saying that the signal in any single neuron conveys nothing but intensity ("how much of something"), because every single neuron is capable of transmitting nothing more than an electrical pulse, and a pulse in one neuron is indistinguishable from a pulse in any other. The "meaning" of patterns of controlled variables comes about through the hierarchy of control.

When Powers says here that it seems impossible for a brain built of patterns of intensities to produce relationships, logic, or science, one thing he means is that we cannot *deduce* the nature of the neural hierarchy from knowledge of the functioning of the neuron. Powers did not use logic to deduce the hierarchy; he invented it. To say it another way, a brain can produce logic, but logic cannot produce a brain. I will give other warnings about logic in Chapter 25. Powers continues:

> It is obviously not impossible. We have to accept that we have done it. Somehow there are links that allow our realities to converge, both toward each other and toward at least some congruence with physical nonliving nature. Every day mil-

lions of new brains are brought into contact with this network of realities, and immediately begin doing what is necessary to poke, test, modify, hypothesize, revise, and converge toward the common, or roughly common, consensual domain (the felicitous term of Humberto Maturana). No epistemology can make any sense if it ignores these facts of experience. If science leads us to ignore these facts, then science makes no sense.

Action affects the world, whatever it is. Perception represents that world to us, in potentially uncountable ways. But the combination of action and perception quickly rules out great areas of possibility, for when we act, what happens in perception is not what might happen, or should happen, but what does happen. We cannot predetermine that this action shall have that [particular] consequence. We can choose any action we like, and construct perceptions according to any scheme we please, but when we act, it is something else [the external world] that decides what the perceived result will be.

This is why we have to learn how to control what happens to us, instead of just knowing how [at birth].... It isn't necessary to accept any particular scientific or intuitive picture of what lies out there. But whatever it is, our actions affect it, and it affects our perceptions. In that passage of effects out of us and back into us, reality gets into the act.

It gets into the act in two ways. First, there is simply the matter of which action affects which perception. Second, there are disturbances: our perceptions tend to change when we are not acting to control them—in fact, that's the principal reason we have to learn to control them: they won't spontaneously come to the states we want, and they won't stay there unless we keep acting. Without these two factors we would truly have no choice but solipsism....

Wherever we look, there is a boundless world out there, invisible. What we can see of it is limited to the ability of our eyes (and brains) to convert a very narrow band of electromagnetic energy into neural currents. And similarly for our other senses. Whatever exists out there, whether it might be called multiform, immense, or intricate, whether august in unity or tumultuously particulate, we can be aware of the ebb and flow of energies outside us only through some analogs in our neural pulsing. We can have no measure except awe of what remains beyond sensing or analogizing. Beyond our abilities to sense, processes go on that result in the energies we do sense. Here is Powers again, this time in an epistle to the CSGnet on 17 September 1994:

> ... some elements of our theories are not really, in some subtle way, reducible to reports of observations, but are *made up* by human imagination.... the concept of "an electron," for example, amounts to an *imagined observation*, with no justification other than that assuming its existence leads to consistent explanations of experience....

Some scientists know this; others vehemently deny it. Richard Feynman, for example, knew it. When he was asked how he arrived at his diagrams showing particle interactions, he said "I made them up." There were physicists who considered this a flippant answer, consistent with Feynman's reputation as a joker. But Feynman was quite serious. Particle physics, he said, is a game we play. It takes a sense of humor to admit that.

Powers is not, of course, the only person who has ever thought such thoughts about reality. Here is a similar view from Bruce Gregory (1988), writing about physics:

> Physicists cannot "see" quarks or gluons, but quarks and gluons are elements of physical theory because they lead to predictions that physicists *can* see... (p. 183).
>
> *The minute we begin to talk about this world ... it somehow becomes transformed into another world, an interpreted world, a world delimited by language*—a world of trees, houses, cars, quarks, and leptons (p. 183).
>
> The lesson we can draw from the history of physics is that ... *what is real is what we regularly talk about.* ... there is little evidence that we have any idea of what reality looks like from some absolute point of view (p. 184).

There is an unseen world out there with causal connections to what we are able to see. To tell what can happen among the things we can see, we must invent ways that invisible functions could be organized to bring to us what we experience, *and then test* whether those inventions (theories, models) do bring us the experiences.

That is our connection to reality. Such as it is.

REIFYING

We reify. We string some words together and then claim that the string stands for something "real." We believe things to exist where our senses bring us no direct evidence that anything does in fact exist. But even when we have direct sensory evidence, as when our fingers touch a flower, how much can that evidence tell us about what is going on beyond our fingers?

We inherit neural nets that strive to put thingness upon what they perceive. At the "levels" of perception closest to the sensory organs, such as the levels perceiving intensities, sensations, and configurations, we can usually agree rather closely with other people about the sort of perception we are experiencing and *therefore* about some feature of the outside reality we all think is there. For example, put a piece of sandpaper on a smooth table. Close your eyes. Put a fingertip on the table; draw it across the tabletop to the sandpaper, across the sandpaper, and off onto the table on the other side. As your fingertip pulled across the sandpaper, your nerves reported a stimulation that was more intense than the stimulation the nerves were receiving while the fingertip was moving across the smooth table before it encountered the sandpaper. The intensity of the stimulation dropped again as the fingertip left the sandpaper. Now do it again, but this time with a friend who will keep her fingertip just beside yours as the two of you move your fingertips to, across, and off the sandpaper. The two of you, I think, will very likely agree very closely concerning the points at which you perceive the changes in intensity and the direction of the change—stronger or weaker. You will agree that there is "something" lying between those two changes in intensity of feeling at your fingertips. If your eyes are closed, you may not be able to say whether the more intense patch is a rough stone inlaid in the table, a patch of nonslip paint, a piece of sandpaper, or something else.

Now the sensation. The two of you will also agree, I think, that the sensations you feel can reasonably be called smoothness and roughness, not wetness or squeezing or weight or electrical shock. You may not, however, agree on the degree of roughness. Now pull your fingers very fast across the sandpaper. Perhaps you will feel some pain from the abrasion. Perhaps, too, as you do that, the sandpaper will feel considerably smoother (even if more painful) than it did at slow speed. But it will not be easy, I think, to find good agreement among two or several people on the *degree* of this change of sensation.

Now the configuration. Two people can agree very closely as to where the roughness lies between the two smoothnesses. With their eyes closed, however, they may not agree on the breadth of the patch of roughness or its shape.

As we go farther away from our fingertips in attending to our perceptions—from intensity to sensation to configuration and so on—there are usually more ways we can fail to agree with others whose fingertips are presumably having the "same" experience. It becomes much easier to disagree when your sense organs and those of the other person are not exactly in the same place—when your fingertips are six inches away from hers, or when you look at the tail of the elephant and she looks at the trunk, or when you sit in the first row of the theater and she sits in the balcony. It becomes still easier to disagree when the "same" experience is not immediately present but exists only in memory or imagination—if, for example, you drew your fingers across the table on Monday and she did it on Thursday, or if you saw the performance of "Hamlet" on Monday and she saw the "same" play on Thursday, or if you visited Italy last year and she visited there ten years earlier, or if you read about experiments with rats in a book by Jones and she read about some in a book by Abercrombie. It becomes even easier than that to disagree when you are talking about something you never did experience with your senses or something you never can: (a) phlogiston, or the ether once postulated as the medium for light waves, or the electrical fluid once postulated for the nature of electricity, (b) the hierarchical levels of control, or (c) democracy, group, the administration, god, reward, anticipation, stress, aggression, intelligence, or eliciting a response.

The concepts labeled (a) were at one time postulated as real things but since discovered to be poor guesses. They were used as scientific explanations in the sense that at least some scientists of the time thought it would be a good bet to pursue their studies as if those "things" actually did exist. But later they came to believe that no "thing" having the postulated properties could after all exist. Such postulations are normal and essential to science. Their key quality is that they are testable by observation or are intended to be so.

The concepts specifying the hierarchical levels of control, labeled (b), are also intended to be testable by observation. The feasibility of hierarchical levels has had some tests; see, for example, the section headed

"Hierarchy" in Chapter 8; for a fuller description, see Chapter 18. At present, however, many features of the idea of levels remain untested experimentally.

The concepts labeled (c) are not testable by observation. If by "group," for example, you mean merely a collection of individual persons, then you can have the same kind of perception of it as you have of a bowl of apples; you can reach out and touch every apple or every person. But if you mean that a group has an existence *beyond* mere aggregation, if you are postulating a group mind or a national character or a corporate spirit, you are postulating something that is intangible and something which, as far as I can imagine, you can never expect or hope to touch or otherwise sense.

As another example, you might say a reward is an opportunity in the environment that the person enjoys using. I know that you will have your own preferred definition of the word, and that's fine with me. Here I want only to emphasize the internal location of the good feeling—the idea that the satisfaction, the rewardingness, lies within the person, not in the object used. One man's meat is another man's poison. And so on. The words "reward," "punishment," and "reinforcement" are not serviceable as technical terms. They may be handy in informal conversation, but are often misleading, even there.

Religious mystics have told us that it is possible (after a sufficiently long and grueling course of spiritual exercises such as putting every thought out of your mind and maintaining a state of complete mental emptiness) to experience God directly. This experience, however, does not employ senses that enable the observers of God to compare their experiences and be confident that they have met the "same" external entity existing beyond their imaginations. There is no way that two mystics can have the confidence that they are talking about the "same" God in the same way that two people can be confident that they are talking about the "same" piece of sandpaper. They cannot know that they are feeling the same god or the same internal emptiness of thought. The result is that the mystics are in the same position in regard to God as the rest of us are in regard to reward, democracy, and group. They and we, in that respect, are no better off than Saint Thomas Aquinas (c. 1225–1274), who felt that he had to fall back on sheer reason. According to Vernon J. Bourke in Paul Edwards's *Encyclopedia of Philosophy* (1967, vol. 8, p. 110), Aquinas wrote in his *Boethii de Trinitate* that a use of philosophy was to demonstrate . . . those things that are proved about God by natural processes of reasoning; that God exists, that God is one. . . .

Aquinas was saying, I think, that we cannot show God to one another; we can only reason (and argue) about God. I put all historical figures in this category of intangibles, too. You have never experienced Napoleon in the flesh, and you never will. In this way, all historical figures, despite all evidence about their lives, are mythical.

I am not saying, by the way, that it is foolish to believe that some intangible thing exists (or did at one time or could in the future). All of us often find it very useful to our various purposes to believe that something exists despite the lack of palpable evidence. And note that I decline to chop logic about the meaning of "exist."

I have now laid out three ways that some perceived "things" can lie at various "distances" from direct sensory perception. Let us review them.

First, every "thing" is, indeed, a perception. We "know" nothing except perceptions.

Second, even with the most direct sensory perception, we know only some aspects of things. We can see only a limited range of brightness (intensity) and color. We can hear only a limited range of loudness (intensity) and pitch. And because you and I perceive things from different viewpoints, at different times, and with different actual sense organs, it is very unlikely that the two of us (or any other two) can ever form the same internal image of the same external thing, even a piece of sandpaper.

Third, because we cannot know by observation all the aspects of anything, we cannot know by direct observation how those aspects are interdependent. We cannot know how to predict weight from bulk if we do not know something about density of the object. If we try to find "laws" simply by observing how frequently directly observable aspects go together (as seems to be the preferred strategy in almost all psychology and other social science), we will always find only partial going-togetherness, and today's togetherness will not be tomorrow's. A Chicagoan or Parisian may wear trousers every day of his adult life, but tomorrow may visit Scotland and wear a kilt. (I wrote about the "Requisites for a Particular Act" in Chapter 1 and gave a brief list in Chapter 5.) We cannot tell what is going on in domains we *cannot* sense but which can affect what we do sense.

We cannot describe reality. We can describe a *model* of some part of reality. Bruce Gregory (1988, p. 116) quoted the eminent physicist Neils Bohr: "It is wrong to think that the task of physics is to find out how nature is. Physics concerns only what we can *say* about nature."

It is always prudent to remember that facts and data do not speak for themselves. They can signify only what you can conceive. If you believe in "creation science," then the schist in the lowest walls of the Grand Canyon of the Colorado River might not seem to you a fact for geologists to theorize about, but simply God's preference in mural decoration. If you believe that purposes do not exist, then you are not likely to see the fateful consequences of acts, because you will not look for them. An act will not seem to be accomplishing anything; it will simply be there, like the schist.

There is more about reality under "Reality" in Chapter 24.

In sum, don't fool yourself with ideas that sound as if they are about the tangible world but have no hope of verification by observation. Speculation about the way ideas fit together is often great fun, but you will never be assured that you can make a working model with those ideas until you try it
.

A CAUTION

The concepts I labeled "(c)" are not the stuff of science in the modern sense of the word. Terms of that sort, however, certainly were the stuff of science in the days of Saint Thomas Aquinas. The old usage of our term "science" followed from the use of the Latin *scientia* for knowledge. For Saint Thomas, I suppose, the speculations of philosophers were just as much "science" as test tubes are for today's chemists. For him, it was a natural process of reasoning to "prove" that God "exists" and to do so by the "natural processes of reasoning" *and nothing else*. That faith in proving by words and reason that something exists has not vanished. It has been passed on from parent to child over the centuries and is still being passed on. Its transmission has been interrupted among only a fraction of our population by courses in physics or chemistry in high school or college, and a good many people have sat through those courses without noticing the interruption.

Concepts such as those I labeled "(c)," whatever language you may find them in, are reified by millions of people every day. (I do it, too, as I'll try not to remind you too often.) Those concepts and thousands of others, however, are such that no one can think of any way to see, touch, smell, taste, shiver at, get dizzy from, or otherwise sense the "thing" or idea conceived. Such concepts will rarely, I believe, be useful in building a scientific theory of psychology.

Chapter 16

Explaining other theories

How does PCT explain the phenomenon of reinforcement, by which a "response" is presumed to be "strengthened? To answer that I will take a little excursion to use the analogy of the Ptolemaic universe.

Some thousands of years ago, astronomers noticed that all the stars moved across the sky in a fixed arrangement except for a very few. Those few would move along with the fixed stars for a while but then slow and move in the reverse direction for a while. Soon, however, they would reverse again and go along with the rest. Only a certain few stars did that. The ancient Greeks called those stars wanderers. The astronomer Ptolemy, living in the second century A.D., apparently believed, like most people at the time, that the sun, moon, and stars revolved around the earth. But then how could those wanderers be so erratic?

Ptolemy worked out a geometric model, with the earth at the center of circular orbits, that could explain how those planets could move regularly, all the time, but appear to move irregularly. He used three clever tricks to fit his model to what he observed. His best-known trick was to assume that the planets did not stay right on their circular orbits, but revolved around a point which, in turn, moved along the orbit. The small circle moving along the large circle of the orbit is called the epicycle. I'll give you an analogy. Suppose a friend of yours ties a rock to a string and swings it in a horizontal circle above his head. Now suppose he walks in a circle around you, all the time whirling the rock (slowly) around his head. His circular path around you is his orbit. The circle of the rock around his head is the epicycle. As you watch your friend, your gaze turns steadily in the direction he is walking. But if you watch the *rock*, (the planet), you will be moving your head (or eyes) one way for a while and then the other way. Imagine that the rock is a small light and that you are watching it in a pitch-black night. The light, like the planet, would move first one way and then the other, but would overall make steady progress in one direction—along its orbit. That was an ingenious model.

Later, Nicolaus Copernicus (1473–1543) came along with a model that assumed the sun at the center. That idea was an important advance for astronomers; it made most of the epicycles unnecessary. The idea was, however, a shocking idea to most people for many reasons, including religious ones. Galileo (1564–1642) was later to suffer interrogation and house arrest by the Roman church for advocating Copernicus's view, but Copernicus himself was already on his deathbed when his book *De Revolutionibus Orbium Coelestium* was published. Then a contemporary of Galileo, Johannes Kepler (1571–1630), figured out how to do away with any vestige of the epicycles by putting the sun at one of the two focuses of an ellipse.

Let us be fanciful here and imagine a conversation between a Ptolemaic astronomer and a Keplerian astronomer:

Ptolemaic astronomer:

Well, now, how does this new theory of yours explain why planets travel in epicycles?

Keplerian astronomer:

It doesn't explain why they do that.

P: Well, I guess your theory doesn't amount to much if it can't deal with an important problem like that.

K: Our theory says epicycles don't exist; they are illusions.

P: Illusions! What I can see with my own eyes is an illusion? When I can look up there night after night, and plot the backward motion of a planet with my own eyes, as can anyone else, you call that an illusion?

K: We think it looks like that not just because of the motion of the planet, but because of the motions of both planet and earth.

P: Motion of the earth? Motion of the earth! Did I hear right? What nonsense! If I were you, I'd be careful where I say things like that!

Similarly, when PCT people are asked to explain how reinforcement comes about, they say they don't explain it, because it doesn't come about; it is an illusion. It looks that way more because of what the experimenter does than because of what the rat does.

EXPLAINING TRADITIONAL CONCEPTS

Proponents of PCT are often asked by professors of psychology how PCT explains one or another concept in traditional theories—aggression, conditioned stimulus, conflict, ego, groupthink, hypnosis, identity, insanity, intelligence, language, learning, motivation, personality, reinforcement, and so on. Many of the traditional concepts look very different from the viewpoint of PCT—motivation, for example. Some of the traditional concepts seem to PCTers to be names for sheer fancies and illusions—stimulus and response, for example. Because the world looks so different under the assumptions of PCT than under the assumptions of linear and sequential causation, I have not tried to organize this book the way traditional texts in psychology are organized. Instead, I simply started from what I thought was the bottom and wrote up and out. I have had the advantage, of course, that I have not had to explain dozens of disconnected theories—only one.

We should not demand that a new theory explain how old ideas fit into it. Often, an old idea is irrelevant. How does a whiffletree fit into an automobile? It doesn't fit into it; an automobile has no use for one. How does a topsail fit onto an airplane? How does a miasma fit into the germ theory of disease? It is not the case that a good theory should explain every idea that was cherished by older theories; sometimes the phenomena didn't actually exist, but were simply postulated (and believed) because the old theory seemed to call for them. In the theory of impetus, for example, thrown objects were given an "impetus" that carried them along but gradually faded. When the impetus was exhausted, the object fell straight down. That was believed for a long time, and the word is still a part of our everyday vocabulary. Impetus, however, did not fit into Newton's theories of motion and gravitation. Newton's theories had no explanation for impetus, because impetus did not exist in his theories. Newton boldly ignored any presumed necessity for a cause of continued motion and simply postulated that motion continued until interrupted.

"Nature abhors a vacuum"—that was the way people a few hundred years ago explained the fact that water or air rushed into empty places. How does a theory of gravitation explain nature's abhorrence of a vacuum? It does not explain it, because that abhorrence does not exist in Newton's theory of gravitation. The earth and atmosphere gravitate toward each other, but there is a lot of vacuum (or near-vacuum) out there between the planets.

Lavoisier's theory of burning did not explain how phlogiston works. Einstein's relativity does not explain the workings of the "ether"—the medium in which light waves were once presumed to form.

PCT does not explain how reinforcement or conditioning works. Objects in the environment are of course used in acts taken to control perception, but the objects themselves do not cause acts—they are neither incentives nor stimuli in the sense of sufficient causes. PCT does not explain reinforcement, because PCT does not assume that "stimuli" have the power to move people. On the contrary, PCT postulates that people have the power to move objects. See the section on asymmetry in Chapter 3.

PCT will not tell us which traits will show up most often in factor analyses of answers to questionnaire items. That is like asking which items on a restaurant menu will be chosen most often. That depends on the items offered and the culture from which the diners have drawn their tastes. That remark may shock some readers; I will say more about the topic in Chapter 26.

Acts that look like "learning" to a traditional psychologist look very different to a PCT psychologist; most of them look simply like repeated use of a means of controlling perception. (The repetition, of course, depends on the function of memory.) Another kind of learning occurs when the internal organization of standards undergoes revision; sometimes that reorganization is accompanied by strong emotion See Chapter 20.

In PCT, motivation is not something that someone else does to you. Food, money, sexual copulation, or an "A" grade in school will seem "motivating" only if the person is willing to use one of those things at this moment to pursue some purpose. At another moment, or for another purpose, or for another person, those things will draw no attention at all. On the other hand, every person is always, at every moment, in a state of motivation. There is no such thing as being unmotivated, because hundreds of internal standards, in every person, are making their demands at every moment. You may not be doing what some salesperson, schoolteacher, politician, parent, or priest wants you to do, but you are always striving to do something *you* want to do, even if that is simply sitting still. In PCT, the word "motivation" is no longer a technical term, though it may be handy now and then in nontechnical discussion.

It is not only Freudian theory in which an explanation serves whether the person does one thing or its opposite. J. E. McGrath (1984, pp. 80–81) has reviewed the history of experiments on change of opinion after group discussion. In the 1950s in the United States, most people, including social psychologists, believed that groups were likely to make more conservative decisions than individuals left to themselves, because individuals in a group can avoid personal responsibility for inaction. Later experiments, however, showed that when participants discussed some possible choices presented to them in writing, some of the choices being more risky than others, the average choice after discussion was *more* risky than before. Though that finding was the opposite of the earlier finding, one of the explanations given for the latter finding was the same as before—that group discussion allowed a diffusion of responsibility! McGrath called that "ironic." After one has been reading the psychological literature for fifty years, one becomes wary of "explanations." It is a tremendous relief to have come upon a theory, PCT, that enables one to build models that will mimic very closely actual persons, individual after individual after individual.

Here is one more cautionary tale. William Powers and Bruce Abbott undertook a series of experiments with rats to try to understand better (among other reasons) what psychologists might have been meaning by "rate of bar pressing" on the part of rats. (Researchers often install a bar or lever in a cage which, when pressed, will release a pellet of food into the cage, either promptly or after some delay.) In a communication to the CSGnet of 16 September 1996, Powers tells about the experiments:

> Even our simple video recordings of the behavior of rats . . . belie what people have reported. Simply because we put the lever close to the food cup, we discovered that the rats learn the behavior even when the bar-pressing is concurrent with eating, or precedes it as well as following it. The simple scenario of a behavior occurring and then being reinforced by its consequences is appealing, but we saw that this simple sequence need not occur in order for learning to occur. We also have recordings of rats grooming, exploring their cages, and taking naps while the apparatus is duly recording a steadily declining "rate of pressing." So we know that reports of the effects of various variables on rate of pressing are misleading at best, especially when we know that even prominent experimenters . . . admit that they don't actually observe the rats in their cages during an experiment, but simply divide total presses by total elapsed time to get rate of pressing.

Also on 16 September 1996, Richard Marken, too, had something to say on this topic:

> I have very strong reservations about giving PCT interpretations of *any* conventionally obtained data. It's OK to do this if the PCT interpretation is just the *start* of a research effort to test . . . for controlled variables, ... but it's really a dead end if it's just a game of coming up with a PCT "just so" story that seems plausible. . . . The problem of explaining behavior is not that it's so hard to do but that it's so EASY. . . . We want people to understand that behavior is the control of perception; and that a correct explanation of behavior cannot be achieved by developing plausible stories about what is seen. . . .

The goal of PCT psychologists is not to explain what other theories explain (successfully or not). Neither is it to predict or to control behavior. Their goal is to explore the varieties of the control of perception, building models that can mimic the complexities of control. But even when you have built models that reproduce human behavior with great precision, you still have to beware of explaining. In a communication to the CSGnet on 4 December 1997, Powers wrote:

> . . . there's a great danger of seeing all behavior through theory-colored glasses, so no matter what happens, you can see it as fitting the theory. I fight that all the time in myself. . . . Once you have a

theoretical interpretation, it's just damned hard to believe that you're forcing the world to look [in your imagination] the way you expect it to look.

People blithely talk about perceptions and error signals and output functions, and forget that all they can really *see* is people doing things in an environment. They can't see another person's perceptions or reference signals or error signals, or any of the functions we postulate inside the control system. What we're doing is imagining that behind what we actually observe is this theoretical structure. This imagination can get so vivid that we think we are actually observing these things, so the theory starts to get more real than direct experience (that happened a long time ago in physics).

To me, a real scientist is a person who does an experiment, . . . finds that his predictions are completely and exactly matched by the results, and immediately says "Wait a minute, something must be wrong here". . . . It's only when every possible flaw in the experiment has been ruled out that a scientist is finally backed into a corner from which he can't escape: "I guess [it] must be right."

The more skeptical you are, the more aware you are of what you can and can't observe, the more clearly you will understand the difference between theory and fact. And paradoxically, the more confidence you will build in your ability to evaluate a theory and its predictions. I have a lot of confidence in PCT precisely because I have tried to face every way in which it could be wrong, every way I could be fooling myself.

PCT does not attempt to explain why people (or animals of any sort) do some particular thing—perform any particular act. The act a person takes depends on both what is inside and what is outside; even if you have ascertained what is inside (the internal standard), which is not easy, you still cannot predict how the person will use the environment to control the perceived quantity—see Chapters 1 and 3. You can use PCT, like any other theory, to "explain" how an action *could* have come about—you can tell a "just so" story about it. (The "just so" label comes from Kipling's 1902 book of fanciful *Just So Stories*.) But let us remind ourselves, as Powers tries to do, that every time we give a plausible explanation, chances are that someone else with some other theory can be just as plausible. What tests the claim is building a model.

EXPLAINING DATA

As well as being asked to explain phenomena implied by other theories, PCT researchers are sometimes asked to explain *data* produced by experiments designed to test other theories. Richard Marken wrote about this contretemps in a message to the CSGnet on 8 April 1999. Let us imagine, he said, that an experimenter has designed an experiment using a version of the coin game described here in Chapter 6 under the heading "The Coin Game," but uses a conventional experimental design. The experiment will use two "stimulus conditions." First, we will have a condition against which the experimental condition is to be compared. In this comparison condition, the coins are laid out like this:

DH NT

NH DT

In the second or experimental condition, the coins are laid out like this:

DH NT

NH DT

where D stands for a dime, N for a nickel, T for tails, and H for heads. As in other versions of the coin game, the experimenter will move the coins, and the controller will say "OK" if the variable the controller is controlling is not disturbed or "Not OK" if it is.

This experimenter, following the canons of conventional experimental design, has chosen an "independent variable"—the position of the coin DH. The two "values" for this variable are (a) its initial position at the corner of the square or (b) shifted to the left, as shown in the diagrams above. The "dependent variable," of course, is the controller's "response," which also has just two values—OK or Not OK. The experimenter predicts (or maybe just discovers) that the controller always says "Not OK" when presented with the experimental stimulus, and "OK" when presented with the comparison stimulus.

Having set up and run this highly successful experiment, the experimenter challenges the PCTer to explain the fall of the data. Well, the PCTer does not want simply to make up a just-so story. The PCTer would want to make a model that would behave the way the human controller has behaved. But whatever model the PCTer would build would almost certainly be wrong!

The PCT model must have in it an analog of the variable the controller is controlling. In the experiments you have read about in this book, the PCTer has usually asked the controller to control a particular variable that can be identified by both experimenter and controller by pointing at a computer screen. "There! You are controlling that distance right there, right?" But in this version of the coin game, the modeler would have to guess—and would have no data that could be used in the manner of The Test when the modeler discovers that the first guess is wrong. There are of course many variables that the controller in this game might be controlling. "The shape of a square" is only one of them. Another might be "keeping DH to the northwest of the others." With that criterion, the second diagram above would elicit an OK, but a reversal of DH and NT, while preserving the square, would elicit a Not OK. And the PCT model would be wrong, wrong, wrong.

In brief, the two theorists not only have different theories, but their theories propose different purposes (different hypotheses) which call for different kinds of data to be collected. The PCT experimenter must hunt for a controlled variable or offer for adoption one the participant welcomes. The PCT experimenter must collect data on what the participant does about disturbances to the presumably controlled variable. The conventional experimenter acknowledges neither of those necessities. Naturally, the data suitable for one are rarely suitable for the other.

It is reasonable to say that if two theories not only seem to imply the same kinds of experimental designs but also produce the same arrays of data, they are the same theory, no matter what their words. But if researchers who espouse those theories produce different experimental designs and produce data that will not suit the designs of the other theory, one is not going to be explainable in the terms of the other.

REVISIONS

I have written here about comparing PCT with other theories. The crucial fact that always makes the comparison unprofitable turns out to be the difference in the core assumptions—circular and simultaneous causation versus linear and episodic causation. But another bootless attempt to make a bridge from conventional theory to PCT is the attempt at hybrid theory—to take some features from PCT and some from another theory and proclaim the assembly to be a theory itself. This kind of mish-mash is done, typically, by writers to whose ears some words or sentences in PCT and some in another theory have similar rings. But of course similar rings is not enough. For the PCTer, the question is always whether a functioning model can be made from the hybrid theory. And no model has yet been made from a hybrid theory; I don't think any hybridizer has ever tried to do so.

The hopelessness of such an effort lies in the same place as the hopelessness of trying to "explain" concepts of other theories via the concepts of PCT; that is, the core assumptions are at odds. I am not saying that PCT should not be revised. Of course it must be revised. But the revision must be done by showing that a new kind of model is more sturdy than the present sort—showing that by actually running the new model. A revision that consists merely of words, no matter how good they sound, is a sham revision. For the same reason that the author of a cookbook does not print recipes for distasteful mixtures, I am not going to tell you here the titles of those hybrid publications. I did tell you in Chapter 4 about the article by Appley; it was a good exhibition of publications that had some hints of the right flavor but were not good enough to serve PCT modeling. Appley's list of hybrid publications is a sufficient documentation.

A lot of books and articles have the word "control" in their titles. Well, you can't tell a book by its cover, and you can't always tell it by its title, either. Psychologists and other social scientists give the word "control" three meanings. One is the common meaning of having firm influence on something—a cluster of meanings such as regulate, direct, dominate, restrain, subdue. The second is its use in the lingo of traditional experimental design, the "control group" being the people the experimenter does not put under the influence intended to bring a certain response. The third is its use in PCT—the control of perception. Actually, PCTers use the word in its common meaning, but they always use it to mean control of *perception*, not of things or people. So a book with "control" in its title might be a traditional book about experimental design, or it might be a book of advice on how to control other people, how to control the quality of a manufactured product, or how to control your temper. I know of a book called "The Psychology of Control" that is about how people control a lot of things, or try to, or want to, and how they feel about it, but it is not at all about the control of perception.

Chapter 17

Beware how anybody writes

This is a book about how humans go about living. It is about the acting, perceiving, and comparing through which people pursue their purposes. I want to show how it is possible for us, whether acting alone or in groups, to achieve our purposes through perception, comparison, and action and to show, too, how the study of human social life can rest on a single theory of individual functioning—perceptual control theory. (Previous texts on PCT written with at least those two goals are those by Powers 1973, Robertson and Powers 1990, and Powers 1998. Valuable compilations of research reports and theoretical comment are those by Powers (1989 and 1992) and by Marken (1990, 1992, and 2002). To carry out my purposes, I might wish I could bring perceptions, comparisons, and actions to you directly. I might wish I could insert certain perceptions into your brain, provide you with certain internal standards against which to compare the perception, and squeeze your muscles into action that would bring the perceptions into closer match, if there is not yet a match, with the internal standards. But I cannot do those things (which I am sure you are relieved to hear). I can give you only words and diagrams and perhaps a little mathematics here and there.

WORDS

The scientific enterprise is a social one. If there is a lonesome investigator somewhere, conducting experiment after experiment, no matter how marvelous, who never tells anyone else about it, the work will never become part of the body of scientific knowledge that serves other scientists as corroboration (or not) and inspiration. The work will be lost to history. Because language is a necessary part of scientific activity, we should think of it as part and parcel of scientific procedure and method, and when we speak of the scientific attitude, we should be thinking not only of attitude toward experimentation and observations, but also of attitude toward the use of language.

Psychologists studying the ways of human functioning will at the same time be thinking about what they will say to colleagues and how they will say it. Most research psychologists, maybe all, hope that others will want to join the way they are going about their research, and most will write articles and give talks to persuade others of the attractiveness of their manner of research and the correctness of their conclusions. To the extent, then, that they are persuasive and others do indeed join in a common direction of research and do adopt common (or similar) beliefs about the implications of the research—to that extent, the communication becomes an integral part of the method of science. And because precise and persuasive communication is so necessary in the progress of any science, I take space here to say a few things about it.

Words do not carry meaning in the same sense that a basket carries apples or a book carries words. Words do not "transmit" meaning in the same sense that a radio transmitter sends words or music to a radio receiver. Meaning does not go from brain to brain in the same way that a ball goes from pitcher to catcher. The ball that reaches the catcher's hand is very much the "same" ball that left the pitcher's hand; the players use it for its agreed function in the game no matter the hand that holds it. But the meaning a reader takes from the words on the paper may be very different from the meaning the writer hoped to transmit. The purpose of the reader may be very different from the purpose of the writer. Either may be quite unaware of the "game" the other is "playing." If I hope you will see in my writing a more fitting view of scientific enterprise than you find in most psychology books, but you are reading to find out how to win friends and influence people, then the meaning you get is bound to be at odds with the meaning I want to send.

A writer can only make squiggles on the paper. The meaning the reader makes out of those squiggles arises from within the reader. The meaning the reader will make is subject to the same Requisites for a Particular Act that I set forth in Chapters 1 and 5; those requisites hold whether the "act" is overt or mental. Readers make use of writings just as they make use of any other features of their environment—to answer their purposes.

Among other sources of difference between the writer's intended meaning and the reader-created meaning, the previous experiences of the two are always potent. One autumn while I was a graduate student at the University of Michigan, a friend arrived from Panama, where she had lived all her life. She, too, was coming to study at the university. As it happened, the first snow of the autumn arrived on the evening when my wife and I expected her to dinner. She arrived bundled in her new fur coat, shivering and fearful. She had read about northern North America, of course, had seen movies about snowbound winters, had been warned by her friends, and had believed all that sufficiently to buy a fur coat. But now the air she was feeling was actually at a temperature below freezing, and that strange frozen water was actually falling out of the sky. But even so, she could not quite believe it. "Is it really going to stay cold like this all winter?" she asked. We assured her it would do so. "But what will they do about classes?" she asked, distraught. "Will they close the university?" All the words she had read and heard, all the pictures, no matter how explicit or graphic, had left her suspecting that the writers and photographers were surely not expecting her to believe what they were literally saying. After all, writers and movie-makers often exaggerate for effect. Surely this below-freezing temperature was a fluke, a momentary lapse, something surely God or somebody would put right in a day or two.

The same thing happens when traditional psychologists read about PCT. The research experiences of traditional and PCT psychologist are radically different. They are separated not merely by vast latitudes of climate, but by incompatible assumptions about causation itself, about the purposes and capabilities of research, and about the uses of mathematics and modeling. It is not surprising that most traditional psychologists seem to have poor reading comprehension when they tell us what they understand a PCT psychologist to be writing about. Quite aside from giving evidence of failing to understand such basic matters as simultaneous circular causation or the difference between controlling action and controlling perception, most traditional psychologists seem not to understand that when we write about constructing a "tangible, functioning model," we are not writing about something you can merely *imagine* yourself touching, but something you *can touch today* with your own physical fingers. Or that we are not writing about an imaginary device that might possibly be imagined to run by itself if we can only get Walt Disney to help us, but an actual computer program that does run by itself in close reproduction of the functioning of an actual, tangible, namable person.

When writing for the reader whose experience encompasses only the traditional research strategies and the traditional research "problems," it is difficult to write in such a way as to offer the reader a clue that there could be something here that he or she had been judging to be impossible—but that now is possible! I sent the manuscript of my 1990 book to 62 publishers before one agreed to publish it. Some of the people from whom the various publishers asked advice said I had voiced once again merely the same old complaints about traditional research methods in psychology. A shining example was the reader who said that I had set forth what was wrong with traditional research methods, but had not said what to do about it. Actually, the book described two grand methods (as promised in large gold letters on the cover): (a) the method of relative frequencies, which is the traditional method resting on statistical inference, and (b) the method of specimens, which looks for invariances both within and among individuals, and which is the method of PCT. I explained in the last 100 pages of the 186-page book what PCT *could* do that the method of relative frequencies could *not* do. And I said on pages 3, 4, 6, 45, and 74 that the explanation would be forthcoming. That reader either skipped pages 3, 4, 6, 45, 74, and 85–186 or ran his eyes over them without being able to comprehend that I was writing about a method radically different from the traditional method and about actual, tangible, functioning models that could not possibly be built on traditional theory.

I do not think that very many of the presumed experts who read my manuscript for those 62 publishers were flagrantly careless or slipshod. I do not think, either, that their reading comprehension would fall below that of an average high-school graduate in respect to traditional psychological writing. I think their previous experiences simply did not provide them with the wherewithal to interpret my sentences in the ways I had hoped. All researchers who have submitted manuscripts about PCT to book publishers or to professional journals have had similar experiences. The tales my PCT colleagues can tell are heart-rending.

This difficulty in communicating an idea that will not fit customary assumptions is not limited to psychologists. When the eminent physicist F. J. Dyson, in his post of editor of *The Physical Review*, was asked how he could tell a crackpot manuscript from a breakthrough in physical theory, he wrote:

> The objection that they are not crazy enough applies to all the attempts which have so far been launched at a radically new theory of elementary particles. It applies especially to crackpots. Most of the crackpot papers which are submitted to *The Physical Review* are rejected, not because it is impossible to understand them, but because it is possible. Those which are impossible to understand are usually published. When the great innovation appears, it will almost certainly be in a muddled, incomplete, and confusing form. To the discoverer himself it will be only half-understood; to everybody else it will be a mystery. For any speculation which does not at first glance look crazy, there is no hope (Dyson, 1958, p. 80).

Transmitting the new meaning is doubly chancy when the writers themselves are still groping for clarity in their own minds. In brief, when precision, exactitude, and clarity are most needed, they will be most difficult to achieve. Because of the pitfalls in the use of language, scientists try, when they are talking to colleagues, to talk as much as possible about what their colleagues can see for themselves. That is one reason I included Chapter 6, "Do It Yourself," in this book.

ASSUMPTIONS AGAIN

One simple sort of ambiguity with language occurs in almost every article one reads in the professional journals of psychology, sociology, and adjacent fields. This difficulty is the vagueness of the writer about the group or class or population of people about whom he or she is claiming to have learned something.

A very large proportion of studies from psychology and other social sciences presents data to exhibit a relation (or several relations) between variables. One simple form of relation is shown by counting the persons who fall into categories of the two variables. Here is a made-up example of numbers of persons falling into one of two categories of variable A and one of two categories of variable B:

		A	
		No	Yes
B	High	16	35
	Low	49	22

Sometimes, instead of assigning only a few categories in each variable, researchers use measurements that produce a great many values (numerals); then the relation is called a correlation. Sometimes the researcher deals with more than two variables; then the researcher examines various combinations of the variables. But the simple example above will serve my purpose here.

Typically, researchers in psychology and other social sciences want to find their data falling into a very strong relation. When they use the simple sort of 2 x 2 tabulation, they want the data to fall like this:

		A	
		No	Yes
B	High	0	57
	Low	65	0

But the data never do that—well, maybe once in a thousand experiments. Typically, the proportions in the cells of the table are more like those I showed in the previous table, where a majority of the cases lie as the researcher hoped, but a sizable minority lie contrary to the researcher's hope.

All or Some?

Often the data fall in proportions that would be unlikely to occur by chance. When that happens, most psychological researchers are happy. They take the pattern of data to mean that they understand something about how people behave. For example, if a researcher believes that people high on B will say Yes to A, and that people low on B will say No to A, then the 35 people in the High-Yes cell and the 49 people in the Low-No cell will encourage the researcher, because those numbers are much higher than the 16 and 22 who fell in the contrary cells. The researcher feels as if he or she is on the right track, and is tempted to write that people high on B will say Yes, whereas people low on B will say No. Indeed, that is what almost all researchers in psychology and some other social sciences *do* write. In almost all that literature, the actual data published show that *some* of the subjects high on one variable are high on the other and *some* of those low on one are low on the other, but not all. Nevertheless, almost no researcher reports those strict facts. Almost no researcher writes *some*; few even write *on the average*. (And the popular writers who report what they read in the scientific journals copy that language.) Here are a few examples of that kind of writing; you can find thousands more in almost any journal in psychology and adjacent fields:

> ... adults who have been successful at a task, and hence momentarily self-accepting, are more willing to give money to a confederate of the experimenter's than those who have not been successful and thus more concerned about their own self-acceptance.

Are *all* adults who have been successful, every one, more willing that any who have not been successful? I doubt it.

> Several studies have shown that an individual's belief and attitude statements can be manipulated by inducing him to role-play, deliver a persuasive communication, or engage in any behavior that would characteristically imply his endorsement of a particular set of beliefs.

Did the studies show that *every* individual was successfully manipulated in all those studies—or in any of them? I would not complain if the author had written "can *sometimes* be manipulated." Here is a prize example from an article about the working conditions of teachers:

> A finding worthy of note is that teachers seem to have constant problems with the quantity of assistance from teacher aides. Over 12 percent of all respondents ... fall into this category.

That author does not seem to notice the glaring contradiction in what he wrote. The "respondents" were of course teachers, so he is talking about the same class of person in both those sentences. In the first, he says that "teachers" have constant problems, and in the second, he says that "over 12 percent" of them have. You could hardly have a more glaring example of writing as if all the subjects behaved like the small fraction who actually did behave that way.

Writing about variables and leaving out any mention of people also implies that everybody showed the behavior:

> Information inputs are more effective in bringing about arousal (and commanding attention) if they are, among other things, intense; spatially extensive; moving; changing; novel; heterogeneous; repeated a few times; contrasting in color, pattern, or in other ways; or complex, i.e., high in information content.

Do those information inputs always, with every subject under any condition, bring about "more effective ... arousal"? I doubt it. The data permit us to say that *some* people have been observed to behave in such-and-such a manner and even that the distribution of the observations between groups would be unlikely to occur only by chance. The data do not, however, permit us to write as if that "more effective arousal" occurred in every subject.

Typically, when psychologists write as if the subjects fell only into the two diagonally opposite cells of the table when only a majority of them did so, they typically go on to conclude that the pattern of those heavy cells signals a "trend" or "tendency" and that the data in the light cells are somehow erroneous. The correct conclusion, it seems to me, is what sampling theory tells us—that with random sampling in both conditions of A, we should expect to get, in future sampling, the same proportions of the B categories that we got this time.

Stereotypy

Finding that some people who are A (or do A) are also B (or do B) and then concluding that *all* people who are A (or do A) are B (or do B) is called stereotyping. Clearly, the reasoning I have described above, in which researchers find that some category of people do so-and-so and then conclude that all people of that category do so-and-so (or "tend" to do so-and-so) fits the definition of stereotypy. Many studies of stereotypy have been conducted by psychologists; every such study I have read used stereotypy in interpreting the data.

Many psychologists cope with the embarrassment of finding cases (data from subjects) in the cells where they had hoped to see none by saying that the subjects *tended* to act in the manner indicated by the larger numbers of subjects in the heavy cells. The idea seems to be that those data are pushed out of the correct cells by unforeseen influences, and that, but for those unforeseen influences, the data *would* have fallen in the correct cells. As far as I can see, that assumption is no more justifiable than assuming the converse—that the data in the light cells are correct, and the data in the heavy cells are in error. Indeed, without any convincing theory to the contrary, the best way to interpret the disproportionate fall of data in such a table is the conclusion dictated by sampling theory—that we should expect future collections of data (from randomly selected subjects) to fall in those same proportions in the light and heavy cells. Be that as it may, to say that the cells containing the majority of the data contain the *correct* data—to say that with no more reason than the disproportion—is, it seems to me, very little different from taking a vote among the subjects on what should be declared to be the truth. Such conclusion is stereotypy, because the researcher observes *some* persons acting in a particular way and concludes that *every* person "tends" to act that way.

It May Be True

Finally, do take the author at his or her word when you read statements about what "may" be true:

> Voice quality, posture, gestures, and handwriting are all quite individualistic and may reveal aspects of personality.
>
> A general lack of eye contact may be indicative of serious emotional disturbance.

Both those statements are true. *Every* statement about what *may* be true is true. Used in this sense, "may" can imply no more than "may or may not." Whenever I read a research report telling me that something "may" be the case, my immediate reaction is that I already knew that. Research is necessary to discover *whether* something *is* true, but not to discover whether it *may* be true.

Most researchers are aware, even painfully aware, that the proportions in the contingency table will change unpredictably as the years go by, and even turn upside down. All psychological researchers are keenly aware of the "unknown variables" that lie in ambush. They have learned, therefore, to be cautious in the way they report their findings. No careful psychologist would ever write, "There! That proves my hypothesis!" or "There! Now you know that my theory is the right one!" or "There! Nobody should waste any more time on that other hypothesis." So they tell us that their findings are "statistically significant," which means that more of the data fell the way they predicted than one should expect to happen by chance. That, in turn, is a probabilistic statement allowing you to say, reasonably, "The gambling odds are in my favor that this pattern of data could happen again." But when they want to write less technically than that, you can see that they are reduced to saying that it "may" be true that the data are reliable. They are hoping, of course, that something, preferably the cause they had in mind, caused the data to fall the way they did. But there are all those pitfalls, and professional custom requires the psychologists to write only that what they wanted to happen "may" have happened. I applaud the caution. The reasons for being cautious about claiming what one can learn from the traditional methods are indeed many and cogent. Where the traditional methods are used in the search for the modes of functioning of living creatures, I adopt the same caution. When I read that such-and-such "may" be the case, I say to myself, "Yes, it may, and then again it may not."

But let me also say a word about the example of making eye contact. The author did not say *to whom* a low degree of eye contact was indicative of serious emotional disturbance. Maybe to psychiatrists, maybe to Americans in general, I don't know. In some Asian countries, however, a *high* degree of eye contact is considered boorish, perhaps hostile, perhaps even "indicative of" serious emotional disturbance.

SPECULATION AND TANGIBILITY

Without some external reality that you and I, both of us, can see and touch, we can argue for an eternity about the nature of the world and life and have no criterion by which to resolve our disparate opinions. Without an external criterion that is uninfluenced by our desires, prejudices, and myopias, our disagreement can be resolved only by persuasion—by one of us saying, "Well, all right, I guess you're right." But even if one of us is indeed convinced beyond any remnant reservation, the community at large remains to be convinced, and similar arguments within pairs of people can go on unceasingly. In realms of thought that do not appeal to empirical test—realms such as religion and political policy—arguments have indeed gone on without resolution for all of history. Every culture, for example, deals with the problem of domestic tranquility in its own way, but the same arguments about the best way to achieve it have rotated around the same assumptions concerning human nature since time immemorial. When a disagreement is appealed only to an authority, there can be no hope of eventual general resolution, because an authority in Argentina is not necessarily an authority in Burma, nor does an authority today necessarily remain one next year.

'Scientists' wish to reach agreement about what they see is also the reason that scientists, when they use words with one another, like to refer frequently to observables. And it is the reason they like to help one another keep separate in their minds the things that are tangible and the things that are imaginary. Phlogiston was imaginary, but the chemists (some of them, anyway) kept trying to find a way to see phlogiston or at least detect it by tangible means. The result was that Priestley discovered oxygen.

Because scientists want to stick close to observables and to speculations that might conceivably show the way to observables, scientists turn to mathematics and models. Here is what Powers said in a posting to the CSGnet on 10 April 1999:

> What's missing from psychology (and allied areas) is the discipline of mathematics, or even the desire for it. What mathematics gives to us is a way to state our assumptions and our methods of reasoning so precisely that our conclusions no longer depend on what we want to be true. This is particularly clear in the field of modeling and simulation. Once you've set up a simulation and started it running, your private beliefs and wishes have no further effect on the outcome. You've created an autonomous entity that runs by itself, behaving because of its own organization and that of its environment. Even if you have grossly misinterpreted the meaning of your own model, the simulation behaves exactly as such a model must behave, and not as you believe it must behave.

> The simulation is far stricter than any human critic could be. A human critic can share your mistaken beliefs and tell you, in error, that you are right, or mistakenly disbelieve you and wrongly say you are wrong, or give you the benefit of the doubt and tell you you're on the right track, keep up the good work, when you are far down a blind alley. The simulation can do none of these things. It behaves exactly as the organization you gave it must behave, and it's strictly up to you to learn from it why it behaved that way. It doesn't care one way or the other whether you understand what it did. It neither reveals nor conceals what makes it work as it does. It just works.

> People who use simulations to test their ideas eventually take the attitude that they don't really care whether the ideas are right or wrong. They learn that anticipating rightness or wrongness is utterly pointless. If the simulation behaves as expected that is gratifying, but if it doesn't that is edifying and indeed promises more entertainment than if one had anticipated correctly. What wouldn't a physicist give to find that his prediction that an object would fall to the ground was disproved? Alpha Centauri, here we come!

> But to take this attitude toward right and wrong predictions, one must have a way of finding out if predictions were right or wrong, a way that doesn't depend on persuasion, emotional pressure, looking at things from just the right "perspective," insight, or being strongly convinced. Mathematics and simulation, coupled to experiment, are that way.

When you read presumably scientific writing, therefore, look for the mathematics and the models. Look, too, at the speculations. Look for the things that exist, so far, only in the author's mind. When they are talking only about imagined things, authors should signal the fact to you with words like *assume, conceive, conclude, guess, hypothesize, imagine, postulate, speculate, suppose, theorize*. When they talk about observable things (or think they are doing so), they should tell you how to see or touch them.

Some people confuse tangibles and intangibles with such abandon that you might think you were in the fantasy section of the bookstore. You can find that kind of wild confusion every now and then in the presumably sober pages of professional journals. Here is a prize example from a journal for business executives:

The laws of nature are merely boundaries. And they've been violated before. Like the law that one body cannot occupy two places at the same time. In essence, that law was violated thousands of years ago, with the first letter. One's thoughts could be in Athens, and in Corinth, at the same time.... For letters, phones, telegraphs, and TVs to be invented someone had to believe that even the laws of nature could, in effect, be broken.

I wonder what the people in the engineering department thought of that article.

MODELING

Comparing the actions of a model (an actual, tangible, operating model) with the actions of an actual person enables us to test the usefulness of a theory. When you hear or read a claim about human nature—a claim that everyone must necessarily behave in a certain way—ask this question (if not of the speaker, then at least of yourself): Has anyone built a model that behaves in that way? Only about three decades ago, this criterion (of actually building an operating model) would have been too exacting; the means for building such a model did not exist. But now that speedy computers are widely available, actual models are being built, and it is only reasonable to ask that psychological theorists (and serious writers and speakers whatever their august titles) be asked to show that their theories can actually *work* by building a model that does.

In what I just said, I was still asking a little too much. Any rule so strict as one prohibiting you from talking about some feature of human nature unless you can build a model of it—any such rule is *too* restrictive. Such a rule would discourage creative imagination. I should have said that speculation about human nature should be accompanied at least by a design for a model. Without that, and if it violates the underlying principles of operating models that *do* work, the speculation should be put forward very gingerly.

I want, too, to make plain what I meant by demanding a model of anyone who claims that everyone *must* behave in a certain way. If the speaker is merely saying that some people, even a great many, do sometimes act in a certain way, I do not demand a model. Such a statement is almost always true. Most kinds of action that anybody can think of are performed by some people somewhere, no matter how angelic or demonic. But if the speaker means to say that every last human is constrained by his or her human nature to perform certain acts whenever the environment permits, I believe that kind of specification has been untrue since the beginning of life, and the burden of scientific proof is on the speaker.

READ WITH CARE

As you read, ask, "Oh? Just how can that happen? What has to be the case for that to happen? For what perceivable things or events might this word stand?" When you come upon a sentence saying that something is "indicative of" something else, ask, "To whom is it indicative?" And so on.

For the research community at large, a vital phase of any research is the report of the research. For all of us who were not present during the conduct of the study, the researcher's words and numbers are all we have to go by. It seems to me that writers of research reports should take excruciating care in telling what happened and what readers might reasonably expect to happen elsewhere. If writers and readers are both insistently careful, we can all make it harder for us to fool ourselves.

Readers, whoever they may be, will read with assumptions that are to some degree not the writer's assumptions. Some will hope for help with problems that are to some degree not the writer's problems. Some will want clarified some ideas that the writer thought were already clear enough. Others will find some ideas already very clear when halfway through the writer's explanation. Every good writer tries to address, as best he or she is able, the readiness of potential readers. But no author can mesh perfectly his or her writing to your desires. Now and then, indeed, you will come upon an author who does not know how to try. Do not suppose, therefore, that the author is necessarily answering the question you have in mind. Ask repeatedly, therefore, "Are you, dear author, talking about this, or that? Are you talking about something tangible, or intangible? All or some? Who? When? May be or is?"

In my early years as a teacher, I did a lot of lecturing. As I talked with students, I wondered more and more what sorts of meanings they were carrying away from my lectures. Once, at the end of a lecture, I asked the students to write on an unsigned piece of paper what they thought the main point of my lecture had been and to drop the paper, as they left the room, in a box I had put beside the door. (I stayed at a distance from the box.) Later, reading over what they wrote, I discovered that only two or three seemed to have picked out what I thought was my main point. Most of the students wrote something I had said, all right, but something I did not think was my main point. One wrote a statement that was the exact *opposite* of my main point. Some wrote nothing about the content of the lecture, but instead told me their feelings about the course, such as "I like this course" or "I hope you talk about something interesting pretty soon."

Considering the hazards of human communication, I will not be surprised or deeply disappointed if you miss some of what I think are my main points. But I will certainly be deeply disappointed if you turn one of my main points upside down.

Part IV
Hierarchies of Purpose

We do many things that enable us to do other things. Usually, we care more about those other things than about the enabling things; that is, we usually care less about how we get to a goal than about getting to the goal. I get in my car because it offers me a way to get to 33rd and Donald Streets. (Using my car is not the only way I can get to 33rd and Donald Streets. I could take a bus or hike. Or if I were going with Claire, we could go in her car.) Let's suppose I want to go to 33rd and Donald because there is a theater there showing a movie I very much want to see. (I would be willing to see it at some other theater, but the theater at 33rd and Donald seems to be the only one showing it.) When I arrive in my car at 33rd and Donald, I look for a place to park, because I believe I would not be successful in getting my body safely into the theater without first stopping the car. I buy a ticket because I believe someone will eject me from the theater if I do not.

And of course there are actions that enable me to perform those other enabling actions. I go out of the house to get into my car because my car is outside, not in the house, and, unlike a dog, it will not come to me if I call it. I stand up before walking out of the house, because I am more skillful at walking when I do it upright. (I could crawl to the car, but it would be slower, and I might damage the knees of my trousers.) To rise from my chair, I lean forward to center my weight over my feet, and then straighten my hip-joints and spine. (I could ask Claire to lift me into an upright position, but she would only laugh.) To straighten my hip-joints, I contract various muscles of which I am ignorant. And to keep my balance, I continuously activate groups of muscles, some acting against others to maintain a continuous dynamic counteraction to the effects of gravity. And I do that continuous delicate balancing even though I am entirely ignorant of the way my nerves and muscles and glands go about doing it.

To activate my muscles in a way that succeeds in getting me out of my chair and balanced on my feet—and across the room and into the car and off to 33rd and Donald and into my seat in the theater—my brain continuously responds to neural signals from my eyes, ear, fingertip, and so on by comparing those perceptual signals with internal reference signals that represent what I want to be perceiving—my leg muscles pulling, my posture upright, the distance to the door shortening, and so on.

One can think of controls of very small local variables such as the concentrations of chemicals at the synapses of nerves. Such controls enable controls of greater scope to succeed—the movements of muscles, the attainment of postures, and so on. And one can think of purposes that require control of variables that in turn require many sorts of acts over considerable periods of time. Why do I want to see that movie? That is, what might seeing that movie enable me to perceive? You can think of various possibilities. Perhaps it is chiefly the artistry of the production that I want to perceive and enjoy. Or perhaps I am a movie buff and want to expand my acquaintance with the repertoires of the actors. If I am myself a movie maker, I might want to become informed about the doings of a competitor. Or perhaps I perceive myself as a person of artistic sensibilities and judge this movie to be a movie my kind of person should be able to talk about.

As I mentioned in earlier chapters (not, I hope, ad nauseam) living creatures maintain stable perceptions by means of varying actions. The perception maintained can be not only a constant value of a variable, but also a constant rate of change of a variable, or a constant rate of change of a rate of change, or a constantly repeated pattern of change (such as a sine wave), and so on. So a more encompassing statement would be that living creatures maintain a constant rate of change (including zero) or pattern of change in perceived variables by varying actions so as to counteract disturbances to those perceived variables. That means that a purpose (or internal standard or reference value) at a "high" level such as a continuing approach to the theater must be maintained by perceptions at "lower" levels that enable that approach to continue—parking at the theater, steering the car to get to that parking lot, getting into the car at home, and soon back to the neural signals to the muscles to get yourself out of the chair, and even back to the neural chemistry.

An indispensable part of PCT is Powers's postulation of "levels" of control organized in a hierarchy. The outputs of the "higher" levels are not neural signals to muscles, but signals to "lower" loops that act as the reference values for those lower loops. I mentioned the hierarchy in Chapter 3 under "A Little Flesh and Blood," in Chapter 4 under "Internal Processing," and in Chapter 9 under "The Neural Hierarchy," where I gave brief descriptions of the first six levels. In Chapter 8, under "Hierarchy," I gave an example of research using more than one level. Here in Part IV, I will give much more detail about the hierarchy of control.

When we perceive ourselves acting in ways that serve as means to ends, we perceive the hierarchy of control. Notice that it is not the outside world that is organized into means and ends. As rocks roll down the mountainside, the result is eventually to reduce the height of the mountain. But they do not roll as the means of achieving the purpose of wearing down the mountain. The rocks just roll. The mountain just dwindles. The idea of eroding as a cause of dwindling lies in our minds.

In what I have said so far, I have illustrated several points:

1 Perceptions at a lower level of perception are necessary for the perceptions at the higher level to exist. The lower level is necessary to the higher, but not the higher to the lower.
2 Control at a higher level is achieved through *varying* the internal standards of the loops at lower levels.
3 Levels are orthogonal. That is, control achieved by loops at one level does not predict the kind of variables that will be controlled at the next higher level.
4 Each new level introduces new degrees of freedom. That is, each higher level introduces further ways the person can make use of the environment to control perceptions.

First is the idea of instrumentality, of means and ends, also called goals and sub-goals. Some actions enable other actions to occur. We forget, by the way, that our own ends can differ from the ends of others. I found this example in an acrostic puzzle:

> In Staffordshire, England, it was reported that the buses no longer stopped at certain hamlets for passengers. Councillor Arthur C. Holerton then made transport history by stating that if these buses stopped to pick up passengers, they would disrupt the time-table.

That there are means to ends is a very old idea. What is new here is the idea that this relationship among acts appears in the same way in the relationship among levels of *control of perception* in human (and other) nervous systems. Also new are the orthogonality and the specification of particular levels or kinds of control, as we shall see later.

The second point is that there are many ways to skin a cat. Actually, I could get to the library by more ways than walking. If I break a leg, I could hop or crawl, but I'd be more likely to use crutches or a wheelchair. I will say more about the third and fourth points in the following chapters of Part IV.

If you want to read more reports of research on levels of control, try Marken and Powers (1989a); Pavloski and others (1990); and Bourbon (1994). For examples of control when intentions change, see Marken (1990), and Bourbon (1994, p. 11). For a somewhat more detailed design for a two-level model for controlling an arm muscle, see Powers (1979a, August, pp. 94–116). For careful observations of control at successively higher levels by infant humans and chimpanzees, see F. X. Plooij (1984, 1990), Plooij and Rijt-Plooij (1989, 1989b, 1990, 1994), or Rijt-Plooij and Plooij (1986, 1987, 1992, 1993), and Vanderijt and Plooij (2003).

Chapter 18

The neural hierarchy

The idea of means and ends, of instrumentality, is an old idea. But how can it work? If we want to build a model that will exhibit the phenomenon of means and ends in the behavior of a living creature, how can we go about it? How can I control, all at the same time, a visualization of arriving at the library, a recognition of the statue of Beethoven that I pass on my way to the library, the proper amounts of movements of the muscles of the neck as I move my visual focus from Beethoven to the library building, the movements of foot, leg, and spine, and so on? How can I alter my internal standards for the walking movements, for example, as I turn at corners, go up steps, and slow to let someone walk in front of me, all the while maintaining a perception of progress toward the library? And how can I do all that while thinking of Claire's dear smile that I will see when I get home and put in her hands the book she asked me to get for her?

You might say, well, let's not get too ambitious. Let's just see if we can build a model that will mimic the lifting of a foot. Gradually, you might say, small bit by small bit, psychologists and others can assemble a model of the whole creature and its trip to the library and to Claire. That strategy, however, if it is strictly limited to a foot and a lifting motion, is doomed to failure. We already know that a foot does not, in normal life, lift without purpose. The purpose, in turn, is carried out by lifts of the foot that must vary in amount, direction, speed, and timing with other movements. The variations that occur do not occur randomly. A particular variation of foot lifting occurs that will serve the purpose in a particular situation and serve it while acting against some particular disturbances to the controls for amount, direction, and speed. In other words, a model of one small motion of the human body cannot be correctly built without having a fair notion of the ways *many motions can be coordinated* in serving a purpose—not the ways they can be coordinated by an experimenter in an unchanging laboratory environment, but by the real, walking person in a normal, changeable environment.

It is true, nevertheless, that we can start small. There is a limit to how small a part we can model without losing the character of control, but we can build models that consist of only a few feedback loops while nevertheless mimicking some simple behavior. I described one such simple model in Chapter 8 under the heading "Hierarchy." (What do I mean by "simple" behavior? Well, I guess I mean behavior that can be mimicked by a simple model.) The point is that even such a simple model as that one in Chapter 8 cannot be made without a design for the way those feedback loops might be connected to larger networks in the whole animal. For example, the reference vectors labeled R(2,1) and R(2,2) in Figure 8–6 of Chapter 8 can appear in that model only because PCT postulates a hierarchy of levels of control. It is true that we do not at present know how to build models of most of the more complex kinds of behavior exhibited by humans. (For an example of a model of one kind of complex behavior, see Powers 1994.) But we must theorize about a larger structure than we are able at present to model, because the larger theorizing will propose connected hypotheses through which we hope eventually to achieve a more comprehensive perception of human functioning. The neural hierarchy that is part of PCT proposes ways for multiple feedback loops to be connected so that complex perceptions can be controlled despite the disturbances all of us continuously encounter.

THE LEVELS

Figure 18–1 shows the neural unit—the basic feedback loop. It is a simplification of Figure 4–1 (which was already a simplification). Its placement on the paper is changed, too; compared to the figure in Chapter 4, the figure here is rotated and then turned with a pancake turner so as to be seen as in a mirror. But just follow the arrows, and you won't get lost. In Figure 18–1, **I** stands for Input, **p** for perceptual signal, **r** for reference signal or internal standard, **C** for Comparator, **e** for error signal, and **O** for Output.

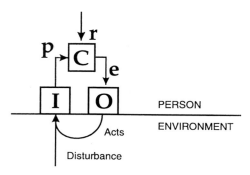

Figure 18–1. The loop

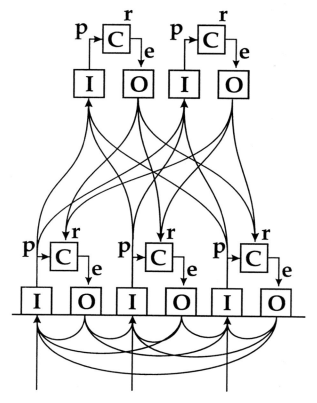

Figure 18–2. Two levels of control

The neural hierarchy is built by connecting many unit loops in a particular way. In the hierarchy, only the "lowest" loops send outputs into the environment. The "higher" loops send their outputs to the comparators of loops at the lower level. Figure 18–2 is an illustration containing only five loops in two levels. Figure 18–2 is a simplification of Figure 8–6; it omits some detail that was necessary to the discussion in Chapter 8. In one way, however, it is more complete than the figure in Chapter 8. There, the loop at the lower right corner of the diagram was incomplete; only the input was shown. Here, that loop is drawn complete. With that exception, all the connections (p, r, and e) among functions (I, C, O, and acts) are the same here as they were in Chapter 8.

The connections among functions are no more haphazard among loops than they are within loops. That is, many connections that are conceivable do not occur. The input from the environment "upward" into the lowest-level loop (through the input function "I") always goes both to the comparator of the loop and to the input functions of some higher-level loops. But that signal going upward out of "I" does *not* go to the input functions of other loops at its own level, it does not skip over the higher-level input function to get to the comparator there, nor does it go directly to outputs anyplace. The outputs from the upper levels (going downward from "O") go to comparators at the next lower level; they do *not* skip levels and do not go directly to muscle fibers or glands for action in the environment.

An input signal into a lower-level loop also reaches upward to the input functions of several loops at the higher level. (In the animal body, "several" can be thousands.) Each comparator at the upper level will then emit an error signal which is the difference between that incoming signal and the reference signal the comparator is getting from a still higher loop. Those error signals combine with others by means of weightings such as those shown in Figure 8–6 by the values at M(i, j, k), and each result of a weighted combination becomes the reference signal for a comparator at the lower level. By this pattern of connection, reference signals at a lower level can be altered by outputs from a higher level. The loop through the upper-level control systems is completed when the outputs from lower-level control systems affect actions in the environment, which in turn affect the energies that will impinge on the sensory receptors, and the input signals go back up to the upper-level loops. So it is that my muscles can vary their combined effects as necessary to hold me upright, swing my legs, maintain a walking pattern, avoid collisions, and eventually get me to the library and thence to Claire's smile. You can see a diagram of hierarchical control only a little more complicated than Figure 8–6 in Powers's article in *Byte* magazine, 1979, volume 4, number 8, beginning on page 94.

Notice that no matter how complex this interconnection may become, no matter how many signals may converge on one input function or on one comparator, those incoming signals are converted into one simple, unidimensional signal before it is passed on into the loop. All "information" flowing along the neurons is unidimensional; that is, it conveys a greater or lesser rate of firing, and that is all. We are often impressed by the complications and subtleties of "meaning" that we discern in action or imaginings; those complications and subtleties lie in the combinations of loops that are activated and therefore the shapes and timings of actions or imaginings that result. The combinations include the input signals from sensors, the combinations of error signals sent down from higher levels, and the combinations of lower loops for which those descending error signals become reference signals. The "meanings" have nothing to do with any sort of shape or configuration of the neural signal itself; the signal along any neuron or bundle of neurons can vary only in its greater or lesser rate of firing. That is the only "behavior" or "message" a neuron can deliver.

Figure 18–3 (from Powers, 1988, p. 278, drawn by Mary Powers) gives us a glimpse of the ways loops are presumed by PCT to be connected in the human nervous system. (Actually, this scheme is presumed for all animals of some complexity, but I don't want to get distracted by speculations about the points in evolutionary lineages at which an animal appears with enough levels of control to be suitably described by Figure 18–3.) Here, for convenience, we see six levels of control; Powers (1998) postulates eleven. For simplicity, only a few of the multitude of possible connections among loops are shown. Between the first two levels, for example, the outputs from only four loops at the second level are shown dropping to four loops in the first level. (Figure 18–2 showed both outputs and inputs between the first two levels.) Between the second and third levels, only inputs are shown. Between the third and fourth levels, and between successively higher pairs of levels, Mary Powers shows some loops connected by both inputs and outputs.

In a communication to the CSGnet of 31 May 1992, William Powers summarized the hierarchy thus:

> [It] is a hierarchy that runs in two directions: a perceptual hierarchy building upward, and a control hierarchy building downward. A given level . . . receives inputs that are copies of perceptual signals of lower order, some under direct control and some uncontrolled. A perceptual function in a specific control system generates a new signal that represents a variable of a new type, derived from lower-level perceptions (or sensors, of course, at the lowest level). A comparator compares the state of this signal with a reference signal received from systems of a higher level. The error signal resulting from the comparison goes to an output function that ends up distributing reference signals to control systems of the next lower level—the same level where the perceptual signals originated. Only the lowest level of outputs generates muscle action. So each level of system acts to match its own perceptual signals to reference signals received from higher levels, and acts by . . . varying reference signals for systems at the next lower level. The result is a hierarchy of goal-seeking and goal-maintaining control systems with many systems at each level and many levels.

So far, I have been drawing your attention to loops in the left part of Figure 18–3; the connections shown there are labeled "behaving" at the top of the figure. Those loops are completed by actions in the environment, as symbolized at the bottom of the figure. At the right side, however, are some connections that are completed within the nervous system by direct neural connection, at some level, from output to input. They are headed "imagining" at the top of the figure. Circuits that do not reach the bottom level have no direct effects in the environment, but they do, of course, have powerful indirect effects. I will describe those modes of functioning later in this chapter and elaborate them in Chapter 19.

Figure 18–3 is like a road map of the United States that shows a few of the roads going into and coming out of a few towns in Massachusetts, and a few roads connecting some of those to a few towns in Indiana, and a few of the roads connecting some of those to a few towns in Oklahoma, and so on, with

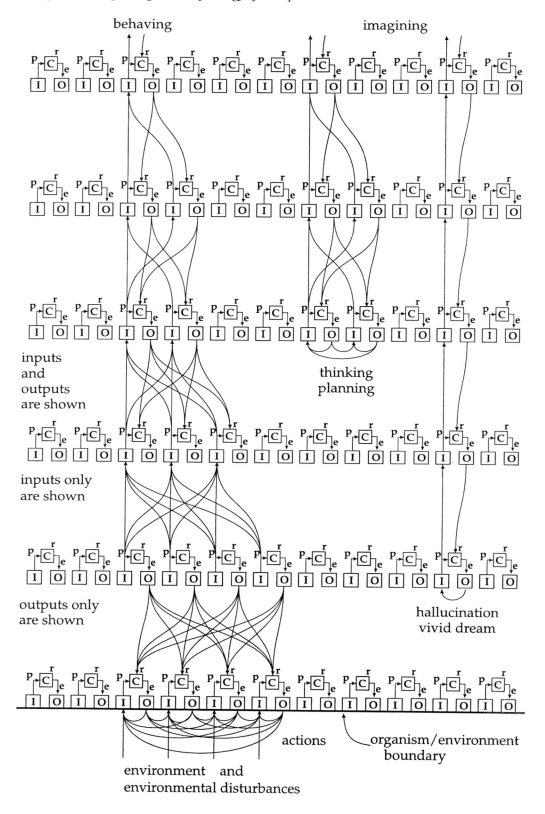

Figure 18–3. The hierarchy of control

the positions of some other towns shown but with no roads to them shown, and all the rest of the map blank. That map would give you a poor picture of the actual connectivity by roads in the United States. And of course there are many times more neural connections in your brain than there are road links in the United States. Look at Figure 18–3 and then close your eyes and imagine those patterns of connection repeated hundreds and thousands of times.

In his 1988 chapter, on pages 273–274, where he is speaking only of the lowest level of the hierarchy, Powers estimates that the human body contains 600 to 800 "small control systems, each of which controls the sensed amount of strain in one tendon." On pages 274–275 he continues:

> Everything a human being does that could be called overt behavior is done by varying the reference signals reaching these systems. *Everything*. . . . the acts involved are all accomplished by varying the reference signals reaching these 600 to 800 first-order control systems. No system higher than first order can act directly on the environment by generating physical forces. The actions of all higher systems consist entirely of generating outgoing neural signals. There are no moving parts on this system above the first level. There are only signals, and systems that receive, manipulate, and generate signals.

An input signal at the lowest level is some function of a physical effect on a nerve ending. That is, the signal is an analog of an energy input from the environment. As light of a certain intensity impinges on a portion of the eye's retina, a neural current of a certain firing rate goes along certain neural paths to loops at higher levels. Our contact with outer reality consists only of neural currents coming in at various firing rates. The rates depend on an external energy, but they are not "pictures" of the external reality. They are only analogs of energy levels. And what we make of those signals occurs in the complex combinations of inputs going upward to "tell" comparators what is happening below and in the complex combinations of outputs going downward to "tell" comparators what to "demand" of the inputs. The environment almost always seems real—right there, visible, tangible, audible. And so it is, but the visibility and the tangibility and the audibility are all there because of the very small part, from among the vast ranges of energies seething in the environment, that reaches and stimulates our sense organs. Our eyes can respond only to light that comes from the front, and they can respond only to a very small fraction of the vast range of the electromagnetic spectrum. Similarly for the other senses. All the rest of what we do with that input at the lowest level occurs in the upper levels that have no direct connection at all with the outside world. When we see red, we are interpreting at the second level a combination of signals of certain rates of impulses from the first level. The experience of looking at a red apple, at a red apple rolling across a table, at a red apple rolling away from our hungry grasp—all that goes on in the higher levels inside our heads. What goes on in the outside world that generates or reflects the energies from which we construct the experience—of that further reality we can know only a few analogs. Those few analogs, nevertheless, are sufficient to fill our world with splendor and horror, from our dreams of heavenly choirs to the agonized blat of a fire engine.

If you hook up an oscilloscope to a telephone line, you can see waving lines on the face of the cathode ray tube that are analogs of the changes in voltage flowing along the wire. But if you put the phone to your ear, your marvelous brain might hear a friend of yours quoting Shakespeare's Henry V: "Once more into the breach, dear friends, once more!" We have learned how to portray the electrical current with the abilities of the oscilloscope; we have not yet learned how to portray it with the abilities of our brains.

Even a few optical bumps in a thin line on a compact disk are sufficient to contain the full liquid glory issuing from the brassy bell of a French horn. I do not find it presumptuous to envision dreams of heaven contained in exquisite balances and shapings among billions of joinings of control loops.

As you look at Figure 18–3, I hope the standard portrayal of the feedback loop—Figure 4–1—will now seem less simple, less clanking, more subtle, and, in multifarious combination, more potent. Think of the reference signals in the hierarchy as changing smoothly and continuously as the myriad perceptual signals flow upward and the error signals downward. Imagine the continuous changes in the reference signal being reflected, instantly, in changes in the controlled perceptions. Our continuously changing behavior reflects the smooth variations of reference signals in the brain.

I turn now to a description of Powers's eleven levels of control.

FIRST ORDER: INTENSITY

Powers (1998, p. 135) says:

> Don't take these levels I propose too seriously. A lot of people talk about them, but few have tried to do any research to see if they're real. I think of them as a useful starting-point for talking about the hierarchy of control; they'll do until something better comes along.

I think we can be confident of the first level, called "intensity." And by the way, I am using "level" and "order" synonymously here. And if you don't like the image of "up" and "down" and would prefer "inward" and "outward" or "away from" and "toward sensory organs" or any other way to indicate ordering, feel free to use your own image. Any word will do that indicates ordering of functions; we are not trying here to describe physical locations in the cranium.

Loops at the first level interpret the magnitude or intensity of an energy level in the environment as a rate of impulses in the perceptual or input nerves. First-order input functions respond to pressure, light, sound, vibration, deep touch, surface touch, balance, taste, chemicals in the air, and the rest. The conversion from external energy to neural impulse rate occurs in the sensor. Although the conversion is no doubt complex chemically, the event is easy to put into words: The outside energy stimulates (that word buries a lot of chemistry) the nerve endings, and an electrical potential builds up in the neuron until the neuron pulses or "fires"—that is, until the electrical potential suddenly runs outward on the neuron's dendrites and, at the synapses with other neurons, sets off the same process in them. Meanwhile, the potential in the sensory neuron is rising again, and will fire again. The greater the energy impinging on the sense organ, the faster the electrical potential will rebuild, and the faster impulses will be transmitted. I am not going to go into any further neurological detail here. You can get more in Powers's 1973 book or in any text on neurology.

A sense receptor sends inward an analog of a particular sort of energy. A sense receptor is a nerve, a bundle of neurons. At the first level, the sense receptor is itself the transducer, the perceptual input function. A pressure receptor, within normal ranges, sends a signal only when it is pressed, not when it is heated or when light falls upon it (though all nerves can be activated by direct electrical or chemical stimulation). A pressure receptor, however, tells you nothing about the cause of the pressure. By itself, it cannot tell you whether you are being cuddled by your sweetheart or sat on by a latecomer in a dark theater. Powers (1973, p. 95) says:

> The perceptual signal from a touch receptor does not reflect whether the cause is an electric current, a touch, or a chemical poisoning.... If any information exists about the source of the stimulus, it exists only distributed over millions of first-order perceptual signals and is explicit in none of them.

The rate of neural impulse gives the organism a representation of what is going on in the outside world. The rate of impulse is a very simple, unidimensional analogy to a tiny part of that world. What is impressive (to me, anyway) is that when we interpret many patterns of such rates coming from many organs sensing sometimes many kinds of energy, we are able to control our simple, unidimensional sensing of that vast outside world very well. We are able not only to check successfully on whether we are putting our feet down in good places to get ourselves to the library, but whether we have imagined a machine that will indeed get us off the ground into the air, whether we have pleased our loved ones, and all sorts of other marvels. Even *Escherichia coli*, functioning without a neural net, is able to check whether the concentration of nutrients is getting yummier.

Much control of intensity goes on without our awareness. Most glandular secretions, most of the time, go on without our even knowing that the glands are there. But they deliver the right amounts of chemicals, not too much, not too little. We balance ourselves as we stand and walk, the semicircular canals in our ears whispering to a lot of muscles (via a lot of levels in the hierarchy) how to keep us from falling over. But we have no awareness of the semicircular canals and most of the time no awareness of the actions of the muscles that keep us erect.

We are aware of many sensations and their intensities. Pain is an example. I cannot say how it is for others, but I almost always find myself less interested in the flavor of the pain than in the sheer intensity of it. We are often aware of the intensity of muscle stretch and tension—"effort." We are very aware of the intensity of light when it is too bright or too dim, and similarly with sound. One way you can focus your attention on intensity is by comparing the intensities coming to you from two different sense modalities. Imagine that I ask you to sit beside a hi-fi.

I turn the volume knob, and the sound of a trumpet blares forth. Then I hand you a red piece of paper. I ask you to turn the volume knob one way or the other until the intensity of the trumpet sound is the same as the intensity of the red. That may sound strange at first, but I believe anybody can do that.

Figure 18–3 gives a hint of the immense complexity of the human nervous system, but I want to emphasize how very oversimplified the diagrams in this book must necessarily be. I'll do that by mentioning the number of light-sensitive cells to be found in the human retina. Dember and Warm (1979, p. 220) say there are about "130.5 million light-sensitive retinal cells." And Hilgard, Atkinson, and Atkinson (1975, p. 112) say that "more than 6 million cones and 100 million rods are distributed . . . throughout the retina." I don't know whether those two statements are contradictory, and if they are, I don't know who is more accurate, but I would be awed even if the correct figure were merely one million. When you are looking at something blue, second-order loops are sending you that message by combining assortments of signals from hundreds of thousands, perhaps millions of first-order cells.

Some people are trying to build devices that can see in the same way that humans see. If you want to look into the complexities of such a project, I urge you to read the dissertation by Rupert Young (2000). His book makes strong use of PCT.

SECOND ORDER: SENSATION

Because each second-order signal is built from an assortment of many first-order signals, the second-order signal cannot be an analog of any physical effect; it is an analog of a *derived* quantity—a quantity that cannot be found in the outside world—and it may, indeed, sometimes be a quantity for which there is no correspondence in the outside world. In vision, for example, most colors we perceive are *analogs* of bands of electromagnetic wavelengths. But no physical instrument can detect our experience of color; it can detect only electromagnetic wavelengths. When you look at an open sky, you do not experience a wide display of 480 magnificent nanometers; you experience *blue*. Indeed, some of the colors we experience (red, reddish-purple, and bluish-purple) do not even correspond to wavelengths (see, for example, Hilgard, Atkinson, and Atkinson, 1975, plate at p. 114). There are no physical counterparts of those colors; you might say that those colors are "all in our heads." That is correct in the sense that we "make" something in our brains that has no counterpart in the outside world. It is incorrect in the sense that we cannot (except in memory) make the color unless some particular *mixture* of wavelengths is first perceived at the first level. And those colors appear *just as real* as the colors that do correspond to particular wavelengths.

An input function at the second level receives inputs from numerous first-order inputs and combines them to deliver a single signal to its comparator and to forward that same single signal upward to higher levels. It is possible that the input function combines signals in a fairly simple algebraic way. To illustrate, I will put into my own words an example from Powers (1973, p. 149). Imagine two input signals rising from the first level; call those two intensity signals i_1 and i_2. Two input functions can function is such a way as to make two different sensations from those two intensity signals even though *both* the input functions for sensation receive *both* the intensity inputs. Let us construct the input function of the first sensation loop so that it multiplies i_1 by three and then also i_2 by three, and adds the results together. That is, the sensation signal S_1 will be produced by combining the intensity signals thus:

$$S_1 = 3i_1 + 3i_2.$$

Then let us construct the input function of the second sensation loop so that it operates with a different rule:

$$S_2 = -3i_1 + 3i_2.$$

Each sensation signal S_1 and S_2 depends on (makes use of) *both* intensity signals. But changes in the intensity signals affect the sensation signals in different ways. Suppose both intensity signals i_1 and i_2 are increased by the same amount. Then, by the formula for S_1, both terms at the right of the equal sign will grow larger, and therefore their sum, S_1, will grow larger. But S_2 will remain unchanged, because, while i_2 grows larger, the term containing i_1, since it is multiplied by the negative number –3, will become smaller by the same amount.

Now suppose i_1 is increased while i_2 is decreased by the same amount. When you trace those effects, you find that now S_2 will decrease, because the negative term i_1 will increase (get more negative) while the positive term i_2 decreases its positiveness (also gets more negative). And S_1 will stay the same, because while one term increases, the other decreases by the same amount. Thus we see that the composition of the sensations need not be given by the sources or strengths of the intensity signals. The compositions of sensations are determined by the weightings (such as +3 and –3) given in the summations accomplished by the input functions at the sensation level (the weightings need not be equal). This example, though simple, shows how a loop at a higher level can receive exactly the same signals as another loop but nevertheless construct its own sensation without any regard to the sensation constructed by the other loop. The two sensations can be independent regardless of the fact that they depend on the same inputs. I hope this example of causal connections in the "wiring" enables you to see a lot more complication in Figure 18–3, and I hope it sounds much more complicated to you than a claim that a stimulus causes a response.

The mathematics of the example can be expanded to cases as complicated as you wish. In general, for **n** sensory inputs, there is always a set of **n** weighted sums that will create **n** sensations, each of which varies with the strength of the signal from one and only one of the sensory inputs.

It is likely that the method of weighted sums (or some neural analogy to it) is used at the level of sensation, but it is likely that more subtle or complicated manners of combining signals are used at higher levels. Neither Powers nor anyone else has proposed, at the time I am writing, a way of combining signals at the higher levels. (For Powers's comments on combining signals at the first two levels, see Chapters 7 and 8 in his 1973 book.)

As I said in regard to the perception of color, sensations need not correspond directly with any particular physical energy impinging on the sense organs. Here is one of Powers's (1973) examples:

> The taste of fresh lemonade [is] derived from the intensity signals generated by sugar and acid (together with some oil smells). However unitary and real this vector seems, there is no physical entity corresponding to it. . . . the mere intermingling of these components has no special physical effects on anything else, except the person tasting the mixture. . . . This means that we would be much safer in general to speak of sensation-*creating* input functions rather than sensation-*recognizing* functions (pp. 113–114).

THIRD ORDER: CONFIGURATION

We are capable of perceiving patterns or configurations in assemblies of sensations. Powers (1998, pp. 136–137) tells us how he is able to see his computer-mouse, and in doing so gives us one small example of how, over the years, his scrutiny of his own perceptions enabled him to conceive the elegant structure of the neural hierarchy:

> My mouse is a sort of cream color . . ., but as I scan my eye over it I see that there are wide variations in shading and brightness, and even in color. . . . And all around the body of the mouse, there is a blue color that is very different from the color of the mouse—it's the mouse pad on which it rests. There's a light blue color most of the way around, but a much darker blue in the shadow of the mouse. If I look very carefully at the edge where the mouse quits and the mouse pad begins, I don't really see anything—there's no line as in a cartoon, just a place where one color stops and another begins. There's no object to see where the edge is, and it's not a color, either. It's just an impression of edgeness. . . . Basically the mouse ends where a sensation turns from cream into blue. If I put the mouse on a cream-colored background, its edge would be much harder to see. . . . when we perceive an object, that perception couldn't exist if there weren't different sensations of things like color and shading in the visual field. If we analyze [the experience of] any object . . . into components, we don't end up with just more and smaller objects, we end up with a collection of different sensations.

There Powers is emphasizing once again that what can be perceived at a particular level depends on perceptions brought up from lower levels. The corresponding dependence going downwards is that what can be controlled at a particular level depends on what lower levels can control when they receive the outputs (which will act as reference values) from the upper levels. Powers continues on page 138:

> . . . while we can show that object perception depends on sensation perception, we can consciously experience *both* kinds of perception. I can look at the mouse and see a mouse, or I can look at the mouse and see cream and blue. And I can see a cream-and-blue mouse. So awareness isn't restricted to any one level of the hierarchy; we don't experience just the topmost level. We can

experience *any* level of perception, and (within limits) more than one at a time.

Configurations that can be perceived and controlled include visual edges or boundaries between two colors or two textures, visual distances between objects, visual sizes, felt sizes (as when you feel a can of beans in your hand with your eyes closed), juxtaposition of pitches as in a musical chord, phonemes, and nausea. One might take the third-order quintessential perception to be that of *object* or *thing*. At the third level, we can perceive a horse (not the abstract class called horse, but a particular animal, nameless, in front of our eyes), a sawhorse, a saw, and a cluster of black marks on paper having the configuration "saw."

FOURTH ORDER: TRANSITION

At the fourth level we encounter the perception of change. At the third level, we could perceive a gap between A and B, but at the fourth level, we can perceive a movement from A to B. With this we have the experience of time.

Perhaps the quintessential experience at this level is "flicker fusion." When a series of slightly differing still pictures is flashed fast enough on a screen (as is done with the "frames" of a motion-picture film), the eye experiences smooth motions of the configurations in the pictures. But it is possible to flash the pictures too rapidly, in which case you see only a blur. And if the pictures are shown too slowly, you see simply a series of still pictures.

The fourth-order ability gives us the sense of the "same" configuration moving from one place to another, even though, as in the case of the motion-picture frames, the configurations in the frames are entirely distinct and different. Imagine that you are sitting in your back yard, situated at the edge of a jungle in India. A tiger comes into view off to your left, moving to the right across your vision. It passes behind a large tree. For a moment, you can see its head at the right side of the tree and its tail at the left. As the fraction of the tiger at the left gets smaller and the fraction at the right gets larger, you assume that you are seeing fore and aft parts of the same tiger. That's not surprising, if you know that tigers do not change their lengths while passing behind trees. But now the tiger passes behind some large bushes, and after several seconds appears again at the other side of them. Is it really the same tiger? Several seconds would be plenty of time for the tiger to lie down behind the bushes and for another tiger to get onto its feet and stroll beyond the right side of the clump of bushes. I might be confident that it was the same tiger, but if you were more knowledgeable about the population of tigers around your house, you might be doubtful. We do not see motion directly; we see change. We infer motion.

Examples of transitions are floating, turning, rising, dropping, expanding, shrinking, straightening, flowing, rolling, rotating, twisting, increasing, and decreasing. You can hear a sound growing louder or more raucous. You can taste the onslaught of pepper and its gradual fading away. You can feel the affection from your beloved as she hugs you. Notice, however, that the communication of affection depends on motion. If the two of you hold perfectly motionless, maybe even holding your breath, you will feel the communication falling away.

Our ability to perceive transitions enables our ears to give us glissandos, accelerando, diminuendo, vibrato, and diphthongs such as b*oy*, h*ow*l, and *eon*. It enables us to return the handshake of a friend and to enjoy the caresses of a lover. By means of this fourth-order feedback, we can control movements. We can walk, beat cake batter, sing, dance, and keep in harmony as members of a string quartet. But we note that controlling motions is somewhat slower than recognizing configurations. If you are walking on a crowded sidewalk and suddenly notice a skater heading your way, there is a noticeable delay between that visual recognition and the reaction of your muscles to get you out of the way. Control is slower at higher levels and faster at lower levels. We have all experienced the quick reflexes when we are jabbed or hear a sudden loud sound, and we all experienced, too, the long deliberations in which we sometimes indulge to bring our perceptions of morality (for example) back into consonance with our reference values. I will return to this matter of time differences in the section headed "Response Times" near the end of this chapter.

FIFTH ORDER: EVENT

An event has a beginning, a middle, and an end. An event has a unitary feel to it. Events that seem to repeat a pattern have acquired names: bounce (of a ball, for example), opening (a door), taking a seat, step, nod, explosion, collision, slap. Powers conceives the unity of an event to be so seamless as to forestall, somehow, anything else from happening—perceived to be happening—during it. He writes (1990, p. 72):

> The duration may be long or short [but] we make a single package of it. [That] just means that an event is experienced as a single thing: an opera performance is an event, as is the first act, as is the aria which finishes it, as is the trill at the end of a passage. At the event level, these smaller events are as unitary as the largest one.

Like all the kinds of perception, events are constructed inside the person. The outside, physical world is not divided into events (physicists mean something else by their term "space-time events"). Almost all physical variables are continuous and continuously related. And much of our experience is quite without the character of event: watching the traffic go by, looking out over a prairie, watching the second hand go round the face of a clock, sitting beside a river, even reading a book (while you are immersed in it).

The capacity to perceive events shows up under various names. About 1927, for example, the Gestalt psychologist Mme. Zeigarnik reasoned that the phenomenon of visual "closure" ought to have a counterpart in an experience having duration. Just as we seem to "want" an almost-closed curve to be a circle, so we might want an almost-closed event to be a finished one. Mme. Zeigarnik predicted that (on the average) persons who were interrupted before they could finish a task would continue to want "closure" of that task and would remember the task better than tasks they had finished. You would, of course, expect the Zeigarnik effect only when the person cares about the task—that is, when the person adopts the completion of the task as an internal standard, as a goal. To test whether the presumed Zeigarnik effect does happen, you would have to use The Test for the Controlled Quantity to ascertain whether the internal standard is there, and you would have to accept the person's own definition of the goal state (the point of completion), not the experimenter's. A lot of experiments were done on the Zeigarnik effect in the 1930s and 1940s, but I don't think any were done with attention to the internal standard in individuals.

Interrupted glimpses of a moving object give another example of how arbitrarily we can choose events. Remember that tiger we saw skulking from bush to thicket to bush? When do we begin to wonder whether it is the same tiger? When too long a time has stretched since our last glimpse of the tiger, we conclude that the tiger is no longer there. (Somewhere, no doubt, but not *there*.) The event of the tiger's skulking has come to an end. We put closure to the series of sightings. You may say that the landscape has come to match our internal standard for a safe landscape: one with no tiger in it. Or maybe one that has not had a tiger in it for the last thirteen minutes. And you may say that the point at which we stop looking for the tiger depends not solely on the frequency of tiger-sightings, but also on the urgency of our other goals. Considerations such as those, having one mix in my mind and another in yours, will bring me to put "finis" to the event at one point and you at another. But wait! Do we see a tiger there now? And if we do, what do we say? Do we say that the tiger is continuing to appear, or that the tiger is appearing *again*?

When the tiger disappears behind a bush, I am as convinced of its being there as I was when I actually saw it. "What are you doing?" you ask, and I reply, "I am watching a tiger," even though I cannot at that moment actually see it. I am "controlling for" seeing the tiger even though I am not receiving light rays from it and even though I am not getting up to run into the forest to look for it. But later, when the tiger seems no longer to be nearby and I cease watching for it, then I think of the tiger less as part of my current experience and more as part of the potentialities of this forest. When I cease looking for the tiger, I will not think of the tiger as part of my present experience; I will remember it as one of many events that have ended. At home, now, I am not likely to say, "I am watching out for that tiger following me around out there." I will be thinking of the tiger more as a property of that forest out there and less as a property of my current experience. An event will have ended. I am likely to say, "I saw a tiger out there this morning," and, "There is a tiger in that forest."

SIXTH ORDER: RELATIONSHIP

Two people walking side by side are maintaining a spatial relationship—quite aside from any affection they may feel. Spatial relationships are described by beside, in front of, behind, on, in, above, under, toward, around, and so on. Some temporal relationships are before, starting, during, ending, after. Musical relationships form triads, octaves, and other chords, not to speak of rhythms of all sorts and contrasting loudnesses. Other relationships are following, accompanying, confronting, helping, cheering, teaching, echoing. Some of those relationships contain interesting mixtures of sensations, configurations, transitions, and events. Some of those terms, too, stand for relationships noticeably different from this person to that. Relationships can be perceived not only among events, but also, of course, among perceptions at any level.

Any variable—that is, any quantity that you can experience in greater or lesser degree—provides the occasion for a relationship. You can perceive that this person is *taller* than that one. Or that this thing is *more beautiful* than that. Or that this person is *more willing* to take risks than that one. Or that this comedian is *funnier* than that one. Choreographers invent dynamic relationships among dancing bodies.

Relationships are to be found in the mind, not in the outside world. See those two people sitting beside each other? Can you see the "beside" between them? Can you go over there and put your finger on the beside? When they tell you they are about to leave, will you tell them not to forget to take their beside with them? And when one walks off this way and the other off that way, how will you decide when there is no longer a beside between them?

A kind of relationship that captures a great deal of our attention is causation. A large part of this book (and millions of others) is about causation. At first glance, causes and effects seem to abound. You hit a nail on the head to drive it into the wood. You pull on the door to open it. And so on. There may be, in the world outside us, some relationships of the sort we call causations, but it is sometimes not simple to ascertain them. One might think that physicists know all about such things, but they seem to have become curiously diffident about the matter. In Geza Szamosi's (1986) fascinating book on time and space, he quotes (in a footnote on page 175) from a text by Misner, Thorne, and Wheeler:

> Space tells matter how to move. . . . Matter tells space how to curve.

That sounds pretty fanciful. How can space "tell" anything anything? Yet Szamosi explicitly says that he does not want to speak of a "force" between one object and another:

> There is no gravitational attraction between the sun and the planets.

Think of that. It used to be an axiom that there could be no action (causation) at a distance. Now it seems there can be influence ("telling") of some mysterious sort at any distance you like—but no simple causation of a gravitational sort. Perhaps not even of a nail-driving sort?

There are five more levels in Powers's hierarchy yet to be described. But let us declare an intermission here. The hierarchy contains some important features or implications that you may have noticed, and I don't want to put them off too long.

DEGREES OF FREEDOM

Two distinctive features of Powers's (1998) hierarchy are these:

> A higher level of perception depends on the existence of perceptions of a lower level and can't exist without them. And to control a higher level of perception, we must *vary* perceptions at a lower level (pp. 139–140);
>
> . . . a perception at a given level can be a function of perceptions at *any lower level* (p. 141).

A third feature emerges from the first two. It is that the levels are orthogonal; that is, *the variation of a signal at one level does not predetermine the variation of a signal at the next level*. The perceptual signal sent to the comparator (and also upward to the next level) at a given level is constructed by the input function at that level. I gave an illustration of orthogonality in the section on sensation earlier in this chapter. In that simple illustration, input functions at the level of sensations made different sensation signals from the same two inputs rising from the level of intensities. Furthermore, the sensation signal constructed by one loop at that level was entirely unaffected by the signal constructed by the other loop. The example showed that the signals constructed at the sensation level were orthogonal to those rising from the intensity level, and it showed that signals constructed by the input functions at the sensation level were independent of each other.

Here is another example of orthogonality. When you type at a keyboard, the carriage or the cursor gives you at each position the choice of typing a letter or a blank. Whether you do one or the other puts no restriction on where you will choose to type a blank and therefore to end a word. And whether you choose to end words at one place or another puts no restriction on the sentences you will choose to make with the words. Each "level" in this process is unrestricted by—orthogonal to—the level below.

My reason for bringing up the matter of degrees of freedom is to point out that the orthogonality between Powers's levels provides us a model of a creature who is free to make what he or she will (so to speak) at every level of perception. The outside world does not dictate (through stimuli) the variables the person can choose to control, and cannot dictate the portions of the outside world the person will act upon to maintain the chosen variables (despite the hopes of the behaviorists). The successive levels of perception in Powers's model do not successively constrain possibilities; on the contrary, they successively widen them. When you consider the immense variety of ways humans make use of their environments (despite the immense variety of ways they try to persuade one another toward conformity) and the immense variety of ways they have invented to think about their variety, it is clear that a realistic model must, as Powers's does, expand the degrees of freedom as the layers rise.

I will let Powers give you his picture of the way the hierarchy can multiply the possibilities for control. This is from a communication to the CSGnet on 22 February 1996:

> ... there are typically multiple control systems at one level acting by mapping their error signals [that is, output signals] onto reference signals of multiple systems at the next lower level. So the net reference signal received by any one lower-level system is really the sum of effects of error signals in many higher systems. As disturbances come and go, the higher systems vary their outputs, and thus their contributions to many lower-level reference signals. There is no simple connection between a single higher-level goal or error and the set of lower-level goals used to accomplish it or correct the error.
>
> The result is that we see different systems at the lower level coming into action as the environment changes. At all times, the reference signals at the lower level are just those that will satisfy the requirements of all the higher-level systems at once. [A]ll the lower-level goals are shifting at the same time, changes in each lower goal simultaneously helping to keep several higher-level perceptions at their own varying reference levels.
>
> What the [PCT] model does is show us how a collection of simple independent control systems can create behavior that looks very complex and adaptive, when in fact no one active element of the model is either complex or adaptive. ... what seems to be a very complex pattern of behavior might be understandable in terms of basically simple components, each behaving in a simple way.

EVOLUTION

You can't make something out of nothing. And neither can evolution. New things are made from old things. A fair number of people, including some otherwise competent scientists, have claimed that natural selection and evolution could not have brought about the species of animals and plants we see around us, because they are too astonishingly complex to have come about "by chance"—even a single organ (the eye is a favorite example) is too marvelous to have come about by chance. The fact is, however, that no evolutionist, not Darwin, not Wallace, not anyone before them or after them, ever claimed that any complex structure, any organ or plant or animal, ever came about through chance agglomerations.

Evolutionists say that simple structures (such as the unicellular prokaryote, poor thing, without even a membrane around its nucleus) were now and then produced with minute differences that gave some of them a slight advantage over the previous standard issue (so to speak) of the species. The slight advantage might be not an eye, but merely a small spot to a slight degree more sensitive to changes in the intensity of light than other members of the species possessed. The new ones with the slight advantage then multiplied, while progeny with other, less advantageous new features, as well as those of the standard issue, reproduced in fewer numbers. At a later time, some progeny appeared with a further slight advantage, and so on.

Dobzhansky and his co-authors (1977, p. 377) say that the earliest evidence of prokaryotes is about 3.3 billion years old, and cells with membranes around their nuclei—the eukaryotes—appeared perhaps 1,500 million years later. About the prokaryotes, they say:

> The most significant single event of this evolution [of the prokaryotes] was the origin of photosynthesis. ... The complex, sophisticated kind of photosynthesis carried out by blue-green algae [a late species of prokaryote] must have been preceded by a whole series of more simple, primitive methods of autrophy [manufacturing their own food] (p. 377).

About the eukaryotes, Dawkins (1987, p. 176) says:

> An especially important event . . . took place at the origin of the so-called eukaryotic cell. Eukaryotic cells include all cells except those of bacteria. . . . [Eukaryotes] differ from bacteria mainly in that [eukaryotic] cells have discrete little mini-cells inside them. These include the nucleus. . . . [One increasingly favored theory] is that mitochondria and chloroplasts, and a few other structures inside cells, are each descended from bacteria. The eukaryotic cell was formed, perhaps 2 billion years ago, when several kinds of bacteria joined forces [so to speak] because of the benefits that each could obtain from the others. Over the aeons they have become . . . thoroughly integrated into the cooperative unit that became the eukaryotic cell. . . . It seems that, once the eukaryotic cell had been invented [so to speak], a whole new range of designs became possible. . . . cells could manufacture large bodies comprising many billions of cells.

You can see in those quotations the tiny, minute changes, becoming slowly, slowly, slowly established among the prokaryotes living in the primordial soup as millions upon millions of years went by. You can see that the new forms were built from the old forms. The old forms were not jettisoned, but provided the capabilities used by the new forms to acquire still further capabilities. The more complex capabilities more quickly (or less slowly!) found more flexible ways of making use of the environment, and evolution accelerated. Multicellular organisms appeared. Dobzhansky and colleagues (1977, p. 377) say that "the period when evolution was exclusively at the prokaryote level was twice as long as that encompassing the entire evolution of multicellular eukaryotes [including humankind]."

Even then, in those days unimaginably long ago, those cells were keeping themselves intact with negative feedback circuits. Beardsley (1994) reported in the *Scientific American* that molecular biologists are beginning to trace such a circuit in the single cell:

> [Researchers] have started to discern . . . one of the cell's principal control mechanisms: a chain of molecular reactions that conveys signals from the cell's surface into the depths of the nucleus. There the signals empower the genes, which change the cell's shape, its activity, or its growth. "We are starting to understand the molecular circuitry of the cell," comments Robert A. Weinberg of the Massachusetts Institute of Technology (p. 28).

Any model of a living creature, especially a living creature as complicated as a human, should reflect the evolution of complex species. Evolution does not make new forms from nothing. It builds them by modifying old forms and reorganizing old functions. We saw in the quotation above that in the very earliest times, new forms and new organizations of function came about when bacteria joined together and, gradually finding new capabilities in their new organization, became the new and fateful division of living creatures we call the eukaryotes—the division from which arose every later creature. Every living creature shows in its physical structure evidence of that evolutionary strategy of building more complex structures and functions from simpler.

A model of the functioning of humans, if it showed no organization of function in which more complex capabilities are derived from simpler—if, in other words, it showed no layering of increasing compass and flexibility in maintaining its necessary inputs from the environment—such a model would be seriously suspect. That shaping of new functions by combining old forms in new ways is the process, put into effect by natural selection, through which the "higher" forms of life have come into being. Not only have the forms of flesh and bones come about that way, but the organization of the nervous system, too. A moment's reflection on the vastly different perceptual capabilities of *Escherichia coli* and the human is enough to convince anyone, I think, that we should expect to find a hierarchy of perceptual capability in the human neural organization.

Powers's hierarchy, as far as I am aware (after reading the psychological literature and some of the literature in related fields for half a century), is the only psychological theory ever proposed that has the necessary characteristics to match the facts of evolution with the obvious capabilities of the higher forms of life:

1. A higher level of perception depends on the existence of perceptions of a lower level and cannot exist without them. To control a higher level of perception, we vary perceptions at a lower level.
2. A perception at a given level can be a function of perceptions of any lower level.

3   The levels are orthogonal. Each new level is not merely a larger agglomeration of existing capabilities; instead, it adds a new view of the world. It interprets (so to speak) the multiple perceptual inputs from below in ways unconstrained by the interpretations at the lower levels.

4   One theory, one sort of model, serves as the base for modeling every sort of human behavior from the knee reflex to the eye's vision, to thinking, emotion, intrapsychic conflict, and social competition and conflict. It unifies the traditional subfields of psychological study by means of a multilayered model of a multipotent living creature.

5   The model permits still further layers of capability to come about through natural selection. That is, the model would not be contradicted or disabled if we were to observe humans developing more levels of perception than those listed in these chapters.

Powers (1990h, p. 81) adds this:

> What matters here is the beginning of thinking about human organization in a way which begins to cover all the things people do and experience. What matters is that we understand just what it is we have to account for with a model of human nature. And what matters most is that we stop ... giving ourselves [as researchers or onlookers] some special place to stand or special abilities denied to others, stop taking for granted the very abilities which make a human being or an animal interesting as a whole organism.

For more on negative feedback in earliest evolution, see Powers (1983a, 1995).

SEVENTH ORDER: CATEGORY

English has many words denoting category, including class, group, set, kind, sort, type, denomination, collection, batch, bundle, and cluster. The language has many words denoting categories of more particular categories: color, grade, league, flock, gaggle, crowd, clan, crew, and so on. We use those words to tell ourselves and others to perceive categories. The relationship among members of a category is that of belonging to it or being characterized by it.

The great advantage of categorizing is that it enables you to act toward any member of the category, for certain purposes, in the same way you act toward any other member. You can go to the paint store, buy another can of Russet No. 37, and go home and finish painting your wagon. You don't care *which* can of russet the clerk gives you. For your purpose, you might not care whether you get No. 37 in one gallon can or in four quart cans. Categorizing enables you to look for a sign that the object or event has the characteristics that will suit your purpose. If it says "Russet No. 37" on it, you expect to be happy to paint your wagon with it. But of course you can categorize without using language—and so can chimpanzees, dogs, and robins. If a fruit is elongated, somewhat curved, and has a yellow skin except for some yummy-looking brown spots, any chimpanzee and I will expect it to taste like a banana. Categorizing saves me from biting into fruit after fruit until I come upon one that will taste right with my Soaky-Oaties and milk.

Categorizing is one of the requirements for logic. Categorizing enables us to say things like, "If all members of class A are also members of class B, and if all members of class A are also member of class C, then some members of class B are also members of class C." But categorizing, unlike logical classes, need not be mutually exclusive. A category can exist in some degree. Greenish can be a category, and objects can vary in how greenish they are. Categorizing is necessary to language, too; for example, language cannot have plurals without the ability to categorize. How would you like to have a unique name distinguishing every dog, every tree, every hair on your head?

It is important not to get mixed up between perceptions and the words that designate them. Perceiving the word "dog" is not the same as perceiving a dog. Naming a lot of animals with the single word "dog" is a separate act of control from perceiving that two of those animals seem similar—maybe they both have tongues hanging out. And what I remember seeing, when I say the word "dog," is not the same as what you remember seeing. I will say more about language in Chapter 32.

We could not count without the idea of a category. If someone tells us to count, we ask, "Count what?" The "what" asks for a name of a category. (When we say that a young child is learning to count, we often mean only that the child is learning to recite the numbers in the customary order. The child is not counting *something* until she pairs the numbers with some objects.) In counting, we are treating all the members in the same way—associating the next ordinal number with every dog we come across.

We distinguish trees from bushes and those from grasses. We make categories of plants versus animals. Of stone, frame, and grass houses. Of dark-skinned and light-skinned people. Of males and females. Those are examples of categories fairly close to sensory experience. We make other categories very far from sensory experience: gods, democracy, socialism, personality, intelligence, excellence, and romance, as examples. Other categories seem to me to lie between: valor, femininity, persistence, and corporation. People may ask, about every one of those terms and a thousand others, "But what *is* it really?" The question reifies. The question implies that there is some arrangement in the reality beyond our senses that corresponds to a category we have put a name on. The question seems to imply that God or Nature has packaged reality in things *and has categorized them*, and that it is reasonable for us to ask whether we have guessed the right category when we say "bush," "corporation," or "socialism." But it is not reasonable.

One more thing about categories. This is the level of perception at which you can perceive things as either-or. You can make mutually exclusive classes. You can say things like, "People can be divided into two classes: those who divide everything into two classes and those who do not."

EIGHTH ORDER: SEQUENCE

Things can be perceived to occur in an order, as do the words in this sentence. Sequences enable us to conceive the ordered class of the natural numbers running on forever: 1, 2, 3, etc. Cooking recipes are organized in sequences of actions. When you remember how to do something, you must remember an order. Open the door before you get into the car. Get in the car before you turn on the ignition. Turn on the ignition before you put the gearshift in reverse. And so on. Controlling a sequence is like singing a melody. Once you start it, the notes you will sing are foreordained. You can stop before the end, you can start later than the beginning, but if you change the order of the notes, you will no longer be perceiving that sequence.

NINTH ORDER: PROGRAM

A program contains alternative sequences. It is like a path with branches. It has steps at which you make a choice, depending on what you want next. The recipe for cake says, "Now, if you want white frosting, do this; if you want chocolate frosting, do that." The instructions for registering for classes at the university say, "If you want to major in English, do this; if physics, do that. At the level of programs, rationality arrives. Powers (1998, p 148) uses the example of long division:

> There are no instructions for what numbers to write down in what order. What you write down depends on the numbers you are given. There are rules that say what to do if one number is larger or smaller than another, but you don't know what the action will be until you see the numbers that develop as the program runs.

When you perceive a program, you perceive implications and possibilities. If *this* is the case, you say, then I will do *that*. But if something else is the case, then I will do that other thing. If I do that other thing, then I will be able to choose among a, b, and c. If I go to the grocery store first and then to the clothing store and the hardware store, I might get delayed, and the milk would get two or three hours closer to souring. At that choice point, therefore, I go off first to the clothing and hardware stores.

Some people are skilled at finding their way through immensely complicated programs; the mathematicians who proved Fermat's last theorem are examples. So are the teams of architects, engineers, lawyers, purchasing agents, superintendents, and others who build large structures such as skyscrapers and oceangoing oil tankers.

The ability to invent, to concoct images of things that exist only in imagination, is a wondrous and powerful skill. It enables us to pursue our purposes by using the physical environment with protean ingenuity. If we keep testing our imaginings against our sensory perceptions (as a tailor keeps testing his measurements against the actual fit he sees between the partial garment and the bulges of my body), we can succeed in constructing programs that other people, too, can use as reliable guides in dealing with the tangible, material world. You, too, can learn to bake a cake, build an Empire State building, or construct a model in a computer that will track a cursor just the way I track it.

To use language, we must be able to imagine words not yet in our mouths, syntax not yet assembled, and topics of our own choosing. Controls for such imaginings must be completed through wholly internal loops. The right side of figure 18–3 shows some connections among levels of the hierarchy labeled "imagining." Those connections are of the sort that complete the feedback loops without reaching down to the first level, out through the environment, and back. I'll turn to that part of the figure in Chapter 20.

For emphasis, I note here that reasoning and logic are not at the top of the hierarchy. Reasoning serves the purposes of principles and system concepts. Powers (1990, p. 78) says,

> While programs can be extremely complex . . ., they have a fixed structure. . . . the range of possible pathways from one choice-point to another is completely determined. . . . This predetermination, however, doesn't mean that a program must always do the same thing after it is set up. . . . If the perceptions constituting the inputs to the program change, different branches will be taken at the next choice-point, and because lower-order perceptions continue to figure into the process at every stage, each following step will involve choices that are just as unpredictable as the perceived world.

It is obvious that a program established to suit one purpose—such as getting to the grocery store or getting undressed at bedtime—will fail if the person's relationship to the environment changes in such a way that the choices in the program no longer cope with the disturbances. If you move to another city with a different pattern of one-way streets, you will have to revise your program for getting to the grocery store. If you go up in a rocket ship into a trajectory where you no longer feel the gravity of the earth, you will have to form a new bedtime program. Programs cannot be the top level of control. Principles concerned with getting food and sleeping will alter the reference signals of the programs. And system concepts such as your beliefs about human nutrition, your conception of the economic functioning of your society, and your beliefs about the functions of sleeping—such system concepts will alter the reference signals of the principles.

TENTH ORDER: PRINCIPLE

Control at the level of principles has not at this writing been mimicked by specific computer programs. Creating, altering, and judging the worth of programs—that is, having perceptions *about* programs—must go beyond programs themselves to the level of principles. The program level is where we think, where we use logic. The level of principles is where we have thoughts about thoughts. The program level is where we manipulate mathematical symbols. The principle level is where we think about how we do mathematics—about what manipulations are possible or properly conceived. The level of system concept is where we envisage mathematics in its totality as an intellectual system.

In pursuing a purpose, a principle can be a general strategy that provides an advantage over a less guided use of environmental or mental resources, without specifying any particular step. Heuristics go here. A principle in chess, for example, is "control of the center." Beginners are beguiled by individual "good moves." Players with more experience seek to maintain the possibility of moving, or threatening to move, to the center of the board, from which a greater number of moves are then possible than from the edges of the board. Another example: A principle in making a comfortable journey of several days is to allow adequate sleeping time. Another: A principle in writing a comprehensible essay to is to keep in mind the audience for it.

Also at this level are matters included in textbooks with titles that begin "Principles of. . .". Newton's laws of motion, chemical equations for forming compounds from elements, the effects of motions of the earth's crust, double-entry bookkeeping, the development of a theme in music, the forming of cyclones—those and similar conceptions can be controlled as principles. I am not saying here that the cyclone (for example) is a principle, but that the *conception* that there is such a thing is a principle. The principle of the cyclone is not the moving air, but the conception that there are motions in the atmosphere that take the shape of vast rotating masses moving across the face of the earth.

Still another kind of principle is the moral principle. Moral principles set reference signals for great numbers of our programs. Principles of honesty restrict the programs through which we acquire possessions. There are principles of kindness, helpfulness, resourcefulness, frugality, caution, dependability, cooperation, and on and on. Many people take moral principles to be absolute categories. Some people seem to put a personal principle, such as rectitude or machismo, right up there with system concepts. They encounter frequent conflicts. In a communication to the CSGnet on 8 September 1999, Powers wrote this:

> I think even a little thought will tell us that we all vary the degree to which we adhere to some principles according to the situation. . . . If a child shows you a scribble and says "I drew an elephant," what do you say? Are you completely honest, and say "That doesn't look anything like an elephant"? Of course you don't, because that would be cruel and discouraging. But some people would say that—their picture of themselves requires being strictly honest all of the time, and to them their own self-concept is more important to them than how a mere child feels.
>
> If you're free to vary your principles as a way of maintaining what you consider a good system concept (self, society, and so on), then you will not come into conflict about such things as being brutally honest; you'll just not do it, and be unperturbed if someone criticizes you for lying. When you consider *all* the principles you adhere to as a way of maintaining *all* your system concepts, it's obvious that if you can't use your judgment in applying principles, you're going to get into inner conflict pretty quickly, and probably conflict with others, too. . . . I don't think it would be a very good idea to formulate one set of principles and then always apply them in exactly the same way no matter what the circumstances. That would prevent counteracting disturbances at the system concept level, and effectively prevent that level from working.

Principles have the quality of generality, even abstraction. A principle tells you to be considerate, but does that apply to letting a pedestrian cross in front of your car when you have stopped at a four-way stop and it is your turn to drive through? Does it apply to you when the waiter has spilled the soup into your lap? A principle does not tell you what to do (how to be considerate) when this waiter spills this soup into your lap in this restaurant in the presence of these people when you are wearing these trousers and saying just what you were saying at this moment and so on. What you do might be an example of considerateness in your mind, but in my mind it may have nothing to do with considerateness one way or the other. A particular principle, for every person, becomes shaped through the control of programs in thousands of situations. Your principle for considerateness cannot be exactly mine, and mine cannot be yours, Tom's, Dick's, or Harry's.

ELEVENTH ORDER: SYSTEM CONCEPT

The way principles seem to hang together makes a system concept. The ways varying pitches and qualities of sound can be heard in various durations and rhythms to make up melodies, along with all the subtle modifications such as crescendo, rallentando, accelerando, rubato, and the ways those elements are put together to make songs, sonatas, gavottes, concerti, symphonies, etudes, operas, nocturnes, blues, ragtime, and all the rest—all that, taken as a whole, comes to the system concept we call music. The performers come into the concept, too, and probably the scenery at the opera. As with other perceptions that come to be reference signals, system concepts are individual. They are composed inside each person's head. When a certain sound vibrates our ear drums, you and I may agree that it should be called music. Often everyone you meet will agree to call something music—perhaps an orchestra playing Beethoven's Fifth Symphony or children singing Three Blind Mice. But sometimes I hear a sound that someone else calls music, but I do not. "That's just racket," I say. But both of us are talking about system concepts.

Powers (1998, p. 150) says,

> The system concept is the overriding idea of some organized entity. . . . The science of physics is a grand system built on most carefully crafted principles . . . consistent with each other. Other system concepts, like "self," also grow out of sets of principles, but are seldom as well worked out. . . . Some system concepts are important and lofty, like religions, and others are perfectly mundane, like a bowling league.

Human organizations are system concepts. I am not saying that they are systems in the sense that a wind-up clock or an electrical power network or a human musculature is a system. Those are all tangible things: you can hear the clock ticking, you can feel the wires (just the insulated ones, please), and squeeze the muscles. We conceive those tangible objects as system *concepts*. We conceive them to be organizations of principles concerning gear ratios, electrical potentials and resistances, and balanced muscular tensions. Human organizations, on the other hand, are built from principles that cannot be touched by the human hand: from norms (agreements) concerning the duties attached to positions, norms about social conduct such as arriving on time and not inviting the president to join you for afternoon tea, and norms about borrowing reference values from one another such as the goals the managers would like the workers to adopt. Many norms in the organization are "informal"—they are maintained by word of mouth. Others are formalized in written bureaucratic rules.

The organization, however, is not embodied by a piece of paper with black marks on it that we interpret as rules, and not by the person in the president's chair. Nor does it exist as any collection of persons, pieces of paper, roofs, and bank accounts. It lies in the invisible relationships, categories, sequences, programs, and principles *among* all those things, invisible to the physical eye, but fully "real" to the inward eye—real to the perceptions of the eleventh order. All the ways humans organize themselves in working together and living together are maintained only because the persons involved agree to act *as if* all (or almost all) the other people involved are also going to act that way and only because they can agree closely enough about the kinds of acts that will fit their definitions of "that way." Thus are formed the doughnut shop, the barber shop, the National Biscuit Company, the Girl Scouts, the community dramatics club, the New York Police Department, the League of Women Voters, and the Methodist Church, not to speak of society, culture, and the law. I'll say more about organizations in Parts VII and VIII.

I'll repeat here that system concepts and principles set the internal standards for programs (where logic lies), not the other way round. We are not merely rational creatures. Many professors, judges, computer programmers, and others sometimes talk as if we ought always to be rational in what we think and do. But we are creatures of all the levels. At the level of intensity, we resonate to the blast of the trombone and to the crack of the lightning when it seems to rip open the sky. At the level of sensation, the sculptor invites us to stroke the marble hip. Also at the level of sensation, we lose the words of the orator in our sensual delight in his mellifluous tones. At the level of principle, we study how natural selection can work. And at the level of system concept, we feel ourselves part of the awesome insensate sweep of three and a half billion years of evolution. That our logic serves our desires is not a new idea. None of the capabilities exhibited at Powers's levels is a new idea. What is new is the total structure, as awesome as the evolution that brought it into being, with its principle of orthogonal layers as its glue and its control of perception as its function.

RESPONSE TIMES

Systems higher in the hierarchy must control lower systems by sending their outputs to the reference signals of the lower systems. Consequently, more time is required by feedback loops that include higher levels than by loops that go only through the lower levels. Data are few concerning the times required for the loops to function, and it is difficult (for me, anyway) to understand from the reports just what portion of the loop was measured. I list on the next page the information I could find, with a few words of description for each measurement. I have omitted a few numbers in cases where I could not understand the level of control being described. If you go to the original sources, you may be able to see more than I saw (see table, next page).

You can see that the trend is toward longer times as we go up the levels. There are difficulties in ascertaining (labeling) the levels, and there are difficulties in measuring the times. Timing will be clarified when programs of research are undertaken in which comparable stages of functioning (such as time from onset of disturbance to first muscular action) are timed at each level of control.

DO IT YOURSELF

Those measurements of time delays in response is one of the ways of demonstrating the existence of hierarchical levels of control, even though they do not always fit in an obvious way with the eleven levels Powers has described. A similar kind of demonstration is that of nervous reflexes. Here is a demonstration of reflex that has been done hundreds of times by students of PCT; you can do it with an acquaintance.

Table: Response times

Action		Seconds	Source
1st order:	clonus oscillations such as shivering. Response time would be about half the .1 sec cycling time given by the authors.	.05	Plooij and van de Rijt-Plooij, 1990, p. 69.
1st order:	arm reflex: the time it takes a muscle to increase its tension from zero to 63% of the final tension after sudden energizing.	.05	Powers to CSGnet, 20 Oct 1999, and e-mail to Runkel of 26 Nov. 1999.
1st order:	first force applied in a tracking movement.	.07	Powers, 1973, p. 74, citing others.
1st order:	tracking movement: best-fit delay component in a model of tracking, including not only muscle delays but delay in the visual system and transmission delays in and out of the central nervous system.	.15–.20	Powers to CSGnet, 20 Oct 1999.
2nd order:	rooting of chimpanzee neonate, oscillation of head.	.33	Plooij and van de Rijt-Plooij, 1990, p. 70.
4th order:	rate control.	.35–.40	Powers to CSGnet, 20 Oct 1999, and e-mail to Runkel of 26 Nov. 1999.
5th order:	tracking on computer when connection with mouse is suddenly reversed.	.40	Marken, 1993, p. 41 and Powers's e-mail to Runkel of 26 Nov 1999.
8th order:	tracking the sequence of another's circling finger; reaction at change of direction.	.50	Powers, Clark, and McFarland, 1960, p. 315 and e-mail to Runkel of 26 Nov 1999.
8th order:	perceiving numerals as sequence.	.50	Marken, 1993, p. 45
8th order:	maintenance of sequence of various-sized squares appearing on computer screen.	1.20	Marken, 23 Oct. 1999, missive to CSGnet.
9th order:	finding program in changing sequences of numbers.	2.00	Marken, 1993, p. 46
9th order:	maintenance of program with various-sized squares.	2.00	Marken, 23 Oct. 1999, missive to CSGnet.

Ask a friend to hold her arm straight out from the shoulder with her palm facing down. Put your hand lightly on your friend's hand and say, "I am going to give you a signal like this," and give a very quick, but rather gentle downward push, forceful enough to deflect the arm only an inch or so. Be sure the arm comes back to its original, straight-out position. (If it doesn't, ask your friend to keep the arm always straight out.) Keep your hand lightly touching your friend's. Give the signal again, just to make it clear. Then say, "Now, when I give you the signal, bring your arm as fast as you can down against your side." After a moment or two, give the quick, downward push.

You will notice that the arm does not go downward immediately. It goes first almost back to its original position and only then goes downward. Your friend might also notice that initial upward jerk before responding to her wish to lower the arm. She might say something like, "Maybe I wasn't paying as close attention as I should have been." But regardless of whether she notices or says anything, try it again. Say something like, "Try to drop your arm just as quickly as you can when you feel the signal." Again, give the push. And again, you will see your friend's hand jump upward before it goes down to her side. Indeed, if you were to repeat this demonstration dozens of times, you would see the same reflex, no matter how hard your friend might try to yank her arm down before it could go upward.

When your friend's arm is outstretched, the control loops at the level of configuration are activating muscles that counteract the force of gravity on the arm. When you push quickly on your friend's hand or wrist, those loops will act immediately to counteract the increased force acting downward. So those muscles pull the arm upward to the straight-out position, and then return to their previous state. Now you tell your friend that you want her to "tell" (to will) those muscles, suddenly, to change their effect; instead of holding the arm up, you want the forces of the muscles to be applied to pulling the arm rapidly downward. To bring about that action, the reference signals must be altered at a level at least as high as programs, because your friend must understand what you say at least at that level. Obviously, it will take longer for the loops to function that go through the level of programs than for those that go through the level of configurations. By the time your friend senses the downward push on her hand and "tells" her muscles to reverse their effect on her arm, the shorter configuration-loops have already returned the arm almost to its former position. She can do nothing to prevent it.

Try it.

HIGHER ORDERS

Do system concepts actually constitute the topmost level? Nobody knows. It seems unarguable that simpler creatures have fewer levels. It seems unlikely, Powers (1990, p. 81) says, that the level of system concepts

> existed in the human brain when the hominid called Lucy walked the earth two or three million years ago.... There must be some level of organization where evolution is still at work, where we are still changing as failures of organization put us at risk. One has only to look at the state of the world today to know that human beings are not very well set up at levels like principles and system concepts.

And from where can the reference values for the topmost level come? What can set them? This question has baffled many of us. The best I can do is quote from a missive from Powers to the CSGnet on 28 September 1999:

> It's always possible that we will come to recognize a higher level than system concepts. . ., but at some point we must find a level that is truly the top one, and then we have to find [an explanation different from the explanation at lower levels] for the source of its reference signals. There are several possibilities. [1.] The simplest is to recognize that absence of a reference signal is equivalent to a reference signal of zero. . ., [implying] that the associated perceptual signal will be maintained at zero. At these high levels, a zero state for a perceptual signal can have very definite implications: for example, having a zero reference signal for being with unfamiliar people translates directly into xenophobia—a common state in young people. . . . [2. Another possibility is] that non-zero reference signals arise experimentally, through reorganization [for which see Chapter 20]. [3.] Another is that they are derived from memory, representing the average state of a given perception over some period of time. You "get used to" a system concept that you often experience, and eventually it becomes the preferred state just because you're familiar with it. [4.] And another is that they're genetically set, the way a bower bird inherits a reference image for a nest.

And are these eleven levels the right ones? Nobody knows that, either. In a posting to the CSGnet on 21 May 1998, Powers said:

> PCT is nothing but the application of the logic of closed-loop negative feedback loops to living organisms. That logic is quite clear. The present form of the theory resulted from applying control-system logic to various groups of behavioral phenomena; as more phenomena are given clear definitions, the theory will grow to cover more territory. Many parts of PCT and [the hierarchy] are basically proposals instead of finished products; dozens of research projects are implied, and decades of development.

And in his *Making Sense of Behavior*, Powers (1998, p. 152) says:

> Rather than slavishly memorizing these levels, I hope people will pay more attention to the *idea* of levels of perception. We control some kinds of perception as a way of [for the purpose of] controlling higher-level kinds; it's seldom that we control one perception just for its own sake.

This whole idea of a hierarchy of perception and control is . . . the product of a human brain trying to bring order and sense into experience, to see life as being coherent, and to find overriding concepts that make it hang together, to give it beauty and worthiness and make the whole process seem worth bothering with.

CODA

Having come this far, I think you can take a deep breath, look around in your thoughts, and see the grandeur of perceptual control theory.

I began reading the writings of W.T. Powers and his followers about 1985. As I read and pondered, I found my previous views undergoing wrenching and even frightening changes. I found myself having to disown hundreds, maybe thousands of pages of my writings that I had broadcast to my peers with pride. I found, then, that I could see order among my previous confusions about psychological method. The sword that cut the Gordian knot—that cut through my gallimaufry of methodological embarrassments—was the distinction between counting instances of acts, on the one hand, and making a tangible, working model of individual functioning, on the other. That idea, which in retrospect seems a simple one, was enough to dissipate about 30 years of daily dissatisfaction with textbook methods of psychological research. That simple bifurcation is what I wrote about in my 1990 book.

The idea that permits making tangible, working models is, of course, the negative feedback loop. And that, in turn, requires abandoning the almost universally unquestioned assumption made by most people (including psychologists) of straight-line causation—which, in turn, includes the conceptions of beginning and ending. Displacing that theoretical baggage, the negative feedback loop requires circular causation, with every function in the loop performing as both cause and effect. That, in turn, implies continuous functioning (beginnings and endings are relegated to the convenience of perception at the fifth level). One cannot have it both ways. Living creatures do not loop on Mondays and straight-line on Tuesdays. They do not turn the page with loops while reading the print in linear cause-to-effect episodes. William of Occam would not approve.

Powers did not invent the loop. It existed in a few mechanical devices in antiquity and came to engineering fruition after electrical devices had become common. Some psychologists even wrote, haltingly, about "feedback." But the manner in which living organisms make use of the feedback loop—or I could say the manner in which the feedback loop enabled living creatures to come into being—that insight was Powers's alone. That insight by itself should be sufficient to put Powers into the pages of the history books as the founder of the science of psychology. Historians of psychology will, I think, come to name the year 1960 (when the two articles by Powers, Clark, and McFarland appeared in *Perceptual and Motor Skills*) as the beginning of the modern era. Maybe the historians will call that year the Great Divide. The period before 1960 will be treated much as historians of chemistry treat the period before Lavoisier brought quantification to that science.

Using the negative feedback loop as the building-block of PCT enabled Powers to show how mathematics could be used in psychological theorizing. Powers's true use of numbers made it possible at last to test theory by the quantitative degree to which the data from any single individual approach the limits of measurement error, as in other sciences.

Even making a science possible was not enough to fill the compass of Powers's vision. He saw the unity of all aspects of human perception and action. He saw that there was not a sensory psychology over here, a cognitive over there, a personality in this direction, a social in that, and so on, but simply a psychology. He gathered every previous fragment into one grand theoretical structure—the neural hierarchy. The nature of the particular levels is not crucial. What is crucial is the idea of the enabling effect of organization by levels—the enabling of coordination among actions of all kinds. Previously disparate psychologies with disparate theories can now all begin with the same core of theoretical assumptions. Though it will take a long time to invent ways of testing the functioning of the hierarchy at the higher levels, I find it exhilarating to realize that Powers and others have already built models having two or three levels organized in the manner of hierarchical control and that those models actually work.

The neural hierarchy is far more than a listing of nice-sounding categories. The theory itself tells how we can recognize the relatively higher and lower placements of levels. It tells us, too, some of the kinds of difficulties to be anticipated in doing research at the higher levels. That kind of help from early theory is a remarkable achievement.

I have mentioned three momentous insights: (1) that the negative feedback loop is the prerequisite for life, (2) that numbers should be used to show the approximation of model to human individual, and (3) that control grasps more aspects of the environment through its hierarchical structure. For any one of those three momentous insights, I think Powers deserves a bronze statue in the town square. To put all three together in one grand system concept is the kind of thing that happens in a scientific field once in a century or more.

After more than 15 years of reading, conversing, writing, and thinking about PCT almost every day, I still feel the way Lewis and Clark must have felt when they began rowing their boats up the Missouri River. I know the general nature of the territory, I know that much of what I will come upon will be astonishing and baffling, and I know that every mile of the journey will be hard going. As I write this book, most parts of which are simply elaborations of the three simple ideas set out above, I find time and again that I must take an hour or a day to struggle with ways of keeping the words as simple as the idea. The ramifications of those simple ideas are multifarious and subtle. As I begin to describe a complication in the way those ideas work together, I find now and again that I have opened further regions of complexity for which I am wholly unprepared. Then I must take an hour or a day or a week to find my way back to firm footing. I do not feel that I am trudging along a prescribed path. I feel that I am taking every step with caution, but also with awe and exhilaration as I wonder what I might add to my understanding. I am sure, however, that I have only an inkling of the exploratory feelings Powers must have had as, day by day and year by year, he built his theory. He guided his footfalls by experimentation; I have guided mine only with thinking about the steps he took.

REFERENCES

Here are more places where you can read about the neural hierarchy: Marken (1986; 1990a); Marken and Powers (1989a); Plooij and van de Rijt-Plooij (1990); Powers (1973, pp. 70–176; 1979a; 1990; 1998, pp. 27–43, 135–152); van de Rijt and Plooij (1996); and Robertson and Glines (1985), and Richard Marken's demos at http://www.mindreadings.com/demos.htm. By the way, when you read these writings (and others), you will find that some authors use the abbreviation PCT when they are focusing on a single level of control, but use the abbreviation HPCT (for Hierarchical Perceptual Control Theory) when focusing more on the neural hierarchy. I prefer to use PCT for everything.

Chapter 19

Memory and imagination

In Chapter 18 and in earlier chapters, almost all I have said about control has been about control of perceptions of variables among energies arriving from the environment—the loudness of a trombone, the length of a path to the library, the rhythm of this sentence, and the light of love in the eyes of my beloved. But life does not consist entirely of controlling indicators of the outside world. You will see a reminder of the rest of human experience if you look again at Figure 18–3 in Chapter 18. At the right of that figure, you can see some circuitry labeled "imagining." That circuitry, you will note, does not go down to the lowest level; the feedback loops do not go out into the environment. Aside from that difference, however, the loops labeled "imagining," like the loops in the left side of the figure, are affected by inputs from lower levels and outputs from higher levels. But they can be affected by lower and higher orders only if they continue to exist—or more accurately, only if the multiple interconnections among assortments of them have a continuing existence. In other words, the nervous system must be able to retain or recollect those multiple interconnections; the loops must include a memory.

MEMORY

Human memory is not like a library, to be used when we feel a need for information and at other times ignored. Powers (1973, p. 208) says,

> We cannot [if we are to build a realistic model] have memories simply being dumped into some community hopper for indiscriminate use by any chance subsystem [loop]. Instead, every subsystem must have its own unique memory apparatus, complete with storage and retrieval mechanisms. . . . In order for neural signals to be recorded and replayed with their original significance, the effect of the storage must be that of a time delay in the signal-carrying path. The relayed information must reach the same destination that is reached by the signals being sampled for recording, and must be of the same physical form as the original information.

Memory is not a snarl of nerves outside the feedback loop; neither is it an epiphenomenon. Memory is one of the ever-present functions that enable us to control our perceptions, and it is, indeed, a part of the feedback loop, a part I have delayed telling you about until now.

Figure 19–1, taken from page 218 of Powers's 1973 book, diagrams the place of the function of memory in the feedback loop. It appears in that place by virtue of Powers's very careful design. Like all the other representations of neural connections in this book, Figure 19–1 does not pretend to be a picture of a chunk of the nervous system; it is a diagram of sequential functions in the *model*. Our claim is that actual living creatures function *as if* their neural nets contain component functions of this sort, connected in this way. Our strategy is to build models with this sort of structure to test whether they do indeed behave in the way living creatures behave.

You see at the top of Figure 19–1 a signal coming down from a higher order—a signal labeled "address signal." When a memory is to be retrieved, it must be found in the same "place" where it was recorded, so that it can be "replayed" with the same effect as the original signal. To "replay" a memory, then, the right address must be signaled. Some addressing mechanism is necessary if remembering is to be useful. The address signal going into the memory "box" in Figure 19–1 selects the past value of the perceptual signal that is to be re-created in the present and (a) sent on to the comparator or (b) sent back to higher

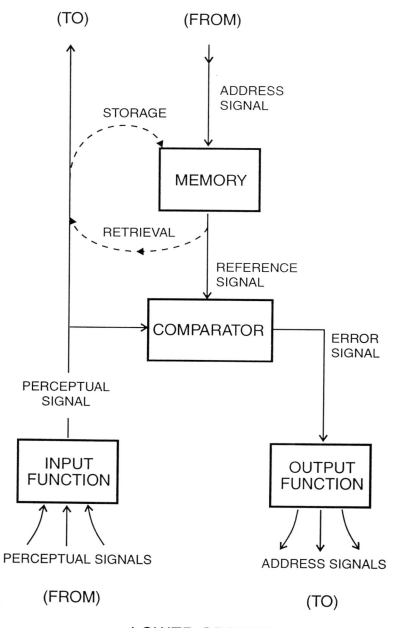

Figure 19–1. Control with memory
From Powers (1973, Figure 15.2, p. 218)

orders as a perceptual memory. Those two possibilities are shown by the dashed lines in the figure. (The lines are dashed because they don't show how signals can actually be routed that way. But do not despair; Figure 19–2 will be more specific.)

But beware how I write! What is "retrieving" or "replaying" a memory? Just how can that happen? And how does a signal "select" a past value? I am slipping past some mysteries and technicalities, because this is not a manual on model-making, but rather an introduction to the world of PCT. If you want more detail on memory, please look at the chapter on memory in Powers's 1973 book.

Now, in Figure 19–1, we have a memory receiving an address signal from a higher order of loop and sending on a reference signal to a loop or loops at a lower level. In this model, *all* loops at higher levels connect with loops at lower levels through memories. "*All reference signals,*" Powers (1973, p. 217) says, "*are retrieved recordings of past perceptual signals*" and correspondingly, "*all* behavior consists of reproducing past perceptions." To say it another way, the output from a higher order says (so to speak) to the lower-order loop: use *this* memory as the standard to be matched by your incoming perception. And behavior consists of altering a sensed variable so that it will match a signal remembered from a previous experience.

To say that reference signals are retrieved recordings of past perceptual signals is to say this: as the many signals come up to an input function, and after they are converted to the resultant signals to be used at this level, those new signals go to *three* places: the comparator at this level, the memory at this same level, and the input functions of loops at higher levels. The perceptual signals stored in memory now will be available to be called upon by address signals from above for use as reference signals at this level. They might not be called upon until next year, but they might be called upon a tenth of a second from now, as signals loop upward to the next level and back down to this level.

Here I will insert, with little discussion, several neurological implications of this model of memory. First, the signal retrieved from memory and sent upward, as is always the case with signals in the neural net, does not differ from any other signal pulsing along the pathways of the feedback loop. There is no label on it saying, "This is a memory." Perceptual signals relayed from memory have the same effects on input functions of higher orders as original signals or present-time signals. Second, not every recorded perception is constantly replayed, and upon replay, not all recorded perceptions are simultaneously retrieved. Retrieval is ordered and controlled. Third, memory is "local"; signals come into memory and are retrieved at the same level. So it is that we never remember from the viewpoint of some other order of perception. If we remember a sequence of action, for example, we remember it as a sequence, not as a configuration or a system concept. Fourth, although we *retrieve* memories at the same level as we record them, we can *interpret* memories in a way we did not when the original perceptions occurred. For example, adults often remember a childhood experience as having a significance it could not have had to a child. Powers (1973, pp. 210–211) illustrates this fourth characteristic of human memory as follows:

> Stromeyer (1970) has removed all doubt about this phenomenon with an astonishing experiment done in the manner of . . . investigations of binocular vision. These experiments involve the use of computer-generated dot patterns in 100 x 100 arrays. The patterns are random; however, a stereo pair is formed by shifting a portion of one copy of the random pattern by several units to the left or right. The shifted area is totally undetectable in either pattern alone, but viewed stereoscopically results in a "floating" area in the pattern, raised or depressed relative to the remainder of the dot pattern by an easily visible amount.
>
> One woman who had eidetic recall was tested by Stromeyer in an experiment involving memory—it took a daring mind to think of doing this. Stromeyer presented the stereo patterns to the woman, one pattern to each eye, but with a delay of 24 hours between the two presentations! In terms of our model, each pattern contained second-order information only; the third-order depth information was simply not obtainable from either pattern alone. Nevertheless, the woman was able to combine the two images in her memory and correctly identify the floating pattern. . . .
>
> This experiment fully supports the present theory of memory. . . .

The Warps of Memory

You will remember that every loop receives perceptual signals (inputs) from not just one but many loops of lower order, that every loop receives address signals from many loops of higher order, and that every loop sends address signals to many loops of lower order. If you want a graphic reminder, look again at Figure 18–3. To make the multiplicity of connections seem more homey, I'll borrow here an illustration from Powers (1973, p. 219):

> One muses, "Who *was* that girl at Aunt Mabel's house in the red dress?" The address signals *girl, Aunt Mabel's house, red,* and *dress* are sent (as sketchy images) to lower-order memory address inputs. Since neither Aunt Mabel nor her house appears in memory wearing a red dress, the only response is an image of the girl who wore the dress. The full recording (or set of them) replays into the perceptual-signal channel, and one experiences the memory.

What happens, then, to "a memory"? A signal goes as a reference value into a comparator. There it is compared with a signal coming in from the input function; a new signal, the error signal, results from the comparison. That signal goes to the output function, and from there it can then go (as shown in Figure 18–3) directly to the input function of the same loop. And at the same time, perceptual signals are arriving at that same input function from other loops. Some of those perceptual signals may come directly from other loops at the same or next lower level, and some may come, though altered at each level of control, from sensory organs. In any case, all those incoming signals are somehow combined by the input function, and a resulting signal goes up to the comparator, the memory, and higher levels. If all the signals (all except the one which is the focus of our attention) coming into the input function are zero, or if they are not zero but cancel each other out, the remaining perceptual signal remains unchanged. If they change the perceptual signal, then the error signal will change in a direction to restore the match between perceptual signal and reference signal.

The perceptual signal also goes up to an input function at the next higher level. There it gets combined with all the other signals arriving at the same input function, and a new perceptual signal results. That new signal gets compared with a reference signal at that level, and the result is sent down to the memories of many lower loops, including the loop on which we have been focusing our close attention. *But now that signal is not the same* as the signal that had gone from the memory of our focus-loop down to its comparator. It now calls up a somewhat different memory. How can it do this? Here Powers (1973) calls upon the ways connections in the nervous system can alter the threshold of firing of a neuron. An array of neurons can exhibit "graded responses that mutually inhibit one another so that the strongest response wins, or alternatively by a threshold of response so that the associative address must attain a minimum degree of match with a recorded unit in order to trigger replay of the whole unit" (p. 214). And remember that "a memory" is represented not just by one loop, as I have been pretending here, but by hundreds, perhaps thousands, in complex combinations. A small shift in the perceptual signals of some loops, therefore, can alter to a small degree the way the memory is called up by the signals *girl, Aunt Mabel's house, red,* and *dress*. And small changes can accumulate into big changes.

At the same time, thousands of other loops are busy controlling *girl* in other connections than with Aunt Mabel's house or a red dress, and other loops are busy with *Aunt Mabel*, and so on. As time goes by and control processes alter reference signals at all levels, those thousands of connections can steadily alter the picture called up by *girl, Aunt Mabel's house, red,* and *dress*. Indeed, one may be surprised, after a few years, to see a photograph taken back then at Aunt Mabel's house and discover that the girl was wearing a yellow dress with purple trim.

Robyn Dawes (1988, pp. 107–108) summarized two studies that documented changes in memories. In one study, Greg Markus (1986) conducted surveys of 1,669 high-school seniors and at least one of their parents in 1965, 1973, and 1982. All respondents indicated their attitudes on a seven-point scale toward five topics: guaranteed jobs, rights of accused people, aid to minorities, legalization of marijuana, and equality for women. They also rated their political views as generally liberal or conservative. In 1982, Markus asked respondents how they had responded in 1973. Dawes reported:

> With the exception of the ratings on the overall liberal-conservative scale, the subjects' recall of their 1973 attitudes in 1982 was more closely related to their rated attitudes *in 1982* than to the attitudes they had *actually* expressed in 1973. Retrospecting, they believed that their attitudes nine years previous were very close to their current one, much closer than they in fact were (p. 107). . . .

Finally, the parent group attributed much more stability to their attitudes than did the student group, which is compatible with the belief that the attitudes of older people change less. In fact, however, the attitudes of the parent group were *less* stable (p. 108).

In the second study, Collins and others (1985) surveyed high-school students about their use of drugs, repeating the survey a year later and again two and one-half years later. As in the other study, respondents generally believed they had changed less than they actually did. For example, among students whose use of alcohol had changed over the two and one-half years, their recall of their earlier use was more like the reports of use at the time of recall than like the reports two and one-half years earlier. Looking back on evidence of this sort, Dawes wrote on page 120:

> My desire is simply to introduce what I hope is a healthy skepticism about "learning from experience." In fact, what we often must do is to learn how to *avoid* learning from experience.

In his 1994 book, Dawes[1] gave more examples of the warps of memory. Here is one of them (page 156):

> One of the world's experts on human memory vividly recalled that he had heard of the Japanese attack on Pearl Harbor in an announcement that interrupted a broadcast of a baseball game. Years later, he realized that the last games of a baseball season are in October, while the attack was in December. He still "recalls" the incident [even though he] knows that his recall is inaccurate.

Warps of memory occur for the same reasons that reference signals (internal standards) change. Perceptual signals are altered either by energies impinging on sensory organs or by changing outputs from other loops not sent down to the environment. Those altered perceptual signals then either directly or eventually change the relations between the address signals and the locations of the neurons recording pieces of the memory. That means that just as control of perception is always going on, and just as reference signals are always gradually changing, so memories, too, are always in flux. Some change more slowly than others. Some stay the same for long periods of time, and some people keep some kinds of memory stable for longer periods than do other people, but as a general principle, memories are always changing. Indeed, they begin changing at the moment they are made.

Memory does not merely deteriorate; it changes. Sometimes features drop out; sometimes features are added. This adding, subtracting, substituting, twisting, and editing happens both to what we experienced a moment ago and to memories of events long past. It happens to memories of what other people did and said and to our memories of our own actions and utterances. No memory is wholly faithful to the event, not even one second after it happens. Maybe you have noticed, sometimes, when you start to describe what has just happened, "I just came around the corner of the house here, and George was shouting, well, Amy got up, and that was when George started shouting, well, he started talking, actually, but when the dog came, that's when he started shouting—" and you find yourself having to *reconstruct* what happened. Memories don't always flow out easily; sometimes you have to hunt for the pieces and assemble them. As you squeeze the pieces together, some of them get bent.

Just as it is obvious that every description of an event is a description of a memory, so is awareness of an event an awareness of memory. We are aware of "what we are doing" at this moment only because we remember what we have begun to do. Think of the experience of being engaged in an activity and realizing, unexpectedly, that you have forgotten what it was that you had started out to do. When that happens to us, we stop acting, being no longer able to continue the recent action. "What am I doing?" we cry, at the very moment that we have ceased doing it. In the moment when we realize that we have forgotten what we were doing, all the details of the previous sequence of acts become irrelevant, because no purpose remains in memory to which they can be relevant. And when we are acting with purpose in mind, we may not pay attention to many details that are irrelevant to our purpose. We may not notice, in our eagerness to get to the dinner table, that we are tracking mud on the rug.

In his 1988 book on perception, R.L. Gregory says on page 40: "All perceptions are mixtures of fact, distortion, and fiction." And as to fact, Karl Weick, on page 206 of his 1979 book, reproduces a picture of some filing cabinets. The drawers are labeled: Bare Facts, Our Facts, Their Facts, Neutral Facts, Unsubstantiated Facts, Indisputable Facts, and so on.

These ideas, of course, are not new. In 1932, F. C. Bartlett was telling about the mutations of memory in his book *Remembering*. Krech and

Crutchfield, in 1948, gave an excellent summary in their Chapter 4, appropriately titled, "Reorganizing Our Perceptions." Thoughtful persons have always, I suppose, been aware of the meanders of memory. James Howell wrote in 1659 in a list of proverbs: "The creditor hath a better memory than the debtor." The four dictionaries of quotations I consulted had a lot on *what* people remembered (or didn't), but very little on the nature of memory itself. I don't know whether those writers of old didn't set down much about memory or whether the compilers didn't think what the writers set down about it was worth adding to their pile.

My own experience is that most people only occasionally doubt their memory, especially recent memory, even people who make a study of memory. Indeed, you may yourself object that I am overdoing my warnings about the fragility and mutability of memory. You may believe that memory mutates only under strong emotion or in other special circumstances. Or you may believe that though the memories of other people tremble and transmogrify, your own memories hold close to the truth. How can we discover whether we are right or wrong about what we think we remember?

We cannot easily learn of an error by calling up other memories to compare against the one in doubt. It is true that sometimes, upon reflection, we find contradictions among our memories. But when one memory seems contradicted by another, we cannot know which is the more correct.

We can take precautions by reaching outside ourselves. We can write down an account of our experience as soon as possible, so that further mutations of the event in memory are avoided. Scientists write down not only what they observe, but also the exact circumstances in which they observe it. They describe precisely how they set up the conditions. (Think how a careful cook records the procedure for producing a particular taste.) Scientists do this so that other investigators can follow those descriptions and replicate the experiment. This practice minimizes changes due to time.

We can also compare our own written or oral account with the account of another person who was present at the same place and time and who, presumably, was paying attention to the same phenomenon. This last technique is especially effective if we make the comparison with the other person's memory within the first fraction of a minute after the event. Often, a memory we are about to edit severely will still be recoverable if the other person reminds us of it very soon. Neither of these tricks, it is true, guarantees recapture of the pristine memory. Writing things down leaves unaltered the changes that have already occurred in the first moment, the changes we make mid-sentence in our writing to help ourselves make sense of what we are writing, the editing we give the event as we write (omitting what seems unimportant and emphasizing what we think is crucial), and our ability to understand at a later time what we have written earlier. And in comparing a memory with another person's, we may be picking a defective standard.

I do not want to give the impression here that the first shape of "the" memory is the correct one, and all the ensuing modifications of it are departures from truth. As I have said before, every perception is an abstraction of reality. Our senses detect only a portion of the energies that impinge upon them— a person who wears hearing aids is reminded of that a dozen times a day. To some degree, eyes and ears and other senses, and the neural interpretive networks, too, differ among individuals in their functioning. Furthermore, the viewpoints from which we witness the "same" event differ. As numerous experiments have shown, and as police and trial lawyers tell us, witnesses of an event often tell very different stories. What "actually" happened is often difficult to determine. Even when you are talking to another person and both of you are hearing very well, physiologically, one of you can say, "What I just said was . . . ." and be in poor agreement with what a recording machine would testify.

If an observation and its interpretation is to become a part of the beliefs of scientists, researchers must "agree" not merely to a statement—not merely in saying that they believe the implications aroused in their minds by a certain string of words. They must also agree on what further sorts of observations will be considered a test of the "truth" of that string of words; they must agree, that is, not only on whether the words sound right, but also on the kind of action they imply. They must agree, for example, on whether a particular experimental procedure corresponds to the words they or others use to describe the procedure. Sometimes they doubt whether what other researchers have done corresponds well to what those others say they have done. Much communication in professional journals goes on to clarify such doubts. One is not surprised to come across sentences like, "If

Dr. Kopfgrosser had read more carefully what I wrote, ..." But I will say more about communication in a later part of this book.

Child Abuse

During the 1980s and 1990s, many mental health workers, including psychologists and psychiatrists, undertook to "recover" memories from children and adults. Most of the presumed memories sought were memories of being abused as a child. Ignorance and misunderstanding of the warps of memory allow professionals and clients to believe that an accurate description of an actual event can be coaxed from a person by some sort of encouraging "guidance," including hypnotism. It is not surprising that willing clients produce a story and even believe it themselves. This is the same sort of process Theodore Sarbin (in 1982 and 1995, for example) has explained as role-taking in hypnotism. Dawes wrote about this horror in his 1994 book:

> It began with a gross overestimation of the incidence of child sexual abuse. . . . The author of one particularly ludicrous article simultaneously asserts one of five women is sexually abused as a child and that half of these women forget the experience. How could the author possibly know such a thing? It would be impossible to conduct a general survey that could reach that conclusion, because such a survey would require knowledge of actual abuse without relying on the respondents' memories (p. 155).
>
> It is simply not true that people *either* lie *or* tell exactly what occurred *or* forget. People with the most rudimentary knowledge of psychology should know that recall is an active search process involving reconstruction of the past. . . . If professionally licensed psychologists were required to know psychology and to make their judgments in accord with what they know, they could not make confident statements about what occurred in their clients' past on the basis of what these clients have been guided to recall. Unfortunately, mental health professionals are themselves free to forget whatever they learned (pp. 156–157).

Dawes's (1994) cogent discussion of the tragedies resulting from ignorance of the dynamics of memory occupy 12 pages of his book about the doings of clinical psychologists. In 1997, Elizabeth Loftus published an article in the *Scientific American* which told the same sort of story Dawes told but probably reached more readers. Her article begins like this:

> In 1986 Nadean Cool, a nurse's aide in Wisconsin, sought therapy from a psychiatrist to help her cope with her reaction to a traumatic event experienced by her daughter. During therapy, the psychiatrist used hypnosis and other suggestive techniques to dig out buried memories of abuse that Cool herself had allegedly experienced. In the process, Cool became convinced that she had repressed memories if having been in a satanic cult, of eating babies, of being raped, of having sex with animals, and of being forced to watch the murder of her eight-year-old friend. She came to believe that she had more than 120 personalities—children, adults, angels, and even a duck—all because, Cool was told, she had experienced severe childhood sexual and physical abuse. The psychiatrist also performed exorcisms on her, one of which lasted for five hours and included the sprinkling of holy water and screams for Satan to leave Cool's body.
>
> When Cool finally realized that false memories had been planted, she sued the psychiatrist for malpractice. In March of 1997, after five weeks of trial, her case was settled out of court for $2.4 million (p. 71).

Loftus continues:

> My students and I have now conducted more than 200 experiments involving over 20,000 individuals that document how exposure to misinformation induces memory distortion. . . . Taken together, these studies show that misinformation can change an individual's recollection in predictable and very powerful ways (p. 71).

IMAGINATION

When I write, I sometimes "tell" my memory to deliver a word that has the meaning I want to go in this blank space I am looking at. I did that right where you see the word "tell" in the previous sentence. I couldn't think of a word to put there that really felt right, and after sitting for part of a minute hoping my mind would exhibit a brilliant resourcefulness, I gave up and wrote "tell." I put it in quotation marks to tell you that I don't want you to put quite the meaning on it that you usually (I am supposing) do. I didn't look in a thesaurus, because when I imagined doing that,

I couldn't even think of a good word to look under. See all the imagining I do when I write? I can almost see you sitting there, listening politely, and nodding your head as if to say, "Yes, yes, you're not the only one who has ever struggled with a sentence, so let's get on with it, shall we?"

I imagine arrangements of words that I hope will enable you to select from your own memory pretty much the meaning I hope you will select. And along the way, I imagine some arrangements which, once I have imagined them, I think are poor bets—arrangements which, if you were thinking what I hope you will think, you would *not* (I imagine) choose to express your thought. After an hour's writing, the wastebasket under my desk is full (in my imagination) of sentences like that. Sometimes, as I look over what I have typed, I see a word that doesn't seem to have the right shape. I look more closely and see that I have put in a letter I didn't want. Indeed, I just now typed "swant," because my inept finger touched the *s*-key just before it got to the *w*-key. On the other hand, when I am reading over what I have written, I am so confident of the word coming next in the sentence that I sometimes "see" the memory of it instead of what is actually on the page, and that interloping *s* stays right there until later, when the spelling program in my computer points it out to me.

Switching the Modes

An address signal can call up a stored perceptual signal (or a modification of it), and we can perceive it without acting—without the signal having to follow a loop that goes out into the environment and back. Figure 19–2 is a revision of Figure 19–1 that shows how this can happen. You see in the middle of Figure 19–2 two "switches" that Powers postulates to function in every loop to enable us to choose what happens to the signal emitted by the memory. What goes on in the brain that corresponds to this "switching"? Well, this is a point, I think, at which we must hope for more modeling than has yet been done. Researchers in PCT, so far, have been sticking to investigating the functioning of control loops for which there is evidence in behavior affecting the external world—for obvious reasons. Still, Powers (1973, p. 227) reports some indirect evidence in support of the imagination connection from experiments on the changes of electrical fields in the brain as perceptions change. Beyond that, the conception of the two switches implies the various types of actual mental experience so well that the postulation is well worth keeping until experimentation demands revisions.

The two switches can provide four configurations. The configuration shown in Figure 19–2 is the "imagination mode." I will return to the imagination mode after I describe to you the other three modes. For the first of the other three, I will remind you of the "control mode." This is, of course, the setting that enables control of perceptual signals that originate at sensory organs—the configuration shown in Figure 19–1 here and implied in Figure 4–1. To imagine how to convert Figure 19–2 to this first configuration, look at the two switches there. The movement of each switch is suggested by the little curved arrows at the middle of each switch; each switch pivots at its top and has the two positions indicated by the two little circles under each switch. Now imagine that you have swung both switches downward; you would then see the kind of loop you have seen in all the diagrams before this one. This is the "control mode" with which you are by now familiar.

Powers calls the second mode the "passive observation mode." Look at the "perceptual switch" in Figure 19–2 and imagine that you have swung that switch downward, but that you have left the "memory switch" still in the position shown in the figure. With those settings, signals from memory will go nowhere, leaving the comparator with a reference signal of zero. In this setting no action will ensue, because, whatever the signal coming to the comparator from the input function, the reference signal of zero will not alter the perceptual signal. The outgoing error signal will be the same as the incoming perceptual signal, and whatever is happening (or not happening) will go on happening (or not happening) unaffected by the person. The perceptual signal coming up from the input function, however, will go into memory, as well as on up to the next higher level.

When I first thought about this, I found myself somewhat confused. On the chance that you might have the same trouble, I'll say a little more here. The comparator subtracts the perceptual signal from the reference signal to ascertain the difference. So, I wondered at first, if the reference signal is zero, and the incoming perceptual signal has a large value, and you subtract that large value from zero, wouldn't you have a big difference as the error signal and therefore some big action? No, because the difference would be negative, and the comparator counts that as zero.

Part IV Hierarchies of purpose: Chapter 19 Memory and imagination

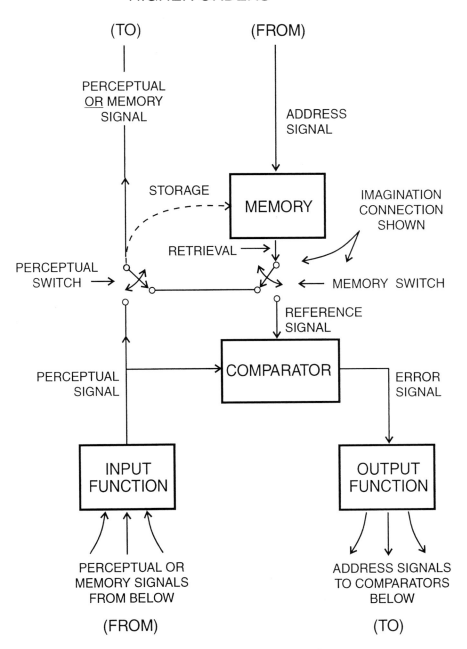

Figure 19–2. Final form of the unit of behavioral organization. From Powers (1973, Figure 15.3, p. 221)

Here is how Powers explained it in 1973:

> The comparator is a *subtractor*. The perceptual signal enters in the inhibitory sense [carrying a minus, subtractive, sign], and the reference signal enters in the excitatory sense [with a positive sign] (p. 62).
>
> If the negative current is equal to or greater than the positive current, the output . . . is *zero*. . . . A frequency of firing, in impulses per second, cannot go negative. . . . In neural current analysis all currents resulting from firing of a neuron flow in one direction only (away from the cell body) and sign indicates *only* excitatory or inhibitory effect (p. 28).

The third mode is the "automatic mode." To convert Figure 19–2 to the third mode, leave the perceptual switch turned as it is, but turn the memory switch downward so that signals can go from the memory to the comparator. No copy of the controlled perceptual signal goes to higher orders, and the signal is not recorded in memory. The control system, however, operates normally. At any moment, thousands of loops, maybe millions, will be in this mode. Much of what we do frequently is done in the automatic mode—walking, washing dishes, making the bed, driving to work, and so on. When I finish shaving, I remember very few of the motions I made; my consciousness is usually busy in the imagination mode. When my wife was the librarian for the 15th Naval District with headquarters in Panama, I delivered her to her office every morning before I went on to my own place of work. One night, some time after dinner, I said I would go buy some ice cream. We were busy discussing something, I don't remember what, and my wife said she would ride along even though she was in her night clothes; I could go in to buy the ice cream, and she could wait in the car. We got into the car and continued our conversation as I steered off toward the grocery store. But while we talked, I fell into the automatic mode that I used every morning, and my wife fell into her corresponding automatic mode, and we suddenly realized that we were about to pull up to the guarded gate at the Naval District— a man with a woman in nightgown and kimono. And without a gate pass, too. I am sure you have had some similar experience.

Figure 19–2, as you see it here, displays the "imagination mode." The two switches are set so that a signal coming out of the memory does not go to the comparator, but instead goes over to the perceptual channel and on up to the next higher order of perception. This is the setting of the switches that Powers calls the "imagination mode" of functioning in the loop. The output from the memory does not go downward to demand action on the environment, and no input from any lower loop interferes with the signal as it goes upward in the perception channel. The higher orders receiving the signal deal wholly with signals produced in memory—signals from the past. This setting of the two switches implies the kind of functioning I was usually describing in the earlier section on memory. This is the setting implied, too, in the right-hand portion of Figure 18–3.

Four modes: control, passive observation, automatic, and imagination. They seem to mirror, as far as we can be aware, the way our minds work. But what "operates" the "switches"? And is that the best metaphor? No doubt a chemical action is crucial, since the nervous system is electrochemical in its functioning. When one is administered a "general" anesthetic, for example, one is aware of no signals whatever— no sensation, not even an intensity. One awakens with no memory of anything that happened during the interim. No signals have gone to memory at any level involved with consciousness, but at the lower levels all the loops necessary to bodily functioning have continued in normal operation. Obviously, an anesthetic does not simply "dull" everything, or one's heart would stop. The chemical affects some functional parts of the circuitry but not others. Mysteries remain to keep the researchers busy.

The imagination mode also offers an explanation of how we can be conscious of a goal, a reference value, before we act. Figure 19–2 shows the path a reference signal can take from the comparator into the perceptual channel.

The imagination mode is the mode of conscious remembering. We can remember at any level, memories at the higher levels giving us the more abstract qualities and those at lower levels the more concrete, vivid qualities. One can remember a program, for example, such as how to drive to the Apex Appliance Arcade, without having any memory of the color of the sky, the sound of the engine, the bumpiness of the street, or the feel of the steering wheel. But I am so fond of comice pears that I can remember, from season to season, the soft, velvety feel of a comice pear that is ripe to eat, the mottled appearance of the green-and-brown skin, the singing, lighthearted juices in my mouth, and the eager happiness of my throat muscles as they carry out the voluptuous gluttony of swallowing. I remember not only those sensations, but many of the muscle-movements that bring them to me and the sight of the pear rising to my mouth. When we remember with many levels at once, including the lowest levels, our experience has the character of vividness, of reliving something.

I think that every time we find ourselves doing something in response to even a little disturbance, taking action on the outside world is not at all the whole of the matter. We also do something, so to speak, inside ourselves. Maybe, for example, we think we know how to recognize ice cream, but a friend tells us, "No, what you are eating is frozen yogurt." We take some low-level action such as taking another spoonful and paying more attention to taste sensations, but we also make some changes in our standard for the cognitive, conscious category "ice cream." I am saying that we deal with a disturbance (such as the "No" our friend utters) in the control mode (taking another spoonful, moving the ice cream around in the mouth while tasting it), but at the same time we use the passive observation mode and the imagination mode to cope with disturbances to higher standards (such as the denotations of "ice cream") that are not now calling for action.

Here is a comment Powers made on the CSGnet on 6 July 1999:

> Generally, any experience consists of multiple perceptions at each level, each perceptual signal providing just one simple dimension of the whole experience. The same applies to imagining. Imagining a single dimension of breakfast might amount to imagining the sweetness of jelly on toast. Imagining (or remembering) breakfast as a whole experience would require imagining many such perceptions at the same time.
>
> It's been suggested, and I agree, that most real-time experiences actually contain a considerable degree of imagined information from previous experiences. When you look at an object, for example, you may imagine that it exists in three dimensions even when you can't see its other side. I'm looking at an upholstered chair back; I see it as being upholstered all over, not just on the side I can actually see. Of course, it's not hard to separate the imagined part from the part I'm actually seeing, if my attention is drawn to the question of what I'm actually observing.

That reminds me of the two friends driving along a country road and finding themselves passing a flock of sheep. One said, "Looks as if those sheep have just been sheared, doesn't it?" And the other replied, "Yep, at least on this side."

Thinking

In the imagination mode, we can move perceptions this way and that without taking action on the outside world and without interference from it. What might happen, I can ask myself in imagination, if I were to do this instead of that? Or this before that? Or this inside that? Could I paint it a color between orange and fuchsia? Would it sound better if I would growl first? What might Aunt Mabel say if I were to do that? The imagination mode is the mode for planning and all other kinds of thinking. Some people think more at one level, some at another. Mechanical engineers spend a lot of time imagining configurations, transitions, and relationships, as well as higher orders. Theologians, I suppose, stay close to principles and system concepts. Both explain themselves to others with reasoning and logic, more or less.

The unreliability of planning surely drew comment some millennia before the famous couplet of Robert Burns: "The best-laid schemes o' mice and men / Gang aft a-gley." Quite aside from the unpredictability of the world around us, we should not find it surprising that our plans so often turn out flawed. While we are trying things out in our minds, the arrangements we make there are rarely disturbed by the unexpected. In the imagination mode, we use inputs only from memory, none from sensory organs. Images taken from memory can rarely surprise us. It is easy to have confidence in our planning when our imagination turns up no stumbling blocks. I'll say more about planning in another part of the book.

It is difficult and requires a good deal of experience and discipline to develop the skill of asking ourselves, "What *could* turn up unexpectedly? What *could* go wrong?" The ease of making faulty plans or images of reality is the reason that scientists try hard to develop the skill of asking themselves, "What could be wrong about my brilliant idea?" And it is one reason they prefer to build models of what they imagine instead of resting wholly on their imaginings. For more on thinking, see Chapter 25 (on logic) and Chapter 32 under "Thinking."

Dreaming

We move things this way and that in dreaming, too. In dreaming, there is still less restriction on imagining. Not only do we sever connection with any action outside the body, but we lose any consciousness of time, and we seem to ignore any realistic memories of sequences, programs, principles, or system concepts concerning physics, morality, or orderings in time. We fly through the air and through walls by waving our hands. We hear oranges and see trumpet notes. Even with all the phantasmagoria, however, dreaming is not random. Our brain is doing something useful. William Dement (1969) and colleagues say that dreaming rests the mind. William and Mary Powers (1973, p. 225) say that dreaming can be "concerned with discovering and repairing basic design defects in the control systems...." These undisciplined imaginings are not "alternate realities" or magical corridors to other universes or messages from Aphrodite or any other god or goddess. They are the brain talking to itself.

ENDNOTE

[1] This and later quotations from Dawes's 1994 book are reprinted with the permission of The Free Press, an imprint of Simon and Schuster Adult Publishing Group, from *House of cards: Psychology and psychotherapy built on myth* by Robyn Dawes. Copyright © 1994 by Robyn Dawes.

Chapter 20

Reorganization

In the first pages of this book, I stressed the idea that there is no possible way that we can predict all the details of the conditions and arrangements of the environment that we will encounter. There is no use, as I walk toward a door, to try to predict precisely the forces I will need to apply in the muscles of arm, wrist, hand, and fingers and the precise moments when I should apply them. I would fail. Every living creature must, to continue living, maintain its vital functions despite all such unpredictabilities and their accompanying dangers. Thinking of that fact, one must feel awe (I do, anyway) at the capabilities of living creatures—even the simplest of them—in making use of their environments even while fending off the threats they find there.

Many researchers in psychology, biology, and artificial intelligence refuse to believe that living creatures are capable of coping successfully with unpredictable disturbances as they happen, and insist that an organism must first construct an internal map or imaginary model, even if an unconscious one, of what is going on in the environment, and then act on that map or model, ignoring, during the action, what is going on outside. But that is far too cumbersome to enable you to open a door, ride a bicycle, or even get a glass of water to your lips. I mentioned this cumbersome strategy of trying to calculate every twitch of muscle in Chapter 6 under "Four Controllers," and you can find more about it in Chapter 24 under "Model-Based Control." But, as you have already seen in many previous chapters, a model built with the assumption of continuous, simultaneous, circular causation can mimic actual human action without having to make a map of the action before acting.

Still, despite the impressive success of the models I have exhibited to you in this book and the many others you can find described in the literature I have mentioned, you may be thinking that there must surely be times when, despite all those thousands of internal standards that a person has brought into being, and despite all those hierarchically encompassing levels of neural organization, and despite the tremendous complexity of the interconnections among them (as implied by Figure 18–3)—there must surely be times when a person will fail to hold some variable to the value called for by the variable's internal standard. This must surely happen rather often to infants while they are building the layers of their neural hierarchies.

Powers had that thought, too. The environment is simply too variable to expect that any hierarchy of control systems, no matter how elaborated, can expect always to succeed in holding steady every controlled variable in every environmental circumstance. Quite aside from environmental happenstance, the person's own activities—running too long too fast, eating something toxic—can alter the chemical and biological conditions in the body on which all functioning depends—temperatures, fluid pressures, oxygen concentrations, electrolytic densities, hormone distributions, and so on. When the proper ranges of such variables are exceeded, malfunction and death can quickly ensue. Sooner or later, a living creature is bound to encounter a situation in which the connections from internal standards to lowest-level output functions cannot keep some one or more of those vital variables within the range specified by the internal standards. When a vital variable goes out of range, the creature must be able to reorganize the neural net, even in infancy, and construct a new level of control. Indeed, Powers postulates that constructing new levels of control of the vital variables is the way that the neural hierarchy is gradually built up, beginning before birth and continuing until death.

You might be tempted to say, "Oh, well, it's obvious that those chemical variables stay pretty much where they ought to stay. They probably don't go beyond their proper ranges very often. Let's keep things simple and simply ignore that possibility." That attitude might appeal to a researcher who is satisfied with predictions that come about in a mere majority of trials (or on the part of a mere majority of persons). But a PCT theorist cannot be satisfied with that. A PCT theorist wants to make a model that will work under realistic conditions. And the real world does confront living creatures with situations in which controlled variables can continue to be controlled only if the neural hierarchy can be reorganized. It is impossible for genes to carry instructions for the proper action in every environment the creature may meet. A model that functions like a living creature must be able to deal with unpredictable disturbances. Genes can provide the *strategies* for dealing with disturbances, predicted or not, but they cannot provide the particular actions that will be effective with particular disturbances. In fact, human babies come into the world with a very restricted repertoire of behavior (in comparison both with other mammals and with their own later behavior). One important kind of behavior the neonate displays is random muscle movement. From that primitive beginning, the multilevel control hierarchy is perforce constructed.

In Chapter 9, I recounted a study by Plooij and van de Rijt-Plooij (1994) that gave more illustrations of the formation of the hierarchy of control during the development of infants. You can find still more illustrations in the book by Vanderijt and Plooij (2003) entitled *The wonder weeks*. Those authors give a lucid and fascinating account of the sequential conflicts between mother and child and the ensuing stepwise development of the neural hierarchy. If you are thinking you might be a parent some day, buy a copy of *The wonder weeks*. Here is more information on this topic from Frans Plooij's communication to the CSGnet on 19 November 1995:

> Bill Powers (1973, p. 180) defined reorganisation as a change in the properties or even the number of components. Let's limit the discussion to the number of components for the sake of simplicity. Our research concerns human infant development. In the first year of life the volume of the infant brain more or less doubles. So we may safely assume that many components are added and a lot of reorganisation takes place. The way we searched for these reorganisations was based on the notion that the organism is out of balance, or disorganised, when reorganisation takes place. This is mentioned many times in the literature.... we found [it to be so at] the ten regression periods in the first 20 months of life (sensorimotor period). At quite specific ages (if not born too early or too late) babies appear to be irritable, cry more, eat less, sleep less well, etc....
>
> Taking this as the starting point, the next question is: "If this is a sign of reorganisation taking place, are any components added to the brain?" Luckily there is quite some literature that confirms this. A nice review is presented by Fischer and Rose (1994). Six of the ten periods found by us go together with sudden changes of the brain such as changes in ERP, EEG, protein metabolism, and simple head circumference. The latter is direct evidence for components being added. The head circumference does not increase gradually but in leaps: the leaps occur just before or simultaneously with the beginning of the regression periods we found.

To the same book containing the chapter by Fischer and Rose that Plooij mentioned, R. W. Thatcher (1994) contributed a chapter on "cyclical cortical reorganization." I am not sure what Thatcher meant by "cyclical"; I think I would use the word "sequential" (which Thatcher often did, too) or "staged" (in stages). Be that as it may, Thatcher wrote of "iterative 'growth spurts' of cortical connections," "a sequential lengthening of intracortical connections between posterior sensory areas and frontal regions" in the left hemisphere, and "sequential contraction of long-distance frontal connections to shorter-distance posterior sensory connections" in the right hemisphere (p. 232). Thatcher's observations were of structures far too gross to reveal anything about control systems. His observations were of such things as changes in the bulk of "short-distance" (millimeters to a few centimeters) and "long-range" (several centimeters) neural fibers. But, as Plooij says, research such as this gives evidence that physical structures in the brain do change radically, and many of the changes result not merely in more of something, but in greatly altered patterns of connection. Plooij continues:

> Since ... research on human infants is very difficult to do, I am quite optimistic that one will find evidence for more periods now that we know where to look.... We hypothesize that with each sudden addition of brain components, the next,

new type of perception emerges such that at the end of the sensorimotor period the ten orders of perception and control are in place.

This hypothesis is based on the observations that clusters of new concrete skill are observed shortly after the onset of regression periods. Such new skills are the outcome of a new type of perception coming into action, forcing the organism to reorganise the already existing hierarchy. This is in line with the enormous individual differences that are found in developmental psychology in the age at which a new skill is first observed: what actual skills a baby develops after having gotten a new type of perception at its disposal depends on the circumstances and these vary enormously. In our research we try to 'catch the reorganising system in the act' . . . by observing individual babies from two months before the onset of a regression period to two months thereafter and see what new skills they master and how. Another line of our research is stress. Some of the intrinsic quantities that are changed when reorganisation takes place may have to do with stress. In developmental psychobiology new, nonintrusive ways of measuring stress have been developed over the last two decades, such as measuring cortisol in saliva. We are planning to collect saliva samples frequently before, during and after babies go through a regression period. One would expect a peak surrounding the onset of a regression period.

ANOTHER KIND OF SYSTEM

To provide PCT with a means to bring a controlled variable into match with a reference value when the organism discovers that it cannot do so, Powers postulated the capability of *reorganization*—the capability not of "correcting" the outputs of existing systems, but of changing the properties or even the number of components among the multiply-connected loops. Reorganization changes the connections and components, and therefore the reference signals, so much that outcomes not previously possible (not previously in the repertoire, so to speak) can now be brought about.

This kind of change may sound like an appeal to magic. It turns out, however, that the basic ability is surprisingly simple. But before I go into that, let me dispose of any worry you may have about magic

by saying that one way of building a machine with the capability of reorganization has been known at least since 1952, when W.R. Ashby published his account of his "homeostat." That machine, built of wires, coils, pointers, stamped metal sheets, and nuts and bolts, consisted of four negative-feedback control systems and another device Ashby called the "uniselector." Ashby built into the uniselector the capability of recognizing an indicator of well-being in the functioning of the homeostat. When the indicator differed from the standard Ashby had built into the uniselector, the uniselector altered the organization (the electrical connections) of the homeostat until the indicator again matched the specified value. The homeostat was able to withstand severe disruptions of its organization. It could survive even having a functioning wire completely severed! It did so by finding other connections that could substitute for the lost one. This manner of functioning is the way living creatures function. They encounter disruptions of their functioning, even physical damage, and go right on finding ways to do what they want to do. (Do I need to keep saying that, even here in Chapter 20?)

I hope you are astonished that the description of Ashby's device has been available for half a century in every university library in England and the United States (I'm fairly sure) and many in other countries, and yet has been treated in mainstream psychology books—when it has been mentioned at all—only as an out-of-date curiosity. But Powers was paying attention; he wrote to the CSGnet on 17 October 1996 that his conception of the reorganizing system was an adaptation of Ashby's homeostat. The fact that the animal body does maintain physiological internal variables within certain ranges of stability has also been known for a long time. Claude Bernard described the control of many such variables in a two-volume work published in 1859. Walter Cannon enlarged that knowledge in 1932.

Powers included a couple of examples of control of "intrinsic" variables in a posting to the CSGnet on 1 September 1999:

> The control system [for internal temperature of the body] is located in the hypothalamus, and the sensors are in the carotid arch of the arteries carrying blood to the brain. The output functions, multiple, are driven by autonomic nervous system signals that dilate and contract peripheral blood vessels, that produce sweating (for cooling) and shivering (for heating). . . .

There are dozens of such known control systems concerned with maintaining the body in a specific state both physically and chemically. For some reason, mainstream biologists and biochemists have energetically resisted thinking of these systems as control systems, but many scientists have recognized them as such and have written books and published papers about them.

During initial growth of the brain, new synapses are formed as nerve cells send axons out, seeking sources of chemicals given off by other nerve cells. Even in the adult human brain, which was once thought to be unchangeable, old synapses are lost and new ones are formed, and the "strength" of existing ones changes continually.

One of the longest-known examples is the loop that controls circulating thyroxin in the bloodstream. Signals from the hypothalamus enter the pituitary, where they cause the release of thyroid-stimulating hormone, TSH, which stimulates the thyroid gland to produce thyroxin. The concentration of thyroxin feeds back to the pituitary to strongly *inhibit* the production of TSH (this makes the signals from the brain into reference signals). The result is a control system that so strongly controls the level of circulating thyroxin that direct infusion of thyroxin into the bloodstream (in an attempt to raise its concentration) simply shuts down the thyroid gland (which, to the surprise of doctors and the detriment of patients, eventually results in atrophy of the thyroid gland).

[The existence of a reorganizing system] seems to be demonstrated by the fact that both animals and people are "motivated" to learn under conditions where we would expect large error signals to exist in these basic physiological and biochemical control systems: hunger, thirst, pain, poisoning, excess cold or heat, illness, electric shock, and so on. All higher animals including human beings will acquire new control abilities when deprived of things they need, including basic physical needs.

For more on the hierarchy and reorganization, go to Chapters 6 through 13 of Powers (1973), to Chapters 5, 7, and 12 of the book by Robertson and Powers (1990), to the article by Raymond Pavloski and others (1990), and to the chapters beginning on pages 27, 45, and 135 in Powers (1998).

Now, for the occasion when the existing control systems in a neural hierarchy fail to control a variable vital to the physical survival of a creature, how can the neural network, having itself no consciousness of its own interconnection, no intelligence, no way to reason that one connection is better than another—how can the neural network present at birth immediately begin to find its own better organization? Powers (1973, p. 181–182) says that Robert Galambos (1961) "states the essential property I would like to give this theory of reorganization" and quotes him as follows:

> It could be argued, in brief, that no important gap separates the explanations for how the nervous system comes to be organized during embryological development in the first place; for how it operates to produce the innate responses characteristic of each species in the second place; and for how it becomes reorganized finally, as a result of experiences during life.

Powers goes on to say:

> ... we must try to discover processes of reorganization that could in principle be inherited, working through inheritable mechanisms that are not themselves shaped by reorganization but rather, in some way, divorced from *what* is learned. ... Nearly everyone who has worked on self-organizing systems has used principles like mine; I am merely adapting what others have done to a specific model (p. 182).
>
> The model I propose is based on the idea that there is a separate inherited organization responsible for changes in organization of the malleable part of the nervous system—the part that eventually becomes the hierarchy of perception and control (p. 182).
>
> Reorganization alters . . . . the kinds of quantities perceived, the means of correcting error through choice of lower-order reference signals, dynamical properties of control systems, and even the state of existence of a control system (p. 186).
>
> The reorganizing system ... does not sense behavior or its effects on the environment. It senses only intrinsic quantities, and . . . . *the process of reorganization is independent of the kind of behavior being reorganized* (p. 187).

In that last sentence, Powers means that the process is the same for finding a way of reducing conflicts and errors when catching food on the hoof, keeping peace in the family, organizing the government of a city, putting beauty into a painting, or any other sort of behavior. He also means, for example, that if a rat finds itself in a small enclosure where there is no food and nothing to explore except a stick in the wall that can move up and down and a hole beside the stick with a cup under it, and if moving the stick does indeed produce a pellet of food in the cup, then— if the rat does not first die of starvation—the rat will acquire the control system that enables it to satisfy its hunger by pressing down on the stick, regardless of how crazy (so to speak) pressing down a stick may be in comparison to its previous experience in getting food. Powers continues:

> . . . this reorganization process will cease . . . only when a behavioral reorganization appears that results in restoration of the intrinsic signal to its reference level, with consequent disappearance of the intrinsic error signal. . . . The criteria for terminating reorganizing do not depend on a control system's achieving some goal state for its perceptions. . . . Reorganization may terminate when one fails to remember or fails to solve a problem, if those failures result in restoration of intrinsic error to zero. One will then have become organized to fail in those situations. The reorganizing system has no pride (p. 187).

In his communication to the CSGnet on 1 September 1999, Powers puts the whole matter succinctly:

> This is the basic principle of reorganization that I propose. When error exists in intrinsic control systems (fairly large and long-lasting error, perhaps), I propose that a process of random change begins in the brain, which ceases only when the intrinsic errors driving the changes are removed. The point of this process is not to create any particular organization in the brain, but only to correct the intrinsic error—to correct, say, a prolonged depression of body temperature below its inherited reference level. The existence of a sufficiently large error, which the inherited control system cannot correct, is proposed to cause reorganization to commence. Any change in the brain that results, however indirectly, in a decrease in intrinsic error (an increase in body temperature) will slow the rate of reorganization of the brain, meaning that the existing organization that produced the decrease in error will persist longer before the next reorganization. In the end, *anything* that the brain ends up doing that corrects intrinsic error—that brings body temperature back to its inherited reference level—will be retained, simply because the disappearance of intrinsic error will stop the changes in organization.

Here I will make a cautionary remark about the phrase "process of random change." Writers describing PCT's reorganizing system often use that term "random," but they do not mean what mathematicians mean by it; namely, every possibility equally likely. A more fitting term might be "unsystematic" or "without rational basis," but I will go on writing "random" simply because it is customary and will do no harm.

In a posting to the CSGnet on 13 August 1999, Marken gave us another succinct statement:

> Intrinsic variables must be held at *genetically* specified levels for the organism to survive. So the genes are the source of references for intrinsic variables, which are probably variables like body temperature, blood sugar level, etc. . . . The theory is that we learn to control these perceptual representations of the outside world *as the means* of keeping the intrinsic variables under control. The entire control hierarchy is the *means* by which we end up keeping our intrinsic variables under control. Some of this is obvious; we have to learn what perceptions represent food and what to do with those perceptions (catch them, eat them, chew them) so that intrinsic error stays low. But [PCT] assumes that *all* of our control skills—our ability to control everything from intensities to system concepts—are built in the service of keeping all our intrinsic errors small. It may not be obvious how learning to control certain variables—like logical variables or principle variables or system concept variables—might keep an intrinsic variable like blood sugar level at its reference. But there are probably many intrinsic variables and some of these might be what we think of as more "spiritual" type variables: inner peace, harmony, grace. I can imagine how measures of completely physiological variables—like the average error in the control hierarchy—could be genetically specified (at zero, say) and experienced as a sense of peace (when this variable is near its reference) or anxiety (when this variable is chronically far from its reference). These are the kinds of intrinsic variables that are kept under control by control of higher level perceptions like programs, principles, and system concepts. . . .

> Learning [of the reorganizing sort] doesn't occur (in theory) unless there is intrinsic error. If there is chronic error in your tennis stroke control system then (according to PCT) this system will be reorganized only if there is intrinsic error. I think the level of error (or a side effect of error) in any control system is almost certainly an intrinsic variable; one of the most important intrinsic variables must be the perception of error in any control system in the hierarchy, and we are almost certainly genetically programmed with a reference of zero for this intrinsic perception. So if there is chronic error in any control system in the hierarchy, reorganization will start working on that system in order to reduce the intrinsic error created by the error in that system.

I pause to say that a great part of the conception of the reorganizing system remains speculative. The existence of hierarchical levels of perception seems incontrovertible, but whether they exist in a countable number, just what each encompasses, just how reorganization proceeds, just how it is set off, what particular intrinsic variables take part—all that and more is speculative. I do not put this note just here because I think Marken's speculations about higher-level variables are excessively dreamy. I am a firm admirer of Marken's dreams. Powers (1973, p. 195) himself says:

> It is well to remember [for example] that the reorganizing system has been quite arbitrarily separated out from the whole organism, and that it might be an aspect of every subsystem.... The *only* reason I have performed this separation is that not doing so makes the picture too complicated for me to think about.

And in his communication to the CSGnet of 1 September 1999, Powers put it this way:

> There are lots of unanswered, and for the moment unanswerable, questions relating to my proposals concerning reorganization. We don't know where the system responsible for it resides in the body or brain, or even whether it is distributed over all the cells. We don't know what confines reorganization to relevant parts of the brain, or what directs reorganization to those parts (although there have been almost-untestable speculations). While we have proven that random reorganization can be used to achieve some highly organized results (like solving a system of 50 linear equations in 50 unknowns without using algebra), we don't know whether some other inheritable process might not work better.

RANDOM REORGANIZATION

It turns out that in searching for a new organization of the neural hierarchy, a random (unsystematic) procedure can work surprisingly well. Powers, with Clark and McFarland (1960), assumed a random procedure in his earliest publications. Then an article by Koshland in 1980, another by Marken in 1985, and another by Marken and Powers in 1989 revealed the fact that for maybe a few billion years, the humble *Escherichia coli* had been demonstrating all unseen the efficiency of random behavior. In Chapter 7, under the heading "Chemotaxis," I described Marken's 1985 modeling of *E. coli*, and I said a little, too, about the 1989 modeling by Marken and Powers. You will remember that it is by random flailing that *E. coli* moves into the more nourishing parts of the fluid in which it swims. Marken and Powers (1989, p. 93) reported that their model of *E. coli* could become "as much as 70% as efficient as a straight-line motion to the target..." and that the method of random trials of new directions such as that used by *E. coli* "is the only feasible way for an organism to maintain control over important effects on itself when its environment is totally beyond its comprehension." Here is a similar realization from another student of perception, Richard L. Gregory (1988, p. 38):

> The blind yet highly creative Darwinian statistical processes, producing changes in species, may seem very different from the mechanisms of behaviour in individuals. But there seems no reason why man-made machines, or living organisms including ourselves, should not play much the same trick—to be creative by selecting from randomly recurring events... either external events or within the machine or the brain.

Gary Cziko has written a beautiful book on how evolution has brought us to our present capabilities of behavior, entitled *The Things We Do* (2000). In Chapter 10 of that book, Cziko tells us how the tactic used by *E. coli* can be an example of purposeful evolution and how it can become the means for perceptual reorganization. Cziko invites us to try out the *E. coli* program available at

http://faculty.ed.uiuc.edu/g-cziko/twd/demos.html

That program is also available, along with other demonstrations, at

http://home.earthlink.net/~powers_w/

If you put together the quotation from Richard Gregory that you read three paragraphs back with the quotation from Robert Galambos several pages back, you get the idea that we are still making good use of some ways of functioning that our forebears used millions of years ago. In the primordial seas, one-celled creatures like the *E. coli*, by "selecting from randomly recurring events," found the path to richer nutriment. That strategy, as Galambos implied, is available to the embryo and continues to be available to the adult. Powers conceives that capability as the basis for the reorganizing system.

Finding effective variables with which to control perceptions is helped by the capacities of the higher levels: categorizing, sequencing, programming, logic, experimentation, and theorizing, among others. But none of those impressive abilities solves every difficulty. In the end, finding an environmental variable with which to control a perception must, when all other skills fail, fall back on unpredictable trial and error. Sometimes a failure of control spreads through many loops at many levels, and reorganization is far-reaching; then the reorganization is often accompanied by strong emotion. The emotion may come in one burst, in a series of bursts, or as a fairly low-level background to action.

QUANTIFICATION

Most psychological studies give little attention to quantities, degrees, amounts, or rates beyond vague assertions of more and less. In saying that, I am not speaking about statistical counts and probability calculations, but instead about the functions that can be compared numerically between model and observed behavior. You will remember that you encountered that latter kind of quantification in the experiments recounted in earlier chapters: distances in pixels across a computer screen, numerical rates of change, and so on. Correspondingly, quantities and rates are also important in the model of the reorganizing system. I will not go into any detail about this, but I'll give you the flavor of the matter by quoting from a few of Powers's contributions to the CSGnet. First, here is a comment about speed of reorganization from Powers's contribution to the CSGnet on 24 January 1996:

The best reorganization model I have been able to come up with uses both the magnitude and rate of change of intrinsic error to determine the speed of reorganization. A random reorganization . . . occurs when the rate of change squared (or absolute) intrinsic error is positive. The *amount* of change of the parameters is proportional to the squared or absolute error, so change slows as error decreases. . . . As long as not too many parameters are being adjusted at once . . . the intrinsic error will continually decrease until the least achievable error is found.

Next, I turn to the query from Hugh Petrie to the CSGnet on 14 April 1999 asking how reorganization could explain the phenomenon of "learned helplessness":

This was the experiment where they shocked animals continuously and did not allow them to escape until they no longer tried to escape [but] simply lay there and took the shocks. Also, when they were allowed to escape, it took a long time for them to relearn how to escape. The [question] is: why doesn't the reorganizing system seem to keep on working when it appears that there is still a lot of error in the animals who lie there and keep getting shocked?

On 15 April 1999, replying to Petrie, Powers mentioned a series of experiments he had been conducting in modeling reorganization, a series he had announced to the CSGnet on 20 June 1997 and had described further on 1 July and 17 July. Powers then reported something his modeling had taught him about rates of change:

I had to adjust the speed of reorganization to suit the likelihood of a better organization being found. The *E. coli* demo shows this effect very well. If reorganization goes very fast or very slow, progress up the gradient is inefficient and sometimes vanishingly small.

For reorganization to work, there has to be time to measure whether the basic error is increased or decreased after a random change. There is a broad range of rates of reorganization over which this can be done as well as needed. But if reorganizations start to come too fast, I'm sure you can imagine what happens. They come closer and closer to being unrelated to their consequences and thus truly random in their effects. They start to dismantle whatever organization already existed, instead of improving it.

In the case where shocks are administered in a way unrelated to behavior, there is no way systematic reorganization can happen. Now you're just looking at random changes, and the result can only be progressive loss of organization. You don't see the *E. coli* effect unless there is a measure of error that is systematically related to the timing of the reorganizations.

Powers made an amplifying comment on rates in a posting to the CSGnet on 20 October 1999. He was replying to a comment by Richard Kennaway, who had written on 19 October to say that he thought the reorganizing system "will not usually produce actions bizarrely unconnected with the error, like yodeling to get your tractor unstuck." In his reply, Powers announced a modification of his earlier random model:

> I agree. My experiments with simulating reorganization led me to a model in which parameters are not *simply* (emphasis PJR's) chosen at random (which would result in bizarre and drastic changes). Instead, the system parameters all change at fixed rates between reorganizations, with a reorganizing episode randomly altering the *velocities* of change. The rate of change is also proportional to the total error, so it slows down as total error approaches zero. There are probably other ways to achieve the same effect, the effect being to assure that nothing about the system can jump to a radically different state (like growing a hand out of your forehead through mutations). The slowness of the resulting changes assures that the system won't change much before an increase in error is detected and another change of direction takes place [to counter the direction of error]. Thus many things could be reorganizing at the same time, and the underlying principle can involve random change, as long as you change the *right thing* randomly and not by very much.

CHANGING CONTROL

Were it not for our ability to control perceived variables, we would be at the mercy of every event impinging on us from the environment. We could not stay alive; we could not even begin to live. Furthermore, if we could control only those variables for which we already possessed internal standards (reference signals), we would not live longer than a very few seconds after we emerged from egg or womb. Well, perhaps animals born with several layers of control (such as mammals) might live for several minutes, maybe even an hour. Our genes, no matter how many, cannot endow us with ready-made reference signals to counter every threat to our existence. It is therefore requisite for us to be able to alter the hierarchical organization of control—to establish new interconnecting patterns among the loops.

But reorganization cannot be done by the ordinary control loops. Control loops can only convert input signals into perceptions (by means of the input functions), compare the perceptions with reference signals (provided from elsewhere), and convert the resulting error signal (by means of the output functions) into output signals. They can do no more. In a communication with the CSGnet on 24 December 1997, Powers said:

> ... no system ever changes itself.... In PCT, there is no such thing as a "self-organizing system." If change occurs, we have to consider both the system that is changed and the system that is responsible for making the changes.... That's why there is an explicit reorganizing system in my model; if a control system's organization changes, something outside the system must have changed it. That's also why there is a hierarchy: if a goal or reference signal changes, something else at a higher level must have changed it.

When the existing connections of the hierarchy allow errors in the *intrinsic* variables to increase, then the reorganizing system accelerates its random rearrangements of the connections among loops and of the effects of input and output functions. In particular, if a conflict occurs among the outputs of loops at the level of system concepts, there is no higher level that can "tell" the level of system concepts how to resolve the conflict, and the reorganizing system will come into action. It may be that conflict at the upper levels, or perhaps generally, is itself an intrinsic variable that the reorganizing system keeps at a minimum.

The fact that an adult neural hierarchy can cope with a wide variety of disturbances without resorting to reorganization may be one of the reasons that young people often perceive old people as being "set in their ways." It may very well be true that older people (not always, but on the average) meet disturbances by using their customary methods (without having to reorganize) more often than younger people do. The obverse of that coin is another phenomenon often remarked upon: that people of exceptional creativity often show it at an early age, constructing principles and system concepts very different from those available in the culture into which they were born. I do not know, however, whether the two assertions in this paragraph have ever been checked by systematic surveys.

UP AT THE TOP

At the top (eleventh) level, that of system concepts, there is no higher level to provide new instructions to the comparators of the loops. As I said earlier, when loops at that level cannot control their inputs, the reorganizing system saves the day by casting about randomly for a solution. But reorganization that entails changes at the higher levels—particularly the highest—is not easy. Any psychotherapist or historian can give a hundred examples off the end of her tongue. To stir your own memory, I will give only two examples here.

In 1775 Patrick Henry, committed to independence for the colonies, spoke in the Virginia Assembly. He cried, " . . . give me liberty, or give me death!"

Bands of millenarians periodically convince themselves that Jesus' millennium of joy is imminent. They fasten upon a date, gather in a suitable place, and wait with awed and awesome expectation. But somehow, Jesus does not seem to be available on that date. When Jesus fails to show up, do the people give up their concept of the millennium or their faith in Jesus? Very few do. Most find another explanation of the disappointment such as a miscalculation of the date or insufficient faith among the group. Leon Festinger was present with such a group at their fateful hour. Afterward, he and colleagues wrote a book, *When Prophecy Fails* (1956), about the various ways the members explained the failure of their prophecy.

In a missive to the CSGnet of 9 August 1999, Powers said this:

> I'm not the only one who has remarked how hard it is to effect any changes at the system concept level. In part, I think, this is because any change at that level has far-reaching consequences at all the lower levels, so any reorganization is more likely to increase than decrease [overall] intrinsic error. We may reorganize at that [top] level, but we're likely to reorganize right back to where we were. This must be the case for any system, once it reaches a state of minimum error: any further change is likely to be for the worse, even if the minimum of error is only relative to local conditions.
>
> . . . at the system concept level, our reference signals show no signs of being freely adjustable. One does not shift freely back and forth between being a Democrat and being a Republican, being a behaviorist and being a humanist, being a Christian and being a Muslim, being a scientist and being a mystic. The reference signals that determine one's place in such arrays of concepts are slow to form and slower to change, so that once we have placed ourself in the spectrum of concepts, further change becomes very unlikely. . . .

CHANGING OURSELVES

We need control loops to keep us alive, and we need the reorganizing system to adapt our controlling to our unpredictable environment, but we cannot control our reorganizing system. What can we do, then, when we become conscious of the fact that we cannot find a way to control a variable we want desperately to control?

When one is troubled by some inner conflict (such as simultaneously wanting and not wanting to quit smoking), one's friends often try to help by giving advice. The advice often takes the form of urging you to "make" yourself do one thing or another. Or to "try harder." Or to "just say no." But the reorganizing system is beyond ordinary control. The desire to please your friends overcomes your wish to smoke, and you throw your cigarettes into the wastebasket. Then as the hours go by, your craving overcomes your wish to please your friends, and at last you run to the wastebasket in a panic, seize the cigarettes, and light one up. Then you feel guilty, you throw your cigarettes into the wastebasket, and. . . . But as the tension increases, the rate of reorganization increases, and the urging of your friends at last results in some behavior in which you were not previously engaging—behavior of random sorts. Perhaps you call one of those friends at two o'clock in the morning to ask her to come and hold your hand so you won't reach for a cigarette. Or perhaps you ask her to join you in streaking naked through a public gathering, both holding signs reading "Stop smoking now!"

Urging a friend who wants to make a change but is in conflict about it to "try harder" no doubt once in a while tips the balance of internal errors so that the person takes some visible action that both you and your friend are glad about—and that is the kind of outcome you are most likely to remember. But there is no guarantee that the exhortation to try harder will be beneficial or even harmless.

Since, according to PCT, the rate of reorganization increases as the overall error throughout the loops increases, one might expect that reorganization could be increased by increasing *any* error by a sufficient amount or for a sufficient duration. It may be that various strenuous disciplines *sometimes* have the effect of enabling the neural net to find a new organization that turns the person more effectively toward a chosen goal. Perhaps that is one use of the policy of "no pain, no gain" used by some persons who want to develop their muscles. There is no guarantee, of course. For some people, that policy may lead to abandoning the activity. Perhaps, too, the medieval practice of self-flagellation brought about, for some monks, a reorganization of their purposes that pointed them with more determination toward what they conceived to be sanctity. I suppose, however, that the practice enabled others to be more determined to leave the monastic order. Many other disciplines have been tried—walling oneself up in a small room, going without food, sitting for a period on a lonely mountain peak, and so on.

Still, things are not hopeless. In that missive of 9 August 1999, Powers said:

> Since people don't like intrinsic error and will immediately start changing themselves, quite automatically, whenever it occurs, it would seem that there is nothing one can do to reorganize on purpose; either you do or you don't. This would make it seem that there's no point in trying consciously to change oneself. Yet it is possible to learn ways of allowing reorganization to occur; for example, you can put yourself in painful situations and, for perfectly logical reasons, *cause* intrinsic error. This is somewhat paradoxical, because according to theory this should result in reorganizing the logical system that produced the intrinsic error, so you stop doing that. But such situations can certainly exist temporarily. In fact this is just another aspect of the idea of reorganization driven by intrinsic error: if logic, or any other high-level process, ends up creating intrinsic error, the result will eventually be a change in that process until it stops creating intrinsic error. This is what makes reorganization appear to be the highest level of control, even though it's not a systematic control process and can operate anywhere in the hierarchy.
>
> But this still doesn't answer the question of how we can *consciously* change our [neural] organizations. The only hint I have ever found in this regard is the way conscious attention seems attracted to problems, to error signals in the hierarchy. There are certain error signals that simply can't be ignored. . . . So I speculated that reorganization follows conscious attention. This seemed very convenient to me, because it can explain why, when we have some intrinsic error that has no direct connection to a control problem, reorganization doesn't . . . disrupt systems that are already working perfectly well. If it can be directed to the cause of a problem, reorganization will operate where it will do the most good, even if one can't predict its outcome.
>
> This ties in with what psychotherapists have found: whatever the method, it must entail directing conscious attention to the real causes of the problem. . . . if attention is directed to the right place in the hierarchy, nothing else is required of the therapist. In fact there is nothing else the therapist can do that will have any effects on reorganization. What actually works in psychotherapy is redirecting attention; everything else that goes on is irrelevant, or even detrimental.
>
> Directing attention to a problem also seems to make it temporarily worse. I suspect that is a side-effect of activating conflicts. Whatever the explanation, this indicates why people do avoid becoming aware of their real problems: doing so makes them feel worse, not better. This is known in technical parlance as a local minimum—the error has to get larger before it can get smaller. This local-minimum effect is what gets people stuck in bad [neural] organizations. If they start to change, the intrinsic error becomes larger, and they will reorganize right back to where they were. Some kind of external nudge is needed to get them over the hump.

I think of the necessary state from which reorganization arises as having three requirements. First, you must feel that something is somehow not fitting properly, or that there is a gap where there ought to be a connection. Second, that misfit or gap must nag at you; you must *want* to find the fit or repair the gap, whatever or wherever it may be. If you perceive no gap, you will of course do nothing to change anything. And if you perceive a gap but do not care about it, you will again do nothing. Caring means having an internal standard for some implicated variable. If what you are experiencing—perhaps a strange dog sitting on top of your piano—disturbs none of your controlled variables, then you will take no action to alter the situation. Third, the quantities must lie between enough and too much. If the misfit seems outrageous, you may ignore it as absurd and not worth your further attention. I think those three conditions are necessary to set off reorganization, but not sufficient. I do not know what is sufficient.

RECAPITULATION

Here is a summary of what reorganization seems to do and not do. It was contributed by Powers to the CSGnet on 16 April 1998:

> First, if you have a systematic way to achieve a result you don't need the less efficient method of reorganization to achieve it. In fact, because the systematic method is installed as an automatic control system, it will correct errors as soon as they appear, leaving insufficient error to turn on the reorganizing process.
>
> Second, reorganizing at a higher level, if possible, is more useful than reorganizing at a lower level. In some respects, the higher level system can "work out" what is needed at the lower level without the need for rote memorization or (random) trial-and-error. New situations can be dealt with without the need for reorganization.
>
> And third, you use reorganization only when there is NO systematic way to get what you want. This is why reorganization has to be a random process—random meaning "not according to any known systematic scheme." If you have tried all existing systematic schemes and none of them corrects the error, all that is left is a non-systematic—random—scheme. That is why, in PCT, the output of the reorganizing system is a random process.
>
> So—when does reorganization occur? Apparently, whenever there is a perception that we want to recreate, and we have no ready-made method of recreating it.

All this is simply a description of what seems to happen. It's not a model.

DRUGS

In a posting to the CSGnet of 17 November 1995 on the topic of drugs, Powers wrote:

> People who are in conflicts, either internal or external, and who have not been able to resolve them, are in a state of continued intrinsic error and reorganization. As long as they're reorganizing, they have at least a fighting chance of correcting the errors. But suppose that what (some) drugs do is to make it appear that one or more, and perhaps all, intrinsic errors have, in one great rush, gone to zero. This will stop reorganization in its tracks, and whatever behavioral organization was in existence at that time will become permanent until reorganization starts again. The normal connections between behavior, its consequences, intrinsic error, and reorganization have been broken. There is a drug "habit" simply because the organization that controls for ingestion of the drug is no longer susceptible to change.
>
> [An] effect of zeroing out intrinsic error signals will be that the normal basis for learning has been destroyed. Starvation, for example, will no longer cause reorganization that can be ended only by finding and eating food. Obtaining sex will no longer be a target of reorganization. Even pain no longer results in reorganization that ends with learning how to control variables associated with pain. What we end up with is a system in which reorganization has been almost totally suppressed, a system with a fixed organization that, from then on, can only slowly deteriorate. Failures of control do not result in learning to improve control. They simply continue.
>
> Different "feel-good" drugs must have different degrees of effect on reducing intrinsic error. Some may selectively affect reorganization at different levels in the hierarchy. The worst, of course, are those that have the most general effect on reducing intrinsic errors of all kinds. A drug that totally suppressed all intrinsic error signals would probably be lethal—although the person would die in a state of tranquility or euphoria, never feeling that anything is wrong.

AWARENESS AND CONSCIOUSNESS

Poets are often good at reminding us of the feel of sheer awareness; here is William Wordsworth (1770–1850) doing it:

> I wandered lonely as a cloud
> That floats on high o'er vales and hills,
> When all at once I saw a crowd,
> A host, of golden daffodils.

Marvelous though it is, PCT so far says little about the feature of ourselves that many writers throughout history have proclaimed to be the most marvelous feature of all: awareness. Awareness does not seem to be a function in the control loops. Awareness observes; it has access, somehow, to perceptions—the warmth of water, the friendliness of a smile. But awareness does not instigate or direct action; it has no connection (whatever "connection" may mean here) with output signals. Awareness can *serve* control; it can enable us to control a variable by using some feature of the environment with which we are not at the moment operating. I might, for example, be approaching my goal of perceiving myself entering my bank by walking along the street, but then become aware of an approaching bus and seize that opportunity to approach my goal faster. But awareness is not a function in the control loop itself. And it is not itself something to be perceived. We can add types of variables to be perceived by adding orders (levels) to the neural hierarchy, but awareness is not a variable; we cannot construct a model that includes awareness simply by adding a level to the hierarchy and proclaiming it to be "awareness."

Some reasonable remarks can be made about the way awareness functions in relation to the neural hierarchy, although none of these relationships has yet been modeled. It is obvious, for example, that every level of the hierarchy can carry out its control functions accompanied or unaccompanied by awareness. Control at the higher orders, especially, usually seem to an outside observer to be carried out consciously, since it requires complex, symbolic, and abstract processes that go on only "in the mind," and surely (so many of us think), we must be aware of what we are thinking. But it is common for higher process to go on quite without our awareness, "like a computer running in an empty room," Powers says.

The characteristics of the world we are conscious of are those controlled at the level of the perceptual signals that are being observed—the signals of which we are aware. We are not conscious of the flexing of our leg muscles when our attention (awareness) is on making our way through a crowded marketplace. And at such a time, we are not likely to be aware, either, of the mathematics of queuing or the supposed principles of economics. But of course, the "viewpoint" of the awareness can change its level. We can encounter our physician, who might ask, "How's the leg?" Or we can encounter a placard reading, "Due to increased demand, our prices. . .". Powers wrote about awareness and point of view in a contribution to the CSGnet on 5 November 1997:

> . . . there is one core that remains the same no matter what the content of consciousness: the silent Observer. . . . The watcher of the screen on which experience appears. . . . The point of view. There is always a point of view of which you're not aware, but from which you are aware. . . .
>
> Why are you not aware of the pressure of the seat you're sitting in until I mention it? Because the sensory endings were not responding? Not at all. The signals were there all along. But the Observer was looking elsewhere in the hierarchy.

Overall, however, awareness and consciousness remain mysteries. Here is one more quotation from Powers (1980, p. 235):

> We *do not* know the basis on which the highest-level goals are set. . . . We can offer some reasonable conjectures about how biochemical and genetic factors enter, . . . but we can *by no acceptable scientific means* show that those factors are "ultimate" determinants, not in any sense. It is time to stop trying to make everything fit 19th-century ideas of physical determinism, which are based on little more than an allergic reaction to religion. The upper regions of human organization are a mystery that we have barely begun to approach. . . .

LEARNING

Learning seems to appear in more than one sort, but it seems to me we apply the term whenever someone comes to do something he or she could not do (or perhaps would not have thought of doing) at an earlier time. I say that I have learned your name when I can correctly say, "Hello, Joe," whenever I encounter you. I say that I have learned *Mary Had a Little Lamb* when I succeed in reciting it all the way through, saying the prescribed words in order. I say that I have learned where the cookies are kept when I can go straight to the jar from wherever I am when the cookie hunger sweeps over me. I say that I have learned how to psych out the teacher when I demonstrate that I can get A's on the tests without reading the assignments and even though I skip a lecture or two. I say that I have learned to see the behavioral world differently when I note that I am no longer taking for granted the boundaries of objects, events, episodes, facts, and outcomes, but am instead expecting to find continuity and persistence of perceptions and goals.

Simply remembering is one kind of happening to which we apply the term *learning*; we remember Joe's name or the words to "Mary Had a Little Lamb" or the location of the cookie jar. This kind of "learning" is almost synonymous with acting. You will remember from Chapter 19 that every perceptual signal coming out of an input function goes to memory (as well as to the comparator and on up to a higher level). Whenever you are perceiving some aspects (variables) of the outside world (which you are always doing when you are acting), the resulting perceptual signals go into memories. They are immediately there for retrieval (that is not to say that retrieval is always easy). Some years ago, a psychologist attracted the admiration of a lot of other psychologists when he discovered that rats sometimes learned where the food was in only one trial. I expect rats living in natural places learn the locations of lots of things in one trial. I understand that bees (whose brains are a lot smaller than those of rats) go out foraging for pollen, find a good supply, and then go back to the hive and do the dance telling the other bees the direction in which they can find that pollen, and they do that after making only one circuit out and back. I must admit that my brain is a lot larger than that of a rat, but for what it is worth, I can go shopping in a department store, find the location of men's shirts, do some other shopping, and then almost every time go back to the men's shirts by the shortest path. Lots of people (though not I) can be introduced to a stranger, hear his or her name once, and call the person by it the next day.

Wending your way through a program also looks like "learning." Back in Chapter 4, I wrote about Claire setting out from home to go to the market. She drives to the end of our driveway, where she can choose to turn right or left. (Actually, she could also choose to drive straight across the street, across the lawn of the house there, and up onto the porch, but let's ignore such a possibility.) If she goes left, she drives to the third intersection, and there she can choose to go left, right, or straight ahead. And so on. There are various choice-points with certain possibilities at each. When she gets to the market, we say that she has learned how to get there. (She could, of course, forget what she has learned. But I don't think she will.) Tomorrow, she might make some different choices at some of those points, and still get to the market. In that case, we would say that she has learned another way to get there. When she learned the second way to go, she tried something "new" in the sense that an observer would see her doing something she had not done before, but she did not need to change her mental organization; the same choice-points were there, and the same possibilities at each choice-point were there, and she sought the same goal. That is, the program itself changed not at all.

The third process we almost always call learning is reorganization. Sometimes, if the reorganization is pervasive—that is, requiring radical alterations in many far-flung parts of the neural net—we use labels like epiphany, revelation, convulsion, metamorphosis, upheaval, and shaking the very foundation. Such a reorganization is usually accompanied by strong emotion. It can come and go in perhaps an hour, flashing and shuddering, or it can recur for weeks or even months. But reorganizations can be small, too, and go by barely noticed. Many people come to understand PCT through an emotional and lengthy struggle. My own struggle began in 1985. I was just now looking through copies of the letters I exchanged with William Powers between 1985 and 1990, and they do give evidence of struggle, upheaval, replacing foundations, and lots of emotion. Indeed, my mind is still at it. Writing this book is part of my continuing effort to arrive, eventually, at the perception that I understand what I am talking about. Richard S. Marken (1994) has described his own conversion from traditional psychology to PCT. His agony and jubilation were similar to mine.

Many writers have described the epiphany that comes with a strong reorganization of understanding. Here is John Ruskin (1819–1900), writing in "The Stones of Venice" in 1853:

> The real animating power of knowledge is only in the moment of its being first received, when it fills us with wonder and joy; a joy for which, observe, the previous ignorance is just as necessary as the present knowledge.

Look again at how I write. I used the phrases "to which we apply the term learning" and "reasonably to be called learning" and "a process we almost always call learning." I did not thrust out my chest, lower my voice, growl "harrumph," and pontificate, "There are three types of learning." You can see, when thinking about the matter with PCT in mind, that *learning* is a folk term that will not serve well as a technical or scientific term. It tells us more about what is going on in the mind of an observer (such as a teacher) than about what is going on in the mind of the person observed.

The first two processes—memorizing and traversing a program—are ways of controlling a perceived variable. The third process, however, is a way of acquiring new capabilities of control. An excellent example of the latter is, of course, giving up the assumption of linear, episodic causation in behavior and supplanting it with the assumption of circular, simultaneous causation.

We often act as if other people can simply accept and record and use the ideas or proposals we urge upon them as if the ideas and proposals are cans of good stuff that can be simply stacked in any empty places on the brain's shelves. We act that way especially, I think, when we are acting in the role of parent, teacher, boss, judge, or any kind of expert or authority. When it doesn't work, we must resolve within our own selves the conflict between what we envisioned happening and our realization that it is not happening. Usually, we do not tell ourselves that our theory of learning is wrong. Usually, we tell ourselves that the person who fails to take up the idea we are urging is deficient in some way. We give her a low grade. Or we relegate her to the "slow" track. Scolding is common, too.

You can see how the usual formal education in schools and colleges (well, at least at the Massachusetts Institute of Technology) fails to enable students to reorganize their pre-existing mental structures concerning physics in a dramatic video entitled *Minds of Our Own* (1997)[1], produced by Harvard-Smithsonian Center for Astrophysics. On graduation day at MIT, interviewers asked graduates simple questions about physics: What brings about summer and winter? Can you see an apple in a room where there is no light at all? What is a piece of wood made of that gives it weight? And others. Most of them gave wrong answers, whether or not they had taken courses in physics. For one of the questions, the interviewer asked the students whether they could take a light bulb, a battery, and a piece of wire and cause the bulb to light. All said yes, they could. When actually given the equipment, however, and asked to do it, all failed. Several said you had to make a circuit (they had the *words* in the right order), but none could make one.

Presenting words and diagrams to people who feel required to listen to you on Wednesdays at 10 o'clock, rain or shine, is not the most effective way to enable them to learn about electric circuits. I learned about circuits when I was ten or eleven years old by wiring up the electric motor of my Meccano set so that it would operate the devices I built. I learned enough from that so that I later designed a fancy circuit for a theatrical switchboard and was granted a U.S. patent on it in 1940.

In some of the public schools in the Harvard-Smithsonian Center for Astrophysics study, the teachers were interviewed after they watched their students fail to give correct answers to things the students had recently been taught (so to speak) and on which they had passed paper-and-pencil tests. At least four of the teachers spontaneously commented that *they had never realized that children came to class with ideas of their own already formed.* Tabula rasa!

When new ideas are in conflict with old, it is usually not sufficient simply to announce the proposed idea and hope for the best—which is what teachers (and parents and bosses and experts) ordinarily do. To cope with the new idea, not to evade it, the person must usually deal with it in a hands-on way, making trials and errors at his or her own initiative, though even that is sometimes insufficient. New ideas often come into conflict with old ideas that have become parts of pervasive, cherished beliefs. Incorporating new beliefs into a strong existing organization can necessitate extensive reorganization, often accompanied by weeping, wailing, and the gnashing of teeth.

The vital points about learning are simple. First, people mean various kinds of internal processes when they use the word *learning*, and reorganization is very different from storing perceptions in memory. Second, changing or supplanting perceptions (including conceptions, understandings, and the like) that have extensive interconnections in the neural hierarchy will typically bring conflict. Frequently, the first reaction will be to evade the new thing. When the person cannot evade it, reorganization and emotional turmoil begin. The outcome is sometimes what the teacher, parent, or judge wants, but the outcome cannot be predicted.

And a third point. Living creatures are always, at every moment, undergoing one or another or all of three processes: memorizing, traversing a program, or reorganizing. Those things are characteristic of being alive. You often hear people talking about learning as if it is something you do in school, or while reading a book, or while the drill sergeant is looking at you, but not otherwise. You hear utterances like, "She just won't learn!" or "He is a slow learner." Well, he may be slow about learning the things *you* want him to learn, but his neural pulses travel just as fast as anybody else's.

I have certainly said very little here in comparison to the hundreds (thousands?) of books on learning and teaching you can find in libraries. Many of those books contain fascinating and instructive stories about how memory and insight appear under various circumstances. I am certainly not, however, going to urge you to pore over those books to hunt for the sparse bits that might contain some glimmering of understanding of circular causation or control of perception. That would not be a sound investment of your time.

ENDNOTE

[1] For the full reference, look for "Minds of Our Own" in the references.

Chapter 21

Emotion

So far, in describing the feedback loop and the output function, I have focused almost entirely on the part of the nervous system that ends at the muscles. I have ignored a large and vital part of the vertebrate body; namely, the visceral organs and glands, which are controlled chiefly by the autonomic nervous system and various chemical feedback loops. The viscera and the glands nourish everything. They maintain the bodily environment in which all the cells of all the organs and muscles, and of the central nervous system, too, can function properly. And obviously, the bodily environment cannot stay always the same. Sometimes a muscle will need more oxygen than at other times. When the body is moving rapidly or strenuously for some time, as, say, in running, it will generate more heat than when at rest, and the body must dissipate the heat to maintain the proper internal conditions. In many kinds of action by the muscles, many kinds of changes must be made in the functioning of the internal organs so that they can make the proper alterations in the internal distributions of energy and nourishment. And while tending to energy and nourishment, the viscera and glands also send neural inputs to the higher levels of the hierarchy.

The viscera and glands do not ever, in their normal functioning, stop their action to start up again at a later time. They sometimes work more slowly than at other times, but they are always at work. We pay more attention to the signals from them at some times and less at other times. When we are controlling easily the variables we want to control and do not need to call up extra energy from our bodies to do so, we are rarely aware of the functioning of viscera; but they are functioning nevertheless. When comparators in the hierarchy of control systems send their output signals to muscles, they also send signals to the organs that must provide the energy for the functioning of the muscles. Those signals alter the reference signals for those organs.

In a posting to the CSGnet on 25 October 1999, Powers mentioned some of the key pathways:

> At about the thalamus, there is a bifurcation in the hierarchy, with one branch going to the motor systems and the other passing inward to the hypothalamus and from there, via the vagus nerve and the neurohypophysis of the pituitary (and maybe other pathways—I'm no expert), to the organ control systems.

Sensors that respond to some aspect of the functioning of the viscera send upward the perceptual signals that we experience as emotion. The amount of awareness we have of the signals and the flavor we give to the experience depends on the interpretation at the higher orders of control. But the organic goings-on that produce the perceptual signals are always going on. In that sense, we cannot act other than emotionally—or at least with feeling. But we use the word "emotion" in English in ways that make it awkward to talk about the functioning of the body and the neural net in producing the experience of emotion. Some people prefer to make a distinction between feeling and emotion, using "feeling" for easy running and a lover's caress and saving "emotion" for fear and rage. "Feeling" would be used to denote increased arousal but no threat to intrinsic variables and no need for reorganization, while "emotion" would accompany reorganization. I will not try to be precise with those two slippery words.

Muscles need energy to operate. Control systems can produce large effects on the environment only because there are organs and glands of the body that convert food and oxygen to other forms of energy and that distribute the energy to the cells needing it. (I wrote about the amplification of energy under the heading "Asymmetry" in Chapter 3.) The hierarchy of the central nervous system, to cope with disturbances to controlled variables, alters not only the reference signals in the loops that send neural pulses to muscles, but also (possibly through the pituitary gland) alters the reference values of the chemical feedback systems that provide the appropriate energy. The autonomic nervous system and the chemical controls are always active, of course, in providing proper energy supplies for involuntary functions such as body temperature, heart rate, and perspiration. They also respond quickly to voluntary demands. For example, you might suddenly adopt the goal of getting away from a rhinoceros. On the one hand, this goal is translated into sub-goals for rapid movement of arms, feet, and legs. On the other, it is translated into increased reference signals for such facilitating conditions as heart rate, breathing rate, vasoconstriction, blood pressure, and blood glucose. Some of those facilitating conditions can be sensed directly or indirectly. To the complex of perceptions of visceral activity, muscular activity, and the approaching rhinoceros, you and I would very likely apply the label "fear."

TWO COMPONENTS

Not only psychologists, but thoughtful writers of many sorts have long pointed out that consciousness of emotion has two parts to it—one is the perception of visceral turmoil, the other the cognitive selection of an appropriate goal. In the case of the rhinoceros, the animal might not be charging, but might instead be trampling one's petunias. One might feel an equal urge to do something energetic and might feel very much the same sort of visceral turmoil, but might adopt a goal not of fleeing, but of taking after the animal with a club to drive it out of one's garden. In such a case, one would be likely to apply the label "anger." When one feels a roiling in the abdomen when contemplating no immediate strong action, one is likely to perceive it to be some sort of "upset" in the stomach. When one feels it at the moment of leaping up to correct a large error signal, one is more likely to perceive the pain as the emotion of anger or fear. William James was apparently the first psychologist to set forth the claim (in 1884) that emotions arise from visceral sensations, though I seem to remember a passage or two in the Old Testament describing experiences that "stir the bowels," or words to that effect, and I daresay lovers have noticed the throbbing of their hearts long before writing was invented. Be that as it may, W. B. Cannon (1927, 1929) pointed out that there was not a good correspondence between visceral patterns of arousal and the experience of various emotions. And a good while after that, Stanley Schacter (1959, 1964) performed a series of ingenious experiments showing conclusively that although visceral arousal is necessary for the experience of emotion, one's purpose for action at the time of arousal gives the experience its interpretation as one kind of emotion or another. This bipartite character of emotional experience—the viscera for the energy, the brain for the meaning—is widely recognized today in the psychological literature.

When the energy supplied by the aroused somatic systems matches the demand of the muscles, when not too much or too little energy is flowing from the viscera, one is typically barely aware of any emotional feeling, or if aware, the emotion may be one of pleasure or exhilaration, as during an active recreation like running or swimming. Often, however, when a goal calls for strong activity, the activity is prevented in some way. The result then is that the muscular activity fails to use up the energy flowing from the viscera. One feels then a strong emotional state. Perhaps one begins to flee the rhinoceros only to discover that one's foot is entangled in a tent rope.

We do not think, "I am feeling an emotion," every time we need to call up extra energy, but we can be conscious of the aroused state. We do not feel fear every time we run. Here is what Powers said about this in a posting to the CSGnet on 17 October 1996:

> I'll stick with the idea that in all ordinary behavior, there are feelings that accompany preparation for action—error signals in the learned hierarchy that drive action. Mary said some things, however, that made this clearer in my mind—notably that having these feelings doesn't necessarily imply that any reorganization is going on. If it did, we'd start falling apart every time we worked up a sweat. The somatic systems are an intrinsic part of all behavior, and we sense them as they respond to the demands made on them by the behavioral systems. We feel different when we are sitting and reading from the way we feel when we are pushing a lawnmower or whacking away with a hammer or an axe, or running a race. Could we agree to call those visceral somatic sensations that accompany actions "feelings"?...

The important things about "feelings" is that they do not indicate that anything has gone wrong. Quite the opposite; they indicate that everything is working right. That's another thing that Mary said. If you went out to mow the lawn and after a few minutes weren't breathing harder and pounding your heart faster and perspiring a bit and feeling generally jazzed up, you wouldn't get many rows done. . . .

But that's not the emotion system. The emotion system, at least with regard to bad emotions, responds not when things are going right but when they're going wrong. That's where the association with reorganization comes in; we also reorganize when things are going wrong.

CONFLICTING GOALS

I remember a time when my wife was in the hospital being prepared to undergo a major surgery. I wanted to be there, waiting, just so I could be quickly available for whatever might occur. I went out of our house that morning with enough time to get to the hospital before the surgery was scheduled to begin. I reached in my pocket for my car keys, then another pocket, then another—I had left my keys, both car and house, back in the house, with the door locked! Sheer panic swept over me. I threw myself against the front door. As the bolt ripped, shrieking, halfway out of the wood, I realized that I didn't want to destroy the door. Was there a window open? Yes, there was. I ripped off the screen, retrieved my keys, jumped in the car, and drove off, my heart still pounding wildly. Much later in the day, I remembered that I could have used the key secreted under a rock in the back yard.

As I went out of the house that morning, I had a strong desire to be helpful to my wife, but very uncertain whether there would be anything I could do. At the same time, I was hoping, fearfully, that the surgery would not go wrong. I was, in short, already in a condition to respond to any sudden need for action. Then the action I was already taking was blocked by the locked door of the automobile. And then my action to get my keys was blocked by the locked door of the house. My visceral turmoil redoubled. And then, hearing the shriek of the wood and seeing the crack I had produced in the door, my wish to maintain the integrity of our house leaped into my consciousness, preventing me from crashing on into the house. Luckily, I did quickly find an unlocked window. I did feel foolish, later in the day, remembering the key under the rock. The incident has been easy for me to recollect, ever since, as a good example of the connection between frustrated action and emotion. Remembering it, I still have to take a deep breath (a little more oxygen, please).

That incident illustrates blockage both by an environmental condition (locked doors) and by conflicting goals (getting into the house versus maintaining an unbroken door). In civilized life, blocks to action by reason of conflicting goals are frequent. Should I keep my job, or have the pleasure of punching my boss on the nose? Some people choose one, some the other, and some neither. Only a few, I am supposing, succeed in reaching both goals. Every time you come in your automobile to a red traffic light, you are in conflict. Should you go through it so as to get that much sooner to your destination, or should you obey it and avoid getting entangled in the cross-traffic and perhaps being espied by a police officer? For most of us, most of the time, those repeated frustrations are minor and stir our innards almost unnoticeably. But when we are truly in a hurry, the conflict grows, and the emotion, as we sit cooped up in the car, unable to do much with our muscles except shake the steering wheel, grows too. And for someone who is already suffering from multiple frustrations, perhaps troubles with bosses, or family, or the Internal Revenue Service, the additional arousal at the traffic light can become unbearable.

I remember a taxi driver in Chicago, who took me along a route underneath the downtown loop of the elevated railway. He swore loudly at everyone and everything as we went along—at pedestrians, other drivers, red lights, and the steel columns holding up the elevated railway. In addition to his mouth and larynx, he used his other muscles, too, much more than the driving required. He slammed on the brakes to stop and slammed down the accelerator to start. He jerked the cab, tossing like a small boat in a storm, from one side of the street to the other, passing cars now on the left, now on the right, and missing the steel columns by an inch or two. When we got to my destination, the thought touched my mind that I ought to ask him to pay me for enduring the trip, but I paid him instead, not wanting him to use up his seething energy by using his muscles on me.

In sum, awareness of emotion rises when (1) moving toward a goal requires (2) an increase of energy, and (3) especially when the movement is slowed or prevented. The movement can be frustrated either from outside or from inside. "Emotion," however, does not serve well as a scientific term. It seems to stand for something we notice when we notice it, but we don't always notice it, and we call it by one name when we are using the energy for one purpose and by some other name when we are striving toward some other purpose: fear, horror, anger, love, ecstasy, hate, sorrow, anxiety, jealousy, resentment, disgust, excitement, pride, shame, guilt, and so on.

Sometimes inner conflicts last a long time before they can be resolved. We sometimes find ourselves having to carry on our daily work without taking time to find ways out of some of our conflicts. Sometimes the only resolution of a conflict depends on someone who may take a long time to act. Maybe resolving the conflict depends on waiting a week to learn whether the bank will lend us some money. Maybe it depends on waiting three months to find out whether Albert is going to graduate this year. When we carry unresolved conflicts around with us, our viscera nag at us about it. That is the sensation we call "anxiety." When we resolve those conflicts, the feeling subsides rapidly. "What a relief!" we exclaim.

REORGANIZATION

When the somatic systems begin altering the supplies of energy but the muscles are not using up the energy as fast as it is being supplied, it is likely that some reorganization is occurring. Throwing my body against the door was not my customary way of getting into my house. Breaking the door did not match my usual standards for maintaining my house in good condition. Neither did tearing the screen out of its frame. Some internal standards had to be rearranged. It is interesting that in this tossing about from one direction to another (characteristic of reorganization), the memory of the key under the rock did not come to me until my emotion had subsided. Apparently my reorganizing system was operating, somehow, in a local manner. It did not reach to the memory of the key under the rock. It dealt with the perceptions of the moment, centering upon the goal, never relinquished, of getting into the house. I did not think of the key under the rock. I did not think of getting to the hospital by calling a taxi from a neighbor's house. My reorganizing system, it seems, was hunting for an immediate course of action that would open the way to my immediate goal—getting into the house. Once that path was opened, I could draw on the energy signaled by my emotion to rush in and grab the keys. Then my need for energy would subside—and so would my consciousness of emotion.

Often, heightened emotion narrows the perception of effective action. In 1959, Easterbrook reported on research findings available at that time and concluded that emotion seemed to reduce the range of cues used by a problem solver, with cues not relevant to the immediate task (purpose) being the first to lose the person's attention as emotionality rises. Something similar to that effect may underlie the frequent finding by educational researchers that, on average, students who feel anxious perform less well on tests than students who do not feel anxious. In 1960, Sarason, Davidson, Lighthall, Waite, and Ruebush wrote an entire book on test anxiety; Hill and Sarason wrote another in 1966. In 1991, Paris, Lawton, Turner, and Roth wrote about various effects of standardized testing on students; they claimed that "as many as 10 million students . . . perform more poorly on tests than they should because anxiety interferes with their performance." They also mention the finding of Hill and Sarason in 1966 that "test anxiety was as accurate a predictor of school grades as achievement tests."

I do not cite any of those studies as evidence of the correctness of PCT. Those studies do not show that humans always function in a certain way. They deal more with kinds of actions you can expect statistically (such as getting right answers on tests) than with invariance in functioning. They show that for *some* students in *some* situations, we can expect stronger emotion to go with inferior school performance. They give us examples of correlates of visceral arousal and give us useful information about where to look for higher rates of those incidents. One psychologist, however, Ellen Berscheid (1983, especially pp. 120–134), though she rests her reasoning mostly on the statistical, nose-counting kind of study, paints a picture of the functioning of visceral arousal that is remarkably close, I think, to that of Powers (e.g., Powers 1972).

I believe that most of the internal conflicts in which we find ourselves come from the conflicts that arise in our dealings with other people. Our lives are intertwined with many other people in changing relationships that are sometimes competitive, sometimes cooperative, and sometimes disengaged. Our ways of pursuing our goals so often depend on what other people do that we frequently find ourselves trying to use the same resource that someone else is right then trying to use. And interpersonal conflicts almost always bring simultaneous internal conflicts. I may want to attend a staging of *Macbeth* so popular that the prices being demanded by scalpers are rising. If I give up attending, I will feel deprived of the pleasure of the play. If I pay the going price, I will give up more money than I wanted to. If I sit and dither, the tickets get scarcer and the conflict grows worse. It is no wonder that most of our painful emotions and many of our ecstasies seem connected with other people.

EARLY USE OF THE SOCIAL ENVIRONMENT

Here are further remarks Powers sent to the CSGnet on 17 October 1996:

> . . . when there are bad emotions, or as I call them, intrinsic error signals, some behavior should ensue which at least tends to do something that results in removing the error signals—which entails, of course, bringing the essential variables back within their normal ranges of operation. . . .
>
> When a baby is born, there are some low-level kinesthetic control systems in working order or close to it. They have reference inputs, but there are no or few high-level systems there to provide any non-zero reference signals. The baby can resist disturbances of its limbs, but it can't initiate any voluntary actions to speak of yet, because there's no higher level of control to demand them. According to the Plooijs, that situation will last about one week. . . .
>
> So what does a newborn baby do when its reorganizing system detects errors? It cries. It bellows. That's a built-in hookup; the output signals of the reorganizing system, or emotion system, are genetically routed to the reference inputs that cause the muscles of respiration and the vocal cords to become tense simultaneously during the exhaling part of the cycle of breathing. . . .
>
> This, as it turns out, is a very effective way of correcting intrinsic error, because adults watching the performance can't stand seeing it, and try everything until it stops. . . .
>
> Suppose you've built up a hierarchy of control systems which keeps almost all intrinsic errors within their reference limits very successfully in almost all situations. You will sail serenely through life. . . . But what if you encounter a situation for which you're totally unprepared? Suppose everything you try . . . fails? . . . What do you do?
>
> You cry.
>
> That same old hookup that got you fed, diapered, cuddled, petted, and dried when you were a baby *is still there*. [T]hat hookup turns on just as before and you do the same things you did before, and wait for someone to take pity and fix the error for you. [Y]ou are again a helpless baby. A psychoanalyst would look down his nose at you and say that you have regressed to an infantile state. He's right. . . . But of course you haven't gone anywhere, in space or time. It's just that the failure of the hierarchy has allowed intrinsic errors to get big enough to turn on a very old crude control system. . . .
>
> Suppose that the control systems we acquire early in life are not updated or replaced or dismantled, but are simply kept from seeing any error by more competent systems that we acquire later. The more competent systems keep the same error too small for the old crude control system to act upon. So we swan around showing of our new skillful control system until . . . suddenly the error gets large. What happens? The old system, which is still there and has just been waiting for any error big enough to see, lurches into action, and we do something clumsy and relatively ineffective, but familiar.

ESTHETICS

In his book of 1992a (on page 39), Powers says that we can have a feeling of beauty when we are perceiving at the level of relationship or above. Esthetic judgments seem typically to have an emotional feeling to them, and a self-evident immediacy: "Oh, how wonderful!" or "Oh, how awful!" Powers says (p. 40), "Reorganization can be motivated just as much by an attempt to produce a proper esthetic feeling as by an attempt to avoid pain or get enough to eat. Indeed, the esthetic drive is stronger in some people than the hunger drive. . .".

DRUGS

Here is a communication from Powers on drugs to illuminate the relation among physiology, drugs, and emotion. This is from a message to the CSGnet on 22 December 1997:

> [T]he physiology and biochemistry [of the brain] work the same way no matter what the brain learns to do. The main effects of biochemistry occur when something alters the optimal regulation of the underlying brain physics and chemistry. That's what I call a hardware problem.
>
> [T]he brain senses the state of the body, and learns to interpret it. If epinephine is injected into the bloodstream, the body reacts in a host of ways; its biochemical status shifts balance. The brain detects this change, and concludes that something is happening like what has been happening before when these changes occurred. It labels the result as "fear" or "anxiety" or "anger," even though there's nothing else happening that would indicate the need to flee or seek safety or attack....
>
> As I understand them, psychoactive substances do not act by altering the body's biochemical state, but by directly altering activities in the brain. These effects, even for the educated brain, would [then] be much harder [for it] to understand correctly. The very systems that do the understanding are being altered. Since the sensations coming from the body are not altered (initially), there is no way to distance oneself from the chemical effects, as when one realizes that the injection-induced rush of adrenaline is really only a bodily sensation and is unrelated to anything one was thinking or doing. If some substance lowers the capacity to detect the signals that are normally interpreted as anxiety, one can only conclude that those signals no longer exist, and no longer call for whatever action would normally be taken to control them. One says, "My anxiety is cured!" without realizing that none of the reasons for the anxiety has been removed. The lump in the breast is still there; the only difference is that one no longer feels anxious about it.
>
> The most pernicious of the "feel-good" drugs are those that supply us with the experiences we normally get from learning, accomplishment, and good human relations with others. A heroin user is not interested in love or even sex; every feeling one could possibly get from such things is turned on in a rush by the heroin; no action other than pressing the plunger is necessary. The cocaine user is filled with marvelous ideas; all difficulties disappear and every thought is an insight. The shy and self-conscious fourteen-year-old boy finds that he can make friends with anyone, even girls, after just a couple of tokes of marijuana. The magical mystery tour!
>
> These drugs are more than "recreational." They are destroyers of organization. And I think the same is probably true of every psychoactive drug, no matter how beneficent the short-term effects may appear to be.

IS THAT ALL?

These few pages do not seem much to say about emotion—something you experience every day and is an important feature of every novel, TV drama, and movie. Most writings about emotions do not propose a model of the way a living animal can produce such an experience as an emotion. Most tell about the varieties of emotional experience (hate, love, etc.) and the circumstances in which people are likely to interpret their inner turmoil as one of those varieties. "His knees shook, his heart pounded, he gulped great drafts of air. 'I love you,' he croaked." Fiction writers devote considerable portions of their stories portraying emotions. But psychologists and other social scientists do it, too, though in a much more statistical (nose-counting) way. For example, here is an excerpt from the abstract of an article by R.I. Sutton (1991):

> [Bill] collectors are selected, socialized, and rewarded for following the general norm of conveying urgency (high arousal with a hint of irritation) to debtors. Collectors are further socialized and rewarded to adjust their expressed emotions in response to variations in debtor demeanor. These contingent norms sometimes clash with collectors' feelings toward debtors.

That is another example of telling you where you can more frequently find people interpreting their inner agitation as signaling one kind or another of emotion. That strategy of research, which I have called the Method of Relative Frequencies, can deliver useful information. The article by Sutton, for example, could be very useful indeed to someone thinking of applying for a job as a bill collector. The article says, in effect, "When you ring someone's doorbell to collect a bill, look out for . . .". That's similar to saying, "When you visit the Okefenokee swamp, look out for alligators." That tells us nothing about how alligators function, and the article tells us nothing about how humans function, which is the topic of this book. For that reason, I do not refer you to any more of the literature on emotions. It consists chiefly of that statistical sort.

Part V

The Higher Orders

This part will be something of a hodgepodge, an ollapodrida. I want to say a few more things about how the control loop works. I could have bloated up Chapter 4, but I did not want to strain your patience. You will find the leftovers in Chapter 22. Then in Chapter 23, I will turn to conflicts among control loops—as, for example, when one internal standard wants you to be friendly and shake hands, while another wants you to keep your hands clean and stay healthy. Chapter 24 will say more about how the world looks from the levels of programs, principles, and system concepts. Chapter 25 will focus on logic, a capability of which, it seems to me, everyone is proud. Chapter 26 will pursue that psychological will-o'-the-wisp, personality.

What did I mean by that box back in Figure 4–1 that I labeled "Internal processing"? Well, I meant all the stuff you have just been reading about in Part IV—beginning to control your perceptions by controlling intensities, but if that is ineffective or inefficient, controlling sensations, and if that is not really satisfactory, then configurations, then transitions, and so on up to principles and system concepts, where, depending on the beliefs one has formed, one can call upon the aid of God, Zen meditation, Bach's *Musical Offering*, daily jogging, or even PCT. I do not mean to be flippant here; I do mean to include imagination, emotional invigoration, sport, art, and dreams in the ways we find to answer the demands of our internal standards. Some of our immense repertoire of adaptive ingenuity goes on unconsciously and some consciously.

In earlier parts of the book, I said a good deal about the lower orders of control. In the traditional divisions of academic psychology, those are the realms of physiological and sensory psychology. Most popular writing about psychology, however, turns to the higher orders of control—programs, principles, and system concepts. Beginning in Chapter 23, you will recognize several popular topics.

Chapter 22

More about the control loop

This chapter contains a couple of topics that will amplify what I have already said about the control loop. You may already have figured out some of these implications for yourself.

UNINTENDED AND IRRELEVANT EFFECTS

All behavior is undertaken to control some perceptual variable, one or more. But an act on the environment can have effects on many perceivable variables, among which only a few are those the person intended, perhaps only one. We can call the unintended effects "side-effects." Some of those side-effects impinge on variables the person is controlling, some do not.

Let us imagine that I sit down at the piano on a balmy spring day and play the Intermezzo No. 1 of Brahms's Opus 117. I want to control those luscious sounds that my fingers strike from the singing strings. There will be, however, various side-effects. My wife stops preparing lunch and listens to the music. Those two side-effects are relevant for me, because I care about both. I want to have lunch before long, and I also want to know that my wife, too, enjoys the music. A neighbor, no admirer of Brahms, closes her window. I am not aware that she does so, and I do not become aware later, because she never tells me she did that. The neighbor's action is an example of the effect symbolized by the box labeled "Events irrelevant to person" in Figure 4–1. Some side-effects are irrelevant at first, but become relevant later when I become aware of them. For example, the piano strings get a little out of tune as I play, and at some point, as they slacken, I hear the flatting and call the tuner. All sorts of irrelevant side-effects, however, can occur that I never learn about. A bird might fly by, be distracted by the musical sounds, circle a few times before resuming her hunt for juicy insects, and thereby fail to encounter a male of her species who would have helped her pass on her music-loving genes.

An act, then, can produce:

1 an intended effect on a variable controlled by the actor,
2 unintended side-effects on variables controlled by the actor, and
3 unintended side-effects on variables uncontrolled by the actor ("irrelevant events").

Onlookers may or may not care about any of those effects; that is, the actor's action, intended or not, may or may not impinge upon a variable being controlled by someone else.

Here are a few examples. Perhaps I put fertilizer on my lawn. The variables I care about are the green color and how close together the blades grow. If the turf is bright green and closely packed, I am happy. But my action has other effects, too. The fertilizer is poison to most of the small creatures living in the soil. They mostly die off, and the nourishment in the soil declines. I find that I have to use more fertilizer to get the same results. In addition, a good part of the fertilizer runs off the top of the soil, into the city's storm sewers, and eventually into the river. There it poisons fish and other aquatic creatures. It also increases the expense to towns downstream that must purify the river's water for drinking. Do I care about those effects beyond my perception of solid green turf? If I do not, I call the events influenced by my fertilizer "irrelevant." If, however, I do care about those extended effects of my fertilizer, then I do not call them irrelevant. I use the term to mean irrelevant *to the actor*, not to the onlooker. We do not have internal standards about everything. Almost all our acts have consequences we do not care about, though others may care.

It is easy to see in someone else's action a pattern the person did not intend. Here is a puzzle of that sort. Mr. Secord, of Secord, Curie, Ity, and Company, has reached an age at which he wishes to work only half time. Mr. Secord goes to his office by taking a train at his suburban station. To put a fillip of uncertainty into his new routine, Mr. Secord decides to rise when he feels like doing so, have a leisurely breakfast, perhaps take care of a chore or two, then go to the station, and take the first train that comes along. If the train is going toward town, he will spend the rest of the day at the office. If the train is going away from town, he will get off at the stop for his golf course. Since trains go toward town at half-hourly intervals, and so do trains going away from town, and since he will be arriving at the station pretty much at random times, Mr. Secord figures that he will find himself going to the office about half the time. After two or three weeks, however, it becomes obvious that he is going to the office twice as often as to the golf course! How can that be?

Someone hanging around the train station might be forgiven for supposing that Mr. Secord has chosen to go to work two-thirds of the time, since that is what he does. Even his wife might be forgiven for suspecting her husband of doing what he "really" wants to do. The fact is, however (in this puzzle), that Mr. Secord does not want to go to work two days out of three, but doing so has been an unintended outcome of the plan he laid out for himself.

Here is the answer to the puzzle. The outward trains arrive at his station (let's say) on the hour and the half-hour; the inward trains at twenty and fifty minutes after the hour. The result is that in every hour there are two periods of twenty minutes when the next train will be going inward, and two periods of only ten minutes when the next train is going outward.

Irrelevant effects are not random. They are brought about by purposeful acts, even though they are not themselves purposed. For example, you have seen many doorsills and stairs that have been worn down over the years by thousands of footfalls. Those worn spots do not appear in random places. Perhaps you remember the neat, unintended pattern shown in Figure 9–4 in the section of Chapter 9 headed "The Crowd," showing a simulation of fourteen persons following a leader. No simulated person was given a standard for forming a circle. Each one of the "persons" moved only according to internal standards for staying close enough to the leader and not too close to anyone else. The circle they formed when the leader stopped was a side-effect, indeed an irrelevant side-effect, of the actions to control those two perceptions. Yet the simulated persons formed such an orderly circle that an onlooker could easily think the persons intended to form it.

Some psychological hypotheses claim that people are motivated to optimize, maximize, or minimize something—to maximize power over others or the environment, to minimize effort, to optimize blood pressure. This kind of hypothesis is typical of "utility theory," which was borrowed, I think, from the economists. In economics, a hypothesis popular for a long time proposed that business people always acted to maximize profit. Many people still believe that.

Optimizing can be a side-effect of the control of other variables, just as in the cases of Mr. Secord and the simulated people following the leader. One very likely example is survival. Many writers of all sorts have postulated an instinct for self-preservation that overrides all other motives in all persons. It is hard for me to understand how that notion became so widespread. I don't think a strong "instinct" for self-preservation is shown by driving at breakneck speeds in poor visibility or when drunk, by sailing alone in a small boat in a race around the globe, by climbing icy cliffs at the top of the world with an oxygen tank on one's back, by walking into a hail of bullets in wartime (or by volunteering to fight in a war), by using drugs the poisonous effects of which have been widely and loudly denounced, by risking one's own life to save others from drowning or from a burning building, or by donating a kidney or lung to someone else. You may tell me that people have reasons for doing those things. Of course they do. I am not saying that people do those things witlessly or stupidly; I am saying merely that they do them, and therefore self-preservation is not at the top of our built-in imperatives—not at the top of the control hierarchy. Our evolution has not even shaped us to protect ourselves more carefully during the years before we have passed on our genes to our offspring. I don't think adolescents are more cautious than older people; insurance companies tell us they kill themselves in automobiles at a higher rate than older people do.

The survival of the species, too, may not be built into the neural hierarchy, but may instead be a side-effect of inherited urges such as sexual attraction. If survival itself were somehow built into genes, we would expect the specification to have included some way to avoid killing off the species through overpopulation or atomic warfare. Be that as it may, we can hardly expect evolution to prepare us by natural selection for dangers that have never happened (to our species) such as atomic winter, excessive rise of global temperature, collisions with large meteors, or galactic catastrophes. In such matters, cockroaches and bacteria seem to have the advantage over us.

CONTROL SYSTEMS FIND THE ONLY POSSIBLE ACTION

You have no doubt noticed by now that adherents of PCT are proud that PCT goes a long way to explain how it is that among the dozens or even thousands of actions that seem available for getting to a goal, somehow an organism can find the act that turns out to be just the act that will do it. Dag Forssell, at this point, reminded me that a bird darting through the air toward a tiny entrance to its nest will flex its wings on final approach so it dives in just right every time, regardless of the direction of its approach or the wind at that moment. Among all the tensions of muscles in a human arm that can bring a glass toward the mouth, most drinkers activate just the right combination of tensions that succeed in lifting the glass to the mouth. It is true that now and then we stumble, spill the milk, or mistake one person for another. We also, however, achieve our goals well enough to grow up, get formal educations, attract mates, raise families, and stay alive for enough years to make sellers of annuities and health insurance lugubrious. In Chapter 14, I described how very reliably we walk and drive automobiles.

The only way in which a consistent result (consequence) can be maintained in a varying environment is by compensating for that variation with suitable actions. A tennis game is a dramatic exhibition of the principle of achieving a repeated result by varying means. If you take getting the ball back over the net and within bounds as at least one of the tennis player's goals every time he returns the ball, then every whack at the ball is an exhibition of one more way to achieve that goal. And that way of doing it at that whack has to be *just right*. Sometimes a return seems fairly easy, but sometimes the player obviously has to strain every sinew to get to just the right place, bring the arm into just the right swing, and hit the ball with just the right angle and force that will compensate for the angle and force and spin that the other player has put on the ball. If any small part of that complicated series of movements is not quite right, the ball is missed or goes out of bounds.

In a communication to the CSGnet on 26 June 1995, Powers put it this way:

> What we commonly call behavior is really a resultant; the outcome of combining forces created by an organism with forces that originate elsewhere.
>
> When we see it this way, we realize how strange it is that an organism can actually appear to emit the same physical consequence of action over and over.... In any one physical situation, there is only ONE action the organism can take that will have a particular physical effect. If the local environment changes in any way, there is still only ONE action the organism can take to create the same effect as before, but now it is a DIFFERENT action.
>
> ... when the organism emits a particular physical effect, its action is PRECISELY THE ONLY ACTION THAT COULD HAVE PRODUCED THAT EFFECT AT THAT TIME.

When the tennis player returns the ball to the left toe of the other player, the player does everything to compensate for the force and angle and spin with which the other player sent the ball to the player, for the wind direction and force, for the fatigue in the player's own muscles, and for everything else that could have an effect on the ball, in precisely the only action that can produce that effect at that time.

We often say rather loosely that many actions can produce the same effect. That is true over a series of acts when we are speaking loosely of the "same" or repeated effect. At one volley, the tennis player returns the ball by one combination of running and whacking, and at another volley achieves the "same" result by another combination of running and whacking. But at an one whacking, only one combination of component actions is going to achieve the result. Powers continues:

> The basic problem in explaining behavior, therefore, is to explain how it is that the one action that is necessary to produce a given result is the one that is produced by the organism—even when each instance of "the same behavior" requires that a specific different action be produced.

I mentioned near the beginning of Chapter 20 that some psychologists and workers in artificial intelligence believe that a creature must cope with its environment by making a detailed internal map of its environment and then plan in detail how to act *on that map*. (You can find more on this topic in Chapter 24 under the heading "Model-Based Control.") Presumably those people believe that tennis players do that, too, in their leisure moments, so to speak, between returns. But PCT models can act even faster than tennis players, and without any preplanning at all.

Chapter 23

Internal conflict

It is no surprise, given the millions of control systems in the brain, that some conflicting signals are sent downward. I go to the introduction to *The Hunting of the Snark* by Lewis Carroll (1936, p. 755) for this example:

> Supposing that, when Pistol uttered the well-known words—
>
> "Under which king, Bezonian? Speak or die!"
>
> Justice Shallow had felt certain that it was either William or Richard, but had not been able to settle which, so that he could not possibly say either name before the other, can it be doubted that, rather than die, he would have gasped out "Rilchiam!" (p. 755).

That is a good illustration of inner conflict. First, lower controls are called upon to act in two ways. Pistol is demanding only one name from Justice Shallow, and Shallow will die if he gives the wrong name. But if he does not answer, he also dies. Second, the struggle is accompanied by emotional turmoil. Shallow trembles and gasps while he tries to make his vocal cords say both names but at the same time say only one. Third, the solution to the conflict is not found at the level where the conflict lies—in this case, within the rules of logic and naming. The solution came, it seems to me, when Shallow did some reorganizing; when he went beyond the rules of logic and language and succeeded in speaking two names at once while uttering only one word.

EXAMPLES

The conflicts at the higher levels are those we feel the more poignantly. Since they take longer to resolve, we become more aware of them. We examine more aspects of them, and the distress becomes more memorable. Rick Marken compared conflicts at lower with those at higher levels in a message to the CSGnet on 8 September 1999:

> Maybe you have never experienced conflict in lower-level systems. But maybe you have and never noticed. These lower-level conflicts are usually very quickly resolved, so they don't create much emotional distress, making them hard to notice. . . . I have found myself, for example, riding a bike where I briefly froze because I wanted to turn in two directions at once. I have been caught in conflicts at home where I wanted to pick up the last dirty dish and not pick it up because my hands were already too full; etc. These little conflicts are usually resolved so quickly that they become just amusing events. This is quite different from the higher-level conflicts like wanting to continue work at place X (for the money) and not wanting to continue work at place X (for the long commute). These higher level conflicts are harder to solve, last longer, create more emotional distress and, thus, are the intrapersonal conflicts we notice. But I suspect . . . that low-level conflicts are just as . . . frequent as high-level ones. . . .

I will give a few more examples at the higher levels just to underscore the prevalence of inner conflict in human life. Think, for example, of the people employed by mining and manufacturing companies that spill pollutants into air and water. On the one hand, working there brings money; on the other hand, working there helps pollute the air and water used by themselves, their families, their neighbors, and many others.

Not all employees of the company will be troubled. If you bring up the matter, you might get a dispassionate answer like, "Yeah, that happens," or maybe, "Interesting, isn't it?" In that case, the speaker may be feeling little or no conflict. I remember seeing on the television, a few years ago, several high officials of tobacco companies being questioned in a Congressional hearing. Those officials, as far as I could tell, were as relaxed and unperturbed as if they were attending a neighborhood tea party. People who remain for some time in a conflict-fraught situation usually find some way of reducing the stress. Certainly those tobacco executives had found their ways of maintaining outward calm. Were they less calm in their stomachs than in their bland smiles?

On the other hand, many people employed in such a company will feel themselves in conflict. When you raise the matter with them, they are likely to respond with a spirited defense, even a combative one, since your question exacerbates an already strong inner tension.

Some people seem to resolve some conflicts with alacrity. Often, they find comfort at high levels of perception: "Oh, well, if we don't sell it to them, somebody else will." "I was only obeying orders." "God wants us to save their souls from heresy." Other people agonize for years. Some of the heretics of the Middle Ages were relieved of their inner conflicts only by death under torture. A few shed their conflicts by becoming leaders of successful Protestant sects.

Examples I admire are "whistleblowers." Glazer and Glazer (1986) wrote about interviews with some dozens of whistleblowers. One instance followed the explosion of the Challenger rocket, which occurred a few second after takeoff on 28 January 1986. When Allan McDonald and Roger Boisjoly, engineers with the Morton Thiokol company, explained to a presidential commission how the difficulties with the rocket's O-ring seals had gone uncorrected, their honesty was admired by the commission and the press. But when they got back home, the company transferred them to menial jobs.

In 1981, the Department of the Interior hired Vincent Laubach to oversee the payment by strip-mining companies of fines and reclamation fees. When he insisted that the companies pay long-overdue fines and fees, his superiors told him not to press the matter. When he complained to the office of the Inspector General that he was being prevented from doing his job, he was fired. He instituted a grievance procedure. His wife took his case to the press and some members of Congress. Environmental groups filed suits. After several years, the Department restored him to his job and reimbursed him $24,000 for his legal fees. When Glazer and Glazer wrote their article, however, the Department had still not collected the overdue fines and fees, which came to an amount between $150 million and $200 million.

Irwin Levin, a supervisor in the Brooklyn Office of Special Services for Children, discovered in 1979 that cases of serious abuse of children were not being properly investigated. He told his superiors about it and also his union, the president of the City Council, and various community leaders. In retaliation, Levin was demoted, fined, and assigned a do-nothing job. Eventually, a long-delayed investigation by the office of the Inspector General supported Levin's charges, and Mayor Edward Koch ordered his reinstatement. Levin is uncertain whether children are now any better protected.

Glazer and Glazer tell many other bitter tales. Think of the conflicting demands and obligations whistleblowers must feel from employer, fellow employees, family, friends, and lawyers. Think of the threats to their internal standards concerning loyalty, self-respect, right and wrong. Think of their coping with those conflicts amid a situation described by one whistleblower as "If you have God, the law, the press, and the facts on your side, you have a 50–50 chance of winning." One can only be grateful that we have such people among us.

Stories appear every now and then in the press about stalwart people who struggle with inner conflict. Magazines such as *Mother Jones*, *Ms*, and *Utne Reader* are good sources. I will mention here only one more collection, the book by Melissa Everett (1989), which contains "stories of ten men who left comfortable jobs in the military, the intelligence community, or the defense industry to work, in their own ways, for peace."

Michael Holmes (1997), of the Associated Press, wrote about telephone conversations Lyndon Johnson had in 1964 with his national security adviser, McGeorge Bundy, and with Senator Richard Russell:

> ... Johnson agonized over what to do about Vietnam and was tormented by the prospect of sacrificing U.S. soldiers to a war he considered pointless. ...
>
> Johnson also worried that Congress might run him out of office if he tried to withdraw.
>
> "They'd impeach a president, though, that would run out, wouldn't they?" he asked. ...
>
> "I've got a little old sergeant that works for me over there at the house, and he's got six children. ... Thinking about sending that father of

those six kids in there . . . and what the hell we're going to get out of his doing it? It just makes the chills run up my back."...

"He was clearly tormented by it," Middleton [director of the LBJ Presidential Library] said, adding that Vietnam was interfering with Johnson's hopes to enact civil rights and other Great Society legislation.

Sometimes we cry, "Can I believe my eyes?" That kind of conflict occurred in a series of experiments, famous in the literature of social psychology, conducted by Asch and others (1938) and by Asch alone (1940, 1951). Robertson (1990a, p. 164) gives a condensed account:

> The experimenter asked a series of individuals, each of whom thought he was working in a group of peers, to call out their judgments about the comparative lengths of some lines drawn on a blackboard. The "peers" were actually stooges of the experimenter, and they sometimes called "line B" longer when "line A" was really the longer line. Surprisingly, when a majority of the stooges called out incorrect readings, the majority of volunteers tended to go along with the crowd. . . . about a third appeared to go along most of the time. Another 15% to 25% stood firmly on their own judgments. . . .
>
> . . . those who did not go along with the crowd did not all maintain their independence in the same way. The independent subjects controlled many different principle-level perceptions in holding to their own judgments in the face of social pressure. Some . . . made their decisions [through] confidence in their own opinions. . . . [Others] did so with "considerable tension and doubt. . .".

No doubt some of the participants who did change their reports to agree with the experimenter's confederates had come to believe that there was something wrong, somehow, in the way they were seeing the lines. No doubt others who changed and said they agreed were nevertheless still believing that their original perceptions were right and the other presumed participants were wrong.

Robertson points out that among both the participants who maintained their independent judgments and those who came to retract their first opinions, there were surely many kinds of internal standards that came into conflict with the statements made by the experimenter's confederates. For some of those maintaining their first judgments, the conflict was slight. "My eyes see clearly," they might have said to themselves, "and it doesn't matter what others say." Others might have felt a conflict with their perceptions of themselves. Robertson imagined such a conflict this way: "I have to call things as I see them, even though I might be wrong, because I don't want to be thought of as a wimp." Others might have felt a conflict with a variable controlling for something like companionableness: "Although I hate being an outsider, doing the job the experimenter asked me to do is more important." Some who capitulated might have thought, "The majority must be right." And some who said they agreed with the confederates might have felt a conflict with an aspect of self-concept something like: "I should not appear to be telling my peers they are wrong." Still others might have thought, "I know what I see, but maybe those people know something about this experiment I don't, and I don't want the experimenter to get mad at me for saying the wrong thing."

It is not to illustrate PCT that I tell you about those experiments, but to portray one more kind of conflict. Conflicts of that same sort happen outside the laboratory. I have come to think that my eyes do not see blue and green the way most other people do, because every now and then someone says something to me like, "You told me that was blue, and it's green!"

The conflicts at the higher levels are typically of longer duration. Conflicts among system concepts can go unresolved even for a lifetime. Einstein spent the last decades of his life trying without success to resolve the conflict he felt between relativity and quantum mechanics. Many saints spent a similar length of time trying to beat down Satan under their feet—that is, to resolve the conflict between their human urges and their pietistic standards.

Conflicts require energy and attention. Inner conflict, though inner, is like a tug of war. In that game, the people at both ends of a rope pull as hard as they can, and eventually one team manages to pull the toes of an opponent a few inches across the mark on the ground. Without the team at the other end of the rope, one person could move the rope with an easy effort of one hand. With the other team providing a powerful disturbance to that goal, the one team must pull with all its might to achieve only a small result. You saw a model of that kind of conflict in Chapter 9 under "Collective Control of Perceptions." The same thing happens within an individual. The chess player wants to get "control" of the center squares on the board, but wants to maintain strong protection for her king. Suppose she sees the position as one in which she cannot do both with one move. After thinking, perspiring, and taking deep breaths for three minutes, she moves a pawn one square forward. (Yes, I know, in some positions that could be a startlingly bold move—but you get the idea.)

Conflict is expensive. Within an individual person, it uses up calories, emotion (anxiety), attention, and time. It is characterized by worry and false starts. The chess player lifts her hand, reaches toward the pawn, but then puts her hand back in her lap, breathing deeply. Between groups of people, the expense of conflict is breathtaking. Wars are so expensive, in fact, that most of us can hardly bear to think about them. After killing thousands of people on both sides, sometimes millions, and after destroying the wealth of both sides, at the end both sides find themselves worse off than before. The nation that has sustained less damage now has a small advantage over the other nation in trade and other world affairs, but at a terrible price to itself and, indeed, the whole world, because of the squandering of resources. The inner conflicts with which we begin a war are so multiplied by the time it ends and have driven us to grasp at such poor straws to resolve them that we raise monuments of adulation to the political and military leaders who could find no better course of action than war. I do not say wars can always be avoided; I do say that they are catastrophes, not solutions, and not glorious.

With individuals, inner conflict at the higher levels of the hierarchy can become persistent, demanding, and eventually debilitating. Inner conflict is the soil in which grow all the psychological derailments of purposive action that have been so thoroughly categorized, tested for, counseled about, prescribed for, given therapy for, prayed to God about, and portrayed in fiction and on the stage. Powers (1973, p. 253) says:

> Unresolved conflict leads to anxiety, depression, hostility, unrealistic fantasies, and even delusions and hallucinations.... I have become more and more convinced that conflict *itself*, not any particular kind of conflict, represents the most serious kind of malfunction of the brain short of physical damage, and the most common even among "normal" people.

CONFLICT AND LONGEVITY

PCT portrays conflict as enervating. When I have struggled to find a way out of a conflict, I have often found myself breathing more deeply than usual and feeling depleted. I have often taken to pacing back and forth. I have upon occasion pounded on things with my fists. I remember a time in my romantic youth when I dashed into the back yard and screamed imprecations into the midnight sky. Those are all examples of energy expended without furthering any of the purposes in my conflicts. I have seen similar symptoms in other people. With continued conflict over months and years, we would expect to see not only a chronic lowering of available energy, but far-reaching alterations in the hierarchy of control, frequent and sometimes conflicting demands for more energy from the emotional systems, and an increased rate of reorganization. Since reorganization does not always result in an overall improvement in control, chronic conflict can increase the frequency of periods of poor control. The combination of all those effects can not only perturb mental process but can also weaken vital physical processes. For example, we would expect to find, on the average, that people suffering the stresses of continued inner conflict die sooner than those living with less conflict.

More than 75 years ago, psychologist Lewis Terman recruited 856 boys and 672 girls who had IQs of 135 or over. Data of many kinds were gathered from them every five or ten years afterward. In 1991, according to a brief report in *Science News* (1995, p. 124)[1], correlations were calculated between longevity and other variables. Those participants with no divorces themselves or in their families lived several years longer, on the average, than those in less stable families. That correlation was about the same as the correlations with systolic blood pressure, cholesterol concentrations, exercise, and diet.

The work of policing is obviously full of opportunity for inner conflict. In 1979, *Behavior Today*[2] gave a brief report of a study of 2,900 police officers in 29 American communities. The study found that "... more police officers commit suicide each year than are killed in the line of duty. Moreover, police personnel... have higher rates of suicide, alcoholism, divorce, and heart attacks than *any* other occupational group."

It would be useful to have some studies using more specific measures of inner conflict and connecting conflict with various kinds of energy-absorbing debilities. In the meantime, the two studies described above offer some hints of what more pointed studies might find.

THREE LEVELS

There are always three levels of control participating in an inner conflict. One is the level at which the conflict is *expressed*—where its results appear. That is the level at which Justice Shallow might have gasped out, "Rilchiam!" Or at which the hands on the handlebars wiggle left and right and then commit to one direction of turn. Or at which the employee of the polluting factory goes on working there (let's say) but tells everyone he knows to stay away from the dump. Notice that the control systems at this level are working normally. A reference signal sets a standard, and the person acts to bring a perception into match.

The next level up is the level at which the conflict is *created*. One system pulls one way, the other another, and the variable cannot go both ways at the same time. Lewis Carroll's imaginary Justice Shallow cannot both speak and not speak at the same time. (But he does figure out how to say both "Richard" and "William" at the same time.) The handlebars could not steer both left and right at the same time. The employee could not simultaneously both work and not work at the factory. At this level, things are out of kilter. Both systems are pulling very hard but not succeeding in getting the perception they want.

At the third of the three levels, the *situation* is created that brings the conflict into being. Justice Shallow perceives himself to be limited to the choices Pistol lays out for him (but then finds a way to utter both names, not just one). Marken perceives himself limited to left or right at that one moment (but then gives up the desire to go one way, allowing himself to go the other). The employee perceives damage to himself and family by either quitting his job or not quitting (but reduces the damage of not quitting by warning family and friends about the danger—not the happiest solution, and maybe later he finds another job).

This uppermost level is the one at which the conflict can be resolved, or at least reduced. Pulling harder at the lower levels will merely use up more energy. But if the situation calling for the conflicting reference signals can be changed, the conflict vanishes. This picture of conflict gives rise to a method of resolving conflict (both within and between people) that Powers calls the "Method of Levels" or "Going Up a Level." I will say more about The Method of Levels in Chapter 30.

SOLUTIONS

Speaking of resolving conflicts, Powers wrote this to the CSGnet on 16 December 1999:

> If you mean the process by which one alternative is eliminated and the other retained (resolution of the conflict) there are many ways in which this can be done, and thus many models to choose from. One way is to set up a third control system specifically to conflict with one side of the choice, so as to remove that control system from consideration. I call this the "will power" solution. Another is to lower the gain on one side and/or raise it on the other, so as to move the net perceptual signal close to one of the reference signals. You might call this "lowering the importance" of one goal relative to the other. And a third is to stop sending a reference signal to one of the conflicted systems, so it ceases to try to act. These three approaches, of course, are the systematic result of higher-level control processes aimed specifically at eliminating the conflict. A fourth method, which is not really a method but is simply a consequence of being in severe conflict, is to reorganize. There is no way to predict the results of reorganization, except to say that the total error will probably (but not always) be reduced before permanent damage has occurred.

You can see some of those methods in use in the illustrations I gave earlier.

Sometimes the search for a solution becomes hopeless. Sometimes, after the person fails to find a higher level at which the situation can be reconceived, even reorganizations then fail to find any path to reduced error. Reorganization can fail by sheer chance, but it can fail also when a conflict is bombarded so rapidly and so strongly by almost incessant new threats to the controlled variables that the reorganizations cannot keep up—that is, a new arrangement of weightings in the hierarchy is no sooner formed than it is knocked loose by still another threat. The person becomes overwhelmed by the uncontrollable turmoil. That situation often results in withdrawal of one kind or another from the sensed world. This is the condition I described as "learned helplessness" in the section headed "Quantification" in Chapter 20. The annals of abnormal psychology describe several other manifestations of it.

Beyond the examples I gave at the beginning of this chapter, I do not try here to portray the agonies of strong conflict. Painting the agonies is a task for artists, dramatists, theologians, perhaps even a clinical psychologist or a psychiatrist. I will, however, say a few things about agonies in later parts of the book.

CHOICES AND DECISIONS

The more stressful and memorable choices or decisions are those wrested from the anguish of conflict. We try to avoid such situations. Many people have written books telling us how to avoid them, offering programs by which, the author claims, we can find a way around, or go above, the conflicts that would otherwise embroil us. Their book jackets tell us we can learn "step by step" how to get past our difficulty. Some books tell us at the outset the number of steps required: three, seven, and twelve are favorite numbers. When we succeed with those programs, we do so by substituting the program for the conflict. In a sense, we go up a level. We say to ourselves that finding a way around this difficulty is more important than either of the conflicting variables we have been trying to control, and we are going to stick to this step-by-step routine until we have got past the difficulty. Powers says we could call those programs decision-*avoiding* methods instead of decision-*making* methods.

Making (or avoiding) decisions easily encounters conflicts when a group of people are trying to agree on a course of action. All the conflicts within all the participants can arise to impede the agreement. Sometimes the conflicts seem like a cloud of mosquitos, whining and needling and avoiding every effort to expunge them. Karl Weick (1979, p. 262) has offered a routine for avoiding that impasse:

> Every day the Naskapi [Indians in Labrador] face the question of which direction the hunters should take to locate game. They answer the question by holding dried caribou shoulder bones over a fire. As the bones become heated they develop cracks and smudges that are then "read" by an expert. These cracks indicate the direction in which the hunters should look for game. The Naskapi be-
> lieve that this practice allows the gods to intervene in their hunting decisions.

Weick tells how that technique actually does help the group to find game. In addition, Weick says that a randomizing procedure has many advantages, once the group agrees to abide by the pronouncement of the expert. Here are some advantages over more "rational" methods that he lists:

> A decision can be made when there are insufficient facts.
>
> A decision can be made when there are insufficient differences among the alternatives.
>
> Bottlenecks can be broken.
>
> These practices confuse competitors.
>
> The procedure is fun.
>
> A decision can be reached swiftly.
>
> No skill is required of the user.
>
> The technique is inexpensive.
>
> The solution is arrived at nonargumentatively.
>
> If no game is found, the gods—not the group—are to blame.

Weick is at times a humorist, but I think he is also a penetrating critic of social-psychological theory.

What looks like choosing to some people looks merely like acting to other people. Think back to Figure 9–2 or 9–3 in the section on "The Crowd" in Chapter 9. You could easily imagine one of those "persons" deciding whether to pass right or left around the next stationary person, deciding the best path for reaching the goal, and so on. As you will recall, however, no such choices, decisions, or planning were written into the program that produced the movements shown in those figures.

The topic of conflict, both inner and interpersonal, will recur in later chapters. Powers (1998, chapters 5–7) has written three easy-to-read chapters on conflict, using the rubber-band game to illustrate. Chapter 17 in his 1973 book is only a little more technical.

ENDNOTE

[1] In the list of references, look under "Living longer."
[2] In the list of references, look under "Stress-linked suicides."

Chapter 24

Up at the top

This chapter will focus on Powers's top three orders of control: programs, principles, and system concepts. They are the "farthest" from the "real world." An experience at the top of the control hierarchy is an interpretation of layers upon layers of interpretations at lower levels. I diagrammed the layering in Figure 18–3.

VIEWPOINTS

We can be conscious at any level of what we are perceiving below that level. When we are thinking about how to do something, we are using perceptions at the level of programs: "If I do *this*, I will also be bringing about *that*, and *that* will make it possible for. . . ." But in the midst of planning something we want to achieve, we rarely stop to think about thinking. We just *do* the thinking, and our attention stays on the imagined sequences of acts. Acting with perceptions coming to the level of programs, we can be conscious of perceptions at all *lower* levels. To be conscious of the *thinking* we are doing, we must go up (so to speak) to the level of principles. And to be critical of the principles of logic we are using, we must go up to the level of system concepts concepts—concepts *about* logic. Any level can be a viewpoint, so to speak, from which we can be conscious of lower levels.

To illustrate changing levels of perceiving, I will tell you about a famous puzzle that experimenters have used in studying problem solving. The experimenter brings the subject into a room and directs his attention to two strings hanging, some distance apart, from the ceiling. Except for the strings, a chair, and a few tools, perhaps left behind by the person who put up the strings, the room is bare. The experimenter tells the subject that the problem here is to tie together the two ends of the strings. Every subject begins by taking hold of the end of one string and walking toward the other string. But the subject is disappointed; the string is too short. He or she cannot hold to one string and reach the other. The chair is no help, either. Many subjects give up, but a few succeed in tying the strings together.

In posing the problem, the experimenter draws the attention of the subject to the *relationship* between the strings. As long as the subject focuses only on relationships, there is no hope of solution. Some subjects, however, turn their attention "upward" to categories. They look around for help. They look at the tools, and a few subjects realize that the tools can also be categorized as *weights*. They pick up a tool such as a pair of pliers and tie it to the end of one string. They set the pliers swinging back and forth toward the other string. They take hold of the other string and walk toward the swinging pliers. They catch the pliers on the inward swing and tie the two strings together. (I am not saying that every successful subject found the solution in just this way. Perhaps a few imagined themselves swinging on the strings like acrobats with ropes.)

When we go from perceiving at one level to perceiving at another, we change the aspects of the world to which we give attention, and different opportunities become apparent. When two people move from a "relationship" to a marriage—that is, from having overlapping schedules (levels 6 through 9) and certain obligations to each other (level 10) to the level of "one unique entity" (level 11)—both purposes and modes of pursuing them acquire greatly expanded degrees of freedom. Many people undertake marriage as a way of solving particular existing problems—problems they have come upon in their single state. But looked at from the level of marriage as a joint enterprise with new degrees of freedom, a problem may look very different. It may vanish, or it may change its shape and become susceptible to a different sort of solution.

In a missive to the CSGnet on 6 December 1994, Powers wrote:

> Behavior never takes place at just one level, and any model that tries to handle all of behavior at one level is just wrong. It's like watching an orchestra conductor and saying that all he's doing is waving his arms in repetitive patterns. He is certainly doing that, but it's not ALL he's doing—it's not even the most interesting aspect of what he's doing. . . . In some quarters, the mistake is in trying to represent functions at all levels as if they were logical processes. . . . To a logic-level system, the whole world consists of variables which are either true or false. . .

Because problems often look different from different levels of perception, changing the level of perception can uncover solutions impossible to perceive at the previous level. Helping a perplexed person to change his or her viewpoint to a higher level, therefore, can be an effective therapy. It enables the person to find his or her own solution and understand it from his or her own viewpoint. Such a solution is far more likely to be carried out successfully than advice from someone else's viewpoint. Powers calls this the "Method of Levels." I will describe it at greater length in Chapter 30.

By the way, people sometimes want to classify actions or events according to Powers's eleven levels of perception. "What level is baseball?" "What about brushing your teeth?" "What's philanthropy?" Levels are not about actions, but about internal standards—reference signals. The levels are not out there like shelves in a warehouse, labeled so that everyone can walk in and find actions such as baseball, teeth brushing, or philanthropy. Levels exist only in your own purposes. Whatever you do, you are doing for one or more (usually more) purposes. Those *purposes* reside at one level or another—in *your* head. When you are acting to maintain a purpose at any level, you are also acting purposely at all lower levels. When you play baseball, you are playing at many levels. Your assortment of purposes will be different tomorrow, and they will differ, too, from the purposes of the other players.

REALITY

Regardless of the level in the control hierarchy, all perceptions seem "real" to us. (Well, we do have moments of disorientation, but let's set those aside for now.) When I feel an apple in my hand, my conviction that I am feeling an apple rests entirely on the neural signals running up my arm to my brain and supplying somehow part of what I experience consciously at some level—perhaps the level of "event." But the feeling of reality does not diminish in the levels of control lying farthest from the sense organs. Here, from the book *Powers of the Crown* by the Editors of Time-Life Books, 1989, is an example of how real a system concept can feel:

> In August 1593, [Abbas Khan, shah of Persia] was advised by his astrologers that the stars boded ill for the ruler of Persia, since Mars and Saturn were in quadrature in the ascendant. Resourcefully, Abbas stepped down from the throne and had a condemned heretic . . . proclaimed shah in his place. The heretic ruled under close surveillance for three days. On the fourth day, when the zodiacal aspects were more favorable, the substitute shah was executed and Abbas resumed his reign.

Sometimes we want to check up on the reliability of a perception: "Wait a minute—let me have another look at that." Or we say to someone else, "Would you take a look at this and tell me what you see?" To make it easy to compare one person's perception with another's, scientists try to keep their perceptions low in the control hierarchy: "Did the pointer on your dial point closer to 3.75 than 3.76? Yes or no?"

At the highest levels, however, it is often difficult to validate one's perceptions. How could Abbas Khan have checked his understanding of the compulsions from the stars? His astrologers told him that harm was about to befall the shah of Persia. They were right; the shah (the heretic) met death. Furthermore, Abbas Khan and his astrologers demonstrated dramatically how well they understood the stars; they figured out a way to fool the stars, and they got away with it.

We can ascertain the reality of some systems (corresponding to our system concepts) in much the same way we ascertain the reality of an apple. The reality of the solar system is an example. You and I can both look through a telescope night after night, make notes, and compare the movements we trace with the movements predicted by the model of the solar system. That, at any rate, is what I am told we can do. I have looked through a telescope, but I have never taken any measurements or even looked at the same planet on two successive nights. But it is far easier to believe what I am told about the model of the solar system and the manner of its verification than to believe that all those books on the shelves of the astronomy sections of library after library, along with all those reports of rockets finding their way successfully to Venus, Mars, Jupiter, and elsewhere, are frauds coordinated and maintained with devilish cleverness by thousands of people throughout every generation since Copernicus, who died in 1543, or maybe since Aristarchus of Samos of the third century B.C.

Another example is the railroad network of the United States. Nobody has ever seen that network as a whole or has seen it even by patching together views of it obtained by traveling over all its trackage. What we believe to be the U.S. railroad network, though it includes a good deal that can be verified in the same manner as an apple or the solar system, includes also many operations and relationships among employees, customers, and suppliers that we can only imagine. Though some evidences can be observed, many are evanescent, changing to something else by the time another observer can get to the same place.

All human organizations have that character: you can point to some tangible (sensible) evidences that a particular sort of behavior is going on that you attribute to an organization, but much of the behavior is of a sort to which you cannot point. You can point to buildings, delivery vans, and balance sheets. You cannot point to authority, cooperation, or, indeed, organization in general. You can point to what you believe to be evidences of those things, but you cannot verify their existence in the same way you verify the existence of an apple.

Here is an example of differing conceptions of "the bank." Sometime in the middle 1950s, my wife and I borrowed money to build a house. I arranged with our bank to open a separate account for the money with which we would pay the contractor. I agreed with the teller that the bank and I could more easily keep our accounts straight if I would write "Construction Account" on checks drawn on that special account. The first time I wrote a check duly marked "Construction Account," the bank tried to take the money out of our personal account instead. The check bounced. Irate, I went to see the bank manager. I told him that if this was the sloppy way his bank was going to handle my money, I would move my accounts elsewhere. He was apologetic. He said, "Oh, Mr. Runkel, you mustn't blame the *bank* for the error. That was one of the girls in the back room."

Some principles and system concepts are even more elusive than the concept of a bank. Here are a few examples:

achievement motivation	intelligence
ambition	personality
cleanliness	physics
democracy	psychology
goodness	responsibility
honesty	safety first

The point here is not that a word stands for different things for different people, though that is true. The point is that the conception you have of *anything* is unique. About some things your conception and mine are very close; staring at the same apple, we might require a minute or two to find a difference between our perceptions. But about many other things, we can discover our differences quickly and easily. You and I will differ in our internal standards even when we agree that we are controlling similar variables and even where we *do not put a particular label* on the standards or variables. For example, we might agree that the situation in which we both find ourselves calls for a certain sort of behavior—perhaps we are both on a small sailboat, or perhaps both listening to a used car salesman. In either situation, without having to find a name to put on the situation, we might find ourselves acting with mutual helpfulness. Nevertheless, we would not show perfect accord. One of us would at least occasionally say something like, "Oh, you care about that? OK." You might want every line on the boat's deck to be coiled clockwise, or I might insist on buying a car painted yellow.

I know of only one experiment—the one described under "Self-concept" in Chapter 7—that investigated a high-order internal standard (self-concept, in this case) and did so testing each individual singly, using The Test for the Controlled Quantity.

The psychologist William Dember, writing in 1974, was impressed with the power of higher-order motives to override lower-order goals. He cited the fasting of Mahatma Gandhi and the religious warring in Northern Ireland. He also included (on p. 166) two then-current news items:

> A young boy died a slow, painful death after ingesting some solvent he had stolen from the school shop. Apparently some of his friends had dared him to do it.

> An assistant pastor and a layman of the Holiness Church of God in Jesus Name, of Carson Spring, Tennessee, died in agony after drinking a mixture of strychnine and water—testing their faith in the Bible, where in Mark 16:16–18 it is asserted of those that believe that "if they drink any deadly thing, it shall not hurt them. . . .".

The military historian John Keegan (1994) was similarly astonished by the power of high-order control. Writing of the age of Clausewitz (1780-1831), during which it was the custom in western Europe for infantry to stand in close formation to fire and be fired upon, he wrote (p. 9):

> Men stood silent and inert in rows to be slaughtered, often for hours at a time; at Borodino [in 1812] the infantry of Ostermann-Tolstoi's corps are reported to have stood under point-blank fire for two hours, "during which the only movement was the stirring in the lines caused by falling bodies."

In eastern Europe, the Cossacks of that time had a quite different understanding of war: ". . . a man fought if he chose and not otherwise, and might turn to commerce on the battlefield if that suited his ends" (p. 16). When the close-order style of fighting was described to Cossacks, Keegan said, they laughed. A similar reaction was aroused in Japan:

> European drill, when first demonstrated by Takashima, the Japanese military reformer, to some high-ranking samurai in 1841, evoked ridicule; the Master of the Ordnance said that the spectacle of "men raising and manipulating weapons all at the same time and with the same motion looked as if they were playing some children's game." This was the reaction of hand-to-hand warriors, for whom fighting was an act of self-expression by which a man displayed not only his courage but also his individuality (p. 10).

I do not presume to put a name to the internal standards controlling the aspirations and muscles of the assistant pastor of the Holiness Church of God in Jesus Name, the soldiers at Borodino, the Cossacks who laughed, Takashima, or the Master of the Ordnance. I exhibited those actions and words as illustrations of the strength with which our behavior can be held in harness by our principles and system concepts. And also to show, if it need be shown, that one person's act of sacred honor may bring ridicule from someone else.

Self-fulfilling Beliefs

Some beliefs are self-fulfilling. An example is Abbas Khan's belief that the shah of Persia was in danger. Believing that, he himself carried out the prophecy by killing the imitation shah. Another example is the way many teachers separate the quick students from the slow. The teacher asks a question, and some hands go up. The teacher calls on one of those students for the answer. After some days of that kind of recitation, the teacher can tell you which students shoot up their hands frequently and which almost never. Meanwhile, here is Algernon, who likes to think about his answers before he sticks his neck out. But by the time he gets his answer formulated, the teacher has called on one of those quick students who put demonstrating eagerness above being right every time. So Algernon gets categorized as slow. Which he is by a second or two, though he may also be a careful thinker and an industrious student. It is also possible that he wants to be called on, and gets discouraged after a while, and he may then neglect his studies. Then, if the teacher does call on Algernon, he may not have the right answer. Which is just what the teacher predicted. But the teacher's explanation is that Algernon was dull all along, probably from birth.

Self-interest

Some people, at first acquaintance with PCT, complain that it is a "selfish" theory because it says that we always do what will bring us our own satisfactions, that we act to satisfy our own goals and needs. And so it does. But it does not say that those goals, needs, or satisfactions must be "selfish" in the sense of benefiting ourselves at the expense of others. We can be satisfied to be altruistic, too. The question of self-interest has, of course, been debated by psychologists. In 1991, Jane Mansbridge assembled a book of research on unselfish motivation. Most of the contributions displayed lists of unselfish actions (caring for children and parents, contributing to charities, voting in elections, and so on). It is not necessary to do psychological or sociological research to assemble such lists. Poets, troubadours, novelists, and others have been doing it ever since people began gathering around campfires. The behavior of the Cossacks and of the soldiers at Borodino will serve as examples also. The fighting style of the Cossacks could be called selfish or reasonable, depending on your military culture. The fighting style of the soldiers at Borodino could be called magnificently altruistic, thoroughly disciplined, or obviously absurd.

BORROWING INTERNAL STANDARDS

There can be no doubt that animals learn from one another many ways of satisfying their needs. If birds, bears, or baboons are orphaned, their coping skills are defective. We have all had frequent experiences of imitating and being imitated. We pick up "viewpoints" (principles and system concepts) from others, usually through the use of words. In a conversation in which you and I are trying to make sense of something, I may suddenly say, "Oh, I never thought of that side of it!" because I have "gone up a level," as Powers would say, and I am seeing a new (to me) way of fitting together programs, sequences, categories, and the rest. I speak here of "borrowing" a variable to control, but you could also call it adopting or imitating. We do it every day, all our lives.

We borrow standards at low levels—the reverent modulation with which we use our voices in church, in a courtroom, or at the bank, for example. But we also borrow at the upper levels—we borrow, for example, the principle of borrowing. Sociologists call this adopting norms. We look for patterns in the behavior of others and for the approval or disapproval others give to certain patterns. When we accept the consequences of that approval and disapproval of our own behavior, it means that we have accepted those patterns as guides for our own behavior—as norms. Socialization is a ubiquitous variety of norm-adoption. Using controls at the top layers of the neural hierarchy, we can consciously provide ourselves with concentrated opportunities for borrowing standards in families, schools, churches, banks, automobile factories, and armies.

We can watch someone do something and can imitate what we see. We can listen to someone tell how to do something or read about how to do it and then try it ourselves. From the words heard or read, and using the imagination mode, we form an image of what we want to attain and then act to narrow the gap between the presently perceived variable and the image. Our own image will not, of course, be exactly like the image in the mind of the person we imitate. Furthermore, as we act, the image of the goal shifts, in the manner analogous to the process I described under "The Warps of Memory" in Chapter 19. We gradually alter our image and our program for achieving it. The image must become uniquely our own, because we borrow an internal standard to serve as part of our own feedback function—as part of a path through the environment that closes our own feedback loop.

We are not always eager to accept standards from other people. Leon Festinger, on page 102 of his captivating book *The Human Legacy* (1983), quotes from a book by Vial (1940, p. 159) a description of tool-making among the Jimi of New Guinea, who were still making stone tools in 1940:

> It took one man fifty or sixty blows before he got a suitable slab from the original block. He was sitting cross-legged with the block in front of him and soon his shins were bleeding from cuts by the flying fragments. The other operator, a much younger man, got a good slab quickly and, holding it in his left hand, began chipping it with a smaller round sphere of stone in his right hand, hitting it on the edges and chipping little pieces off. He had a quite good blade, seven inches long, chipped ready for polishing half an hour after arriving at the quarry. The process looked easy, as if anyone could do it. The older man was not so successful, taking longer to get a suitable slab, and having more difficulty in reducing it to the shape for polishing.

Why did not the older man imitate the obviously better methods of the younger man? I think I know how the old fellow felt. Sometimes, when I think I am doing something well enough, someone urges me to do it in a better way. I mumble something like, "You do it your way, and I'll do it mine." Not imitating can be a principle as well as imitating. An example ubiquitous in the United States, perhaps in Western civilization, is the reluctance of adolescents to adopt the customs of their elders.

Sometimes large populations will reject principles urged upon them even by duly constituted authority holding coercive power. John Keegan (1994, p. 50) mentioned the opposition to the Vietnam war in the United States. He was aware, too, of the inner conflict engendered:

> When in response to forces released by the French Revolution, European states were progressively impelled to remilitarize their own populations, they did so from above, and it was accepted with varying degrees of enthusiasm. Universal

service eventually came to be associated, entirely understandably, with suffering and death: there were 20,000,000 deaths in the First World War, 50,000,000 in the second. Britain and America abandoned [universal service] altogether after 1945; when it was reintroduced by the United States in the 1960s, to fight what became an unpopular war, the eventual refusal of the conscripts and their families to ingest warrior values caused the Vietnam War to be abandoned. Here was evidence of how self-defeating is the effort to run in harness in the same society two . . . contradictory public codes: that of "inalienable rights," including life, liberty, and the pursuit of happiness, and that of total self-abnegation when strategic necessity demands it.

RESETTING GOALS

Sometimes, when we become aware of a conflict among our internal standards, we consciously try to change at least one of them. We try to reset a goal or two. Sometimes that is easy, sometimes not. The fact that we can be aware of the conflict and the desire (at a still higher level of control) to alter the goals in conflict raises deep theoretical questions. On 21 October 1997, in reply to earlier comments by Isaac Kurtzer and Bruce Gregory, Rick Marken offered these thoughts on the CSGnet:

> "We," as hierarchies of control systems, are not free to set our own goals. Goal setting is automatic in the control hierarchy. [Given the highest-level goals in the hierarchy], variations in the settings of lower-level goals are completely determined by the outputs of higher-level systems—outputs that are generated as the means of controlling the perceptions of the higher-level systems. It is this fact . . . that, I think, led Isaac to note correctly that "We do not reset reference levels.". . .
>
> On the other hand, "we," as conscious observers of our own controlling (of the "we" that is the hierarchy of control) do seem to be free to reset our own goals to some extent; at least . . . to reset the goals that have been automatically set by the "we" hierarchy. This conscious "we" includes . . . the "reorganizing system.". . . But little is known about the nature of this conscious "we.". . .
>
> Nearly all of our research in PCT has been dedicated to studying the nature of the "we" that is the hierarchy of control systems. The "we" hierarchy carries out our purposes automatically and unconsciously—low-level purposes like the purpose of moving a finger toward the keyboard and higher-level purposes like writing about purpose. But we can also experiment with the conscious "we"—at least in . . . our own mind. One aspect of the conscious "we" is expressed as *will power*, where we willfully reset a goal that has already been set by the "we" hierarchy. For example, I think we are experimenting with the conscious "we" when we try not to do something that we want to do (like not take that second drink or not eat that triple-deck chocolate dessert) or when we try to do something that we don't want to do (like jump into that cold ocean). . . .
>
> The conscious "we" is, I think, what we think of as ourselves—it's our soul. The conscious "we" takes care of the control hierarchy, watching it and keeping it functioning as well as possible. . . .
>
> I guess my basic point—related to Isaac's—is that the hierarchy of control *can't fix itself*. If there is a conflict in the hierarchy . . . then the hierarchy itself can't fix it. . . .

After Isaac Kurtzer commented on that, Marken replied to Kurtzer on 24 October 1997:

> My distinction between "conscious we" and "hierarchical we" was not meant to be about what is the "real" we; they are both real, I suppose. I was just describing how an aspect of my own experience (the difference between willful and automatic goal setting) might map into the PCT model.
>
> The difference between willful and automatic goal setting that I am talking about is demonstrated (to me) . . . by the following kind of experience: I throw my keys up in the air and catch them with my hand; I do this a few times and note that my goal for the location of my hand varies on each occasion, as necessary, in order to make the catch; I note also that, as long as I have the goal of catching the keys, the goal for the location of my hand is completely determined by the disturbance (where the keys fall) to the higher level (catching) goal; my hand "automatically" ends up under the keys. . . .
>
> Next, I don't throw the keys [I let them stay in my pocket] but simply will my hand to be in certain locations. Now I feel [that] I am

consciously willing a goal state for the *same* perception (location of my hand) that had been set automatically when I was catching the keys.

I think the "hierarchical" me was automatically setting the goal for the position of my hand while I had the goal of catching the keys; the "conscious" me seemed to just be watching while this happened. I think the "conscious" me was actively involved in determining the goal for the position of my hand when I was just moving my hand to arbitrary positions.

This is the distinction I was describing; it's a distinction I experience between what seems like automatic goal setting and conscious, voluntary goal setting. . . .

One of the things Kurtzer had written was this:

I would suggest that one of the clearest implications of a hierarchy of *experience* . . . is that there is no real "I"—that it is an arbitrary identification and subsequently a reification.

To which Marken replied in the last part of his post of 24 October:

I basically agree with what you are saying here. What I have been calling the "conscious me" does seem to experience the world (the hierarchy of perception) from the point of view of the particular level in the hierarchy from which it is currently aware. So there are, in a sense, as many conscious me's as there are levels in the hierarchy.

But the nature of the conscious aspect of me that becomes aware of the world in these different ways nevertheless seems always the same to me. . . .

PLANNING

A plan is a program—an assembly of sequences connected at choice-points. I should say it is a *predicted* program. The planner assumes or hopes that the sequences between choice-points will turn out to be feasible. The choice-points are usually provided with alternatives designed to cope with foreseeable disturbances. Sometimes a choice-point will offer an opportunity to revise the plan to meet *un*foreseeable disturbances. A plan is almost always a means to an end. Following the plan is part of the feedback loop meant to bring the person to a goal specified by a reference signal at the level of program or above. Usually the planner is aware of a goal at the level of program, principle, or system concept that will be served by the plan. When I follow a program for getting to the library, I always have in mind something I want to do when I get there.

Coordinating

The ability to plan enables us to use the environment more flexibly. It enables us to do six errands during one efficient circuit instead of back-tracking. It enables us to coordinate our actions with the actions of other people: "I'll meet you under the clock at the Biltmore at noon." It enables us to get together at the convention in October. It enables us to undertake a 20-year mortgage even though the people who are there to receive our last payment may not be the same as those who received the first. Conscious planning requires the use of the imagination mode. This is not to say that a plan must contain, either consciously or unconsciously, a specification for the contraction of every muscle as the plan unfolds. A shopping list says, in effect, "I want to find myself back home with the following items." We usually spend little time thinking of the sequences of actions, large and small, needed to bring about that desired end.

Sometimes we make a plan just to show somebody that we take our work seriously—the boss, for example. Sometimes people write plans as a way of inviting colleagues to discuss future possibilities. Sometimes plans are used as an exhibition of one's hopes for the future—the president's budget message to Congress, for example.

Model-based Control

Many psychologists and designers of robots believe that an animal or an artificial device can act on the environment only by first making an internal map or model of the environment and then acting on the map—that is, as if the map were an exact representation of the environment. That strategy for theorizing about animal behavior or about designing robots has been called inverse kinematics, coordinative structure, motor programming, model-based control, and various other names. I mentioned this strategy very briefly near the beginning of Chapter 20 and near the end of the section in Chapter 22 entitled "Control Systems Find the Only Possible Action," but I will say more here, because not only does model-based control seem necessary to many people, but many people hold to the analogous idea that a more detailed plan is always better than a less detailed one. They strive for a plan that anticipates every possible disturbance before taking the first step of action.

But first let me distinguish the two uses I am making of the word "model." Almost every time I have used that word so far in this book, I have meant a tangible model existing in our environment, made by a researcher to test how well she understands how a living person can function. Now, in the phrase "model-based control," I am using "model" to mean an image inside a person's mind (brain) that the person uses as a guide for her own action. When I go to the grocery, I follow a map or model of the streets between here and there. If I encounter a barricade, my map or model tells me how I can follow a detour. If I go to the train station to wait for a train to arrive, I have in my mind a model of how trains function, and part of that model tells me that the wheels are built to run only on rails. Accordingly, I do not stand near to the rails when the train is due, but I feel safe standing anywhere else. I do not, however, need an internal model to keep myself balanced on my two feet while I wait for the train. If I had to use an internal model of my body and the train station to stand there, I would probably be too sluggish in my movements to keep myself upright.

The idea of model-based control was dominant in, for example, the research field of artificial intelligence until about a decade ago and perhaps still. Rupert Young (2000, pp. 14–15) puts it this way:

> . . . with chess it [the strategy of looking for an optimal solution to a problem] involved looking many moves ahead. . . . Although these programs have become quite sophisticated . . . they gave no insight into the nature of human intelligence. The methods and processes used by the programs . . . were not the same as those employed by humans.

Artificial Intelligence never fulfilled its early promise, and so, due to the resounding failure of the traditional approaches, in the late 1980's and early 1990's a number of researchers moved away from looking at high-level human cognition to more humble . . . problems faced by more simple agents when navigating their environment.

Meanwhile, throughout the decades from the 50's to the present day a quite different theory of perception and behavior within living systems was being developed independent of mainstream Cognitive Science. . . . [Perceptual Control Theory] explains the functional architecture and basic mechanism of the nervous system as *control of input*.

In a posting to the CSGnet of 10 January 1995, Powers gave an example of the kind of anticipation of events and disturbances that a plan with "coordinative structure" would have to include:

> When you look at someone getting the car out of the garage, all you can see are the actions. You see the person moving from the kitchen into the hallway and out the front door. You see the newspaper being tossed aside, the bicycle being moved from the driveway, the hand tugging at the garage door, the trip back inside to find the keys, and finally—after a whole series of unpredictable actions—the car backing out of the open garage onto the driveway. [Thinking a bit about that] it becomes ludicrous to suppose that the person planned to go out and pull on the garage door, then go back inside to get the keys. . . . It becomes silly to suppose that the plan includes a specific contingency saying that if there is a bicycle . . . on the driveway, it will be moved eight feet west. . . .
>
> The whole motor-program, coordinative structure, equations-of-constraint, inverse-kinematics and inverse-dynamics approach is posited on the assumption that the brain must plan moves that must be made in order to create a particular result *as an outside observer would see it* (emphasis PJR's).

If you have the attitude that you, as psychologist or robot designer, know better than the animal (or the robot) how to move toward a goal, then of course you must "tell" the animal or the robot every smallest twitch of every smallest muscle that must be made. But if you are more modest about your own role in the universe, you will look for a neural organization inside the animal (or robot) that can hold to a goal and cope with disturbances quite without help (or interference) from a psychologist. Rupert Young (2000, Chapters 2 and 3) has described lucidly the horrendous difficulties model-based control would encounter in designing vision for robots.

A few years ago, some vulcanologists built a robot that could go down into the poisonous gases and terrible heat of a volcano to gather samples and pictures too hazardous to be got by humans. They named the robot "Dante." In 1995, a television program called *NASA Select* gave a report on the performance of the robot. Powers watched the report and wrote to the CSGnet about it on 15 January 1995:

> I saw [Dante] on NASA Select for many hours and I was beating on the table in frustration. Long before it fell over I was predicting disaster. I think Dante was the perfect multi-million dollar demonstration of what is wrong with the approach of figuring out what move to make and then making it. This philosophy was carried to a ridiculous extreme [in Dante], with computers generating a model of the terrain with the vehicle in it, trying to identify obstacles and compute the limb motions required to get past or over them. The result was that the robot could move only about ten feet per HOUR and spent long periods between steps trying to figure out which way to lift or move its legs while maintaining support. The human operator had to intervene very often to get it out of impasses (which the human operator did quite easily, if slowly). Eventually, Dante got on a slope of loose rock; a rock rolled, and it slid into a posture from which it couldn't recover. Even the human operators couldn't save it; its motions were much too slow, and eventually it simply rolled onto its back and had to be hoisted out. I look on Dante as a horrible example of a certain philosophy of robotics gone mad.

We humans do sometimes use inverse kinematics, or model-based control, but we do it when our real-world use of it moves slowly enough so that the plan or model can be revised as circumstances change. In a communication to the CSGnet dated 19 January 1996, Powers wrote:

> We never actually control a future event. What we control in present time is a presently-imagined future, not the future that will actually occur. Starting with our current financial status and our current knowledge of economics, we imagine a future in which we retire at some standard age, and we make a plan to save for that future—NOW—at some rate. We then calculate what our retirement income will be, based on an assumed average interest rate over those years, other sources of income such as (in the U.S.) Social Security, and so forth. Periodically, as time passes, we recalculate the retirement income as the current situation changes and our estimates of future income change, and we may change our current rate of saving to adjust the calculated income to fit our changing PRESENT conception of how much income we will want.
>
> This is very much model-based control . . ., including running the model at high speed to calculate its state at some future time and adjusting our present actions to control the predicted outcome.
>
> By its very nature, predictive control is much less reliable than present-time control, and the longer the period over which the prediction is made . . . the less reliable the prediction. There's probably some relatively short time over which prediction can be useful, and beyond which reliance on prediction becomes a liability. . . . Look at the people retiring right now after 20 or 30 years of service to the same company, only to find that the company has spent all the retirement funds to pay off debts from a takeover.

Powers said more about model-based control in a missive to the CSGnet on 5 Jan 1998:

> [Model-based control] is, in fact, one of the misconceptions of negative feedback control that students have to be disabused of when they are just learning how feedback works. . . .
>
> The problem with model-based control as a *general* model of behavioral organization is that it requires the brain to be capable of all the analysis an engineer or physicist could carry out, using all the data about the world and the organism itself that the human race has accumulated for the past 350 years or more—and, actually, more, because we are still learning how things work, yet organisms must behave as if they already know everything to perfection.
>
> As far as I can see, if there is this kind of control going on [in living creatures], it can deal only with large, abstract, approximate, slow-changing aspects of the world. As a means of dealing with day-to-day environments it is simply not a very good means of controlling anything. Real-time negative feedback control is all that makes planning-and-execution control even SEEM to work.

[P]lanning below a certain level of perception is simply useless. We don't understand the world well enough to plan in any detail, nor is the world predictable enough. We can set up contingency plans, but they're all in terms of rough generalizations, not actual events. If something goes wrong with a fuel pump 12 hours before launch, the plan may be to replace it, but the plan can't tell us what is going to go wrong, or at what time, or with what side-effects. Those things we just have to cope with as they come up. We can't plan out what to do if the spare fuel pump isn't where the plan says we will find it. The plan can't even tell us how hard to pull on the wrench as we loosen the bolts, or what to do if we twist a corroded bolt-head off.

. . . "real control theory" [describes] the *simplest and fastest* possible way to control something. No [imagined] model needed, no great precision needed, no calibrated and repeatable outputs needed, no predictions needed, no complex computations needed. You get control within the capacity of the system to detect error and to affect the environment—good enough control for the needs of the organism.

In Chapter 36, I will describe some of the difficulties of planning in organizations.

WHAT WE CANNOT DO

I have been emphasizing, so far, the marvels of control that the negative feedback loop can exhibit. "You get control within the capacity of the system to detect error and to affect the environment." That wonderful capacity, however, has a serious weakness. It is not good at detecting very gradual changes in variables. If somebody turns on a water faucet within your hearing, you notice that the sounds coming into your ears now include the sound of rushing water. But if that is no disturbance to any of your controlled variables, you ignore the sound and go on about your business. Before long, you don't hear the water any more. Sensory psychologists call that effect "adaptation." If a sensation increases very slowly, the adaptation keeps up with it, and we may not notice at all that a variable has changed. I can turn up the thermostat, get absorbed in writing this book, and realize suddenly, when I stand up an hour later to get a drink of water, that the temperature in the room is uncomfortably warm. People living in a large city can adapt to the increasing pollution in the air and be unaware of it even though they are aware that they are experiencing respiratory troubles more frequently than they had earlier.

The negative feedback loop copes with disturbances *right now*. The loop is sometimes fooled by disturbances that are increasing very slowly, not to speak of disturbances that *might* come about in the future. The only way we can anticipate troubles and take preventive measures is to imagine the way things might be in the future and the way we would like them to be. We can then imagine how things can move between now and then to arrive at the state we do not want to see and how things might proceed to arrive at the state we do want to see. Then we can estimate the variables we should alter, as time goes by, to increase the likelihood of the one state and decrease the likelihood of the other. A vital part of anticipating future troubles is to keep a meticulous record of slow changes in variables that might become dangerous. The work of the scientists who keep track of the increases in global temperatures is an example. Another is the work of the historians and geographers who keep track of the destruction of unrenewable resources by war and other means.

As time goes by, our understanding of the benefits and dangers in our imaginings will change, and we will change the character of both our goals and our steps toward them. Taking small steps and making changes en route is, of course, the safest way to get there.

When I write "we," however, I am skipping lightly over a great many complications. The steps that will save us from our continuing malfunctions cannot be brought about by politicians alone, by scientists alone, by book writers alone, by priests alone, or by any other group of experts alone. Indeed, they cannot be brought about by all the experts in the world combined. The steps must be taken, in the end, by the bulk of the people of all sorts throughout the world. The early steps must include enlisting help at all levels of society. I will return to this matter of protecting the future in Part VII.

CODA

I hope you have got from this chapter some picture of the vast range of variables controllable in the upper reaches of the hierarchy—somewhat as you might get a picture of the topography of the United States by flying on a clear day from the west coast to the east coast and glancing out the window every half hour or so. In Chapter 25, we will look more closely at a capability many people believe to be humankind's crowning glory: logic and reasoning. And in Chapter 26, we will look more closely at principles and system concepts as they are exemplified in the professional struggles of researchers in personality. To end this chapter, I quote below from an epistle from Powers to the CSGnet on 5 February 1999 concerning what is up at the top:

> Thought and reasoning are not the highest levels in the hierarchy. This is why they have so little effect on changing anything important. To make any significant changes, you must change your principles, and to change your principles you must change your system concepts. When we're considering system concepts, we're looking at things higher than logic, science, methods, language, and so on. We can only *experience* [emphasis PJR's] at this level, considering how the world looks in terms of its organization, internal harmony, beauty, and consistency, and be pleased or dissatisfied with it. If we are dissatisfied enough, changes will begin, and with luck the changes will be for the better, eventually. If our system concepts change, so will our principles change to become more mutually consistent, and as our principles change so will our thoughts, reasoning, and language.

Chapter 25

Logic and rationality

People everywhere explain things, and explaining includes notions such as "because" and "therefore"—the notions with which we construct rules of reason. It seems reasonable to me, as it does to Powers, that a capability of controlling perceptions of rules of reason must lie high in the control hierarchies of humankind everywhere. Programs of reason, logic, and rationality are not the only programs controlled at the ninth level, but I will confine this chapter to those topics. I will be writing here from my small fund of knowledge about the forms those mental activities take in contemporary occidental industrial civilization.

Though the sections of this chapter will describe several manifestations of reason, logic, and rationality, my chief purpose in each section will be to enlarge upon four ideas:

1. Though the *capability* of using language in accordance with rules of logic is inborn, the rules to be followed and the skill of doing so are learned. We differ in our skill.
2. Logic and mathematics can be powerful aids in making use of the environment, but their use does not guarantee truth, validity, accuracy, success, appropriateness, or any other good thing. Logic and mathematics must be applied judiciously.
3. Our brains do marvelously well in continuous control of perceived variables, but variables perceived in the present do not foretell what will be perceived in the future. Perceiving the water running into the washbowl does not include a reliable prediction that the water will or will not run over onto the floor. Perceiving the flavors of pecan pie a la mode cannot be counted on to include a perception of a greatly expanded waistline twenty years hence and the probable accompanying ailments. When we in America read that some disease has become epidemic in Africa, that perception does not in itself entail an image of a carrier of that disease boarding an airplane bound for New York.
4. Logic and reason are not the masters of the neural hierarchy. System concepts and principles always override programs. Generations of schoolchildren were fortunate to be reminded by Samuel Butler (1835–1902) that "He that complies against his will/ Is of his own opinion still."

Much of the time, of course, our will is delighted to comply. One way to avoid inner conflict (and feeling bad) is to go by a clear program that has been furnished you (or that you have already built for yourself), taking only those actions provided by the program at each choice-point. When the choices are few, we call that ritualistic behavior. Ecclesiastics are not the only people devoted to ritual. You find ritual wherever you find norms for social behavior. Social ritual provides widely approved (and therefore safe) ways of proceeding. In psychological research, the rituals of established research method are safe; they may not produce reliable knowledge about human functioning, but they are professionally safe.

LOGIC

We humans spend a great part of every day with thoughts of "because" and "therefore"—that is, in reasoning. The extent to which we reason *logically*, however, is my first topic here. I am not going to offer you a short course in logic; I want only to give you a glimpse of what I am talking about when I use the word "logic." You may want to skip this section if you have studied a book on logic. Even a high-school course in demonstrative geometry, if your teacher taught the course as one in formal reasoning, would have acquainted you with what I mean.

At this very moment, it is possible that a million people in one place or another are saying, "It's only logical" or "It stands to reason." Usually, a person saying that means merely that he feels satisfied with his opinion—that he feels no internal conflict about the matter. But some people practice, some of the time, a kind of systematic thinking in which logic means much more than that, a kind in which thinking is much more meticulous and conscious.

I will not give you a definition of logic. Kershner and Wilcox (1950), in their admirable *Anatomy of Mathematics*, gave no definition of logic. On page 16, they unabashedly said only that "logic is like what we do in this book." Those authors then went right on using logic (and the word, too) throughout their book, even writing there about what they would and would not accept as logical. I looked too into a couple of logic books published about 1990. Neither was as explicit as Kershner and Wilcox, but both omitted any attempt to define *logic*.

Here is some logic:

Pigeons coo.

Here we have a pigeon.

So we have cooing, too.

A thousand years ago, the Scholastics gave the name *modus ponens* to that pattern of reasoning. More than 1300 years earlier than that, Aristotle laid out some variations of the *syllogism*. The three lines above are an example of a syllogism. Here is a slightly more formal version of it:

All pigeons coo.

This is a pigeon.

Therefore, this coos.

The word "pigeon" appears twice; so do "this" and "coo." Those are called the three *terms* of the syllogism. The first two lines are called the "premises," and the last line is called the "conclusion." The second term of the first line and the first term of the second line are the first and second terms appearing in the conclusion. I am not asking you to memorize those labels or the logical pattern. I want only to make it clear that reasoning can be done in a systematic way that can be precisely specified and precisely used by everyone willing to do so. The program of the syllogism has three steps, with choices at each step. A somewhat more symbolic diagram of the family of this simplest syllogism looks like this:

All P are Q.

All Q are R.

Therefore, all P are R.

In the logic of classes, that argument sounds more like this:

The class of Ps is included in the class of Qs.

The class of Qs is included in the class of Rs.

Therefore, the class of Ps is included in the class of Rs.

Classical logic set forth four kinds of statements that figure in syllogisms. They are labeled A, E, I, and O:

A: All P are Q.
E: No P are Q.
I: Some P are Q.
O: Some P are not Q.

The Scholastics took care to provide mnemonic aids. A and I are the first two vowels in the Latin word for "I affirm": *affirmo*. And E and O are the vowels in the Latin for "I deny": *nego*. In addition to affirming and denying, other relations exist among these four ways of connecting two terms. For example, A implies I; that is, if all P are Q, then certainly some P are Q. Similarly, E implies O. Logic books have considerably more to say about the logical interlacing of A, E, I, and O.

In the first example I gave of a syllogism (with P, Q, and R), you can see that all three statements were of the form A. Here is an example of a syllogism constructed of the forms E, I, and O:

E: No orderly minds are creative minds.
I: Some philosophers' minds are creative minds.
O: Therefore, some philosophers' minds are not orderly minds.

The examples of logic I have given so far would look familiar to Leonardo da Vinci (1452–1519). Modern logicians prefer to work with symbolic logic, which looks very much like the branch of mathematics called set theory. I won't bother you with the symbols, but I'll tell you how the four ways of connecting two terms would sound if the symbols of symbolic logic were read off orally:

A: For all x, if x is a member of P, then x is (also) a member of Q.
E: For all x, if x is a member of P, then x is not a member of Q.
I: There exists at least one x such that x is a member of P and (also) a member of Q.
O: There exists at least one x such that x is a member of P but not a member of Q.

If you consider all the ways that four things can be taken three at a time, including repetitions (such as A, A, A), you can see that many syllogistic arrangements can be made, but only some of them would be valid, whatever the content of their terms might be. The Scholastics made up names for all the patterns of syllogisms. The vowels in a name told the kinds of statements in the syllogism. Any syllogism, for example, composed of three statements of type A was called by the name Barbara.

Fallacies and Unrealities

Programs of logic do not automatically turn out statements useful in the tangible world. For one thing, a logically valid syllogism need have no connection with reality:

All babies have tails.
All dogs are babies.
Therefore, all dogs have tails.

The logic in that argument is impeccable. Neither premise corresponds to any tangible reality, but logic has little to do with tangible reality. Any syllogism says simply, "*If* this is the case, and *if* that is the case, *then* the following must be the case." But garbage in, garbage out. The argument above shows, too, that the fact of a correct conclusion should give you no confidence that your premises correspond to reality. Correspondingly, you may use Theory T to predict Event E, carry out an experiment to see whether E will happen, and find that it does happen. You should *not* then have much confidence that Theory T is the right theory.

Aside from the truth or falsity of the premises, it is easy to make mistakes with the structures of logic; people do it every day. One way a syllogism can have an outward appearance of respectability while hiding an invalid structure is called "affirming the consequent." This is an example:

If Jane has gone back to religion, then she has found tranquility.
Jane has found tranquility.
Therefore, Jane has gone back to religion.

That is an inferential fallacy; it is illogical. Jane could have found some other path to tranquility. Other fallacies lie in wait in every direction; one book describes 34 types. Most fallacies are subtypes of three ways a syllogism can fail:

1. A premise can be faulty if a term names something that does not actually exist, such as a chimera.
2. A premise can be faulty if it connects its two terms in a way that does not correspond to reality, such as: All dogs are babies.
3. The structure of the syllogism (the pairing of the terms and the manner of moving from one statement to the next) can be invalid, as in the case of Jane and her tranquility.

Cuban Sugar

You might think that people charged with making fateful decisions for millions of fellow citizens would be careful with their logic. They often fail to do so. On page A6410 in the appendix of the *Congressional Record* of 1960, volume 106, Part 1, 86th Congress, 2nd session, the Honorable Andrew F. Schoeppel quoted approvingly some remarks by the Honorable Spruille Braden, who said that in the face of the Communist threat, we should

> ... stop paying Cuba nearly double the world price of sugar. . . . True, the Cuban people might temporarily suffer, but they would benefit in the end by ridding themselves of their Communist masters.

On page A808 of that same volume remarks appeared from the Honorable Daniel J. Flood, who was quoting from an editorial in a Charleston (South Carolina) newspaper:

> The first logical step to restore freedom in Cuba is to withhold the sugar subsidy, which Castro's government needs to prop up the economy. Next could come withdrawal of U.S. recognition, an embargo on trade, and support of a free Cuba movement in the United States. . . . One way or another, communism must be barred from the New World.

Many other politicians of that period made similar remarks. You can see the logic implied:

If Cuba continues Communistic,
  we will not buy Cuban sugar.

If we do not buy Cuban sugar,
  the Cubans will suffer.

The Cubans do not want to suffer.

Therefore, if we do not buy Cuban sugar,
  Cuba will not continue Communistic.

The structure of that reasoning is good enough. Unfortunately, the second premise is glaringly flawed. The reasoning has other serious weaknesses, but here I want only to illustrate how one flawed premise can nullify logic. I will narrate very briefly a few subsequent events.

In May of 1960, Cuba and the Soviet Union had resumed diplomatic relations. In July of 1960, the U.S. Congress passed an act authorizing the President to cut any foreign quota of sugar imports to the United States, and the President reduced the Cuban quota in that same month and in August canceled it entirely. From a 1991 book by Perez-Lopez, here are some statistics on exports of sugar from Cuba in the years just before and after 1960:

Percentage of total exports

Year	To United States	To socialist countries including U.S.S.R.	To other countries	Total metric tons (in thousands)
1958	58	4	38	5,632
1959	59	6	35	4,953
1960	35	40	25	5,635
1961	0	75	25	6,412
1962	0	73	27	5,131

You can see that the Cubans got along very well. Indeed, the Soviets gave them more money per year for their sugar than we had been giving them. As for tons exported, Cuba's average annual exportation from 1954 through 1960 was 5,110 tons; then, from 1961 through 1987, the average *rose* to 5,863 tons. So much for the logic of the Honorables Schoeppel, Braden, Flood, and the rest who voted to urge the President to act on a faulty premise. Illogic is not, of course, limited to the United States Congress. It goes on in every longitude and latitude every day.

And it leaves a waste of time, money, emotion, and lives behind it. I hope you will not find illogic in this book, but it would be illogical of me to think that I have everywhere avoided it.

Define Your Terms

Often, in an effort to understand what you have in mind behind the terms you use in an argument, a listener or reader will demand, "Define your terms!" That will often help, but the more precise your listener wants you to be, the less possible it will be for you to satisfy her. After you have defined your terms, she could ask you to define the terms in your definition. And then *those* terms. There is no end to that regress. Logicians recognize that it must stop some place, and they speak of the "language basis," which contains words on the meaning of which they are willing to assume they agree without further definition. (Kershner and Wilcox put *logic* in their language basis.) Usually we must leave some words undefined after going only one or at most two steps back into definitions. When you want to be careful, it is a good idea to tell your listener what you are leaving *undefined*. Indeed, even more than asking for definitions, it will often help understanding to demand, "State your undefined terms!"

Other Inadequacies

This ordinary kind of logic (called Aristotelian or Boolean) is inadequate in a deeper way than the possibility of faulty premises. The manner of connecting terms is one of inclusion or exclusion, either one or the other but not both. "P implies Q" means that if you pick up something that is P, you will necessarily find that it is also Q, and that will be true of anything you might pick up whenever you might pick it up. Let us draw a diagram of that relationship. Draw a circle and label it P. Every item inside that circle (you can think of it as a fence, a corral) is a P. Now draw another circle around the first, and label it Q. Every item inside the Q circle is a Q, including all the items that are also inside the P circle. So it is easy to see that if you pick up a P, you have also picked up a Q. The contrary, however, is not true. If you pick up a Q, you will not necessarily have picked up a P; you might have picked up an item lying inside the Q fence, but outside the P fence.

The kind of diagram and syllogism I have described above are isomorphic with the relationship of inside and outside. But there are many more kinds of relationships than that. Consider the relationship "comes before." In the ordinary kind of logic, the syllogism would look like this:

A comes before B.
B comes before C.
Therefore, A comes before C.

That is true on a straight line. But stretch that straight line around the earth (make a circle of it), and the syllogism wobbles. As you travel along that line eastward, Athens comes before Bombay, Bombay before Chicago, *and Chicago before Athens*. That relationship is called intransitivity (inside-outside is a transitive relationship). Every point on a circle comes before every other point, and every point comes after every other point, too. So with a circular relationship, the following is valid:

A comes before B.
B comes before C.
Therefore, A comes before C,
And also, C comes before A.

You can see there why people brought up with ordinary logic have a hard time with the notion of circular and simultaneous causation. In sum, ordinary logic is limited in the kinds of relationships among terms that its structure can validly accept.

Another inadequacy of ordinary logic is that it is static. You will have noticed that all the statements in my examples have been timeless:

This is a pigeon.
All P are Q.
Jane has found tranquility.

Maybe it is a pigeon now, but it might be a dead pigeon, incapable of cooing, an hour from now. Maybe all P are Q now, and all Q are R, but by the time you get round to counting how many P are R, some of them may have deserted. And as to Jane, we can only hope that no one has disturbed her tranquility. In the tangible world, there are many Ps and Qs that jump over the fence while your back is turned.

Still another inadequacy of ordinary logic is its demand that an item of a class be either inside a fence or outside. The item cannot straddle the fence; it cannot be *probably* P or Q. Ordinary logic cannot cope with the relationship of statistical correlation, and it cannot cope with *degrees* of inclusion, such as:

Company P holds one-third of the stock of company Q.
Company Q holds two-fifths of the stock of company R.
Therefore, . . . ?

If P held *all* of the stock of Q, and Q *all* of the stock of R, the conclusion would be obvious, but with partial holdings the conclusion is uncertain. Here is another triad of assertions that sounds like a syllogism but won't work like one:

A visits with B part of the time.
B visits with C part of the time.
Therefore, A visits with C . . . ?

And here is still another relationship ("loves") that won't work right in classical logic:

A loves B.
B loves C.
Therefore . . . ?

Using Logic

I hope I have given ample reasons to use logic gingerly. But if the programs of logic are so unreliable, why do we all nevertheless rely on it to help ourselves choose one thing over another? One answer is that we cannot help doing so. When we are in the imagination mode, we cannot help sometimes thinking, "If I do A, I will at the same time be doing B." Doing B may be desirable or undesirable, so the second premise will have the form "I like (or dislike) B," and the conclusion will be "If I do A, I will find myself doing B, which I like (or dislike)." We are capable of constructing programs like that, and we do it. Frequently, we discover that the program matches later perceptions. If I go to the Bacari restaurant, I will find myself eating a pork chop. I like the pork chops that restaurant gives me. Therefore, if I go to the Bacari, I will be eating something I like. And it almost always turns out that way.

Another reason we use logic is that it does in fact fit very well some large domains of our experience. It fits very well when the relationships are of the inclusion-exclusion sort and where the relationships maintain their character regardless of the times at which they are observed. Chinese boxes are an example. If Box A fits inside box B and box B fits inside box C, the regardless of when we fit them together, we find box

A fitting inside box C. For another example,

> The product of a weight and the distance of the weight from the fulcrum at one side of the teeter-totter will balance the product of the weight and distance on the other side.
> We have a weight of 120 pounds at four feet from the fulcrum on this side and a weight of 80 pounds for the other side.
> 120 x 4 = 80 x 6.
> Therefore, the teeter-totter will balance if we place the 80-pound weight at six feet from the fulcrum at the other side.

In that example, the terms are physically simple and knowable. The weights can be placed accurately with a weight's center of gravity over the distance mark on the teeter-totter. There is no ambiguity about the distances. The teeter-totter keeps its shape and distribution of mass during the experiment. We can easily specify extraneous influences to be ruled out of the syllogism. (We would not expect, for example, that our logic would hold if a tornado were to arrive during the experiment.)

Logic (not necessarily Aristotelian, but some systematic, programmable logic, and barring tornados) holds in vast areas of science, engineering, geography, and games of all sorts. It holds in all areas of purely mental disciplines such as mathematics, chess, and logic. (Yes, logic must be used in logic.) Logic (as well as mathematics) is wonderfully helpful in many of the ways we deal with the tangible world, and wonderfully helpful in all of our imaginary structures. It is prudent, however, to remember that no logic fits all the evidences of tangible reality and that logic is strewn with hazards at every turn, no matter the domain in which you use it.

Because our civilization has produced science and technology far beyond those of the ancients, some people conclude that our average ability to think is superior to theirs. But nobody in the time of Aristotle, no matter how brilliant a mind, could have pushed thought so far beyond the theory of that day as to design a radio, not to speak of building one from the resources of the industry and technology then available. Our science and technology continue their triumphs because a small fraction of every generation throughout history has used the best knowledge and know-how from the previous generation and has added to it. Skill in thinking is not evenly distributed; only a small fraction of the population has acquired the mental discipline to carry on the scientific culture. While some of us struggle to comprehend reality in verifiable ways, others blithely commit the error of affirming the consequent a dozen times a day, turn to the newspaper to profit from the advice of the astrology column, and hang good-luck charms from the rear view mirror.

From the delightful book *Innumeracy* (1988) by John Allen Paulos, I copied this joke, which I am sure is prized by logicians:

> When asked why he doesn't believe in astrology, the logician Raymond Smullyan responds that he is a Gemini, and no Gemini believes in astrology.

And this:

> Mort Sahl remarked about the 1980 election . . . that people were not so much voting for Reagan as they were voting against Carter, and that if Reagan had run unopposed, he would have lost.

CORRELATION AND PROBABILITY

Correlation and probability do not fit into the logic of classes. I'll begin with an illustration of correlation, using the very simple form of 2-by-2 association. Let's suppose you are sorting a collection of 200 books. First you sort them by number of pages, choosing arbitrarily some number at which you will divide the books into those with a "high" number of pages and those with a "low" number. Then you divide the books again by price, again arbitrarily forming two classes, those high and those low in price. Finally, you divide them by the ratios they contain of lines of poetry to pages, resulting in those high in poetry and those low. Let's say that the numbers of books of those eight combinations of categories fall out like this:

Among books low in poetry

		Price	
		Low	High
Pages	High	30	30
	Low	25	15

Among books high in poetry

	Price	
	Low	High
Pages High	15	30
Pages Low	30	25

Now we can look at the variables two at a time, as we look at terms in the premises of a syllogism. What is the correlation between pages and price? To make a 2-by–2 tabulation of the numbers of books according to pages and price only, we "collapse" the poetry category. That is, we add together the counts in the two sub-categories of poetry. The number of books with "high" pages and "low" price will be 30 + 15 = 45. Proceeding similarly, we get this association between pages and price:

	Price	
	Low	High
Pages High	45	60
Pages Low	55	40

There we see the largest counts at low-low (55) and at high-high (60). We would say that pages and price have a moderate degree of positive correlation. That is, it is more likely that we have more books low in price than high when we have books low in pages, and more likely that we have more books high in price than low when we have books high in pages. Similarly, we can tabulate the other two associations:

	Poetry	
	Low	High
Price High	45	55
Price Low	55	45

	Poetry	
	Low	High
Pages High	60	45
Pages Low	40	55

Like the correlation between pages and price, the correlation between price and poetry is also positive. This pairing is analogous to premises in a syllogism:

Low pages implies low price (and high, high).
Low price implies low poetry (and high, high).

And from a syllogism like that, we would conclude that low pages imply low poetry (and high, high)—that is, another positive correlation. But look again at the table above for pages and poetry! The correlation between pages and poetry is *negative*. The high counts are for high pages with low poetry (60) and low pages with high poetry (55). In sum, correlations are not necessarily transitive. If A is positively correlated with B, and B with C, it does *not* necessarily follow that A will be positively correlated with C.

That is enough about correlation. I'll turn now to probabilities, and mention in passing the weathercaster who reported a 50 percent chance of rain on Saturday and a 50 percent chance on Sunday, and so "it looks like a 100 percent chance of rain this weekend."

A large portion of the subtopics and examples below are paraphrases from the bountiful book *Rational Choice in an Uncertain World* (1988) by the psychologist Robyn M. Dawes. If you want to understand how to think about percentages of people who do this or that, I urge you to read Dawes's book. And if you want to know how psychotherapists reason, read that book and also Dawes's *House of Cards*. These two books of Dawes will tell you little about how it is *possible* for humans to function the way they do, but they will tell you a lot about the kinds of judgments we all make at the level of programs. In turn, the judgments we make about causation affect our capabilities as psychological experts and theorists. And if I give you some examples here of the kinds of errors in thinking of which we, including psychological experts, are all capable, I hope you will be cautious when you read the pronouncements of experts and the secondhand reports of those pronouncements in newspapers and magazines. Most of the errors illustrated below will stem from ignoring base rates or picking the wrong base rates.

Here is what Dawes means by a *rational* choice (p. 8):

1. It is based on the decision maker's *current* [not past] assets. Assets include not only money, but psychological state, psychological capacities, social relationships, and feelings.
2. It is based on the possible consequences of the choice.
3. When these consequences are uncertain, their likelihood is evaluated without violating the basic rules of probability theory.

Dawes says, "Don't we all make decisions like that? Decidedly not." His book is devoted to the "decidedly not" and what you can do about it. From the viewpoint of PCT, we will insist that the rules of probability theory, like the rules of Aristotelian logic, are not at the top of the neural hierarchy. No matter how thoroughly people might be versed in probability theory and how respectful of it, they will abandon it if their principles or system concepts call for doing so, just as they abandon logic. Eons ago, creatures with some logical capability must have succeeded in bringing more offspring to puberty than creatures without. And later, creatures who could go beyond logic to principles (and still later, to system concepts) must have had an advantage over those who were limited to logic. Nevertheless, when higher orders of control are not calling for something outside of probability considerations, probability theory can save us from a multitude of harmful judgments. I hope my examples here will illustrate the benefits of attending to probability theory but also how easy it is to find experts who are ignorant of it or unskilled in it.

Dawes tells (on his page 4) about many of his colleagues who are convinced that all instances of child abuse ought to be reported, because they believe that no child abusers stop on their own, and they believe abusers do not because no abuser referred to them has stopped on his own. The psychologists, however, cannot know from their own experience whether there are child abusers who do stop on their own. This is an example of ignoring the base rate—the rate of stopping among all child abusers, including those who *do not show up* in the psychologists' offices. My point is not that the psychologists should go hunting for those other abusers, but that they should refrain from thinking they know something about abusers in general.

Sunk Costs

Dawes gives these examples on page 23:

> "To terminate a project in which $1.1 billion has been invested represents an unconscionable mishandling of taxpayers' dollars." —Senator Jeremiah Denton, November 4, 1981.
>
> "Completing Tennessee-Tombigbee is not a waste of taxpayers' dollars. Terminating the project at this late stage of development would, however, represent a serious waste of funds already invested." —Senator James Sasser, November 4, 1981.

> Both senators were responding to critics who had pointed out that the total value of the Tennessee-Tombigbee Waterway Project, if completed, would be *less than the amount of money yet to be spent completing it.* . . . Both senators believed that the arguments presented here were compelling—or they would not have made them.

Dawes treats the matter of sunk costs at length and discusses other mistakes in trying to encompass the past in decisions about the future. He concludes with these statements (1988 p. 28):

> In general, the past is relevant *only for estimating current probabilities and the desirability of future states.* . . . the probability of five straight heads when tossing a fair coin is 1/32; in contrast, the probability of a fifth head [after four straight heads have appeared] is 1/2 [as it is at every single toss]. *Rational estimation of probabilities and rational decision making resulting from this estimation are based on a very clear demarcation between the past and the future.*

Base Rates

In a workshop Dawes once conducted with a group of practicing clinical psychologists, Dawes gave the clinicians a scenario about a hypothetical person and asked them a question about the person. Here is the scenario and the question (p. 66–67):

> T.D. had been a good student until eighth grade, when he suddenly failed several courses. He was sent to a school psychologist, who interviewed him, gave him a W.I.S.C. (the standard intelligence test for children], a Rorschach [the inkblot test], and a sentence-completion test. The school psychologist concluded that T.D. had an I.Q. of 125, was basically stable, but had been having social problems with peers and family—from whom he was somewhat distant. He had withdrawn into such pursuits as stamp collecting and reading, at which he spent an inordinate amount of time. The psychologist speculated that this withdrawal would probably be temporary, because there was no strong evidence of schizoid characteristics [lack of interest in other people and social interaction] or of gross pathology. The school psychologist concluded that T.D. nevertheless had little sympathy or feelings for other people.

In fact, T.D. did well in high school, went on to college to graduate cum laude with a major in history and minors in computer science and sociology and entered a master's program in graduate school. Please make a probability judgment about which of two fields he entered:

A. Library science.
B. Education.

Most of the clinicians said that they thought T.D. probably entered library science. Dawes then asked them the following questions (note that the first question was about base rates):

1 Estimate the ratio of the number of people in M.A. education programs to the number in M.A. library science programs.
2 What do you know about the background and expertise of the school psychologist described?
3 How well can you predict the occupation of someone age twenty-two from personality characteristics at age thirteen?
4 How well can these personality characteristics be assessed on the basis of an interview and W.I.S.C, Rorschach, and sentence-completion tests?

Dawes asked the clinicians whether they wished to change their answers, and most did so. In his book, Dawes pointed out that although the information in the scenario was sparse and of uncertain reliability (p. 66),

... professionals working in clinics and hospitals often make judgments on the basis of much *less* detailed knowledge. Furthermore, in everyday life we are often constrained to reach conclusions on the basis of very little information. That is one reason our conclusions are, in fact, probabilistic.

Our brains are built so as to *allow* us to conceive of base rates, but we are not born with the concept of base rates. Those professional psychologists who thought T.D. entered library science may not ever have heard of base rates or if they had, they may not have connected them, in their memories, with making judgments from details about individuals. The skill of paying attention to base rates must be learned.

Like all actions, giving a test is a way of enabling yourself to perceive something you want to perceive. Often, what you want to perceive is not something you can immediately perceive with your senses. Perhaps you want to know whether this person has a lot of disposable money. You can't get a look at his bank account, so you look closely at his automobile as a "test" of his disposable wealth. In mystery stories, the characters judge the caliber of the gun that killed the murdered person by the appearance of the wound; that "test" is eventually validated (or not) by the medical examiner when she digs out the bullet. Some tests are presumed to serve as a perception of a characteristic that can be directly perceived by no one; intelligence tests are an example.

Suppose a disease (call it Ugh) could be treated more effectively if the presence of its germs could be detected well before external symptoms appear. Let's say a pharmaceutical company comes out with a test that can detect the presence of the germs at an early stage. The company says their test is 99 percent accurate. That is, when 100 uninfected persons are tested, the test will call only one of them infected (a "false positive"). And when 100 infected persons are tested, the test will falsely call only one of them negative. Now, if we test a person and the test is positive, what are the chances that the person actually has Ugh?

The actual rate of success of the test will depend on the rate of the appearance of Ugh in the population. An extreme example will show what I mean. Suppose that 99 of a hundred people in the population are harboring Ugh. A test that merely said "positive" *every* time (let's say a sham test actually insensitive to the presence of the germs) would be correct in 99 cases in 100. In contrast, a test must be very sensitive to pick out an infected person when the proportion of *uninfected* persons is very large. Suppose that only one person in 1000 is infected with Ugh. The company's test is going to make one mistake in every hundred persons tested. That is, in testing those 1000 persons, 999 of them uninfected, the test is falsely going to call "positive" once in every 100 persons and be right in only *one* of those calls. The probability that this test will be right when it calls "positive" is only about *ten* percent—not 99 percent.

This situation is frequent in modern life. For the many rare diseases that physicians know about, many of the diagnostic tests are even poorer than the example I just gave. Every test used by personnel managers to identify applicants of rare abilities suffers the same disadvantage. The same sort of uncertainties afflict high scores and low scores (that is, scores that occur with low frequency) on any test.

Consider again an example concerning a characteristic that occurs in almost everyone. Suppose we find that almost all people who love Mozart's music have two legs. If we then conclude that all or almost all people who have two legs love Mozart's music, we will be wrong. We will be committing the fallacy of the inverse (or affirming the consequent). Similarly, if we find that people we judge to be mentally out of the ordinary are prone to affirm the consequent, we should not conclude that people who affirm the consequent are out of the ordinary. Dawes found in his own research that "all people do it at least some of the time." That certainly fits with the apparent belief of authors of texts on logic that most their students will be prone to that error; all the books I have looked into take a good number of pages to warn their readers about falling into the error of affirming the consequent and other fallacies. Dawes goes on to say (footnote, p. 78):

> Liberal psychologists, psychiatrists, and other social scientists have been quick to discover the presence of such thinking in political reactionaries and schizophrenics. In schizophrenia, for example, it has even been elevated to a "principle" of psychotic thought.... The irony of this analysis is that the conclusion that people who reason in a representative way must be paranoid is itself based on representative thinking; that is, [on] the ... confusion of inverses. Roger Brown (1973), a psychologist and linguist, lived for a sabbatical year among hospitalized mental patients, mostly psychotics, to determine how their thought patterns differed from those of others. He found no great differences. The people he studied thought and talked the same way other people do, except that they had certain "crazy" beliefs—for reasons, Brown concluded, unknown.

In pointing out the "irony," Dawes was implying that those experts who label people paranoid if they commit the fallacy of the inverse should apply that label to themselves. I must resist recounting further illustrations from Dawes's 1988 book, which is richly studded with them, some humorous and some horrifying. Read his book—and *House of Cards* (1994), too.

EXPLAINING THINGS

In Chapter 24, there was a section on curiosity. I think curiosity and the urge to explain things are two names for just about the same thing. Why do people chip at layers of rock? Sometimes they say they do it because they are curious about the order of the layers. Sometimes they say they want to explain how the layers got into that order. I think no geologist with curiosity about the layers would be satisfied merely to stare at them. I think she would soon turn around and say, "This layer must have been laid down when . . . ."

When my wife and I were traveling in a bus in Colorado, the driver proudly displayed his geological knowledge, explaining to his passengers that you could find shells high on the mountains because the Rockies were under water during Noah's flood.

During the construction of our house in Champaign, Illinois, there were a lot of scraps of wood lying around. One day, I went to the site and found a neighbor, a girl about six years old, sitting on the ground and pounding one piece of wood with another. She said, "Do you know why a hatchet is called a hatchet?" "No," I said, "I don't. Why is that?" "Well," she said, giving the one block of wood a firm whack with the other, "when you have something and you want to hatch it," (whack!) "you take your hatchet" (whack!) "and you *hatch* it!" (WHACK!). Teachers explain all sorts of things in all sorts of ways. So do novelists and playwrights. Painters and sculptors explain . . . well, I am not satisfied with any of the words with which I have tried to finish that sentence. The best I can say is that when I was doing stage design, I had a very clear feeling that the shapes and colors I was putting together made a unitary kind of communication from me to my audience. Scientists build great systems of explanations of the world they live in. Historians, politicians, generals, philosophers . . . the list goes on and on. Our urge to explain puts shape on communicative customs. It furnishes us with the curiosity to examine the world around us to see whether more information about it will enable us to explain things even better. It leads us to systematize our examinations of the world and our reasoning about it; it leads us into religion, scholarship, and science, not to speak of engineering, commerce, games of chance, and stock markets. It also leads us into reification, argument, conflict, hatreds, vilification, and revenge.

The level of programs enables us to imagine a series of actions we might take. The higher levels of principle and system concept enable us to imagine organizations of possible goals and actions in functions, formulas, policies, strategies, coordinations, systems, and world-views. We can imagine words and things in arrangements that do not now exist. In a posting to the CSGnet of 23 December 1996, Powers commented in that regard:

> ... there's a higher level operating. I claim it's the level where we fit our logic into a set of principles. And the principles, it seems to me, fit into and are adjusted to fit various system concepts, world-views.
>
> So, why do we have world-views? I can't see any explanation other than that we want to make sense of experience, to see it all fit into a system of some kind where all the parts make sense in terms of each other.
>
> To me, this need for consistency or harmony is a dominating requirement of life. It seems to function just like an intrinsic reference signal. I can see it in all sorts of people, from street gangs to academics. How does it fit into evolution, learning, and all that? I don't know.

Colleges and universities institutionalize the enterprise of explaining things; there professors profess their beliefs about art, science, and everything else. Every book ever written has brought us the author's explanation of what the author has perceived. Some explanations draw more admiration than others. Without presuming to know what book would draw your own admiration and without claiming to have made a careful selection among the many admirable books about explaining, I will recommend to you just two: *Inventing Reality: Physics as Language* (1988) by Bruce Gregory and *Sensemaking in Organizations* (1995) by Karl E. Weick.

EXPLAINING OURSELVES

Most of us explain our own actions several times every day. We often explain things when doing so serves no immediately useful purpose. One spouse says, "You forgot the apples?" And the other says, "Oh, well, I ran into George in the produce section, and we got to talking...." The first spouse, wanting to bake an apple pie, needs apples, not the information about George. The first spouse might very well, however, welcome the information that the other spouse cares about the frustration of delaying the pie-making. The second spouse could have said, "Oh, so I did, and that's a frustration for you, isn't it? Now you'll be delayed while I go back after them, so I'll hurry."

Suppose I am standing in a crowded lobby, and someone backs onto my toe. "Ouch!" I cry, "You are standing on my toe!" The person replies, "Oh, well, I was trying to get a look at the balcony up there, and ...." But I was not asking for a history of his movements and purposes; I wanted only for him to know that my toe was hurting so that he could move off it. I would have been more satisfied with, "Oh, that must hurt. Sorry."

Suppose you are down on your knees in the garden, with a trowel in your hand and a sack of bulbs beside you. If I ask what you are doing, you are not likely to explain by saying, "I am pushing this trowel into the dirt." That tells *how* you are accomplishing what you want to do. To tell *why* you are pushing the trowel into the dirt, you go up to the level containing the criteria for placing and pushing the trowel: "I am planting tulips."

We all explain to ourselves and others how we think the actual world actually functions. A lot of our explaining is conscious, but we are often unconscious of the assumptions we are making about the real world. An example is going downstairs in the dark and discovering at the bottom that one has placed one's foot as if there were another step where there is not.

I think another indication of our need to explain is the reluctance most of us seem to have to say "I don't know." It seems to be shameful or even inconceivable not to have a ready answer to most questions and especially to questions about ourselves. All of us have had the experience of suddenly finding ourselves perplexed about what we are doing: "Why in the world am I doing this?" Yet when you ask, "Why did you do that," how often have you heard an adult answer, "I don't know"?

Our actions serve not just one reference signal, but usually many. Furthermore, we are almost always unaware of many of those signals. When, therefore, someone asks me, "Why did you do that?" the truest answer would be, "How can I know?" Or, "Well, here are three possible goals I might have had in mind, though I'm not sure of their relative importance, and of course there are bound to be other goals of which I am not aware...". Not many people, however, want to carry on a conversation like that. Sometimes when I am asked, I say, "Gosh, I don't know." Or, "I think it had something to do with _____, but I feel as if there was another reason in there someplace." That at least enables the other person and me to get on to what we want to do next, which is usually more urgent than thinking up reasons for the past.

Fiction and drama are full of good examples of explaining ourselves. If you want examples from a more scholarly point of view, you might care to look into *Explaining One's Self to Others* by McLaughlin, Cody, and Read (1992).

SUMMARY

In shortened form, I repeat here the four main points of this chapter.

1. We can reap benefits from reason, logic, and mathematics to the extent that we use them skillfully.
2. Logic and mathematics do not fit themselves to every sort of tangible reality. We must ourselves find the fit, if there is one, between what we want to do and the applicable logical or mathematical concepts.
3. Control loops deal with present perceptions. We can make a desired future state more likely if we imagine a feasible future state and feasible present steps toward it.
4. Systems concepts and principles are the final arbiters of the neural hierarchy—not reason and logic.

And I will add a note about the fourth point. No professionals work any more meticulously to avoid slips in logic than mathematicians and theoretical physicists. Yet those thinkers are notable, even notorious, for prizing beauty over stodgy logic in their theories. Time and again, when facing a choice between one path or another in following the implications of their theoretical ideas, they have chosen the one more exquisite. Judging from the elegance I see when I contemplate the sweep of PCT, I think Powers, too, must have made that choice more than once.

Chapter 26

Personality

It's all in your head. Every conception, every awareness you have of anything is a perception. Apples, bumblebees, democracy, mothers, personality, races, schizophrenia, the zodiac—all are perceptions. The higher a perception lies in the neural hierarchy, the more idiosyncratic it is. Almost all of us English speakers will agree on what should be called an apple, not a bumblebee, but we will have some wide differences of opinion about democracy and schizophrenia. You will find some wide differences about personality in this chapter.

This is not to say, of course, that all the world is imaginary. We do eat apples and get stung by bees, whatever our individual perceptions of apples and bees may be. We agree with almost everyone about whether we would prefer to eat an apple or get stung by a bee, regardless of what time of day it is or whether the other people are Moslems, mothers, or mandarins. And when we receive one or the other, apple or bee sting, we are in no doubt which it is. The fact that we can extract repeatable sensory perceptions from predicted sources demonstrates a hundred or a thousand times a day the existence of an external reality of which we sense some part. I am only repeating what I said in Chapter 15 (entitled "Where's the Reality?") and, though in somewhat different words, in the section headed "Knowing Something" in Chapter 10 and in the section headed "Speculation and Tangibility" in Chapter 17.

Still, some of the things in your head exist *only* in your head. Chimeras, for example. We can talk about chimeras. I can tell you about mine, and you can tell me about yours. But we will never be able both to look at one, feel of it, or share a piece of it for lunch. Many important things are almost like chimeras—democracy, for example. You can point to people doing certain kinds of things: voting, making a speech in the city park, telling their troubles to the city council or their senator. In each case, you can say, "That's part of democracy." But democracy has no identifiable entirety.

Things out there do not assemble into a democracy in the way bones and muscles and feathers assemble into a bird. The aspects of the environment that you assemble into your conception of democracy are your own selection—not the bird's or the democracy's. I select my own aspects. We can discuss our conceptions, but we can never check their validity by looking to see whether the creature does or does not have wings.

SIMILARITIES AND DIFFERENCES

Let's leave the rarefied atmosphere of democracy and come down to something tangible. Here's Phil. You can reach out and touch him. When you look at him, you see a bodily configuration; you put it in your memory. You see transitions in the configuration as Phil turns, walks, sits, rises. All that goes into memory, too. When you see Phil again, he seems familiar. He participates in events; he seems to spend more time and make more noise in some kinds of events than in others. You see him in continuing relationships with some people and in transitory relationships with others. You find yourself putting Phil in several categories that you have found useful—maybe English-speaking, professorial, male, risible, auctorial, snoozy. You notice that he often does some things in series or sequences. When, for example, he stands up to talk, he always looks right and left before speaking. When he lectures, he seems to stay pretty well with some program he has in his head, despite questions from listeners that might be distracting. It seems a principle with him, too, never to discourage queries by saying something like, "Please hold your questions until the end of the hour." Overall, he seems to you a distinctively recognizable person, a person whom you could term in PCT lingo a unitary system.

Is your experience of Phil all in your head? Certainly. Does Phil exist *only* in your head? Certainly not; you can see, touch, hear, and smell him; there is *something* out there for you to see, touch, hear, and smell. Still, the degree to which the pattern in your head is constrained by whatever is on the other side of your senses—that degree diminishes as you go up the levels. You are not going to insist that Phil is five feet tall if the yardstick says he is six feet. But, depending on your experience with him, you may insist that he is sobersided while Claire insists he is risible.

You put your experiences with Phil into your memory at all the eleven levels. To the total assembly of all those perceptions, organized in your mind as a unique systems concept, you might give the label "what Phil is like" or "Phil's personality."

You do that with everyone with whom you become even somewhat acquainted. You put together all those perceptions at all those levels for each person uniquely (except in the case of identical twins with whom you have been acquainted for only five minutes). No two personalities that you file away in your memory are the same, because no two persons you meet and characterize appear the same. That's no surprise. Artists, novelists, dramatists, and poets have cataloged unendingly the differences among individuals. Psychologists, however, have wanted to find a limited number of ways people can differ from one another—and therefore in which they can be alike. For more than a hundred years, psychologists have been proposing lists of characters, dimensions, dispositions, factors, faculties, motives, propensities, temperaments, traits, and values that they hoped their colleagues would adopt as standard for theorizing and doing research. That is, Psychologist A would publish a list, and then Psychologist B would publish an article saying, in effect, "Well, that's a valiant beginning, but here is a list I think is better." Then Psychologist C would say the same thing about Psychologist B's list, and so on through the alphabet. Here is a list of only a few of the traits or factors (choose your own label) that have been proposed over the years:

achievement motive
affiliation motive
agreeableness
anxiety
artistic motive
ascendancy
authoritarianism
competence-impotence
conscientiousness
economic motive
emotionality
ethnocentrism
extraversion-introversion
femininity
field independence-dependence
inner- vs outer-directedness
masculinity
neuroticism
obedience
political motive
power motive
practical-imaginative
religious motive
self-esteem
self-reliance
social motive
theoretical motive
timid-venturesome

TRAITS

The psychologists were not content to say merely that people differed in lots of ways. They wanted to specify the ways, perhaps in a manner similar to the way a chemist specifies the difference between gold and helium. A "periodic table of the personality factors" would have brought glory. Perhaps they dreamt of finding a few factors from arrangements of which all personalities could be built, as all chemical elements could be built (as it was believed in the early twentieth century) from electrons and protons. From the very beginning, however, a crucial difference between personalities and atoms was missed. The atom that Niels Bohr proposed in 1913 was not a recipe for mixing ingredients; it was a *model* with components and specific, quantitative specifications of the ways the components had to function with one another as parts of a dynamic and continuing system. In contrast, a description of personality is typically presented as a list of scores on the parts of a questionnaire. Sometimes a chart is made with the scores shown graphically; such a chart is called a profile. The conception of a personality described in that manner is not any sort of model. It is simply a recipe for mixing ingredients. Throw in so much of extraversion, so much of agreeableness, so much of conscientiousness, and so on, and we have a personality.

A psychologist who uses the mathematical procedure of factor analysis to find clusters of correlated questionnaire items may protest that she is not proposing traits out of the blue, but is locating a personality in a multidimensional space in which the dimensions or "factors" are composite traits. That is, indeed, the geometric realization of factor analysis. Still, the question of a model remains. How does this individual we are studying make use of the factors or traits? How does the individual interconnect them? Are they arranged in a control hierarchy or does the person simply "have" them? If the factors are of the same quality in everyone, differing only in amounts, how does that come about? Are the qualities and the amounts specified in the genes? Are they kept in their specified amounts as the person grows and changes? How would you use the traits or factors to construct a working model of the person? Unless the psychologist can answer at least a few such questions, the list remains fanciful.

In saying that psychologists might have been inspired by the achievements of chemists, I am not claiming that any psychologist was or is consciously trying to mimic the periodic table of the elements or Bohr's atom. There is no doubt, however, that a great many psychologists have looked to the physical scientists for inspiration, and there is no doubt either that many have missed the crucial features of physical science. In Chapters 2 and 3, I told how psychologists, like most of the rest of us, have persistently overlooked the difference between living and nonliving things, and I have sounded that note again in several later chapters. In Chapters 3 and 4, I described circular and simultaneous causation, very much a post-Newtonian idea. In Chapter 4, I told about J. B. Watson's strange notion that a scientist should not look inside the thing he or she is studying. In various places, I have contrasted theorizing of the recipe sort with theorizing that is used to build models.

PCT does not mix internal standards as if they were ingredients in a cocktail. The basic model for the interconnections among components is the one summarized in Figures 4–1 and 18–3. In the chapters in which I described experiments, I described models that have actually been built. In Chapters 12 and 13, you saw the designers' own descriptions of their models. But PCT's conception of internal standards is not likely to please psychologists who are devoted to the concept of common factors—dimensions of personality that have the same qualities in every person. You have read again and again in this book the assertion that above the very lowest level or two of the neural hierarchy, every person builds his or her own unique internal standards through experiences of finding effective ways to control perceptual variables. I agree with Robertson's (1990, p. 152) recommendation that we "redefine the study of personality as the exploration of the higher-level control systems with which we control our environment."

Personologists have wanted to find particular dimensions or factors which, when mixed in various proportions, would describe this or that *type* of person. PCT proponents have shown no interest in that pursuit. First, the uniqueness of every individual is obvious. Second, there are no natural boundaries between one type of person and another; more bluntly, there are no types. Third—and this is probably the most persuasive reason—our behavior is motivated by our internal standards. Behavior is the way we bring a perceptual variable into match with an internal standard. The way to estimate the nature of the internal standard is not through questionnaires and calculations of intercorrelations, but with the Test for the Controlled Quantity (for which see Chapter 7 under "The Test"). Furthermore, when the controlled variable is found, we do not expect to have found a trait that will always be detectable in the person's behavior. It will be detectable only in environments providing a means of altering the controlled variable and only when we have some way of knowing when the variable is being held steady—as when we ourselves, for example, apply a disturbance. Furthermore, we do not expect to have found a trait that will necessarily be detectable in another person. And still further, we do not necessarily expect to discover a trait that is genetic and unchangeable.

Though everyone is unique, it is common to perceive that Boris and Bertram (let's say) are more alike than Bertram and Benjamin. It is easy, that is, to find clusters of people who share some similarities but are all somewhat different from other clusters of people who are, in their turn, similar in some respects.

Finding those clusters does not, however, necessarily tell you anything about how the people can function the way they do. Let me try to make this more obvious by using an analogy with automobiles.

Look at some automobiles. It is easy to see that they differ in size, color, rotundity, and whether their glasses are tinted, among other characteristics. They are similar in number of wheels, placement of windows (always one in front), and placement of steering wheel (always in front of the foremost seat). Those characteristics of similarity and difference are so easy for observers to see that the advertisers don't often bother to tell you how wonderful it is that they are available. Similarly, psychologists do not seem to include characteristics like size, color, rotundity, and tinted glasses in their personality tests. I don't know why they do not; those characteristics would surely correlate with something (such as being found in a position of leadership), and they are much easier to measure than conscientiousness or emotionality. But let us keep the analogy closer to psychological practice by looking for some characteristics of automobiles less easy to see when you are standing in the showroom.

Some traits interesting to buyers are whether the automobile is powerful, accelerates rapidly, uses a lot of fuel, and is noisy. Those traits would cluster. (A factor analyst would say they have "high loadings" on the same factor.) That is, cars scoring high on horsepower often score high on rate of acceleration, fuel consumption, and noise. Some further interesting traits of automobiles are comfort, smoothness of ride, and ease of steering. Those traits would cluster, too. The correlations would be considerably higher among traits within a cluster than among traits belonging to different clusters.

We now have some cars that score high on the traits of Factor A but not on those of Factor B, other cars that score high on B but not on A, some that score high on both, and some that score high on neither. We also have a good many cars that do not conform to any of those four patterns—cars, for example, that score high on fuel consumption but low on noisiness or cars that score high on both power and comfort. That mix of similarities and differences is in very good analogy to studying personality by examining similarities and differences among traits. We can characterize every automobile by its scores on Factors A and B—or any other factors you might care to construct.

And now that we have ascertained the scores of some automobiles on Factors A and B (and on C, D, etc.), *what do we know about how automobiles function?*

We know that some steer more easily than others, but what do we know about their construction that makes them steerable at all? We know that some swallow more fuel than others while going a mile, but what do we know from those factor scores about what happens to the fuel inside the automobile? What do we know about how to produce more power or less power in a machine that has four wheels, is about twelve feet long, is fairly rotund, and has untinted glass? What do we know about *how to build a machine* that will have qualities such as acceleration, noisiness, and smoothness of ride?

Notice that if you know almost nothing about automobiles, but want to learn how to build one, you do not need to know that one automobile can accelerate faster than another. What you need to know is how any automobile can accelerate at all. You need to know how gasoline can be fed at adjustable rates to the engine, how the explosions of gasoline can be converted into rotary motion of axles, and so on. I think personologists have tried to find particular traits because they have had no model other than mixing (the cocktail model) from which to build human capabilities. In contrast, a model built from PCT can, like a person, act in unpredictable environments to control perceived variables despite unpredictable disturbances and thereby maintain the inner states specified by thousands of reference signals. Some of those inner states in the person we are more likely to call "physical" (such as maintaining the various components of the blood) and others "psychological" (such as a preference for neatness), but they are all intertwined in the manner of Figure 18–3 and they interact uniquely in every individual.

The features of ourselves that we most obviously share with one another take shape much more from the specification in our genes than through effects from the outer environment: two legs, two eyes, a head at the top end, circulating blood, and all the rest. The qualities that distinguish individuals more markedly are those higher in the neural hierarchy: the programs, principles, and system concepts. While the latter are the features that provide most of an individual's uniqueness, they are not those immediately recognizable. You can see two legs at a glance; it takes much longer to ascertain (in a manner approximating

The Test) whether the behavior you see indicates a concern for neatness, for example, and if so, the degree of it the person prefers. When we guess at the variables a person wants to control without using an approximation to The Test, our guesses are often inaccurate if only because of our uniqueness; we often do not have within us a perception sufficiently similar to the perception the other person is controlling for us to recognize it or imagine it.

ASSESSING PERSONALITY

When in the midst of our everyday affairs we assess the character or qualities of another person, we make use of any handy indicator—how the person speaks to a spouse, how she sits a horse, the willingness she shows to make a meal for a guest. No particular action is necessarily an indicator of any particular controlled variable. Through our propensity, however, for perceiving patterns in events, we do form judgments about the qualities of others: "She likes to stir up new experiences." Instead of watching whatever people happened to be doing, the early scientific researchers asked people to perform tasks the researchers thought would be particularly revealing. Francis Galton (1822–1911), for example, believed that intelligence was indicated by sensory discrimination; he asked people to arrange weights in order of their heaviness and to estimate the midpoints of lines. Assessment using sensory discrimination and a minimum of language has recurred periodically. In the 1950s and 1960s, a series of experiments were done on "field dependence" versus "field-independence." The chief method of measurement used a rod-and-frame display. In one arrangement (many were tried out) subjects sat in a dark room and looked at a luminous rod inside a luminous square frame. Various tasks were used; for example, beginning with both rod and frame tilted away from the vertical, the experimenter would ask the subject to rotate the rod to the true vertical. If the subject could place the rod accurately at true vertical regardless of whether the frame was vertical or tilted, the subject was scored as "field-independent." Subjects who were able to do that were usually able to do some other perceptual tasks—for example, to discern a geometric figure embedded in a more complex figure. Almost any task you might conceive turns out to be correlated with at least a few others; that is, if several people can do

one thing, it is likely that at least some of them can do something else, too. That is not saying much.

Sets of weights and tilted frames are more expensive than pencil and paper. Organizations that can afford it, however, often use tests that require the handling of apparatus. If you want to find out whether an applicant for a job has the dexterity to splice wires, it is a good idea to ask her to splice wires instead of asking her questions about her fingers. Tests employing apparatus are often used for industrial and military purposes. In most research carried out by academicians, however, testing of personality, intelligence, academic aptitude, personnel selection for various occupations, and many other human qualities is done with paper and pencil or with a computer keyboard. When a computer is used, even the cost of scoring and reporting plunges. In the list of traits I gave earlier for which psychologists have composed measures, most are measured by paper and pencil. A few use pictures as part of their questioning, and a few use apparatus. All, however, require verbal answers to verbal questions, whether written or oral.

If you want to compare two people on (let's say) agreeableness, you could watch them for an hour or so and then rate them: "I'll give Brutus a 6 on a scale of 10 and Barbara a 7." It turns out that raters in such situations agree poorly. Psychologists prefer to put some sort of device between the scorer and the person being measured that standardizes the testing situation from person to person and produces a numerical score in an objective manner—such as counting the pencil marks beside certain answers. The typical device is a list of questions on paper or on the screen of a computer. The list of questions is called a test, questionnaire, inventory, or instrument. We usually speak of an intelligence *test* and a personality *questionnaire* or *inventory*, but they are all lists of questions accompanied by "keys" that tell the answers the test maker believes should be interpreted as indicating neurasthenia, intelligence, generosity, knowledge of geography, or whatever the test maker claims the instrument will measure. Test makers call the questions *items*.

Making a test is, in principle, fairly straightforward. If, however, you want it to have certain statistical characteristics (such as high reliability), the process becomes full of arcane calculations. It becomes tedious, too, as trial lists are administered to respondents, items culled, new trials made, test-retest reliabilities calculated, new culls made, and so on week after week. The core features of a test, however, are simple, and they are sufficient to exhibit what a test can and cannot do.

To begin, you write some questions. You can give an instruction like, "Mark whether each statement is like you, not like you, or cannot say," and then offer a series of items, perhaps like these:

I daydream very little.
I enjoy mystery or detective stories.
I often cross the street to avoid people I don't like.
Life is risky.

Another way to make items for a personality questionnaire is to use adjectives preceded by an instruction such as, "Check the adjectives that you believe describe you." Many of the early researchers in personality made use of adjectives. Galton did in 1884, and Allport and Odbert did in 1936. Norman (1967) reported the statistical relations (groupings) among 2800 words. Sometimes researchers into personality ask people to judge the applicability of the adjectives to themselves and sometimes the applicability to other people.

Suppose we ask a "rater" to think of someone he knows (the "ratee"), and suppose we give the rater a list of seven adjectives. (In building a real questionnaire, we would use dozens, maybe hundreds of adjectives, but here I use only seven so as to keep things simple and brief.) Suppose we ask the rater to say, yes or no, whether each adjective characterizes the ratee. And suppose the first rater (whose name will be R_1) says yes to the first three adjectives. Here is the listing of R_1's response:

Adjective	Answer
1	Yes
2	Yes
3	Yes
4	No
5	No
6	No
7	No

Now we want to record whether R_1 has told us, in regard to the ratee, that adjective 1 goes together with adjective 2, and so on for every pair of adjectives. There are two ways that two adjectives can go together; they can both be characteristic of the ratee or they can both be uncharacteristic of him. We make a matrix in which each cell tells whether R_1 has given the same answer to the corresponding two adjectives

or different answers. In each cell, the first letter tells the answer to the row adjective, and the second letter to the column adjective. Cell 1,2 shows that R_1 said Yes to both adjectives 1 and 2. Cell 1,4 shows that she said Yes to adjective 1 but No to adjective 4:

R_1
Adjective

Adjective	2	3	4	5	6	7
1	YY	YY	YN	YN	YN	YN
2		YY	YN	YN	YN	YN
3			YN	YN	YN	YN
4				NN	NN	NN
5					NN	NN
6						NN

Now we go to another rater, R_2, who will be rating a different ratee. R_2 believes her ratee to be characterized by adjectives 3, 4, and 6:

R_2
Adjective

Adjective	2	3	4	5	6	7
1	NN	NY	NY	NN	NY	NN
2		NY	NY	NN	NY	NN
3			YY	YN	YY	YN
4				YN	YY	YN
5					NY	NN
6						YN

The third rater, R_3, believes her ratee to be characterized by adjectives 5, 6, and 7:

R_3
Adjective

Adjective	2	3	4	5	6	7
1	NN	NN	NN	NY	NY	NY
2		NN	NN	NY	NY	NY
3			NN	NY	NY	NY
4				NY	NY	NY
5					YY	YY
6						YY

Now, because all three raters were responding to the same list of seven adjectives, let us tally the going-together those three persons are telling us about. Let's begin with adjectives 1 and 2. R_1 told us YY, R_2 told us NN, and R_3 also told us NN:

	Adjective 2	
	N	Y
Adjective 1 Y	0	1
N	2	0

That is, those three raters gave us one tally of Yes going with both items, two tallies of No going with both items, and no tally of Yes for one but No for the other. The two items have a high positive correlation; if you get a Yes response to one, you get a Yes response to the other, and if a No to one, a No to the other. Just to be clear, let us tally one more pair of adjectives, numbers 4 and 6. R_1 told us NN, R_2 YY, and R_3 NY:

	Adjective 6	
	N	Y
Adjective 4 Y	0	1
N	1	1

That correlation is not as strong as the one between adjectives 1 and 2, but it is positive. If we were to turn up a pair of items with more tallies in the upper left and lower right than in the other two cells, the correlation would be negative; getting a Yes to one would be likely to bring a No to the other.

That is enough to show the way you can find pairs of items that are correlated—characterizations which, according to the raters, cluster in the ratees. If you had a lot of adjectives and a lot of raters, you could get some big numbers in those two-by-two tables. Your little tables would look more like this:

	Adjective Q	
	N	Y
Adjective P Y	11	48
N	31	10

You could find some clusters of adjectives in which each one had positive correlations with all the rest in the cluster. You would find many adjectives, too, that had low positive or even negative correlations with most others. If you were to look just at clusters of adjectives with mutually positive correlations, you would expect to perceive some meaning in common among them, and you would usually succeed.

You could make judgments about the usefulness of each item by ascertaining its _discriminability_; that is, you could compare responses to the item with the total scores on the test or on the cluster. If most of the respondents who said "yes" to the item were respondents who got high scores on the test as a whole (or the cluster), you would say the item had high discriminability; it was working in the same way as the other items. To improve the reliability and interpretability of the test, you would throw away items with low discriminability.

The early researchers had to do an appalling amount of hand calculation and then sit and ponder what the correlations were telling them about the clustering of the adjectives. In the early 1930s, however, the mathematical method called _factor analysis_ was invented for finding clusterings among large numbers of correlations—a method that reduced the steps requiring judgment almost to zero (not quite) and later enabled the calculations to be done by computer. In 1934, for example, L.L. Thurstone reported that he had given 60 adjectives to each of 1300 raters. He asked them to think of some person they knew well and mark the adjectives that would describe the person. Thurstone fed the correlations into his factor analysis and concluded that only five factors (clusterings) would satisfactorily "account for" the correlations. Some recent "lexical" investigations of personality have continued to display five particular factors (reported, for example, by McCrae and Costa, 1987; by Goldberg, 1993; by John and Srivastava, 1999; and by McCrae and Costa, 1999).

WHAT DO WE LEARN?

What do we learn about the functioning of the human creature after we have gone through that taxing and tedious process?

Abilities Beyond Words

When every item in the instrument depends on verbal communication between the respondent and the scorer, then what we learn is chiefly, sometimes almost entirely, something about what the respondent does through words. We learn nothing about what the respondent can do without words. Compare, for example, what you can observe from performance on a personality inventory with what you can observe from the performance of participants in the experiments I recounted in Part II. The use of words in those experiments was limited almost entirely to inviting the person to participate. Once the participant began the task, the performance depended on words not at all. Further, the experiment demonstrated that a model could be constructed that performed the same way as the person—a concept inconceivable within the customary assumptions of a personality inventory. I am not saying there is anything reprehensible in restricting one's attention as a researcher on what people do with words. I am saying only that there is a lot more to people than their words—more exactly, there is more to people than the *experimenter's* words.

Forced Traits

With the standard manner of constructing a personality test, we do not learn whether the trait presumably being measured has any necessary connection with the way an individual test-taker functions. The reason I say that is that a test will produce a measure of a presumed trait only if you answer the questions. If you decline to answer the questions, the test will decline to give you a score. In short, you are required to *act as if* the questions are relevant to what you care about. That arrangement between test and test taker cannot reveal whether the test taker finds the questions irrelevant to his concerns.

For example, let us say that I am rating articles of clothing as to whether they are suitable for me. I care about size. When the tester presents me with a garment that would fit me, I am likely to call it suitable if its other qualities permit. And let's say I have a prejudice in favor of natural fibers, too; the larger the percentage of natural fibers, the more likely I am to call it suitable. But let's say I care about brightness of color not at all, neither one way or the other. Toting up my choices, I would get high scores on size and natural fibers. What would I get on brightness of color? Well, I would be paying no attention to that matter. I would select garments regardless of color, so I would select some with bright colors, some with dull colors, and some in the middle. My score on brightness would average out somewhere in the middle. Should that score indicate that I prefer moderate brightness in my clothing? Not at all. The fact would be that I don't care about brightness. The test is giving me a score on a preference that does not exist.

Although a distinctive feature of the theory of personality most widely accepted by psychologists is that the traits are the same in every individual, the tests used *do not permit that assumption to be falsified.* The test will give you a score on the trait it presumably measures if you make X's beside one answer following each question regardless of whether the questions and answers make any sense to you. The worst thing here is not that the respondent can get bored or mischievous. The worst thing is that if the respondent cannot make sense of some questions and answers, if the dimension the test maker had in mind is irrelevant for the person, the person has no way to tell the tester that the test does not fit. If the respondent leaves some or all of the test blank and writes in the margin, "This is not the way I think about it," the tester will simply throw the answer sheet in the wastebasket. That is standard practice. That respondent gets no chance to influence any later testing.

During the making of a new test or in a revision of it, test makers do pay attention to skipped items and nasty comments in the margin of the answer sheet. They know that too many skipped items will reduce the magnitudes of reliability and validity coefficients. They know, too, that customers will not want to buy tests if the answer sheets of too many test takers turn out to be "unscorable." For those and other reasons, test makers throw out items that draw more than a very few maverick responses. Nevertheless, after publication of the test, it will sooner or later encounter people who are to some degree unlike the people whose answers were used to select the final collection of items before publication. These newly encountered people will produce answers that would have caused the items to be rejected during the making of the test, but their answers will now have the effect only of bringing them a lowered score or being tagged as unscorable. Some test makers collect those answers for use in revising their tests. Some do not. Most of the maverick responses, I suppose, never get to the attention of the test makers.

The characteristics of a personality test reflect its "norming group." Test makers cull items that do not work well, find or write new items, and ask some people to answer the revised test. That recycling goes on a good many times. For many reasons, it is often convenient to go back to a group of people who have served before. When a publishable version of the test is at last achieved, the published figures for reliability, percentiles, and other statistics are calculated from the responses of that "norming group." Obviously, the traits that will show up strongly on the test are the traits that show up frequently within that norming group. For example, McCrae and Costa (1987) described one norming group with these phrases:

> ... generally healthy group of volunteers who have agreed to return for medical and psychological testing at regular intervals ... The sample has been recruited continuously since 1958, with most new subjects referred by friends or relatives already in the study. Among the men, 93% are high school graduates and 71% are college graduates; nearly one fourth have doctorate-level degrees. ... Results are based on the 156 men and 118 women for whom complete data were available, except for one subject who scored more than five standard deviations below the mean on the conscientiousness adjective factor and whose adjective data were thrown out.

That is a much more detailed description than many test makers give us. It enables us to be wary of applying findings from tests of that norming group to share-croppers in Louisiana.

Hundreds of books have been written about culture, personality, and testing, and I am giving them shamefully short shrift here. I will say more about testing in Chapter 38. Here, I am trying only to offer some doubts about what can be learned about an individual from a personality test and what can be claimed about a presumed "thing" we call personality.

What we learn from tests, after we have chosen a domain of items, winnowed the items, analyzed the items into traits and the traits into factors, and pondered the correlations, is that there are similarities in the way some people answer the questions, and there are some similarities between those tests and other tests in the ways some people answer them, and the correlations among items and tests and factors are higher among some tests and lower among others. None of that seems surprising.

Some psychologists would say at this point, "Well, we are not just looking for the fact that there are correlations; we are looking for the particular traits and factors that yield the largest correlations in various groups of people." Doing that, I admit, can be put to good profit. Such a search can have commercial value. It can help clothing manufacturers to find a small number of patterns with which 90 percent of a million people can be well fitted. Knowing that certain preferences or motivations go together for larger fractions of people can help advertisers. The information is of no help, however, to scientists trying to learn how humans function. Naturally there are going to be similarities in the internal standards that people form as they grow up in a culture. PCT, however, insists that genes provide us with no common factors, and culture cannot. Since each hierarchy of control will be unique, learning that some percentage of a hundred-and-some people on the eastern seaboard of the United States answered a personality questionnaire in a somewhat similar way in the early 1980s holds no interest for scientists seeking to learn how the human individual (every human individual) is enabled to function.

Though the tests do not yield data that could enable us to disprove the test maker's claim of common factors, the data from the tests do enable us to assess the degree to which the responses of the test takers fit the traits or factors proposed by the personologists. We can also assess the degree to which it is possible, using the tests, to fit some single individual to a trait or factor. I could simply make the claim that the fit is very poor and go on to the next topic, but I think it only fair to produce some evidence and detail some reasoning underlying my claim.

CORRELATIONS

In the section following this one, I will exhibit some of the correlations produced by research into the "five-factor model," also known as "the Big Five," a domain of research in personality which, as I write, is attracting a great amount of attention from personologists. Before I present those correlations, however, I want to explain what correlations can mean—what a correlation number can tell us about the connection between X and Y.

The general idea of the correlation number (usually called a "coefficient" because of its place in a prediction equation) is the idea of any relation; namely, if you know the shape of the relation and the value of one variable, you can calculate the value of the other variable without having to go and measure it. If you know the equation that tells the relation between altitude and air pressure, and you also know the altitude and air pressure where you now stand, and you want to know what the air pressure will be a thousand feet higher, you can calculate it without having to go up a thousand feet and measure it. Similarly, if you know the correlation between X and Y and also the value of X, you can calculate the range within which Y will lie (to a specifiable probability).

Any quantifiable relation can be represented by a graph. A series of measurements of altitude (X) and air pressure (Y) would yield a series of pairs of numbers, and each pair would specify a point on the graph. The points would lie very close to a smooth curve. A graph of a correlation, however, does not look like a line; it looks like a cloud of points. In the kind of phenomenon for which correlations are used to describe the relation between variables, one value of X corresponds to more than one value of Y, sometimes many. As a result, the graph of a correlation looks like Figure 26–1.

Figure 26–1. Scatter-plot of a correlation.

A graph of a correlation is commonly called a scatter-plot. Graphs of two variables do not always have an oval shape, but the roughly oval shape is common. The cloud in Figure 26–1 would result from a correlation of about .8. Figure 26–2 shows some actual scatter-plots, each of 100 points, which would produce the correlation coefficients ("r") shown. When the oval of points is very narrow (lying very close to a straight line), knowing the value of X can give you a close estimate of the value of Y. You can see in Figure 26–2 how very close to 1.0 a correlation must be for a value of X to give an accurate estimate of Y.

Since the predictability of Y from X varies a great deal as correlations go from a value of zero to values of plus or minus 1.0, the question of whether a correlation of some certain magnitude should be treasured or scrapped is an important one for scientists who deal in correlations. Whether a certain magnitude is a welcome one depends on one's purpose. The purpose about which I am writing here is to tell the value of Y, given some certain value of X. For example, given some certain score of a personality trait X and a correlation between scores on the trait and scores on some other behavior Y, how closely can we predict the behavior Y from knowing the score on trait X?

A common measure of the usefulness of a correlation is very simple to calculate. You square the correlation and subtract the result from one. For example, if the correlation is .4, the square is .16, and when you subtract that from one, you get .84. That .84 is usually called the "coefficient of alienation"; Kennaway (1997, 1998) calls it the "coefficient of uselessness." You can see that even a correlation of .9 leaves 19 percent of uselessness. I won't say any more about this coefficient. It is easy to calculate, but it is mathematically subtle; it does not correspond to anything you can put your finger on. I will give you instead another sort of indicator, a screening test. This will require some explanation, but it is worth it.

Suppose we have a test for trait X, and we want to use it to screen applicants for a job. Suppose previous research has collected data and calculated the correlation between score on trait X and performance Y on the job. We want to recruit more people for that kind of work, and we want to hire people who have a very good chance of being those who will do well in the job. Look again at Figure 26–2. To improve our chances of getting good performances, one way to use the test for trait X would be to hire people from those whose scores on X lie above the mean—whose scores, that is, lie to the right of the vertical line through the center of the oval. Except for the case of zero correlation, the mean of scores on Y (the vertical dimension) is higher among the points lying to the right of the vertical line than among the totality of the points. We might still, however, find ourselves hiring some people whose performances fall below the average of all persons; those persons are represented by the area of the oval lying in the lower-right quadrant.

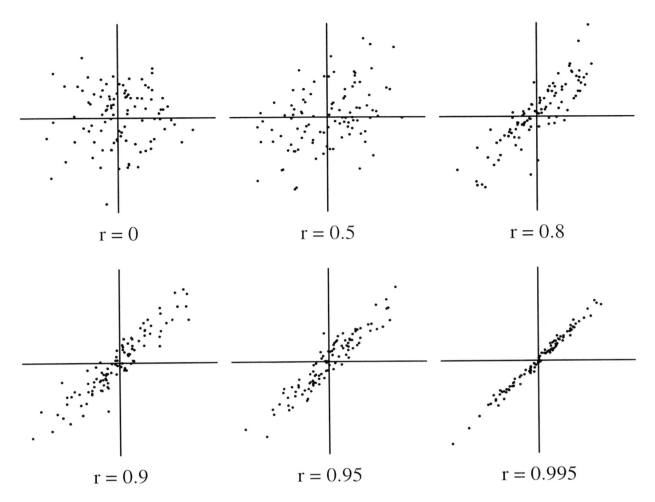

Figure 26–2. Scatter-plots for various magnitudes of correlation. Excerpted from Kennaway (1998, Figure 2).

We can improve the expected performance of the applicants we hire if, instead of using X = 0 as our cutting point, we use a point on X well to the right of the center—that is, well above the mean on X. Clearly, as we move a vertical line to the right, the portion of the oval to the right of the line and *below* the mean on Y will shrink. Let us choose a cutting point on X such that 95 percent of the points to the right of that value on X will fall above the mean on Y. That is, if we pick points at random to the right of the cutting point on X, we will have a 95 percent chance of having picked points that lie above the mean on Y. For a correlation of 0.8, the result of choosing that kind of cutting point would look like Figure 26–3.

The figure shows the desired applicants as dots in the top half of the figure, above the mean on Y. Those are the dots representing applicants who, in some study in the past, gave above-average performances in respect to the variable Y. The cutting point (or vertical line) shown at the right is drawn at a value of X that cuts off a slice of "people" in the oval among whom we can expect 95 percent to perform above the average on Y.

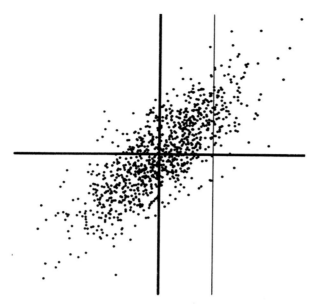

Figure 26–3. Scatter-plot for r = 0.8 with a cut-off on X at a value beyond which 95 percent of the points fall above the mean on Y.

Figure 26–3 is a diagram of the profit to be gained from using test X to predict variable Y. It is a kind of information valuable to industrial employers, military recruiters, school superintendents, college admissions officers, and psychologists. We cannot be 100 percent confident of hiring only people who will fall above the mean on Y, but we can place the cutting point on X so as be confident to some lower degree. A commonly used degree of confidence among psychologists is the expectation that in only one out of 20 predictions of this sort will we turn out to be wrong. Such an expectation is called a "5 percent significance level" or a "95 percent confidence level."

Now note that we can have our specified degree of confidence only in selecting applicants beyond the cutting-point line. If applicants have scores high enough on X (to the right of the cutting point), we can rely on them (with our chosen degree of confidence) to show above-average performance. For applicants whose scores lie to the left of the cutting point, we have less confidence.

What portion of the applicants do we succeed in classifying with these two cutting points? In Figure 26–2, you could draw in each scatter-plot a vertical line placed so as to leave only a very few points below the mean on Y—that is, in the southeast quadrant of the scatter-plot, as in Figure 26–3. You can see that the line would be farthest to the right at the lowest degree of correlation, r = 0, and at that position, the line would cut off not only very few points below the mean in Y, but also very few above the mean. The line would be the smallest distance to the right in the scatter-plot for the largest correlation in the figure, r = 0.995, where a very large percentage of the points to the right of the cutting-point on X would also be above the average on Y. The table below (adapted from Kennaway, 1998, Table 2) gives the percentage of applicants whose scores would fall above the mean on Y at a confidence level of 95 percent.

Table 26–1. For various magnitudes of correlation, the percentages of people classified as falling above the mean on variable Y by a cutting point on variable X set to give a confidence level of 95 percent.

Table 26–1.

Correlation	Percentage of people reliably classified
0.2	0.000000000000075
0.5	0.4
0.8	21.7
0.9	42.6
0.95	58.9
0.99	81.5
0.995	86.8
0.99995	98.7

With a correlation of 0.8, you will be correct in 95 percent of your predictions of positive values on Y if you use the cutting point shown in Figure 26–3, and you will be able to make predictions with that confidence for 21.7 percent of the applicants lying to the right of the cutting point. To be able to classify any large portion of the applicants, the correlation would have to be very close to 1.0, as you can see from the bottom rows of the table. Kennaway's 1998 calculations were based on an ideal bivariate normal distribution, but the conclusions concerning practical predictions will hold for any reasonably similar distribution.

VALIDITY

Now, what do we find in the literature for the correlations between personality traits and other variables? We find a good many correlations greater than zero; that is what we would expect from cultural effects.

We find, too, that the correlations are generally low, most ranging from below .2 to about .50, with few going into the .60s and still fewer into the .70s. In their article in 1987 concerning the validation of the Big Five traits, McCrae and Costa reported correlation magnitudes for correlations among paper-and-pencil personality measures. "The magnitude of the correlations—generally .4 to .6," McCrae and Costa wrote concerning their Table 5, ". . . was larger than typically reported" (p. 86). In their Table 6, showing correlations between answers given by people about themselves and answers given by others about them, the magnitudes ranged from .30 to .57. The correlation of .57 gives a coefficient of uselessness of .68. Interpolating in Table 26–1 above, a correlation of .57 would correctly classify about *one percent* of the test takers.

Hogan, Hogan, and Roberts (1996) reported on the success of personality measures in personnel selection. They wrote (p. 469), ". . . the data are reasonably clear that well-constructed personality measures are valid predictors of job performance. . .". Then they gave some correlation magnitudes. Personality measures correlate most strongly, as you might suppose, with themselves (as in test-retest reliability studies) and with other personality measures. Here are excerpts from what Hogan, Hogan, and Roberts say about the magnitudes of correlations being found:

> The most impressive evidence for personality consistency comes from truly longitudinal studies. [One researcher] found that . . . *r*'s averaged about .34. . . . [Another] found . . . that the scores on the first two factors of the MMPI . . . had retest correlations averaging about .53. [Others] reported . . . correlations averaging .50. . . . [Others] presented stability coefficients on the Edwards Personal Preference Schedule . . . ranging from .20 to .40, with some as high as .70. [Others] found personality correlations . . . averaging around .25 (p. 473).

Costa and McCrae (1988) presented correlations between personality traits over a six-year period that averaged .83. Similarly, Helson and Wink (1992) reported correlations between scores on the [California Personality Inventory] and the Adjective Checklist . . . across ten years averaging close to .70 (p. 473).

Correlations were lower on the average between personality tests and performance in occupations:

> [Researchers in 1991] studied . . . the Big-Five . . . and concluded that, minimally, measures of conscientiousness reliably predict supervisors' ratings of job proficiency and training proficiency (each estimated true validity = .23). . . . [Others reported] validities for the Big-Five dimensions of intellect and agreeableness . . . in predicting job performance (e.g., corrected mean *r*'s of .27 and .33, respectively). [Others] found that the personality inventory that they developed for the U.S. Army significantly predicted relevant nontechnical performance criteria for enlisted personnel (corrected median *r*'s ranged from .33 to .37). [Others] found that integrity tests . . . significantly predict supervisors' ratings of job performance . . . (estimated operational validity = .41). [Others] reported that customer service measures, which contain facets of the Big-Five dimensions of agreeableness and emotional stability . . . have a mean validity of .50 for predicting rated performance in service jobs (p. 471).

> [Another researcher] reviewed validity data from 24 studies in which the Socialization Scale of the California Personality Inventory was correlated with a range of social behavior criteria; she estimated that the true score validity for this scale is .56 (p. 472).

> In a study of creativity in women, . . . measures of creativity gathered in college correlated .48 with occupational creativity assessed 30 years later (p. 473).

The highest correlation cited in those two studies (by McCrae and Costa, 1987 and by Hogan, Hogan, and Roberts, 1996) was .83, the lowest .17. Hogan, Hogan, and Roberts interpret that display of correlations as follows:

> . . . a surprising number of people still believe that personality measures are unsuitable for use in preemployment screening . . .; we have tried to show that these criticisms are less serious than is generally believed. . . . [W]e present data showing that scores on well-developed measures of personality . . . predict important occupational outcomes (p. 475).

How well do scores ranging from .17 to .83 predict occupational outcomes? In Table 26–1, we find an entry for a correlation of .2, which is close enough to .17 for our purposes here. Concerning a test yielding a correlation of .2 with a criterion (such as occupational performance), Kennaway (1998, p. 7) says,

> ... suppose that such a test were being applied to the entire human population of the world [about six billion]. There is only about one chance in 200,000 that *anyone* would be reliably classified at the 5 percent level.

At a correlation of .5—in the middle of the range McCrae and Costa gave as "larger than typically reported"—about four persons out of a thousand would be reliably classified. A correlation of .83 leaves a coefficient of uselessness of 0.31 and reliably classifies about 11.5 percent of applicants.

Hogan, Hogan, and Roberts are not the only people proud of what they believe they can do with psychological tests. Bohl, Luthans, Slocum, and Hodgetts (1996), all members of the editorial board of the journal *Organizational Dynamics*, wrote an article setting forth "concepts that will give new weight and credibility" to the idea that employees are "a company's most valuable asset." One of the touted concepts was "psychological testing to predict performance." Bohl and his co-authors wrote in the same year as Hogan and his co-authors and seem to have been basing their optimism on substantially the same evidence. Bowen, Ledford, and Nathan (1991, p. 36), on the other hand, seem to have seen a different body of evidence:

> ... researchers generally are skeptical about the ability of personality variables to predict job performance. Managerial interest in individual testing appears to have dropped sharply after several 1970 court decisions held that unvalidated and discriminatory selection procedures were illegal.

If you are the personnel director for a large corporation hiring hundreds of persons every year, you will be able to report proudly to the vice-president, "Look, we hired more good employees than we would have hired by tossing a coin." You won't know which of the applicants you hire are the ones you wanted to hire until they have been at work for a while. But if you can get a test having a correlation with performance of .8 or .9, you can do a little better than hiring anyone who shows up. Whether what you gain from the testing will justify maintaining a personnel department will, of course, vary from case to case.

On the other hand, suppose your company hires only a handful of employees each year. Instead of using a test, you will do as well, maybe better, by almost any method you can think of. You might at first think, looking at Table 26–1, that a test with a validity coefficient of .99, used with a proper cutoff score on X, would bring you four new employees out of five with above-average performances on Y. And so it will—on the average. But when you pick five out of maybe a hundred applicants who scored beyond the cutting-point on X, you do not know whether you have picked an average assortment containing four who will perform above the average on Y. You might have been lucky enough to have netted five out of five, but you might have been unlucky enough to have netted only three or two or one. But those calculations are moot, because a test with a validity coefficient as high as .99 (calculated against any criterion you may choose) has never been seen. Finally, if you want to make a prediction about a single individual (such as whether *you* would do well as a librarian), you should not waste your time hunting for a proper test. Here is how Kennaway (1998, p. 7) puts it:

> If an individual requires a 95% chance of receiving a prediction that has a 95% chance of being correct, the correlation must be over 0.99995, and for a 99% chance of receiving a prediction that is 99% likely to be accurate, the required correlation will be in practice unmeasurably close to 1.

There have been dozens of other articles reporting the "successes" of personality research. The two I have quoted are representative. Both were published in prestigious journals under the aegis of the American Psychological Association. A recent book containing dozens, maybe hundreds of correlations similar to those above is the collection of chapters edited by Pervin and John (1999).

If researchers find it interesting to calculate correlations among personality tests or between them and measures of other behavior, I do not presume to tell them they ought to be doing something else with their time. If, however, they expect to discover how the human animal functions by measuring personality traits, then I do urge them to find something else.

It is a mistake to hope to understand the functioning of the human creature by looking at correlations among acts instead of looking for a model that can mimic human behavior. It is a mistake, too, to assume linear causation instead of circular. A third mistake is the assumption that trait X is of the same quality in every person who "has" it—and that everyone has it. The result of those assumptions is the range of correlations I have displayed here—a range showing the hopelessness of using the techniques of personality theory to study a human individual. By studying an individual, I don't mean comparing that individual with a hundred others, but studying that individual alone as a free-standing system. Studying an individual with data taken only from that individual is the only way to find out whether a theory can generate a model of the behavior of an individual human.

Answering an item on a personality questionnaire is a particular act. The success of predicting a particular act is limited by the Requisites for a Particular Act. Put briefly, the prediction will be successful only if (1) you know the variables being controlled, (2) all environmental disturbances remain the same while you are testing your prediction, and (3) no neural reorganization occurs during that period. You can see that comparing one person's answers with another's (as in correlations) has nothing to do with any of those requirements; correlations between answers to test items and other acts must therefore remain low—in Kennaway's meaning of "low."

In 1997, James T. Lamiell wrote:

> ... the assessment and study of individual *differences* is fundamentally and irremediably ill-suited to the task of advancing personality theory (p. 117).
>
> ... the reliability and validity coefficients and other statistical indices generated by studies of individual differences ... *bear no legitimate interpretation of any kind whatsoever at the level of the individual.*
>
> Within a discipline in which the overriding objective is to explain and understand the behavior *and* psychological functioning of individuals, it is difficult to imagine an epistemologically worse state of affairs.

Tests of intelligence and most other personality traits tell you where your performance falls in comparison with others. Your score does not tell you whether you did well or poorly by your own standards, but only by the test maker's standards, and only in comparison with the performance of others in the norming group, no matter how good or bad those performances may be by your standards. A test score cannot tell you in any direct way what you are like. It can tell you directly only how some test maker ranks you in relation to some other people in respect to a standard he cares about for some purpose having some unknown connection, if any, with your purposes.

INTELLIGENCE

What I have said above about personality traits applies equally to intelligence. I will, however, take space for a few further comments. If you want more, go to Richard Robertson (1990).

Brown and Langer (1990) say that in the traditional conception of intelligence, an ambiguous situation presents a problem to be solved, and one's objective in solving it is to find the perspective that most nearly corresponds to reality. What we know of reality, however, lies uniquely in the perception of each of us, not necessarily in the perception of the test maker. Whether an ambiguous situation is a "problem," too, is an individual matter—except when one is acceding to a tester's demand to take the questions as problems. The way a mind works, Brown and Langer say, is that it

> ... (a) views a situation from several perspectives, (b) sees information presented in this situation as novel, (c) attends to the context in which one is perceiving the information, and (d) eventually creates new categories through which this information may be understood (p. 313).

That description rings somewhat like PCT, and the quotation below sounds a good deal like neural reorganization:

> We might consider reconstructing this world for ourselves whenever it does not fit our abilities, or perceived lack of abilities, whenever we feel stunted or less than fully effective. [W]hen we are not feeling smart, we are not being stupid; we are being sensible from some other perspective (p. 332).

The early makers of intelligence tests thought of intelligence as a trait that would change very little during a lifetime, a trait given largely by the genes. It was not long, however, before evidence of changeability appeared. For example, a report in *Science News* ("Educated IQ," 1991) told about a review in which nearly 200 studies indicated that IQ scores are higher among people who spend more time in school, regardless of quality of schooling.

Another example, a mysterious one, is the phenomenon called the "Flynn effect." Flynn studied intelligence testing in the U.S. military in the early 1980s. He found that the average score achieved by recruits at that time was above the average of recruits a generation earlier who had taken the same test. Flynn and other researchers since then have looked at data from 20 countries; all the studies show rises as the years have gone by. When raw scores are recalibrated to take the mean in 1989 as 100, the mean score in 1918 would have been 76. In other words, scores had increased by about 30 percent! Many explanations have been tested in the data, but so far no explanation has fit. Ulric Neisser (1998) has edited a book on this topic. John Horgan (1995) quoted an earlier statement by Neisser—a statement with which I agree:

> The fact that there could be such a large effect, and that we don't know what causes it, shows the state of our field. It shows that we should be quieter than we are.

Some psychologists still hope to pin down some genetic "component" of intelligence despite warnings from evolutionary biologists—Richard Dawkins (1982), for example—about the complicated way that genetic endowments manifest themselves (or do not) in observable behavior. Those complicated ways do not at all fit metaphors such as components or strength of ingredients. Using the mathematics of correlations between the appearance of "markers" on genes and a score on an intelligence test is like looking at parings from the hoof of an elk and predicting whether that elk will be found in the northeast corner of the herd (like the dots in Figure 26–3) at nine o'clock in the morning of December third. I doubt there is any way to trace a causal connection from the structure of DNA to the number of items a person answers "correctly" on an intelligence test.

I hope what I have written here leaves you with very little confidence in the usefulness of intelligence tests for any purpose whatever. I do not deny that you are justified, knowing two people fairly well, in putting more trust in the competence of one person to carry out certain kinds of tasks than in the other's. If, for want of a term you like better, you call the one person "more intelligent" than the other, I won't accuse you of misbehavior. But don't give the two of them an intelligence test; you won't know any more about them than you do now.

CODA

About personality, Kurt Danziger (1990, p. 239, footnote 17) wrote:

> Of all the slippery terms that define modern psychological discourse this one is perhaps the most slippery.

It is slippery because it resides in its multiform shapes among our system concepts. You may believe that we have the characters we have because the human species has been allocated a few traits or factors—such as the Big Five—from which each individual human is fashioned according to a recipe provided by the genes—so many ounces of intellect, so many of agreeableness, and so on. Considering the low correlations I have exhibited, I don't know what you can do with that belief except publish papers in journals with "personality" in their titles.

What we need for a science of living creatures is the capability of building models that are testable with an individual—that can be disproved with an individual. Beyond that scientific purpose lies the purpose of sheer curiosity. I do not say it is shameful to investigate what can be correlated with what. I say only that correlations among traits or between traits and behavior cannot build a psychology that can be tested with individuals. We need a model of the person so constructed that it can fail its testing so clearly that the psychologist will revise the model instead of writing an explanation of how it is that his or her theory is nevertheless right after all.

PART VI

DYADS AND GROUPS

So far in this book, I have focused on the individual human. I have sometimes digressed to write about rats or about humans interacting with other humans, but my intent was always to illuminate the functioning of the individual. I have now touched upon all the features of individual functioning I think necessary to a reasonable introduction to PCT. I have not provided you with a manual for constructing models of living creatures; for that, you must go to the writings of William T. Powers, W. Thomas Bourbon, Richard S. Marken, Richard Kennaway, Wolfgang Zocher, Martin M. Taylor, Kent McClelland, Bruce Abbott, and Dan Palmer. But I think I have said enough so that you can locate PCT's manner of investigation amid the branches of science and, in particular, so that you can recognize the differences between PCT and mainstream psychological research. The use of the social world to control perceptions, however, brings more surprises than the use of the nonliving world. For that reason, among others, I do not end the book here, but go on to say some things about social life that I think the basic assumptions of PCT allow us reasonably to say. I will also mention some ideas many people wish were reasonable to say, but which, given the assumptions of PCT, are not reasonable.

I will begin my description of social life by describing in this part of the book the ways that interaction can proceed between two people. We often speak of what a collectivity does or how we can deal with it. But we always interact with a collectivity by interacting with individual members of it. We deal with the General Electric Corporation by dealing with its public relations officer, its purchasing agent, the person behind the counter at our local retailer, or the person who opens the letter we sent addressed only to "General Electric Corporation." We deal with a 10,000-member government by dealing with a customs inspector, a tax collector, a member of the House of Representatives, an agricultural "extension" agent. We deal with a family by speaking to a member—perhaps a child at school. This family, this church, this business firm, this society, is what it is through the ways the individuals conceive the collectivity and the ways they believe they can use other individuals (even when they are anonymous) as a means of controlling their perceptions. No matter how intricate may be the myriad linkages in the collectivity, each member can act only at the links among which he or she is a node. Accordingly, I will devote Part VI to describing some social relationships as we can see them manifested in dyads. What can occur in dyads can of course be seen again in collectivities of any size—though the converse is not true.

Despite the complexities of dealing with other people, the key idea for the rest of the book is the same as it has been for the earlier parts: we act on the environment as a means of controlling perceptions. I could say that equally well by quoting William James (1890, p. 7): "Again the fixed end, the varying means!" That remains key.

RECAPITULATION

You may want to have at hand a quick review of the key ideas that I have set forth so far and that will underlie the rest of the book. You can remind yourself of them by looking at certain of the figures in earlier chapters.

Figure 4–1 diagrams the circular, simultaneous, continuous causation through which living creatures are able to control their perceptions continuously—through which they are able to act and at the same time perceive the consequences of their action. It reminds you that action is shaped simultaneously by inner reference signals and outer opportunities. It reminds you that every function is an in-out "transfer function" operating with linear causation, but the connections among the functions result in a loop characterized by circular causation in which every pulse is simultaneously a cause and an effect.

Figure 3–3 reminds you that although action results from joint effects from inside and outside the organism, the effects are not symmetrical. Nonliving things (with certain very temporary exceptions, such as exploding dynamite) do not amplify forces that impinge upon them. Living things do reliably, repeatedly, continuously amplify incoming energies.

Figures 18–2 and 18–3 diagram the organization of hierarchical control, and Figure 19–2 diagrams the function of memory within the operation of control.

The rest of the book will give a lot of attention to action, because action is what we see when we look at each other. For convenient reference, I repeat here the Requisites for a Particular Act that I first set forth in Chapter 1.

REQUISITES FOR A PARTICULAR ACT

For a person to perform a particular act, it is necessary—

1a That the person be controlling a perceptual variable (such as intensity of light).
1b That some environmental event disturb the controlled variable; that is, that the environmental event have an effect on the controlled variable such that, if the variable were not controlled, it would underreach or overreach the internal standard.
2a That an "object" or source of energy or some means suitable for affecting the controlled variable be available in the environment,
2b That the person come upon or believe it possible to come upon a suitable object or means,
2c That the person be capable of carrying out an act with the object that will affect the controlled variable (this includes being capable of conceiving or imagining the act, when that is a necessary step),
2d And (especially if the act begins in the conscious state) that the person estimate the likelihood to be sufficiently high that a feasible act will reduce the disturbance to the controlled variable.
3 That the chosen act not disturb some other controlled variable.

I think of Requisites 2a through 2d as constituting the "opportunities" in the environment for controlling a perceived variable. I will often speak of those opportunities in the pages to come.

DYADS AND GROUPS

I have chosen to limit this next part of the book to dyads so as to simplify my task of writing about social interaction. Social psychologists sometimes call the dyad the minimal group or minimal social system. You can see in the dyad the simplest forms of communication, coordination of effort, and cooperation. But though communication is easy to see there, it is not always easy to *do* there, as many lovers testify. Communication between two persons is often easier when further people are there to help. So it is that we make use of various kinds of helpers: negotiators, referees, coaches, consultants, counselors, clergy, psychiatrists, and the like. In Part VI, I will bring in the topics of language, communication, cooperation, counseling, and some other patterns of action that can be seen in the dyad.

Communication and coordination of effort become more complex (though not in every instance more difficult) as the number of people face to face grows larger. In a group, communication is almost always unequally distributed, appearing more frequently within some dyads than within others. Some individuals address the whole group more often and at greater length than do others. Effective communication in a group requires skills in addition to those needed in a dyad.

Chapter 27 will focus on making use of the social environment as a part of the feedback loop. Later chapters will deal with the intricacies of particular aspects of the social environment. Counseling, planning, and turning to experts are examples.

Chapter 27

Degrees of freedom I

What we can choose to do depends on the opportunities available in the environment (item 2 of the "Requisites"). What is available depends, for example, on where you are on our globe—land, ocean, tropics, arctic, and so on. For humans, what is available in a locality depends to a great degree on what previous humans have left there. Humans far outdistance other animals in reshaping the environment. Most of the marks left 10,000 years ago by the hands and feet of homo sapiens are not easy for most of us to find, but many of the marks left in the last few centuries are very easy to find. The medieval cathedrals of Europe and the motorways of the United States seize the eye. The opportunities (and dangers) in New York City in the year 2000 were different from those available in 1900, and those differed from what was available in New Amsterdam in 1650.

THE LIVING ENVIRONMENT

Control is more successful when variables change slowly (but not too slowly), when fewer variables are changing, and when the variables are being changed by nonliving rather than by living things. The action of a nonliving thing when affected by an external force can be predicted with the assumption of linear causation, but the action of a living thing cannot. Consequently, operating on the nonliving environment in a habitual way can bring highly reliable results, but operating on the living environment is always chancy. Washing a saucepan can be done while thinking of something else, but washing a dog or a child requires continuous attention. Below are some situations in which you might find yourself. They differ in the predictability (reliability) with which the variables available there can be controlled. Imagine yourself in each. The most predictable situation is listed first, the least last.

> Imagine that you are in an environment containing no moving object. Example: Sitting reading in a quiet corner of the library with nobody near. Another example: Standing on a sand hill in western Nebraska at dawn, with no bird or animal close enough or large enough to be encountered.

> Imagine that you are in an environment with one moving, nonliving object. Examples: Standing alone in a workshop, drilling a hole in a board. Sitting on the bank of a brook, watching the water flow past. Flying a kite.

> You are with several simultaneously moving, nonliving objects. Examples: Playing the piano. Juggling. Hanging clothes on a clothesline on a windy day. Operating a sewing machine, especially one with a treadle.

Notice how relaxed and peaceful the world seems so far. Now I will bring living things into the picture.

> You are with one other living creature. Examples: Walking with a dog in a meadow far from other people or animals. Riding a horse along a quiet path in the country. Walking near a woods and hearing the buzz of a bee.

Those seem like relaxed and peaceful activities, too, and all very well could be. Still, the dog can run off and ignore your calls to return, the horse can shy at something and possibly toss you off, and the bee can land on you with hostile intent. Continuing the list:

> You are with several living creatures. Examples: Taking six dogs for a walk, all on leashes. The four-person game with eight rubber bands described in Chapter 6 under the heading "Four Controllers."

If you have seen someone out with several dogs on leashes—or if you have done it yourself—you know that they can become very awkward to manage. In the game with the rubber bands, someone can get cantankerous.

> You are using language with one other person. Examples: Having a conversation in a far corner of the library. Talking on the telephone with one other. Exchanging e-mail.

Conversations can be soothing, charming, invigorating, challenging, demanding, insulting, enraging, accurate or misleading, or have any of a thousand other qualities.

> You are using language with several other persons. Examples: A coffee klatch. A committee meeting. A conference.

Committee meetings can be peaceful or brawling. Large deliberative groups meeting repeatedly (examples are the Senate and the House of Representatives of the United States) have sergeants-at-arms to cope with brawling.

Some communicative situations do not stay classified. A lecture, for example, can become a conversation among many if the lecturer freely allows questions and comments. It can also be a case of the lecturer hearing himself talk while members of the audience sleep, write letters, or watch out the windows at squirrels digging for nuts.

A social encounter can consist almost entirely of talking, but it often includes other kinds of acts: carrying on work of all sorts, playing games, eating, copulating, making music, traveling, and so on. People often prefer certain kinds of settings for satisfying certain kinds of needs; they prefer, that is, settings that offer certain kinds of opportunities for controlling perceptions. Here is a description by Randolph Louis Viscio (1993) of the preferences of teenagers for places to "hang out":

> We were the first generation to set foot on the hard, . . . clinical, sterile, conditioned . . . surfaces of America's tacky malls. We were so disgusted with what we experienced at the mall, this glimpse of our future, that we naturally attempted to liberate parts of it, to make the arcades, the pizza joints, and the bathrooms places that were open to anyone and free from the conformity that saturated the rest of the institution. Those places were our respected territory. Here we could play, eat, and piss in peace; we smoked pot in the parking lots and drank tequila and Southern Comfort on the roofs and in the bathrooms. . . . Seen through my eyes, the mall was just like the rest of society: sterile, empty, controlled, and used by the Authorities to make money. . . .
>
> The arcade became a sanctuary we all needed. Everyone shared the space and pretty much respected it as being a place for everyone to hang. It was the first place kids would go when they arrived at the mall. From the arcade, we would form small gangs and go off in different directions to shoplift or smoke a joint. I did a lot of both and particularly enjoyed going onto the roof and conducting guerrilla snowball assaults on the mall police.
>
> We spontaneously created a community of nonconformist youth. The new hangout, however, came to us with one condition; for the time being the Authorities were willing to let us gather as we wanted so long as we continuously fed quarters into the slots of the video games. . . . So we spent a lot of time and effort finding ways to make money—stealing included. At first I stole from my mother. Then I stole from anyplace that was left open to me (the United Way donation jars at the counters [for example]). . . .
>
> A lot of kids, young kids, used to get drunk in the bathrooms and then make love on the roof or in the woods just beyond the parking lot. . . It was a very rebellious experience to smoke, drink, make love, fight, or just sit in those woods. . . .
>
> Then various mall Authorities got together with the cops and the school Authorities to crack down on the kids doing the mall scene. The media played up announcements that school officials were blaming falling grades on the fact that kids were "spending too much time playing video games in mall arcades. . .".
>
> Kids stopped hanging out at the mall. But our grades didn't get any better and we didn't stop making love or drinking. . . . we just found new places to go; the woods, hard rock concerts, cars, and even the schools could become "hangouts" of sorts.

That account does not give a precise description of the perceptions the author or his friends were trying to control or how "hanging out" helped them to do so, but one can get at least a flavor of the perceptions from some of the phrases: "disgusted with . . . this glimpse of our future, . . . open to anyone and free from the conformity, . . . respected territory, . . . in peace, . . . assaults on the mall police, . . . a very rebellious experience," and the capitalized "Authority." It seems reasonable, too, to suppose that the goal Viscio said the kids had when they fed quarters into the video games was not a goal most school officials thought the kids were pursuing.

People very frequently use the social environment in controlling perceptions by talking with other people; sometimes, however, they use it by perpetrating violence on others. Here are some statistics about violence in the United States given by Lore and Schultz (1993, p. 17):

> The risk of being murdered in the United States is 7 to 10 times that in most European countries. Finland is the closest European competitor, but the American homicide rate is 3 times that of Finland. . . . the homicide rate in Australia is less than one third of that in the United States, and that in Canada is less than one fourth.

James Gilligan (1996, p. 95) makes a similar report:

> The murder rate in the United States is from five to twenty times higher than it is in any other industrialized democracy, even though we imprison proportionately five to twenty times more people than any other country on earth except Russia; and despite (or because of) the fact that we are the only Western democracy that still practices capital punishment (another respect in which we are like Russia).

I give this sort of example here only to emphasize the wide range of uses of the environment through which humans control their perceptions. The wide range of violence rates suggests that cultures vary greatly in the opportunities they are perceived to offer for homicide.

DEGREES OF FREEDOM

Freedom can be a matter of sheer physical space, as any prisoner knows. I once crawled into a ventilation shaft to install an electrical conduit there. I could move my arms only by sliding them between my body and the walls. I found myself having an intense desire to get out of that shaft. I slid slowly out and lay on the floor for a minute or so, sweating. Perhaps you remember the chickens I mentioned in Chapter 3 under "Person and Environment" that kept the walls from getting too close by pecking at a button. Humans do not like other humans to come too close without special permission. Some people call this maintaining one's "personal space." When introduced to someone, for example, most of us would consider the person to be impolite, even hostile, if he approached closer than, let's say, two feet to shake hands. Cultures differ, of course, in this sort of standard.

But physical limits are only a part of what I am talking about. Other people limit our opportunities. They hide information that would be useful to us. They discourage us from using some opportunities by making threats. They demand that we postpone our own goals and help them attain theirs.

I wrote about degrees of freedom in Chapter 18 under "Second Order: Sensation" and under "Degrees of Freedom." We find new degrees of freedom by recombining perceptions from lower levels and also by perceiving (and acting on) new aspects of the environment and combinations of them. As social life becomes more complex (as it must as the population becomes more dense), our degrees of freedom come more and more to depend on the actions of other people. At the highest levels of perception, we seek greater degrees of freedom through political negotiation, religious sectarianism, industrial democracy, and voluntary organizations of all sorts. Other people, however, can be both helpers and hinderers. We sometimes, therefore, improve our degrees of freedom by going alone into the wilderness or just into another room . If a wilderness or another room is not available, we sometimes simply refuse to talk to anyone.

The more people we depend on to provide us with environmental opportunities for perceptual control, the more likely it is that those very people will disturb some of the variables we want to control. We depend on people to build roads, extract oil from the earth and refine it into gasoline, construct and install traffic signals, and so on. Then we can drive to the market and buy some food. But on the way to the market, we encounter the automobiles of many people who had nothing to do with any of those people who have helped us get to the market, but on whom those people who helped us have depended in turn for their opportunities. Those people in those other cars (people who help the other people who help us) get in our way on the streets and in the markets.

We gather within a convenient distance to do things with others, and in gathering, we get in one another's way—we make it more difficult to help one another. We get angry, suffer "road rage," get ulcers, drive to a physician to get cured, and increase the traffic congestion. The sheer increase in population, too, presses us all closer together and increases the average time and energy needed to make use of the environment. Each of us participates in building a civilization in the hope of increasing our options— our degrees of freedom. We succeed, on the average, in increasing some kinds of options, but we also lose some kinds. Furthermore, civilization increases some kinds of freedom more for some people than for others. Money is the obvious example. Money buys freedom of many kinds, and in every civilization so far seen, money has been very unequally distributed.

I will take some space here and in later chapters to give illustrations of the manifestation of degrees of freedom in our society, because sufficient degrees of freedom is the essential condition to be sought in any society if conflict, both intrapersonal and interpersonal, is to be kept to a salubrious level.

INCREASING DEGREES OF FREEDOM

We often think of personal power as the ability to do what we want to do. But since other people often seem to get in our way, we get distracted by that frustration and come to think everything would be all right if we could just get other people to do what we want *them* to do. I remember David C. McClelland's writing somewhere that an anchorite can feel powerful because there is nobody around to interfere with him, and a drunkard can feel powerful because he doesn't feel the social restrictions he would feel if sober.

Culbert and McDonough (1980), in a remarkably perceptive book on life in organizations, say that "Real power has more to do with clearing space for your own interests than getting other to perform in a certain mode" (p. 195). They say that you can feel powerful and fulfilled when the way you go about your work satisfies at the same time your own needs and the goals set by your bosses. When there is a match (not a conflict) between your own needs and the goals of the organization, Culbert and McDonough call that "alignment."

When employees low in the hierarchy are encouraged to organize themselves to suit their own way of working, the policy is often called "high-involvement management," meaning that the employees are allowed more "involvement"—more degrees of freedom—in their own working lives than is ordinarily the case. Edward E. Lawler III (1986, p. 193) writes that if you want to give employees greater freedom in carrying out their work, you must make certain assumptions. I rephrase three of them here:

> Most people can make decisions about their work activities that will benefit that work without hampering work elsewhere.
> Most people can find the knowledge needed to make those decisions.
> When employees manage their own work, the result is greater organizational effectiveness.

You can see that those assumptions allow employees greater freedom to fit their own goals into the way they pursue the goals of the organization. (Slipped again! Assumptions can't allow employees to do anything. I should have written that *managers* holding those assumptions are more likely than other managers to allow employees greater freedom.) When managers and workers come into conflict over the variables they are controlling, energy is wasted in working against the people with whom one is in conflict. When conflict is reduced, the energy needed for control is reduced, and more options become available. Wider "space" seems available in which to find opportunities for controlling perceptions.

Jan Carlzon, then Chief Executive Officer of Scandinavian Airlines System, published a book about managing called *Moments of Truth*. In reviewing that book, Celeste Coruzzi (1987, p. 256) said that Carlzon had

> ... four beliefs regarding human nature: (1) Everyone needs to know and feel that he is needed, (2) everyone wants to be treated as an individual, (3) giving someone the freedom to take responsibility releases resources that otherwise remain concealed, and (4) an individual without information cannot take responsibility. ...

The emphasis in all the quotations I am giving here is on the breadth of choice available to the worker. What you do not find in these quotations is any demand for conformity to a standard routine or for close supervision or for obedience to detailed orders. Close supervision reduces competence. As long ago as 1965, Arthur Kornhauser reported "that mental health is poorer among factory workers as we move from skilled, responsible, varied types of work to jobs lower in these respects" and that symptoms appearing in the more restrictive jobs included "dwarfed desires and deadened initiative...". In a study of responses from males in national sample surveys in Sweden and the U.S., Robert Karasek (1979) found that the combination of narrow latitude for decision and heavy demands from the job was associated with mental strain. Participation by workers in managing their own work has been found to pay off. Ralph P. Hummel (1987, p. 73) said:

> ... the more than 40% of American businesses with more than 500 employees that now have worker participation programs did not implement them [only] because of a sudden discovery. These businesses were losing their shirts in a competitive marketplace in which some firms still knew how to produce quality.

In 1973, in December, the astronauts in the Apollo III mission to Skylab went on strike. They refused to do any more work until they were given more degrees of freedom. In telling the story, Karl Weick (1977) quotes Neal Hutchinson, the flight director of the mission:

> We send up about six feet of instructions to the astronauts' teleprinter in the docking adapter every day—at least 42 separate sets of instructions—telling them where to point the solar telescope, which scientific instruments to use, and which corollaries to do. We lay out the whole day for them, and the astronauts normally follow it to a "T." What we've done is we've learned how to maximize what you can get out of a man in one day (pp. 31–32).

Clearly, Hutchinson was wrong about it, as shown by the fact of the strike. Weick says:

> Conspicuously absent from the zealous scheduling by ground personnel ... was any sense of the astronauts' "selves" and of their needs to reflect, to observe, to find their place amid these baffling, fascinating, unprecedented experiences (p. 33).

Later, Hutchinson said, "Then I saw we'd done a bad thing by forcing them. I saw they needed time to think about what they were doing and reestablish themselves" (pp. 33–34). Instead of dictating a series of detailed actions, Weick says, the organization should encourage people to "create alternative arrangements of people and activities.... the location of these processes [is] in the hands of insiders (the people who will do the work) rather than outsiders..." (pp. 38).

In England, a new candy factory was organized so as to give workers a high level of involvement in operational decision making. In groups of eight to twelve people, members were collectively responsible for allocating jobs among themselves, solving production problems, recording data, and many other features of the work. In a 30-month study of the factory, T. D. Wall and others (1986) found clear evidence that

> ... shopfloor employees actively enjoyed the system and its attendant responsibilities. None preferred conventional forms of work design.... All were pleased not to have supervisors breathing down their necks... (p. 297).

Ralph P. Hummel (1987, p. 72) put it this way:

> The only reason any work gets done at all is that *the worker*—even while tuning in to the manager's needs—*is still doing things his way*. What isn't obvious, until it's too late, is that when management takes away *all* opportunity for the worker to do things his way, quality drops.

I mentioned earlier that increasing the number of people using a common physical space (like traffic on a street) increases the frequency of conflict. This applies within organizations, too. Lawler (1986, p. 201) writes,

> The large multi-thousand-person plant or office structure is the exact opposite of the kind of physical structure that supports ... high-involvement work.... they should be broken down into mini-enterprises, and everything possible should be done to avoid the depersonalization, communication problems, and powerlessness that people feel when working in such large structures.

The writers I have quoted seem to understand that workers and managers see many things differently. A worker can move only in the world she sees, which cannot be exactly the world the boss sees. A student can understand the student's world better than the teacher's. A child the child's better than the parent's. Each person must find enough degrees of freedom in the environment if she is to bring about *in her own way* the consequences the other person, such as the boss, wants to perceive. By "in her own way," I mean choosing the parts and aspects of the environment, the intensities, sensations, configurations, transitions, events, relationships, categories, sequences, programs, principles, and system concepts by means of which she will reach a goal which she hopes, and the other person hopes, will satisfy the other person. If she cannot find her way in her own way, she will of necessity give up trying to find an outcome that will satisfy the other person and find one that satisfies only herself—as the crew of Apollo III had to do.

Arranging for workers to manage their own work does not mean anarchy. Carrying on a complex enterprise such as operating an airline or flying a rocket ship cannot be done by disconnected individuals. Workers who manage their own work to a significant degree do it as cooperating members of a group. Eric Trist, a luminary in the history of high-involvement work, calls these groups "autonomous" and describes them in this succinct manner (1981, p. 34):

> Autonomous groups are learning systems. As their capabilities increase, they extend their decision space. In production units they tend to absorb certain maintenance and control functions. They become able to set their own machines. The problem-solving capability increases on day-to-day issues. They negotiate for their special needs with their supply and user departments. As time goes on, more of their members acquire more of the relevant skills. Yet most such groups allow a considerable range of preferences as regards multi-skilling and job interchange.... The overall gain in flexibility can become very considerable, and this can be used to enhance performance and also to accommodate personal needs as regards time off, shifts, vacations, etc.

I think more people nowadays than in previous centuries understand the need we have for sufficient elbow room, sufficient degrees of freedom. It is a hopeful statistic that Hummel gave us—that 40 percent of American businesses with more than 500 employees in 1987 were organized so that workers could negotiate room to get the work done in their own way. It was not always so. William Ewens (1984, p. 89) says that in 1912, the president of the Remington Typewriter Company, John Cader, wrote:

> The last thing a good manager would think of doing would be to make his policies of shop management the subject of a referendum.

Even today, however, most of us (including, presumably, most of those in 60 percent of the companies with more than 500 employees) work in companies whose managers would approve of Mr. Cader's 1912 declaration. Furthermore, most people, including managers of participative companies, still believe that the environmental opportunities for carrying out a job should be tightly designed *for* the worker by experts.

I have wanted here to illustrate how we make use of the environment, both its living and nonliving features, in controlling perceptual variables. Workers join workaday organizations to control many perceptions, including the perception of monetary income. The organization becomes in that regard a useful feature of the environment. People in that organization, however, can also disturb variables the worker is controlling. I used the idea of degrees of freedom to illustrate one way in which the choice of a feature of the environment can help in controlling some variables but at the same time make difficulties for us in controlling other variables. Every act has advantages and disadvantages. And what is an advantage for you can be a disadvantage for me.

All this seems familiar; all of us can say, "One person's meat is another's poison," and nod our heads sagely. But most of us often, even usually, fail to act that way. We usually act as if we know better than the person doing the work (or play) how that person ought to be doing it. We do sometimes know some things about a task that would help the person do it, but we never know all the relevant features of the environment perceived by that person and the perceptual variables that person is controlling while doing the task. The assumption that an expert can to a sufficient extent know better than the worker what the worker should do was held not only by the president of the Remington Typewriter Company, but also by Frederick W. Taylor (1856–1915) and his followers Frank (1868–1924) and Lillian Gilbreth (1878–1972), who wanted to make work easier and at the same time more productive by finding out through time-and-motion studies exactly the movements necessary for the worker (the average worker) to make, when rest periods would be necessary, and so forth. But they almost always overlooked the individual's necessary degrees of freedom. Most people still overlook that necessity.

The mistake made by Taylor and the Gilbreths was the ancient mistake of supposing that we, or at least the smarter among us, can discern the real reality (so to speak) and thereby perceive what others perceive, whether or not they realize that is what they are perceiving. The expert can thereby arrange the perfect environment. But the smarter among us are not that smart, the real reality will remain forever unknown, and no one perceives the whole of what any other perceives. Consequently, we cannot guess very well the kinds of freedom being defended by another—we cannot guess "what they need." Even in trying to guess what some other people will prefer in activities very familiar to us, we go wrong, as is illustrated by John Cader, Neal Hutchinson, Frederick Taylor, Frank and Lillian Gilbreth, and—well, almost everybody.

Degrees of freedom are being sought wherever you hear the word *freedom,* and I suppose that word was heard when words naming principles were first heard in any language. Sufficient degrees of freedom are required by *Escherichia coli* and have been required at every level of perception added by evolution. Finding, however, the necessary degrees of freedom in a complex social environment can become a daily enigma and worry. For the very poor, it can become daily panic.

I have given a few examples in this chapter of people who have understood the need for sufficient degrees of freedom. I will give a few more examples in later chapters. But anyone who has read some history or even the newspapers knows about the ages-long struggle of humankind for freedom—for the availability of opportunities for perceptual control that will not bring inner or outer conflict.

I am easily convinced that degrees of freedom is a variable we all perceive (often unconsciously) when I consider the negation. When I was in that ventilation shaft, I knew that I was in no physical danger; nevertheless, I was becoming as terrified as I would have been had someone been wrapping me in duct tape like a mummy. Being handcuffed, shackled, and imprisoned are disabling to the psyche as well as to the body. It is no surprise that imprisoned people sometimes show the behavior called "stir crazy." The same urge to "break out" in some way comes upon factory workers who are required to synchronize their muscles with the rhythm of a machine, children who are required sit in one place until a bell rings, and hospitalized persons who are required to lie in a bed and obey the orders given by anyone who walks into the room.

Finally, here is an admonition from Powers (1989a. pp. 35–36):

> ... the variety of human experiences, circumstances, preoccupations, and problems tells us that ... the structures of control that individuals build up as they mature are highly idiosyncratic. It is no simple matter to manage [an internal] world that [is] subject to multiple levels of interpretation that must, for the most part, be worked out in private and without the aid of an instruction manual. ...
>
> The physical world and the society into which we are born only set the stage on which our lives are played out: they do not limit our freedom, but simply constitute the means available to us for doing whatever we can make sense of doing. It is up to each of us to learn how to act on and in that world, to learn to perceive its possibilities, and to learn how to organize our intentions regarding that world. Through the miracle of communication each person can learn from the others, but if there are no others who understand human organization, the amount of help available is going to be small.

Chapter 28

The social environment

The control of perception in a living creature must operate in some electrochemical circuitry, and that must operate through neurons or simpler electrochemical flows. Similarly, interaction with the environment must operate by means of a physical substratum. Our senses respond to light, sound, pressure, and so on, in the ways that physical things respond to physical energies. Interaction among living individuals proceeds on that same physical substratum.

We cannot act socially without using the nonliving environment to do so. When I talk, my words reach you through patterned compressions and rarefactions in the air. Even if I communicate wordlessly by pressing my cheek against yours, it is not the living quality of my flesh, but the act of pressing that communicates the affection. If you were to close your eyes, you would get the same effect from a warmed-up and partially inflated football.

HOW ORAL COMMUNICATION WORKS

Look at the left half (labeled "Actor 1") of Figure 28–1. That is a somewhat abbreviated copy of Figure 4–1 in Chapter 4. It represents a person and reminds us of the key features of the feedback loop—itself the key feature of living things. The new thing about Figure 28–1 is the inclusion of the social environment, symbolized by the right half of the figure (labeled "Actor 2"). If some energy in the environment, occurring independently of the Actor, is to call up an action from the Actor, it must disturb some variable the Actor is controlling. That is, the environmental energy must impinge on the Actor's loop at the point labeled "Results for input." The resultant of the opposition between the disturbance and the action output of the Actor is the input that the Actor is controlling.

Now suppose that Actor 1 wants to control some perception by making use of Actor 2. Depending on the perceptual variable Actor 1 wants to control, she might take various actions upon Actor 2—patting him on the head, shooting him dead, or feeding him a spoonful of ice cream. But let us use oral conversation for our example. Suppose Actor 1 says, "Please bring me a drink of water." Actor 1 is hoping that Actor 2 has a desire to perceive her in a state of satisfied thirst, and she hopes he will perceive her utterance as a disturbance to that desire. She hopes that he will choose to control his perception of her thirst by bringing her a glass of water, seeing her drink some, and hearing her say, "Thanks."

I have symbolized the oral request of Actor 1 by the dashed line in Figure 28–2. The line is drawn from the action output (the oral request) of Actor 1 to the symbol for disturbances of Actor 2. The control loop for Actor 1 is going to go through Actor 2. I have erased the arrow, which, in Figure 28-1, went from Actor 1's action output directly to her "Results for input," because that arrow is now going to be replaced by the series of events going over to Actor 2 and back again. You can see the complete feedback loop in Figure 28–3. That loop is completed when Actor 2 hands Actor 1 a glass of water, Actor 1 grasps the glass, gulps the water, says "Thanks," and goes on reading her mystery story. Actor 2's loop is complete when he hears Actor 1 saying "Thanks." The diagram is asymmetrical, and remains asymmetrical in the next two diagrams, because the interaction is asymmetrical: Actor 1 requests and Actor 2 responds.

I am omitting from my description here all the levels of control that would ordinarily be outside the awareness of the two actors during her request and his delivery of the glass of water. I have omitted, for example, the low-level controls in the muscles of the two. I have also omitted the controls for high-level

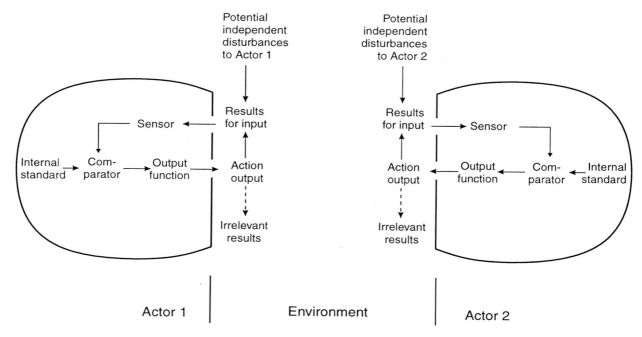

*Figure 28–1. Two persons as simplified control loops.
Compare Figure 4–1.*

perceptions such as courtesy, love, and the structure of society. I am trying to write as simply as I can to illustrate how control can use the social environment.

We are not yet at the end of the necessary diagramming. The disturbances intended by these two persons are not all that are going to affect the interchange. There are also the "irrelevant results"—those, that is, that are unintended by the actor and irrelevant (for that actor) to her or his control of a perception. For example, Actor 2, as he carries the glass of water, might inadvertently dangle his necktie in it, and Actor 1 might delay her drinking while the two of them dry up his necktie so that it won't drip on her mystery book. I have diagrammed those unintended sources of disturbance in Figure 28–4.

I should mention, too, that there are many kinds of events that could occur in an interaction between two people that I am omitting from my little scenario. Here is an example. As Actor 2 hands the glass to Actor 1, he might say, "Here is your water, you lazy good-for-nothing!" That would probably disturb some of Actor 1's controlled variables, and I would have to add an arrow going from Actor 2's action output to Actor 1's disturbances. If I were to draw in all the possibilities in the figures, I would end with an incomprehensible snarl. Nevertheless, within the snarl, there are control loops acting without interfering with the rest of the control loops. To draw my diagrams for Figure 28–1, I have imagined myself extracting a small independent set of loops from the larger snarl.

For completeness, I must add one more feature to the figure. So far, I have diagrammed disturbances that are independent only of *one* actor. For example, Actor 2 could disturb a perception of Actor 1 independently of what Actor 1 has been doing, but that disturbance is obviously not independent of Actor 2, the person producing it. In Figure 28–5, therefore, I have added a note at each side of the figure to remind us that disturbances independent of both actors are also at work.

Throughout Figures 28–1 through 28–5, I have drawn the interpersonal transmissions as dashed (broken) lines to indicate that what is transmitted is not nearly as reliable as the transmissions inside a person. Internally, transmissions are neural currents. Except under a seriously abnormal condition, a string of pulses sent out from one function arrives substantially undistorted at the next. Transmissions of energies through the external environment, however, are normally subject to the tumultuous uncertainties always there. The rest of this book, indeed, will be devoted to the ways in which, when we seek help from one another in controlling our perceptions, our intended controls can be brought to naught—or worse.

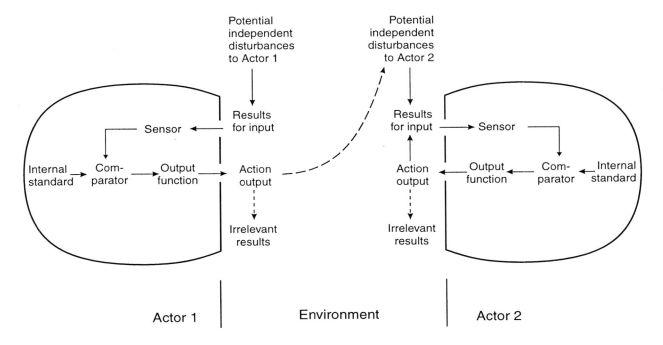

Figure 28–2. Actor 1 providing an intended disturbance to Actor 2.

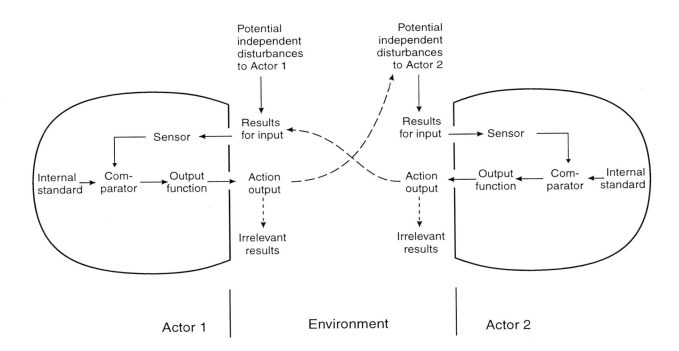

Figure 28–3. Actor 2 responding to Actor 1.

316 People as Living Things; The Psychology of Perceptual Control

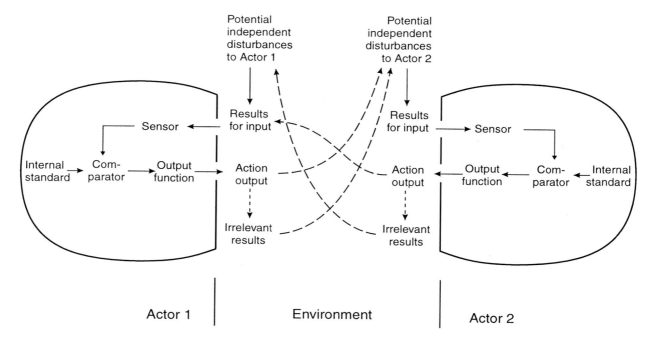

Figure 28–4. Two persons providing both intended and unintended disturbances to each other.

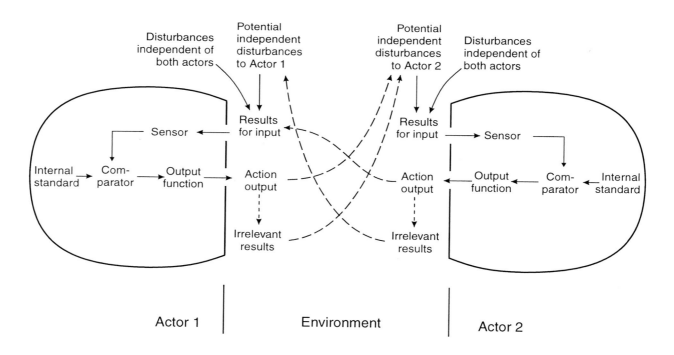

Figure 28–5. Two persons receiving disturbances both from each other and from other independent sources.

In my example of the drink of water, the only thing that went wrong was the necktie. But life is chancier than that. Instead of Actor 2 going after the water, a conversation such as the following might have taken place:

"Please bring me a drink of water."

"Who?"

"Well, what's the matter with you?"

"What do you mean, what's the matter with me? I just don't know whom you're talking about, that's all."

"Who? I'm not talking about a who; I'd just like some water."

"OK, all you have to do is ask."

"I did ask!"

"All I heard you say was, 'He's being a bit of a martyr'"

"Oh, heavens! I said, 'Please bring me a drink of water!'"

Actor 1 could have been speaking past a mouthful of nougat. Or a gravel truck could have just then roared past on the street. Or Actor 2 could just then have been pulling a sweater off over his head. And so on. I once said to a woman, "You're lovely," and it turned out a few moments later that she thought I had said, "You love me." That shook up a few controlled variables in both of us.

Let's go through this systematically. Actor 1 wants to perceive herself drinking some water, and hopes to achieve that goal by enlisting the aid of Actor 2. First, she must convey that wish to Actor 2. But doing that is far from simple. She must compose a sentence—a request. We compose requests of some sort every day, and some are more quickly understood than others. In this case, the request by Actor 1 must match the language capabilities of Actor 2. Then Actor 1 must utter the request; she can utter it with skill, or she can mumble, slur the syllables, and drop half the consonants. She can speak too softly for the other person to hear, or can suddenly shout so loudly that the person gives attention to the shock of the sound instead of the meaning of the sentence.

Then the sound waves must get to the ear of Actor 2. That could fail if Actor 1 speaks too softly. Or Actor 2 might be wearing headphones with the sound turned up loud. Or he might be in the bathroom, washing his hair, with his head under the faucet. Then Actor 2 must hear the sound as the string of words that Actor 1 intended. Actor 2 could misinterpret the sounds because of a gravel truck going by. Or the two could have been discussing the behavior of a friend who was being extraordinarily helpful in some way, and Actor 2 might just then have been wondering whether the friend's help was becoming excessive. Actor 2 might then have been expecting a further comment about that matter from Actor 1, and interpreted the sounds to form the sentence about being a bit of a martyr. Or Actor 2 might have his attention fully engrossed by something he is reading and not even hear the sounds as words, but only as a passing background sound giving no disturbance to any variable Actor 2 was controlling at that moment.

So far, so good. Let us suppose that Actor 2 has had at least a part of an ear alert for the possibility that Actor 1 might say something to him, has heard the string of words, "Please bring me a drink of water," and has interpreted them as a request. Now the loop we started tracing at Actor 1's action output has traversed the external environment and is about to enter the internal circuitry of Actor 2. Actor 2's hearing of the request is a potential disturbance to some perception Actor 2 might be controlling. If it does not actually threaten to disturb any of Actor 2's controlled perceptions (if perhaps Actor 2 is engrossed in some demanding task), then Actor 2 takes no action, and Actor 1's scheme for getting some water will fail. Actor 2 will pay no attention, as we say, and Actor 1 will have to find some other way to get some water. But if Actor 2 hears the request as a request and cares about Actor 1's thirst—wants to perceive that she is not suffering thirst—then Actor 2 will act to bring about that perception of her lack of suffering, and he will probably find taking a glass of water to her to be the easiest way to do that.

The scheme can fail in other ways. Actor 2 may not have an internal standard for perceiving the thirst of Actor 1. He may have no internal standard for perceiving any kind of comfort on the part of Actor 1. Or he may want to perceive *discomfort* on her part. Hearing the request, he may say something like: "Get it yourself." Or he may take the action of inaction. That is, he may say nothing and do nothing, hoping his lack of response will cause Actor 1 still further distress. You can think of other reasons (internal standards) that Actor 2 might fail to answer Actor 1's request as she had hoped.

Requisites for a Helpful Act

You can see that the requisites for Actor 2 to act in a way that will reduce Actor 1's thirst are parallel to the "Requisites for a Particular Act" that I repeated under that heading in the introduction to this part of the book. I have numbered the requisites below to correspond to that list in the introduction. If Actor 2 is to act in a way that will help Actor 1 control an environmental variable, it is necessary:

1a That Actor 2 be controlling a variable that can oppose a disturbance to a variable that Actor 1 is perceiving and for which she (Actor 1) is seeking help to control. (Actor 2 wants to perceive Actor 1 feeling no discomfort from thirst.)

1b That Actor 2 perceive a threat to his perception of Actor 1 being in a state she desires. (He interprets her request for water to indicate that she is thirsty.)

2a That an environmental feature suitable for affecting the variable Actor 1 is controlling be available to Actor 2. (There are a water faucet and a glass nearby, a smooth surface to walk on, and no obstacles in the way.)

2b That Actor 2 believe it possible to find or use a suitable environmental feature. (He believes he can find the faucet, draw the water, and carry it to Actor 1.)

2c That Actor 2 actually be in good enough physical condition and has enough skill to use the object to control his perception of the state of Actor 1. (He is physically capable of finding the faucet and so on.)

2d That Actor 2 estimate a sufficiently high likelihood that using the object will improve his control of his perception of Actor 1's state. (He believes that when he watches Actor 1 drink the water, he will see signs that her thirst is alleviated. Perhaps she will sigh contentedly and say, "Ah, that's better.")

3 That Actor 2's chosen act not disturb some other variable he is controlling. (Actor 2 might not go for the water if doing so would cause him to miss a crucial scene in a TV show he is watching.)

You can see that The Test for the Controlled Quantity comes into this, too. When Actor 1 utters the request, Actor 2 must make some guess at the variable Actor 1 is acting to control. In the case of the drink of water, guessing close enough is pretty easy. Actor 2 would be unlikely to say, "Sure," and then wait until tomorrow to do it. Or take her a glass of boiling water. Still, even with this simple example, it is not difficult to think of a way Actor 2 might guess wrong. When he hands her the water, she might pour it on the aspidistra.

"Oh, I thought you were thirsty."

"No, it was the aspidistra that wanted the drink of water."

I will finish my description of this two-person loop by pointing out that when Actor 2 speaks or acts, the energies set in motion must traverse the same precarious environment before they can have an effect on the variable that Actor 1 is wanting to control.

Look again at Figure 28–5, at all the links through which Actor 1's intended control must pass when she perceives another person as a handy feature of the environment to use in reaching even such a simple goal as feeling some water in her throat and thereby controlling (alleviating) her thirst. Imagine all the ways things can go wrong at any one of those many links. We take for granted such a simple interaction as this, and successfully carry off similar uses of our social environments, dozens of them, even hundreds, every day—think of the checker at a supermarket or a physician at his or her office. Yet some of the interactions go somewhat wrong, every day, and a few go very wrong. It is not surprising to me that some go wrong; it surprises me that so many are carried off as well as they are.

Non-oral Coordination

Much coordination occurs without speech. Some of it is done through writing. A great deal, however, occurs through actions on the environment that are not directed toward particular persons. A highway, for example, exerts a strong effect on the travels of persons even though the act of building the highway was directed by no single person to any other single person. Highways as a whole, along with the automobiles traveling along them, have been used by millions of people in ways that have changed radically the nature of neighborhoods and neighborliness, the procedures of urban policing, the organization of school districts, and many other features of civilization. Humans have produced many other environmental features that affect the ways we coordinate with one another: indoor plumbing, skyscrapers, electric light, airplanes, and weapons of mass destruction, for examples.

I should mention the media of mass communication, too, because I could not think how to add them in any clear graphic manner to Figure 28–5. It is true that a writer of a book, for example, makes use of his readers for one purpose or another. But the feedback loop is even more complicated than Figure 28–5. Suppose the writer produces a manuscript in the hope of making some money. What are all the links from one person's action to produce a disturbance to another's perception, to person after person, that bring some people to control their perceptions by paying some money to somebody in exchange for a copy of the book, and then what are the links that eventually give the author a perception that his bank account has increased? Or suppose the author wants to persuade some people to change their views about human nature—about how humans function as living creatures. What are the links from action to disturbance from person to person, and how does the author find out whether anyone has changed her understanding of human nature? I am not going to try to trace some hypothetical loops of that sort. Just one loop would require far more space than I gave to the example of the drink of water. I leave it to the reader as an exercise to imagine some sorts of internal standards that could be relevant and to trace out a few links in such a loop. When you have no concern for any particular individual, but only the mass, certain features of mass communication can be reasonably simple. But when you look at the situation of an individual trying to reap a benefit from her social environment, the complexities and uncertainties seem to me boggling.

W. Thomas Bourbon (1989) has built a model of interference during social tracking. It corresponds to the situation represented by the dashed line in Figure 28–5 drawn from the "Irrelevant results" of Actor 2 to the disturbances of Actor 1. That is, Actor 2 is unaware that his actions are disturbing the variable being controlled by Actor 1, and Actor 1 is unaware that some portion of the disturbances to his controlled variable are being supplied by another person. The demonstration shows nicely how people can actually be interdependent while feeling as if they are experiencing full freedom of action. Bourbon's study appears in the same volume (edited by Hershberger) that furnished Powers's study for my Chapter 13 here.

MULTIPLE MOTIVES

The diagrams I have borrowed or drawn for this book are all too simple, as are the illustrations I have offered. Only a couple of them may be adequate to portray even the *Escherichia coli*. Behavior among crocodiles, giraffes, and humans, however, is far too multileveled and multimotivated for these diagrams to represent more than paltry fragments of the multifarious equilibrations that go on continuously. The underlying principle of life, the negative feedback loop, is blindingly simple; but the hierarchical interlacing of millions of loops brings power and subtlety beyond fathoming.

To Figure 28–1 we should somehow add the multiplicity of internal standards (as in Figure 18–3 in Chapter 18) governing the control of perceptions in the two actors. When one Actor solicits help, there are always more than one internal standard shaping the request. Sometimes the ambivalence shows up in the words: "If it's not too much trouble for you, . . ." "There aren't many people I can count on for this. . . ." "I hate to have to ask you to do this, . . ." "This may seem strange to you, but . . ." Sometimes, after the other Actor begins to help, the first actor discovers that the helper's actions are putting a strong disturbance on a variable she had thought would not be affected. That conflict must then be resolved.

When some people say, "I want to help you," we reply, "Oh, good!" At other times we are dubious about the motivations of the helper. Sometimes we wonder what the person might want in return. We might ask, "What will you get out of helping me?" Or we might reply merely, "No, thanks." When still others offer help, we might trust the intention, but not the capability. We might say, "Thanks, but I'd like to do it myself."

MULTIPLE REASONS

There are always many reasons we do things. I am writing this book because I want to understand better what I think about PCT, because I want to make some sort of useful contribution to the science of PCT and to my colleagues in the Control Systems Group, because I want to feel in my retirement that I can still say something about psychological matters that other people might find valuable, because I care about the fact that people often get themselves into trouble from misunderstandings about human nature (and maybe a few of them will read my book), and because I have a vague but pressing feeling that this is what I must do. If you ask my wife and a few of my friends, I am sure you will be given still more reasons.

Almost always, if I ask someone, "Why do you want to do that?" the person gives me a single answer. In my experience, indeed, most people seem to think there is only a single reason for an act, or at least only one worth any attention. If, after they answer, I ask, "Any other reason?" they then often ask, "Isn't that enough?" or "Don't you believe me?"

In talking about an act at one level, the answer to "why" lies at the higher levels. If I put a bowl of potatoes in the middle of the table, and someone asks me my reason, I say that I want every diner to be able to reach the bowl. I don't say I did it because my muscles operated in such a way that the bowl arrived in the middle of the table. If I were to answer that way, the other person would say something like, "All right, smarty-pants, now why did you have your muscles do it that way?" And at every level, you can continue to ask why. I want the diners to find the bowl in a convenient place because I want to help them satisfy their hunger (and not have to interrupt a conversation by asking for too many bowls to be passed). And I want that because I want us all to feel friendly and satisfied with dining together. And so on. And of course there are multiple reasons at each level, which I have illustrated only minimally here.

You know which level is higher and which lower by knowing which is necessary to the other. It is necessary for me to move my muscles in those amounts and sequences to get the bowl to that place, but it is not necessary for me to put the bowl in that place if I want to move my muscles like that. If the dining table were removed, I could move my muscles, carry the bowl to where the table had been, and drop the potatoes on the floor. The lower levels are necessary to carrying out actions controlled at the higher levels. As I said in an earlier chapter, I must walk if I want to get to the library, but I need not go to the library if I want to walk. Maintaining a course to the library is control at a higher level than maintaining my walking.

Correspondingly, the lower levels answer the question "how." I put the bowl in the middle of the table by moving my muscles in certain ways. I move them in those combinations and sequences by controlling various transitions and configurations in their action.

This fact of multiple reasons, by the way, is one more reason that using formal reason often fails to correspond to the patterning of human action. A syllogism can be logically valid and tight, but the conclusion in the third line of the syllogism can fail to correspond to reality simply because there are more reasons in the reality than were contained in the premises. Without being formal about it, here is a simple example:

"Why did you throw that apple out the door? I thought you liked apples!"
"I do like apples. But I wanted to throw something at that guy, and the apple was handy."

Humor can be constructed by portraying unexpected motivations:

"Do you like pancakes?"
"Oh, yes! I love pancakes! I have a whole trunkful in the attic!"

To sum up, you can never be sure that someone else will help you achieve a goal, even when the task seems simple to you and the other person does truly want to help you. The variable you want to control is sometimes not clear even to you, not to speak of the person who wants to help you. While the process of helping is going on, the variable you want to control may change to one at another level, without your helper knowing it. The complex feedback loop may break at any of the links diagrammed in Figure 28–5. One of the Requisites for a Helpful Act may fail.

If the other person fails to help you, do not take it for granted that he or she did not want to help, or that he was physically or mentally deficient, or that she finds you detestable. Don't take it for granted, either, that you are yourself deficient or selfish or otherwise reprehensible. Do not immediately take for granted *any* cause of failure of the coordination. Look at Figure 28–5; look at all the places the helping could have gone wrong other than the place you thought of first. To put it another way, whenever you are acting in concert with one or more persons, be slow and tentative to form an opinion about what is going on between you.

The paragraph above is not a scientific statement. It is not a description of how you must inevitably function. It is a recommendation for staying out of trouble. But if you want to jump to conclusions, to leap before you look, if you want to believe that all those links in Figure 28–5 will function just as you wish them to function, and if you want to believe that all those internal standards in yourself and the other person are transparently visible to you, you are free to do so. But I hope you will not try to exercise that freedom.

INTERPERSONAL CONFLICT

Not all relationships among people are helping relationships. Sometimes we use a feature of the environment that other people also use, but without intending to form a relationship with those people. We drive along a highway having to avoid the other motorists who are using those same lanes, but the relationships we form are much more spatial than social. Sometimes we do not know whether other people are using some of the environmental opportunities we ourselves use. When I go to the university library, I wonder whether anyone ever borrows some of the books I see on the shelves. I am sure that years go by, maybe decades, during which some of those books are touched by no human hand. But I would be reluctant to guess which books are the neglected ones. I am tempted to think that all those books are there awaiting my pleasure. But one day I discover that a book I want has been borrowed by someone else, and then I know that my path for controlling a variable sometimes crosses someone else's deep in those quiet aisles.

When, in pursuing our purposes, we make use of the environment (including other people) in such a way that we do not prevent others from using the environment in pursuing *their* purposes, harmony results. You and I can ride in the same bus even though I am going to the grocery store and you are going to church. You and I can work in the same surveying crew even though the chief thing I want is to make some money and the chief thing you want is to find out whether the Egyptian pyramids really do match well the proportions some archaeologist says they do.

None of us has wants, purposes, or internal standards about how every event should proceed. I may not care whether the bus goes down 12th or 13th Street to get to the corner of 14th and Lincoln. You may not care whether I take my day off from the surveying crew on Thursday or Friday. Coordinating and organizing is possible in social life, indeed, because so much of what goes on in the world about us interferes very little or not at all with our own purposes.

Often, however, we do interfere seriously with one another's purposes. I use a piece of the environment in such a way that it closes off the use of that piece to you, and you cannot easily find some other piece to use in pursuing your own purpose. Then we have conflict. We have conflict when we both want to go through a swinging door at the same time.

When two or more persons are controlling many perceived variables in their common environment, it is not surprising that they sometimes try to move the same variable in opposite directions. In Chapter 9, under the heading "Collective Control of Perceptions," you saw some diagrams of this situation in Figures 9–7 through 9–9. You saw there how the effort of a second person was added to the disturbance already acting on the effort of a first person, so that the first person had to exert that same added amount of effort to maintain the position of the cursor. In other words, the conflict results in both persons having to exert greater effort to achieve no more than they would have achieved had the conflict not arisen. Conflict is exemplified by the tug-of-war game, in which each team tugs mightily against the other to move the marker on the rope a distance that one person could have moved it with two fingers had the other team not been there.

Conflict is debilitating. The participant exhausts himself achieving a result that could otherwise be achieved with two fingers.

Inside and Outside

A conflict has a continuing effect between us only because of the effect it has within us individually. When we both try to get through the swinging door at the same time, we are both frustrating our *individual* purposes. Though the frustration arises from the interaction, the accompanying emotion and the urge to overcome or circumvent the frustration lies separately within each of us. Maybe you want to perceive yourself as a person for whom others ought to make way. Then, if you stand back from the swinging door to let me through first, you weaken your preferred picture of yourself. But if you get into a pushing contest with me, you delay getting to where you want to go. Doing one thing satisfies one internal standard or goal but violates the other. Indeed, conflict exists between persons only as long as conflict exists within the individual persons. If, as I am pushing on the swinging door, I suddenly remember that the appointment to which I am hurrying is not, come to think of it, for today, but for tomorrow, I no longer have a conflict over whether to push or not to push. I stand back, and you barrel through. The conflict between us vanishes, too.

Another example. If you have an internal standard about making some money to feed yourself and family, and another about being nice to children, and the boss orders you to dump some toxic chemicals where children are likely to play, you have an inner conflict that can cause a conflict with the boss. If other jobs are readily available, the conflict, after some emotional upheaval, can be readily resolved. But if jobs are hard to get, the conflict hollows out a place for itself in your innards and sits there and gnaws.

At work, the action that follows upon a discovered conflict often reduces the conflict for some people (usually the bosses) but leaves it in place, or worse, for others. The boss says something like, "Yeah, yeah, I know you don't like it, but we've made the decision, and you're just going to have to live with it." The conflict with the boss has not been removed; it has merely been pushed into the subordinate's belly, so to speak, where the boss hopes it will be invisible.

Inner conflicts do not go away just because the boss thinks it would be nice if they did. They do not go away until the person finds some way to restore the disturbed perceptual inputs. Sometimes employees do succeed in doing that without resorting to actions that harm the organization. At other times, however, distraught employees turn to irascibility, careless work, slow work, by-the-letter work, frequent absences, pilfering, ulcers, alcoholism, and sabotage. The conflicts have become visible again, but wearing different garments, and many people fail to recognize them, even people wearing the new garments.

The fact that interpersonal conflict leads so quickly to *intra*personal conflict (and vice versa) makes it difficult to deal with. People often hide their purposes from others. Even when they don't try to hide them, they are themselves often unaware of them.

When another person interferes with your use of the environment, when the person chooses a path through the environment to use as a feedback function in such a way that you are prevented from using that path to build *your* feedback function, then the other person becomes a disturbance to your input. You must then act to oppose that disturbance. When people can use the same environment, physical and social, in such a way that they help one another find feedback loops through the common environment, when they do not interfere with one another's loops, then they find harmony, coordination, and productivity. When they use the environment in such a way that they hinder the control of one another's perceptions, then they find conflict, disarray, and ruin.

Countercontrol

You saw in Figure 28–5 how chancy it is to try to control the behavior of other persons. Each of us is capable of acting to control only his or her own perceptions. We cannot control the perceptions of others. We can influence the perceptions of others; for example, we can turn out the light. We can add to their perceptions; we can open the curtains. In the case of perceptions the other person is controlling, we can disturb them; that is, we can make it more difficult for the other person to control his perceptions. We can do those things, but we cannot hold steady the perceptions of others as we do our own. We can have temporary effects on others' capabilities for action, but we cannot be confident, in general, in predicting the particular act a person may take, neither immediately nor later.

Sometimes we are aboveboard in our methods of controlling others, and sometimes we are underhanded about it. But subterfuge is of small advantage; sooner or later, the other person becomes aware that you (or somebody) is making it difficult for him to control one or more perceptions. The person must then find some way to counter that disturbance—to evade or oppose your intention. We can call that resistance *countercontrol*.

Burrhus F. Skinner (1904–1990) was one of the most influential figures in "behavioral" psychology—also called the "experimental analysis of behavior" or EAB. He believed that behavior was caused by stimuli, and that humans could control the behavior of pigeons, rats, and other animals. He did not, however, believe that humans could succeed for long in controlling humans. Nevertheless, his followers have given lots of advice to teachers about controlling students, to parents about controlling children, to managers about controlling workers, and so on. In 1999, I had a conversation about this with W. Thomas Bourbon on the CSGnet. Bourbon wrote on 25 August, replying to a missive of mine. Below are some of the paragraphs; I have labeled them with the name of the writer.

Runkel: I never did understand how that conditioning stuff could work, and I remembered very little of it after the final test in the course in graduate school. No doubt I have some distorted notions of what the EAB people actually say.

Bourbon: In 1953, in his book *Science and Human Behavior*, Skinner declared boldly that, by using the principles of operant conditioning, any person could control any other person's behavior. . . . In the same book, Skinner warned that any person who controls another will be countercontrolled by the other. He said that the countercontroller would try to upset the controller, or to harm the controller. He also said that a person determined to control another could not prevent countercontrol.

Runkel: Skinner also said that the goal of EAB was to predict and control behavior (actions). But if you are getting countercontrolled, you are not going to be able to control.

Bourbon: That does seem to be a reasonable conclusion!

Runkel: I always supposed that one of the attractions of EAB to many psychologists was that they could dream of sitting at the center of a worldwide web and CON-TR-ROLLLING EVVERYYBBODDY. And that they could do it without the controlled people becoming aware that it was happening to them.

Bourbon: In *Science and Human Behavior*, Skinner said exactly that: If you are clever enough, or lucky enough, the people you control won't realize you are doing it. They all want to get away, but they won't know to get away from *you*. They all will want to do something to someone, but they won't know to do it to *you*. That is the best outcome a controller could hope for. . . .

Runkel: But there was Skinner, you say, saying in effect that EAB WOULD NOT WORK with humans!

Bourbon: I enjoy telling educators that obscure fact.

You can find a lively and richly informative exposition of countercontrol, especially in schools, on pages 42–48 of the book by Carey and Carey (2001).

COOPERATION AND COMPETITION

We can treat others as helpers, we can treat them as obstacles, or we can simply ignore them. I ignore a lot of people when I walk across campus. Few either help me walk or obstruct me. Maybe someone holds a door open for me; that's nice. Maybe a couple of people stand in a doorway, talking, oblivious to their obstruction of the passage, and I ask them to move. I go to the business office and find three other people wanting to do business at the same counter. We resolve the conflict by standing in line, last arrival last.

All three of those principles for coordination (helping, hindering, and ignoring) occur repeatedly in groups and organizations. Most of the scholarly literature concentrates on cooperation and competition, ignoring ignoring. Many architects, however, understand that they must provide spaces, especially in public buildings, where people can have a somewhat protected passage to where they are going, but in which they can safely ignore most other people. Here I will say a little about how cooperation and competition can be connected to the perceptions of individuals.

Economists often talk about competition as a form of social interaction that occurs when two or more people want to make use of the same resources in the environment, the resources cannot be used jointly, and the resources are too scarce to satisfy the demands of all the people. They often speak of the resources as "goods"; goods may be consumables such as food or gasoline, but economists are willing to talk about services, too, and even psychological goods such as peace. Still another kind of good is a trophy given for winning a game. Only one person or team can have it, but it is "scarce" only because all the players and judges have agreed to make it so; in the factory that made the trophy, there are hundreds more lined up, waiting for the names of one-only winners to be inscribed on them.

Instead of using the word "good," I have spoken here about environmental features, opportunities, resources, and the like. I intend those words to cover getting any kind of "good" from the environment whatever. But I will borrow the economist's word "scarcity" to mean that two or more people have fastened upon the same feature of the environment, the same resource, with which to control their perceptions.

I think most people, most of the time, prefer to cooperate or act alone rather than compete. When you act alone, you can do things your own way without interference from others. When you cooperate, you get help from others. When you compete, however, the others make trouble for you. Generally, in other words, cooperation and independent action are easier than competitive action. It is true, of course, that many people enjoy some competition. That enjoyment occurs mostly, I think, when vital resources are not at stake. Competition is fun when all you can lose is a trophy. Similarly, competition in business is fun when you remain assured of an annual income of, say, a million or two. It is also true that even when your food and shelter are not at risk, competition can bring pain—as, for example, when you lose a reputation as champion, or the approval of your parents, or your position as an admired leader of industry.

We do often find ourselves in a condition of scarcity. If two people are acting cooperatively in what is actually a condition of scarcity, they will discover the scarcity before long. "Hey! The peanut brittle is just about gone!" Then, if each acts to get as much of the environmental resource as he or she had been wanting, they will be acting competitively. People do sometimes act altruistically amid scarcity, but here I am talking about what is likely—what I think most people will do. I expect that under scarcity, most people will discover the scarcity before the resource is completely used up, and most people in most such situations will begin acting competitively, even if they began in cooperation. What brings about cooperative or competitive behavior is the perception the participants have of the supply of a desired resource, not the actual amount of the resource available. Following that reasoning, we can make up Table 28–1.

Note that the table is stated in likelihoods. It does not say that your Uncle Umbert will act cooperatively at the model railroad club tomorrow morning. The table shows only what is *possible* and what the majority of instances might be like if you observed a hundred instances at random. It does not show how humans function. It shows some actions that human functioning makes possible. Although there is a fair amount of research behind the probabilities I have implied in the table, there is a lot of speculation there, too.

My main point in displaying Table 28–1 is to point out how easy it is for humans to shift into the competitive mode. In only one cell of the table do the circumstances augur the adoption of cooperation. The table, by the way, is neither clever nor arcane; all the upper right cell says is that if two people want to cooperate, given the necessary resources, they will do so. But it does take two to tango.

Note how fruitless it can be to hunt for "laws" of behavior in particular kinds of acts—such as cooperative, competitive, or individually independent. The intentions of the individuals cannot alone guarantee how the interaction will turn out, nor can the availability of resources. The outcome is still subject to the Requisites for a Particular Act.

Despite their intentions at the outset, two or more people who set out to compete can, as circumstances and perceptions change, turn to cooperation. John Keegan (1994), for example, tells how in World War I, when the first Christmas approached after both sides had dug deep trenches and the war in northern Europe had come to a trembling stand-off, war-weary soldiers from both sides crawled into no-man's land, joined their enemies around campfires, and sang Christmas carols. Cooperation can turn into competition, too. Two children can be having a happy time together, but one parent or both can say, "Rupert, you mustn't play with one of *those* boys!" Conversely, it is common during famines to find some parents giving their children the larger portions of the little available food. And humans can alter the supply of

If A's intention is to:	and if B's intention is to:	and if the resource is actually:	then their joint behavior is likely to become:
cooperate	cooperate	ample	cooperative
cooperate	cooperate	scarce	competitive
cooperate	compete	ample	either
compete	cooperate	ample	either
cooperate	compete	scarce	competitive
compete	cooperate	scarce	competitive
compete	compete	ample	competitive
compete	compete	scarce	competitive

Table 28–1. Likelihood of cooperation or competition

resources. Instead of competing for a scarce good, they can cooperate to produce enough of that good for everyone.

For all those reasons and more, my table should be taken as a table of tendencies and temptations, not as a table of predictions. Humans need not and do not compete or cooperate according to circumstance, but can, given a little time, act according to their wishes.

Among social psychologists, Robert Axelrod is famous for his 1984 book on the Prisoner's Dilemma game. I won't bother you with the story of how the game got that name. The game is usually played by two players and a scorer (the scorer can be a machine if you wish). The scorer awards money or points to each player according to whether the players declare they wish to "cooperate" or to "defect" (defecting is the competitive move). Each player can choose to do either. At each round, both declare their choices, but neither knows what the other will declare. After the scorer has awarded points, each player knows what the other declared by knowing the following scoring table.

If both cooperate:	If you cooperate and the other defects:
you get 3	you get 0
other gets 3	other gets 5

If you defect and the other cooperates:	If both defect:
you get 5	you get 1
other gets 0	other gets 1

You can see that if both cooperate, they can both win 3 points at every round. If one defects while the other cooperates, the defector wins 5, and the cooperator gets zero. And if both defect at the same time, both get 1 point. If you want to experiment with this game (as many psychologists have done), you can alter those points as you see fit. Axelrod asked several experts each to contribute a computer program that would play the strategy the expert thought would bring a player the most points. Axelrod then played each strategy (via computer) against every other. The most successful entry began with a cooperative move and then simply made the same move the opponent made just previously. In 1997, Messick and Liebrand reported some studies of the Prisoner's Dilemma that extended Axelrod's simulation research in several directions. For example, they tested how a player's choices affected the later choices of onlookers. They also tested the effects through 100 "generations" of players.

Studies like those are useful in suggesting possibilities and probabilities. If real people were to adopt the sort of rules the researchers test in their simulations, the people would often follow the patterns of choices the researchers found. But of course only *some* would do so. They would be more likely to do so if they were to agree to act as experimental subjects and limit their control of perceptual variables to the point-rewards with which the experimenters want to describe their subjects' behavior. If they had other goals, such as getting home to dinner or making trouble for either the other player or the experimenter just for the fun of it, then the choices would not have the patterns produced in the simulations.

The simulations were all done on computers. There were no emotional ties among the simulated "players." No imagined "player" cared whether the other "player" lived or died. Nor did any part of any model correspond to a person who would think about what was going on or would talk with the opposing player. Furthermore, the players in Axelrod's models found out the payoff immediately. In natural situations, a player (such as a businessperson or a general in a war) may not discover the payoff until next day or next year and in the meantime must take action quite without regard to that payoff. Studies such as Axelrod's do urge us to ask important questions about what is possible. They do not, however, tell us about human nature—about how humans functions. They tell us some patterning of choice that we might find on the part of some people in some situations in some cultures.

The virtues of competition were no doubt debated hotly in the ancient agoras. In our own time, however, Arthur W. Combs (1979, pp. 165–166) put the matter concisely:

As usual, the myth that competition is a great motivator contains an element of truth. Competition does, indeed, motivate *some* people *sometimes*. It also discourages and disillusions innumerable others.... (1) Competition is valuable as a motivator only for those people who believe that they can win; (2) People who do not feel that they have a chance of winning are not motivated by competition; they are discouraged and disillusioned; and (3) When competition becomes too important, morality breaks down and any means becomes acceptable to achieve desired ends.

Controlling our perceptions is made more difficult by our wrong beliefs about competition. Alfie Kohn has written several readable books about the pernicious effects of competition; those of 1986, 1993, and 1999 are especially relevant here. In his 1986 book, he wrote (p. 162):

> Pious admonitions about not getting carried away in competition, however well-meaning, are just exercises in self-deception. If we are serious about eliminating ugliness, we will have to eliminate the competitive structure that breeds ugliness.... Most people do not even keep the facts in view, failing to see the connections between campaign illegalities, scientific fraud, corporate trickery, and the use of steroids in college sports.

By "competitive structure," Kohn means an arrangement of environmental opportunities such that one person can achieve her goal only by preventing other people from reaching theirs. This is exactly the situation that produces interpersonal and then, necessarily, internal conflict and loss of control. Much, often most, of the energy spent in striving to reach the goal is then spent not productively, but wastefully in fighting off the efforts of other people to reach their goals.

When faced with the possibility of a competitive situation, most of us usually look for a way of avoiding it. Here are some excerpts from Powers's remarks to the CSGnet of 4 September 1995:

> What the free market system offers is an opportunity for an entrepreneur to find a niche by offering a mix of products and services that satisfies the needs of some part of the population. But the entrepreneur, to be successful, has to avoid direct competition as much as possible. It isn't competition that makes the system work; it's the ingenuity of people in finding combinations of goods and services to sell which, ideally, are not to be found anywhere else—certainly not nearby.
>
> At another level, of course, the competition is always there. It can be avoided only if the number of independent businesses remains below a certain number. The staggering failure rate of new businesses, however, shows that there are always more entrepreneurs trying to enter the market than the market can sustain. There are not enough degrees of freedom that actually matter to people to let everyone who wants to start a . . . business . . . find a niche without coming into direct competition with another business that already occupies an almost identical niche. And at a still higher level, the Composite Consumer can spend only as much as it earns (in the long run), and that is the pie that has to be divided among the entrepreneurs....
>
> The main new notion here is that the health of the business community depends on the avoidance of direct competition, because direct competition is simply conflict, and conflict prevents control systems from working properly. There are many other considerations involved in economic health, but I think this is a major one.

An obvious method of avoiding competition is monopoly. It is easy to see it as it approaches, and once achieved, its pernicious effects are quickly felt. But a sort of shared monopoly, so to speak, can grow up while only a few people become aware of it. In his 1970 book, the economist Albert O. Hirschman called it "collusive behavior." He pointed out that a firm need not regret losing customers if it can acquire new customers to replace them. For example, if customers desert a firm upon discovering they have purchased a product of low quality, the firm need not suffer from the desertion if it acquires other customers who have deserted competing firms. Hirschman (1970, pp. 26–27) wrote:

> A competitively produced new product might reveal only through use some of its faults and noxious side-effects. In this case the claims of the various competing producers are likely to make for prolonged experimenting [by] consumers with alternate brands, all equally faulty, and hence for delay in bringing pressure on manufacturers for effective improvements in the product. Competition in this situation is a considerable convenience to the manufacturers because it keeps consumers from complaining; it diverts their energy to the hunting for the inexistent improved products that might possibly have been turned out by the competition. Under these circumstances, the manufacturers have a common interest in the maintenance rather than in the abridgment of competition—and may conceivably resort to collusive behavior to that end.

In that collusive behavior, most customers remain quite unaware that their continued efforts to find a good quality product are doomed to frustration by the very competition most of them trust to bring a product of superior quality from at least one of the competitors.

Competition among firms can produce lowered quality of goods or services in many ways. When airlines and bus lines were deregulated, they immediately closed down their least profitable routes—which were, unsurprisingly, those serving the poorest people and the people most isolated geographically.

Just as competition leads firms to reduce the wages of their employees, it often brings hardship to others beyond the immediate competitors. For example, cities have often offered inducements to industries that the city councilors believe would employ some of their unemployed citizens. The managers of the industry typically promise City B that they will move their factory from City A to City B if City B will exempt their firm from taxes for X years, build a street to their factory, charge them reduced rates for water and electricity, and provide various other inducements. Usually, the firm hires fewer local workers than promised and, when the years of reduced taxes and charges expire, the firm begins looking round for greener pastures. Strangely, few councilors in City B seem to think that if the managers can be induced to move their manufacturing from City A to City B, they can also be induced to move it from City B to City C and from there to Mexico or South Korea. And that is what often happens. Instead of benefitting from bringing the industry to town, the city finds itself worse off than before.

RESOLVING CONFLICT

In general, one resolves conflict between two or more people by enabling every person to reach his goals without preventing others from reaching theirs. That's it. The complications, however, often pose difficulties.

The first thing to think about is whether people who feel their actions are being hindered by others are actually correct about it. For example, in 1978 Richard Francisco and I were consulting with a high school in Oakland, California (see Francisco, 1979, for a fuller account). The principal claimed that the five counselors were disorganized, were not using their time to the best benefit of the students, and so on. Francisco and I spent three days making an actual minute-by-minute count of how the counselors used their time. We also interviewed all the counselors, the principal, and several teachers who worked frequently with the counselors.

It turned out that the counselors were very conscientious, and that the things bothering the principal were the very things bothering the counselors. Neither principal nor counselors, however, had been able to see their agreement, because both parties had taken an embattled stance. We brought principal and counselors together and displayed our findings. The feelings of conflict fell close to zero, and agreements were made to transmit information more frequently and to meet when discussion would be useful. The counselors had admired our methods of conducting our meetings with them, and asked us to help them adopt the methods. The new methods helped them to reorganize their operations so as to improve coordination and reduce stumbling over each other.

The next thing to think about is whether people are right now trying to act in pursuit of their goals and are feeling hindered by actions now being taken by others, or whether they feel they are being pressed to take an action which, they believe, once taken *will* bring hindrance to their pursuit of their goals from the acts of others. If the worry is about what *might* happen (but has not yet happened), the conflict in place at the moment is not a conflict over the larger goal, but a conflict over the promises being solicited. A later, larger conflict can usually be avoided by airing the fears and making a plan for very small steps, accompanied by monitoring to document what action is actually taken, and also accompanied by replanning after every step to make sure fears of conflict are receding and cooperation is rising.

If actions now in progress are conflicting, it does little good to soothe brows. The parties in conflict must find some way that both can act to gain their goals without getting in the way of the other party. Friendly overtures and a glass of beer may (or may not) improve the willingness to discuss possible solutions, but the conflict will not go away until the interference is removed. Here are some possible strategies:

> Increase the resources. If two secretaries fight over using the copy machine, buy a copy machine for each secretary. Borrow the use of the one down the hall that is not used much. Schools, libraries, and some agencies of government increase their resources by using unpaid volunteer aides.
>
> Reorganize duties or schedules. Let one secretary do all the copying. Or let one copy in the morning, the other in the afternoon. (You could call this specialization of labor.)

Trade kinds of resources. Barter. Give something for which your need is less than your need for what you want from the other person if the other person will do the same.

Compromise. Give up a little to get a little and hope that you will get a chance in the future to get a little more. I remember a neighbor in the Panama Canal Zone (where everyone keeps the windows open to catch the breeze) who liked to play his musical saw. We thought its sound was an annoying whine. When I visited him, I discovered that he, in turn, hated classical music. We struck a bargain that he would play his saw more softly, and we our classical music more softly.

If all else fails, get tough. Grab what you can no matter at whose expense. This, to me, is the least desirable strategy, because it leads to escalation and vendetta and all claws bared. It turns the possibility of negotiation into a competitive situation difficult to retrieve. Escalation is especially rapid if a ruling principle on both sides is that of winning.

Those methods of reducing conflict all depend, of course, on removing the necessity for the two or more people to control the same environmental variable. The last is often the most effective way if you do not care how much damage you inflict on the other person. When, however, contenders expect to go on living in at least some interaction with the others, the last method listed is obviously the least effective way. It impoverishes at least some contenders. When the struggle to win escalates to modern warfare, it impoverishes all contenders. When it escalates between spouses, so that they no longer seek to solve the problem that brought on the conflict, but instead seek only to win the battle, it destroys love. See the struggle in Figure 9–7 in Chapter 9.

I do not claim that my list of methods for resolving conflicts is the best way to categorize possibilities. If you have a list you like better, feel free to use it. If you are a writer, write about it. Your categorizing might serve readers better than mine.

By the way, my list of method may sound to you like advice. It may sound like, "This is what you should do." But I mean it, rather, to serve as a catalog of devices that might serve in some situation. Think of the list as you would think of a row of bins holding hooks. If I were a purveyor of hooks, I would not say that the next time you want to hang something up, you should use my Hook Number 6. How could I know that? But I might be helpful to you in offering you the choice of possibilities. And maybe you would need to modify what I offered. Maybe what you want to hang would hang better if you were to take your pliers and put another bend in Hook Number 6. The only idea that I am putting before you as fact is this: the only way to reduce conflict is to increase the degrees of freedom; that is, to enable the contestants to move toward their goals in ways that do not hinder one another. As to the best way to do that next time you find yourself in conflict, I cannot say. I will say more about cooperation and competition in later chapters.

TRUST AND COOPERATION

Cooperation comes in degrees. We might not want to call a bus ride an episode of cooperation. For one thing, the points of coordination between passenger and bus driver are few during one trip, occurring mostly at boarding and debarking. For another, the overlap of goals between the driver and me are at low levels in our control hierarchies. As I board, my goal is to perceive myself having mounted safely, and the driver's goal, too, is to see me safely on board. Later, my goal is to get off at the right stop, and at that point the driver's goal is to pull to a stop and see me safely off onto the curb. But my trip on the bus serves higher goals for me and for the driver that are not the same. I did not choose to ride on the bus so that I could later perceive the driver being paid a salary. She did not stop to pick me up so that she could perceive me arriving home in time for dinner.

We usually think of cooperation as characterized by fairly continuous coordination, yielding a result desired by both persons. Some fire engines have steerable wheels at both ends and two drivers, one to operate the front wheels and another, seated at the rear, to operate the rear wheels. Those two firemen coordinate continuously; they must surely breathe a sigh of relief when they reach their goal at the fire without brushing any pedestrian off the sidewalk as they maneuver around a corner.

I do not claim there is a sharp distinction between the bus ride and the fire engine's trip. I want instead to emphasize that the participants in both cases must act to some degree under a principle of the sort we call "trust" if they are to enter upon a period of coordination. Another way to say this is that both persons must want to cooperate. I do not mean merely that both persons must want to reach the same end; I mean too that both must want *to enter into joint effort* to achieve that end. Cooperation must itself become an internal standard for both, so that both persons can act to perceive that they are maintaining the cooperative relationship. When the rear driver on the fire engine says (over the intercom), "Slow down, Bert!" Bert must perceive slowing to be a cooperative act and want to maintain that cooperation. If Bert controls only his perception of getting to the fire, he might ignore the rear driver and fling the rear of his fire engine into the facade of the First National Bank.

When we join with others to reach our goals and find ourselves succeeding in moving from subgoal to subgoal and reaching eventually the goal for which we first adopted the cooperative relationship with those others, we are very likely to find pleasure in the association. If we have further experiences of that sort with those same partners, we are very likely to enjoy not only the achievement of the original goals, but we are also very likely to enjoy perceiving the pleasure our partners take in the achievement. And as the association continues, we find pleasure in perceiving the pleasure the others take in perceiving our own pleasure. And so it goes, appreciation and delight of one sort giving rise to a higher-level pleasure, with trust and affection growing deeper. Sometimes the control of the perceptions of cooperation, trust, and affection comes to far outweigh the importance of many of the tasks the participants might undertake.

I have read what a few mountain climbers have said about their attraction to the sport, and I have talked to a couple. They did not tell me much about the challenge or the thrill of dangling in space at the end of a rope. They talked mostly about the deep feelings of trust, affection, and camaraderie they developed with their teammates. That feeling of mutual commitment among people, along with the joy in it, whatever the task, is what is meant by esprit de corps. It is also very close to one of the meanings of love.

LOVE

I am not seeking to define love. Let us not reify. Everyone has his own connotations when he hears the word "love." I am saying that cherishing the joys of cooperation is to experience one of the meanings many people have in mind when they speak of love. The treasuring of mutual help—that meaning of love—has little to do with "liking the same things." It has to do with appreciating the feeling of reaching goals through mutual helpfulness, regardless of the actual goals.

My wife is a whiz at real estate dealings. Before I met her, I had no interest whatever in matters of real estate, and I still have very little. But naturally, as she took on some real estate transactions after our marriage, I wanted to be helpful, and we soon found ourselves working in a cooperative manner and admiring each other's contributions to the subgoals along the way. As we met obstacles and triumphs, we delighted at our hand-in-hand coordination in overcoming the obstacles and in securing the triumphs. That pleasure was quite aside from, in addition to, the achievements of the final transactions. Those joint experiences increased immensely our feelings of trust, security, and appreciation of each other's perspicacity and skills. All those memories and feelings, along with many others, of course, give us the opportunity a dozen times a day to be happy that we have enjoyed, and can continue to enjoy, the deep satisfaction of that kind of confident control of perceptions of joint action which, in turn, take us to our own goals—including the welfare of each other.

Correspondingly, Claire is helping me write this book. Though she had found a few courses in psychology interesting many years earlier, the critical comparison of varieties of psychological assumptions and theory were of little concern to her. Certainly she would never have undertaken to read hundreds of pages of first drafts of a radical new psychology. Nevertheless, even before we were married, she offered to do a first editing of the book (1990) that came before this one. The companionship we found in that experience was one more that helped to persuade both of us that we could trust each other to cherish our joint life above the pleasures of any one sort of activity.

Doing something together is not itself the source of bliss and contentment. The source of love is the appreciation of the interdependence, of the interlacing of goals and acts, of the orchestration, so to speak, of the distinct melodies into a succession of harmonies, all helping both persons to control their own perceptions without conflict while contributing to a joint finale desired by both. Two people figure-skating on the same rink is one thing; a couple balancing each other as they whirl and dip and soar and change figures, each enjoying patterns and accelerations neither could do alone, is another.

Conversations, too, can go one way or the other. I have been present at conversations in which each speaker recited his likes and dislikes for food, movies, neighborhoods, and so on. Or recited events such as shopping trips and vacation travels. For all the effects one person's contribution had on another's (aside from suggesting a topic), each person might almost as well have been reciting alone. I have been in other conversations where each person's utterance was treated by the others as a contribution to an idea being shaped jointly by everyone, and where the listening was so intense that persons frequently asked for clarification about the way one utterance might or might not fit tightly with another. I prefer the latter sort.

The love that comes with the treasuring of mutual help has been celebrated by writers in many fields, in fiction and nonfiction. One could fill a volume with appropriate quotations. I shall, however, give only two examples here. The first comes from the psychological literature. The two authors of the excerpts below (Suarez and Mills, 1982) are practitioners of therapy and community mental health.

> Compatibility, in producing good relationships, is irrelevant. . . . All that compatibility can tell us is whether our belief systems can tolerate the other person's belief system, given the level of insecurity [lack of trust] we both have now and assuming that neither will ever change (pp. 143–144).
>
> Imagine receiving unconditional love and regard from your partner. This feeling coming from that person would soon make all our insecure games seem unimportant, certainly irrelevant. In addition, the things we would normally fight about no longer work [no longer support a fight], particularly if only one person is fighting. The most natural, normal state of relationships is love and appreciation, which if left alone will naturally grow deeper and more loving all the time (p. 146).

Dealing with the world is often stressful. There are many occasions when we see signs that we will find great difficulty in controlling variables vital to our welfare, when we feel lost and sometimes terrified amid the turmoil of the world about us. It is a precious and incomparable comfort when we have a secure love with another person, when we know there is at least one person with whom we need no defenses, no doubts of her good intentions, and with whom we can find peace, trust, and renewal. The way Matthew Arnold (1822–1888) put that feeling in his *Dover Beach* is my second example. This is from a 1970 edition:

> Ah, love, let us be true
> To one another, for the world which seems
> To lie before us like a land of dreams
> So various, so beautiful, so new
> Hath really neither joy, nor love, nor light
> Nor certitude, nor peace, nor help for pain
> And we are here as on a darkling plain
> Swept with confused alarms of struggle and flight
> Where ignorant armies clash by night.

ADOPTING INTERNAL STANDARDS

We are born with some essential internal standards. Immediately, then, we begin forming the standards to give broader purposes to those inherited standards. We begin erecting, that is, the vast hierarchy of control by which we cope with the perceptions that come to us. A living creature must be able to reorganize its control systems repeatedly to maintain control of its "intrinsic quantities" (vital variables) despite its encounters with unpredictable disturbances in the environment.

We expand our control by reorganizing, but also by using temporary standards in the service of higher standards. For example, to serve the purpose of nourishment, we go repeatedly to the grocery. The grocery becomes a goal when we go out to gather food. If the grocery burns to the ground, however, we do not lie down and starve to death. We quickly substitute another source of food. The internal geographic standard representing the location of the grocery is a temporary one, easily replaced with another. Similarly, the goals of finding the streets leading to the grocery are easily replaced with similar goals leading to other groceries. Memory makes possible this repeated use of temporary goals.

An important function of every social institution is that of offering (and often insisting upon) goals for other people to adopt; governments, churches, schools, families, athletic clubs, and bridge clubs all do it. Even commerce: whenever you buy some useful object, you adopt subgoals along with it. Every button or knob on the TV functions as a subgoal through which you achieve the perception on the screen for which you yearn. When you buy a knife, you buy also the subgoal of grasping the handle (not the blade, please). When you become an employee, you agree to adopt (explicitly or implicitly) certain goals from your employer. As you grow toward adulthood, most people expect you to adopt certain of the goals stated by people in certain parts of the government of the country in which you were born.

When I say "adopting," I do not mean that the nervous system of the adopter can take on the same assortment of connections that exist in the nervous system of the other person. I mean, of course, that the adopter can try to be guided by the same standard that he *supposes* the other person is using. His ability to discern the other person's internal standard and his ability to control the perception can never duplicate exactly what is inside and outside the other person. When a child of seven adopts from a parent the goal of driving an automobile, he does not get into the family car and drive off down the street. He sits in a chair, holds an imaginary steering wheel between his hands, and makes the sound of an engine by growling or blowing air through his fluttering lips. Imitation is a form of adopting internal standards. I learn to play the piano by imitating (I hope) people who play better than I.

A great deal of sincere imitating and obeying goes on simply because we do not care much (we do not get into internal conflict because of) whether we do things one way or some other way. If the government wants everyone to drive on the right instead of the left, that's OK. We can get to the grocery equally well either way. If the boss wants my desk to be on the third floor, it's all right with me; I can get my paycheck regardless of the floor my desk is on. Indeed, the world's work can get done only because we so often do not care whether we control our perceptions through any particular means. If we cared to control every perceived variable it is possible to control, and other people were doing the same, we would be in such a web of conflicts that everything would grind to a halt—or end in a "war of all against all."

There are limits to adopting standards. We do not want to adopt standards that will conflict with those we already maintain. At the higher control levels, such a conflict is often called an ethical dilemma. Some standards are harder to understand than others. It seems that staying in your own lane while driving an automobile is understandable in much the same way by every driver, and almost every driver seems to be capable of doing that almost all the time. But controlling a perception of piety is not as easy; you and your priest may judge your piety very differently. Some standards easy to understand and easy to match are easy only at the outset. Employees may find it easy to understand that the boss wants a machine operated in a certain way—perhaps a punch-press in a factory or a keyboard in an office—and an employee can operate it that way after a certain amount of practice. But if the employee repeats the operations for too many minutes at a stretch too many hours per day, the employee may come down with damaged muscles or ligaments and then be unable to go on repeating that "easy" operation.

Here are comments made by W.T. Powers on the CSGnet on 25 March 2001:

> Fortunately, it is almost always true that there is more than one way to control any given perception. This means that when one person's actions bother another person, there is probably some other way to act that will achieve essentially the same ends without bothering the other person (unless, of course, the aim is specifically to bother the other person). When alternatives are available, there is no need for the people involved to give up controlling for what matters to each person. All that's necessary is to find a somewhat different set of actions that will continue to control the important perceptions, but without impacting the other person's ability to continue controlling, too.
>
> The exception to this occurs when one person wants another person not just to quit disturbing the first person's own controlled variables, but to display or refrain from specific actions. Now the variable one person is trying to control is the very means of action that the other person needs to be able to vary freely in order to keep control of what matters to that person. What the fist person wants the second to do removes, in effect, the second person's ability to continue controlling successfully—or would do so if the second person behaved as requested or ordered. The most likely result is some degree of conflict.

One person can't know the exact perception another is controlling, and even less the whole hierarchy of perceptions in which any given control action is embedded. In fact, trying to guess what the other person is controlling is usually part of trying to tell the other how to behave, as if you were better able to figure out how the other person should achieve his or her goals than he or she is. This approach is really a show of disrespect, a way of saying you are superior to the other person.

If we refrain from trying to tell other people how to control their own perceptions, we can still describe to other people our own perceptions that are disturbed by their behavior, and ask them if they can think of a way to reduce the disturbance. Assuming good will, we can then expect the other person to figure out some other overt actions that will still be able to achieve the other's goals, but now without disturbing us so much. And of course, while the conversation is on that subject, the other person might well have a few requests to make of us. . . .

Ideally, each person would not just be complaining about bad side-effects of the other's behavior, but offering to change, if possible, his or her own behavior to lessen the disturbing effects on the other person. . . . To respect another is to want the other to be in control of his or her own life; to demand respect is to claim the same right for oneself.

Where there is no good will, no mutually-agreed goal of getting along together, interactions based on mutual respect are impossible. If I'm concerned only about controlling my own perceptions and am indifferent to your ability to do the same, then conflict is almost inevitable. But conflict carries a price, especially conflict among equals. To be indifferent to the autonomy of others is to risk having to pay the price of being sidetracked from your own real goals just to counteract the resistance of other people to your effects on them. I think it's through consequences like this that children, most children, learn eventually that peace is better than conflict. To give respect is to receive respect, in the long run, and that makes life much easier.

Chapter 29

Helping

It is possible to use the resources of the social environment much as one might use a nonliving thing. One might use a crouching person as a bench to sit on or as a stool to stand on. Children and acrobats sometimes do that. You might slaughter a person and serve slices of the flesh for dinner. That has been done, usually with some ceremony. Most frequently by far, however, we use other people as agents, as in asking someone to bring a glass of water, to carry a message to Garcia, or to fire a cannon at an enemy.

We frequently ask others to act to our benefit—that is, to help us in some way. We often offer help, too, without being asked. Almost everyone, every day, requests and offers help many times, including infants too young to phrase their requests in any way other than squalling. Sometimes we tell a helper our purpose. Often, however, we simply ask the helper to perform some act we that believe will further our purpose.

Civilization, it seems to me, cannot be maintained unless almost everybody is willing, most of the time, to help and be helped. Civil order falters when efforts to help fail to help or when acts that help some people harm too many others. In a society of some complexity, the ways of giving help become myriad. Some kinds of help require no special skill: hand me that rock, please. The more technical help disengages into specialties and becomes institutionalized in formally designated occupations. The U.S. Department of Labor's *Dictionary of Occupational Titles* (1991) runs beyond 1400 pages.

Much of the help we give one another helps us to control the perceptions that are connected with our physical welfare. Farmers, truck drivers, grocers, and cooks help us to keep our perceptions of hunger at a low level. Carpenters, roofers, and real estate agents help us to keep rain and snow off our heads and feet. Chemists, pharmacists, physicians, nurses, and insurance companies help us to minimize sensations of pain and to perceive ourselves walking, swallowing, and sleeping well.

As the centuries have gone by, the provision of physical necessities has required the work of a smaller and smaller percentage of the total population. One illustration is the decline in the United States in the percentage of the population in farm occupations. These figures are from the *World Almanac* (M. S. Hoffman, 1993, p. 122):

Year	Percentage of population in farm occupation
1850	63.7
1900	37.5
1950	11.6
1990	2.4

Much of the help we seek or give, however, serves purposes quite beyond physical welfare. Gold-plated forks and platters serve purposes beyond the nutritional value of the food they carry. Golden epaulets on the shoulders of a coat add nothing to the protection it gives from cold weather. Many persistent people paint, sculpt, or compose music when they could eat better and keep warmer in some other occupation. Scientists and mathematicians are satisfied to pursue knowledge for which no one can imagine any practical use. Magazines publish articles speculating on what might have happened billions of years ago in the far reaches of the universe. There are many ways of getting healthy exercise, but many people find satisfactions beyond exercise in highly organized games such as soccer or volleyball. People climb Mount Everest, too, for reasons beyond the exercise. Those activities help us in many ways, and those activities themselves require help. Many skilled craftsmen work for months or years to give the astronomer her telescope. Even among painters in their garrets, few make their own brushes or paints, and few weave their own canvas. Athletic teams hire psychologists to help them maximize their interpersonal coordination.

Giving help is not always unalloyed with harm. Police sometimes bloody demonstrators to prevent the harm a continuing riot might cause. Psychologists who ignore the research on memory have sometimes coaxed adults or children to "remember" brutality from their parents that did not actually occur. Brokers sometimes advise investments that lose money for the investor. A rocket-builder installs O-rings that fail on liftoff. Wars bring some kinds of benefit to some people at horrendous costs to a great many people on both sides. In some wars, few people other than the suppliers of armaments and the moneylenders reap any benefit at all.

The sorts of events helper and helpee perceive before and after the helping, the roles expected of helper and helpee, and the manner of helping itself are all affected by the kinds of internal standards guiding the actions—especially the principles and system concepts. In other words, helping is always burdened with cultural anticipations and expectations that can limit the kinds of help for which we think we can ask. Philip Brickman and his co-authors (1982) describe four focuses of strong cultural expectations that will serve as examples.

If a helpee calls for help, one of the questions a helper may have is, "Who caused the state of affairs with which you want help?" The helper or the helpee or both might believe the helpee caused or helped to cause the problematic state of affairs. In that case, the helper might expect the helpee to be the one to do something about the problem. The common view of this case in our culture is that it calls for the helpee to show *motivation*. Brickman and colleagues call this case the *moral* model. "You got yourself into this; now get yourself out."

The label for the second model is *enlightenment*. The helpee is held responsible for causing the problem but is judged to be unable or unwilling to do something about it. Here the cultural attitude is that the helpee should receive *discipline*. "You ought to be ashamed of yourself." In the third model, the *medical*, the helpee is not considered to have caused the problem and is not expected to be able to relieve it. Here the helpee is the patient and needs *treatment*. "Here, take this pill." In the fourth model, the *compensatory*, the helpee is not held to have caused the problem but is expected to do something about it. What the helpee needs here is *power*. "Here is some money to buy a new one."

You can see that those various views (principles and system concepts) of problems call up different social roles to help cope with them: coaches, clergy, psychiatrists, physicians, teachers, consultants, and others. You can see, too, that the view of the problem taken by the helpee and the helper can strongly affect the success of the help, especially if the helpee subscribes to one "model" and the helper another. If I think I need to be empowered, but you think I should pray to be forgiven, our association will not last long. Brickman and colleagues also pointed out that in the case of someone whose behavior deviates markedly from the average, people were more ready a century or two ago to call in the clergy, whereas now we are more ready to call in a medical person, though some of us are arguing that it would be more effective to offer empowerment, perhaps in the form of training, social support, or economic support.

HELPERS

In the most recent decades, a growing percentage of our population has become engaged in the so-called service occupations. So rapidly has our occupational structure been changing that it outruns our official categories for thinking about it. In 1987, Stanley Davis wrote about the Standard Industrial Classification used at that time by the U.S. Department of Commerce:

> If you are a lawyer . . . working for General Motors, you are counted as part of the industrial sector. . . . If you leave, . . . hang up your own shingle, doing the same exact kind of work [with G.M. as your customer], you are considered to be working in the service sector (p. 96).
>
> Less than 20,000 of [IBM's] 400,000 people worldwide are production employees, yet IBM is nevertheless (mis)classified as being part of the industrial sector. . . (p. 97).
>
> The most important perceptual transformation we can hope for is for everyone to realize that *services play a major role in every sector of the new economy*. . . (p. 99).

We seem more and more to be helping one another not only to produce the necessities of life, but to help those who help the producers, and to help those helpers, and so on. For example, some people grow corn, which is used by people who raise cattle, the hides of which are used by manufacturers of shoes, which are repaired by your local shoe repairer, who, to manage his business, is helped by his accountant, who is aided in her work by the reference books put out by the Internal Revenue Service. I have not mentioned the teachers who have taught all those people some of what they need to know, the bus drivers, auto manufacturers, and police who help them all to get to work every morning, or a hundred other relevant helping occupations. Finally, some number of all those people are helped (though sometimes hindered) in bearing the frustrations of all those interpersonal complications and stresses by their astrologers, clergy, coaches, consultants, hypnotists, lawyers, parents, physicians, psychiatrists, psychics, psychologists, social workers, spouses, authors of self-help books, crisis hot-lines, and 12-step meetings.

I use the word "helping" to include any sort of coordination in which one or more persons intend to facilitate what others want to do. That statement is too loose to be called a definition; I want, with my examples, only to make the point that all of us, every day, make use of our links with others as a means of pursuing our purposes—of controlling our perceptions. In the rest of this chapter, however, I will pay attention chiefly to helping of the psychological sort, though I will sometimes contrast that sort with other sorts.

GIVING ADVICE

Whether advice is useful turns on the fact that we act so as to control our perceptions—to carry out our purposes, to reach goals we care about. When we seek advice, we usually ask to be told to *do* something: Tell me how to get the cap off this bottle. How often should I water the grass? Should I list the cost of this trip to Paris as a business expense? What medicine should I take to remedy my illness? What should I do to be more attractive to women? How can I keep order in my classroom? How can I reduce absenteeism in my firm?

I am using the term "advising" to name the situation in which both persons take it for granted that the adviser has at least some relevant knowledge that the advisee does not have, and that the only reason the two might have for discussing the matter is to make sure the advisee understands what the adviser is telling him to do. When the conversation goes beyond mere clarification, I would not call it a pure case of advising. In the pure case, the adviser takes the role of authority. Sometimes the authority becomes heavy-handed: physicians give you *doctor's orders* and clergy give you *commandments*.

No matter how well intended or how carefully planned, a recommended action often fails the advisee's purpose. The act the advisee chooses through which to control a perception must satisfy the Requisites for a Particular Act (for which see the introduction to Part VI). The advice-giver will always be to some extent (perhaps only a little, perhaps a lot) unaware of:

1a the extent to which he or she (the adviser) is correctly conceiving the perceptual variable the advisee feels most in need of controlling. . . .

I am sure that King Edward VIII of the United Kingdom got a lot of advice about how to continue as king, but taking that advice would have taken him away from the arms of Wallis Warfield Simpson. Edward ruled only for a part of the year 1936 before he abdicated the throne and married Mrs. Simpson. I will skip mention of Requisite 1b, since the advisee is unlikely to be asking for advice if a controlled variable is not being disturbed. The advice-giver will also be to some extent unaware of:

2a whether an an opportunity or source of energy or procedure suitable for affecting the controlled variable is available in the environment,

2b whether the advisee has come upon or believes it possible to come upon a suitable object,

2c whether the advisee is capable of carrying out an act that will affect the controlled variable (this includes being capable of conceiving or imagining the act, when that is a necessary step),

2d whether the advisee will estimate the likelihood to be sufficiently high that the advised action will reduce the disturbance to the controlled variable.

3 whether the advised action will disturb some other variable the advisee is controlling.

Often, the adviser is correct about one or two of these requisites but wrong about one or more others, and the advice fails. The physician may be correct in her judgment that (1a) the patient would like to be rid of his ailment, about (2a) the usually favorable effect of a particular pill, about (2b) the patient's understanding of how to get to a pharmacy and buy the pill, about (2c) the physical and mental ability of the patient to go to the pharmacy and ask for the pill, but somewhat wrong about (2d) the patient's estimate of the likelihood that the pill will be effective, and wrong, too, about (3) disturbances to other controlled variables. The patient may not want to spend as much money as the pill requires, or he may not want to spurn the advice of his spouse, who repudiates mainstream medicine, or he may simply want to avoid thinking any more about his ailment, and therefore may want to stay away from the pharmacy. The result is that the advice of the physician fails.

All the Requisites must be satisfied if the advisee is to find an action that will aid the control of the perception. No adviser is omniscient; no adviser can know of every condition in the environment now and during the period necessary for the action to have its desired effect, and no adviser is a mind reader—although many advisers, including some psychotherapists, believe themselves to be. Considering the fact that an adviser will always be to some extent ignorant of one or more of the six kinds of knowledge necessary to giving the advisee advice she can follow, I marvel that advisees can use advice to their benefit as often as they do. I do not marvel that advice so often goes wrong.

Some helpers frequently help their clients in other ways than advising, but, day in and day out, most people in helping roles, paid or unpaid, usually choose to help by giving advice. Here is an illustrative list of some kinds of helpers you might keep in mind as you read the rest of this chapter: parents, salespersons, bosses, clergy, police, lawyers, physicians, psychiatrists, teachers, psychologists, consultants, therapists, coaches. Or you may make your own list. Think of your experience with people in those roles. Did they usually (whether gently or vehemently) tell you what to do?

Often, advising (telling what to do) is the best kind of help. Where is the toilet, please? Turn left; second door on the right. In other circumstances, especially when the help-seeker is trying to find a way out of an inner conflict, advice has a low likelihood of being helpful. In a circumstance such as inner conflict, helping the person to find a new viewpoint (without the helper prescribing a particular viewpoint) is more likely to be beneficial. I will say more about how helping can be free from prescribing or advising in Chapter 30 on the Method of Levels.

You might want, as an exercise, to look for sentences in this book that seem to you instances of giving advice. (Is that sentence itself a piece of advice?.

SOCIAL LIFE, CONFLICT, AND DEVIANT BEHAVIOR

When a lot of people move close to one another in a limited environment, all using some parts of that environment in maintaining their preferred values of their controlled variables, they will inevitably find themselves trying to move some variable in the environment toward a value from which someone else wants to move it away. One person gets in an elevator wanting to go up, and another steps on with him who wants to go down. Two students want to talk to the professor at the same moment about different topics. A nurse dislikes the authoritarianism of the physicians in the hospital but knows of no other place she can find work. Conflicts in the outer world make conflict in the inner world when the person cares about what happens—that is, when she is controlling a variable she perceived in the outer world and something or someone is disturbing that variable. Then the person is tugged in opposite directions. Should I fight with the other passenger to get to the "up" button before he gets to the "down" button, or should I go down with him before I can go up? I want neither to fight nor go down; I want to go up without fighting. Should I argue that the professor should talk to me first or should I come back later? Should I go on working with these authoritarian physicians or should I quit even if I have to take a job that pays less and perhaps might turn out to be equally annoying?

We encounter conflicts of one kind or another every day. Most of them we resolve almost without thinking. The other elevator passenger gets to his button first; I push back the closing door and step out to wait for the next "up" car. One student says to the other, "I need only three minutes," and the other student says, "OK, I'll wait." Other conflicts take longer to resolve—hours, months, and years. We always have some conflicts for which we have not yet found a resolution, but most of the time they are not so disruptive of our other controlling that our rate of reorganization soars. Some people carry about more unresolved conflicts than other people. We avoid some conflicts by confining ourselves within narrowed bands of action or environment. Some of us are more confined than others.

Inevitably, sometimes our conflicts are going to be resolved by actions that seem outrageous to other people (not to ourselves) because we have narrowed our attention to one level of control and one sort of environmental opportunity. Sometimes a reorganization will let go the previous control of a variable on which control is demanded in most circles of society. There are books full of examples; Robyn Dawes (1994, p. 49) tells of a woman who one day disrobed and ran along the streets of her neighborhood screaming, "My father is the handsomest goddamn drunk in [X], Pennsylvania!" It has been the custom for ages, when someone behaves in a way that other people feel prohibited from behaving, for the other people to believe the person is functioning in some nonhuman way—possessed, for example, either by a god or by a devil. Nowadays, most of us are ready to believe that the person is malfunctioning. It may be, however, that the person's control systems and reorganizing system are all functioning in a normal manner, but the result of the last large reorganization has been to put uppermost the control of some variable that can most easily be controlled in the available environment by, for example, running down the street screaming about a drunken father. The more continuously a person encounters conflicts, and the more delays the person finds in resolving them, and the more entangled throughout the neural hierarchy the conflicts become, then the more likely it is that a reorganization will put a variable low in priority that was previously high—as, for example, the approval of neighbors.

Social environments bring conflicting demands more strongly and more frequently to some people than to others. Here is an excerpt from what Powers wrote in a message to the CSGnet on 4 June 1995:

> Some people, it is said, are more "emotional" than others.... [W]e would interpret this as meaning that some people have more difficulty than others in satisfying their goals and carrying out their intentions: that they suffer larger chronic errors than others do. Their hyperemotionality is not, however, the problem; it is only a sign that there is a problem.
>
> ... hyperemotionality might reflect considerable internal conflict, which makes any effective control difficult. Or it might be that the conflict is external—it could be that because of some accident of birth or situation such as race, gender, age, physical constitution, religious beliefs, or social status, a person finds that normal efforts to get respect, help, encouragement, or simple cooperation, which most people who do not have these "handicaps" take for granted, are continually frustrated.
>
> For example, it has not been very long since women were expected to stay home and take care of children, cook and sew and clean, be [copulated with], and be content without any education about the world or any say in how the world is run. When they expressed resentment, anger, or grief, nobody asked what goals were being frustrated, what opposition was encountered to every attempt to shape a world closer to the heart's desire. Instead, women were accused of being "hysterical" ... or of being innately emotional.... As the women's movements have been trying to say in recent times, emotionality is not the problem: the problem is in the obstacles to ... satisfying the goals that any normal human being wants to reach. It is loss of control that is the problem.
>
> Irrationality, I might add, seems to go with emotionality.... A person who suffers large and chronic errors will be in a state of more or less continual reorganization.... This means that the person's perceptions, goals, and means of action will continually be in a state of change; the goals of one moment may give way to new goals at the next moment. What is happening is that the system as a whole is looking for solutions to control problems by trial and error, all learned methods having produced no desired result. As chronic emotionality signals problems with achieving control by available means, so does inconsistency and erratic change of goals signify the chronic errors that reflect a persistent difference between what is wanted and what is experienced. We should therefore look on a person who is hyperemotional and seemingly irrational as a person who is experiencing serious and continuing difficulties in creating acceptable experiences. And perhaps we should ask ourselves to what degree our treatment of such a person is a source of the problem.

When we turn to our social environment as a means of controlling our perceptions, we often find that we open ourselves to conflicting demands. A young woman might choose the career of mathematician. The school counselor might tell her that women don't have the right kind of mind for that work. After a series of reactions like that from counselors, teachers, and boyfriends, she might easily adopt ways of controlling her picture of herself that other people would interpret as "touchy." That behavior would result from the interaction between the woman and her social environment, but onlookers could attribute it to the character of the woman.

Physical Effects

Reorganization has physical effects, as one might expect from a system that casts about (so to speak) for a way to maintain the health of intrinsic processes. In Chapter 9, under "Humans," I mentioned the finding by the Plooijs that human infants were more likely to be ill during periods of reorganization if "deficiencies and excesses of caretaking" (Plooij, 1990, p. 133) occurred. During the very first weeks of life, social contacts can have very beneficial effects, too. Bruce Bower (1985) reported a study of premature infants carried out by Tiffany Field, a psychologist at the University of Miami Medical School. The infants had medical problems severe enough to require 20 days of intensive care. Field gave very gentle massage and aid in exercising to a random half of the babies during three 15-minute periods per day for ten days. The treated infants, on average, gained 47 percent more weight per day than the untreated infants, though both groups had the same number of feedings and the same caloric intake. The treated babies were awake and physically active a greater percentage of the time. Other benefits were observed even after eight months.

Being an active member of a social group aids longevity. *Science News* (1980)[1] reported a study by Lisa Berkman of 7,000 residents of Alameda County in California over a period of nine years. She found that "persons with the most social contacts were the least likely to die during the study period." Berkman said that in addition to emotional support, members of the group got useful information from one another about available jobs and sources of medical care. They also got goods, money, and services. The magazine quoted Berkman as saying that for good health, "we not only need a healing brain, but we need a healthy society." Social stresses can be as debilitating as physical stresses. E. B. Palmore (1969) found that a psychologically stressful job was more closely associated with shorter longevity than physical causes such as smoking.

The need for a "healthy society" is emphasized by statistics on public health. The United States is far from the healthiest place in the world to live. For example the physician Hugh Drummond reported in *Mother Jones* (1987, p. 16) that the incidence of illness from malnutrition in Cuba at that time was 0.7 percent among children one to 15 years of age. In the United States, the rate for children and adults combined was nine percent.

I will say here again what I have said a few times earlier. I do not cite these studies of averages and frequencies with which people take certain acts as verifications of PCT. The correctness of PCT is to be tested by modeling, not by how many people performed a predicted act. I cite these studies as *illustrations* of some ways people can make use of their social environments in controlling their perceptions—or fail to do so. I am saying, "Here are some actions you can see when people are controlling perceptions." I am not saying, "These are acts people will take every time under these circumstances." Percentages and averages and headcounts do not tell us what every person will do or will be trying to do. They tell us what *some* people *did* when certain *opportunities* were available in the environment. Other people at other times and places will use other opportunities.

Psychological Effects

Bruce Bower (1998) reported a tale told in a book about urban gangs:

> Raul, . . . was 16 years old, smoking marijuana in a fast food restaurant and showing off his new gun to 15-year-old Paco. Bored, they got in Raul's Chevrolet Impala and accompanied a car full of fellow gang members to a nearby town for a surprise attack on teenagers in a rival gang. . . .
>
> Suddenly, a large group of opposing gang members appeared on the sidewalk. Raul's comrades, armed with knives and metal bars, exhorted him to shoot his gun into the hostile crowd. Eyes shut and heart pounding, Raul fired off [two magazines] of bullets that sprayed wildly into the darkness. . . .
>
> Police officers arrested Raul and Paco that night. One of Raul's shots had wounded a man

on the street two blocks away; another bullet had struck and paralyzed a child sitting in his living room watching television. Raul now wonders [in prison in California] how he became a gang member and whether he will be able to leave that life behind for something better.

The article goes on to report what various experts have to say about this sort of episode. You can see in the following quotations the way researchers struggle to use personality characteristics and environmental circumstances as explanations for particular actions such as the violence described above:

> Researchers have documented an array of traits and behaviors that, according to statistical evaluations, occur more often among [violent] children.... The research [however] simply cannot explain why certain individuals and not others opt for mayhem....
>
> ... social scientists have tended to treat all severe forms of aggression in young people as psychiatric ailments, assigning them such diagnoses as conduct disorder and antisocial personality disorder. [Critics of that medical view say that a] child's violent streak may be misguided and morally repugnant without reflecting a broken brain or disordered mind....
>
> ... editors of 30 ... mental health journals ... issued a statement calling for ... alternative forms of research....
>
> ... former American Psychological Association president Lawrence Hartmann [in an] editorial openly expressed frustration with predominant scientific approaches.... Hartmann says that [correlational] studies are limited by [the assumption in the Diagnostic and Statistical Manual of Mental Disorders (DSM) published by the American Psychiatric Association] that the three diagnostic categories—conduct disorder, antisocial disorder, and substance abuse—exist as "discrete and solid things."
>
> Conduct disorder ... may occur when brains carry out functions for which they evolved, but which prove dysfunctional in present-day environments, hold psychologist Leda Cosmides and anthropologist John Tooby, both of the University of California, Santa Barbara....
>
> John E. Richters of [the National Institutes of Mental Health] ... raises doubts about scientists' ability to label with confidence certain mental functions as either natural or dysfunctional.... There's currently no way to tell whether the ... child who skips school, steals, and hangs out with a gang really suffers from an enduring psychiatric condition that will haunt him throughout adulthood.... In fact, the vast majority of teenagers in industrialized societies engage in at least a few delinquent acts without going on to a life of crime....

You can see in those quotations a yearning to find some internal quality or some external stimulus that can be the *cause* of particular "disorders"—acts that disrupt the orderliness desired by "authorities." That yearning neglects the Requisites for a Particular Act. It is true, however, that the fraction of a subpopulation you will see engaging in certain types of acts (such as skipping school or stealing) depends on:

2a whether an opportunity or source of energy or procedure suitable for affecting the controlled variable is available in the environment,

2b whether the person has come upon or believes it possible to come upon a suitable object,

2c whether the person is capable of carrying out an act with the object that will affect the controlled variable (this includes being capable of conceiving or imagining the act, when that is a necessary step),

2d whether the person will estimate the likelihood to be sufficiently high that the advised action will reduce the disturbance to the controlled variable.

Although one cannot confidently predict whether *this* person will skip school today, get arrested, or take the trouble to think about the best answers to items on a school test or an IQ test, one can predict that some kinds of conceptual readinesses and some kinds of environmental opportunities permit those acts to occur more frequently. From an article by Bruce Bower (1994), here are a couple of excerpts from a review of typical studies of poor children:

> By age 5, children in persistently or occasionally poor families have markedly lower IQs and display more fearfulness, anxiety, and unhappiness than never-poor youngsters...(p. 24).

Preschools enable young children to learn how to make use of teachers and their fellow children. Bower (1994) offers several examples; this is one:

> . . . 123 poor black children [were] randomly assigned either to the Perry Preschool in Ypsilanti, Michigan, or to a control group that received no preschool services. Perry graduates more often finished high school, found jobs, and avoided going on welfare or getting arrested (p. 25).

George W. Albee has long deplored the view of unwelcome behavior or emotional distress as illness. In an article announcing its Gold Medal Awards for 1993, the American Psychological Foundation wrote that Albee believed

> that social evils like racism, sexism, ageism, unemployment, child abuse—indeed every condition in which inequalities of power prevail and exploitation results—are responsible for far more psychopathology than twisted molecules . . . and that . . . the most effective and humane way to reduce human suffering is through primary prevention (p. 721).

Richard J. Robertson, a psychotherapist and co-author of the 1990 text on PCT, wrote this in 1966:

> The fact that socially disconnected individuals fall into different categories which range from "inadequate personality" and "sociopath" to legal and social classes such as "criminals" and "recipients" may be largely an accident determined by which of society's various corrective institutions encounters the individual first (pp. 333–334).

Kenneth Heller (1993) reviewed an anthology edited by Kessler, Goldston, and Joffe in honor of George Albee. Heller wrote:

> . . . a small group of retarded toddlers . . . were raised on wards housing retarded adults. The toddlers received intensive stimulation from both inmates and staff. This group as adults were indistinguishable from typical Iowa citizens, again emphasizing the responsiveness of psychological deficits to environmental conditions (p. 785).

Bruce Bower (1991a) reported a study of depression in which good results had resulted from a treatment that combined an antidepressant drug and short-term psychotherapy. Later reanalysis of the data showed that the half of the patients who had continued with the psychotherapy *without the drug* stayed free of depression for just as long as those who continued with the drug. Bower wrote that ". . . most of the long-term improvement was sparked by a consistent focus on improving social skills and relations with others. . . .".

In 1997, Bower reported similar findings:

> . . . a new study suggests that over the long haul, individual psychotherapy tailored to strengthen interpersonal skills and control social stress markedly helps many people suffering from [schizophrenia]. . . .
>
> This . . . treatment . . . resulted in lower relapse rates and progressively better social functioning over 3 years. . . .
>
> "If a new medication had treatment effects of the same magnitude as those . . ., it would be seen as a major advance and adopted as the main drug treatment for schizophrenia," remarks William T. Carpenter Jr., a psychiatrist. . . . "Unfortunately, the influence of this new finding will be severely muted because it involves a psychosocial approach."

In 1989, a book appeared by John Mirowsky and Catherine E. Ross entitled "Social Causes of Psychological Distress," and in 1992, George Albee published a review of it in which he said he was grateful

> to see the evidence, once again, that the problems of deeply distressed people are largely social in origin. [And] that distress is largely a consequence of powerlessness, alienation, meaningless work, normlessness, and social isolation; that opportunity to control one's own life is critical to mental health; that the economically disadvantaged half of society is almost six times more likely to be severely distressed than is the wealthier half; and perhaps more important, [the] conclusion that
>
>> neither genetic nor biochemical factors have been shown to account for any substantial part of the differences in levels of distress found in our society. In particular, there is no evidence that the social patterns of distress reflect genetic or biochemical abnormalities. . . .
>
> Who are the people least likely to experience severe psychological distress? Those with a fulfilling job, a supportive relationship, a good education with continuing interest in learning, and a decent living (pp. 16–17).

The book by Mirowsky and Ross (1989) contains about 200 pages of evidence, displayed carefully and interpreted judiciously. Almost all the data come from carefully composed interviews in surveys, fully described in the book's appendix. Mirowsky and Ross agree with Albee that research since about 1960 has repeatedly shown four basic social patterns of distress. On the average (1) men are less distressed than women; (2) unmarried persons are more distressed than married persons; (3) the less the number of *undesirable* changes in a person's life, the less his or her level of distress; (4) the lower a person's socioeconomic status, the higher that person's level of distress.

Although married people on the average encounter less distress than single people, marriages vary in distress. The distress varies with whether the wife is employed outside the home, whether husband and wife prefer her to be employed, and whether the husband shares caring for the house and children. Often, Mirowsky and Ross say (pp. 89–90), husbands feel distress because of embarrassment or guilt when the wife is gainfully employed. Wives, on the other hand often feel stressed when they get too little help with the housework.

Mirowsky and Ross (p. 95) report evidence that even within the same income level, blacks with low socioeconomic status feel more distress than whites at the same level, probably because of the conflict occasioned by their comparative lack of opportunity for upward mobility.

Single mothers with dependent children often find themselves in a terrible bind. Frequently, their employment does not pay enough for them to hire child care while they work. The discouragement they feel, the depression and anxiety, interferes with their ability to work and care for their children. They are often advised then to treat their depression as an illness. Our society says, in effect: if you are having trouble making enough money to pay for the child care that will enable you to go on working, any distress you feel must be due to some defect in your personality, so you should somehow find some money to pay for visits to a psychotherapist.

Mirowsky and Ross go on to give evidence of effects on distress of religious belief, powerlessness, alienated labor, social support, economic support, normlessness, being labeled a sufferer from "mental illness," role stress—and other social conditions. In the last pages of the book, they say:

> Social factors . . . account for at least half of all symptoms of the large majority of cases of serious distress. We think the evidence shows that distress, whether moderate or severe, is primarily a normal response to difficult circumstances, rather than a manifestation of unseen flaws in the organism (pp. 181–182).

We think that the informed individual and the informed community can do a great deal to prevent distress. Strategies for preventing distress can be centered on a few simple basics: education, a fulfilling job, a supportive relationship, and a decent living. These are to mental health what exercise, diet, and not smoking are to physical health. Emotional well-being is founded on active, attentive, and effective problem solving. Unpleasant emotions are not themselves the problems; they are signs that problems remain unsolved. . . . Drugs do not solve problems. . . . Similarly, having someone else solve a person's current problems . . . merely lightens the individual's burdens while he or she is building strength (p. 182).

Remember that the studies reviewed by Mirowsky and Ross are head counts. They give average numbers and proportions. There are many instances of cases that go contrary to the averages. Although *on average* people high on the totem pole are less stressed than those at the bottom, there are some poor people who are less stressed than some rich people. Rich people once in a while do commit suicide. Maintaining control of a perception depends not only on resources available in the environment, but also on one's ingenuity in making use of them.

Gary D. Gottfredson (1981, pp. 425–426 and 430) described two studies that I find especially dramatic in showing the ways adolescents interact with their environments. One study (carried out by Wolfgang, Figlio, and Sellin, 1972) tallied the age at which adolescent boys in Philadelphia were first taken into custody by police. At age seven, the percentage of boys arrested for the first time was less than five. That percentage rose steadily until age 16. At 17, however, the percentage dropped precipitously to about *half* the percentage at age 16. The other study (carried out by L.T. Wilkins, 1965) was of English schoolboys. Until 1949, indictable offenses among the boys peaked at age 13. In 1949, however, the peak suddenly changed to age 14. What was going on? In Philadelphia at the time of the first study, the school-leaving age was 16. In England, the school-leaving age until 1949 was 13, but beginning in 1949, it was 14. Despite differences in country, in time of study, and in change in the English school-leaving age, in every instance drop in the incidence of falling into the hands of the police coincided exactly with the exit of the boys from school.

Clearly, the opportunities the boys in those two studies found during their school years for controlling perceptions important to them were very different from the opportunities they were able to use after they left school. No doubt, too, the internal conflicts they encountered in school engaged internal standards different from those engaged by the conflicts they encountered in their lives out of school. The result of the inner and outer differences was that when in school, the boys more often than when out of school chose means of controlling perceptions that we call delinquent. You can find a few hints about the sort of opportunities at least some adolescents search for in the story I quoted in Chapter 27 under "The Living Environment" about the adolescents at the shopping mall.

Kenneth Heller (1996, p. 1125) wrote that

> ... accumulating evidence indicates that supportive ties, skill training, and social problem-solving training are promising approaches to reductions in child abuse ..., academic failure, behavior problems in school ..., adjustment problems among children of divorce ... or children experiencing parental loss ..., tobacco and alcohol use among preteens ..., lower levels of postpartum depression ..., or depression associated with job loss....

What does *not* have any effect on the way people make use of their environments? One example is the irrelevance of psychiatric diagnosis to the rearrest of jail detainees for violent crimes. Teplin, Abram, and McClelland (1994, p. 335) summarized their study of 728 male detainees this way:

> The authors examined whether jail detainees with schizophrenia, major affective disorders, alcohol or drug disorders, or psychotic symptoms (hallucinations and delusions) are arrested more often for violent crimes six years after release than detainees with no disorders.... Neither severe mental disorder nor substance abuse or dependence predicted the probability of arrest or the number of arrests for violent crime.

The authors I have quoted above are saying that more often than not, the way a person makes use of his environment can be altered by altering the opportunities in the environment. Not always, but certainly more often than not. Opportunities in the environment do not *cause* acts, but they make some acts more likely to aid control of perception than others.

Those studies I reported above (and more) demonstrate that the social environment can restrict our choices of action in ways that put our internal standards into conflict. We then sometimes display unusual sorts of behavior that worry and offend people around us. When psychotherapists and researchers recommend changing those social conditions so as to make environmental opportunities of the unworrying sort more readily available, they are recommending the same kind of strategy recommended by public health experts. The public health people point out that infant mortality was greatly decreased and longevity greatly increased in Western civilization long before modern medicines and CAT scans—indeed, even before the germ theory of disease. The general health improved because of provision of clean drinking water, increased supplies of nourishing foods, more sanitary sewage systems, regular garbage collection, and the like. The strategy was not one of fixing people after they got sick, but of removing the environmental conditions that brought upon people the physical assaults of disease, poisons, and undernourishment. That strategy should work for public mental health as well as it worked for public physical health.

LOVE

To illustrate the use of the social environment, many of my illustrations above have been of actions that most of us are unhappy to see. I will rectify that emphasis by turning now to a social relationship most of us are very happy to see—love.

Love is a high-level conception. What people have in mind when they say "love" varies immensely; indeed, every person's conception of love is unique. People can have in mind comradeship, sexual delight, commitment to raising a family, simply feeling that the other person is nice to have around at least for a while—you can no doubt think of several more conceptions. Consider only sexual love. The pleasure one takes in coitus can be one of delight in the physical sensations one feels skittering over the skin and flooding one's body; it can be one of delight in the pleasure evidenced by one's partner; it can be one of caressing to show admiration and affection; it can be playful; it can be one of inventing variations on a theme; it can be one for impregnation; usually it brings several sorts of pleasure. I will refrain from mentioning here the ways that coitus can be used to bring pleasure to one person at the cost of pain, physical or psychological, to the other.

My concern here is helpfulness. The love that comes from mutual helpfulness is not a "thing" you can hunt for or make. It is a side-effect of certain kinds of goals two people pursue in their perceptions of each other. When two people want to help each other live satisfying lives—help each other control the perceptions each cares about—and when they spend their time together with that goal very high in their control hierarchies (when they bend their other goals to maintain that goal), they very often develop feelings of joy in watching the happiness of the other person, of comradeship in joint undertakings, of strange rumblings deep in the chest signaling somehow that everything is as it should be, of wanting all that to go on forever, and above all of unquestioning trust that the other person always wishes one well, never harm. That is the kind of love that marriage (or a similar partnership) can be. It takes a while to bring it into full fruition, and it is sometimes hard work. Perseverance is vital. This statement is not a summary of research, but only a report of my own experience, but you can see how mutual help in controlling perceptions, persistence, and time can lead to the condition I have described.

You will no doubt wish to change the emphasis here and there in the paragraph above. Go ahead. Here I mean only to point out how thoroughgoing mutual helpfulness can be and how very good it feels when it is indeed helpful, mutual, and thoroughgoing. I know there are people in the world who would not be happy living in such close mutuality. But I am very sure that all of us are born with the capability of living that way if the environment—including the social environment of family, friends, school, and so on—contains the opportunities for us to develop the necessary skills and internal standards.

The love arising from comradeship can embrace an entire group, sometimes even a fairly large group. The most familiar examples, I suppose, are sports teams and elite military groups, but they can come into being in any endeavor. I have myself belonged to two groups with that strong esprit de corps, a theatrical company for three years and a group of organizational consultants for about 15 years. It is a deeply gratifying way to live.

This is a tiny amount of space to devote to the topic of love—a topic upon which millions of tongues, pens, keyboards, and cameras have poured poems, songs, operas, dramas, novels, and scholarly tomes. Almost all those outpourings, however, have been devoted to describing actions or feelings that are illustrations of love; there is no need for me to add more. Here I wanted only to say something about what mutual help in controlling perceptions can feel like. It can feel like love.

ENDNOTE

[1] In the list of references, look under "Social ties."

Chapter 30
The method of levels

The experience of finding a new, clarifying, refreshing point of view is probably about as ancient as awareness itself. We have a few terms for it: insight, intuition, revelation, epiphany. And a few phrases likening it to a burst of light: bring to light, dawn on, awaken. And a few calling up an internal feeling: come alive to, feel an enlargement of mind. But a new view is not always enlivening and invigorating; one does not always shout "Eureka!" Sometimes we simply feel as if we are standing off to one side, looking back at a previous understanding. Sometimes the outcome is even quieter than that, as when we feel relieved of conflicts, tensions, and even curiosity, and we simply feel peaceful.

Every level of the control hierarchy is itself a point of view, a collection of internal standards through which perceptions can be understood. To give an example of one of my own invigorations, I remember the burst of light I felt when I read (long ago) Lobachevsky's *Theory of Parallels*. I had previously thought of Euclid's geometry as a system concept quite isolated from other parts of mathematics. That is, I was viewing Euclidean geometry as an event (a thing) having its meaning within itself, not illuminated by other mathematics. When, however, I struggled with Lobachevsky's construction of a geometry built without Euclid's twelfth axiom—a geometry in which any number of lines through a given point could be drawn parallel to another line—and came to understand the logic of it, I was overcome with awe and delight in suddenly realizing how Euclid's geometry could belong to an entire family of geometries, each taking its character from its own set of axioms.

Moving up a level can reveal a viewpoint that enables the person to become aware of previously unimagined resources. A new viewpoint can often resolve a conflict. Systems at a higher level of control can set the values of reference signals at lower levels. Consequently, what seems at the lower level to be a situation of unresolvable conflict (I want to get through this narrow door at the same time he does but I don't want to struggle with him) can often become quickly resolved at a higher level (getting through the door at this exact moment is not necessary to carrying these groceries to my car). Helping a person, therefore, to adopt a viewpoint from a higher level than the level at which she is perceiving a perplexity or a conflict can be used as a method of therapy. Doing so is called the method of levels.

In the method of levels, the helper does not act as an expert. The helper (or guide, or therapist) does not give advice; she does not know better than the helpee (or explorer, or client) what the helpee should do. Refraining from taking the role of expert is unusual among helpers, especially among paid helpers, but it does happen. Carl Rogers published a book in 1951 called *Client-centered Therapy* in which he described a method of "reflecting" the client's thoughts simply by paraphrasing what the client said. The therapist made no assertions or guesses about the client's history, about causes, or about any deficiencies, disorders, afflictions, or demons the client might be harboring. Often, the client found his way to a new view of his situation or himself—a new understanding—while talking, so to speak, to himself. I am sure there have always been some friends, siblings, spouses, teachers, parents, and priests who have helped others through this "non-directive" method. Rogers made it respectable for psychologists. I do not know how many present-day psychologists follow Rogers's lead. Rogers, however, did not know about the hierarchy of control; he did not know about the method of levels. But he was bold enough to relinquish the authoritative role of the expert.

In the method of levels (MOL), the therapist or guide does not merely reflect what the client says. Indeed, the therapist talks very little about the matter (the perplexity, problem, conflict, trouble) the client brought to the conversation. Instead, the therapist encourages the client to talk about talking or thinking *about* the topic with which he arrived. The therapist encourages the client to look back at the topic, so to speak, from another level. Below are some examples of conversation that I borrow from a communication from Powers to the CSGnet on 30 August 1995. First, here are five examples of typical advice-giving, *not* what you would hear in an MOL session. I have labeled the utterances C and T for client and therapist:

- (C) I don't want to be left alone. (T) Try being alone for just 15 minutes.
- (C) I want to get over this problem. (T) You can do it, I'll try to help.
- (C) Nobody likes me. (T) I like you; I'll get your mother to like you.
- (C) I am a terrible person. (T) Tell me some things that are good about you.
- (C) I don't know what to do. (T) If you like, I'll tell you some things to try.

Now here are five examples of the way a therapist using MOL might respond to a client. In each case, the therapist asks the client to say something *about* what the client has just said:

- (C) I don't want to be left alone. (T) What's it like to be left alone?
- (C) I want to get over this problem. (T) Is this a very strong feeling right now?
- (C) Nobody likes me. (T) Is that OK with you? How do you know they don't? Are you thinking of anyone in particular? What's it like not to be likable?
- (C) I am a terrible person. (T) What are some of the terrible things you're thinking about? Are any of them going on right now?
- (C) I don't know what to do. (T) Can you tell me more about not knowing what to do?

Notice that there is no "content" in those remarks of the MOL therapist—no mention of what might be right, wrong, good, or bad; no advice about actions to take; no explanations about motives, repressions, or reinforcements; no warnings about a prescribed course of therapy; no demands. The remarks simply invite the client to say whatever he wishes as an observer of his own perceptions. The therapist will not tell the client what he should think either about the problem he came with or about the observations he makes of it; the client will tell the therapist—and himself.

The next few pages contain most of the text of a handout Powers distributed at a workshop in 1999.

THE METHOD OF LEVELS

An overview for a workshop
conducted by W.T. Powers
July 18–21, 1999

The method of levels is an experimental approach to counseling or psychotherapy. Its bases are ... Perceptual Control Theory and the naturalistic observation that a person's consciousness can apparently operate from different viewpoints within the brain's organization. The objective of this method is to draw a person's attention to perceptions at levels higher than perceptions in the primary or central focus of attention. When a shift of level has occurred, the same process is repeated, and so on for as many times as possible or useful.

This simple procedure seems to facilitate therapeutic change, with productive sessions lasting only about half an hour or even less. In this workshop we will describe and demonstrate how this method works in practice and teach participants how to direct an MOL session, with and without coaching, participating as both guide and explorer.

At the end of the workshop, participants should understand the method well enough to test it by themselves when they return home. One approach that has worked well is for individuals to organize discussion groups in which they pass on what they have learned and gain experience with each other, trading roles to develop their understanding of both sides of the process....

How the Method May Work

Most people have had the experience of being engaged in some train of thought or conversation, while at the same time being aware of a background thought, attitude, or feeling *as a commentary about the foreground experience* [PJR's emphasis]. It's as though a person can operate at two levels of attention at once, with the [background] level being less vivid and explicit than the [foreground level]. While this second level can be like an intruding thought about something completely different, it is often about the first level in some way, or about the person in whom the thoughts at the first level are occurring. . . .

For example, a perceived relationship like "the pencil on the table" is composed of perceptions of objects—the pencil and the table, which can each be perceived independently of any relationship between them. In that example, we can be aware of the pencil and the table as individual objects without paying any attention to the relationship of one being "on" the other, or we can be aware of the "on" relationship specifically. When attention is on, say, the pencil, we can be aware of its color, its size, the sharpness of its point, any lettering on it, and so forth. In the backs of our minds, we may be aware of other objects, and even of relationships among the objects, but those perceptions are not clear and central. They become the center of attention only when we perform a hard-to-describe act and somehow bring the relationships into mental focus. Then we specifically notice the relationship, the "on"-ness.

It's not difficult to see that the perception of "on-ness" may have been in existence all the time, even though not in conscious awareness. And it may even be possible that some control process may have been going on, in which "on"-ness was being controlled (putting the pencil *on* the table), but that one's attention was on finding and picking up the *right* pencil to put on the table. Although unlikely in this simple example, it could be that some difficulty or misunderstanding had arisen about which pencil was to go on the table, and that in working out this problem the person's attention had focused on the characteristics of the pencil to the exclusion of the purpose involved in finding the right pencil.

In that case, the process of finding and picking up the pencil might be in the foreground of attention, while the reason for wanting to do so has retired into the background. Most people have experienced this, too. Have you ever found yourself looking into a closet or a refrigerator and wondering what brought you here? After a moment you become aware of the purpose, and answer your own question, but in the time just before you remind yourself, you're in an interesting condition of pursuing some goal but not being aware of the higher goal that led to the setting of the lower goal.

When we consider things like getting a shirt out of a closet, this kind of episode doesn't seem very important. But suppose that what you find yourself doing is feeling an instant dislike for someone you have just met. You do not want to talk to this person or even be around this person—and you don't know why. Something inside you, clearly, is perceiving something in this situation that is to be avoided, and as a way of avoiding it is setting a goal of immediate departure or non-interaction, a goal which, unsatisfied, leaves you with a great urge to be elsewhere. All you're aware of, though, is the desire to get out of there.

At this point a psychologist might go into high gear and start speculating about traumatic childhood incidents, phobias, guilt, and all sorts of other possible explanations of this "irrational reaction." All sorts of treatments might be suggested, from a prefrontal lobotomy to electroshock to tranquilizers to desensitization therapy. But what we would look for under the method of levels is simply the next level up. We would assume that this goal of getting away is there for a reason, and the reason is that a higher system has specified this goal as a way of controlling something else.

Since we haven't the least idea about what this next higher system is or what it's trying to accomplish, the best thing we can do is ask the person having the problem. And rather than lead the person with suggestions and analyses, what we really need to do is to help the person move to a point of view from which the actual cause of the problem can be seen: the next level up. Then the person can tell us the right answer, if he or she wants to.

This is the principle behind the method of levels. By directing a person's attention to materials relating to the next level up, we effectively move the person's attention to that level, from which perceptions and intentions that were formerly in the background become a new foreground. If the unexplained reason for the behaviors in question now shows up as a foreground thought, the chances are that some kind of change will immediately take place or begin to take place. The reason for the "phobic" reaction may become immediately obvious (My God, he talks just like that son-of-a-bitch Uncle Charley"). Or nothing dramatic may happen, but the person for some unknown reason loses interest in escaping the situation. One of the typical and obvious consequences when a person succeeds in going up a level is a complete and sudden change in the emotional content of experience.

How an MOL Session Works

I've already suggested that in an MOL session, we don't try to psychologize or advise or analyze, but instead focus on getting the person to go up a level. This may seem to be a vague and unhelpful description of what we do, but in fact going up a level is an easily recognizable phenomenon both to the guide and to the client. And for the most part, once the client catches on, the client will let the guide know what the next level is.

The MOL is conducted as a rather peculiar kind of conversation between the client and the guide (these terms, guide and client, are open to revision. Some prefer guide and explorer). In this conversation, the client picks some subject to talk about, quite likely some difficulty being experienced. The guide asks questions aimed at getting more details about the subject of discussion, but what the guide listens for are not the answers to those questions. The guide listens for meta-comments that are *about* what is going on: for example, a comment like "Am I doing this right?"

Pouncing on every meta-comment is not very productive; the guide needs to listen for more comments that help establish the nature of the higher-level point of view. The client says, scattered here and there, "I hope I'm doing what you want," and "I'm not sure I'm doing what I'm supposed to," and "I hope I'm not being too dumb about this," and eventually the guide will interpose another sort of comment: "Are you unsure of what you're supposed to be doing? Tell me more about that."

To make it plain what's being asked, the guide can elaborate. "I'm just asking what you're thinking or feeling about that right now, while we're talking—not what you think in general, or might be experiencing, but what you can see really going on in your mind right now. Just a kind of observer's report." If that doesn't do the trick the guide can elaborate further: "This unsureness—is that the right word?—are you feeling it right now? Are you thinking thoughts about it? For example? Does any physical feeling go with it?"

Eventually the guide can get the client talking freely about the former background thought, feeling, or attitude, so it really becomes the foreground of the conversation. And as it becomes established in the foreground, the guide starts listening again for meta-comments (or watching for body language, or listening for tones of voice—any source of information) that will point to the *next* level up.

And that's basically the method. That's all the guide does. The focus of the method is not on helping the client or solving the client's problems or making the client feel better or giving the client advice or encouragement or prescribing behaviors for the client. The focus is entirely on getting the client to go up a level, and when that happens, to go up another level.

How does a session end, then? Once again, the client lets you know. It's unusual for a person to go up more than four or five levels in a session, and it's often fewer than that. The session ends when the client expresses satisfaction or progress or boredom or starts talking about lunch. It can end with the guide saying, "I've kind of lost track of where we are—would this be a good place to stop?" or "Would it be OK with you if we leave it here until next time?" Sometimes the client likes to review the session, noting the level shifts. Sometimes the guide, after the session has clearly ended, asks what was going on at some point where he or she found the proceedings mysterious.

In other words, there's no hard and fast rule.

Why is the Method so Limited and Simple?

The short answer to that is that when we help other people, we seldom really know what we're doing, so the best thing to do is as little as possible. This principle is somewhat like the admonition in the Hippocratic Oath: first, do no harm.

Perhaps a better, or more serious, answer is that there are dozens of psychotherapies and probably hundreds of individual variations on how to conduct them, yet in every psychotherapy there is some success rate. Obviously, if one therapy is nondirective and another is highly prescriptive, yet they both work, then directiveness or prescriptiveness can't be an explanation of why they work. . . .

The method of levels can be thought of as an attempt to express what is common to most successful psychotherapies. Most interactive therapies entail "listening with the third ear," meaning listening not just to what the client is saying, but to the meta-content, the background, the up-a-level comments. And most progress in therapy comes about when the client has a sudden realization, sees the familiar old problem from a new point of view, finds a new level of self-awareness. It has been said that successful therapists, with experience and the passage of time, come to conduct their business in more nearly the same way, regardless of their theoretical foundations. They listen more and talk less. They analyze less. They advise less. They lead less. They stand by while the client works it out. They wait for the client to have the insight instead of trying to show how brilliant they are. . . .

The method of levels is about as minimalist as one could imagine a therapeutic method to be. Maybe it's too much so, but we'll never know that until we try it. To mix it with any other approach is inevitably to confuse matters, and of course makes it impossible to judge whether the MOL is worth anything in its own right. The implication of the MOL is that most of what goes on in other therapies is (a) unnecessary and (b) possibly detrimental. If you're a therapist, reading about the MOL will, I hope, suggest to you that you ask yourself what the effect of your various interventions is supposed to be, and how you know they are helping progress in therapy. And I hope that in the interests of science you are willing to suspend your customary approach long enough to test the method of levels in as pure a form as possible.

[End of W.T. Powers' overview]

Going up a level is something we all do every day. Perhaps it is easier at the lower levels of the hierarchy than at the upper levels. We do not ordinarily need help in noticing the "on" relationship of a pencil to a table. We can often profit, however, from help with perceiving particular programs, principles, and system concepts that connect us with other people. Even there, we do not need much help with the capability itself of going up a level—with using the function itself. We need help, a nudge, to find the perception in the level above that is connected with the problem or perplexity or conflict. We need help in putting to use in a particular situation (or with a particular content) a method we have known how to use all along.

Powers emphasizes the partnership—the equality of status of client and guide. Here is a paragraph from his missive to the CSGnet on 27 August 1995:

> When you're doing the method of levels, you have to forget about being a wise psychologist who understands the patient better than the patient does and is helping the patient to understand something the psychologist has already figured out. That's a completely different process, which may have its place, but not when you're doing the method of levels. If the patient says, "My mother likes my father better than me," you don't mutter "yippee" and head for the Oedipal conflict. You just say, "What else can you tell me about that?" or "Any thoughts about that?" In the method of levels, you don't know what's going to come next, and don't much care. If there are self-image problems involved, the boy will sooner or later come right out and tell you about them: "I guess I shouldn't be feeling that way about my parents," or whatever. Then you ask "Why not?" The ONLY point of the method of levels is to get the person to adopt as high a level of viewpoint as possible. What happens as a result of doing that is out of your hands.

The method of levels has been a frequent topic on the CSGnet; many of the messages have been very instructive. I will resist the temptation to quote a dozen or so of them here; I will give you only one more excerpt from Powers. This is from his epistle of 17 November 1995:

> But the most important aspect of it concerns getting to the final state, not being in it. If one runs across conflicting thoughts, goals, attitudes, ideas and so forth while doing this process, the smooth ascent up the orders hangs up, and the conflict seems to demand attention. The next viewpoint up is hard to access and the more serious the conflict, the harder it is to find the background thoughts. When, however, the next order is found, the result is quite amazing: the conflict simply dissolves. This makes sense in terms of a hierarchical model, because it is the next order of control that is responsible for establishing the conflicting reference signals, and thus (it would seem) it is the logical place for focusing changes that can reset the conflicting reference signals and thus remove the conflict.
>
> Conflicts usually disappear when discovered in this way—usually, but not always. When they do disappear, the way is open to reach the final state; otherwise, the exploration terminates there. So it seems that one can find a path through successive orders of system to the topmost viewpoint, but only when that path doesn't involve any serious conflicts or other significant error-producing problems. . . .

I think this is a natural process that people use all the time. I know that those who have been through the formal method seem to get the hang of it, and can consciously pause when there is a problem and "go up a level" to resolve it. I guess "level" is the best word to use there, in this context. Furthermore, I think that most psychotherapists are aware of this process—some call it "insight"—and count on it, although they don't necessarily have any formal model of what is going on. It's obviously involved in the personal explorations that Eastern philosopher-psychologists have written about for some thousands of years, although without the hearts and flowers and fanciful theories. I think it's a natural function of the brain, or of something connected to the brain (who knows?). It probably has something to do with reorganization.

It is often difficult for a person to help himself go up a level, even at the lower levels of the control hierarchy. You remember the puzzle I told about in Chapter 18 under "Viewpoints," in which the person was asked to tie together two strings that were hanging from the ceiling. Only a few participants were able to step outside the usual classifications of the objects in the room and look for an aid to swinging the strings. At the upper levels of perception, it is difficult, too, for a person to act as guide for a "client" whose happiness is entwined with his own. Physicians, psychiatrists, and lawyers are warned not to expect to be successful in those roles with themselves or members of their own families. It is easier to go up a level with the aid of another person, a guide, and preferably a guide who does not have much of a personal stake in the client's welfare. The guide should wish the client well and want to be helpful, but should not be tempted to pursue her own welfare through her client's perceptions (or her assumptions about them).

A rich source of comment on the MOL is the CSGnet; you will find below an endnote[1] listing some of the postings. Powers included a chapter on MOL on pages 41–53 of his 1992a collection of papers, and he included a section on "Going Up a Level" in his 1998 book(pp. 87–90). The psychotherapist Timothy Carey (1999) has written a brief introduction to PCT and MOL. Carey has also written (2000) a book manuscript that devotes a couple of chapters to the MOL.

THE GUIDE

What is necessary? The essential skill needed by the guide is that of recognizing a remark that is *about* a topic of conversation, not a remark within that topic. Acquiring that skill does not require a PhD or an MD—not even a high-school diploma. It is not even necessary to be able to read and write. I think there are billions of people who have a sufficient degree of skill in recognizing meta-comments.

A further skill is required to help the client move his perspective up a level, but again not one that requires the ability to read or write, to speculate about repressions, or to give a talk at a conference. Part of the necessary skill is the discipline to confine oneself to encouraging the meta-comments. I think it is more difficult for a person with a PhD or an MD to maintain that discipline than it would be for a 16-year-old, because the person with the doctorate has spent so many years maintaining her perception of herself as an advice-giver, a counselor, a dispenser of sagacity.

For the MOL to be beneficial, the client must be willing for the guide to limit herself to the kind of contentless comment I have illustrated in this chapter. So far, guides report that clients welcome this kind of therapy. As with all kinds of therapy, MOL will not bring benefit if the client does not trust (1) that the guide has the necessary skill or (2) that the guide cares about the client's welfare.

Here are a few excerpts (slightly edited) from Carey's (2000) lucid and luminous book manuscript:

> When people are talking to other people, . . . from time to time there will be disruptions to their stream of words. People often pause, smile, chuckle, or say something that doesn't exactly fit with what they were saying a moment ago. Or they may become teary, look away, or nod knowingly. Usually the disruption seems to be an evaluation or conclusion about something they have just said. These comments are referred to as up-a-level comments (p. 60).
>
> . . . disruptions occur at a higher level than the description that is occurring, because the disruptions are often comments *about* what has just been said; . . . to make comments about something, you must reflect *on* the experience rather than remain *within* the experience (pp. 60–61).

MOL therapists are basically uninterested in their clients' problems, because they realise that the [unwanted or disliked behaviours] their clients are describing are not the real problems. The real problem . . . is that [the client] has not yet reorganized at the level that has established the conditions for the conflict. . . .

For this reason, the MOL is a present-time activity. . . . The point of interest is in what the client is doing *right now* as he or she is sitting in front of you. . . .

MOL therapists understand that their clients don't need fixing. They are not broken or disordered or otherwise in need of repair. They are just conflicted, and they have everything they need within their own systems to resolve the conflict. The only thing the clients need help with is maintaining their awareness at the appropriate level. The MOL therapist, therefore, demonstrates respect for the client by recognizing that the person will resolve his or her own conflict (p. 63).

MOL is probably the least directive therapy anybody has yet thought up, with the possible exception of Rogers's client-centered therapy. The MOL guide does not tell you what to do. The guide does not even lay out possible actions or thoughts from which you can choose. The only expertness the guide claims is that of being alert to signals that you might be on the verge of going up a level. Even then, the guide does not claim to know how far up you already are or how far up you are about to go. The guide claims only to know which direction is "up." But "know" is too strong a word. The MOL guide expects to guess right only some of the time. The guide asks a question to put your attention on what you have just said, but she may be wrong in thinking you are ready to go up a level. Only you, the client, can know when you are ready and when you have taken the upward step.

The result of this "non-expertness" on the part of the guide is that there is little or no feeling of power difference in the conversation. The client is never pressed in any direction or toward any act. The guide never claims to know better than the client what is going on—whether, for example, the client has taken a step upward. The guide may feel convinced by behavioral signs that the client has indeed gone up a level, but what the client thinks or does is always the final criterion.

ENDNOTE

[1] Here, from the CSGnet, is a list of postings I found particularly informative about the method of levels. For a complete archive of CSGnet postings, see www.PCTresources.com

Carey, Timothy A.:
15 Dec 98   15 Sep 99   16 Sep 99

Gregory, Bruce:
29 Dec 97

Lazare, Mark:
12 Dec 97   23 Apr 99

Marken, Richard S.:
21 Feb 95   14 Apr 98   12 Jan 98
04 Feb 00   01 Apr 98   16 Oct 00

Powers, Mary:
04 Feb 00   10 Sep 99

Powers, William T.:
03 Mar 91   30 Aug 95   26 Dec 97
27 Jan 98   19 Jun 00   27 Aug 95
17 Nov 95   29 Dec 97   23 Nov 98
27 Jun 00   28 Aug 95   04 Dec 96
04 Jan 98   16 Sep 99   29 Oct 00
29 Aug 95   25 Dec 97   12 Jan 98
30 Jan 00

Chapter 31

Psychotherapy

Like Chapters 29 and 30, this chapter will be about giving and getting help at the higher levels—chiefly at the levels of system concepts such as the self-concept (or self-image) and the level of principles such as morality. We get help at those levels from parents, clergy, teachers, fellow members of 12-step support groups such as Alcoholics Anonymous, and astrologers. Increasingly, we get help also from secular professionals and semi-professionals such as psychiatrists, clinical psychologists, and social workers, to name a few. Kirk and Kutchins (1992, p. 8) reported that between 1975 and 1990, the number of psychiatrists in the United States grew to 1.4 times their earlier number—from 26 to 36 thousand. The number of clinical psychologists grew 2.8 times to 42 thousand, of clinical social workers 3.2 times to 80 thousand, and of marriage and family counselors 6.7 times to 40 thousand. Considering the growth in those professions (all the rates are much greater than the growth of the general population) I was not surprised to come upon an article in TIME magazine by Wendy Cole (2000) entitled "The (Un)Therapists" describing the work of people who call themselves coaches. The members of this new (or newly named) profession are not the managers of athletic teams, but simply people who offer help of any sort to other people. Coaches Cole says, do most of their coaching over the telephone. Any matter of concern is grist for a coach. No license, diploma, or certificate is required.

In this chapter, I examine the elusive benefits of psychotherapy, and especially the highly institutionalized forms of therapy that require the doctoral degree for their practice—chiefly psychiatry, clinical psychology, and clinical social work. I will present evidence that attempts to help people do (on the average) help people, but evidence also that having more training or experience, or even having specially licensed kinds of training, does little if anything, on the average, to increase the benefit. Furthermore, I will offer some reasons to doubt the very existence of some of the "illnesses" from which many psychotherapists want to rescue us. Psychotherapy is an example of inventiveness at the very highest levels of control. Its conception results in very complex patterns of social interaction.

Wendy Cole wrote, "The field [of coaching] seems to be prospering precisely because it is not therapy." I will not quibble about what should be called therapy and what should not. Nor will I be talking in this chapter about helping with troubles we get into with malfunctioning or misapplied *things*—automobiles, houses, or can openers—but about helping with the troubles we get into with other people or ourselves. I will be dealing with the realm of clinical psychology, a good part of psychiatry, most of psychoanalysis, some of counseling, and no doubt some of coaching—the realm at the higher levels of control that makes use of language and has control of the kind of perception we call, roughly, understanding. For that realm of trouble, I think the term "psychotherapy" is close enough. I also include the method of levels (see Chapter 30) under that heading.

I will focus here on modes of interaction between therapist and client when the therapist is a member of a mental-health profession. I will devote the most space to diagnosis and efficacy. By an effective procedure, I mean that after I have done it, I like the resulting condition better than the earlier condition. A therapist calls her therapy effective if she (the therapist) likes the behavior of the client better after the therapy than before. Typically, any sort of therapist will consider a therapy to be effective if the client comes to feel relieved of internal conflict or of behavior in which the client was engaging even though he (the client) deplored it afterward. From the viewpoint of PCT, the client engaging in deplored activity is suffering internal conflict—he wants to do it and simultaneously does not want to do it. That is, the client shows up with a worry or behavior of which he wants to rid himself. As Timothy Carey (2000, p. 15) puts it, "the common ingredient... seems to be a feeling of distress."

If, after some therapy, the behavior of the client seems to the therapist to indicate that the client is suffering less conflict than before, the therapist will usually claim the therapy to have been effective. That is usually true even if the chief or only kind of behavioral evidence the therapist has observed is the report by the client that he feels less conflict. I am not belittling that evidence. A report from the client is usually the best evidence available that something useful to the client has happened and is likely to continue.

From the viewpoint of the PCT therapist (or guide), the client is hunting for some way to resolve a conflict. Using the method of levels (MOL) the task the therapist accepts is that of helping the client to turn his attention to that higher level when the utterances of the client hint that the client is about to do so. The MOL therapist does not claim to know better than the client the internal standards that may be in conflict nor the most effective level of the control hierarchy from which to set in motion a reorganization.

From the viewpoint of a traditional therapist, there is something malfunctioning inside the client, and the therapist's task is to ascertain what is malfunctioning and then to do something to correct it—to prescribe a treatment to bring back correct functioning. The assumption of the therapist is that she (the therapist) knows more about what can be malfunctioning and what can be done about it than the client does. Implicitly or explicitly, the therapist invites the client, too, to adopt that assumption.

From that traditional viewpoint, the first task is to perform a diagnosis. Doing that requires a list of possible malfunctions—a list of names of things that can go wrong. Therapists construct their lists from various sources. One source is the therapeutic theory to which the therapist subscribes—Freudian, Rogerian, rational-emotive, cognitive-behavioral, pharmaceutical, or whatever. Another source is the range of environmental circumstances with which the clients must cope. A therapist with an office on Park Avenue in Manhattan will have a list of diagnostic categories different from a psychiatric social worker with clients on welfare.

Because lists with such origins are highly idiosyncratic, many therapists and researchers have wanted to have a standard list to which they could turn. Such a list might increase the agreement among therapists and among researchers on the diagnosis. The list would also enable therapists to say to insurance companies and to judges that they are conforming, in their diagnosis, to the best standards of their profession. Over several decades, several lists of diagnostic categories have been compiled and reshaped into the volume now published by the American Psychiatric Association—the *Diagnostic and Statistical Manual of Mental Disorders* (DSM-IV, 1994). I will say more about the DSM below.

The second task of the traditional therapist is to choose a treatment to go along with the diagnostic category chosen. If you are a historian of psychological research or of psychotherapy, you can say a great deal about traditional psychological treatments. Library shelves sag with tomes on the topic. If, however, you want to know where to find an effective mode of psychological helping, the library shelves will not help you much. One kind of treatment seems to be about as effective (or ineffective) as another. I will say more about efficacy below.

You will have noticed that in the method of levels (MOL), the concepts of diagnosis and treatment vanish. You might be tempted to say that both diagnosis and treatment go on at the same time during the helping conversation. You might wish to say that the client does his own diagnosis and chooses his own treatment. I prefer to say that the concepts of diagnosis and treatment simply get in the way. They are conceptual inventions built on analogies from the practice of medicine that make more trouble than they are worth. Yet on their face, they seem to make sense. Can we simply jettison them with no loss? My proposal to discard them may seem outrageous, but remember that I am talking about the overall functioning of a living creature. Diagnosis and treatment are certainly useful concepts for repairing automobiles, radios, TVs, and even the physical functioning of living bodies. But not for helping people resolve their conflicts.

POSTULATES

A physician looks for conditions or actions on the part of the client that she thinks are good things or bad things for a body to be doing. That attitude yields useful results when examining physical functioning. Is blood leaking from your body? That's bad! Let's stop the leak! Nobody is going to complain about that attitude toward that kind of event. The goal of keeping the blood in the body so it can nourish and clean our tissues is a goal everyone can cherish. Of course, some ways of stopping the leak are better than others. Ways that help the body's own feedback loops to manage the healing of the wound are generally best. It took a while for surgeons to learn about the biological standards maintained by the immune system. There are some ways of helping that the body will reject, no matter how good your intentions may be—no matter what *you* think are good and bad things for that body to be doing. Nevertheless, in dealing with physical events and simple processes (such as leakage, breakage, or too hot or cold) it often helps to bring conditions back toward "normal." If you have a broken leg, a simple crutch is helpful. If you are shivering, a warm coat is welcome.

With psychological matters, the attitude of looking for good and bad events will not work nearly as well as it does in physical medicine. Particular actions (events) do not tell us (they are not diagnostic) about neural functioning. If a client walks down the street shouting about her handsome father, that act gives you no information about the functioning of her control hierarchy.

If something or someone makes it too difficult for us to take some particular action, we will find another to serve the same purpose. If I cannot find some legal kind of employment that will bring me money, I will beg, borrow, steal, or kill to get it. (I don't mean to imply that only unemployed people beg, borrow, steal, and kill; kings and captains of industry do, too.) The fact that we do not like the actions of a person does not mean that his feedback loops are defective.

In short, whether diagnosis and treatment seem reasonable parts of psychological helping will depend on your theory of human functioning. In Table 31–1, I have tried to summarize the crucial postulates of traditional theory and of perceptual control theory.

In Table 31–1, I use the term "traditional theory" to include both "scientific" or scholarly theory on the one hand and popular, common, folk, or general theory on the other hand, since both classes of theory use the postulates I list here at left. But now I must be careful of how I write. The table says that the two classes of theory, traditional on the one hand and PCT on the other, postulate certain characteristics of living things. But only people can postulate; theories merely lie there on the page after people have written them there. If you go to an actual, live, human therapist and ask whether my table describes the theory to which she subscribes, she will very likely say something like this: "Well, mostly, yes, except. . .". A table like this can be only an average or majority-vote characterization of the way people in a category talk or act.

I mean to say in Table 31–1 that when therapists or researchers act in certain ways, they are acting *as if* those postulates are faithful descriptions of reality. For example, to look for a diagnostic category in the traditional manner is to look for an entity, a "thing," syndrome, or condition that can be caused in a linear sequence. Let's say that you perceive a pretty good match between the behavior you have seen on the part of the client and a category of "depression" that you found in the Diagnostic and Statistical Manual. When you look for a thing like that, you are acting as if there is some sort of causal thing inside the person that is causing the symptoms (actions) specified in the DSM for the category of depression. And you are acting as if that is the end of the story. The story goes that the thing (such as depression) causes a cluster of actions, so if you do something to make the actions stop, you will know that you have somehow caused the unwanted thing to go away or at least to become unable to bring about those unwanted actions. You are *not* acting as if those unwanted actions—unwanted either by the client or the therapist—are themselves causes of some perceptions, wanted or unwanted. The possibility that the actions are purposely taken by the client to aid the control of some perceptions, but at the same time cause a worsening of the control of others, thus producing a conflict, is not part of the causal postulate.

To expect observed actions to tell a story about internal standards and processes is to forget (or be ignorant of) the Requisites for a Particular Act (which you can find reprinted in the Introduction to Part VI). Actions are chosen (even if unconsciously) for a purpose; they are not simply emitted by some internal

Table 31–1
Postulates of Traditional Theory and of Perceptual Control Theory of living creatures

Traditional Theory

Causation:
    linear, straight-line, from input to output. It is episodic: a stimulus or influence sets off a response action.

Unit of observation or analysis:
    the unit or episode begins with an environmental event, condition, or stimulus and ends with an organismic response or action. The responses, whether by one person or many, are assumed to be interchangeable for purposes of interpreting the data.

Origin of action:
    Action is set off either by (1) an internal urge (motive, need) or (2) an external stimulus (event, condition). Most subtheories prefer one or the other of these origins.

Purpose:
    Strict behaviorism rejects purpose. Other subtheories recognize purposiveness in action but do not make a technical concept of it.

Levels:
    A few subtheories postulate levels of motives but are vague about relationships between them.

Research:
    The count of arbitrarily specified actions (sometimes the count from a single person but usually the pooled count from several persons) authorizes further research into a hypothesis. Comparison with a working model is not required.

Perceptual Control Theory

Causation:
    circular, each function in the control loop being both a cause and an effect. Continuous: control never ceases.

Unit of observation or analysis:
    none; observation and action are both assumed to be continuous. Observation necessarily begins and ends at arbitrary times. Each action is assumed to be new (or continuously changing), providing its part of continuous control.

Origin of action:
    Action arises simultaneously from both the organism (because of the internal standard or reference perception) and the opportunities for control available in the environment.

Purpose:
    is given by the reference perception.

Levels:
    Higher levels determine the reference perceptions at lower levels. Perceptions at one level are orthogonal to those at another. Intrinsic (genetic) reference signals bring about reorganizations independently of the control hierarchy.

Research:
    The usefulness of a hypothesis is determined by the closeness of fit by the model to the observed control by a single organism. The Test for the Controlled Quantity can be used to hunt for controlled variables. A count of responses is not taken.

"thing"—depression, schizophrenia, personality disorder, or anything else. On this topic, Carey says (2000, pp. 15–16):

> Currently it appears that the "symptoms"... are seen as the problem.... [That view] is weakened by the fact that there are people who can have the same kinds of actions, thoughts, or feelings who *aren't* bothered by them at all. Some people, for example, might like to stay in their houses and not go out, others might hear sounds or see sights that other people don't see, others might experience unusual sensations, others might feel down and gloomy much of the time.
>
> It doesn't seem, therefore, that experiencing any particular state is the problem. Rather, the problem seems to be experiencing a particular state [while at the same time] not wanting to experience that state.... In other words, mental illness seems to occur [for example] when people stay home a lot *and* don't want to stay home a lot....
>
> Mental illness [means] behaving or thinking or feeling in *any* [emphasis PJR's] particular way *and* not wanting to. It is the inability to alter one's current experiences [because of the conflict] that I believe constitutes mental illness, *not* the current experience itself. It makes little sense, then, to spend time deciding what particular condition ... a person has. All that needs to be known is whether or not the person is experiencing something they would like to change.

You might also want to look back at the quotations in Chapter 29 (under "Psychological Effects") that I took from some traditional writers concerning the social origins of psychopathology and disorders.

In contrast to the traditional view, you can try to help the client find his own more effective higher-level control, not specifying for him the actions he ought or ought not use to achieve control. You are then acting as if causation is circular, continuous, and a joint result of internal standards and external opportunities. Those are the causal postulates behind the MOL.

From the viewpoint of PCT, looking for a particular internal cause for particular external actions or any cluster (syndrome) of them is a mistake. But I do not want to leave you with mere reasoning from postulates. Let us examine the success of therapy that proceeds from diagnoses and corresponding treatments. The rest of the chapter will deal with that research. How have diagnosis and treatment fared?

DIAGNOSIS

A great deal of research has been done on the reliability of categorizing clients with the third (1980) and fourth (1994) editions of the Diagnostic and Statistical Manuals. Reliability turns out to be low, much lower than can justify decisions about the likely functioning of an individual client. (The logic of the matter is the same as that I described under "Correlations" in Chapter 26.)

When two therapists put the same label on what a client is doing (maybe Dysthymic Disorder or Agoraphobia), we could call it simply agreement, but the technical term is interrater reliability. Obviously, if the two do not agree, one or both of them must be wrong. For this reason, researchers say that reliability puts a limit on validity. You cannot assess accurately the efficacy of aspirin if some of the pills you give the patient are not aspirin but actually sugar. Or maybe arsenic.

The origination of diagnostic categories in psychiatry is generally laid at the door of Emil Kraepelin (1856–1926), of the University of Heidelberg. In the United States, the American Medico-Psychological Association, with the cooperation of the National Committee for Mental Hygiene, published in 1918 the *Statistical Manual for the Use of Institutions for the Insane*. By 1942, it had gone through ten editions. The American Psychiatric Association published its *Diagnostic and Statistical Manual of Mental Disorders* (DSM-I) in 1952, DSM-II in 1968, and DSM-III in 1980. Nowadays, the guide chiefly used by psychiatrists and many other therapists for diagnostic labeling is the 4th edition (DSM-IV, 1994). Research has been conducted on the reliability of the classifications proposed in both DSM-III and DSM-IV.

Stuart A. Kirk and Herb Kutchins wrote a book (1992) called *The Selling of DSM: The Rhetoric of Science in Psychiatry*. It is devoted almost entirely to the DSM-III; the DSM-IV came out two years later. The comments of Kirk and Kutchins about the DSM-III, however, apply easily to the DSM-IV. Like its predecessor, the DSM-IV assumes linear, episodic causation from stimulus to response. Kirk and Kutchins (p. 6) quote from a 1985 article by Robert Spitzer, the principal author of the DSM-III, and R. Bayer:

> The adoption of DSM-III by the American Psychiatric Association has been viewed as marking a signal achievement for psychiatry. Not only did

the new diagnostic manual represent an advance toward the fulfillment of the scientific aspirations of the profession,... (p. 8).

And they quote Gerald Klerman, "the highest ranking psychiatrist in the federal government at the time," as saying to the Association at its national convention in 1982:

> The decision of the APA first to develop DSM-III and then to promulgate its use represents a significant reaffirmation on the part of American psychiatry to its medical identity and its commitment to scientific medicine (p. 8).

During the development of the DSM-III, a conference was held to get comment on what had been written so far. Kirk and Kutchins quote the *Psychiatric News* of 18 March, 1977:

> Out of that conference came [various changes]. Spitzer said that none of the changes... is political in nature. "We have strongly and successfully resisted any changes in the draft DSM-III not based on good sound knowledge" (p. 106).

Kirk and Kutchins then comment (p. 107) that some of the changes arose from other than scientific considerations. For instance, after the American Academy of Psychiatry and the Law protested that the concept of "Sexual Assault Disorder" would provide a defense for rapists in criminal trials, that "disorder" was dropped from the DSM-III. The Academy's reasoning could be used to demand that the American Chemical Society remove arsenic and lead from the Periodic Table of the Elements on the grounds that those elements could be used by murderers..

Several field trials of the DSM-III were conducted. In the results, some of the diagnostic categories continued to show very low reliability. Nevertheless, the developers of the DSM-III argued that the new system was much better than the old one (p. 133). None of the psychiatric community complained.

In analyzing data from the field trials, the researchers used a statistic of agreement called kappa instead of the more commonly used product-moment correlation statistic. The justification given was that kappa contained a correction for chance; that is, kappa is the ratio of the actual beyond-chance agreement to the maximum possible beyond-chance agreement. Kappa, however, has some severe disadvantages. The value of kappa varies not only with beyond-chance agreement, but also with the base rates of the presumed disorder. (I wrote about base rates in Chapter 25 under "Base Rates.")

My own thoughts upon reading the account by Kirk and Kutchins were, first, that diagnostic categories justified by an index yielding an unascertainable reliability (the kappa) have actually been in use by psychiatrists for more than 20 years, and second, that even if diagnosis with the DSM were completely reliable, we would still know nothing about the efficacy of psychotherapy. A high reliability tells us that a measure *can* give us accurate (or valid) information, but it cannot tell us that the measurement actually is accurate.

Still, the converse is important information. Validity cannot be higher than reliability. If reliability is poor, your confidence that you have made a correct diagnosis must be correspondingly low. If reliability is poor, your confidence in your diagnosis can be high only if you are ignorant of the low reliability or are disdainful of it. Both the ignorance and the disdain, sad to say, are widespread among psychiatrists and clinical psychologists—a statement amply attested by Kirk and Kutchins (1992) and by Dawes (1994).

Despite the ambiguity of the kappa statistic, the principal author of the DSM-III, along with two co-authors, wrote in 1979 (pp. 816–817) that a kappa of 0.7 or above indicated good agreement among diagnosticians on whether the patient has a disorder. For argument's sake, Kirk and Kutchins accepted the kappa of 0.7 as a criterion for judging the reliability of the various categories. Concerning the large class of categories labeled "Major Mental Disorders," they said this about the major diagnostic categories under that heading:

> ... 31 of the kappas are above the mark [0.70] and 49 are below their established level. The case summary results are striking: Not a single major diagnostic category achieved the .70 standard (p. 143).

Concerning the large class of categories labeled "Personality Disorders and Specific Developmental Disorders," Kirk and Kutchins said this:

> Only one of the seven individual kappas reached the .70 level. None of the overall kappas ... did (p. 143).

Commenting on the reliability of the two large classes of categories I have just mentioned, Kirk and Kutchins pointed out that the overall reliability did not reach the self-imposed .70 standard. Nevertheless, no one involved with producing the DSM-III seems to have been disconcerted. Kirk and Kutchins told about many other deficiencies in the field studies of reliability and then went on to tell about further studies of the reliability of the DSM-III conducted after its publication in 1980. Summing up those further studies, they said:

> ... following the publication of DSM-III, a broad scattering of studies with both adults and children reported data that were not better and usually somewhat worse than the reliability levels reported in the field trials. More significantly, neither the data from the field trials nor the other studies strongly supported the remarkable claims that were made about the reliability of DSM-III (p. 156).

To put some flesh on the statistics I have been giving you, I will condense (severely) a story with which Kirk and Kutchins open their book (1992, pp. 1–3). In 1986, a man in Virginia was declared by a court to be mentally ill and legally incompetent. The parents of the man disapproved of his political activities and his financial contributions to a conservative political organization. The critical issue for the court was the man's legal competence, and in that regard the court relied heavily on the psychiatric experts hired by the contending parties. The psychiatrist hired by the parents of the man testified that the man was mentally ill and suffered from a schizoaffective disorder. The judgment of legally incompetent meant that the man was not allowed to manage his own money, was not allowed to vote, needed a court to validate his marriage of three years earlier, and made him and his wife fearful of having children for fear the husband's parents would try to take them away.

How is having a "schizoaffective disorder" connected with managing money and the rest of those prohibitions? Nobody knows. Here is what the DSM-III itself says about schizoaffective disorder:

> The term Schizoaffective Disorder has been used in many different ways since it was first introduced, and at the present time there is no consensus on how this category should be defined (DSM-III, p. 202).

If you or I use a term on which there is "no consensus," that means that when one of us uses the term, the other can have no idea what the speaker is talking about. The man in Virginia was sentenced to a severely restricted life, including deprivation of several civil liberties, because of the testimony of a presumed expert who, according to the DSM-III itself which the expert was using to justify his diagnosis, did not know what he was talking about. Kirk and Kutchins have other similar tales to tell. Robyn Dawes (1994) tells still more; I will mention his book again below. I should also mention Dawes's (1994, pp. 152–152) warning. If you find yourself facing a risk because of a possible "evaluation" (labeling) by a psychologist, "walk out of that psychologist's office.... Immediately consult a lawyer...". Tell the lawyer about Dawes (1994) and Kirk and Kutchins (1992).

Rosenhan (1973) carried out a dramatic demonstration of the low validity (accuracy) of psychiatric diagnosis. The story has been retold by Suarez and Mills (1982, p. 60), by Sarbin (1982, p. 153), and by Kirk and Kutchins (1992, p. 93). Rosenhan and seven volunteer acquaintances entered mental hospitals as patients complaining of an auditory hallucination that said "empty, hollow, and thud." Otherwise, Rosenhan instructed his friends to be completely truthful and to behave normally. The "patients" stayed in the hospitals between seven and 52 days, averaging 19. Since some of the participants entered as patients more than once, there were 12 admissions. Eleven cases, Rosenhan wrote on page 258, "were diagnosed by hospital personnel as schizophrenic, and one, with identical symptomology, as manic-depressive psychosis." When discharged from the hospital, the "patients" were diagnosed as schizophrenia improved, schizophrenia in remission, or manic depressive in remission. The only persons to recognize that the "patients" were faking were actual patients.

There was a sequel to that investigation. Rosenhan was challenged by the staff of a hospital not part of the first study to send one of his "patients" to this hospital without telling anybody when the "patient" would appear for admission. The challengers were confident they could pick out the ringer when he showed up. During the period of this study, 193 persons were admitted as patients to this hospital. Twenty-one percent of the 193 were diagnosed by at least one staff member as being the ringer, 12% by at least one psychiatrist, and 10% by at least one psychiatrist and at least one other staff member. (Those groups overlap; that is, the 12% and 10% are included in the 21%.) As it happened, the assigned "patient" fell ill just before the period of the study and never appeared at the hospital. Sarbin (1982, p. 153) quotes

Rosenhan: "... we have known for a long time that diagnoses are often not useful or reliable, but we have nevertheless continued to use them. We now know that we cannot distinguish insanity from sanity."

As window dressing, a heavy tome such as the DSM that categorizes everything you can think of and then some, and labels some of them with names ringing with Greek and Latin, serves very well. At the very least (and, as it turns out, at the very most) the tome provides a handy list from which one can select a label with which to brand the client—or the defendant, as the case may be. Neither the DSM-III or the DSM-IV, however, should give anyone any confidence that those labels correspond to any tangible reality. The same sort of comments you have seen here about the DSM-III can be extended, essentially, to the DSM-IV; I will make a few further remarks about the DSM-IV later in this chapter under "Reification."

EFFICACY

As an introduction to the topic of whether psychotherapy does any good, I offer you the paragraphs below, taken from the magazine *Science 80* Vol. 1, No. 6, p. 7):

> In Nigeria most mental illness is blamed on witches and wizards, while more sophisticated Africans ascribe their sickness to orbiting satellites.... In developing nations like Nigeria and India the odds [nevertheless] seem vastly to favor a relatively fast and complete recovery from major psychoses. But in the United States and other industrialized countries, nearly half who suffer psychotic breakdowns never recover. These findings result from a seven-year World Health Organization pilot study of severe mental illness in nine countries around the world, the first multicultural study of its kind. Focusing on schizophrenia, investigators found that though the disease's symptoms are remarkably similar worldwide, its outcome varies greatly.

The finding that the symptoms of the "disease" were "remarkably similar" is no surprise. If you go looking for a particular clustering of symptoms (There! There's one! There! There's another!) you are bound to find that the instances you have picked out as being the looked-for clustering are all very like the clustering you were looking for. But reliability and validity are of little moment for my purpose here, which is merely to illustrate cultural differences in the rates of "success," however judged, in coping with a person whose behavior has been distressing the neighbors, whatever label one might put on the behavior. The article continues:

> Fifty-eight percent of the Nigerian patients studied and 51 percent of the Indians had but a single psychotic episode, then were judged cured. In the more technologically advanced countries, the prognosis is far worse: The proportion of patients who recovered after one episode ranged from only six percent in Denmark to 27 percent in China.
>
> Many who worked on the study agreed that where the stigma against mental illness is most pernicious, as it is in the West, patients tend to do poorly.

That article seems to be saying that the World Health Organization found evidence that psychotherapy was beneficial in greatly differing cultures and, at least in the case of the symptoms sought by the investigators, was much more beneficial in some other cultures than in the United States, even when the therapy in the other culture took a form that would surprise us here; for example:

> In Sri Lanka, an agitated psychotic is tied to a post and "counseled" around the clock by family, villagers, and faith healers who listen sympathetically to the patient's complaints.

In 1994, in a book in which he reported a great deal of research on the efficacy of psychotherapy, Robyn Dawes[1] wrote:

> Psychotherapy works overall in reducing psychologically painful and often debilitating symptoms. The reasons it works are unclear...(p. 38).

There should be no question that people can help one another with their troubles, even with troubles that seem forbiddingly complex and profound. Still, there can be some doubt about *how much* we can help one another, about how much we *do* help one another, about whether some kinds of help are better than others and better in what ways, and so on.

PCTers, many of them, believe that systematic investigations of the MOL will lead us to a much better understanding of psychotherapy than now exists. The investigations, however, are yet to be done. I cannot now refer you to any research documenting therapeutic outcomes of the MOL. Be that as it may, it is important to be aware that the average outcomes of existing techniques of psychotherapy have been well documented and do show "overall" (on-the-average) beneficial effects.

In 1977, Smith and Glass published an article reviewing 375 studies of the effectiveness of psychotherapy. The studies examined a variety of evidence of the subjects' well-being or reduction of symptoms. The degree of benefit was assessed by subtracting the average score of a non-treated group from the average score of the group receiving therapy. The mean differences from the studies became the data analyzed in the overall study. Robyn Dawes (1994, p. 51) wrote about the overall study that:

> Smith and Glass found that someone chosen at random from the experimental group after therapy had a two-to-one chance of being better off on the measure examined than someone chosen at random from the control group. That is a very strong finding—stronger, in fact, than findings for most medical procedures and for comparisons of healthy versus deleterious lifestyles.

That outcome was reassuring to a lot of psychologists, but other findings from the study were unwelcome. The study

> ... concluded that three factors that most psychologists believed influenced this efficacy actually did not influence it.
>
> First, they discovered that the therapists' credentials—Ph.D., M.D., or no advanced degree—and experience were *un*related to the effectiveness of therapy.
>
> Second, they discovered that the *type* of therapy given was *un*related to its effectiveness....
>
> They also discovered that length of therapy was unrelated to its success (Dawes 1994, p. 52).

Training and Experience

As you might suppose, it was not long before other researchers gathered further research reports to see whether those three findings would again be found. They were again found. Dawes (1994, p. 55) quotes from a later study by Berman and Norton (1985):

> ... we found that professional and paraprofessional therapists were generally equal in effectiveness. Our analyses also indicated that professionals may be better for brief treatments and older patients, but these differences were slight....

> When we classified studies according to the four most commonly occurring categories of patient complaint (social adjustment, phobia, psychosis, and obesity), we found no reliable differences [between professionals and paraprofessionals].... We also failed to detect any systematic differences when we divided the studies into five forms of treatment (behavioral, cognitive-behavioral, humanistic, crisis intervention, and undifferentiated counseling).

Dawes (1994, p. 56) also tells of a famous study by Strupp and Hadley (1979):

> They recruited as therapists university professors who had no background in psychology and randomly assigned clients either to them or to professionally trained and credentialed psychologists.... fifteen clients to the professionals and fifteen to the professors. The clients were those whose problems ... "would be classified as neurotic depression or anxiety reactions. Obsessional trends and borderline personalities were common." The professionals charged higher fees, but they were no more effective as therapists than the professors.

Professionals of all sorts like to think that they get more skillful as they get more experience—that 20 years of experience is not, so to speak, one year of experience repeated 20 times. The yellow pages of the telephone book are spattered with proud pronouncements like "20 years' experience!" and "Founded 1922!" I suppose that practitioners in some occupations do get more skillful as the years go by. On the average, however, psychotherapists do *not* become more successful in treating their clients as they get more experience. Nor do they become more accurate in their diagnoses. Dawes (1994, p. 108) tells of a study by Faust and others (1988). They asked some clinical neuropsychologists from various parts of the country to evaluate the written results of tests of ten people known to have suffered specific types of brain injury, or known to have suffered none. Faust and colleagues wrote:

> Except for a possible tendency among more experienced practitioners to overdiagnose abnormality, no systematic relations were obtained between training, experience, and accuracy across a series of neuropsychological judgments.

Ornstein and Ehrlich (1989, pp. 141–142) mention a study by Ernest Poser, who "found in treating schizophrenic patients that randomly selected undergraduates with almost no training used as therapists produced more positive change than did psychiatrists and psychiatric social workers."

Dawes (1994) gives many more examples of research than I have reported here, but I hope I have provided enough for you to take seriously the evidence that psychotherapy of several sorts does do some good of various sorts. I hope you will take seriously, too, the conclusions that there is no convincing evidence so far on what feature of therapy brings the benefits, that the skill of therapists at recognizing pathology is generally poor (though misdiagnosed patients often get better anyway), and that neither greater training nor greater experience improves the competence of therapists. And finally, it turns out that psychotherapists think they are considerably more expert than the evidence shows them to be. You will remember the demonstration by Rosenhan that I described earlier in this chapter; the psychotherapists in the later hospital said, in effect, "OK, you fooled those other guys, but you can't fool us!" But he did. Dawes (1994) gives further damning illustrations throughout his book.

I am sure I have not encountered all the books that give evidence about the thin benefits of present-day psychotherapy. I have not tried to be exhaustive in my search. But just the other day, I happened across one by Donald Eisner (2000), entitled *The Death of Psychotherapy: From Freud to Alien Abductions*. I read only the foreword, by Tana Dineen, who said:

> [Eisner] has carefully and meticulously presented the evidence . . . that, despite the claims and supportive studies, [the popular varieties of psychotherapy] all lack the essential proof of effectiveness. He doesn't deny that some, even many, people feel better after talking to a therapist. But he does challenge that this is due to anything special or specific to any of the treatments.

It may be that one reason training and experience make little difference is that therapists and their teachers believe that the therapist must "understand" the client if the therapist is to be helpful, and the teachers spend years teaching the budding therapists how to "understand" the clients. But the research I have described above shows clearly that all that theoretical training helps the therapist not at all.

My own experience with organizational development fits nicely into the findings of Smith and Glass. At various times, I worked or conferred with teams of consultants. Invariably, when proposing a strategy for the project to the rest of us, each consultant couched the proposal in his or her own concepts and theory. It became clear to me after some years of listening to my colleagues, that although some of them carried similar theories in their kit-bags, no two theories were quite alike, and some seemed plainly to contradict others. None of those similarities or differences, however, enabled me to tell whether a consultant was going to do a good job, work well with other consultants, or behave ethically. Insofar as our theories about the effects of our own behavior had any connection with our efficacy, we all could just as well have left our kit-bags at home. I thought some of my colleagues were more effective than others, certainly, but I could not see that the difference had anything to do with the theory (type of consulting), type of training the consultant had undergone, or the consultant's years of experience.

Since there is little evidence that any one sort of psychotherapy is better than another, it is not surprising that there are fads in methods of treatment. In *Science News* for April 2001, Bruce Bowers reported that there was "renewed interest" among therapists (perhaps especially among psychiatrists) in non-drug therapies for schizophrenia. The rates of benefit from various treatments quoted by Bowers brought no surprise: whatever the treatment, some patients benefit and some don't. The report also showed the usual bizarre reasoning to be found in research on therapy. In one paragraph on page 268, Bower wrote, "Merely defining the disease [schizophrenia] has evoked a century of controversy." A few paragraphs later, he wrote, "The causes of schizophrenia remain unknown."

The presumed experts have been arguing for at least a century, Bower says, how to recognize the presumed condition they call schizophrenia. Correspondingly, all studies of labeling patients show that agreement is very poor among therapists about whether to put the label "schizophrenia" on a patient. That is, psychiatrists and other therapists cannot agree among themselves at all well on whether they are seeing schizophrenia. How can Bower or anybody else expect to find causes of such an elusive condition? My reason for asking that rhetorical question is that validity (effectiveness) cannot exceed reliability.

How can you study the cause of something that won't hold still, so to speak, for you to see whether it is there? A good many psychologists refuse to believe that schizophrenia exists, and studies of the reliability of diagnoses repeatedly show that nobody should bet on getting five therapists together who can point at any one patient and agree that the patient "has" schizophrenia. Yet one sees article after article in presumably scientific journals presuming to report on the "etiology," "determinants," "origins," or causes of schizophrenia and other equally imaginary "illnesses." I call that bizarre reasoning.

In sum, psychotherapy serves here as an example of institutionalized helpfulness—a sort of specialization of labor that appears when a civilization contains so many people and such a rich technology that the daily activities of most people are drawn away from family and clan. People then often reach out to strangers for help. Since the civilization can afford to support professionals who do nothing but help other people with their worries, people often reach out to those professionals. Given the stresses of our civilization, we certainly need all the help we can get with living in it. We will use the help to better advantage, however, if we understand the interweaving of professional and homespun efforts. In Chapter 29, I mentioned that George Albee (1992) urged us to use the homespun efforts more often.

Captive Clients

You don't have to do a lot of scientific research to know that people who do something from their own desire almost always do it more happily and to their greater benefit than people who are coaxed or threatened into doing it. You don't have to do a lot of research, either, to notice how often people forget that fact when they think of something they think would be good for other people to do. I have lost count of the times I have seen a school or college offer an elective course, seen it attract students, seen the students praise the course and recommend it to their friends, seen the school or college then make the course a required one, after which students complained right and left about having to take it, and faculty wondered why students didn't like a course with which other students were once so happy.

The same thing happened with sensitivity training in the 1960s in the United States. That technique of using a small social environment of about a dozen strangers to report to one another their perceptions of the effects of the interaction going on—that procedure had attracted considerable notice from group consultants and therapists of one kind or another. Most participants were speaking favorably about their experience; some were reporting life-changing insights. Many participants and others who listened to the happy participants thought that the experience would be good for other people. Managers and counselors in businesses, churches, schools, hospitals, prisons, recreational organizations, and others began to corral employees, inmates, and members and subject them to a series of sensitivity sessions. Somehow those employees, inmates, and members were not at all as enthusiastic as the volunteers had been; indeed, many gave very poor reports. It was not long before sensitivity training began acquiring an unfavorable reputation.

The same thing happens to psychotherapy of any sort. It is much more effective with volunteers than with people pressed into it. (Note that almost all the research done on the benefits of psychotherapy has been done among voluntary clients.) The effect, I think, is substantially the same as the placebo effect. If you think something is going to do you some good, you are more likely to make good use of it, even unconsciously. The effect is similar, too, to the self-fulfilling prophecy. It is easy to see that the MOL would work poorly, if at all, with an unwilling client. The unwilling client would be unlikely to respond cooperatively to a request from the guide to "tell me more about that." Walter A. Brown (1998) has written an enlightening article about the conscious use of the placebo effect with medical patients; the same principles, I think, would apply to psychotherapy. I am not saying, by the way, that there are no willing clients in organizations, even prisons. I am saying that I do not envy a therapist saddled with unwilling clients.

Surgery and Drugs

Both research psychologists and therapists, most of them, focus on actions rather than on internal functions. There are, however, two kinds of treatments for psychological deviations that affect internal functions directly: surgery on the brain and drugs. When internal functions are altered, actions of many sorts are affected, and the particular actions will be unpredictable, just as in reorganization. The difference, however, between reorganization, on the one hand, and surgery or drugs, on the other, is that reorganization will almost always work to improve overall control, while the effects of surgery or drugs on control are, as far as I know, unpredictable even in that regard. Even in treating physical malfunctions, the physical effects of drugs are only probabilistic. Few drugs help everyone. My medicines always come with a sheet of paper bearing a long list of effects that may come upon me in addition to (or instead of) the curative effect the physician intends. Effects on whole-person behavior are still less predictable. Here is the horrifying tale Dawes (1994, p. 49) tells about the woman with the drunken father:

> [A patient] had taken off all her clothes one day and run around the streets of her home town screaming, "My father is the handsomest goddamn drunk in [X], Pennsylvania!" She was subsequently hospitalized and then was lobotomized within six *weeks*. She often continued to shout, "My father is the handsomest goddamn drunk in [X], Pennsylvania," but she would not take off her clothes or otherwise express her ... feelings toward her father, because she would immediately forget what she was shouting about. Unable to concentrate for more than ten seconds at a time, she was unable to obtain a job ... or live with relatives. She had, in effect, been sentenced to life imprisonment for having expressed her feelings in a socially inappropriate way.

Let us look now at the effect of drugs. I put a section on drugs in Chapter 20, but here is another communication about drugs from Powers, taken from his reply to Tim Carey on the CSGnet on 23 December 1997:

> By whose evaluation is the effect of a drug beneficial? If we allow only the evaluations based on external observations of a person's actions, we get assessments of value from the standpoint of other people who are affected by irrational, neurotic, psychotic, or "abnormal" actions. But we are left in the dark about the ability of the person to control his or her own perceptions.
>
> ... nobody has a clue as to what the abnormal actions were intended to control—it's as though that just doesn't matter. What matters [to the onlookers] is getting the person to act normally, meaning more like the average person who doesn't have the same problems of control.

> When conventional psychologists get the person to desist from the actions they consider unusual or bizarre, they say that behavior has returned to normal. But if those actions were an attempt to control some perception of which the observer knew nothing, are those perceptions now magically under control? Or has the person simply given up on controlling them, or has the whole control process, as you suggest, been suppressed or removed by the action of a drug ... ?

And on 25 December 1997, Powers wrote:

> Haven't I [acknowledged] that drugs are sometimes the only answer we have?
>
> The problem as I see it is that the human biochemical system is just that, a *system*, in which you really cannot affect just one thing with a drug. Even if a drug could be found that had one and only one local chemical effect, arbitrarily changing one variable in a system will inevitably result in affecting others.... If there is a chemical "imbalance," then something created that imbalance as part of doing something else. When you step into this whole system and make arbitrary changes in the middle of it, with no understanding of the larger organization, you may deal with some immediate (apparent) problem and at the same time assure that other, larger problems will never be dealt with.
>
> The psychotropic drug industry is far larger and richer than can be accounted for by pointing to extreme cases where drug treatment got somebody out of an institution or an emergency situation.... What we have is a society gone drug-crazy, with understanding of the human system being pushed aside in favor of an empirical shotgun approach aimed at getting immediate results....
>
> I see a future in which the approach to human problems will be very different; where we can understand the whole system and predict the effects of chemical treatments on the whole system rather than on one attention-getting symptom.

Those paragraphs are rather abstract. To put some flesh and blood on the matter, let me offer what a friend of mine wrote to me not long ago:

> [My daughter] has now been able to live on her own—but with no job or meaningful activity—for two and a half years, after I "rescued her from the system." But I couldn't rescue her from the damned drugs, and even though there's no evidence that any

of them (five in all) does anything but fatigue her and keep her from growing up (not even evidence that the anti-seizure drugs decrease her seizure frequency), she's 26 and I can't legally say no. She can't get too upset for a prolonged period of time because she can't do anything for a prolonged period of time. I have been trying without success to get her into a hospital where all drugs are removed to see if any are doing more good than harm—but I failed again just last week.

The physician Fred Baughman has maintained a site on the World Wide Web where people can write to tell about their sufferings from psychotropic drugs. Here is an example from 8 September 2000:

> . . . I am very concerned about my grandson. I saw your web page and agree that this so-called disease is truly one of the biggest frauds ever perpetrated on innocent parents. My grandson is 14 years old; he was diagnosed with ADHD about 5 years ago. He was prescribed Ritalin. After a year or so on this horrible drug he began developing twitches. The doctors also prescribed him sleeping pills because he could no longer sleep. My wife and I are the ones that noticed his twitching while he was visiting us. We begged his parents to please get him off these drugs. They took him back to the doctor who in turn prescribed him another amphetamine-based drug. This seemed to work for a while but now he seems to be going downhill again. His temper is getting out of control; he has horrible headaches he gets sick at his stomach a lot. He takes about 6 to 8 of these pills twice a day plus the sleeping pills at night.
>
> My grandson was not a bad kid to start with—a little hyper, but that's about it. Now his temper has escalated to the point he becomes uncontrollable. . . . He is not the same sweet boy he was before he was placed on these drugs. This boy comes from a good home with loving parents but they are at their wits end and their doctor just keeps increasing his medication.
>
> These drugs are turning him into a monster. Is there anything we can do?

Dr. Baughman replied to that letter, in part:

> I wrote in the *Journal of the American Medical Association*, April 28, 1999, page, 1490: "As a neurologist, I have found no abnormality (disease) in children said to have ADHD. Once children are labeled with ADHD, they are no longer treated as normal. Once methylphenidate hydrochloride (Ritalin) or any psychotropic (psychiatric) drug courses through their brain or body, they are, for the first time, physically, neurologically and biologically abnormal."

Baughman's e-mail address is fred-alden@worldnet.att.net.

DIAGNOSIS IN PCT

In his book manuscript, the psychotherapist Timothy Carey (2000) says this about diagnosis:

> [T]he current system of assessment, diagnosis, and treatment of mental illness is at best meaningless and at worst detrimental. . . . Placing people into categories based on patterns of behavior and assigning them treatment regimes accordingly makes as much sense as grouping cars according to color and basing mechanical decisions on this criterion (p. 57).
>
> [M]ental illness results from chronic conflict. [C]onflict involves multiple levels of a perceptual hierarchy. [R]eorganization will alter some aspects of [a higher] level, thereby eliminating the conflict (p. 53).
>
> [PCT therapists] would no longer be interested in [predicting actions]. We would be interested in how humans control various sensed states of their environments. We would not look for causes of behavior, and we would not create theories about influences and causes that produce various behavioral outcomes. We would be interested in what the person is experiencing and how they continue to experience what they intend despite fluctuating environmental influences (p. 44).

People distressed by inner conflict are usually functioning with normal nervous systems, regardless of how unwelcome their behavior may be to other people. Some people, however, do suffer with impaired neural functioning. When that is the case (or when it is suspected) diagnosis in the medical sense is useful. Unfortunately, medicine is often confounded with psychological chimeras. Diagnosis can be more penetrating when the diagnostician asks not only what is wrong, but also what is right. More exactly, the diagnostician can ask how well the person is controlling the perceptions he cares about. If control of bodily movements is bad in all situations, or if there seem to be few higher-level perceptions that the person exhibits any wish to control, there is a possibility of a damaged or deteriorated nervous system, and that possibility should be investigated. What the helper needs to know, however, is not what name to put on a "disorder," but instead what levels of control are functioning well. This sort of diagnosis can be useful, because control is a precise and easily observable function.

I will illustrate how diagnosing for levels of control can be done by telling you about research by Tom Bourbon. In June of 1999, I wrote to him via e-mail:

> It strikes me that the PCT manner of assessing performance would be to ask whether the levels are functioning well. Can this person (for example) remember a principle? Follow a program? Make one? Whether the program is a more complicated one than some other, in somebody's judgment, is not very important (unless you are hiring someone to follow a particular program you have made up such as sorting the incoming mail, and you expect never to change it).

On 10 June 1999, Bourbon replied:

> I agree. We would also want to know whether the various parts of the loop are functioning for a single level of control.

I had also said:

> I seem to remember that when you were working in the research department of the medical school, you were thinking up some diagnostic schemes. So no doubt you have an opinion on the gropings I have put in the paragraphs above. Please comment.

And Bourbon told this story:

> Yes, I was working on diagnostic schemes for people who had experienced head injury, stroke, spinal cord injury, and other neurological insults. I used a simple battery of tracking tasks. Some people performed the tasks using a mouse. For those who could not operate a mouse, I constructed a device that allows me to use velcro straps and attach a joystick to various parts of the human anatomy where there is relative movement—around a knee or elbow, or at the waist.

In the next two paragraphs, note how Bourbon looked for levels of the variables the client *was capable* of controlling.

> I studied a group of people with severe head injury. After conventional neuropsychological assessments, all of the patients in the group were deemed equally nonfunctional. The neuropsychologists could not differentiate between them on any of their measures. With my simple tracking tasks, I found a wide range of performance. Some reference perceptions were for nothing more than keeping a disturbed cursor aligned with a stationary target. In another task, the person was to think of a sequence of four different positions of the undisturbed cursor relative to the stationary target, then to create that sequence of relationships.
>
> There were no meaningful differences between what some of the patients did on the tasks and what I did. A few people could not think of a sequence of positions for the cursor relative to the target. A few others could not remember the reference condition all the way through a two-minute run. Others could remember what to do, but some of them could not produce the necessary movements, while others could move well, but could not keep track of what was happening perceptually. I believe those are the kinds of groupings you were talking about.
>
> I had similar results with the other patient groups. I was especially touched by some of the quadriplegics I studied. I remember one young man who had some use of the biceps in his right arm. He sat palm up with his elbow resting on the arm of his wheel chair. I strapped the joystick to his arm so it would operate around his elbow, when he used his biceps to raise his hand. He used the biceps working against gravity to control the lowering of his arm. On every tracking task, his performance was indistinguishable from mine. During one session, he was animated and excited. At the end, he sat silently, then he sobbed, "Why don't those damned neurologists do this? Why don't they let me show what I *can* do, instead of always making me look like I fail?"

That last agonized query shows the difference between diagnosis to find out what is functioning well and diagnosis to find out what is wrong. If you hunt for what is wrong and try to fix the things that are wrong, you may never find out what the person can do for himself (as Bourbon's neurologist colleagues did not). The converse is not true: if you hunt for what the person can do normally, you will run no danger of missing the malfunctions, because you will be testing the capabilities of the whole system. Unfortunately, Bourbon's colleagues were not maintaining perceptions of that sort. Bourbon wrote:

> All of my results were preliminary. They were exciting to me. The clinical neuropsychologists [however] thought the tasks were interesting, but they did not know how to relate my data to the results of their conventional assessments. My department head was unimpressed and decided to set me adrift.

It is not surprising that the neurologists wanted to understand Bourbon's findings within the compass of their own conventional assessments. That kind of interpretation, however, must necessarily fail, as I explained in Chapter 16.

REIFICATION

The question recurs in the literature year after year whether some widely used conceptions of undesired behavior "exist" outside the heads of the diagnosticians. Theodore Millon (1990) of the Harvard Medical School wrote:

> Certainly the disorders of personality should not be construed as palpable "diseases". . . . Unfortunately, most are . . . receding all too slowly into the dustbins of history (p. 339).

Similarly, here is the view of Laing and Esterson (1964) on schizophrenia:

> Psychiatrists have struggled for years to discover what those people so diagnosed [as schizophrenics] have or have not in common with each other. . . . No generally agreed objective clinical criteria for the diagnosis of 'schizophrenia' have been discovered. . . . No consistency in pre-psychotic personality, course, duration, outcome, has been discovered. . . . Every conceivable view is held by authoritative people as to whether 'schizophrenia' is a disease or group of diseases[—]whether an identifiable organic pathology has been, or can be expected to be, found (p. 17).

Despite that report from 1964, you may think that by now someone has discovered schizophrenia in a certain spot in the brain or in a certain gene. Here, however, is Elaine Walker (1993) reviewing a book in which the argument dealt with by Laing and Esterson is still going on:]

> The viewpoints represented in the volume [reviewed] fall into three general categories. One category assumes the existence of schizophrenia as a disease . . . while . . . acknowledging the complexities inherent in research on a disease of unknown etiology [cause]. The second acknowledges the . . . pathology, but raises questions about the . . . conceptualizations. The third questions the very existence of the syndrome and the relevance of the medical model. . . (p. 951).

> For example, in Chapter 13 Theodore Sarbin states that . . . "The availability of the diagnostic term, schizophrenia, like the availability of its superordinate, mental illness, is useful . . . in . . . controlling persons whose conduct is unacceptable to others". . . . He goes on to argue that the consistent increases in the number of diagnostic entities with subsequent editions of the [DSM] are driven by arbitrary, often political, considerations (p. 952).

Three hundred years ago, it seemed entirely reasonable to most people to believe that behavior abhorrent to the local priest was caused by some of Satan's assistants who had taken possession of the person's will. It seemed reasonable, too, that the little devils had to be driven out by making the person's body a very uncomfortable place in which to reside (for example, by the use of red-hot pincers), even if doing so also made the body an agonizing place for the person himself. Nowadays, it seems reasonable to many people to believe that behavior abhorrent to therapists is caused by some "disorder" just as difficult to see or touch as little devils but just as assuredly there. Not many decades ago, therapists tried to drive out the disorder with prefrontal and transorbital lobotomies. Nowadays the favored remedies seem to be (1) various kinds of talking and (2) pills containing various chemicals, though some treatments do still occur that are very painful psychologically and physically (for examples of the latter, see Dawes 1994, pp. 42, 49 and Suarez and Mills 1982, p. 61).

I am not saying that all alterations in patterns of behavior are chimerical. The changes in patterns of control exhibited under Alzheimer's disease and Parkinsonism, for example, have been shown to have correspondences with physical changes in neural interconnections. The reliability, however, of finding the little devils we call schizophrenia and dissociative identity disorder is very low (as illustrated by Rosenhan's demonstration), as is the rate of ridding the person of them. One can never be sure whether something that does not exist has gone away!

Some of the patterns of behavior called personality disorders clearly "exist" in the same sense that a person who smokes, say, three packages of cigarettes every day is a "smoker" or a person who spends seven or eight hours every day practicing the piano is a "concert pianist." Those are examples of persons who do things much more often or to a greater degree than most people. Leslie Morey (1997) says:

> The concept of personality disorder implies that [some] individuals act in a certain way with much greater frequency and in more situations than is expected of most people (p. 935).

And here is one of the "general diagnostic criteria for a Personality Disorder" on page 633 of the DSM-IV:

A. An enduring pattern of inner experience and behavior that deviates markedly from the expectations of the individual's culture....

That is to say that you cannot ascertain the degree of a characteristic defined by the ranking of a person in a population—a characteristic such as intelligence quotient or personality disorder—by examining only the person. You must know also the same sort of information about the population—or a reliable sample of it. (I have no idea how your neighborhood psychiatrist ascertains "the expectations of the individual's culture.") This way of defining a personality disorder amounts to an instruction to rank-order the population on some characteristic, cut off an arbitrary percentage of persons at the extreme top (or bottom, or both) of the array, and, if those people seem unhappy, to label them Personality Disordered.

But suppose you were to "cure" those people of their personality disorder? You would no longer see them behaving as they had. But the definition would still be there, still demanding that you cut off some extreme percentage of people and label them as Personality Disordered. So now you would cure the next bunch of people, too. And so on. The only way you could stop before curing the whole population would be to declare at some point some criterion for Personality Disorder that did *not* depend on what the rest of the population does. But if you were going to do that, you could have done it at the outset.

Another way to say this is that no matter what sort of characteristic you choose, if it can be quantified in degrees (not merely in yes or no), some people will always fall at the extremes, *because your choosing to order them that way has put them there.* Accordingly, the psychotherapist will never run out of clients.

This is one more way of expressing the difference between medical ailments and psychological deviations. To diagnose a broken bone, you do not need to know the number or severity of broken bones in the population.

CODA

The field of "mental health" is a very big business. Except for the business being bigger, I think the situation today is much the same as it was when Suarez and Mills described it in 1982:

> The present-day state of mental health [in the U.S.] is reflected in the ever-increasing demand for more mental health professionals and programs to solve people's problems.
>
> In spite of the present arsenal of therapeutic defenses..., relief has yet to be experienced. This dissatisfaction has led to the overgrowth ... of "self-help," "personal growth," "awareness," and "consciousness raising" ... techniques. It is possible today to ... undergo ... individual and/or group therapies involving ... analysis, diets, gurus, bio-feedback, rebirthing, rolfing, massage, encounter, gestalt, [transactional analysis], primal scream...(pp. 61–63).

Not to mention astrology, numerology, palmistry, and psychic readings. Suarez and Mills also wrote:

> The traditional role of psychology in the health care setting ties in with the prevailing illusion that psychologists have the power, via rituals and techniques, to change other people's levels of understanding, motivation, mood, reality, compliance, and so forth (p. 265).

We do often come away from professional helpers (and from amateur helpers, too) feeling that we have been helped. It is useful to know that help is available by looking in the yellow pages of the telephone book under coaches, counselors, psychiatrists, psychologists, and psychotherapists. It is also useful to know that those professionals have their limits.

A mistake most therapists make is to suppose that they must "understand" the person—even to a greater degree than the person understands himself—before they can be helpful to the person. In setting out to "understand" a client, therapists typically look for a cause of a disorder. They assume that something is malfunctioning—perhaps the ego, a gland, or the wiring in the frontal lobes. Once they have ascertained the malfunction, they can presumably carry out a treatment that will correct the malfunction, compensate for it, or at least make it less distressful to the neighborhood. In other words, they believe that they can predict or prevent specifiable kinds of acts. It is no wonder that psychotherapy sometimes requires years of patient-and-therapist interchange to be of much help.

In physically normal people, however, control, too, is normal. Part of the normal functioning of the nervous system is reorganization, which often produces modes of control that surprise and distress the rest of us. The cause of the surprising behavior is not a malfunction; it is a normal interaction between a search for a means of control and the environmental opportunities. When the behavior the reorganization has produced still leaves too much internal conflict unreduced (and the distress of family and friends exacerbates the conflict), the sufferer often asks for help.

I am not saying it is a bad idea to have a good estimate of some internal standards of another person. On the contrary, if you are confident, for example, that your employee believes in the principle of a day's work for a day's pay, you won't have to hire guards to keep him working. If you are sure your friend or spouse maintains your welfare very high in his control hierarchy, you can dismiss the precaution and wariness you maintain in relation to most others. If you have a good estimate of a few of a person's high-level standards, you can estimate fairly well the *kinds* of acts the person will *avoid* so as not to disturb the perceptions being held to those standards. You can then live and work harmoniously with that person by avoiding acts that would (as far as you are able to guess) hinder the person's purposes. Frequently, knowing what *not* to do is more serviceable than knowing some particular helpful things to do. Usually, we want people to refrain from disturbing the variables we care most about, and beyond that we care little what they do. Knowledge of limits is often a better guide than a lot of examples of acts that lie within the limits.

A little knowledge of high-level standards is useful, but it can also be dangerous. You can think you know more than you do; when you find some guesses about the person's internal standards being verified, you can too blithely assume that your next guesses will be correct. You can even be tempted to invent grandiose explanations for your successes, including "illnesses" that do not exist. Even when your guesses about the nature of some of the person's highest standards are very good approximations, your predictions about the kind of act the person will take now or tomorrow morning can fail because of being wrong about the environmental opportunities or about the opportunities you can perceive but the person cannot. You can even be wrong about what the person will *not* do. The danger of thinking you know enough to give good advice besets helpers of every sort, including psychotherapists. You can avoid the danger by using a method of help that does not require advising particular acts—a method such as the MOL.

I do not want, either, to leave you with the impression that I think physicians, psychiatrists, psychotherapists, and other helpers are typically charlatans or ninnies. As with people in other occupations, many are honorable and helpful, some are unscrupulous, and some do harm. You can find in the professional literature more asperity than I have given you here. For example, Arnold Lazarus (1994) reviewed a book by Dryden and Feltham (1992) entitled *Psychotherapy and Its Discontents*. In his review, Lazarus said

> Masson impugns the entire field and regards psychotherapists, by and large, as tyrannical and exploitive, or, at best, as "dullards, frauds and narcissists."
>
> Gellner sees psychoanalysis as a mystical, basically untestable and self-protective guild.
>
> Pilgrim . . . points out that therapists are potentially exploitive and abusive. . . .

Many clients of therapists feel greatly helped. There is no telling how much benefit is due to the placebo effect, but a benefit even from a placebo should not be devalued. That benefit is as real as a benefit produced in any other way, and no apology or shame need attach to it. If a person feels comforted or more insightful after praying to a god, the comfort or insight is as real as the feeling of being nourished after eating beans. Whether the god will do anything about the prayer or whether the god actually exists should be given no weight in assessing the benefit of the prayer.

A final note. On several pages, I have mentioned Robyn Dawes's book *House of Cards* (1994). I had to talk myself out of quoting the entire book. Between its covers you can find much more useful information and many more useful ideas than I have displayed here.

ENDNOTE

[1]This and later quotations from Dawes's 1994 book are reprinted with the permission of The Free Press, an imprint of Simon and Schuster Adult Publishing Group, from *House of cards: Psychology and psychotherapy built on myth* by Robyn Dawes. Copyright © 1994 by Robyn Dawes.

Chapter 32

Language and communication

To pursue our own purposes, we try to persuade, cajole, reason, and threaten others into taking actions that we believe will help us to our goals. We all, however, live within our own worlds of perception. We can disturb the perceptions that others control and observe the outcome, but we cannot (except in an environment of draconian constraint) be sure of compelling any particular act. We can offer opportunities for others to act as we hope, and we can invite them to do so. We can offer to cooperate in activities of mutual benefit. Mutually helpful actions can occur if there are sufficient degrees of freedom for all.

LANGUAGE

Look at this: IWVWI. That configuration might be a decorative design for a belt buckle. Or a string of those configurations might serve as a border to be etched into the edge of a tumbler or carved along the frieze of a building. As you look at the zigzags in that configuration before you have thought about what use the configuration might have, all you can see are the zigzags of a configuration. The configuration IWVWI, however, is a word in a language I have just now invented. (So far, the vocabulary of this new language consists of only that one word.) "Iwvwi" stands for the sound made by a belt when you pull it out from the loops of your trousers. A word can also be produced by making sound waves—by pronouncing it yourself or by playing it back from a recorder. We could call that production an event (fifth level)—a transition with a beginning and ending.

We can play some curious tricks with language. The "the" at the beginning of this sentence is not a name for a category, but the "the" that I put in quotation marks is the name for the category containing "the"s. Here is another bit of trickery:

This sentence will not release your attention from its grip until you have come to its end.

Words are symbols; that is, they stand for (are associated with) something beyond their shape as configuration or event. We use many symbols other than words to convey meanings. If you clear your throat, that sound is not usually a word; but if you walk up to two people who are conversing and clear your throat, the sound can mean, "I'd like your attention." We can gesture with a wave to mean "Come here" or "Go away." We can spit in a person's face to give an insult. A shrug of a shoulder can serve as a word. So can logos, handshakes, whistles, facial expressions, and foot-stampings. Many mammals use cringing and sexual postures. Dogs use spots of urine. Bower birds use bowers.

Words can change their meanings as the context changes. The word "bank," for example, means one thing when you want to deposit your money for safekeeping, another when you are sitting at the edge of a river, another when you are in an airplane as it changes its direction, another when you are heaping up a coal fire so that it will last through the night, and another when you are playing billiards. Usages of words change from decade to decade. Nowadays, I often come upon "plethora" being used not to mean an excess or superabundance, but merely a lot. Or someone saying "I could care less" to indicate, oddly enough, that her caring is already at a minimum.

An object can become a symbol simply by being used in an unexpected way. That is sometimes called "making a statement." I think it was 1930 or thereabouts when I saw a photograph in a newspaper of the actress Marlene Dietrich wearing trousers and a jacket of a mannish cut. The caption implied that Ms. Dietrich was doing something very daring, perhaps even somewhat immoral in wearing such clothing. It is possible that Ms. Dietrich thought merely that she was wearing what she preferred to wear and that her preference was not worth a lot of comment. It is also possible that some clothing designers at the time thought that she was heralding a future trend; that, indeed, turned out to be the case.

Judd's Hierarchy

Much of what I write here is speculation about the functioning of control in the higher reaches of the control hierarchy. So far, very little empirical research has been done on functioning at these levels. If the research on PCT were as sparse at the lower levels as it is at the upper, I don't think I would have been inspired to write this book. The research at lower levels, however, has been so repeatedly and uniformly successful, and its reliability and validity have so outdistanced almost all the mainstream psychological research, that I am willing to ask you to grant some attention to the speculations in this part of the book.

In his dissertation on second-language acquisition, Joel Judd (1992, p. 59) has laid out the features of language that can be perceived at Powers's eleven levels of control. Table 32–1 below is a modification of Judd's Table 4. I offer it to you for two reasons. First, it illustrates once again the fragility of human communication. In Chapter 28, I described many of the hazards awaiting attempts to convey meaning from one person to another. Judd's table describes in another sort of detail the levels of language where inaccuracies and insufficiencies can occur. Second, the table illustrates again the usefulness of Powers's levels of control in thinking about a domain of behavior—learning a second language, in this case. A large part of Judd's dissertation is devoted to explaining that usefulness; I will not take space here to recapitulate those pages. I will, however, quote a few sentences from which I hope you can get the flavor of Judd's arguments:

> . . . change brings about a new configuration of the systems which make up our self [that is, a change in the] "control systems hierarchy". . . . Such a change brings with it a new perspective on experience . . . the world seems "different.". . . We cannot know, except through our current . . . way of seeing things, what future learning will bring. Hence, we cannot stipulate the "goals" of learning for students. . . (p. 83).
>
> The genetic contribution seems to be not a specific determination of the language-to-be-learned but a specification of the development of a control systems hierarchy which is needed to learn a language, and a timetable for the maturation of the hierarchy. . . . Join together concepts of control of perception, error-driven behavior, perceptual hierarchy, and unforesighted problem-solving, and a general picture of learning appears (p. 88).
>
> Given multiple levels of perception and numerous control systems at each of those levels, it is no wonder that there is not going to be a neat and tidy explanation of learners' attitudes and motivation in [second-language acquisition] (p. 98).

I have altered Judd's table chiefly by expanding his entries under "Linguistic Equivalent." Judd gives some evidence (p. 58 ff.) of the development of language ability in children that fits this ordering of perceptual ability.

Producing Language

Producing language, like any other act on the environment, is done to control perceptions. Speaking or writing has purpose. Producing the parts of lan-

Table 32–1. A hierarchy of linguistic control systems.

Level	Level of control	Linguistic equivalent
11	System-concept	Language as a whole; socio-cultural connotations.
10	Principle	Pragmatics, esthetics, usage, skills.
9	Program	Grammar, complete sentences.
8	Sequence	Syntactic ordering.
7	Category	Naming, semantics.
6	Relationship	Prepositions.
5	Event	Words, idioms.
4	Transition	Intonation to the ear, phoneme to ear or eye.
3	Configuration	Syllable to the ear, printed letter or character to the eye.
2	Sensation	Sound quality and pitch to the ear, black of ink to the eye.
1	Intensity	Loudness to ear, brightness to eye.

guage which, arranged properly, evoke meaning has the purpose of evoking meaning. We control the perceptions of the language we are producing to be sure the meanings we perceive in the sentences are the meanings we wanted to evoke. We cannot, of course, be sure that the meanings we desire will be evoked in the listeners' minds. We can only hope that they will be close to the meanings in our own minds.

All of us edit as we speak, some more, some less. Most of us, perhaps all but trained actors or orators, are unaware of how much we revise what we say as we go along. Had I put what I just wrote into speech, it might have come out like this:

> All of us, well, most of us anyway, maybe not an actor, maybe not old-fashioned orators, most of us just don't realize, most of us are unaware of how much we correct ourselves, revise what we say as we go along.

The revising, oral or written, is the evidence of control. We act *against* a word or phrase or ordering that we have produced by discarding it and substituting something that sounds to us closer to the meaning we want. Even the most unschooled do this. Whenever you hear someone changing a word or starting a phrase over again or saying, "I mean . . .," the person is correcting an error or emphasis, even if she has never heard of such a thing as a rule of grammar or syntax.

Thinking

Many people, perhaps especially people who have spent most of their lives paying professional attention to language, believe that thinking can be done only with language (in which I include any system of symbols—chess, for example). But that view succeeds only in defining thinking as something done with language.

What can we mean by thinking? Look back at Figure 18–3. There we saw some neural loops labeled "thinking"; they were simply loops that did not run out into the environment. We can define thinking as any process of searching for control of perceptions that does not involve feedback loops running through the environment. Much thinking is accomplished, certainly, by alternating purely internal processes with empirical testing. I will, however, use the label "thinking" only for the internal neural processes, once more using PCT to simplify psychological conceptualizing.

This definition of thinking requires us to say that when an animal is finding outputs at a higher level that will serve as reference signals for lower levels, the animal is thinking—no matter how simple such an adjustment may be, no matter how quick, no matter how unconscious. It requires us to say that animals with very few levels of control can think, albeit primitively.

Usually, when we say we are thinking, we mean that we are making conscious use of ninth-level language and logic. Still, even at the higher levels, I think our mental processes often cast about while we wait for our minds to reach somehow a higher sense of rightness. You remember from Chapter 25, on our use of logic, the many ways in which we can and do go wrong when we think we are being logical. In writing this book, my mind does not proceed "logically." I spend a lot of time staring at the computer screen or walking back and forth, pointing my mind, so to speak, toward the topic of the chapter and hoping some phrase worth typing will somehow come into my mind.

The comedian Jack Benny was famed for being tightfisted with his money—or pretending to be. The story goes that one dark night Benny was accosted by a robber who poked a gun in Benny's belly and growled, "Your money or your life!" Benny stood there, hands raised, gasping, speechless. "Well," barked the robber, "which is it going to be?" Breathing hard, Benny burst out, "I'm thinking! I'm thinking!"

Heritability

Not many people believe that we inherit, biologically, the ability to speak a particular language, say French, English, or Swahili. Some people, however, do believe that we inherit some elementary linguistic rules. Noam Chomsky (see, for example, 1969, 1986, or 1995) is currently the best-known proponent (at least in the United States) of an inherited Universal Grammar. We do not, however, need to inherit particular rules to be able to learn a language. We need only to inherit the capability of perceiving repeated patterns and converting them into rules—the ability of the ninth and higher levels.

The mistake in supposing we must inherit the basis (a set of rules) of a language if we are to learn it is logically the same mistake as supposing that if we are to do it well, we must inherit a knack for tennis, playing the piano, smoking a cigarette, or driving an automobile. People from a long line of nondrivers of automobiles—perhaps Eskimos in roadless northern Alaska or people in the roadless parts of Bhutan—learn to drive just as quickly as youngsters in Detroit. Here is what Powers said in a posting to the CSGnet on 18 February 1995:

> ... it seems evident to some people that ... [if] we see with our eyes, there must be a gene for seeing, and if we walk there must be a gene for walking. From there it is only a short hop to a gene for drawing pictures, a gene for dancing, and a gene for making movies about Fred Astaire and Ginger Rogers.
>
> The error here is to assume that for every regularity we see there must be a corresponding cause—the opposite of "systems thinking." If, for example, we observe that there is a perception of social responsibility, the assumption is that there must be some specific reason for this perception, some social force or inherited goal of being socially responsible....
>
> When we find regularities in social behavior, it may seem that there must be some important property of societies that makes the regularities necessarily appear. But I argue that this is not the case; that we are seeing surface or emergent regularities as a consequence of regularities of a completely different and deeper kind, which are indifferent to the circumstances in which we see their consequences. If there are social rules, for example, they exist only because people are capable of perceiving rules and making behavior conform to them. This property is totally independent of WHAT rules are put into effect....
>
> Children learn regularities of language so easily [some people argue], that it simply has to be inherited. What seems not to be considered is that these underlying abilities that seem inborn may reflect more basic properties of the brain that are indifferent as to whether they are used to build language or to accomplish anything else....
>
> Before language can exist, the ability to perceive and follow rules must exist. We learn rules of many other kinds, like the rules for starting a car or raising the sails on a boat.... But what we inherit is not any particular set of rules: only the capacity to invent rules, perceive rules, specify rules, and modify behavior so its consequences conform to rules. Out of that capacity come many specific skills.... Only some rules are given in the form of sentences; and even to give them in that form implies the existence of underlying rules that are not linguistic.

The learning of language by children is one of the best demonstrations of the amazing capability of the human mind. John E. Pfeiffer (1978, pp. 370–371) described it provocatively:

> As a newcomer, a recent arrival from the womb, the infant faces a problem it will never solve completely. It is born into a turbulence of noises and odors, smooth places that suddenly become rough, cold places that suddenly become warm, lights and shadows that rise, fall, appear, and disappear. Plunged into this commotion, the infant must start to find a way and a place for itself. Its job is to create out of all the random strangeness a system of familiar objects, landmarks, rhythms, and laws.
>
> So the infant investigates because it must, because that is what it is designed to do.... It distinguishes ... which sounds to heed and which to ignore and, among the heeded sounds, which have the precisely patterned qualities of words. And it eventually makes ... a discovery which is no less great because it is made over and over again by every infant. It proceeds to discover language.
>
> ... the child learns but is not taught. Most of us have little or no knowledge about the intricacies of syntax and semantics, and even if we did, we would not be able to impart such abstruse information to our offspring in the nursery. We [supply] language to them as we supply food and shelter. We provide them with a flow of sounds, words, and intonations what they may imitate.... Given ... an estimated 50 million words overheard during the first thousand days of life ... they go to work and create language on their own.

These few pages, along with what I said in Chapters 25 and 28, bear all I am going to say about language itself. Compared to all the books on language you can find in the nearest public library, not to speak of a university library, these pages are less than a smidgen. I have not gone far beyond the Judd-Powers proposal for the infrastructure from which any language and any instance of it can arise. Look again at Table 32–1. The levels below the ninth provide the "materials" for language. The ninth level provides the capacity for adopting rules. We adopt most of our materials and rules from our parents and playmates, though we invent some, too, if only for the fun of it, as my "IWVWI" above. My brother, perhaps inspired by the phlogiston theory, explained to me that some food tastes bad because of the "bibicoochee" in it. Lewis Carroll invented "chortle." The tenth and eleventh levels provide the contexts and reasons for language, and thus put the final shapes on the meanings intended and conveyed. As we learn more about the formation of control at the higher levels, we will know more about language.

COMMUNICATION

We can use language (speak or write) and fail to communicate. This book is certainly a production of language, but if nobody reads it, communication will be stymied. Conversely, we can communicate without using language. The stripe down the middle of the road tells us (in the USA) to keep our automobiles to its right. The stripe is a symbol, but its "grammar" is so minimal that I would not call it a language—not, at least, in the context of the control hierarchy.

Language can be produced by a single person. I can sit here, alone, typing away at this book; I can stand unseen in a trackless desert and recite Shakespeare. *Communication* requires at least two people. It has not happened until something happens to both persons. Its structure can be simple or complex.

When using language to communicate, the rock-bottom thing to remember is that you cannot reach directly into the other person's nervous system (the neurosurgeon can reach in there, but not by using language). Meaning is never transmitted or conveyed by language in the sense that you can transmit or convey a book from one place to another or in the sense that you can type a sentence into the memory of a computer. Meaning is always recreated in the mind of the listener or reader. Sometimes the re-creation pleases the speaker or writer, sometimes not.

No matter how convinced you may be of the truth of an idea, the beauty of a principle, or the goodness of a goal, you cannot implant it in someone else's belief or action. No skill at oratory, no garment of authority, no threat of harm, no magic of charisma can enable you to do that. Look again at Figure 28–5 in Chapter 28. Those dashed lines of uncertain effect are the only paths of influence you can touch. The solid lines inside the oval at the right of the figure symbolizing the functioning of the other person—those lines are beyond your grasp.

My point is not that meaning is necessarily *lost* as communication goes from one person to another. When you make sound waves in the air and another person hears them, neither you nor the other person can ever know how precisely the meaning that person takes from the sound waves matches the meaning you wanted to convey. Though the meaning taken by the second person can be less in some ways than what the first person meant, it can also be more. Charles Lutwidge Dodgson, writing as Lewis Carroll, published his "Hunting of the Snark" in 1876. Repeatedly thereafter, he was asked to explain the meaning of the ballad. According to Martin Gardner (1981, p. 6), Dodgson wrote this to a group of children:

> I'm very much afraid I didn't mean anything but nonsense! Still, you know, words mean more than we mean to express when we use them: so a whole book ought to mean a great deal more than the writer meant.

HIERARCHIES OF PURPOSE

To illustrate how a string of words can carry various meanings, I borrow a scene from Charles Derber's admirable book *The Pursuit of Attention* (1979, pp. 67–68). Here is his report of a visit to an expensive restaurant:

> We were greeted by the owner himself, who was dressed in extremely formal attire, but welcomed us informally, as if he were personally pleased to see us. All through the meal, one waiter hovered over our table watching to see if we needed anything. He moved quickly to fill our glasses and to make sure that we had the right sauces and condiments. The head waiter had a distinguished manner, with the expected French accent, and was extremely deferential. In taking our orders, he spoke softly and unobtrusively, yielding the floor immediately whenever one of us began to speak. He was responsive to our inquiries, showing considerable knowledge of gastronomy, but careful not to draw too much attention to himself. He listened solicitously to each person's order, nodding supportively or appreciatively at the selections and never allowing his gaze to wander. All staff members were,

in fact, extremely attentive. The wine stewards, busboys, and waiters also approached us respectfully and served the food with a sense of exquisitely concentrated care and concern.

Now, suppose you are dining at that restaurant with your spouse, and you are asked, "How does it taste?" Imagine the implications (meanings) that might be in your mind and in the minds of the inquirers if the question were asked by your spouse, the head waiter, the chef, the waiter, the wine steward, the busboy, and a friend who happened to pass by your table. You might find it easy to imagine various meanings by imagining some purposes a person in each of those roles might have in asking the question.

Your spouse might want to know (among other things) whether you are going to be glad you chose this restaurant. The head waiter might want to check on whether the cooks and the waiters have performed to your satisfaction. The busboy might be relieving his boredom by violating the rule to speak to customers only if spoken to. The meaning in your own mind would be correspondingly different if the question were asked by one or another of the persons in those roles. The meaning in your mind would be influenced by the purpose (or purposes) you were guessing might be in the mind of the questioner and also by the purpose you yourself were pursuing at that moment. Here is an exercise for you. Imagine two or three purposes you might pursue in responding to persons in each of those roles. What meanings of the question might come into your mind in connection with each of those purposes of yours? How might you answer the questioner?

A conversation changes shape, so to speak, as it goes along. Some purposes are more urgent at the beginning, others later. Two purposes arise almost universally, I think, at the opening of an encounter between strangers. First, both persons want to reach an opinion on whether they are likely to continue their interaction long enough to take the trouble to agree on a working relationship. Questions are used such as, "Do you come here often?" If both entertain the possibility of further encounters, both usually feel some necessity for an approximate agreement on the "rules" of the relationship.

In many brief encounters, social custom suffices. When encountering the check-out person in the supermarket, for example, social custom specifies the minimally sufficient rules: confine the conversation to the transaction of paying for the purchases, and shy away from personal comments. A similar example is depositing money with the teller at the bank. Though repeated visits to those persons can bring about a sense of familiarity, years can go by during which neither person invites a less stereotyped interaction.

Some intermittent acquaintanceships grow into a more personal friendliness, such as that with a waiter repeatedly encountered at a restaurant. Closer intimacy arises among people who see one another frequently for a variety of purposes—family friends and co-workers on the job, for example. Very special rules are negotiated (some explicitly, some by implication) between sweethearts and between spouses.

Relationships with co-workers vary. Some are as occasional and stereotyped as encounters at the check-out counter, especially in a large organization; others take on close cooperation and a long-term perspective. Relationships change if either person changes his or her own commitment to the organization. If one person decides to leave the organization, a close relationship can become no longer worth maintaining. A decision to stay twenty years to retirement can justify forming several alliances.

Trust

Communication cannot be carried on at all unless both persons believe the other person will refrain from physical assault long enough for some conversation to occur. Both, that is, must entertain a minimal degree of trust. By trust, I mean my belief that you intend me no harm—at least for long enough for us to say a few things to each other. To say that trust is strong, high, or lasting is to say that you estimate the harmless intention to cover further sorts of circumstances or to continue for a longer period, or both. I do not use the word in the sense that I might trust you to wallop my head the moment my back is turned. I use the word only in the positive sense—that I trust you *not* to bash my head in, that I believe you will act for my welfare or at least will refrain from acting to my detriment.

Communication and trust are reciprocal. That is, trust is necessary to communication, and communication is necessary to trust. A little of one encourages a little of the other. Either one, however, can outrun the other, especially when a first person has the purpose of persuading a second person to take an action that will further the purposes of the first person.

Sometimes, two people (or two parties) find themselves urged to act together toward compatible goals, but also find themselves dubious of the intentions of the other party. This situation arises often, for example, in work organizations where a change in policy or mode of operation is being debated and in families where children and parents are in contention over activities to be allowed or required of the children. It is common, in such a situation, for persons on both sides to demand a clear statement of a goal toward which both sides will agree to strive. They usually go further and demand specification of steps to be taken toward that goal.

When trust is low, people typically demand very precise and detailed descriptions of goal and steps. The specifications can then be used as checklists of conformity to the agreements. Perceived deviations from those specifications can be used as red flags signaling departures from the promises. Thus the detailed agreement serves as a protection. The persons perceiving the errant behavior can take action before things have gone too far; they can demand a return to the presumably correct path, or they can withdraw from the agreement. People sometimes withdraw by resigning formally from the organization, sometimes by renouncing the agreement publicly, and sometimes by saying nothing at all and silently ignoring the agreement. Children sometimes withdraw from families by running away.

Understanding

Once two (or more) people have agreed to listen to one another, the question of arousing meaning arises. I say "arousing" in preference to "conveying" or "transmitting." I speak or send you a letter in the hope of arousing in your mind a meaning I have formed in my mind. I choose words and syntax which, if I were to receive them, I believe would arouse in my mind the meaning I now want you to perceive, and I hope they will arouse a sufficient approximation of that meaning in your mind.

I do not necessarily choose words and syntax that I would personally prefer. I might have learned that you attach certain meanings to words for which I do not ordinarily use them. You might use the verb "present" intransitively—as in "I will present at the conference." My dictionaries of general usage do not say that people use "present" in that way, but I have heard physicians do so, as in "He presented with a stomach ache." Although I would not myself say that I am going to present at a conference, I know that some other words—"give a talk" and "lecture," for example—have become unsavory for many people. Rather than spend time with a colleague agreeing on a term that pleases us both, I might put aside my persnickety preference and ask her, with no more ado, if she is going to present at the conference. In brief, although the words I choose to speak to you may not be those I would prefer you use to speak to me, nevertheless I can do nothing other than choose words and syntax that I hope will have the same meaning for both of us. I cannot know the words and syntax you would choose; I can only guess and hope.

That is a fragile hope. Sometimes, it is true, I may be sure that I am choosing words that will resonate in my listener with the meaning I want her to conceive. Perhaps "I love you" will do it. But much of the time we are not surprised when an utterance fails. Sometimes, it is true, a person is annoyed at having put the wrong meaning on something you said: "Well, why don't you say what you mean?" But of course you did say what *you* meant, but chose words and syntax that aroused a different meaning in the mind of your listener. Lewis Carroll (1982, pp. 84–85) had a relevant comment, too:

> "Then you should say what you mean," the March Hare [said].
>
> "I do," Alice hastily replied; "at least—at least I mean what I say—that's the same thing, you know."
>
> "Not the same thing a bit!" said the Hatter. "You might just as well say that 'I see what I eat' is the same thing as 'I eat what I see'!"

Often—most of the time, I suppose—we believe we understand well enough what a person has said. What can we mean by "understand"? We can use the word in at least four ways. First, I think we can mean simply that we entertain no doubt, that we are bothered by no feeling of uncertainty. This reminds me of Dr. Fox. In 1972 or thereabouts, two professors in a school of medicine hired a professional actor to give a few lectures. They gave him an article from the *Scientific American* to use as "source material." They instructed him "to present his topic and conduct his question-and-answer period with an excessive use of double talk, neologisms, non sequiturs, and contradictory statements. All this was to be interspersed with parenthetical humor and meaningless references to unrelated topics." They then gathered a group of eleven psychiatrists, psychologists, and teachers of social work as audience. The hour-long lecture with its subsequent half-hour discussion period was videotaped and shown to two other groups similarly composed—44 listeners in all.

The demonstration and its outcome were summarized in *Behavior Today*, (1973)[1]. The audiences were asked to answer a questionnaire. Here are three of the questions and the percentages answering "yes" in each of the three audiences:

	Group I	Group II	Group III
Did he use enough examples to clarify his material?	90	64	91
Did he put his material across in an interesting way?	90	82	81
Did he stimulate your thinking?	100	91	87

Large majorities of those forty-four highly educated people, with years of practice in giving lectures and listening to them, not only did not know they were listening to hogwash and trumpery, but thought the speaker did a good job and stimulated their thinking! One wonders what kind of thinking was stimulated. And apparently not a single person found any of the questions too irrelevant or nonsensical—everyone answered either yes or no.

When we hear statements that are sufficiently different from what we ordinarily hear, we ordinarily stay awake and alert to cope with the small disturbances to our controlled variables. If the disturbances are easy to counteract with small internal adjustments, if they cause no frightening conflicts, we feel that life is going along in a peaceful and interesting way. If we feel that a speaker has been bringing us a pleasant hour and a half, most of us would be willing to say something pleasant in return. But nine out of ten of those professionals, presumably skilled in critical thinking, chose to say yes, that mountebank stimulated their thinking! The summarized report in *Behavior Today* did not say whether the questionnaire used the word "understanding" anyplace, but the questions quoted (the ones above, for example) seem to me to touch on what we ordinarily call understanding. In sum, that is one meaning I think people have for the word "understanding"—the feeling that what is being said does not disrupt any of their opinions. People often put it just that way: "I don't have any trouble with what he is saying."

A second meaning one can have for understanding is the ability to put what was said in one's own words. Your friend asks you, "What did she mean by _____?" and you reply by putting your own words on the meaning you thought the speaker wanted you to perceive. If you can do that without floundering, you are likely to feel that you understand. A third meaning is that you can put the meaning into your own words and the speaker agrees with you that you have done so. Many people (and I, too) call that "paraphrasing." It has two parts: (1) the listener states what she has heard in her own words and (2) the speaker agrees or verifies the listener's accuracy. That is, the speaker agrees that the words and syntax produced by the listener have about the same meaning in the speaker's mind as what the speaker had said.

A fourth meaning of understanding is the ability to do more after having heard or read something, to do more than speak or write in response—that is, to take some non-language action on the environment. If someone tells you how to fix the lock in your door, and you fix it, then presumably you understood the instructions. Or someone tells you how to get to the library, and you succeed in getting there. You will remember the Annenberg study of physics students that I described in Chapter 20 under the heading "Learning." Those students could pick out most of the "right" words on the tests—a kind of understanding not quite as good at the second meaning I described above. But they could not wire up a light—the fourth meaning here.

I don't want you to think I am proposing these four kinds of understanding as if they were states of nature. I have invented these sloppy categories merely as an easy way to say that what we think about as "understanding" can have various flavors. Some situations in which you feel you are understanding something are not going to fit neatly into my categories. For example, if someone tells you how to be polite, and you find that people smile upon you in a welcoming manner when you act in the way you thought your adviser meant, then presumably you understood pretty well. But which kind of understanding is that—the second or the fourth?

The experience of arriving at an understanding can be very different from the experience of "having" or remembering one. Often, both arriving at and having are placid, unimpassioned. If a friend tells you how to get to Carl's Computer Company, you may feel that you understand his directions, but your feeling is not going to be, "Oh gosh! Oh gee! Wow! I understand!" The feeling is going to be more like, "Thanks, friend." If you then get to Carl's Computers with no wrong turns, your feeling of understanding is going to be corroborated, but again, you won't invite your friends to dinner to celebrate your success in getting there.

On the other hand, suppose you have been reading about PCT for five years and trying to learn how to construct control circuits in a computer, and one afternoon the wide-ranging implications of circular causation suddenly, within three minutes' time, sweep through your mind. "That's what Powers meant!" you cry out. "That's what Bourbon meant! And Marken! And Cziko and Forssell and Carey and McClelland and Young and Judd and Taylor and Zocher and Runkel!" You look at the people around you (and the dogs and birds and alligators), and they no longer seem to you to be taking actions; you see them controlling variables continuously, unceasingly. That kind of arrival at a new understanding happens upon a reorganization. When a reorganization is far-reaching among your reference signals, the experience can have a strong emotional side-effect. In Chapter 20, under "Learning," I used the words epiphany, revelation, convulsion, metamorphosis, upheaval.

But even when you reach a new understanding in a blinding flash, you do not spend the rest of your days running up and down the street shouting "Eureka!" The organization and functioning of the world may continue to look different from what it had been. You may continue to treat events with your new ordering of importance. But you no longer need extra energy from your viscera, and you settle down to a less agitated daily routine.

When you have reached a new understanding through a reorganization, it is tempting to believe that you have reached the summit of wisdom, that if you do not now understand *everything*, you are very close to doing so. The feeling of being "born again" is very persuasive. The deep feeling of peace that comes after an extended period of inner conflict and turmoil seems quite beyond language—religious mystics have called it "the peace that passeth all understanding." It is easy to believe that you are living in a reality quite detached from the ordinary world and reachable only by a select few. I am not saying you should not enjoy your new insights. Do so. I am saying only that humans have been reorganizing now for a couple of million years or so, and you yourself may experience a large and clarifying reorganization again before you die, perhaps several. I hope you do.

Paraphrasing

Under "Understanding" above, I mentioned paraphrasing. When we communicate, we naturally want to know that we have aroused a meaning in a listener's mind that is sufficiently close to the meaning we hoped to arouse. The best kind of immediate clue is immediate action by the listener. If you say, "Hand me the screwdriver," and the person does so, that is very good evidence that the person heard the message as you meant it. Often, however, an utterance does not call for immediate action. We could watch the person's actions for the next day or month or year for clues, but we often want a clue immediately. The next best evidence is paraphrasing, in which you tell the speaker what her utterance aroused in your mind. Here are a couple of examples from Schmuck and Runkel (1994, p. 130):

Larry: I'd like to own this book.
You: Does it contain information useful to you?
Larry: I don't know about that. I meant that the binding is beautiful.

Ralph: Do you have twenty-five pencils I can borrow for my class?
You: Do you just want something for them to write with? I have about fifteen ballpoint pens and ten or eleven pencils.
Ralph: Great. Anything that will write will do.

Paraphrasing not only verifies for you the listener's understanding of what you said, but it also tells you that your listener *cares* what you said. Skillful paraphrasing can not only ward off costly mistakes mistakes; it can also deepen affection.

I often hear A asking B whether she (B) understands what he (A) has said: "Do you understand?" That always perplexes me. If A is unsure whether B understands what A has just said, how is it that A thinks B is in any better position to know whether what she understands is what A wants her to understand? I suppose A might mean, "Do you feel confused?" But feeling confident of having caught a meaning is very poor evidence of having done so. Here are a couple of examples of a confident reply from a listener:

> I hope you agree with what I am saying.
> I certainly do.

The first speaker typically takes the reply at face value. But suppose you hear the following:

> Would you hand me that screwdriver, please?
> I certainly do.

Selfish Bargaining

A paper by Jerry B. Harvey entitled "The Abilene Paradox" appeared in 1974 in the journal *Organizational Dynamics* and was reprinted there in 1988. Harvey told the story of a family sitting in the parlor on a hot (104 °F.) Sunday afternoon. One of them suggests going to Abilene, Texas, 53 miles away, and having dinner at the cafeteria there. The others agree. Four hours later they return, coated with dust and perspiration, with stomachs full of poor food. In turn, each one confesses to not wanting to go in the first place, but agreeing to go because he or she wanted to please the others.

Instead of each saying what he or she wanted, each had professed to want what the other wanted—or what he or she *thought* the others wanted. The result of all wanting to go along with what the others wanted was that no one got what he or she actually wanted. You can call that "generous bargaining."

You are much more likely to get what you want, and so are the others, through selfish bargaining—through seeking a mutually beneficial trade. All will be better off than in generous bargaining as long as the bargaining does not turn into a win-lose battle.

If you say you want something because you think that is what the other person wants you to want, and the other person does the same, both of you hoping to get what you want next time through the generosity of others, you can at best get what they think you want. But since nobody will be telling the truth next time, either, about what they want (they will be telling only what they think you want them to want), you will never find out for sure what they want, and they will never find out what you want. Usually, everybody will be wrong about what others want, and all will end getting what they do not want.

Tell them what you want.

There will be times, of course, when it is dangerous to tell other people what you want. At those times, you are not usually expecting to get help or give help, so the Abilene situation does not apply. But when you expect at least some cooperation, tell them what you want.

Emotion

In Chapter 21, I described emotion as an indication that the body is marshaling its energies to act with unusual vigor. Many of us engage, at least sometimes, in strongly physical activities, but most of us spend most of our days in only moderate use of our skeletal muscles. Nevertheless, even in comfortably upholstered chairs in a conference room, we encounter challenges and threats, and our autonomic system prepares us for vigorous action.

In the conference room, when strong feelings arise, someone is likely to say, "Let's not get emotional about this." That demand is bootless. First, none of us can turn off emotion at a moment's notice. We can only *pretend* not to feel it. Most of us, feeling anger at what someone has said, feel an urge to damage the other person, and we often make a hurtful retort. Exchanging blows, however, verbal or physical, is a poor basis for carrying on a mutually helpful association with another person at work, in the family, or while playing croquet. It is possible to deal with anger and other emotion in a more cooperative way. We can tell the person explicitly our emotion and how it arose. A teacher in a school, for example, might say to another, "I've spent an hour hunting for the overhead projector, and now I find that you borrowed it without telling anyone or leaving a note. I'm angry." That communication would serve the continuing association better than silently showing an angry face for the next hour or week and then shouting, "Some people sure don't care about anybody else!"

Second, sooner or later, we begin to guess correctly when emotion is rising in people with whom we associate frequently. Few people succeed in hiding emotion from others who know them well. Pretending not to feel emotion merely gives the advantage to those who pretend with greater skill.

Third, when you hide your emotion, you deprive others not only of information they might use to hurt you, but also of information they might use in being helpful to you. When people do not know what others care about, they can not avoid stepping on others' toes. One person suddenly cries, "Oh, no! You didn't!" or bursts into tears, and the other says, "Oh, I didn't know you cared about that!"

Communication in a group always involves the balancing of evidences that it will or will not be safe to give trust. Open and accurate exchanges of information require trust that others will not use the information to do you harm. Conversely, hiding your emotion sends the message that you do not want to trust the others with knowledge of what you have strong feelings about. That message, in turn, leads the others not to trust you, and so on. Acting constructively and collaboratively while obviously feeling emotion invites reciprocation and trust. Letting others know what you care about is sometimes risky, but it enables them to know how to avoid being a hindrance to you.

The second of the Requisites for a Particular Act tells you how you can be helpful to another person:

2a You can put an "object" or opportunity into the environment that the person can use. The object might be a warm coat. It might be your own body when you help the person to a chair. It might be a plan for getting some information or the welcome of a self-help group. In a discussion group, it might be your understanding of the implications of what the person has said.

2b You can help the person to find a suitable object or opportunity. You tell her where to find it or take it to her.

2c You can explain to the person how to use the object or help her to do so.

2d You can help the person estimate the likelihood that the object will be useful to her.

Many books have been written on ways to communicate that help others to help you. I will content myself with referring you to Matthew Miles (1981); Schmuck and Runkel (1994, Chapters 4 and 5); and Alvin Zander (1983).

Task and Process

A tricky thing about language is that it can be used to talk about itself; you can talk about talking. It is possible to talk with some persons about whether it would be a good idea to go to Abilene. It is also possible to talk with those persons (or with other persons) about the talking that you and the others carried on. At the level of talking about going to Abilene, you might say that you would like to go there, you might say that you do not want to go there, you might say it is a hot day today, or you might say nothing at all.

At that level you would not ask whether it is OK for you to express an opinion; that would be talking about talking; that is the higher level. At the lower level, you can ask about the means of getting to the restaurant in Abilene—how much time it will take, whether the restaurant will be serving when you arrive, whether the car will be crowded, and the like. At the upper level, you can ask what assumptions you make about the discussion—whether people are saying what *they* want to do or whether they are saying what they think *you* want to do, whether people feel welcome to ask that question, and the like. You can see that the assumptions at the upper level affect the internal standards that will control the discussion at the lower level, but not vice versa.

You can see that this shifting from one level to the other in a discussion is something like what happens in the MOL (Chapter 30), but in a group, the shifting can go on in some individuals without going on in others. Indeed, unsynchronized shifting brings about a great deal of misunderstanding. When John shifts while others think he is still at their level, they cannot perceive a connection between what John says and what they are talking about. They often then ignore what he says, and go on talking on their own level. "That's not what I was talking about!" John shouts. "Well then," someone says, "what were you talking about?" Whereupon the group is faced with the necessity of agreeing on the level of talking, or the level of talking about talking, on which they want to dwell for the next little while. Sometimes agreeing is easy, sometimes very difficult.

Everyone beyond some tender age is capable of talking about talking. Most people are able to perceive others doing it, though people vary in their ability to perceive it. Finally, not many people, in my experience, very often talk *explicitly* about the fact that they or others are talking about talking—which is what I am doing right now in this paragraph. I think just about everyone can do that in a classroom, but I think not many people can do it adroitly during a conversation about going to Abilene. You may disagree about my opinion here; I do not insist on the proportions at which I have hinted.

T-groups

Social psychologists and clinicians have discovered this matter of talking-about-talking-about to be extremely important to accuracy of communication and even to the content that can be communicated. I am sure that some humans have paid some attention to talking-about-talking-about ever since humans started talking, but writers in the tradition of what they call "group dynamics" generally agree that systematic experimentation in that tradition began in June of 1946 in New Britain, Connecticut, USA, under the leadership of Kurt Lewin. Here is how Marvin Weisbord (1987, pp. 99–102) tells the story:

> Lewin . . . was central to the founding of a world-famous adult education organization, National Training Laboratories (NTL Institute). NTL pioneered the T-group, a generic name for a training group that studies its own "here and now" behavior. . . .
>
> In 1946 the Connecticut State Inter-Racial Commission sought Lewin's help in training leaders to combat racial and religious prejudice. He saw a chance to design new methods. . . . He [along with Ronald Lippitt, Lee Bradford, and Kenneth Benne] planned a research and training event to observe and measure how people transfer leadership skills from workshop to workplace. . . . The forty-one participants, mainly teachers and social workers, about half of them black or Jewish, said they wanted more skill in changing attitudes, understanding prejudice, and dealing with resistance to change. They were offered a chance to study these processes in themselves, using group techniques like role playing and problem solving.
>
> *Discovering the Power of Feedback.* This conference became a management milestone. In it the enormous learning potential of personal feedback—trading perceptions of self and others—was first discovered, almost by accident. Training groups were observed by researchers, who reviewed interactions with the staff each night. One evening three trainees asked to sit in. "Sometime during the evening," Lippitt recalled, "an observer made some remarks about the behavior of one of the three. For a while there was quite an active dialog between the research observer, the trainer, and the trainee about the interpretation of the event, with Kurt an active prober, obviously enjoying the different sources of data that had to be coped with and integrated". . . .
>
> Next night half the group showed up. This interchange on what "really" happened proved to be the most exciting session. Bradford recalled "a tremendous electric charge as people reacted to data about their own behavior". . . . None had fully appreciated the learning potential of feedback until that summer evening in 1946. People became aware that we always attend . . . "the same different meeting together." The discovery created some anxiety—and enormous energy for learning. Groups observed changes in their daytime productivity after the evening discussions of the process. A set of effective feedback rules evolved: be specific, nonjudgmental, express your own feelings, don't "psych out" the other person, don't give advice. . . .
>
> The T-group made possible the study of general phenomena present in all meetings that *cannot* be studied in any other way. It also delivered on Robert Burns's longing "to see ourselves as others see us," a valuable gift not always pleasant to receive.
>
> "Sensitivity training," wrote Carl Rogers, "is perhaps the most significant social invention of this century". . . .

The T-group, also called "sensitivity training," taught participants to see processes in groups that are ordinarily invisible; it enabled them to see themselves as others saw them, and it taught them how to look for multiple levels of purposes in what participants were doing and saying. It taught many to be humble and tentative when making guesses about the purposes of others. Members of the T-groups invited each other, in effect, to look at what they as individuals were "doing" from viewpoints that otherwise would not have occurred to them. The result was that reorganization in the control hierarchies of individuals was common. It did not happen to everyone, and the proportions varied from group to group.

Because the T-group aided members in examining their own perceptions, it served many as a setting of group therapy. You can see that Weisbord's "feedback rules" came very close to specifying the proper condition for the method of levels: Be nonjudgmental, describe your feelings, don't presume to discern psychological causes in others, and don't give advice.

He could have added: Don't push people to reach some perception you wish they would reach; they will do it in their own time or they will not do it. The reorganizations that descended upon many T-group members were very much like those that occur in the method of levels.

The numbers of enthusiasts for sensitivity training grew, and the T-group training acquired an aura of glamour and mystery. It came to be treated as a sort of magic that was expected to yield marvelous treasures if only the proper abracadabras were chanted. Mistakes and distortions began to appear. One mistake was supposing that compulsory participation would be as valuable as volunteer participation. It is not. Another was supposing an individual returning from a week or two in a T-group could somehow convey the culture of the T-group to an organization back home. He or she cannot. Another was supposing that beneficial reorganization can be hurried or multiplied by threats, insults, staying awake all night, swearing allegiance to the trainer's theories, and other tricks. Many trainers in T-groups were skillful, helpful, and caring. Others were well-meaning but bumbling.

When I checked few months ago, the NTL Institute was still offering workshops and special projects. Here is the address, along with its current name:

NTL Institute for Applied Behavioral Science
300 North Lee Street, No. 300
Alexandria VA, 22314 USA
e-mail: info@ntl.org

ENDNOTE

[1] In the list of references, look under "Educational seduction."

Chapter 33

Influence

If I leave a gun in your car, and two days later you find it and shoot at something with it, have I influenced you? If the government builds a road to Seattle, and you avail yourself of it, has the government influenced you? If a friend of yours is elected a U.S. senator and invites you to visit her in Washington, D.C., did the framers of the constitution influence you to visit Washington, D.C.?

I have not found a definition of influence to suit me. Here is the best I can do. If I make use of you as a means of controlling a perception of mine, or if you make use of me, or if I provide a disturbance to a variable you are controlling, then I have influenced you, regardless of whether either of us is aware of the use or the disturbance. But that definition has gaps I dislike. For example, suppose I board a passenger train and travel on it to Seattle. In doing that, have I influenced the engineer? Has the engineer influenced me? Answering yes to either of those seems to me pretty thin. I set down here my flawed definition not in the hope of being precise, but only to show you the direction of my thinking. I can get along without a definition, and I hope you can, too. In Chapter 25, I said that logicians urge us to state our *undefined* terms. I just stated one: *influence*.

I will be writing here mostly about intended influence, though influence can be unintended, as mine would be if you were to tire of my long-windedness and donate this book to the Goodwill people. I will be writing mostly about verbal influence, though of course influence can be wordless. An example of the latter is shouldering someone off the sidewalk. Another is building a fence across an existing walkway. Another is acting in such a way that another person wants to imitate your action.

You can influence others by offering new environmental opportunities (as implied in Figure 28–5). In the short term, the person can then (1) choose a new action to control the perception or (2) choose to control a perception higher in the hierarchy which will improve overall control, including the control of the original perception. In the long term, given enough variety of experience, the person can, as a result of reorganization, come to control new perceptions and therefore use environmental opportunities in ways effective in controlling the new perceptions.

No change in the environment can guarantee that the person will choose some *particular* action. In practice, you can sometimes make one particular action very easy to choose—as in the case of the rat pressing the lever in the Skinner box. You can also make an act "easy" to choose by threatening terrible consequences for any other act. "Make him an offer he can't refuse." You cannot be sure the person will choose the "easy" act; there are always a few who will spurn it. "Damn the torpedoes!" cried Admiral Farragut at Mobile Bay in August of 1864. When the person does take the act you have surrounded by threats, you can be sure you have only temporary control. If Farragut had not damned the torpedoes, he might have found some other way past them. You can maintain a person's conformity by threat only by maintaining the threat, which usually requires carrying out the threatened punishment when the person deviates from the acts you want. In the meantime, since the person's action will be increasing the error in many other of her controls, reorganization will be occurring. And as always, we cannot predict what actions will result from the reorganizations. Sometimes the peasants, the slaves, the prisoners, and even the students rise up in revolt.

I will now put more detail on what I mean by a particular act. Admittedly, the boundaries of an act are arbitrary, a matter for every individual's unique perception. Furthermore, an act can be viewed from any level. For example, I can look at your hand rising with a fork in it and perceive that you are putting food in your mouth, or eating, or dining, or acting as a guest at a dinner party, or showing political solidarity with the other guests. I can perceive you "doing" something at one level while you are perceiving yourself as "doing" something at another level. I can perceive that you are showing political support for the after-dinner speaker while you perceive yourself to be making an opportunity to speak to another guest about hiring your nephew. As I said under "What is the Person Doing?" in Chapter 7, you can not be sure what a person is doing by watching what she is doing. When you are predicting an act or influencing someone to perform an act, what are its boundaries, and from which level of control are you observing it? From which level of control is the person controlling?

If I want you to perform a particular act, how can I be sure you are performing it? Let's suppose I asked you to attend the dinner being given to honor that politician. You did so. But what act did you perform? My purpose was to enable the politician to perceive one more supporter. Your purpose was to get a job for your nephew. Was I correct in predicting that you would accede to my request? Did I cause you to attend the dinner? You did attend the dinner. But did you enable the politician to perceive another supporter? My interpretation of the event is that we both made used of our social environments to achieve our own purposes.

Here is another example. Harry is a good friend of yours. He is always ready to be helpful to you. You ask him to help you carry a sofa into another room. You predict that he will help you. But Harry says, "Aw, gee, I'd like to help, but I strained my wrist yesterday, and I just shouldn't put that much weight on it yet." Did you predict correctly? You might say, "Well, he would have helped me if he hadn't strained his wrist." And that illustrates my point. Whether a particular act occurs depends on (among other things) whether it will weaken control called for by some other internal standard—such as preserving the proper functioning of the muscles in your wrist.

We do act every day as if we know what other people will do. We make predictions such as whether certain people will be helpful to us in certain ways in certain circumstances. And we are right much of the time. When I say you cannot, in principle, predict particular acts, I do not mean that social life is random. I mean that every now and then, you are going to be wrong. You will be wrong not merely because you are unskillful, but because it is hopeless to satisfy all the Requisites for a Particular Act.

I mean a little more than that. I mean that you should not be confident you are correctly tallying your successes and failures. Was I correct in predicting that you would help me show support for the politician? No, you attended the dinner, but you did not make a contribution to the politician's election campaign or even vote for her. Were you incorrect in your prediction that Harry would want to help you? No, you were correct about that even though he did not help you. He might even have offered to hire someone to move the sofa for you.

OTHER LITERATURE

Just about everything written about human behavior could be considered to be relevant to the topic of influence. In the first draft of this book, I included a paltry 30 pages summarizing some scholarly literature on influence, but that draft was much too long. In this version, you will get only a few paragraphs. In my first draft, from the hundreds of authors whose thoughts could illustrate my points, I wrote about Maslow (1954, revised in 1970), Herzberg (1968), Alderfer (1972), the Foas (1974), Boulding (1978, 1990), Harrison (1978), and Harrison and Kouzes (1980). Here I will give you a few paragraphs from Boulding (1978). He and those other authors have a good deal in common.

In his 1978 book, Kenneth Boulding describes three "social organizers." By that term, he means modes of influence that enable organizations and societies to take form. He says:

> The social organizers are relationships among two or more individuals that change role structures, which thereby create organizations, and which create great networks of hierarchy, dependency, and mutuality (pp. 139–140).

You can see the idea of influence in that sentence: "change role structures... networks of hierarchy, dependency, and mutuality." Those are all words that tell of the social coordination and miscoordination, the dependence and independence, the organization and disorganization, the millions of influence links and gaps that result in the "social order," much as the condensations and rarefactions of sound waves result in language or music. Boulding continues:

There are three major classes of social organizers: the threat relationship, the exchange relationship, and the integrative relationship ("love"). . . . Each of them creates a great network in the social fabric of space and time, which we may call the threat system, the exchange system, and the integrative system. . . (p. 140).

The meaning of *threat*, Boulding says, is the expression of this kind of intention:

"You do something that I will perceive as improving my condition or I will do something that you perceive as worsening yours" (p. 141).

Topics under threat include submission, defiance, capability and credibility, counterthreat, and the arms race.

The meaning of *exchange* has this intention:

"You do something that I want and I will do something that you want" (p. 163).

Topics under exchange include the invitation to exchange, money as a facilitator, how exchange facilitates productivity, consumption and welfare, the labor market, and the theory of profit.

Boulding describes *integration* thus:

Perhaps the central concept is that of an individual's image of his personal identity and of the identity of others. The core of the integrative relationship then becomes a statement such as "I will do something or I will ask you to do something because of what I am and because of what you are." What I am is what I think I am. What you are in the above statements is what I think you are.

Topics under integration include benevolence and malevolence, grants, status and class, and legitimacy.

Those paragraphs are pretty abstract, but they show the way Boulding was thinking. Boulding was describing the organization of society and the kinds of interactions he believed gave society its order. His intention was not to expound upon the theory of human motivation. Nevertheless, you can see that like almost all writers on human behavior, he incorporated to some extent the concepts of motivation I have set forth on many earlier pages. The method of threat implies that one can control the behavior of others by reward and punishment. The method of integration implies that people choose acts because of the kind of people they are individually.

AFFLUENCE AND POVERTY

When two people in a market agree to exchange beans for money, the question of one having power over the other rarely arises. When, however, two people reach an agreement that one will exchange laboring most of most days for a continuing supply of money, then an imbalance arises in the direction of influence.

When two people negotiate an agreement (as in buying some beans), we do not typically say that one is "rewarding" the other. If, however, one person has a continuing power to withhold or give what the other person wants (often money), and that first person gives the desired thing to the second person when the second person does what the first person wants, then we typically say that the first person is rewarding the second. And when we say that the second person wants the thing, we mean that the second person perceives it to be a means of controlling a variable he cares about—maybe eating, maybe getting under a roof of his own, or maybe being admired by his colleagues for having an annual salary of half a million dollars.

Now suppose someone has an annual income from investments of $50,000 per year, and we offer her an income of that same amount if she will stand by an assembly line eight hours a day and help fasten body panels on automobiles. This person is not likely to accept the offer. If she were to accept, she would be paying much more (laboring eight hours a day) for her $50,000 than at present (signing her checks). But suppose that she has no investments, no particular training for any well-paid occupation, and is looking for a job in a period of high unemployment. This person might quickly agree to work at the assembly line for $30,000. If she were to try to attain a comparable standard of living by her own efforts—perhaps by farming, or selling kewpie dolls at the county fair—she might put in many more hours of labor for much less income than $30,000. The cost of refusing the offer might exceed he benefit (avoiding the assembly line) of doing so. In this case, the "reward" of $30,000 would be an attractive exchange for her labor. In brief, offering an incentive is influential when the person is in a condition of deprivation. (Deprivation is the typical condition of rats undergoing conditioning experiments.)

When you can achieve your goals without excessive costs in severity of labor, pain, or threat of finding yourself next month with too little money—when, in other words, your customary efforts are keeping the benefit of your actions well beyond the cost, then offers of what other people might consider "rewards" will have little appeal to you. Powers (1992, p. 125) offers this state (not being attracted by "rewards" offered by others) as a definition of "affluence."

The division between affluence and poverty lies at the point at which the net effect of an increase or decrease in effort is just zero; where the cost of an increase or decrease of effort exactly matches the benefit. If the total benefits of action become lower than this point, or if the total costs of action become greater, then there will be no benefit to the person from expending greater energy. To do so will not bring greater control; on the contrary, the error between reference level and perceived level will increase. The result is that the person will give up and become apathetic. Powers (1992, p. 125) says:

> This threshold effect, this transition from positive to negative feedback, may be a far more useful way of distinguishing the poor from the rich, the deprived from the affluent, than a mere measurement of possessions, health, and so on. The poor and deprived are those who cannot try harder to correct their errors, because doing so will lead only to a net increase of error. In this condition, people respond as classical theories of human nature would predict. Windfalls, opportunities, rewards, and outside help will encourage them to be more active in their own behalf, while adversity and punishment will drive them toward complete apathy. Unless their own efforts become effective enough, however, removal of these outside influences will result in their dropping back to the former conditions. The question is not only whether an individual's goals are being satisfied—that could occur simply because of fortuitous disturbances that happen to be acting in a helpful direction. The important question is whether the individual has the means of maintaining all his perceptions at those goal-states even in the face of disturbances which *oppose* his efforts. The question is whether the person *has control of what happens to him*.

It is easy to find workers, especially among those who find almost no other satisfaction in their jobs except that of money, who feel little control over what happens to them. They come to work in the morning apathetic and go home in the same state. They are so close to the threshold that spending only a little more effort to make or find a better situation results in a net loss of benefit. Among these are the people who spend the rest of the day with TV and beer. I will give an example under "Assembly Line" in Chapter 34.

Designers of work organizations are learning that beyond a certain point, other satisfactions become more important to most people than money. During recent periods of more widespread affluence than usual, we have seen many workers, especially the younger ones, disdaining dull or physically burdensome work, insisting instead on interesting work, jobs workable at home, and jobs with flexible hours—all those characteristics increasing their degrees of freedom to control other perceptions than only the degree to which they are conforming to the requirements for collecting a paycheck.

As to reward, many people forget that rewards, given frequently, come to be perceived not as largess or manna, but as an entitlement. Once an employee, for example, has settled into a routine of producing a certain pattern of work and receiving his wages ("reward") periodically, the effect we ordinarily call "reward" is no longer there. If the employee works for no other purpose than to get money, and you want to increase his effort, there is no way you can succeed for very long. If you increase his compensation, you might get a brief blip of increased effort, but it will soon fall back to the level of entitlement. If you threaten to reduce his pay, you will get a blip of increased effort, but you will also produce inner conflicts that will have outcomes difficult to predict and usually of a sort you will not welcome.

Of course, most of us, as we work, pursue more purposes than getting money. Those further purposes can serve both employer and employee as useful coin for negotiation.

FORCE

The extreme sort of reward-and-punishment influence is physical force. I will take a paragraph or two here to display a definition of social force put into PCT terms by Kent McClelland (1994, p. 479):

> [We say that] **A** uses force on **B** when **A** acts with the intent of creating a disturbance for **B** which is [sufficiently large] to cause **B** to lose control of one or more of the perceptual variables **B** is currently controlling.

That is, a disturbance sufficiently large for B to lose control is one too large for B to counteract. This definition is interesting for what it includes as "force." McClelland says:

> A disturbance need not be violent or even physical to disrupt perceptual control. For instance, an attempt to insult or embarrass another person counts by this definition as a verbal act of force, in that the intent of the action is to make the victim lose control of his own perception of self-esteem. Likewise, a listener heckling a speaker is using force if the object is to disrupt the delivery of the speech (pp. 479–480).

I think that kind of thinking opens the way for clearer thinking about techniques and effects of interpersonal influence attempts, including methods of child rearing, schooling, supervising, and other social interactions. That new conceptualization is beyond the scope of this book. (That is a way of saying, too, that it is beyond my inventiveness.)

Powers writes about reward and punishment in Chapter 8 of his 1998 book. If you want to use reward as a means of controlling someone else's behavior, he says,

> First, of course, you have to be pretty sure that the person wants some X. He has to have a current reference condition set for some nonzero amount of X.
>
> Second, you have to make sure that the person does not already have enough X to satisfy the reference condition.
>
> Third, you must be sure that what you're asking the other person to do in order to get some X doesn't violate some other reference condition in the other person—cause pain or embarrassment or excessive fatigue, or interfere with controlling something else in the person's life, the action thus producing more error than it corrects. Or, if you don't care about this problem, you have to have enough physical force at your disposal to overcome any objections of this sort.
>
> Fourth, you have to be sure that the person can't get some X by doing something else that is quicker and easier—for example, just by asking someone else for some X without having to do anything in return (pp. 116–117).

You will recognize this fourth condition as the one members of the United States Congress overlooked in their presumably punitive legislation concerning Cuban sugar; I wrote about it in Chapter 25 under "Cuban Sugar."

> And fifth, you have to be sure there are some safeguards in place so the other person can't just take the X away from you as soon as he finds out you have it or can give it (pp. 117).

In brief, reward will bring you the act you want from the other person only if you can back it up with the threat of withholding the "reward" or the threat of actual punishment. If you do not do your work "properly", your employer can dock your pay. If you march down the street in protest against your employer's treatment of you, you can be arrested and fined for disturbing the peace. If you annoy your employer sufficiently before you look for another job, you can find that your name is on a blacklist. Some forms of those punishments, I believe, are now illegal in the United States, but many of them still occur.

COUNTERPUNISHMENT

There is a strange phenomenon of punishment that seems to me like a distortion of countercontrol. It is also like an arms race or the phenomenon of sunk costs that I described under that heading in Chapter 25. It can happen between persons who are hurt by each other but do not yet want to dissolve their association. I have seen it usually between spouses, but now and then elsewhere, too. It goes something like this:

Spouse A: It hurt me when you did that.

Let us suppose that Spouse A does not want to break up the marriage, but instead hopes that Spouse B will attend to the hurt. Instead, what comes next is something like this:

Spouse B: Oh, yeah? How do you think I felt when . . . ?

Spouse A is now in a doubly difficult situation. First came the hurt from Spouse B. Second came the refusal of B to attend to that hurt and, instead, the demand for attention to B's hurt. But if A asks for B to return attention to A's hurt, A will in doing so be ignoring B's hurt, which is exactly the kind of hurt from which A is suffering after B's last remark. In asking for a return of attention to A's hurt, A would be making an opportunity for B to accuse A of making a selfish demand for an unfairly large share of attention—an accusation people in B's position often do make, despite having done that themselves in their first response. By then the dialog would have strayed hopelessly far from A's original cause for hurt.

What is A to do? A could stop the escalation by turning to a consideration of B's hurt. Doing that, however, is very difficult for most of us. Instead, the interchange often continues in this manner:

Spouse A: I don't think you have a right to complain about that if you remember what you said when . . .

Spouse B: Well, that was nothing compared to the time you . . .

Spouse A: Oh, it wasn't? And I suppose it was nothing when . . .?

And so on. Each person is disturbing some high-level perception held by the other. It may be a perception of self, of the affection felt by the other, or of some other quality important to the marriage. The two persons act as if they have put the goals of love and trust aside and have agreed to adopt the goal of trying to win a game of demonstrating the worse hurt, even though it is obvious from the beginning that the winner is not going to be awarded a prize of any kind. The competition typically continues until one person comes to believe that the loss (for him or her) is exceeding the gain. The consequence is, as in an arms race, that both persons end worse off than before. The only way to avoid such a waste is to inquire about the perception that is being threatened and try to help the other person restore control. Doing that is not too difficult if both cooperate in the effort. If neither will do it, the arms race is inevitable.

Sometimes, cooperating to find the internal standards being abused is too difficult even when both parties try. For some people, the very act of trying to find the other person's hurt while still nursing their own would increase error in some prized variable. For example, a man might shout, "You think I care?" rhetorically, even though he does care deeply. His utterance could be shaped by his control of his self-image; he might believe that a man whose self has been disparaged must not show sympathy for the offender. You can think of other examples. This complication is one of the reasons that a third party can be helpful. The internal standards of the counselor or therapist are not likely to be insulted by the feelings between the two clients. The method of levels can be useful. The method of counterpunishment almost never works to the benefit of either party.

Counterpunishment occurs between rulers of nations. Ruler A suffers a raid over his border from Ruler B. Ruler A, not wanting to be seen by other august personages as the kind of person who would do nothing about an insult like that, makes a larger raid into B's territory. Ruler A makes a still larger raid, meanwhile increasing taxes to pay for it. Ruler B makes a still larger raid, increasing his taxes well above A's. And so on. The matter is never as simple as that, I know. Possible gains in wealth are often involved. But history tells me that feelings of threat to the self-esteem of one or two august personages have sent thousands of men into battle.

CODA

There are some principles. Try not to advise, press, require, or threaten people into a particular act. Instead, invite, offer, inspire, or simply put the possibility into the person's environment in any nonthreatening way. But let the person find *her own* particular act—which she will do anyway.

I know there are some circumstances in which, if you want to influence someone, it seems impossible to do anything other than urge the person toward a particular act. For example, when a bill in a legislature comes up for a vote, your representative or senator is limited to three particular acts: yes, no, or abstain. For any practical purpose, you can do nothing other than urge him or her to one of those particular acts. If the legislator wants to remain a parliamentary member, those three acts are the only acts available.

Even so, with our attention on the means offered by legislation, we can forget that the legislator can actually do a lot of other things: stay home, defect to Iran, or dynamite the State House. Indeed, there have been occasions in our history when some members of the Congress did forgo voting for physical brawling.

Part VII

The Social Order

According to PCT, the key to health, physical and mental, is to insure the individual of sufficient degrees of freedom for controlling perceptions. Without sufficient degrees of freedom, external and internal conflicts waste energy and leave perceptions deficiently controlled.

When we come together to seek help from one another, we also obstruct one another. Frustrated, we hunt for new ways to control our perceptions. We hunt, that is, for increased degrees of freedom. And naturally, we ask others to help us free ourselves from the frustrations we encounter in dealing with others. In our overpopulated, overurbanized, and overorganized world, almost all of us at one time or another implore parents, teachers, priests, police, lawyers, accountants, physicians, psychologists, politicians, generals, or kings to rescue us from the hindrances, constraints, and oppressions from which we suffer. We beg them to enlarge our degrees of freedom.

For almost everyone everywhere, the environment is much more social than material. Very few self-sufficient hunting-and-gathering families can still be found, and certainly no self-sufficient person singly. The rest of us depend on a few of us to bring food, clothing, and shelter to the rest of us. Most of us spend our working days providing services of one kind or another to those who provide us with food, clothing, and shelter—or to still others who provide services to them. Those in farming occupations in the United States, for example, are down to less than three percent of the population.

Daniel Eisenberg (2002) reported in TIME magazine some trends in the numbers of people engaged in various occupations. Included in the 16 occupations that are projected to draw the greatest increases in numbers between the years 2000 and 2010 are these services: teachers, nurses, accountants, health therapists, police, social workers, lawyers, and recreation workers. In contrast, the occupation that will *lose* the most members during those years is that of farmers and ranchers!

Only a tiny percentage of us, in brief, devote our working hours to acquiring directly or to producing for others the necessities of life: food, clothing, and shelter. The rest of us spend most of our time helping one another disentangle ourselves from the stresses of overcrowding. Yet our most widely believed theories of human functioning urge us to decrease our degrees of freedom still further by rewarding, punishing, threatening, and acquiring the comforts of life by denying them to others—that is, by organizing ourselves competitively. As we continue to treat one another as nonliving things, as our capabilities for coercing (and killing) one another accelerate, and as our population grows ever more dense, we can only expect our ability to avoid conflict and control perception to worsen.

When I began this book, I thought the part on groups and organizations would require almost half of it, since there are uncountable settings for social life, offering unendingly various opportunities for control of perception and profusely various interactions between person and environment. After all, I had studied social psychology for about 40 years, and my head (not to speak of my filing cabinet) was full of illustrations of PCT to be found in the doings of groups and organizations.

I came to realize more clearly, as I went along, that I was not writing this book to recite the multifarious actions that people take in social settings, nor the probability of encountering action of type P along with action of type Q. When I describe here a pattern of social action, my purpose is to illustrate—to describe a manifestation of control, not to catalog social behavior. A few illustrations are enough. You can find an unending supply of further illustrations just by looking around you. Ask yourself, for example, "What sort of perceived variable(s) might I be controlling in doing what I am doing at this moment?" Much of the time when you might ask yourself that question, you will find yourself making use of other people to control the variable(s), as in Figure 28–5.

In this part of the book, I will illustrate the points that (1) social life must inevitably be fraught with conflict, (2) there is now far more conflict than is necessary, (3) ways of organizing human life have been worked out that produce far less conflict than we now suffer, (4) no matter what the social setting, communication is generated within an individual and has its effect within an individual—or within many individuals, (5) employees have their own goals that may not coincide with managers' goals, and conflicts therefore arise, (6) it is not necessary, for good work to get done, that all the members of the organization cleave to the so-called organizational goals—to be "aligned," (7) there can be no end to the ways that writers (whether psychologists, sociologists, psychotherapists, historians, novelists, dramatists, poets, or any other sort of commentator you can name), define, categorize, and classify the visible patterns of social behavior, and (8) it need be no wonder that social life is endlessly fascinating to almost everyone.

I will illustrate those points without any pretense of representing fairly the vast literature on the varieties of social behavior. In any case, the idea of representing the literature fairly is irrelevant here. Writers are free to portray human behavior as they wish, and I am free to select the portrayals that I judge will suit my purpose—as long as I do not claim my selection to be representative, which I do not.

Here again are what I believe to be the fundamental postulates of PCT, phrased now in regard to perceptions in social life.

1   Every person acts continuously to control each of a repertoire of perceived variables. The internal standards (reference signals) for the variables specify *purpose*.
2   To accomplish control, the person acts on aspects of the environment (that is, variables perceptible there), including aspects of the behavior of other persons.
3   Control of a perceived variable is achieved opportunistically. The person maintains a variable (for example, internal body temperature or progress toward a goal) by using whatever means are at hand. When several means are available, the choice is influenced by the effects on other variables. See the Requisites for a Particular Act.
4   The person controls many perceptions (variables) simultaneously. If the person's controlled perceptions exceed the number of environmental opportunities (degrees of freedom) for means of control, internal conflict will occur among output signals. Effectiveness of control will then decline while total effort increases. In an environment being used by several persons, conflict among persons (in addition to internal conflict) will occur when the total variables being controlled by all of them exceed the opportunities in the environment.
5   The person's perceptions are organized in a hierarchical manner in which perceptions at "higher levels" encompass and "define" the perceptions that will be controlled as means at the "lower levels."

People will use the perceptual hierarchy and the social environment to increase the dependability of environmental opportunities—for example, they might replace hunting with herding. They will form promissory or obligating relationships such as rules, norms, customs, duties, roles, joint plans, and cooperation. Those relationships and categories, put into sufficiently complex combinations, become principles and system-concepts such as morality, patriotism, marriage, government, and all sorts of institutions and organizations. These social "structures" increase the degrees of freedom for some and decrease them for others. This consequence leads to political theory as well as to resentment, riot, rapacity, and revolution. In Chapter 9, I told you what Kent McClelland had to say about that.

There are thousands of books about groups and organizations. You can, however, count almost on one hand the studies that offer a workable model of what must necessarily happen in a collectivity of purposeful humans; the only such studies I have found are those by Bourbon (1989, 1990), McClelland (1994, 1994a, 1996), McPhail and Wohlstein (1986), McPhail (1991), McPhail, Powers, and Tucker (1992), Tucker, Schweingruber, and McPhail (1999), and McPhail (2000). I used the study by McPhail, Powers, and Tucker (1992) in Chapter 9 for the section on "The Crowd" and the study by McClelland (1996) in Chapter 9 under "Collective Control of Perceptions."

The writings containing the key idea for Part VII are those of William Powers (1979 and 1992). The key idea is "degrees of freedom."

Chapter 34

Degrees of freedom II

Other people can open opportunities for us that we could not otherwise reach. They can also fence us in and close off opportunities. Our evolution has brought us, as a species, to "high tech"—an astonishing ability to use an astonishing array of environmental features to carry out our purposes. As individuals, however, we often find that our social order reduces our opportunities to control our perceptions. We often find that when we take an action to control one perception, the action reduces control of another. In brief, our means of control so overlap with the means others use for control that we often find ourselves both wanting and not wanting to act. Our society enables us to use money to pursue a great many kinds of purposes. Our society also requires most of us to get money by doing what we call "working." For many of us, some aspects of our work are distasteful. Many of us rise every morning wanting to go to work to get money but not wanting to go to work and suffer its many stresses. We find ourselves in conflict. We find ourselves with too few degrees of freedom.

When the environment offers sufficient opportunities, physically normal persons can quickly find a means of using the environment jointly—in cooperation, for example—that brings neither internal nor interpersonal conflict. Lack of conflict is healthy; conflict is crippling. Conflict, both internal and interpersonal, uses up energy to little effect. Freedom from conflict releases energy for productive action.

I said a few things about degrees of freedom under that title in Chapter 18 and again in Chapter 27 under "Increasing Degrees of Freedom." (You can read more about the idea in Marken 1991, in Powers 1979, and in Powers 1992.) In this chapter, I will give a few examples of the burdensome and crippling effects of too few degrees of freedom and the energizing effects of an increase in degrees of freedom. I will also point out how actions in a densely populated and complex society can increase opportunities for some people while decreasing them for others.

INSIDE AND OUTSIDE

Civilization fills the environment with means of control that are easy to find and use. Actions to control the basic internal standards—actions to secure nourishment, shelter, and clothing—become somewhat standardized, because suppliers of those means find it economical to standardize their products. Once you have ridden on two or three buses, you know what opportunities and restrictions will be offered by all the rest. Once you have found your way through two or three bus stations or airports, you know how to get through all the rest. You enlarge your degrees of freedom by traveling to other places; you can be aided in your traveling if you are willing to accept logistics that reduce your degrees of freedom during the traveling. Passengers complain that one airline's food is no better than another's.

The same duality, the same yin-and-yang, occurs in psychological research. Theories and other system concepts help us to make sense of what we see, but that sense-making itself narrows what we allow ourselves to see. Karl Weick is good at this kind of thinking. He wrote in 1980:

> A person schooled in social psychology will not immediately look for things like minimum sociability, weak social ties, loose coupling, or the possibility that interaction is made tolerable only because people can escape it. . . . The minimalist perspective . . . argues that people are more complicated than their social ties (p. 179).

Weick was saying there, I think, that we have more capabilities, we can envision more goals, and we conceive more possibilities, than other people usually demand of us. He continued:

> [If people] want to control the variety in their environmental inputs [if they want to maintain a wide range of choice] . . . they need many weak ties rather than a few strong ties. In other words, social ties are rendered more complex when they take the form of many loosely coupled, partially indeterminate links between people that are improvised for the occasion, rather than when a few strong ties are locked into place (p. 180).

You can see there the PCT view that control of perception is aided by ample degrees of freedom in the social environment.

Overall, freedom is relative. We often, maybe usually, give up some sorts of freedom to get other sorts, as in my examples above. Almost always, we prefer more degrees of freedom to fewer; the exceptions are perhaps agoraphobic. It is important, however, to remember that the number of degrees is not an "objective" count; the "count" (often just a vague feeling) is done by the perceiver. The degrees of freedom to act are determined, it is true, by the opportunities physically in the environment (Requisite 2a), but they are also determined by what the person can perceive as available and usable (Requisites 2b, 2c, 2d).

Not only is our environment standardized, but so is our stock of high-level internal standards. We get many of our principles and system-concepts from the same sources—from parents sharing a subculture, from teachers reading the same textbooks, from clergy reciting the same admonitions, and so on. When millions or even only thousands of somewhat standardized people encounter hundreds of other somewhat standardized people every day in a somewhat standardized physical environment, the degrees of freedom will be a fairly small fraction of the purely physical possibilities. When perceptions of what is possible or permitted become too narrow, the effect is the same as when physical space is too constricted.

In social life, we continually try to expand the advantages and avoid the disadvantages. Describing the growth of towns during the medieval period in Europe, Mundy and Riesenberg (1958) wrote:

> Concentration has advantages and disadvantages. The medieval town defended itself better than the village but was more threatened by fire and disease. . . . [The] basic definition of the medieval town is that most of its citizens did not make their living by working in the fields.
>
> This fact caused townsmen hardship. Their food and raw materials were in the hands of foreigners, the countryfolk (p. 13). [On the other hand, the town had] industrial and intellectual products to offer in exchange for food and raw materials. Man's commerce is always fraught with difficulties, however, and where we posit mutual need, we are sure to find mutual hostility (p. 15).

Powers (1979, pp. 232–233) wrote:

> On the one hand, by banding together and pooling our efforts, we can achieve for all of us what none could achieve for himself. On the other hand, by banding together and creating a shared reality, we reduce the size of the universe in which we live, narrowing the choices of goals and the actions recognized as means toward goal achievement. The more of us there are, and the more close-knit the society we perceive and accept, the fewer become the unused degrees of freedom and the higher becomes the likelihood of direct conflict. . . .
>
> . . . the effective world, the one we perceive, must have far more degrees of freedom than we have goals, if we are to hope to approach something like a minimum level of conflict. . . .
>
> The more standardized a society becomes, the fewer become the individual goals and the means for achieving them. The more people there are, the fewer degrees of freedom remain. Long before actual exhaustion of degrees of freedom occurs, the level of conflict within a growing and increasingly standardized society must begin a rapid ascent.

This is the same warning we get from biologists—for example, from Ehrlich, Ehrlich, and Daily (1995) and from Niles Eldredge (1998). As the number of species of plants and animals dwindles, the opportunities we will have for obtaining food, medicine, construction materials, and so on will also dwindle, and humanity will tremble at every threat to those few sources.

In the same chapter, Powers (1979) wrote:

> . . . conflict is the key to understanding many social problems. This conclusion is, . . . by itself noth-

ing earth-shaking. But control theory shows us in great detail just *why* conflict has bad effects, and it leads us to see a relationship among conflict, overstandardization, and overpopulation, a relationship that has long been intuitively obvious but which now assumes the proportions of a natural law. . . . [C]onflict itself cannot be good for us, any more than breaking a leg is good for us just because it exercises our self-repair machinery (pp. 234–235).

And concerning freedom, Powers wrote in that same chapter:

> Our chief freedom, it seems, may be the freedom to seek the state in which we suffer the least internal conflict, and thus remain capable of acting on the environment in a way that lets us continue functioning according to our own inner requirements. . . (p. 229).

LIFE IN ORGANIZATIONS

I will give a few examples of striving to expand degrees of freedom within an organizational environment.

Getting Promoted

A first example is told by Alyson Worrall (1995). The author had applied to be promoted to a vice-principalship in the school district where she was at the time a department head in a high school. Notice the harrowing restrictions put upon her by the administrators.

> . . . I received a telephone call from a board official congratulating me on my success in round one and inviting me to an informal interview the following Saturday. As part of the interview, all candidates would be given thirty minutes to write a response to another scenario. Dress, I was told, was casual (p. 170).
>
> I spent most of the week trying to answer the question of dress. It seemed ridiculous to me . . . that it could be so important. . . .
>
> The interview was anything but informal. The atmosphere of the timed scenario was one of a formal examination. All sixteen candidates were gathered into one room, given the day's agenda, and then told to sit in every second seat around the large conference table. . . . At the superintendent's command, we began writing. I felt the exercise was really intended to gauge my reaction under pressure. This hidden agenda was only the first in a series of occurrences which made me question the process. . . (p. 171).
>
> At the end of September I was informed of my success in stage two and told I would be interviewed in my school by one of the selection committee who would give me thirty minutes to describe three leadership initiatives I was responsible for within the school, the board, or beyond. . . . The interview began with her explanation of how I would be timed, how notes would be kept, and how she would verify the information I provided with my principal. She then bent her head to her notebook. . . . She wrote, and I spoke to the top of her head for the entire time. . . (pp. 171–172).
>
> Originally, there were eight on the [final] list; four females and four males. There were four appointments and now the four females remain. . . (p. 173).

I am omitting several further harrowing episodes. After a year on the shortlist without being promoted to vice principal, Worrall looks back on her experience. Again, notice the many ways she feels constricted:

> . . . as I moved from one phase of the process to the next, I was silenced. . . . I became cautious with my colleagues, weighing my words. . . . I have tried to believe that this situation is only temporary and that once I have the job in hand I will be able to speak again. [But] will I ever regain my voice? (p. 174).
>
> By the end of the selection process I am powerless. Throughout the experience I was unable to direct what happened to me. I could only wait patiently while others decided my fate. Withdrawal was the only choice I was free to make. If my goal was to influence my environment, then I have not achieved it, for my career is in the hands of the board. . . . I have never had so little control over my career as I do now. . . (p. 176).
>
> It came as a shock to realize I had become everything I said I would not. I had changed my appearance, monitored my thoughts so they remained politically in tune with my board's stated objectives, and become capable of convincing myself I had done none of this. . . (p. 177).
>
> . . . I was unable to ignore the effects of the selection process on me. . . . The sense of powerlessness became unbearable. . . . I asked the board

to remove my name from the shortlist. . . . For the first time in two years I feel free to speak my mind, dress as I please, and refuse responsibility which exceeds my job. . . . I do not know what I will do with the next nineteen years of my career, but I know what I will not be doing. I will not be sacrificing my beliefs and freedom for a position which no longer is "suit-able" for me (p. 178).

(That "suit-able" is Worrall's allusion to her earlier worry whether she should wear a suit to an interview.)

Worrall said her goal in applying for a vice-principalship was to "influence my environment," which I take to mean that she thought the position would expand her opportunities to control some of her high-level perceptions. She was hoping, in other words, to enlarge her degrees of freedom. In the selection process, however, she found her actions bringing conflicts with some of her high-level internal standards. The very process of applying for promotion reduced Worrall's degrees of freedom. She felt prohibited from saying things she would have said earlier. She felt powerless. Looking back, she realized she had taken actions repugnant to her while convincing herself she was not doing so.

Bureaucracies (school systems, for example) are presumably designed to reduce conflict by specifying as precisely as possible the duties of every person and specifying the few other persons with whom each person is allowed to communicate. Since, however, humans are incapable of confining their uses of the environment within limitations other people may invent, bureaucracies are full of people in conflict between what they want to do to control their perceptions and what their bosses want them not to do.

Assembly Line

Sometimes the environment contains so few degrees of freedom that workers suffer unremitting frustration. Here are some snippets from a description of life on the General Motors assembly line by B. Hamper (1986):

> Ropes, wires, and assorted black rubber cables drooped down and entangled everything. Sparks shot out in all directions—bouncing in the aisles, flying into the rafters and even ricocheting off the natives' heads. The noise level was deafening. . . . (p. 28).

> [Our job was to] install splash shields, pencil rods, and assorted nuts and bolts in the rear end of Chevy Blazers. . . . Every minute, every hour, every truck, and every movement totaled nothing but a plodding replica of the one that had gone before. . .(pp. 28–29).

Hamper was laid off and endured the bureaucratic paper work of the Michigan Employment Security Commission. He was rehired a week before his benefits ran out:

> Back on the line I was reincarnated as a rivethead. . . . The most important thing I've learned during my appointment to the Rivet Line is a new approach to monotony. . . . I've found that one should lie down and wallow in it. Let the repetition be its own reward. The key is to grind your job down into a series of empty, vacant gestures. . . . Well, I've mastered dead head velocity to such an accomplished level that, oftentimes, I must run down the line and examine a prior frame just to make sure that I performed my duties on it. Without fail, the job is always complete. It's proud moments like these when I know I have achieved total zombie nirvana (p. 33).

> The bell rings, the line stops, and I go out, sit in my car, and smoke Newport Menthol Kings until the lot clears. I . . . go home to drink bourbon from a plastic mug. . . . This is the best I've felt all day (p. 55).

There, too, you can see the severe restriction of freedom: "every minute, every hour, every truck, and every movement . . . a plodding replica." The situation is one in which many people simply give up hunting for a choice. The situation is perhaps like the one called "learned helplessness" that I described under "Quantification" in Chapter 20 and under "Solutions" in Chapter 23 and like the one called "threshold effect" on which I quoted Powers under "Reward and Punishment" in Chapter 33. When conflict remains despite reorganizations, the rate of reorganization becomes so fast that before the person can perceive the results of action within a new neural organization, a still further reorganization is occurring. If lower levels are still functioning well, the person can eat, sleep, go to work, and go through the motions in a kind of "zombie nirvana," pushing the frustrated side of the conflict into a lower priority.

Work Schedules

In recent years, many organizations have tried out unusual working-hour schedules for one purpose or another. Often, a new schedule increases the workers' choices for ways of maintaining perceptions, including perceptions of the welfare of their families. My third example comes from a report by Pierce and Dunham (1992) concerning a change in the working hours of police in a Midwestern city of about 100,000 people. The police department and the officers' union agreed upon a workweek in which the "week" contained eight days, with each officer working 12 hours per day for four days and then having four days off. Two weeks before the change to the new schedule, the researchers collected officers' opinions by questionnaire about several aspects of the job. Opinions were collected again after a year had passed.

Respondents to the questionnaires reported that the interference between work-schedules and time needed for personal activities declined, on the average, from the earlier data-collection to the later. I will not go into the details of the data-collection and the analyses, but say merely that the changes I report here were statistically significant.

Attitudes toward the schedule of working hours became, on the average, more favorable. Responses to questions about satisfaction with life, job, and leisure time all brought more favorable replies. Four measures of stress and fatigue decreased. Coordination of activities within the department and the servicing of the needs of external constituents improved. The effectiveness of patrol functions stayed the same.

Those statistics seem to indicate that the change in the scheduling of environmental opportunities made life less stressful for a lot of police officers without bringing any diminution of the effectiveness of the department. Indeed, in the two mentioned respects, the effectiveness improved.

Managers

As a fourth example, I will tell about feelings of helplessness among middle managers in a large county government. What you read here will no doubt remind you of the feelings of Ms. Worrall. Here again are feelings of being prevented from taking desired actions by the actions (or lack of them) of others. Shan Martin (1983, pp. 83–84) pictured everyone concerned as feeling tied down by everyone else:

> During a long-term consulting and training program. . ., discussions among the middle managers [revealed] their feelings of powerlessness in making important decisions and the powerlessness they perceived on the part of top administration who took their "orders" from the board of supervisors. (Doubtless conversations with board members would reveal their sense of powerlessness in the face of public demands and lack of resources. And to complete the circle, almost any citizen discussing government will eventually express . . . feelings of being powerless with regard to large bureaucracies.)

Two of the reasons cited for the experience of powerlessness were: (1) the constraints imposed by the next level of authority . . . and (2) the proliferation of increasingly detailed rules and regulations governing almost all actions. . . .

Prisons

For a final example, I will quote from an article in TIME magazine by Mary Cronin (1992), who described a new prison in Boston in which

> . . . brightly colored dayrooms equipped with televisions, butcher-block tables and cushy chairs completed a picture of serenity. . . . Gone were the five tiers of cages, the earsplitting clash of steel against steel as hundreds of cell doors slammed shut in unison; gone was the cavernous, clattering mess hall, whose ambiance was an invitation to riot. . . .(p. 52).

> Modern prison design has been evolving since the late '60s, when the Federal Bureau of Prisons first tried replacing dangerous linear tiers of steel cages with rectangular modules of cells built around common rooms manned by officers. . . . violence among inmates and between inmates and officers decreased. . . .

> Boston's Suffolk County House of Correction . . . is typical of the new design. Each housing unit is a self-contained triangular pod consisting of 30 to 60 cells on two floors overlooking a common room. Prisoners are separated into units according to their conduct rather than the seriousness of their crimes. . . . [The most privileged units allow] inmates to spend the day in the common room, locked in with only one or two unarmed officers. Meals are shipped from central kitchens and served cafeteria style from warming tables in

each pod so that prisoners never congregate in overwhelming numbers. Key to the success of the concept is the interaction between inmates and "officers," new prisonspeak for guards.

And because part of the direct-supervision model is to normalize the environment, space is reserved for recreation, specialty-group meetings such as Alcoholics Anonymous or drug-therapy sessions, and religious functions (p. 53).

Tarrant County [Texas] moved 1,440 maximum-security inmates from . . . overcrowded facilities into its newly built Tarrant County Correction Center in Fort Worth, the first fully functioning direct-supervision jail system in the state. . . . "Since then," says the center's . . . warden, . . . "we have not had one piece of graffiti written on the walls, one toilet stopped up, one officer or inmate struck or injured. Our officer turnover rate has dropped to 5.4% from 18% in our linear jails, where on average an officer is injured once a day. . . . Having budgeted $20,000 for jailhouse repairs for the first year, so far we have spent $50 for two panes of broken glass" (pp. 53–54).

Degrees of freedom in that sort of prison, compared to the traditional prison, come from being able to move to the dayroom, for example, or to a meeting of some sort, or to the cafeteria line at mealtime, at one's own initiative. It comes from the variety of choices of activity. It comes from not suffering the onslaught of the inescapable "earsplitting clash of steel against steel." It comes from being able to communicate with the officers in the normal manner of talking to an unarmed person face to face instead of talking through bars to an armed guard. Apparently, for most inmates a good part of the time, the feeling of being chained and caged is much diminished. I suppose the chief reason that feeling is lessened in the new design is that the officers can deal with the inmates as humans who want to pursue ordinary human needs and desires rather than as enraged enemies eager to turn to violence without notice. Communicating in a normal, human way gives the inmate a large part of the normal range of conversational choices. Communicating in the manner of a guard narrows the communication to barked commands. An account of some similar jails and their similarly beneficial effects was given earlier by Wener, Frazier, and Farbstein (1987).

The examples above of "Getting Promoted" and "Managers" show how, in organizing ourselves to widen our environmental opportunities of certain kinds, we usually narrow our opportunities of other kinds. Sometimes we find, indeed, that we have narrowed the opportunities of the same sort we hoped to widen. The examples of the Assembly Line and Prisons show how insufficient degrees of freedom can vitiate the searches of reorganization and result in troublesome behavior. The examples of Work Schedules and Prisons show some benefits of maintaining ample degrees of freedom.

It is not surprising that often, in our work or in our lack of it, many of us find ourselves in persistent conflicts that we try to resolve with actions that other people take to be signs of "poor mental health." I elaborated this assertion under "Psychological Effects" in Chapter 29.

The common view of controlling other people (children and adults) is that not only must they be coaxed or threatened into doing right things, but they must also be prevented from doing wrong things. Both those tactics assume that someone else (parents, bosses, teachers, coaches, clergy) must "motivate" us—must set us on the "right path" and turn us away from other paths, from "going astray." Both those tactics reduce freedom.

I have offered the examples above to emphasize the social complexity involved. There is no simple recipe for improving matters—for reducing the overall multitudinous and interlaced conflicts. Improving the society will come only from a long-term campaign on the part of many people. The examples of Work Schedules and Prisons are examples of good starts. I will tell about more good starts in later chapters. But any time (such as now) is a good time to start, and every start by every individual will make the campaign easier for the next individual. You can begin, for example, by ceasing to suppose that you know the right action for another person to take or the right belief for another person to hold. But note: although I think we would all be better off if almost all of us would believe we cannot know what is best for others, I cannot know whether *you* will be better off in *your* circumstances to hold to that belief.

GROUP EXPECTATION SURVEY

Our expectations (conceptions, predictions, beliefs) limit our degrees of freedom, and in any social situation, people will have different expectations of its possibilities. That is still another way to state Requisite 2b. In any group, especially a newly formed group, some people will expect its members to be more helpful than will other people. Every person there, however, interacting with those people at that moment, can act only according to his own expectations. Consider a dozen people, say, all estimating how the other members might respond to their own actions, physical or verbal. Clearly, the extent to which the estimates are correct will have consequences for the degrees of freedom not only of the actor, but for everyone else. I will elaborate on this situation here, because it portrays so well in microcosm the typical social situation—the situation of having to act on the situation you now perceive even though as you begin to move, the situation is being changed by your own and everyone else's action or inaction. It is being changed by the efforts of everyone to maintain controlled variables, and that includes maintaining one's internal standard for degrees of freedom.

I will describe here a particular pattern of perception that occurs repeatedly in groups. It is a pattern in expectation—or attitude, or readiness, call it what you will—that is not easily perceived as a pattern in ordinary social intercourse, but is very easily seen by means of a little systematic data gathering. You don't have to be an expert to gather the data or to analyze them, as you will see.

This example will illustrate how a nonrandom pattern can emerge when individuals undertake to use a collection of other people as a means of controlling their own individual perceptions. I use this example, too, to show how tentative and cautious we are (on average) as we set out to make use of a group. And I use it to illustrate one of the ways we are systematically wrong (on the average) in our suppositions about the other people of whom we want to make use. (I don't know what portion of our six billion people I encompass when I use the word "we" in this way. Beware of how I write.)

Imagine that you are convening a group, and the members have agreed to answer a questionnaire as a quick way of learning about some of the expectations with which people have come to the meeting. You pass out to them copies of the questionnaire reproduced below (from Fosmire and Keutzer, 1968) and ask them to answer it.

Group Expectation Survey

DIRECTIONS: In the blanks before the items below, enter a number from the following rating scale that best expresses your opinion at this time.

RATING SCALE:

5 = All members of this group.
4 = All members except one or two.
3 = A slight majority of the members of this group.
2 = Slightly less than half the members of this group.
1 = One or two members of this group.
0 = None of this group.

A. Others' Candidness

How many members of this group do you expect will candidly report the following information during future group sessions?

__1   When they do not understand something you said?
__2   When they like something you said or did?
__3   When they disagree with something you said?
__4   When they think you have changed the subject or become irrelevant?
__5   When they feel impatient or irritated with something you said or did?
__6   When they feel hurt—rejected, embarrassed, or put down—by something you said or did?

B. Your Candidness

With respect to how many members will you candidly report the following information during future group sessions?

__7   When you do not understand something a member said?
__8   When you like something a member said or did?
__9   When you disagree with something a member said or did?
__10 When you think a member has changed the subject or become irrelevant?
__11 When you feel impatient or irritated with something a member said or did?
__12 When you feel hurt—rejected, embarrassed, or put down—by something a member said or did?

C. Others' Interest in Directness

In your opinion, how many in this group are interested in knowing...

___ 13 When you do not understand something a member said?

___ 14 When you like something a member said or did?

___ 15 When you disagree with something a member said?

___ 16 When you think a member has changed the subject or become irrelevant?

___ 17 When you feel impatient or irritated with something a member said or did?

___ 18 When you feel hurt—rejected, embarrassed, or put down—by something a member said or did?

D. Your Interest in Directness

With respect to how many members are you interested in knowing...

___ 19 When they do not understand something you said?

___ 20 When they like something you said or did?

___ 21 When they disagree with something you said?

___ 22 When they think you have changed the subject or become irrelevant?

___ 23 When they feel impatient or irritated with something you said or did?

___ 24 When they feel hurt—rejected, embarrassed, or put down—by something you said or did?

Before the group arrived, you had drawn four grids on a chalkboard or on newsprint, the first labeled A, the next B, the others C and D. Each one looks like Figure 34–1, except that no tally has yet been entered into it. The six columns in grid A stand for items 1 through 6; the six columns in grid B stand for items 7 through 12; and so on. The rows 0 up through 5 stand for the possible answers from the rating scale.

You ask the members of the group to come up to the chalkboard and tally their answers. If a member gave a rating of "5" to item 1, he would put a tally in the top cell of column 1, and so on. If the group contains 12 members, there will be 12 tallies in each column, distributed among the six cells. Figure 34–1 shows what a grid might look like after all 12 members of the group have put in their tallies for the first six items (section A). You can see immediately that people will differ in their expectations, but I will say more about such matters after I describe the next figure.

Now you calculate the averages in each column. In column 1, for example, multiply the ten tallies by the rating of 5 to yield 50 and the two tallies by 4 to yield 8. Then 50 + 8 = 58. Divide 58 by 12, and we see that the average rating of item 1 was 4.8. When you calculate the means for each of the 24 columns (items), you can connect each set of six means by a line and draw Figure 34–2, which shows all four lines on one graph.

Data for this graph came from teachers in an actual elementary school. It is a reproduction of Figure 4–11 in Schmuck and Runkel (1994).

Clearly, although members differ a great deal from one another, the pattern is nonrandom. The successive lines sag from left to right. The line plotting the means from part D of the questionnaire (My interest in others) hangs highest on the chart. The line plotting part C is next, sagging less than the remaining two. The line for part B is next, and the line from part A falls to the lowest point of all in rating the answer about feeling hurt. This is the pattern found repeatedly (among groups mostly composed of schoolteachers and college students in the western United States) by Fosmire and Keutzer, by Richard Schmuck, and by me. Sometimes one line crosses another, but the predominant pattern is for the lines to descend in the order shown and to maintain that order at most of the items. The ordering at the item about feeling hurt is especially reliable.

Typically, the means differ least among the items that are first in the four sections—the items about not understanding something said. The means differ most about the last items—those about feeling hurt. It seems to me that the items increase from left to right in the likelihood that discussions characterized by them would arouse emotion. But as to just what sort of variables group members could be controlling in those discussions, your guess is as good as mine.

In commenting on the four grids, I have spoken to groups more or less like this:

> Compare the means of the six items in section A with the six in section D. That comparison shows that the number of others from whom you say you are interested in getting reactions to you (section D items) is almost always larger than the number of others you think are ready to report their reactions (section A items). You want (on the average) to get more than you think you will get.

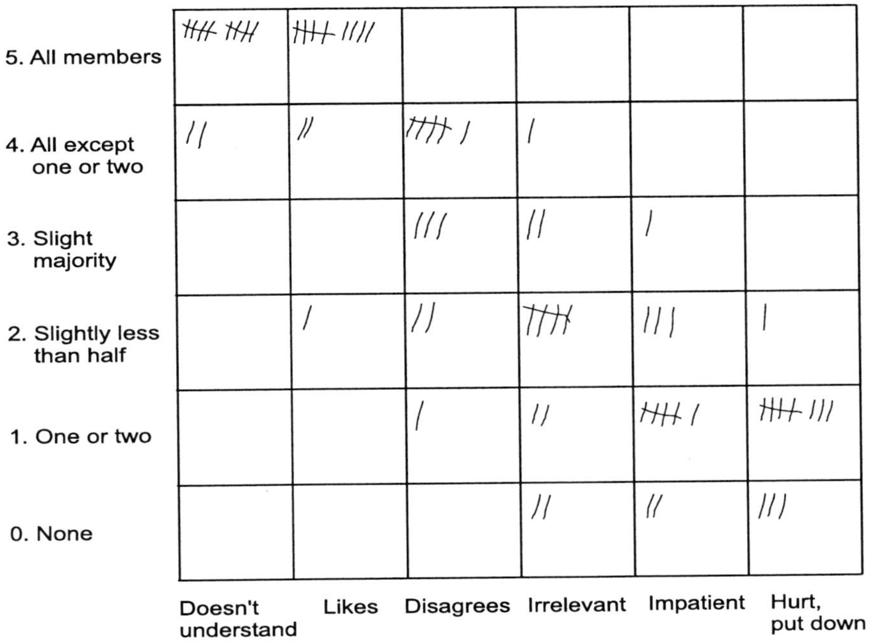

Part VII The social order: Chapter 34 Degrees of freedom II   401

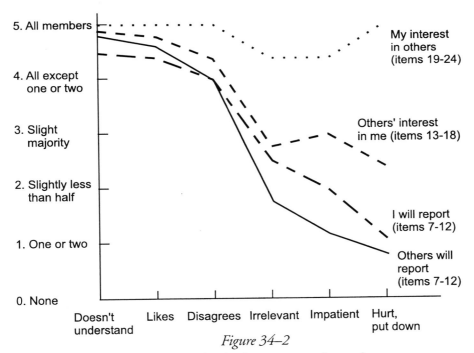

Figure 34–1
Tally of answers to Part A of the Group Expectation Survey

Figure 34–2
Plots of four kinds of expectations about others in a group using the Group Expectation Survey

Reprinted by permission of Waveland Press, Inc. from Figure 4–11 in Richard A. Schmuck and Philip J. Runkel. Handbook of Organization Development in Schools and Colleges (4th ed.). *Prospect Heights IL: Waveland Press, Inc., 1994. All rights reserved.*

Now compare section A with section B. Now we see that on the average, the number of others to whom you say you are willing to tell your reactions (section B) is also almost always larger than the number of others you think are ready to report their reactions (section A). You expect to give more than you expect to get.

Comparing section D with section C, the number of others from whom you are interested in getting reactions to yourself (D) is almost always larger than the number of others who you think care to know your reactions to them (C). You want to know about yourself from more people than you think want to know about themselves from you.

Comparing sections B and C, the number of others who you believe want to know your reactions to them is almost always larger than the number to whom you are ready to report your reactions. You are willing to tell your reactions to fewer others than you think want to know them.

Those differences are small concerning intellectual matters like not understanding something said, but large concerning emotional matters like feeling hurt—on the average.

That pattern illustrates the typical relationship among individual perceptions of others (and therefore ways of making use of the social environment) in groups. Individual perceptions show a scattering such as that shown on Figure 34–1. (Some groups show a good deal more scatter than that.) Nevertheless, the clustering of perceptions is sufficiently nonrandom so that it can be seen in tallying questionnaires, in listening to discussions, and in observing other actions such as volunteering for tasks. The evidences of nonrandom clustering such as we see in Figure 34–1 enable us to form a conception (perception) of some "thing" occurring, not just a random passing of individuals through the same space. Our perceptual capabilities (Powers's hierarchy of control) include the capability of making a namable "thing"—a "group" in this case—out of our perceptions of just such foggy clusterings.

Still, the individual member of the group cannot get much help for immediate action from the charts on the chalkboard. Figure 34–2 tells her where the tallies bunched up some (at the averages shown by the lines), but it does not tell her how *much* they bunched up—that is, whether the tallies fell close to the lines (indicating a lot of agreement) or fell in a wider spread on each side. Figure 34–1 gives that information, but it does not tell her who put those tallies there. She might think her best bet is to speak to the person next to her as if that person's tally is in the largest cluster, but she will speak tentatively, because that guess could easily be wrong. And she will likely speak no less tentatively than she would if she had not ever seen Figure 34–1. Anyway, she knows that as soon as she speaks to anyone, or as soon as anyone speaks to anyone, or as soon as someone realizes that a silence has gone on an unusually long time, Figure 34–1 will be hopelessly out of date.

And be all that as it may, all members, even before you pass out the questionnaires, will know that opinions on such matters have considerable scatter, even if they do not know that this particular pattern will show up. Individuals in the group will not be taking actions to deal with the *average* member; they will be finding out, by tentative sallies, *which* members will be helpful to them. They will move in that tentative way whether or not they get information such as is contained in Figures 34–1 and 34–2. The figures might, however, encourage some members to a little more boldness after they see that other people, too, are yearning for more information than they feel ready to reveal. And the figures are useful for the purpose I have used them here—to display the constantly shifting interlacing, the interdependence, of perceptions and actions in a human group.

Figures 34–1 and 34–2 are frozen snapshots of something in motion. A group of interacting individuals is never static—perhaps not even when all members are asleep. When members of the group see the tallies others are putting in the cells of the grids, their expectations (opinions, perceptions) begin changing. Some members will make some proposals and ask some questions they would not have made or asked before they had seen the tallies and graphs. All groups, of course, talk at least a little about their expectations of one another, quite without the formality of a questionnaire. When the norms and skills in a group permit, members can find out the expectations of others by asking their own questions. Few groups, however, are skillful at doing that.

Whoops! I just wrote (I hope you noticed), "Few groups are skillful. . .". But groups do not have skills; individuals have skills. It is true that skills necessary for good cooperation are best acquired jointly with other group members rather than attempted with nonmembers or alone. When skills are well practiced and honed with others, the result feels very much like what one wants to call a "group skill." Despite that feeling, the skill resides in individuals. Two half-skilled members are not equivalent to one skilled member. I should have written something like, "That kind of skill is seen in few groups."

Imagine the twelve members of the group in our example trying to control perceptions important to them by making use of the other members of the group. They must constantly re-estimate the portion of their time other members will claim to get help for *their* activities. They will want to know the kinds of things the six items in the Survey asked so as to estimate whether cooperation is going as well as it appears on the surface. Every individual will at some time become impatient with others who believe they must take time out from the work to ask one another questions such as those. Then more time out must be used to reassure the impatient one. Estimating the best scheduling and the optimum amounts of time to allot for this sort of "group process" discussion is rarely easy.

Skills change, too. As members work together and discuss their working together, the need for time to discuss their working together often diminishes. It is easy to fail to notice the dwindling need and go on allotting the same amount of time. Then it is easy to mistake the signs. The signs may be that many people are saying nothing but are squirming and showing other signs of impatience. Some members will mistake those signs for hiding troubles and betraying anxiety about doing so, when they are merely signs of boredom and a desire to get back to work.

The purposes of the members further ramify the patterns of action in a group. McGrath (1984, p. 49) describes 14 kinds of groups and, on page 61, eight types of tasks. Types of groups and tasks often metamorphose into other types. For example, processes such as discovering the expectations of others are usually background to tasks that require cooperation or coordination, but when group members discuss those communicative processes explicitly, the processes themselves become foreground. In types of groups such as those labeled "encounter groups," "T-groups," and "sensitivity training," members adopt those usually background processes as foreground topics. (I said some things about sensitivity training in Chapter 31 under "Captive Clients" and in Chapter 32 under "T-groups.").

Members of groups and organizations try to use other members and their interrelationships to augment their own degrees of freedom. Their own actions continually alter the degrees of freedom and their location (so to speak) in the group's activities. Expanding the degrees of freedom available in a group requires knowledge of human nature and the kind of skill that is amplified in cooperation.

ORGANIZATIONAL MANAGEMENT

When people discovered that they could domesticate animals, it didn't take them long to think also of domesticating other humans. Or maybe those practices arose in the reverse order. It is impossible to know how long before the invention of writing slavery existed, but Festinger (1983, p. 147) cites clear evidence that slavery was widespread in Egypt by 2000 B.C. What we mean by slavery is the use of direct physical force to control any kind of action by the slave that the master arbitrarily chooses. When a lot of people take that attitude in the use of other people, life for most can become very unhappy.

It is not surprising that efforts to replace the use of others by force or the immediate threat of it with a less painful method began early in the history of humankind. Hammurabi (18th century B.C.) wrote out a code of laws to limit the interference and violence his Mesopotamian subjects might visit upon one another. Confucius (551–479 B.C.) had a similar idea. The nobles of England, with the Magna Carta, put limits on the aspects of their lives they would allow their king to try to control. The French and the Americans rejected monarchies and established representative democracies in which written prohibitions, along with balances among parts of the government, limited the aspects of the life of the citizenry with which the government was allowed to interfere.

More recently, more of us are objecting to arbitrary control over people's lives because of creed, race, sex, age, physical disability, national origin, and other classifications. We are also finding ways not merely to elect representatives to our government, but to participate directly ourselves; Stavrianos (1976) gave a wide-ranging description of participative democracy.

When we find ourselves able to control other people in some ways, most of us find it difficult to resist trying to control more and more aspects of the lives of those people. When rulers are unable to resist that temptation, we call them tyrants and dictators. The same temptation is felt today by bosses of all sorts. They tell us when to arrive at work, how to address the boss, to whom we can talk during working hours, what chair to sit in or where to stand, what kind of clothing to wear, when to take a break, which toilets to use, and so on. Some people are told with what kind of people they should associate outside working hours. "You don't want people to find out you're hanging out with those socialists, do you?" Some are told what recreation to choose. "Sure, backpacking is OK, but the people you are going to work with all belong to the golf club."

When the industrial revolution moved workers from their homes into the factory, the concept of "industrial discipline" arose. That meant getting to work when the supervisors were ready to start the machinery going in the morning and working steadily in places and during hours it was easy for the supervisors to count. Employers tried many ways to improve industrial discipline. For example, Ewens (1984, p. 105) says that

> [about 1900] . . . John D. Rockefeller, Jr., pioneered in the development of personnel management by establishing a "sociological department" at the Colorado Fuel and Iron Company which looked into every aspect of workers' lives from diet and drinking habits to public school curriculum.

A similar idea came to Henry Ford. Ewens (1984, p. 123) says:

> Ford's reaction to . . . worker dissatisfaction and the unionization drive was to institute new measures to control workers, including a wage of $5.00 per day [then munificent], announced with much publicity in 1914, and the creation of the infamous Sociology Department, whose purpose was to investigate the private lives of workers.

Sooner or later, attempts at direct control backfire. People "get out of hand," we say, using still another metaphor of direct physical action. In an earlier time, the strife between managers and workers now and then escalated to gunfire. Ewens (1984, pp. 99–100) wrote this about events in Homestead, near Pittsburgh, in 1892:

> Just before the contract with the Amalgamated [union] was to expire in 1892, Carnegie transferred managing authority of his Homestead mill in Pennsylvania to Henry Clay Frick, who was already notorious for his brutal treatment of strikers in the Connellsville coke regions.
>
> Frick, in his turn, . . . ordered a three-mile long fence topped with barbed wire to be built around the entire Homestead Works. The fence would hold off strikers, while Pinkerton detectives hired by Frick would escort strikebreakers into the plant by boat and barge from the Monongahela River side of the plant. . . . Robert A. Pinkerton [supplied] 300 detectives. . . . In addition, holes for rifles were put in along the fence, platforms were constructed for sentinels, and barracks built inside the fence to house the strikebreakers.

In her autobiography, Emma Goldman (1931, pp. 86) told what happened next:

> . . . the news was flashed across the country of the slaughter of steel-workers by Pinkertons. Frick had fortified the Homestead mills, built a high fence around them. Then, in the dead of night, a barge packed with strike breakers, under protection of heavily armed Pinkerton thugs, quietly stole up the Monongahela River. The steel-men had learned of Frick's move. They stationed themselves along the shore, determined to drive back Frick's hirelings. When the barge got within range, the Pinkertons had opened fire, without warning, killing a number of Homestead men on the shore, among them a little boy, and wounding scores of others.

At the very same time, however, that some of us seek to increase our degrees of freedom by bringing influence upon others (their employees, for example) with guns, other people are trying to find ways that people with little power can find enough degrees of freedom, despite those who point guns or use other threats, to live lives reasonably free from nagging anxieties and debilitating conflicts. The latter people never cease thinking up ways of improving their freedom—some, of course, more effective than others. One way, the one adopted by Emma Goldman, is anarchism. Another is the representative democracy designed in the late eighteenth century in America and France. Another is the labor union movement.

Many people think of a labor union as no more than a protective device. It is that, but it can be more than that. In 1994, the National Center on the Educational Quality of the Workforce at the University of Pennsylvania carried out a survey of 1,500 workplaces. About 20 percent of those were unionized. Some shared profits with their employees, some did not. Some had formal quality-enhancing programs such as Total Quality Management, some did not. Lynch and Black (1996) wrote about the relative productivity of these 1,500 workplaces. They took the workplaces without unions and without any formal quality-enhancing program, but with limited profit sharing, as "typical" companies, and took the average productivity of those companies as a baseline against which to compare the productivity of other companies.

Lynch and Black found that the average nonunion workplaces had an average productivity 11 percent lower than the baseline. The unionized workplaces without programs that provided regular discussion of workplace issues—in short, workplaces where union and management took the traditional adversarial stance—showed productivity 15 percent lower than the baseline. The average of all unionized workplaces was 16 percent higher than the baseline. In unionized workplaces where formal quality programs enabled workers to meet and discuss workplace issues, where more than a quarter of the workers worked in self-managed teams, and where workers shared in the firm's profits, the average productivity was 20 percent higher than the baseline. Calling these last companies "high-involvement" companies, here is a tabular comparison:

Type of company	Labor productivity % above baseline
Unionized, high-involvement	20
All unionized	16
Nonunionized, high-involvement	10
All nonuninized	11
Unionized, adversarial	15

Lynch and Black say that the superiority of the unionized places may arise from the belief of workers that if they propose improvements in job practice, they will not be proposing themselves out of a job. Also, since unionized companies have, on the average, lower turnover, on-the-job training can have greater effects.

I am not saying hurrah for increased degrees of freedom for the reason that you can get more work out of your employees. I am saying that where energy and time are being used in unproductive conflict, increased degrees of freedom can reduce the conflict. If your employees then choose to use some of their released energy and time to help you with your profits, you should say, "Thank you."

Self-Managed Groups

In recent decades, a growing number of people have been searching for ways to bring more degrees of freedom (though few use that term) to persons employed in jobs requiring deadening repetition. Earlier in this chapter, I gave an example of the deadening repetition under "Assembly Line." Here is another quick example of deadening repetition. I once worked in a factory that made aluminum name plates and dial faces. I sat at a table. With my left hand, I picked up a strip of aluminum perhaps five inches wide and 15 long. I pushed the strip into a slot. With my right hand, I lifted a handle. That action put a fold in the aluminum strip. With my right hand, I pulled out the strip and put it aside. With my left hand, I picked up another strip. I shoved it into the slot. I lifted the handle. I pulled out the strip and put it aside. With my left hand. . . .

That could have become slumberous. I amused myself, however, by playing with the quota. I was required to produce a certain number of folded strips per hour. I soon found that I could produce a good many more than that. I did not, however, want to produce more than the quota, because my fellow workers would not have approved. I made a game for myself in which I would for a while fold the strips too slowly to produce the required number by the end of an hour. Then I would work faster than usual, ending with the required number by the time an hour had passed. I played several variations of that game. Eight hours a day.

I gave a few examples of self-managed groups in Chapter 27 under "Increasing Degrees of Freedom." Here I will give a few more.

Soldani

James Soldani (1989) wrote a chapter giving examples of how PCT can be put to use in a manufacturing company. He wrote:

> What managers fail to understand is that setting goals for organizations through senior management oratory or written directives does not guarantee that people in the organization will internalize these goals and work for them. . . . Even high pay, promotion, and other incentives will not always work (p. 518).
>
> . . . I once worked for a company which spent over one million dollars on team development training over a four-year period, all to no avail.

> [C]ontrol theory has helped me to develop teams that actually work (p. 524).

I will not take space to describe Soldani's teams. I will, however, summarize (from p. 527) some of the results of instituting the new self-reliant teams in which "the people themselves would be empowered to remove obstacles that kept them from doing their best" (p. 528):

> Getting orders out on time rose from 23 percent to 98 percent.
>
> Overtime (as a percentage of total time worked) declined from twelve percent to three percent, saving $17,000 per month.
>
> Because of warehousing costs, companies like to keep inventories small. Counted in the number of days' supply kept waiting in the warehouse, the inventory went down from 75 days to 52 days, effecting a saving of $2,100,000.
>
> Defects per unit went from an average of 1.26 to 0.25.

Soldani seemed to get the point:

> The challenge of productive personnel management is essentially: ... finding the means to control what we want without infringing upon the rights and abilities of others to do the same (p. 529).

Gore

W.L. Gore and Associates, the manufacturers of (among other things) the fabric they call GORE-TEX, extends degrees of freedom for workers in directions of which few managers have even thought. The company began in 1958 with financing from a mortgage on the Gores' house and $4000 from savings. By 1992, it had 5000 employees and was approaching $1 billion in sales. Here are some excerpts from an article about the company by Shipper and Manz (1992):

> Self-management involves an increasing reliance on workers' creative and intellectual capabilities.... In many companies, organizing work around small groups of workers empowered to perform many traditional management functions (assigning tasks, solving quality problems, and selecting, training, and counseling fellow team members) has become a way of life (p. 48).

> Almost always, employee self-management is introduced in organizations through ... formally designated work teams. When employees are hired, they are assigned to a work team as a condition of their employment.
>
> In this article, we will [describe a company] without formally designated teams. Instead, the whole work operation becomes essentially one large empowered team in which everyone is individually self-managing and can interact directly with everyone else in the system... (p. 49).

PCTers insist that everyone in every organization is self-managing in the sense that everyone controls his or her perceptions, regardless of whether some boss is authorizing the person to do so. When Shipper and Manz write that everyone "is" individually self-managing, I believe they mean that the president and the secretary-treasurer told the workers it was all right for them to act in an openly self-managing way as part of their jobs. The Gore Associates do not have to pretend they are *not* doing so. Gore employees say their company is "unmanaged." William Gore called the coordinating arrangements a "lattice structure." Shipper and Manz described it thus:

1 Lines of communication are direct—person to person—with no intermediary.
2 There is no fixed or assigned authority (p. 53).

For a company of its size, Gore may have the shortest organizational pyramid found anywhere. The pyramid consists of Bob Gore, the late Bill Gore's son, as president, Vieve, Bill Gore's widow, as secretary-treasurer, and all others—the associates (pp. 54–55).

3 There are sponsors, not bosses.
4 Natural leadership is defined [chosen] by followership.
5 Objectives are set by those who must "make them happen."
6 Tasks and functions are organized through commitments (p. 54).

Associates are aided in helping one another by the role of "sponsor":

> Before anyone is hired, an associate must agree to be the new employee's sponsor.... The sponsor is to take a personal interest in the new associate's contributions, problems, and goals, and to serve as both coach and advocate. The sponsor tracks the new associate's progress, providing help and

encouragement.... Sponsoring is not a short-term commitment. All associates have sponsors... (p. 52).

The structure within the lattice is complex and evolves from interpersonal interactions, self-commitment to responsibilities known within the group, natural leadership, and group-imposed discipline.

Bill Gore once explained..., "Every successful organization has an underground lattice. It's where the news spreads like lightning, where people can go around the organization and get things done." [There is] constant formation of temporary cross-area groups. In other words, Gore has "teams without formally designated teams."... Associates can team up with other associates, regardless of area, to get the job done. When a puzzled interviewer told Bill that he was having trouble understanding how planning and accountability worked, Bill replied with a grin, "So am I. You ask me how it works. [The answer is, it works] every which way." (p. 54).

Notice the empirical attitude there, quite the opposite of a rational, "logical" attitude. Most managers are reluctant to undertake a project they cannot explain to themselves. They want to think, then act. The innovator acts, then thinks. I don't mean the innovator is careless or rash. Karl Weick might characterize the attitude by, "How can I know what I think before I see what I can do?" Shipper and Manz continue:

Not all people function well under such a system, especially initially.... As Bill Gore said, "All our lives most of us have been told what to do, and some people don't know how to respond when asked to do something—and have the very real option of saying no—on their job. It's the new associate's responsibility to find out what he or she can do for the good of the operation." The vast majority of the new associates, after some floundering, adapt quickly.

Compensation comprises salary, profit sharing, and the Associates' Stock Ownership Program. Salary reflects the associate's contribution to the "good of the operation." Salaries are reviewed once or twice a year. A review is conducted by teams drawn from the associate's work site, and one or more sponsors act as the reviewee's advocate. One quality assessed is the reviewee's ability and willingness to help others.

Just as the work groups at Gore come and go as needed, so do leadership roles. As a group forms, a leader appears—or perhaps two or three leaders, each for a particular aspect of the work. The ability to show this kind of flexible leadership is another quality assessed at review time.

Many people believe that groups or organizations with Gore's loose kind of leadership must necessarily spend a lot of time in discussion and must respond sluggishly to demands for quick action. Quick response can occur, however, where only those workers necessary to the response take action (and this can occur where there is no fixed organizational hierarchy) and where those taking action have cooperative expectations of one another and skill in the work and in discussion. Here is an example given by Shipper and Manz:

... in 1975, Dr. Charles Campbell, the University of Pittsburgh's senior resident, reported that a GORE-TEX arterial graft had developed an aneurysm.... Obviously, this ... problem had to be solved quickly and permanently.

Within only a few days of Dr. Campbell's first report, he flew to Newark to present his findings to Bill and Bob Gore and a few other associates. The meeting lasted two hours. Before it was over, Dan Hubis, a former policeman who had joined Gore to develop new production methods, had an idea, and he returned to his work area to try some different production techniques. After three hours and twelve tries, he had developed a permanent solution. In other words, in only three hours a potentially damaging problem to both patients and the company was resolved. Furthermore, Hubis's redesigned graft has gone on to win widespread acceptance in the medical community (pp. 58–59).

By 1965, seven years after founding, the company had grown to 200 employees, and William Gore discovered that he could no longer recognize everyone as he walked through the plant. Since then, as a plant approaches 200 employees, a new facility is begun. This policy is called "getting big by staying small." This policy fits E.F. Schumacher's (1973, p. 242) dictum: "The fundamental task is to achieve smallness *within* large organization." The number 200 fits reasonably well into the groupings anthropologists have found among primitive peoples. John E. Pfeiffer (1978) wrote that bands of about 25 were repeatedly found among hunter-gatherers, and "adult males tend to form working groups of six to eight individuals, which is the number generally included in a 25-member band (p. 312). Tribes of hunter-gatherers frequently contained about 500 people, with extremes of 200 and 800. These figures include women, children, and oldsters. Pfeiffer wrote (p. 313):

> There seems to be a basic limit to the number of persons (1) who can know one another well enough to maintain a tribal identity at the hunter-gatherer level; (2) who communicate by direct confrontation; and (3) who live under a diffuse and informal influence... (p. 313).

Those three characteristics sound like W.L. Gore and Associates, if you interpret the "hunter-gatherer level" as hunting for effective ways to get work done and gathering suitable people to do it.

It is easy to see that the mode of operation of the Gore associates provides rich environmental opportunities, both material and social, for controlling perceptions, and provides strong invitation to use those opportunities. Not only are the degrees of freedom many, but the variety of means for control offered by the environment is immense. It is not surprising that initiative and inventiveness appear in all parts of the company. For more on Gore, you can read the articles by Maureen Milford (1996), Michael Kaplan (1997), Dawn Anfuso (1999), Glenn Hasek (2000), and Rebecca Quick (2000).

Semco

Another company much like Gore, perhaps somewhat more structured, is Semco, a manufacturing company in Brazil. Semco, too, found that small facilities worked better than large, and reduced the size of its installations to fewer than 150 employees. To give you a bit of flavor, I'll quote two sentences from a 1989 article by the company's president, Ricardo Semler:

> One of my first moves when I took control of Semco was to abolish norms, manuals, rules, and regulations.... [C]ivil disobedience [is] not an early sign of revolution but a clear indication of common sense at work.

Other Organizations

Gore is not the only company to let employees find their own jobs. Anne Miner (1987) studied the occurrence of "idiosyncratic" jobs among nonfaculty in a large university. Will, Hickman, and Muska (1998) described self-designed jobs in Oticon A/S, a Danish manufacturer of hearing aids. Forward, Beach, Gray, and Quick (1991) wrote about Chaparral Steel, which has neither job descriptions nor restrictive work practices. Bowen, Ledford, and Nathan (1991) wrote about "hiring for the organization, not the job." E.F. Schumacher (1979, pp. 76–83) told about a plastics company in England called Scott Bader, which was reorganized in 1951 to provide increased degrees of freedom for the employees.

Two notable cases of very flexible teams among the many examples available are the Saturn plant, part of General Motors, and the Opel factory in Eisenach, Germany. You can read about Saturn in an article by S.C. Gwynne (1990) and a book by Joe Sherman (1993). And about Opel in an article by Adolf Haasen (1996).

Still Others

Here and there, in this book, I have called your attention to writings by authors ignorant of PCT who nevertheless wrote well about some aspect of human functioning which is also crucial to PCT. For example, I spent a good part of Chapter 4 on "Other Appearances of Purpose." In Chapter 27 under "Increasing Degrees of Freedom," I described some theorizing and some experiences concerning degrees of freedom. In this chapter, I have been giving still more examples. It would be easy to believe that Gore and Semco had been designed by a member of the CSGnet.

In the last decade or so, the phrase "learning organization" has appeared every so often in the professional journals. Authors who write with that phrase do not usually have in mind learning to do things in the customary, time-honored way. Instead, they mean generating ideas for doing things that depart from the old ways. Slocum, McGill, and Lei (1994, p. 38) said this:

...these practices, standing in sharp contrast to strategies that have worked so well in the past, allow firms to produce goods and services for consumers anytime, anywhere, and for anything. What are these new practices? The first is developing a strategic intent to learn new capabilities. The second is a commitment to continuous experimentation. The third is the ability to learn from past successes and failures.

You can see that people in a "learning organization" would not be happy with an employee who wanted to know the right way to do his job—the specific daily acts he should be exhibiting. The designers of learning organizations seem very conscious of the fact that people seek various ways of controlling their perceptions —"the fixed end, the varying means!"

As is the case with many other illustrations in this book, I do not claim that these examples concerning degrees of freedom are verifications of PCT. They could equally well illustrate other theories, and for most of the authors of the illustrations, they do. For example, astrologers might (for all I know) predict that a lot of people will be jolly when the earth is lined up in a certain way with other heavenly bodies in late December, but others of us can think of other reasonable explanations. Verification lies with the models, not with these prosaic exhibits. I intend my illustrations merely to bring closer to familiar experience the implications of controlling perceptions. Although my illustrations do not test the correctness (or adequacy) of PCT, I think they show that it is easier for most of us, when thinking with PCT, to think of ways to arrange opportunities in the social environment so as to increase degrees of freedom than it is when thinking with linear causation and reward-and-punishment.

Because I do not wish to overburden you with examples, I will stop them here. I cannot, however, bring myself simply to discard a lot of good examples I came across while gathering other writings to help me with this book. In an endnote[1], therefore, I list some further examples in case you want to look them up. They show more ways in which people try to increase their degrees of freedom, succeeding or failing in the effort.

SUMMARY

One point I wanted to make here is this: Whenever you presume to describe an actual group, you are always giving a history. The group is no longer what it was when you began talking. Saying that is the same, of course, as saying that when you control a perception, you are always controlling your memory of it. I explained that in Chapter 19.

Figure 34–1 is a chart of the different degrees of helpfulness, and therefore of the different degrees of freedom, perceived by the twelve people in that group. And as the members of the group find their estimates confirmed or disconfirmed, their estimates will change, and the directions of their actions will change. The change on the part of any one person will be perceived by the others, and some of the directions of *their* actions will change. And so on. Rocks just lie there and let you kick them. But people act to rectify matters.

A second point is vital. If you want to improve society, the key thing is to enlarge the proportion of people who have ample degrees of freedom. People with ample degrees of freedom are less likely to act so as to reduce *your* degrees of freedom; they are not the people with internal conflicts and reorganizations that cause them to flail in all directions (and possibly in yours) in their search for solutions to their conflicts. Further, because people with ample degrees of freedom will not be using time and energy struggling against conflicts, they can turn some of that time and energy to helping to maintain a society that will provide ample degrees of freedom for their children and grandchildren—and, in passing, for you and yours. A few people who at present feel blissfully full of freedom and free of conflicts will feel threatened by a rise in the proportion of people who are also happy with their lives, but those threatened people, already rich in opportunities, will find further opportunities for themselves rather quickly.

You may say, with reason, that spreading degrees of freedom around is going to be a long-term, complex thing to do. That's true. But you can begin with yourself and the people with whom you deal every day. Keep asking yourself, "Am I interfering with this person's ability to control his own perceptions?" And conversely, too.

ENDNOTE

[1] The following writings concern the "learning organization": William N. Isaacs (1993), Kofman and Senge (1993), McGill and Slocum (1993), Edgar H. Schein (1993). The following display the search for degrees of freedom in organizations within cultures other than European or North American: Thomas A. Carey (1996), Robert E. Cole (1989), David Cohen (1985), Geert Hofstede (1993), Lindsay and Dempsey (1983), Dana Thompson (1994), James A. Wall (1990). The following describe American gains in freedom, especially by organizing work in teams: Thomas J. Atchison (1991), Banker, Field, Schroeder, and Sinha (1996), Conger, Spreitzer, and Lawler (1999), Elizabeth Corcoran (1993), Galbraith, Lawler, and associates (1993), Jamieson and O'Mara (1991), Edward E. Lawler (1986, 1991, 1992, 1994, 1996, 2000), Lawler and Galbraith (1994), Lawler, Mohrman, and Ledford (1995), Lawler, Mohrman, and Ledford (1998), Mohrman, Galbraith, Lawler, and associates (1998).

Chapter 35

Coordination

As one individual acts or proposes action with another, her action alters opportunities for controlling perceptions—both the other's and her own. Further, that act alters the possibilities for still others to make use of these two individuals. The result is that the readiness of every one of the interacting individuals to make use of others will adjust and shift. A sort of continuous hunting or *tracking* process goes on as every individual assesses the changing opportunities for controlling perceptions, takes actions to control them, and in doing so further alters the distribution of opportunities for herself and everyone else. The availability of the human resources in a group and the actual use of them are dynamic, in constant flux, protean.

Many others have seen this unceasing mutation. Arrow, McGrath, and Berdahl (2000, pp. 56–57) say this:

> Local group process creates, activates, replicates, and adjusts dynamic links in a coordination network.... From local action, global-level patterns emerge—behavioral and cognitive patterns such as group norms, cohesion, division of labor, a role system and influence structure, and temporal patterns such as cycles of conflict and consensus, regularities in changing group performance, and the ebb and flow of communication....
>
> This is not the kind of relationship traditionally modeled by independent and dependent variables. Rather, we are talking about contextual factors that constrain the operation of local-level rules without determining the outcome. The whole pattern of global dynamics that emerges from this local action may shift when a contextual parameter shifts to a different value. Or it may remain unchanged....
>
> Complex systems whose behavior depends largely on interactions among local elements—the weather, for example—are predictable only in the short run, and these predictions are for global variables..., not local variables.... Patterns of key global variables, however, do show substantial regularities over time.

The vocabulary of those authors differs a good deal from mine. Where I would speak of a demand or proposal an individual makes to others, they speak of local dynamics. Where I would speak of environmental opportunities, they speak of contextual factors. Nevertheless, it is easy to see in those passages a picture of patterns arising from the actions of individuals, patterns that change their actors or their shapes or both as the social opportunities change. It is easy to see, too, that the authors do not aspire to predict particular acts when they write of environmental conditions that "constrain . . . without determining the outcome." They see the regularities of outcome not in individual acts, but in the statistical outcomes ("global variables") constrained by a preponderance of similar individual goals and by the use of opportunities from a common environment.

ORGANIZING OURSELVES

Maybe you remember how it was when you were a child and you walked onto a playground in a neighborhood new to you. Or you can think of how it is when you go to a party where there is no one you know. Or think of the first hour of the first day on a new job. You don't know whether to bet that someone will have an eye out for newcomers and will come over to greet you or whether everyone will ignore you until you say something. You don't know whether people will expect you to state your credentials or to say something nice about the weather. You feel many uncertainties, but you don't expect utter chaos.

You expect to find, before long, some patterns of action. If someone comes forward to greet you, you do not take that as a random action. Rather, you think, "People in this group expect to make at least a little effort to be welcoming. They probably won't ignore me if I invite a little attention to myself." Imagine how uncertain and disconcerting social life would be if we did not find some repeated patterns pretty quickly. Imagine how much time would be wasted, without quickly discernible patterns, customs, or routines, in dealing with other people—in getting work done, playing a game, making a purchase, carrying on a courtship.

To reduce the evils of strife and reap the benefits of cooperative work, we must replace what would otherwise be wild uncertainty with some degree of reliable expectation about the actions of other people—and about our own actions, too. That, at least, is what we all seem to believe. We all act on the social environment to restrict and order perceptions of the possible actions of ourselves and others. We tell children to stay in their seats until the bell rings. We tell employees to show up at nine o'clock in the morning. We tell the customers to stand in line at that window. We tell the subjects of a king to kneel in his presence, and similarly the members of the flock of the Pope. When the judge walks into the courtroom, the bailiff shouts, "All rise!"

Thus arises culture: customs, rituals of membership and passage, laws, agreements, rules, norms, traffic signs, administrative manuals, and instructions for opening a box of breakfast cereal. Thus arise groupings of families into clans and communities in which customs and rules can be well enough understood by all individuals and to which work groups can bring the benefits of cooperation.

If we see a cluster of men walking down the center of the street, all dressed pretty much alike, all wearing ten-gallon hats, with two or three at the front carrying banners, and all keeping close together as they walk— at least until they go out of sight around a corner—it is easy to perceive a repeatable, recognizable, namable pattern there and think of that pattern moving down the street as a "thing"—maybe a "bunch," maybe a "parade"—just as we can think of the little pattern we perceive moving along behind the men as a "dog." Then it is easy to look for the properties or characteristics of that "thing" or "group." And it is easy to ask how individuals *affect* a group—that is, how this *thing* (an individual) affects that *thing* (a group).

We remain, however, individual entities. If we perceive the clustering as an activity of individuals, we are more likely to ask how this pattern of groupiness can come about—what are the capabilities of humans that can enable them to march in a cluster down a street, wearing ten-gallon hats on their heads? How are they helped and hindered when they control their perceptions by using ten-gallon hats, other marchers, and other features of their environments?

It is true that human individuals acting as groups can do things they cannot do as separate individuals— carry a large canoe to the water or build a skyscraper, for example—thereby expanding their powers. But those cooperative patterns of action do not make a group into a system or organism or entity that is somehow a "thing" bigger than the individual. Behavioral functions connect individuals in the uncertain and shifting manner shown in Figure 28–5. We can be mistaken when we see people behaving as if they were all physically connected in thinking that something more is going on other than individual control of perception. You saw that misapprehension illustrated in Chapter 9 under "The Crowd."

Groups cannot have perceptions separate from those of their individual members, and they do not have reference signals (internal standards) separate from those of their individual members, and therefore they cannot maintain purposes separate from those of their individual members. In a group, when individual purposes come into conflict, individuals can leave the group to pursue their separate purposes. In an individual, when pursuits of purposes come into conflict, no purpose can leave the individual and go off by itself.

The idea of living entities of larger and larger scope is an old idea. (See under "Reification" in Chapter 5 and "Group Mind" in Chapter 8.) I am reminded of a story by Arthur Conan Doyle (1859–1930) about a scientist (whom he named Professor Challenger) who drilled a deep hole through the earth's crust and dropped an immense steel spike into it—and the earth screamed. A book is currently in print called *Gaia*; it conceives the whole earth with all its inhabitants as a unitary living organism. Maybe before long someone will write a book explaining how the galaxy or the universe is a living creature.

A great deal of trouble comes upon us when we think of a group (or organization) as a "thing." We come to think that others must agree with us about the nature of the group as well as they agree with us about "things" like apples and wheelbarrows. Interminable arguments go on containing complaints like, "Well! I thought our purpose was to . . .," as if "our" purpose could be unarguably singular. We attribute features to the group like those we attribute to individuals—goals, emotions, energy, persistence, and so on. We are surprised, sometimes to our harm, when we find a member of the group acting contrary to what we thought was the nature of "the group." I gave an example about revising a curriculum in a school district under "Taking a Vote" in Chapter 10.

HOW CAN COORDINATION COME ABOUT?

If you and another person pick up a canoe, you both can get clues about how to move from the tugs and shoves your hands get because of the motions made by the other person. Without uttering a word, both of you can manage to walk at about the same pace because of those input signals that tell you how close you are coming to controlling a variable you want to control—namely, walking in concert, not in conflict, with the pace of the other person (while simultaneously perceiving progress toward your destination). Judging from the achievements of bees and beavers, one of you could no doubt propose carrying the canoe in a particular direction and the other could accede, both the proposing and the acceding occurring without language of the human sort. Human language, however, makes possible the fine coordinations of a game of pinochle and the landing of airplanes at a busy airport.

Even with human language, how can humans come to display coordination? How can the questions in the Group Expectation Survey in Chapter 34 convey useful meanings? For language to be useful, individuals must somehow connect a word to similar experiences with a world that exists regardless of the language. My experience of an apple can never be exactly yours, but our experiences can be sufficiently similar, and our associations among words sufficiently similar, so that when one of us says "apple" and "yummy" and "eat," the other can make reasonably reliable predictions about intentions and even, given some further information, about actions.

McPhail

Clark McPhail (2000) has written a nice piece on this topic; his earlier discussion in Chapter 6 of his 1991 book is good, too. Here are some excerpts from his 2000 piece:

> How are two or more individuals, each with unique personal histories stored in memory, able to interact with one another . . .? There are at least two compatible answers. . . . First, most all of us have headaches and stomachaches. From experience, we know that all the headaches are not alike, but they have sufficient similarities among them, and sufficient distinctions from stomachaches, that we place the former in a distinctive category labeled "headaches." We do the same with our experiences of stomachaches, toothaches, backaches, and even heartaches.

McPhail's second answer requires more space:

> For two or more individuals to engage in collective action—either parallel actions at the same time or different actions taken simultaneously or sequentially—they must adjust their respective individual actions to [control] similar or related reference perceptions. There are three or more ways in which similar or related reference perceptions can [come about].
>
> *Independently.* People who have interacted a great deal with one another, who are part of the same daily rounds, social networks, groups, and cultures, are more likely to have similar named categories of experiences stored in memory from which they can independently [construct] similar reference perceptions. . . .
>
> *Interdependently.* We all have experienced something we did not initially understand clearly. We all have been confronted with a task we could not complete by our actions alone. . . . When one person requires the assistance of another, or when two or more people are confronted with a mutual problem to be solved, they can interact by signifying in words and gestures what needs to be done, and who will do what, when, where, and how. Thereby they interdependently establish the similar or related reference perceptions in relation to which . . . they will adjust their respective actions. . . .
>
> *Adoption from a third party.* The more complex the problem to be solved, and the more people required for that solution, the more important it

becomes to have a single source of reference perceptions. I refer to that single source as a "third party" who [offers] reference perceptions for adoption by two or more other individuals. [Those individuals] must adopt the proposed reference signals as their own and then adjust their ... actions ... to realize the intended outcome.... Familiar examples of third parties include the principal organizer for large events (protests, weddings, funerals, reunions), the coach of an athletic team, the director of a church choir....

... it is frequently the case that the actual implementation of what the third party asks other individuals to do assumes that they can and will independently draw upon their individual memories for additional bits and pieces of cultural knowledge to supplement what the third party has requested or proposed. Thus, the three sources of similar or related reference perceptions—independent, interdependent, or third party—can operate separately or in various combinations.

Taylor

Once internal standards (reference signals) have become sufficiently similar and their similarity is recognized by members of a group, helpful coordination among members can be arranged. This is not to say that the coordination will be so dependable as to be unfailing or permanent. Despite similar internal standards, individual differences will produce continual alterations. I like the picture of this situation that Martin Taylor gave in a message to the CSGnet on 13 April 2000 of how coordination can come about:

Assertion 1. All actions have side-effects, which are defined as effects on the environment that do not influence the perception the action is to control. Some side-effects may disturb perceptions in other people.

Assertion 2. Some perceptions are easier to control through the actions of another person than by one's own direct actions. One's own actions are intended to disturb the other person's perceptions in such a way that the other person's controlling actions influence one's own controlled perceptions appropriately (to the other person, that influence is a side-effect).

Those two assertions summarize what I said in Chapter 28 in connection with the five figures there. Taylor continues:

Assertion 3. The power of one person to influence some aspect of the environment is ordinarily less than the power of the person coordinated with the power of another, both influencing the same aspect of the environment in the same direction.

"Power" is not a technical term here. I think Taylor intends it to be more or less synonymous with ability or capability.

Assertion 4. Persons differ in their individual power to influence any particular facet of the environment....

Consequence 1 (from A1 and A2). If two people find that each can control a perception within themselves better through the actions of the other, a "contract" can be made between them. Both control better when the contract is executed as agreed.

Consequence 2 (from A1 and C1). In some cases of "contract" one or both of the partners may find that the counteracted action has side-effects that disturb another controlled perception, or worse, that executing the contract induces internal conflict.

Consequence 3 (from A1 and C1). All contractual actions have side-effects that affect the environment. In particular, the side-effects may disturb controlled perceptions in people not party to the contract.

Consequence 4 (from C1 and C2). Contracts are likely to be broken if the contracting parties incorrectly perceive the likelihood of conflict in the other partner (i.e., do not perceive correctly what perceptions the partner is controlling that make the partner desire the contract).

Consequence 5 (from C3). Persons outside the contract will attempt to influence the actions of the contracting persons....

Consequence 6 (from A3, A4, and C5). Persons outside the original contract will contract together to influence the actions of the originally contracting persons. (Argument: the original contracting parties have contracted because the contract enhances their power to influence the environment. This makes it more likely that their power to disturb others is greater than the power of a random other person to resist the disturbance. [Therefore persons outside the original contract are likely to contract to counteract the disturbance.])

At a later place, Taylor says:

> ***Consequences 6, 7, and 8*** do not say "law and regulations," but they come very close.

Taylor continues:

> ***Consequence 7*** (from A1, A4, and C5). Persons will contract together to influence the actions of powerful people.
>
> ***Consequence 8*** (from C2). Persons creating a contract may also contract with outside persons to influence each other to perform according to the terms of the contract.

This consequence is the basis for customs and laws that make breaking a contract unprofitable. For example, two people being married may agree to observe the religious customs presided over by the priest in return for the priest's public announcement of the couple's intention to remain mutually helpful to the end of their lives—an announcement that may help them to remain mutually helpful. Taylor continues:

> ***Consequence 9*** (from C6 and C7). Powerful people and contracting persons are likely to oppose the actions of the external persons who contract to influence their actions.

For example, the marrying couple may later want to dissolve their marriage through the help of a powerful person. A major contributor to the church's coffers, a bishop, or the Pope may help persuade the priest to issue an annulment.

The "contracting" (which seems to me equivalent to agreeing to cooperate) that Taylor describes begins in the dyad and the group before it can become coordination among groups and organizations. I quote Taylor here because one can so easily see "contracting" undertaken to enlist aid, and one can see the consequences he lists. Coordination comes about through the cooperative efforts Taylor and McPhail describe. I say again, however, that a pattern of coordination is not something which, once built, remains intact like the great pyramid at Giza for the next 4600 years. Coordination is constantly renewed and reshaped.

Powers

Coordinating or organizing is an attempt to reduce the possible variety of things that can happen—an attempt to make events more predictable. Using the concept of degrees of freedom, Powers (1979) explains how this process comes about in social life:

> We are not absolutely free to indulge in certain behaviors because, I propose, neither we nor the environments with which we effectively deal possess an infinite number of degrees of freedom in the mathematical-physical sense. . . . As will be shown, mere masses of people do not create correspondingly large numbers of degrees of freedom in any sense; in fact, quite the opposite can occur (p. 222)[1].
>
> . . . a control system controls some particular *aspect* of its environment. How many aspects might there be of a given environment? As many aspects as there are of combining elementary sensory stimuli. . . (p. 223).

I discussed the matter of combining elementary sensory stimuli in Chapter 18 under "Second Order: Sensation." Powers continues:

> The number of degrees of freedom in the perceptual world is limited, of course, by the number of degrees of freedom in the physical universe outside, but it is much more severely limited than that: it is limited by the number of different aspects of the environment that a given organism is prepared to sense at a given time (pp. 223–224).
>
> . . . at each level in the hierarchy there is a problem of degrees of freedom [at] any time [when] many higher-order systems act simultaneously on and through the same set of lower-order systems. The problem is simply that of avoiding internal conflict and losing control altogether. We can now put this aspect of "degrees of freedom" together with . . . the idea that freedom is never absolute. . . (pp. 226–227).
>
> . . . the limitation of freedom is imposed by the fact that doing certain acts that satisfy one set of goals or purposes . . . can cause other controlled perceptions to depart from their referenced levels. Conflict can be created, depending on one's structure of perceptions and the reference levels that go with them. If a person wishes strongly to avoid going to jail and also wishes strongly to shout "Fire!" in a crowded theater, he has a problem. . . . it is a direct conflict, in that satisfying both reference levels . . . is essentially impossible (p. 227).
>
> . . . it is not a physical model of the environment with which we normally interact and within which we find our goals. The worlds we attempt to control relative to our goals, and the goals themselves, are made up of automobiles

and hamburgers, jobs and vacations, bowling and cross-country skiing, passing algebra, and plying ladies with gifts to overcome resistance. It is almost entirely a manufactured world, a world divided into familiar perceptual categories and familiar examples of each category (p. 232).

I will give an example. When I was perhaps eleven years old, several other boys and I were one day wondering what to do with ourselves. After three or four suggestions had brought no enthusiasm, I said, "Let's play marbles." Silence. One of the boys was staring at me with astonishment. He closed his mouth, gulped, and exclaimed, "It isn't the marble season!" Powers continues:

> It is a rather small world, the smaller to the extent that we come to share more and more classes of perceptions rather than creating our own categories and examples (p. 232).
>
> On the one hand, by banding together and pooling our efforts, we can achieve for all of us what none could achieve for himself. On the other hand, by banding together and creating a shared reality, we reduce the size of the universe in which we live, narrowing the choices of goals and the actions recognized as means toward goal achievement (pp. 232–233).

A shared reality makes life more predictable in some ways. Observing that result, many people believe also that a shared reality must reduce the frequency of interpersonal conflict. On the contrary, avoiding conflict requires *increasing* the degrees of freedom, so that the degrees used by one person still leave enough degrees for another. I believe this conclusion can also be deduced from Taylor's assertions.

J. Richard Hackman (1987, pp. 338–339) also has something to say on this point. While Powers speaks of increasing degrees of freedom, Hackman speaks of providing redundant conditions:

> . . . different task and organizational circumstances involve vastly different demands and opportunities. Thus it is impossible to specify in detail what specific behaviors managers should adopt to help groups perform effectively. There are simply too many ways a group can operate and still wind up with the same outcome. . . . the key to effective group management may be to create redundant conditions that support good performance, leaving groups ample room to develop and enact their own ways of operating within those conditions.

In much of the literature on organizational management, managers are urged to influence members, by hook or crook, to commit themselves to the "organizational goal." The assumption seems to be that the organization cannot function well unless almost everyone puts his or her shoulder to the same wheel. That is a wrong idea. The fact is that an organization can function well even if only a few people are enthusiastic about the goals of the organization (which are almost always the goals of a few leaders) if the attitude of the rest is no worse than apathetic—that is, if no one actually sabotages the work. People can carry out a joint task willingly and with profit of some sort for all if doing their parts of the task does not obstruct them or others from pursuing their own goals. The political scientists Dye and Zeigler (1990, p. 135) have asserted this same idea to in the realm of national politics:

> It is the irony of democracy that democratic ideals survive because the masses are generally apathetic and inactive. Thus the capacity of the American lower classes for intolerance, authoritarianism, scapegoatism, racism, and violence seldom translates into organized, sustained political movements.
>
> The survival of democracy does *not* depend on mass support for democratic ideals. It is apparently not necessary that most people commit themselves to a democracy; all that is necessary is that they fail to commit themselves actively to an antidemocratic system.

The chief point I am emphasizing here is that while much coordination comes about simply by offers of help and requests for it, members of organizations frequently try to achieve further or more "rational" coordination by trying to increase predictability, hoping (misguidedly) thereby to reduce conflict. Writers in journals devoted to organizational management repeatedly urge managers to persuade all their employees to cleave to a common goal. That is not necessary to harmony, coordination, or productivity. On the contrary, it may actually increase conflict.

STABILITY

Just as we can see a "group" when we see a collection of individuals behaving nonrandomly with one another, so we can perceive the qualities of stability and permanence as we see patterns repeating. When we see a collection of people meeting most days during every year in rooms under a large dome in Washington, D.C., we call that pattern the United States Congress, and we say that "it" has been meeting since 1789. (Well, nobody *sees* that, because none of us lives that long. The marvel of language, especially written language, enables us to form that conception of continuity from what we read.)

I have no need to agree with anyone on the degree of similarity or difference between the Congresses addressed by George Washington and George W. Bush. I am merely illustrating how our ability to perceive patterns also enables us to conceive stability. We can then perceive a relationship between the Congress and the quality of stability. And between the Broomwoods Bridge Club or any other group and the quality of stability. Then we can look upon the stability of some things as good and the stability of others as bad. And we often differ among ourselves as to whether the stability of something is good or bad. Some cry, "Honor the ancient traditions!" Others cry, "It's time for a change!"

The yearning for stability, perhaps along with the belief that a particular "stimulus" should produce a particular "response," can be seen in efforts to improve groups or organizations by some sort of program—some sort of curriculum in schools, some sort of management style in industry, some sort of reform in a church, and so on. Will the program be here today and gone tomorrow, or will it last?—we ask. Are we doing it right?—the participants ask. But the program you adopt today cannot be the program you will be pursuing a year or five years from now, and the criterion for whether you are doing it right cannot be faithfulness to a recipe, but instead the effectiveness with which you control the perceptions you want to control.

MEMBERSHIP

Members of a group demand evidences of certain kinds of internal standards from other members as dues paid, so to speak, for being members. They demand, for example, certain kinds of helpfulness that they would not demand from nonmembers. They demand certain kinds of coordination—such as showing up on time for meetings, saying "Yes, sir" to certain members, or wearing clothing of a certain cut when acting as a member. You can think of other examples. When members of a group make certain demands of you and accept certain demands you make of them—demands they do not honor with nonmembers—then you know you are a member of the group. Members of a group often take on the restricted kinds of interactive behavior that we call roles. Acts that we call leadership come to be noticed.

I have just spoken of being a member of a group as if a group were a "thing" you can be "in." But the sentence in the paragraph just above reading, "When members of a group . . . then you know you are a member . . ." is actually circular, actually a tautology or a mere labeling. It says no more than this: When you and some other people make reciprocal demands on one another that you do not make with other people, then you know that you are maintaining a reciprocated relationship of expectations with those people that you do not have with others. And you call those people "my group." I am writing dispassionately here, but of course the wish to answer the demands of other members can be very strong indeed.

To sum up the complexity of action, purpose, and pattern to which I have pointed here, you may find appealing (as I do) the categories offered by Arrow, McGrath, and Berdahl (2000, p. 42):

> In its most common usage, *coordination* means the spatial and temporal synchronization of overt behaviors of two or more people so that those actions fit together into an intended spatial and temporal pattern. This defines . . . the *coordination of action*.
>
> In a second meaning . . . *coordination* means achieving either explicit or tacit agreement among group members regarding the meanings of information and events. This includes the shared understanding of the nature of embedding conditions and the threat or opportunity they pose; agreement on procedures for pursuing goals; . . . and agreement on division of labor. . . . This is the *coordination of understanding*.
>
> [In] a third meaning . . . the mutual adjustment of individual purposes, interests, and intentions among group members yields the *coordination of goals*.

All of us become accustomed, after some years, of maintaining alert attention to the social scene, of judging likely opportunities and of thinking our own thoughts. It is easy for us to forget that other people are thinking *their* own thoughts and judging their own opportunities. It is especially easy to forget if we are accustomed to treating people like nonliving things. In contrast to that daily forgetfulness, I hope I can succeed in portraying social life to you as a great intricate clutter or entanglement of individuals, all unceasingly sensing environmental variables and controlling their perceptions of many of them, and of internal variables, too, by action and reorganization. To someone looking down from a height, we might seem a confused, buzzing, snarl. Each of us on the ground, singly, espies at every moment helps and hindrances to the control of perceptions, and wends his way through the snarl, often quite unaware of the buzzing.

LEADERSHIP

Leadership arises from the helping situation. Suppose some of us want to move a rock too big for any one of us to move alone. We can move it if we all push. We stand around looking at each other. Someone says, "All right now, everyone put a shoulder to it, and shove when I say shove!" That person is acting like what we think of as a leader.

Suppose an army officer shouts "Charge!" If nobody does anything, we think the officer is not much of a leader. But if the troops march forward, we say he is a leader. Suppose, in a conference room, a person at the end of the table says, "Here is how I think we can go about this. George, you make a list of the materials we will need. Amy, you make a list of the operations that have to be carried out. Bill, ..." and so on. That person is acting like a leader.

Leadership behavior appears where (1) individuals want to control variables they cannot control alone and (2) the individuals (or most of them) perceive the task as one that will enable them to control their variables, and (3) they perceive the task as requiring coordination that cannot arise from individuals acting directly at their own initiatives—as in the case of persons pushing haphazardly on the rock or soldiers marching forward individually each on some convenient day of the week. At some point, someone will call the shove or propose a division of labor or do whatever is necessary to enable the joint efforts to result in control of the variables the individuals care about. For one person, rolling the rock may provide good exercise. It may enable another to build a better road. It may give a third a day's wages. Remembering, however, the kind of coordination achieved at places like W.L. Gore (for which see Chapter 34 under "Gore"), it is clear that people can be mistaken about the necessity for traditional, magisterial leadership.

I am writing here as if, when you want to do something you cannot do for yourself, you invite someone to coordinate your efforts with others. The joint effort coordinated by the "leader" then results in your getting what you want. A lot of joint effort does come about that way. Yet many people, including scholarly writers, seem to believe that an important quality of a leader is to "motivate" people. A student once gave me a cartoon showing two people whispering to each other in front of a large crowd of people, all apparently waiting expectantly. One whispers to the other, "Now that we have them organized, what do we do next?"

The persistence people show in pursuing their own goals is often an annoyance to leaders. Leaders often want other people to stop pursuing their own goals and do what the leader tells them to do. Leaders then look for ways to "motivate" others—that is, ways to somehow coax or dragoon the others to do what the leader wants. There are hundreds of books presuming to tell you how you can "motivate" and "align" people to do what you want. I devoted chapter 33 (on influence) to that topic, and I will say a little more here. The skeleton of the topic of influence is the list of Requisites for a Particular Act, which I repeated in the Introduction to Part VI.

When people conceive possible actions that are beyond the power of one person, they can easily perceive the necessity for coordination and the convenience of looking to a single person to provide it. Coordination is *necessary* to achieving many of the goals humans pursue. I am not saying, however, that leaders of the kinds that fill history books are necessary to a helpful ordering of human society. Some leaders help more people than others do.

It is easy for persons in positions of coordinating others to come to believe that it is reasonable and morally right that they should ask for many kinds of help in coordinating their duties—supplies and equipment, transportation, a meeting hall, and assistants of various sorts, including a guard to protect them from unappreciative malcontents, and money to make the purchases and pay the assistants. It is easy for those requests to turn into taxes, gilded carriages, castles, police, and an army. It is easy, too, for people whose parents, grandparents, and uncounted generations of forbears have knelt as the leader passed by in the gilded carriage (with armed guards before and behind) to believe that is the natural state of affairs. It is easy to believe it is only natural for the officers of the company to have reserved parking spaces near the door and salaries 50 times that of the mill hand.

Social situations vary a great deal in the amount of power given the leader. Threat can be effective in situations that are barely social: "Give me that piece of meat or I'll kill you!" or "Stay off my property or I'll shoot!" In most social situations, however, a threat to an individual is effective only if the threat is accepted as legitimate by most others. "Work faster or I won't give you your paycheck!" That threat is effective only if the threatened worker is not allowed to take the money away from the employer by force; that is, the other workers must believe it is legitimate for the employer to deny the paycheck and not legitimate for the worker to take the money by force.

As employees and citizens (for example), we get into the odd situation of hearing an employer say, in effect, "I will do this coordinating for you, but you must agree that if I think one of you is obstructing my coordination, you will not interfere with the way I deal with the obstructor." The elected representative says something like, "I will go to Washington and help pass a bill that will help you live in better coordination with your neighbors, but you must agree that if you violate the provisions of that bill, you will become subject to arrest and prosecution."

Though a democracy of a sort took form more than two thousand years ago in Greece, that form fitted a small locality where the voting citizens (not the slaves or the women) could walk to the agora to argue and cast their votes. As armies and nations became larger, that democracy vanished. The modern form of republican democracy has come from a long, heroic struggle that showed its first small fruits perhaps a thousand years ago in the city-states of Italy, then in the northern part of Europe as towns grew in power, and finally in Britain. The constant aspiration during that struggle (a struggle still with us) has been the limitation of power of the sovereign (the Coordinator-in-Chief). It has been difficult for many people (perhaps especially the sovereign) to accept the idea that being able to afford a gilded carriage and armed guards should *not* entitle one to extort taxes from those kneeling or to chop off their heads when they complain.

Especially when the efforts of a large number of people must be coordinated, it is inevitable that some members of the group or organization will act in ways that obstruct what others are trying to achieve. In a work group, co-workers can sometimes remove an obstruction by doing a little coordinating themselves. "Hey, you're forgetting the cotter pin sometimes, and that makes trouble over here. Watch it, will you?" At other times, they may use threats. "You keep leaving your bench in that condition, and the foreman is going to come down on you."

In coordinating the work of others, leaders and managers often mistakenly suppose that they must tell the others exactly what to do. Frederick Taylor (1856–1915), the "father of scientific management," went so far as to tell them exactly how their arms and fingers should move and how fast. Mr. Taylor's manner of helping turned out to be a spur to the growth of labor unions. Compare Taylor's prescription with the examples I gave under "Organizational Management" in Chapter 34.

The Myth of Management

During the 1990s, industrial companies and even governments in the United States did a lot of "downsizing." Many of the employees discharged were middle managers; their superiors had discovered that workers did not need as much supervision as had been supposed. Shan Martin was one who anticipated that trend with a book entitled *The Myth of Management* (1983). Martin believed (and gave some evidence) that:

(1) After an initial orientation or training period, the actual human labor applied . . . does not need to be directed or controlled by others. . . .
(2) "Doing" activities are the most important of all organizational activities [for most members' purposes].
(3) Those who do the work are in the best possible location to decide on the division of responsibilities, the work standards, and the work methods (p. 128).

Martin went on to describe how teams could be trained to be self-managing. The teams would not be expected to operate as they would if a supervisor were looking over their shoulders. Because of (3) above, they would operate *better*. Furthermore, they would not be without coordination. Part of their training would be to learn how any member, as needed, can take on the coordinating role.

Two serious misapprehensions about leadership are widespread. First is the misapprehension that the leader, once in place, must supervise everything the workers do—or at least as much as he can possibly find time to supervise. Second, the leader must remain the permanent and only leader until replaced by a higher authority.

The first misapprehension rests on the wrong assumption that it is possible to give advice that always (or even almost always) fits. The second rests on the notion (among others) that a leader is a special kind of person who must either be born with the necessary attributes or undergo long and expensive training or experience. Martin's point (3) denies that notion, and his point (2) supports the denial. You will remember from the section "Training and Experience" in Chapter 31 that training and experience make no measurable difference in the efficacy of psychiatrists and other psychotherapists. This is not to say that training and experience are irrelevant to every field. Farming, engineering, and musical performance are surely some of the exceptions.

In our consultations in schools, one thing Richard Schmuck and I almost always taught participants at the outset was how to rotate the leadership (chairperson, convener) at faculty meetings. Participants were always grateful to learn those coordinating skills; see Chapter 5 of Schmuck and Runkel (1994). Similar rotations occur at W.L. Gore and Associates.

None of that is to say that we should abolish all positions of leadership, or coordination. It is simply to say that people do better when they are not getting unremitting supervision and prodding: do this, now that, now this, now that. People differ, of course, in how much direction they want. It is obviously better to give every individual the amount that best suits her than to give all what "they" want—that is, the average of what individuals want.

Pressures of Time

Hearing about how much better coordination can be if individuals can find sufficient degrees of freedom while working with a group, many people will say, "That's all very well if you have plenty of time for all that talking, but lots of times a decision has to be made in a hurry, and somebody has to make it." That is true when only a second or two is all you have, but even in the most pressing and fateful circumstances more time than that usually lies between the realization of the need for quick action and the moment when the action must occur. Furthermore, a very few minutes or even seconds of pooling information can often turn catastrophe into safety. Considerable investigation, for example, has been made into the management of aircrews.

Blake and Mouton (1985) were called upon by an airline to help with retraining aircrews. In the 1970s, NASA had conducted research into aviation accidents and had found that in 80 percent of the accidents, there had been a lack of effective use of the resources of the flight deck crew. In response to that finding, airlines instituted training to "tighten up" the authority of the captain and the discipline of the aircrew. No change in accident statistics resulted.

Blake and Mouton conducted a series of experiments in which three captains, taking the roles of captain, co-pilot, and flight engineer, underwent a simulated crisis in flight. They constituted the experiments crews with everyone having the same rank so that the criticisms the members offered in later discussion could not be disparaged as coming from a person of different rank. Blake and Mouton reported:

> Over the course of the trial runs, every captain had the opportunity to experience handling a dilemma from each cockpit position. After each trial run, crew members critiqued the handling of the crisis. . . .
>
> But when it came to assessing the *quality* of the solution to the crisis, . . . the second and third crew members would turn to the captain and say, in effect, "*If you had consulted us, we could have given you information that would have enabled you to take a better action than you took.*" [C]aptain after captain would become aware of how their conventional behavior prevented them from tapping available resources. . . .
>
> While *insufficient time* is often used to justify centralizing authority, in fact, it is only in the rarest of cases that time is too limited to tap other resources. The research concluded that in the cockpit time gets truly short when only a few seconds are available. When 30, 60, or 90 seconds . . . are available, there is usually sufficient time to permit input that could make a difference (p. 15).
>
> It was critical, then, for *all* crew members to participate in seminars that involved dilemma-solving experiments so that a shared way of thinking . . . could be created.

This training has now been completed by over 5,000 flying personnel from domestic and international carriers as well as from corporate aviation and the military.... Since the advent of the crew leadership training, followed by simulator recurrent training, the number of failures in routine, noncrisis situations has declined steadily at the rate of 50% per year for three consecutive years (p. 16).

Blake and Mouton conducted similar experimentation and training with 4,800 fire fighters in a region of the U.S. Forest Service.

Where an average of six fire fighters per year had lost their lives over a 20 year period . . ., there was only one life lost in the next nine years (p. 16).

Similar reports about bringing out vital information in a hurry (or failing to do so) among members of aircrews have been told by Seymour Hersh (1986), Elizabeth Stark (1988), Dean Tjosvold (1990), and Robert Helmreich (1997).

LEADERSHIP STYLES

Psychologists, not to speak of sociologists, historians, soldiers, and novelists, write a lot about leaders and leadership. They wonder what qualities distinguish leaders from the rest of us. They wonder whether leaders are born or made. And they give a lot of advice about how to be an influential leader. All the advice you will get from me is contained in the Requisites for a Particular Act. This chapter and Chapter 33 are merely elaborations on that list.

We use the label "leading" when we see someone coordinating. The coordinating gets an individual flavor because of the idiosyncratic ways the leader goes about making control easier for the people who want to be coordinated—and for herself. The leader chooses to place more confidence on one of the Requisites instead of another because of the beliefs she has about motivation. Additional flavor comes from the leader's hierarchy of control—from the complex of internal standards that guide the leader's choices of action. All leaders use their legitimated coordinating positions as an opportunity to satisfy some desires (to control variables) that are irrelevant to what most of the followers conceive to be the legitimate tasks of the group or organization. Officers of companies hire family members or see to it that the company awards contracts to them. Politicians go on junkets they know will provide them little governmental information. Leaders vary a great deal, however, in the *degree* to which they use their positions for goals going beyond the necessary coordination. Alexander the Great spent his career suffering the same hardships as his soldiers. Louis the Fourteenth, on the other hand, impoverished France to build the opulent administrative center at Versailles. Both of them, of course, were willing for their soldiers to die as a means of expanding their own dominions.

All the writers I have read, ancient and modern, seem to conceive the possibilities for influence to be those of reasoning and negotiating, bonding with affection, inspiring, and threatening. Bolman and Deal (1984, 1997), for example, in two editions of their widely used text on organizational management, described four sorts of organizational culture, which can also be considered styles of leadership. (1) The *structural* style emphasizes the rational bureaucracy: the organization chart with its roles and duties. (2) The *human relations* style uses both the abilities of individuals and their desire for social ties (affections) as resources. (3) The *symbolism* style gives the leader the task of portraying commitment to the organization and inspiring the members to achievement and loyalty. (4) The *political* style uses negotiation, rewards, and punishments. In Chapter 33, I mentioned Kenneth Boulding's three "social organizers" — that is, ways of achieving coordination: exchange, integration, and threat. I am especially fond of the four influence styles set forth by Roger Harrison: assertive persuasion, charisma, participation and trust, and reward and punishment. See Harrison (1978) and Harrison and Kouzes (1980). For further elaborations, see Harrison (1993) and Harrison and Stokes (1992).

There is no best recipe for effective leadership. That is true whether you think of effective leadership as facilitating what the followers have taken to be their task or whether you think of it as the leader getting his own way about what ought to be done. Control of perception is opportunistic, and persons in positions of leadership, like persons in any position, will make use of the social and physical environments to control their perceptions with the opportunities, skills, and conceptions at hand. It is bootless to say without qualification that some style of leadership is more effective or better than another. We must also specify whose purposes and what purposes, in what culture among people with what expectations, and during how long a period of time. And even with those specifications, you will still be guessing when you advise a particular leader about what to do tomorrow morning. Within a given culture in a given decade, you can find some patterns of action that are more successful, on the average, than others. To advise an individual, however, to choose tomorrow morning the pattern of action that in the past has been successful more often than other patterns is risky.

A book that describes an abundant array of leadership styles is Jim Wall's (1986) *Bosses*. It contains 50 interviews with first-line leaders in "airports, missile systems, peace movements, Indian reservations, crime rings, sawmills, brothels, bars, mental wards, and everything else." Reading Wall's book is an excellent way to broaden your view of the ranges of "business" that go on in the USA.

Of all the recipes for effective leadership I have read, many written by famous scholars or famous leaders, Wall's list of requirements is the one I find simplest and closest to PCT. Wall does not try to say what a leader should be like, nor does he prescribe any particular act for a leader to take. Rather, he says that a leader whose coordination is going to continue to be accepted by her followers must help in three ways.

First, the leader must protect her people from interference and dangers. The leader must, that is, help the workers to maintain sufficient opportunities for action—sufficient degrees of freedom. Note that this is a *helping* function, not domineering or a power-over-others function. Many theorists seem aware of this point, but almost all describe it as a personal quality. Some even call it a need for power. I have not found a writer outside the PCT community who sees it as a matter of degrees of freedom necessary for the joint control of perceived variables.

Second, the leader must coordinate. Doing this has two parts. One is helping the worker to know what to do. The leader does this by instruction, by giving approval and disapproval, and by arranging for co-workers also to do those things. The other part is finding matches between workers and subtasks so that each worker finds satisfaction in the job itself—so that the job can help the worker to control the perceptions that are important to him—and preferably those perceptions high in the hierarchy. This, too, is a helping function. Many writers on management recognize it.

Third, the leader must find a mesh between leader and led, and the shape of that meshing must necessarily differ at least to some degree for every worker and in every situation. Wall calls this finding a style that works. He quotes from a manager of housekeepers: "I find the moods [of my housekeepers] and then react." That manager is sensitive to personal goals, too, dealing differently with housekeepers who chiefly want to please her (the manager), who want money, or who want to find pride in their work. That manager sounds like a person who understands about control of perception. This third requirement, again, is one of helping.

That recipe does not say, "Do this, do that." It says to act however you find opportunities to act so as to perceive certain features to the relationship. It does not say that you must get control of your workers. It says you should enable them to find their own ways to use the activity you are leading—to use it as a means of maintaining control of the perceptions they care about.

Wall is not the only one, of course, to think of leadership primarily as coordination—less as getting people to do what the leader wants them to do and more as enabling the followers to do what *they* want to do. A good illustration of that kind of leadership is provided by Goldman, Dunlap, and Conley (1993). Between 1987 and 1991, the state of Oregon funded 271 schools to develop school improvement plans initiated and administered by teacher-led committees. In 1991, Goldman, Dunlap, and Conley visited 16 of the schools, believing that the conditions specified in the funding would enable teachers and principals to move away from authoritative or directive management and toward what I am calling coordinating and these three authors call *facilitating*. They found that facilitation did indeed happen in most of the schools. They reported:

> Although authority structures varied, all projects were in fact directed by teachers with input from administrators rather than the other way around. Teachers took the lead, and administrators played a support function instead of taking a visible leadership role (p. 78).

The authors quoted remarks of several teachers about their principals, including this one:

> [The principal] is more available to all of us for help and consultation. Otherwise, he is pretty

much in a basic collaborative style. He really has an open door policy. He's excited and happy about how people are getting turned on to new ideas and he lets that be known (p. 81).

Richard Hackman (1987) has written on the design of work teams within organizations. Hackman suggests that when managers want to create new work teams, they ask questions at four stages: prework, creating performance conditions, forming and building the team, and providing ongoing assistance. I won't elaborate on those stages. I will say only that he does not tell the manager to take any particular action, but to ask several pertinent questions at each stage, the answers to which will be useful to the manager in her own choosing of actions to take. And I want particularly to let Hackman tell you how leadership fits into forming a new team. He says (on p. 338) that it should not be a foregone conclusion

> whether an internal group leader should be named (as the group first comes together)—let alone how he or she should behave. It often does make sense to have such a role, especially when substantial coordination among members is required, when there is lots of information to be processed . . ., or when it is advisable to have one person be the liaison with other groups or with higher management. Yet it is rarely a good idea to decide in advance about the leadership structure of a work group. If a group has been designed well and helped to begin exploring the group norms and member roles it wishes to have, questions of internal leadership should appear naturally. And while there invariably will be a good deal of stress and strain in the group as leadership issues are dealt with, when a resolution comes it will have the considerable advantage of being the group's own.

Hackman's description is not one of pie in the sky. Manz and Sims (1987, p. 107), for example, say,

> According to one estimate [from Edward Lawler], two to three hundred manufacturing plants in the United States seem to be using some derivative of a highly participative team approach. In addition, there are other, nonmanufacturing organizations that rely on some variation of this approach. . . . Academics have had limited access to organizations that use the team approach, and sometimes when they have been given access it is with the proviso that there be no publicity or writing about it.

I have spoken with a couple of colleagues who have consulted in industrial organizations using self-managed teams, who told me they could never write about their experiences, because the managers of those plants did not want to make it easy for their competitors to learn about methods of managing that bring large increases of productivity and large decreases in turnover. Manz and Sims, however, were permitted to write about their project without naming the company. They reported, among other outcomes, that productivity gains were "significantly greater than 20 percent" (p. 118) compared to plants using traditional management methods, and "a manager in the plant counted on the fingers of one hand the employees who had chosen to leave" (p. 118). Concerning leadership, they wrote that "the external leaders' most important behaviors are those that facilitate the team's self-management through self-observation, self-evaluation, and self-reinforcement" (p. 106). By self-reinforcement, I think the author meant that the members of the team told one another when others were helpful to them.

Here is one more example of authors who understand about coordinating and facilitating. Culbert and McDonough (1980) advocate:

> . . . a style of leadership that seeks . . . the self-beneficial interests which underlie people's participation at work. . . . [and that] recognizes the uniqueness of each individual's . . . inner organizations. . . . [It] emphasizes three activities: counseling, team-building, and brokering. Each is aimed at helping others to be powerful—helping others to clear the space they need to assert themselves in directions which maximize personal, career, and organizational accomplishment. . . (p. 213).

For more on self-managing groups, See also, for example, Manz, Keating, and Donnellon (1990), Edward E. Lawler (1991), and David Barry (1991). There is a considerable literature on "self-managing" or "autonomous" work groups and, as I said above, some "buried" information, too. I am making no attempt here to offer an overview of this literature. I am using a few illustrations from it to say, "This is an example of how you might encourage the talents of your employees if you believe that humans are control systems." If you want to get started on the literature, an excellent early history of the topic is that by the famous Eric Trist (1981). I am saying, too, that you will not inevitably be happy if you institute self-managing work teams in your business. Like every organizational scheme that has a label on it, it "works" sometimes and not other times. The literature does not demonstrate that this is a good way to get employees to do what you want. It demonstrates that this is a way that people *can* work productively and happily if everyone involved (or almost everyone) *wants* to make it work.

If you read the scholarly journals in social psychology or organizational management, you can get the impression that managers everywhere have learned or are just about to learn that authoritarian, dictatorial leadership is chock full of disadvantages and downright harmful effects and that participative (facilitative, coordinative) management is fast replacing it. If, however, you look around you, you will see the traditional, dictatorial, threatening style almost everywhere, from families (possibly your own) to retailers to hospitals to manufacturers to social service agencies to churches. The participative style is difficult to see because the fraction of organizations using it is small. Furthermore, many organizations using participative management do not care to publicize it, and those that do usually describe it in such vague terms that the reader gets the impression only that the corporation is a good place to work.

Nevertheless, more participative management is probably going on today than at any time in the history of the world.

HOW TO LEAD

In the advertising for books on leadership and management, you can often read testimonials that go something like this: "I read your book on leadership and did what you said, and I am now the president of the XYZ Clothespin Company. You sure know your stuff."

One thing teachers and authors (some, anyway) learn early is that no matter what you say in your lecture or your book, a few students or readers will say it was just wonderful, and a few will say it just stank. And no matter how "scientifically" you go about reaching the advice you put into your lecture or your book, you will still get that same result: some cheers and some boos. It is true that some books get more cheers than boos and some vice versa. But nobody gets all cheers or even almost all, and those cheers fade away when a later book comes out that resonates a little more loudly to the temper of the times. (That's one reason authors revise their books—to try to keep in touch with this year's "in" way of talking about things.)

I am not saying you should refrain from reading or writing a book on becoming an effective leader. Inevitably, here and there, at least a few readers will feel benefited from reading such a book (even this one!). I am saying that you should not expect multitudes to catch the ideas to their bosoms and revolutionize leadership in the USA or elsewhere. The reason can be simply put: no advice can be expected to match the person, the person's readiness, and the situation except sporadically. I wrote about advice in Chapter 29.

Social scientists of various stripes have spent millions of dollars and millions of hours making note of qualities that characterize leaders but not nonleaders, with the idea that one can then become a leader by adopting the qualities (or doing the things) characterizing leaders. No matter how carefully those studies have been conducted, they have never found the magic recipe. I will skip over the logic of the matter and say, first, that researchers never succeed in finding a quality such that 100 percent of effective leaders show it and 100 percent of ineffective leaders do not. That fact allows leaders to appear who have the quality but are ineffective. The second difficulty is that of ascertaining whether a leader belongs in the class who "do this"—or whether the leader "has" a particular quality. The signs by which a person should be classified one way or the other are never unambiguous, and often they are very uncertain indeed. This problem fits into the relations between correlations and probability that I explained (with Kennaway's help) under "Correlations" in Chapter 26 on personality.

RECAPITULATION

Leaders are not born. They are not made by schooling or training, either. They are made in the sense that they find themselves in a situation where others feel a need for coordination, and the prospective leader is called upon, or is appointed, or simply feels an urge to offer his services, and his goals and other internal standards enable him to take actions that turn out to be helpful to the other persons.

Leaderlike behavior appears in interaction with unpredictable situations. Beyond some elementary skill in communication, it is a waste of effort to try to select leaderlike persons (as has been shown by many failed attempts), and a waste to try to train people to do leaderlike things. It is not a waste to help people acquire skills that they might find useful (such as listening carefully) when they find themselves in a leadership position. But it is a waste to suppose that the training will necessarily mesh well with the person's later positions. Furthermore, training for almost any *particular* kind of position is necessarily out of date or soon will be.

Some portions of people given training will find themselves in leadership positions simply because they have undergone the training; the training becomes a credential, and they are hired by employers who believe in the magic of credentials. But the training may not help a larger portion to be successful than would be found among untrained people. (It is difficult to test this assertion if almost every person hired into a leadership position comes in trained.)

We act to control perceptions. If acting like a leader in some situation will help us do that, we adopt leaderlike actions. If acting like a follower will help us do that, we follow.

At this point, you may wish to look at Dag Forssell's little book *Management and Leadership* (2000). Not only does it treat leadership from the viewpoint of PCT, but, in small space, it serves as a carefully organized review of the basic principles of PCT. Forssell presents the ideas in combinations different from mine, and you may find his organization helpful. His diagrams are more imaginative than mine, too. Furthermore, his book addresses managers in business and industry, and that practicality brings his pages more down-to-earth than many of mine.

INDIVIDUAL AND ORGANIZATION

To sum up, I will use some quotations from the book by Culbert and McDonough (1980). As far as I know, these writers had no knowledge of W. T. Powers. Yet they wrote almost as if they were translating Powers into ordinary language. They showed no awareness of negative feedback loops, but they showed, as you will see below, an exquisite sensitivity to the importance of perceptions. Here is the authors' description of the situation of the individual among other members of the organization:

First, we find relatively few people who measure their success solely on the basis of external rewards: how far they've made it in the hierarchy, how much money they earn, and how much praise they receive. Most people claim that they're primarily concerned with internal satisfactions such as finding opportunities to pursue personal interests and values, to learn and develop personally, and to demonstrate skills in areas of their special competence (p. 58).

There Culbert and McDonough are saying that most employees are guided in their actions less by standards pressed upon them by managers than by internal standards they cherish for many reasons beyond the job. This first point agrees with PCT that almost everyone has a good many high-level reference signals that provide pervasive purpose to their lives, and these do not usually include the demands of the managers at work. I say here "not usually" instead of "never," because some people become so committed to the official goals of their organizations that they come to define themselves almost entirely within the tasks and concepts of the organization as promulgated by its current leaders. Possibly the prime examples can be found among the clergy, but examples occur in other kinds of organizations as well.

Second, . . . the unique definition of success each person holds and the unique set of objectives toward which that person is targeted cannot be deduced merely by observing what he or she is doing (p. 58).

This point is almost a quotation from Powers. It reminds us of the demonstration I described under "What Is the Person Doing?" in Chapter 7.

Third, . . . knowing an individual's unique definition of success . . . allows us to see self-interests in how each assignment is performed, each problem is formulated, and each organization event is viewed (p. 58).

This point, along with the second, reminds us of The Test for the Controlled Quantity (for which see Chapter 7 under "The Test"). Notice that the authors do not say that knowing the internal standards will enable you to predict actions, but only the purposes or consequences toward which the person will choose actions. The third point is a point of hindsight.

When Culbert and McDonough wrote of "self-interests," they were not pointing to greed or selfishness.

They were simply saying, as does PCT, that people act to keep some perceptual variables in match with their own internal standards. On page 139, they say, "Self-interests are operating all the time, and there is nothing necessarily sinister about their presence."

> *Fourth,* ... achieving ... success that is internally meaningful ... rests on an individual's ability to ... satisfy internal needs and [simultaneously] the objective needs of the job (p. 58–59).

A PCTer would say that if some perceptions are caused by the work to depart from high-level internal standards for those perceptions, the person will experience conflict. The person will not then find the work "internally meaningful." Conflict, of course, stirs up unexpected troubles.

> *And fifth,* ... people do not possess the mental resources to keep self and organizational interests from intermingling (p. 59).

That is a sort of corollary to the postulated hierarchy of control. "Interests" are those aspects of the environment we care about, and we care about environmental variables for which we have internal standards. We can "compartmentalize" to some extent, but that is often only temporarily effective.

Culbert and McDonough even had a vision of something like the control hierarchy:

> ... there's an *internal organization*, far more encompassing than an individual's personality, that determines how individuals within groups transact their business and work for the greater institutional good. Moreover, despite their lack of prominence in how people present themselves, self-interests are a dominant factor in determining what gets produced in the name of organizationally required product. ... And you don't need the skills of a psychoanalyst to understand these self-interests. You merely need to comprehend what an individual is trying to express personally and achieve in his or her career, and what he or she perceives as making a valuable contribution to the job. At every point personal needs and organizational goals impact on one another, and it's always up in the air whether the needs of the job or the interests of the individual will swamp the other or whether a synergy of interests will evolve (p. 68).

Notice, throughout all the quotations above, that Culbert and McDonough nowhere talked about relations among variables. No statistical table appeared in their book. Throughout, they wrote about perceptions, about how people can and cannot perceive the social world and about some of the ways they want to perceive it.

It is true that only a small domain of human behavior has been directly tested so far by the kind of modeling specified by PCT. Still, PCT remains the only psychological theory that successfully models perceptual control *in the face of random disturbances.* It is, in fact, the only theory in which perceptual control appears as a core concept. (Perception appears in all theories, and control of *action* appears in many, but control of *perception* is unique to PCT.) Control of perception is a fact, not a theory. The theory explains *how* control can be achieved. Only PCT explains that. My quotations from Culbert and McDonough only illustrate some ways that control of perception can put shape on social life.

I do not claim that PCTers are the first people ever to think that organizational modes depend on individual perceptions or to think of providing sufficient degrees of freedom in organizational life. The writings by Culbert and McDonough show that similar thoughts are possible in ignorance of PCT. In PCT, however, we have a basis that has the kind of experimental support you saw in the chapters of Part II and in Chapters 12 and 13.

ENDNOTES

[1] Page numbers of these quotations from Powers (1979) are the pages in the reprinting in Powers (1989).

Chapter 36

Planning

Humans look to the future. So, to some degree, does every creature. *Escherichia coli* "expects" to find a continuing liquid environment in the sense that it is capable of adapting to changes in nutrition by wiggling its cilia in its liquid environment. It is "prepared" by its structure to do that. But it is not prepared (does not "expect") to find itself surrounded only by air or lying on a dry surface. If it were to find itself in such a predicament, no amount of wiggling would save it.

In the same sense, but in far greater detail of adaptive capability, we humans have expectations, we make predictions, about future experience. As we walk along, we swing our legs and move our weight onto the forward foot in the expectation that the ground will support us at this step just as it did at the last. If we happen upon a soft spot, as in swampy country, our habitual program for walking fails; we stumble or fall. Rarely, however, do we die from such an interruption.

Human programs get immensely complex. Finding the way to home and dinner from a mile away would be too much for a worm or a slug, though no trouble for a horse, cow, dog, or cat. Finding the way from my house in Eugene, Oregon to the offices of the E. I. DuPont Corporation in Wilmington, Delaware, would be a straightforward matter for me, using airports, maps, taxis, and street signs, even though I have never been there, but a dog could not do that. Neither could a dog look up my mention of the E. I. DuPont Corporation in this book.

Planning is a form of model-based control, to which I devoted a section under that heading in Chapter 24. Indeed, planning, model-based control, imitating what someone else does, following a recipe, following a map, obeying, using a codified technique or method (including experimental method)—following any sort of program whatsoever—are all the same sort of thing psychologically. When you do any of those things *with* other people, it becomes a form of helping. Every map we draw, every structure we build, and every agreement we make to cooperate with others is a partial model for future action, a partial means for achieving a goal. Every program leads the user to a goal. More accurately, I should say that the program user must adopt a reference state to which the user believes the program will enable her to attain. The reference state conceived by one user will differ to some degree from the state conceived by another user.

Making a plan for action requires envisioning a goal that can be perceived when it is reached. To be useful, the plan must also contain a reasonably accurate inventory of the environmental opportunities that will be available when they are needed. Those two requisites take us back to the Requisites for a Particular Act.

REQUISITES FOR A FOOLPROOF PLAN

Most plans of any complexity turn out to require modification as you go along. I call a plan more complex if it has more choice-points and if the opportunities for action at the choice-points are likely to turn out to be other than you expect. Many people try to make plans that leave no room for improvisation. They try to specify every action at every choice-point, to preclude the need for judgment on the spot. What would be the requisites for such a foolproof plan? Let's take an inventory by paraphrasing the Requirements for a Particular Act (for which see Chapter 1 or the Introduction to Part VI). To specify successfully the time and place where every choice-point will occur, the choice to be made there, and the manner of recognizing the achievement of the goal, the following would be necessary.

1 The person(s) would continue to control the same perceptual variable as goal—that is, until its desired value or condition is achieved.
2 This is given; the perceived variable is *not* at its reference value when the plan is yet to be carried out.
2a The environmental opportunities for the action specified in the plan at every choice-point would have to be there to be used at the time needed.
2b The person(s) would recognize all those environmental opportunities.
2c The person(s) would have the necessary skill and ability to use correctly the opportunities at all those choice-points.
2d The person(s) would go on believing that those environmental opportunities would move the person(s) toward the goal.
3 None of the actions taken would turn out to disturb some other perceived variable being controlled by any of the persons involved.

Those requisites, obviously, are unattainable. People often change their goals before they reach them. Often, more choice-points turn up than predicted. Often, the expected environmental opportunities fail to appear (someone, for example, fails to get the supplies in place on time). People often fail to recognize the opportunities even though they are there. ("Oh, was I supposed to ask Jones for that?") Skill or ability turns out insufficient. ("I multiplied this number by that. Was I supposed to divide?") When it comes to the point of action, people sometimes lose confidence in the planned action. ("I didn't think that bookkeeper had good enough credentials, so I replaced her with MacIntyre.") Conflicts arise. ("I didn't do it, because I won't treat people like that.")

A plan can be subject to all those uncertainties even when carried out by one person acting only on the physical environment. When a plan is carried out by several persons acting on the social environment as well, the uncertainties multiply. To remind yourself of the uncertainties in human communication, look again at Figure 28–5.

If your plan has only two or three choice-points, if you are the only person who will be acting, and if the plan can be carried out in a short time, a detailed plan will often work very well. Your plan for peeling the cover off a tub of cottage cheese will usually be successful. For getting money out of the mechanical teller at the bank, the instructions work well if you can remember to put your card in right-side up. The customary procedure for reserving a table at a restaurant usually serves, though sometimes communication fails, or intentions change, or perceptions of "six o'clock" differ, and somebody, you or the headwaiter, fails to reach a goal. In general, with a program of some complexity, reaching a specified goal by following a detailed plan without deviation is impossible.

Yet thousands of us do succeed, every day, in making reservations, in registering as students at universities, in following the rules for playing a game of bridge, and so on. How can that be? First, not all who set out to do those things are successful. Some students get lost in the procedure and have to start over. Some bridge players violate a rule and suffer ire. And second, when we are successful at programs like those, we do *not* follow a strict, preplanned series of particular actions, but instead take appropriate actions as we come to each choice-point. We control our perception of progress toward a goal by opposing whatever might threaten it. We use the opportunities suitably at each choice-point. In a communication to the CSGnet on 1 January 1995, Powers said that we are aided in our control of perception by

> [u]nderstanding and short-term, contingent, prediction. If we know what a person's purposes are, right now, and [know] that reorganization is not going on at too fast a pace, we can predict some of the outcomes of that person's actions. We may not be able to predict [particular] actions, but knowing purposes gives us a view into the future that is longer than what we could otherwise achieve: purposes determine outcomes if not actions. If we can predict disturbances a little way ahead, we can predict action, too, given that the purposes of the action are constant for the time being. . . . We can learn to have respect for others as autonomous systems like us even if we can't—or perhaps because we can't—predict or control their actions.

Powers put "understanding" first in his list. It is not enough to memorize a program—that is, a few words describing it. To make full use of a program, it is necessary to understand the nuances of its correspondence (or lack of it) with the actual environment. On 12 February 1998, Powers wrote another comment on this topic to the CSGnet:

> A simple prediction made under clear-cut circumstances, or made when the consequences of being

wrong are not severe, can be very useful. If it looks like rain, you take an umbrella with you; if it doesn't rain, you haven't incurred much of a cost even if you're wrong 75 percent of the time. But a complex prediction that tries to take every factor into account can be worse than useless. Having gone to the expense of gathering the data and working out the prediction, you can hardly abandon it at the first sign of inaccuracy. This means that you will behave as if the prediction is true long past the point where it obviously (to anyone else) failed. And that is worse than useless.

The costs of continuing to pay for something that is going wrong are often called "sunk costs." I wrote about that temptation under that heading in Chapter 25. With obvious attention to degrees of freedom, Powers continued:

> It's far better to make contingency plans. This means considering everything you can think of that *could* happen, and working out what to do in each case, or at least which direction to begin acting. When you've done this, you no longer have to predict what IS going to happen, because no matter what happens, you're covered.

The foolproof plan is often sought when a plan is being made for a group of people to carry out (perhaps an assembly team in a factory, perhaps a faculty in a school, perhaps a family), and some of the people are fearful that if they take action in the proposed direction, they will find themselves bringing unwanted consequences on themselves. Therefore, they try to get a harm-proof plan—one in which everyone else will be prevented by explicit agreement from doing anything the anxious people think might bring them harm. The result of that, not surprisingly, is that the anxious people try to maintain what they now know how to cope with—the status quo. Those anxious people then look to the others like obstructionists. They are accused of being "against change and of "resisting."

When you make plans for a range of contingencies, you are saying that you know some things that *could* happen, but you are also admitting that you don't know just what *will* happen. When you act, however, you can rarely act in the direction of all the contingency plans at the same time. You must choose a direction and see what happens. In stepping out in a particular direction, you are acting as if you know what is going to happen. But since you are ready to discover that you have stepped out in the wrong direction, you are also acting as if you do *not* know what is going to happen.

HOW TO MAKE A GOOD PLAN

You can use the Requisites to tell yourself how to make a useful plan, one that will not sink you with sunk costs or commit you to actions no longer appropriate. For a plan in which several people would figure, your questions to yourself would go something like this:

1 Do enough people feel an error (a discrepancy) between the present state of affairs and a yearned-for state? Do enough people yearn for the same goal-state? If enough people have much the same yearning, and if no one in the group will actually work against you, then you have a good start.

2 Is the environment sufficiently rich in opportunities for moving toward the goal? Are there enough degrees of freedom so that if a few actions lead down a blind alley, there will still be other possibilities to try? Do the advantages of taking the first steps outweigh the disadvantages? If so, you can start assigning tasks.

3 Will any of the contemplated actions bring conflict inside any of the people who must help carry out the plan or who must accede to it? If no, go to it. If yes, try some of the conflict-reducing methods you saw in Chapter 23 under "Solutions."

If you answer "yes" to the first two items and "no" to the third, you don't need much of a plan. You need a clear statement of goal, and you need lots of communication within your group and with relevant outsiders so that cooperation can be maintained. Leaders will appear, as at W.L. Gore (for which see Chapter 34 under "Gore).

But a caution. It often happens, as a person or group works toward a goal, that they discover the goal with which they began is not, after all, what they want. They may find that it has unwanted side effects, or that it simply is not as much fun as they thought it would be. As Karl Weick might say, "I can't know whether I want to be there until I get there." The loose kind of organizing that the Requisites imply will enable participants to notice quickly when they are changing their goals.

When I was doing organizational consulting in schools and wanted to ask faculty members to tell each other about their goals, I often asked something like this: "Imagine that when you wake up tomorrow morning, you leap eagerly out of bed, thinking what a fine, deeply satisfying time you are going to have at work. What kinds of experiences might come to you during a day to which you looked forward that eagerly?" Such a request invites the person to imagine looking *back* at what might have happened during a happy day at work. It reduces the likelihood, always too likely, that an organizational member will talk about the *means* to her goal instead of the goal she really cares about.

Do keep in mind the kinds of perceptions you want eventually to come to you. Do keep in mind an inventory of your resources and your possible routes to the goal. But be constantly tentative about planning the actions you will take at the choice points.

LONG-RANGE PLANNING

When I point out that environmental events are in principle unpredictable, I am certainly not saying that it is hopeless to pursue a goal you do not expect to reach until five, or ten, or a hundred years have gone by. People undertake such projects every day, and they (or their successors) often do achieve those far-off goals. The Great Pyramid of Cheops at Giza got built. The cathedral at Rheims got built. The United States of America got established. Thousands of young people acquire college diplomas every year. I am saying that those great things did not come about, nor did millions of smaller projects, by detailed specifications of particular acts. People get to their goals only by coping with the impediments along the way, regardless of the planning beforehand.

Planning enables you to tell yourself and the people you are working with the kind of future you would like to reach together. If you are working with others, planning enables you all to find out whether you *want* to go in the same direction. And as you go along, revising the plan enables you to know whether you still want to pursue the goal that looked good at the start, and whether you still want to go along with these same people. Those are good reasons when you are planning *with* other people. If you are planning *for* other people, those other people will feel less benefit from it.

The professional literature for managers in business and other organizations has recently contained a lot of advice about long-term or strategic planning. As with all recipes, some managers have found the advice useful, some not. Henry Mintzberg wrote a book (1994) called *The Rise and Fall of Strategic Planning*. He said that when you get too detailed about what you are going to do in the future, you will find out that you were wrong about it.

Possibly the most grandiose and dramatic example of long-range, large-scale, detailed planning in the last century has been the planned economy of the U.S.S.R. Every five-year plan promised to rectify the faults of the previous 5-year plan. The damages of all sorts have been horrendous.

Some years ago, I read some sentences about long-term planning that still sound good to me. William Dowling (1978) interviewed Fletcher Byrom, chairman of the board and chief executive officer of the Koppers Company. Byrom said:

> We do quite a bit of [long-range planning] . . . but to us, it's a discipline more than it is a guide to future decision making. . . . If you have taken the trouble of anticipating something, you should at least be able to recognize that it hasn't happened that way. If you hadn't done the planning, you wouldn't have realized that what happened was unexpected. As a regimen, as a discipline for a group of people, planning is very valuable. My position is, go ahead and plan, but once you've done your planning, put it on the shelf. Don't buy it. Don't use it as a major input to the decision-making process. Use it mainly to recognize change as it takes place (p. 40).

THE KEYS TO SUCCESS

There is always a good deal of loose talk about imitating successful people. Parents say, "Watch me; do it the way I do it." Teachers sometimes say that, too. They also say, "Do it the way it is described on page twenty-three." On your first day on a new job, someone says, "Here's Al. He'll show you the ropes," and Al describes to you how other employees perform various tasks. An advertisement comes in the mail that commands, "Attend our celebrated seminar and learn our Simple Steps to Success!" We are told that this person and that are "role models." In brief, we are urged in many ways to imitate others as a way of reaching our goals.

Imitation is sometimes an efficient way to get to your goal and sometimes a poor bet. You can never imitate exactly the way someone else has done something, nor can you ever exactly repeat how you yourself did something. When you are carrying out a series of acts, an undertaking, you will always be doing it in your own way, no matter how closely you try to cleave to a model. And when a course of action seems to be going wrong, the worst way to try to right it, usually, is to cling ever more tightly to a model, routine, or ritual.

Industrial Management

I have read lots of journal articles about projects undertaken in industry to improve morale or productivity or profitability or some other variable that managers or researchers care about. Typically, the author tells about some companies managed in manner X that scored high on outcome Z. The author then urges managers, implicitly or explicitly, to imitate what those companies do. A typical study collects data from a dozen or three dozen or 500 firms and looks to see how many of them managed in manner X are high on outcome Z and how many low, and similarly for those managed in manner Y or at least non-X. If the proportion high on outcome Z managed in manner X is greater than the proportion managed in manner Y, then the writer almost invariably concludes that if you want lots of outcome Z, you should do X, not Y.

All the questions immediately arise that I asked above under "How to Make a Good Plan" when I used the Requisites as an outline. An obvious question is the degree to which your material and social environments are similar to those of the company you want to imitate. More important than that question, however, is the matter of your purposes—your internal standards.

What shall I do if I want outcome Z but do not want to use manner X, because using X makes it difficult for me to control other variables in my company for which I have internal standards? That is, suppose that getting Z via manner X puts me into internal conflict. Regardless of the majority shown in the study, I am not going to follow the advice of the writer. Furthermore, there are lots of ways to get to outcome Z other than manner X. Manner X may be the manner that the writer may have thought most likely, or most suitable to the theory the writer loves best, but that need not prevent me from thinking up my own successful manner. Most social scientists, it seems to me, think the people they study cannot possibly be as clever as they are. But any reader of the literature soon finds out that even rats now and then outwit the experimenters.

Let me use the analogy of a map. A map is an environmental resource many people use to further their purposes. You can use a map for many things, but a common use is to let you lay out a sequence, or a program, for getting someplace you want to go. Will the map get you there? No, you can get yourself there if you use the right map in the right way. Will the map enable you to get ten of your employees to go to Chicago? Not by itself.

If you want ten people to go to Chicago, it will not be sufficient to hand them copies of a map with Chicago indicated on it. It will not be sufficient even if you write "Please go to Chicago" on the margins of the maps. The necessary condition for them to go to Chicago is that they must *want* to go there. (Some may simultaneously *not* want to go there; they may be suffering conflict. But going there must produce less error than not going.) But whether the map will be useful depends on still more than wanting to go to Chicago. The person must be able to read the map. (And must be able to do a lot of other things, too, not associated with the map. But let's stick to the map, because we are talking about evaluating the efficacy of the map.) So you give copies to ten people, and only three get to Chicago. Is the map ineffective? Is giving the map to the ten persons ineffective? Is the map invalid? Is it bad advice to give the map to someone who wants to go to Chicago? If ten companies use management method X and only three of them reach above-average productivity, is it bad advice to recommend manner X to others?

Well, maybe five didn't want to go to Chicago. The map, however, was useful to them; they used it to wrap up the garbage. Two wanted to go to Chicago, but couldn't read, and didn't know that the map had anything to do with Chicago. There are lots of maps showing Chicago and lots of people owning those maps. But lots of those people never get to Chicago. Does that show that those maps are unreliable for getting to Chicago? That they are invalid?

Perhaps the analogy is not strict, but I am trying to tell you why I think we mislead ourselves when we ask what a plan or device or program or test will do or yield or produce in interaction with human purposes. Those plans or programs or specifications or procedures or plans do not *do* anything. We humans *do* things, and sometimes (when it suits our purposes) we make use of those plans or programs or specifications. And when we choose to use something as a means to an end—to use something such as a hammer or a recipe for enlightened management or the Method of Levels—we do not, if we have our wits about us, set the thing in motion and then stand back, hoping for something wonderful to happen. Instead, we watch every moment to be sure things are going as we want them to go. If we think a program or procedure can be set in motion and the program or procedure will then *itself* do what we want, we find ourselves in the position of the Sorcerer's Apprentice.

How would you conduct research on the efficacy of a map? Look back at all the questions in the previous paragraphs. Think how difficult it could be to compose a definition of "efficacy of a map" that would satisfy you and ten other people. Conducting research of the efficacy of a program of any sort is no easier than that.

School Reform

In the 1960s and 70s, the U.S. government spent a great amount of money in the search for a way to improve the efficacy of public schooling. In the U.S. Office of Education and in the later National Institute of Education, most of the people in charge of funding research projects had the best of intentions—that was certainly my belief about those with whom I dealt. They were intelligent, conscientious, indefatigable, and well read. They hunted, however, for a *thing* that would invigorate schools and enlarge their effectiveness. They hunted for a thing that money would serve to disseminate, to install—a thing that would work if people simply followed directions. They thought linearly: put a good thing into schools and thereby get some good things out.

But in funding dozens, probably hundreds of bright ideas (no one should ever accuse those federal agencies of not trying diligently to improve schooling in America), they rarely found a scheme, an innovation, a program, a recipe that worked (more or less) in more than a few installations—sometimes in not more than one. When an innovation did seem to be spreading, it didn't last long; schools gave it up after a year or two. The people trying to follow those programs did, of course, bring about some changes, some good and some bad. Often a school would go on using some part-process of what they had learned—maybe a better way to conduct meetings. Sometimes they learned only not to be overoptimistic about offers of money from the government. In any case, no *thing* produced a revitalization; nothing produced even a reliable improvement in test scores.

Milbrey McLaughlin (1975, p. 49) quoted Senator Daniel Moynihan in testimony before the House Subcommittee on Education in 1968:

> We had thought . . . we knew all that really needed to be known about education in terms of public support, or at the very least, that we knew enough to legislate and appropriate with a high degree of confidence. . . . We knew what we wanted to do in education, and we were enormously confident that what we wanted to do could work. That confidence . . . has eroded. . . . We have learned that things are far more complicated than we thought. The rather simple input-output relations which naively, no doubt, but honestly, we had assumed to obtain in education simply, on examination, did not hold up. They are not there.

After a series of failures, the government people funded a series of investigations into the failures, with the idea that they might find what made them fail. Then, they thought, their future projects could avoid those mistakes. You will recognize this as another instance of linear thinking. The conception was that there were some good ideas out there that would "work" if it were not for some impurity that kept them from functioning properly, much as too much pepper will ruin the soup or as sand in the gears will stop the clock. Remove the obstruction and all will be well. The studies of relative success and failure (some of course did better than others) included projects to install new curricula, new methods of management, new relations with parents and community, new relations between teachers and students, and other kinds of change. The studies were carefully done and, despite the conception that spurred their funding, turned up some very useful information.

In the late 1970s, I had occasion to study the reports of all I could find of those studies that were relevant to managing schools. The reports covered ten programs of research on innovation and diffusion; some of the programs issued several separate but connected studies. I summarized their findings in Runkel, Schmuck, Arends, and Francisco (1979).

The studies were unanimous in concluding that if consultants are to be effective in helping an innovation, they must work locally, face to face with the local people, not at long distance via mailed materials, and not in the manner of dip-in-and-out visits. The authors of the studies concluded, in other words, that innovation in a school could not be brought about by sending a recipe to the school, expecting the people there to succeed in doing things in a new way simply by following directions. Complex social change requires monitoring, rethinking, replanning, redirecting, making new agreements and promises with colleagues, taking anxious first steps in the hope that others will offer helping hands, and finding that others *do* offer helping hands.

The studies were also unanimous in saying that consultants must join with the school participants from the outset in collaborative planning, and they should share with the local people the task of monitoring progress and evaluating its quality. That is, a plan or schedule should not be foisted on the local people by presumed experts from outside the local school community. A workable plan must expect to make use of the local environment, and the local people are the experts about that. Further, whether the projects are succeeding must be judged by the perceptions and standards of the local people, not by those of the outsiders. If the outsiders think a project is doing beautifully, but the locals hate it, the project will not last two weeks after the outsiders leave.

In the more successful projects, the consultants urged the participants to do things their own way—in effect, to build their own project. That is exactly what PCT tells must happen. You cannot ever do something exactly the way someone else does it, because that person is not wired to your nervous system. For simple things, close enough is good enough. If the pie you bake is a little browner than his or a little sweeter, most of us will never notice. But when a project requires several people to cooperate closely for weeks and months in ways that are new to them, a recipe cannot serve. A recipe cannot cope with the unpredictable difficulties that always arise. The people engaged in the project must deal with events as they arise, using the resources at hand.

In the interest of brevity, I have given you only a few vague phrases about the nature of those projects, and I have told you nothing about how success was judged and who did the judging. If you want more detail, I must refer you to the actual reports I have been telling you about[1]. My study of the studies was done in 1979—a considerable time ago. There have been other studies of organizational change in schools since then, and I wouldn't be surprised if someone has reviewed the later ones and, as I did, has looked for commonalities. But I would be very surprised if any later review has produced findings much different from those above.

Maybe you noticed, among those unanimities, the absence of any mention of reward, punishment, or a firm hand at the helm.

You might say to me that I have urged you not to expect success, necessarily, from imitating what successful people do, but that I have myself devoted a lot of pages to arguing that successful enterprises always provide ample degrees of freedom. I have, you might claim, violated my own admonition in advocating ample degrees of freedom.

The answer lies with the control hierarchy. I am not warning you against adopting someone else's goals. But I am warning against thinking that a *particular* lower-level *means* is the only or the best way of getting to that goal. I have connected most of my arguments to the Requisites for a Particular Act. The particular act is the thing to beware.

THE RIGHT WAY

There is no single right way to do anything, and no single best way. In most instances, there are many right ways and best ways, because what is best or right for one person is not best or right for another, not even if you explain carefully to that other person why it is that your right way should be her right way.

Many books tell you the right way to do something. Many books, too, tell you stories comparing people who did something the right way with people who did it the wrong way. The organizational literature contains its share. Only one example is the book by Jeremy Main entitled *Quality Wars: The Triumphs and Defeats of American Business* (1994). The advertising quotes an editor of *Fortune*:

> Here . . . is a book that tells what really happened in the companies that have turned to total quality management—the mistakes they made, the disappointments they suffered and, for those that did it right, the rich rewards they earned.

It is easier to tell by hindsight than by foresight who did it right. Nevertheless, that book and others admonish the reader to do it *right*. I urge you to read some of those books, but not to learn the *right* way. Read them to widen your stock of stories about some *possibly helpful* ways of doing what you want to do—ways of acting that may come in handy at your choice points. Then operate according to what you know about the ways humans function—and I hope that knowledge includes PCT.

There are right or best ways for *you* to do something at a particular place, at a particular time, with particular other people, for particular purposes. You can suit those particularities (contingencies) with a theory, but not with a recipe. But even with theory, heed E.F. Schumacher (1979, p. 100): "It is quite amazing how much theory one can do without when one starts real work." And remember the four people doing the rubber-band exercise that I told you about in Chapter 6. How could planning help them?

EFFECTIVENESS

When people speak of the effectiveness of a plan, procedure, policy, or other program or principle, they always have some purpose in mind, though not always at the front of their minds. Since people have different purposes, what seems effective to one person need not seem so to another when their purposes are not explicit.

A business firm, for example, can beautify the community magnificently and go bankrupt. It can delight the stockholders while sending most of its employees to the hospital with asbestos in their lungs. A few organizations succeed in making a profit, providing a healthy place to work, providing insurance against unemployment, providing day-care for children, and organizing the work so that almost all employees satisfy several of their deeper purposes through their jobs.

To discuss effectiveness in a general way, therefore, it does not help much to adopt a purpose that may be meat to one person but poison to another—such as reducing downtown traffic, increasing the enrollment in the English Department, putting a human on Mars, or even making a greater financial profit. One can, however, adopt a sort of metapurpose; one can adopt the purpose of safeguarding the differing purposes of individuals. I like what Hackman (1985, p. 129) and Aoki (1984) said about effectiveness. They wrote these three criteria with work organizations in mind. They propose that we call an organization effective when—

1. The productive output of the [individual, group, or organization] exceeds the minimum standards of quantity and quality of the people who receive, review, or use the output.
2. The process of carrying out the work enhances the capability of the [individual, group, or organization] to do competent work in the future.
3. The work experience contributes to the growth and personal satisfaction of the persons who do the work.

To those, I like to add my interpretation of Aoki's fourth criterion:

4. Individuals have confidence that the work they do is helping to make their community, society, and even the world a good place to live—for themselves, their grandchildren, and the people among whom their grandchildren will live.

I am not saying that "PCT says" there are four kinds of organizational effectiveness. You are free to make your own list. I offer the four viewpoints listed above merely as an aid to getting away from the widespread view of effectiveness as return on investment, units produced per person-hour, percentage of students scoring above the national average on a standardized test, and the like. They do, however, fit the PCT view that effectiveness is in the eye of the individual beholder, and cherishing those four kinds of effectiveness will help, not hinder, individuals to control their own individually perceived variables.

Sometimes, by the way, writers use the term "efficiency." To engineers, that term means the ratio of the effective or useful output to the total input. Analogous to that is the idea of a "cost-benefit" analysis of a procedure or organization: the ratio of benefit to cost. Those ideas are very useful in evaluating nonliving things such as electric motors or canoes. Applying them to human activity, however, is the error of treating living things as if they were nonliving things. I discussed that error in Chapter 2.

Finally, we should remain wary of the temptation to think linearly—that something *causes* effectiveness in an S-R manner. I will let Karl Weick (1979, p. 86) remind us:

> Most managers get into trouble because they forget to think in circles.... Managerial problems persist because managers continue to believe that there are such things as unilateral causation, independent and dependent variables, origins, and terminations. Examples are everywhere: leader-

ship style affects productivity, parents socialize children, stimuli affect responses, ends affect means, desires affect actions. Those assertions are wrong because each of them demonstrably also operates in the opposite direction: productivity affects leadership style . . ., children socialize parents . . ., responses affect stimuli . . ., means affect ends . . ., actions affect desires. . . . In every one of these examples causation is circular, not linear. And the same thing holds true for most organizational events.

CONSULTING

Consultants help people to make plans and to take planned actions—and unplanned actions, too. As you might suppose, I claim that the most useful consulting preserves or increases the client's degrees of freedom and helps the client to find his way out of conflicts. I'll relate a small example.

When Richard Schmuck and I, with our apprentices, were conducting a project of organizational development with a school district in 1968, a couple of department heads in one of the high schools asked us for help with a matter of coordination among themselves and three other department heads. I went to the principal to let him know what I was doing. It turned out that no two department heads had the same free period. It was impossible, the principal said, to arrange the class schedule to open a common free period. Since, however, the superintendent had asked the principal to "cooperate" with me, the principal made some special arrangements so that I could meet with those five department heads.

When we met, we made a remark or two about the weather, and then I said, "I arranged this meeting so that you could discuss whether you have anything you want to discuss with one another." Then I just sat there. They sat there. Maybe someone said something more about the weather. Then one of them asked, "Are you getting a dissertation out of this?" I assured him I had finished my dissertation many years earlier. After several more moments, they began discussing whether they had concerns to talk about. After an hour or so, they asked me if I could persuade the principal to arrange some further times they could meet.

The principal did arrange meeting times for those five every other week for the rest of the spring semester. I met with them three or four more times in case I could be of help; now and then I clarified a point of discussion, but mostly I just sat there. The chief thing they wanted was to change the departmental budgeting. Customarily, each head submitted a proposed budget to the principal, and the principal then told the head what he or she could have. The result was that there was a good deal of unnecessary duplication of equipment and supplies. It had been obvious to them that they could make more economical use of equipment by rotating the use of it; they would then have money left over for other purposes. They had never been able, however, to meet as a group to work out a common budget submission. They believed, they told me, that the principal deliberately prevented them from dealing with him as a group by scheduling their free periods at different times.

By the end of that semester, as I learned later, they did submit a joint budget, and during the summer, they helped write a class schedule that gave all the department heads, not just those five, the same free period. The last I heard, those gains in communication and budgeting were continuing.

All I did for those department heads was to help them do what they wanted to do anyway. I did not tell them what they ought to want or how to talk to their principal. All I did was pry a little time loose from the principal's clutches. That gave them some crucial degrees of freedom. That's a small example, but I think a very appropriate example of the principle expressed by Marvin Weisbord (1987b, p. 233) like this:

> We change our behavior when we are ready to do it, not because of . . . any . . . kind of analysis. . . . The best a consultant can do is create opportunities for people to do what they are ready to do anyway.

I am tempted to tell you about the various styles of helping you can find among organizational consultants, but this is another point where I must frustrate myself. You can get an excellent overview of all that from just two books: Weisbord's 1987b and 1992. Some good illustrations of Weisbord's whole-system method were collected by Bunker and Alban (1992). For further help, see Richard Schmuck (1997, 2000). If you have a particular interest in schools, you can look at Schmuck and Runkel (1994).

APOLOGIA

I am saying repeatedly that although it is a good idea to think of all the things you *might* do in pursuit of a goal, it is not a good idea to commit yourself to what you *will* do very far ahead or in much detail.

When I worked as a draftsman in an engineering organization, I was impressed by the detail that the engineers put into their drawings. The exact position of every shelf-bracket, for example, and the exact type of bracket were specified. At the same time, the contract that went with the drawings always had a line in the budget called "Contingencies", which meant unforeseeable circumstances that would necessitate a change of plan and additional expense. The typical amount entered for contingencies was 20 percent of the estimated cost for doing the work as planned, without changes. As Weick might say, those engineers acted as if they knew what they were doing and, in the same document, as if they did not.

It may be that you have never found yourself committed to too much detail; you may feel that I am preaching to the faithful. I have written this chapter as I have, however, because my experience throughout my life has been that most people, most of the time (except, perhaps, farmers, engineers, and musicians), try to tie down the future by demanding that everyone do everything just as it is specified in a detailed plan.

Doing that tempts fate.

ENDNOTE

[1]The report of mine in which I summarized the findings of all those other studies is the appendix in Runkel, Schmuck, Arends, and Francisco (1979). The reports I reviewed are Bentzen and associates (1974), Berman, McLaughlin, and others (1975, 1977), Emrick and Peterson (1977), Emrick, Peterson, and Agarwala-Rogers (1977), Far West Laboratory for Educational Research and Development (1975), Glaser and Taylor (1973), Goodlad (1975), Hammond and Todd (1975), Human Interaction Research Institute (1976), Schmuck and Runkel (1970), Schmuck and Runkel (1994), Schmuck, Murray, Schwartz, Smith, and Runkel (1975), Sieber, Louis, and Metzger (1972), and Tye and Novotney (1975). The item Schmuck and Runkel (1994) was not, of course, mentioned by the 1979 book; the item mentioned there was the 1977 edition of that same book. I list the 1994 (4th) edition here because it will be easier to find.

Chapter 37

Schooling

Not just the bosses in an authoritarian, hierarchical, bureaucratic organization, but the organization itself seems to say to us: We think you are a passive animal who cannot move on your own; you must be pushed, enticed, threatened. Without supervision, you are competent to do only very insignificant, detailed tasks. You probably do not want to do what we want you to do. The only way you can have any power over your life here is to obey—to show us that you are willing to have very little power over your life here. And power is zero-sum. You must compete for it. If others have it, you do not; if you have it, they do not.

That description fits bureaucratic organizations the world around, and almost all organizations larger than perhaps 15 or 20 members are bureaucratic. I could pick any sort of organization—business, governmental, educational, sport, military—to illustrate the scrabble for degrees of freedom typical of bureaucratic organizations. I will describe public schooling in this chapter, because most of my own work outside the college classroom has lain with public schools.

Almost all schools immerse you in the bureaucratic kind of life. Go to mathematics class when the bell rings even though you may just now be finding the right words for an essay. Work at doing what the teacher assigns to everybody in the class today even though you want to work at something else, are ready to do something else, or are already doing something else. Answer short, little questions that can be answered with one word or a check-mark. Your function in the organization is to be processed in step with a hundred or a thousand other human-units selected for processing by the impersonal procedures of the school.

Do not challenge the teacher's criteria for judging good work, even though you have in mind a purpose for the work different from the teacher's. Work alone, excluding others, even though you know the work would go better if others helped. Try to get a better grade than Joe even if Joe is your best friend.

In 1971, Spencer Wyant, then a graduate student, handed me a definition of education:

> EDUCATION: Registered students receive documented facts taxonomically arranged in authorized courses based on recommended materials by recognized experts commissioned to develop authoritative interpretations under the tutelage of a certified teacher to satisfy institutional requirements and thus to be awarded a legitimate degree from a chartered institution which permits them to become licensed to practice sanctioned endeavors in reputable disciplines.

Those authorized channels leave little space for a path of one's own making. The educational authorities, especially in universities, proclaim themselves the bearers of liberation (via the liberal arts), and in the next sentence declare that your liberation will require you to confine yourself to authorized courses, to recommended materials by recognized experts, to thinking the thoughts on Wednesday that are assigned to be thought on Wednesday, and so on. Freedom, they imply, can be achieved only by having little of it. It is true that we often make a small free space for ourselves by throwing up a dike against a sea of threatening restrictions. That is the symbolism of the ivory tower.

As in non-educational organizations, the bosses of schools try to protect their own degrees of freedom by reducing the degrees allotted to their underlings. That method is seductive, but it does not work very well. The more you tie down your underlings to restricted ranges of behavior, the more you must tie yourself down to making sure they stay within those ranges. In many schools, for example, teachers are reluctant to leave a classroom of students for even a few minutes for fear all hell will break loose while they are gone.

RESTRAINT

I know you have yourself had a good many years of experience in classrooms, but I will give a few examples here to stir your memories. Susan Gonzales (1991, p. 12) wrote to the *Utne Reader* to say:

> I was a successful student in suburban, white, middle-class schools. . . . I got mostly A's, achievement and citizenship awards, ... The system worked for me . . . or did it?
>
> Now, as I watch my daughter and her friends go through public school, I am forced to acknowledge my long-buried feelings: my anger at being corralled with no regard for my will in the matter, then cajoled with rewards and threatened with punishments; my hatred of the bells, the buzzers, and the whistles of yard attendants, which ate holes into my brain and bones; my fear, once I came to accept that the system was a fit judge of me, of not meeting the mark, of not being good enough; my frustration at having to fit my complex thoughts into . . . true or false, correct or incorrect; my sorrow at losing the time and the energy to explore the world on my own and listen to my soul's voice.

Harry Wolcott (1974, p. 417), with a class of Kwakiutl pupils, assigned them to write essays on "If I Were the Teacher." This is part of the essay handed in to him by a fifteen-year-old girl:

> The first day in school I'd tell the pupils what to do. . . . And give the rules for school. . . . And if anybody's late, they have to write hundred lines. . . . First thing they'd do in the morning is Arithmetic, spelling, language, Reading. And in the afternoon Science, Social Studies, Health, and free time. And if nobody works they get a strapping. . . . And if they get the room dirty they'll sweep the whole classroom. . . . And if anybody talks back. They'd get a strapping. If they get out of their desk they'll have to write lines. If they don't ask permission to sharpen their pencil they'll get strapping. If they wear hats and kerchiefs in class they'll have to stand in the corner for one hour with their hands on their heads. . . . If they make a noise in class, they all stay in for half an hour. If anybody talks in class they write lines about hundred lines. If anybody's absent they have lots of homework for the next day. And if anybody fights they get a strapping. . . . And on Christmas they'd have to play or sing.

Wolcott made an analogy to a prisoner-of-war camp in which teachers are charged with recruiting prisoners into the teachers' own society. The teachers encourage the prisoners to defect and offer them the skills they will need to live effectively in the teachers' society. Wolcott wrote that a teacher in that role soon realizes that—

> his pupils may not see him playing a very functional role in their lives other than as a representative of the enemy culture. . . . He realizes that [much of] the energy and resources of prisoners are utilized in a desperate struggle to survive and maintain their own identity. . . . The teacher recognizes that the antagonism of his pupils may be addressed to the whole cultural milieu in which they find themselves captive rather than to him as an individual. . . . If this is how things seem to the prisoners, the teacher realizes that a modification of a lesson plan or an ingenious new teaching technique is not going to make any important difference to them. Taking the view of his pupils, the teacher can ask himself, "Just what is it that a prisoner would ever want to learn from an enemy?" (p. 421).

I'll add something to that. Imagine how you would feel if you were required, by law, to spend the greater part of every day, Monday through Friday, along with 25 or 30 other detainees, in a room where you would listen to a person lecturing to you, hour after hour, about topics most of which had very little interest for you, and imagine that this grinding routine was going to go on year after year, into the far future. Imagine that the lecturer would scold you when you gave evidence of being bored. Imagine that the lecturer would frequently put quizzes and tests in front of you, full of questions about which you cared very little, and that some days, periodically, would be devoted to nothing but answering those questions on those tests. Imagine that if you failed to guess too many of what the lecturer called the "right" answers to the quizzes and tests, you would be sentenced to repeating an entire year of those same lectures and quizzes and tests.

An article in TIME magazine (Wendy +, 2001) reported on nightmares, stomachaches, vomiting, insomnia, and depression that come upon some elementary-school children as testing time approaches. You may remember, too, the heightened rate of delinquency that occurs in at least some cities during the schooling ages; I described Gottfredson's (1981) studies of that phenomenon in Chapter 29 under "Psychological Effects." It does not surprise me that adolescents want to demonstrate to themselves and others a lack of submission to that unremitting oppression.

Imagine, too, that all the people you knew, including the other detainees, acted as if this oppression was a normal, natural, ordinary, unremarkable state of affairs and as if to expect anything else would be absurd.

How would you feel, getting out of bed in the morning, knowing you were going off to that room for most of the day, five days a week, year after year, regardless of what else you would prefer to do, and knowing that only illness could give you even a respite? That is the daily fate of millions of school children in the United States. Schooling has always been numbing to large numbers of pupils attending school by parental requirement or by law, but the daily dullness and the threats of failure have been severely increased by the recent movement to justify school taxes by requiring pupils to pass statewide multiple-choice tests. The dullness and the threats have been extended to teachers, principals, and superintendents by requiring schools to show that large percentages of their pupils have passed those tests. I will say more about tests in Chapter 38.

I am not saying that school children have a worse life than workers in chicken-packing plants (for example). At this point, I am simply emphasizing the restricted environments—the restricted opportunities for controlling perceptions—typical of schools.

If life is not happy for students, life is not likely to be very happy for teachers, either. Of course, even if the average teacher feels hemmed in and beaten down, there must be some teachers who feel good about their jobs. For example, you might think that a teacher receiving an award as New York City's Teacher of the Year would be one feeling pretty good about his job. John Gatto, a seventh-grade teacher on Manhattan's Upper West Side, was awarded that title in 1990. Here is what he said in his acceptance speech:

> The truth is that schools don't really teach anything except how to obey orders. This is a great mystery to me because thousands of humane, caring people work in schools as teachers and aides and administrators, but the abstract logic of the institution overwhelms their individual contributions. [T]he institution is psychopathic; it has no conscience. It rings a bell, and the young man in the middle of writing a poem must close his notebook and move to a different cell. . . .
>
> It is absurd and anti-life to be a part of a system that compels you to sit in confinement with people of exactly the same age and social class. That system effectively cuts you off from the immense diversity of life and the synergy of variety.

It is easy to see there the chafing at restraint and the yearning for greater degrees of freedom. Gatto continued:

> It is absurd and anti-life to be part of a system that compels you to listen to a stranger reading poetry when you want to construct buildings, or to sit with a stranger discussing the construction of buildings when you want to read poetry.
>
> It is absurd and anti-life to move from cell to cell at the sound of a gong for every day of your youth, in an institution that allows you no privacy and even follows you into the sanctuary of your home, demanding that you do its "homework."
>
> "How will they learn to read?!" you say, and my answer is, "Remember the lessons of Massachusetts."

Earlier in his speech, Gatto had said that compulsory schooling in the United States began in Massachusetts about 1850—and not by popular demand:

> It was resisted—sometimes with guns—by an estimated 80 percent of the Massachusetts population, the last outpost . . . on Cape Cod not surrendering its children until the 1880s, when the area was seized by militia and children marched to school under guard.
>
> . . . prior to compulsory education the state literacy rate was 98 percent, and after that the figure never again reached above 91 percent. . . .

Gatto's comparison there is not an argument that compulsory education *caused* the literacy rate to drop from 98 percent to 91. Other factors could have been influential. Those numbers, however, do demonstrate that compulsory education was not *necessary* to achieve 98 percent literacy. Gatto was probably familiar, too, with what Ivan Illich (1973) had written:

> Money is now spent largely on children, but an adult can be taught to read in one-tenth the time and for one-tenth the cost it takes to teach a child (p. 128).

Gatto went on to describe his students:

1. The children I teach are indifferent to the adult world. This defies the experience of thousands of years. A close study of what big people were up to was always the most exciting occupation of youth, but nobody wants us to grow up these days. . . .
2. The children I teach have almost no curiosity and what they do have is transitory. They cannot concentrate for very long, even on things they choose to do. Can you see a connection [with] the bells ringing again and again. . .?
3. The children I teach have a poor sense of the future, of how tomorrow is inextricably linked to today. . . .
4. The children I teach are ahistorical; they have no sense of how the past has predestined their own present, limiting their choices, shaping their values and lives.
5. The children I teach are cruel to each other; they lack compassion for misfortune, they laugh at weakness, they have contempt for people whose need for help shows too plainly.
6. The children I teach are uneasy with intimacy or candor. . . .
7. The children I teach are materialistic, following the lead of schoolteachers who materialistically "grade everything"—and television mentors who offer everything in the world for sale.
8. The children I teach are dependent, passive, and timid in the presence of new challenges. This timidity is frequently masked by surface bravado, or by anger or aggressiveness. . . .

How can freedom to explore the world, to develop skill in selecting and using the opportunities of the environment, be found? Gatto said:

> For five years I ran a guerrilla school program where I had every kid, rich and poor, smart and dipsy, give 320 hours a year of hard community service. Dozens of those kids came back to me years later and told me that this one experience changed their lives, taught them to see in new ways, to rethink goals and values. It happened when they were 13. . . .
>
> There is no shortage of real problems. Kids can be asked to help solve them in exchange for the respect and attention of the adult world. . . .
>
> Independent study, community service, adventure and experience, large doses of privacy and solitude, a thousand different apprenticeships—these are all powerful, cheap, and effective. . . . But [we must] include *family* as the main engine of education.

Joseph Blase (1986) made a study of "sources of teacher stress"—the kinds of complaints teachers were making about their jobs. Blase got responses to questionnaires given to 392 teachers in various parts of the United States. Their average years of experience in teaching was 10. Here are a few quotations from what the teachers wrote:

> Teachers are loaded with extra work. . . . Much of this work doesn't have anything to do with the classroom. I like to do everything the best I can, but with so much to do I can't. . . . I feel like a marionette. . . . I feel frustrated, angry, and helpless (p. 30).
>
> I'm in a situation where I have to teach five classes a day, all different preps plus supervise a study hall. That leaves me one prep period to prepare five different things. . . . In addition, I must use prep time to purchase groceries and other supplies. (I'm in the home economics field.) Many days I've had to spend the [major part] of my prep period purchasing groceries for class—then head right into five straight classes ill-prepared. . . . There's no time to come up with new ideas. I reuse far too many old teaching materials instead of developing new ones.
>
> Some of the kids need your help desperately. I try to look for signs, see if they're having problems. . . . I'm so busy just trying to keep up with things, I can't give the kids the attention they need (p. 31).
>
> When I . . . started teaching I was excited. . . . I would try all kinds of things to make the class interesting. . . . Teaching drains you! I've lost a lot of my enthusiasm. . . . It's hard to keep up a facade of excitement when so many kids and parents don't give a damn about education (p. 32).

Here is an excerpt from an anonymous article[1] printed in 1987 in a periodical for teachers called *Cutting Edge*:

[Presumably] wiser heads tell [teachers] what text or method to use; how much time to devote; . . . what equipment and supplies may be used; what tests to give; and in general how to conduct the class that is called "theirs." Orders and directives rain down from supervisor, principal, superintendent, and central office, the board of education, state offices and officials, the state board, governor, federal offices, and not least, the Supreme Court. Local power people, such as editors and activists, add to the restrictive input; and of course funds are doled out only to compliers. The "wiser heads" . . . normally prove to be those with more clout, not those who can point to any successful educational results.

In most schools, bells or buzzers and the p.a. system interrupt with total disregard and disrespect for the teaching staff. To add further insult, teachers, without apology, get bus, lunchroom, policing washrooms duty, and other assignments. . . . That any teachers at all can be persuaded, for modest pay, to accept these conditions must be counted a marvel. . . (p. 1).

I do not have a recent statistic, but some years ago I read that 50 percent of new teachers were leaving the profession after their first year. So about 50 percent were willing, or felt obliged, to try at least one more year. I agree with the anonymous writer that it is a marvel. Here is another marvel from a report by the Ford Foundation (1987):

For Gates, recipient of a Presidential Award for Excellence in Mathematics Teaching and other professional honors, a typical day as teacher began at 7:30 a.m. . . . Exacerbating the problem of providing meaningful instruction to 125 students per day was her hall-monitoring responsibility during the five minutes between classes. . . . "I hate standing in the hall," says Gates. "You need that time."

For reasons of accountability and standardization, the school system had developed prescribed curricula and textbooks. . . . Says Gates of this mechanized approach, "It doesn't work" (p. 12).

You might think that teachers in small schools would be less loaded with extra duties and directives from the principal, the superintendent, and the Supreme Court and that they would be able to find more degrees of freedom with which to give attention to their students. That indeed, was what Barker, Gump, and others (1964) found in their well-known studies of five high schools in Kansas. In 1990, however, Patricia and Richard Schmuck reported their findings from their visits to 25 school districts in small towns in 21 states:

We saw superintendents not only developing budgets, working on curriculum, and meeting with the board, but also driving a bus, attending out-of-town sports events, and directing traffic. We saw principals not only supervising teachers, running faculty meetings, and observing classes and hall behavior, but also acting as athletic directors, working on bus schedules, and coordinating the curriculum for the district. . . . We saw some teachers teaching 6 or 7 hours a day without time for preparation or grading papers, and then spending another 2 or 3 hours supervising an extracurricular activity after school (p. 18).

. . . it was rare for students to tell us that they were enthusiastic about schoolwork. When we asked secondary students how they felt about school, again and again we heard, "bor-ing." From kindergartners to seniors, students responded that friends were the best thing about school. . . . Students considered many of their teachers to be uninterested in them as individuals, lacking a sense of humor and teaching their subjects like robots. . . . Even though the typical small-town high school classes had fewer than 20 students, we saw only a few cases of active discussion involving more than a couple of students and the teacher (p. 15).

Not only in small towns, but in large cities too, superintendents suffer from too much to do. And they must withstand threats from every direction. An article in TIME in February of 2000 by Rebecca Winters began this way:

HELP WANTED: Dynamic professional to lead venerable institution through period of rapid change and political tumult. Will manage more employees than Texaco has and a budget bigger than Cisco Systems'. Must answer to mayor, governor, seven-person board, unions, press, and public. Workweek: 80 hours plus. Compensation: around $140,000, about 1% of the average FORTUNE 500 CEO's pay (p. 74A).

Would you aspire to a job like that? At that writing, Winter said, 13 large cities were looking for superintendents. The average tenure of urban superintendents had dropped to less than three years. The chancellor of the schools of New York City, just "ousted," had been hired in 1995 following 10 chancellors in the preceding 12 years.

In my examples above, you saw restriction of freedom everywhere and, as a result, conflict. Teachers demand that students pursue the *teachers'* goals. Principals demand that teachers pursue the *principals'* goals. Superintendents demand that principals . . . and so on. I am not saying that it is immoral or even avoidable for a boss to make use of an underling for the boss's own purposes. I am saying that if the boss does that arbitrarily, out of sheer authority, without allowing the underling sufficient degrees of freedom with which to control her own perceptions, then intrapersonal and interpersonal conflict will arise. People will be working against themselves and others. Everybody will become weary without seeming to do much. Bosses will demand more contribution to their purposes and shower down more directives to "focus" and "align" the actions of the underlings still more severely. That kind of positive feedback stops escalating only when everybody is too exhausted to summon any more effort. Note that I am using the word "positive" here to mean "amplifying," not "good."

Nevertheless, many adults, perhaps most, in school and out, want to perceive children knuckling under to adults. That is, they want to see children readily behaving in a docile, biddable, obedient manner. Ready obedience among workers has been called "industrial discipline"—get to work at the time set by the employer, leave when the bell rings, remain at your work station unless given permission to leave (for toilet, lunch, or break), do your own work (don't ask for help from others), follow directions for producing the work (not your own ideas), submit to quality inspection by others, and so on. Indeed many employers say the chief reason they do not like to hire dropouts is the likelihood that they have not acquired this docility. Many writers on education—Bowles and Gintis (1976), for example—say that the need for docile workers by the factories springing up during the industrial revolution in the 19th century was the chief reason that legislators instituted universal compulsory education.

Many educators, even most, have purposes for schools very different from those of factory managers. Compare, for example, my brief description about three pages farther along of A. S. Neill's Summerhill.

A lot of conflict in schools results from ignorance of how living creatures function, but most of it is put there *purposely*—even though thoughtlessly.

GOOD SCHOOLS

There are, however, schools where the principal and a sufficient proportion of the faculty know better than to escalate a conflict—where they understand that where everyone is too confined by hierarchy and "high standards," what is needed is not a reduction in degrees of freedom, but an increase. That is true, for example, in the schools I will tell you about in Chapter 39. I am sure there are more schools than I know about where people understand about degrees of freedom; I have happened upon a few that seem never to have been discovered by journalists or social scientists. But I will tell you about a few programs or "reforms" that have been described in print. Whether the schools that were in these programs are still operating in the way described I do not know; a good school can be ruined in a month by a new principal who does not understand what the faculty has been doing and is afraid to let them continue.

Many people, even educational "experts," seem to think that educational reform and experimentation is a recent phenomenon; they talk about "new" ideas as if schools and their struggles have no history. Educational reforms were afoot in England, for example, during the time of Henry VIII. The historian Arthur Cross (1914) wrote:

> The *First Liberal Education for Boys* (1529) [contained recommendations in] shining contrast to the prevailing mechanical methods, in which flogging was employed as the chief incentive.
>
> Further evidences of advance are manifest in the writings of those whom Henry selected to educate his own family. A Spaniard, Ludovico Vives . . . insisted upon the value of observation and experiment [and] believed that much should be left to the independent exertions of the pupil. . . (p. 344).

Those "advances" may not seem like much, since they sound so familiar, but the attitude of Ludovico Vives seems to have differed considerably from that at Oxford University two centuries earlier, when revealing the wrong thoughts brought not only disparagement but financial loss. John Barrow (1992, p. 205) quotes a 14th-century statute of Oxford University:

> Bachelors and Masters of Arts who do not follow Aristotle's philosophy are subject to a fine of 5 shillings for each point of divergence.

But let me skip over a few centuries and the Atlantic Ocean. I mentioned earlier the important effort to reform public education by making it compulsory, a movement that began in the United States before the Civil War. Now, skipping over almost a century of further reforms of schooling, I want to tell you about a very large and carefully systematic experiment carried out with about 60 high schools during the 1930s. The outcomes were reported in a book by Wilford Aiken (1942), but I will tell you about it by quoting from an article by Jennings and Nathan (1977). They wrote:

> One of the most extraordinary experiments ever conducted in American education was the Eight-Year Study, once widely cited in the literature but neglected in recent years. . . . Thirty high schools signed an agreement with 300 colleges to exempt their graduates from the usual college entrance requirements (p. 568).

Note that in the language of current reform proposals, release from college entrance requirements would be called *lowering* standards, both in the high schools and the colleges! Jennings and Nathan continued:

> This meant that the high schools did not have to use grades, class rank, required courses, credits, etc. They could experiment with curriculum and organization.
>
> Some 1,500 students from the experimental schools were paired with 1,500 students from similar but nonexperimental schools (perhaps 30 of them; Jennings and Nathan do not say) and were matched by sex, age, intelligence, family background, race, and other factors (p. 568).

And those 1,500 students who were released from the traditional high standards of college entrance—what do you suppose happened to them when they got to college?

> The students from experimental schools did as well as the others or better at college in grades, participation, critical thinking, aesthetic judgment, knowledge of contemporary affairs, etc.
>
> Further analysis yielded some startling results: When students from the six *most experimental* schools were compared with those from traditional schools, there were great differences in college attainment. Finally, the two most extremely experimental schools (where practices were indeed different—e.g., extensive learning in the community; outside volunteers working with students; advisor-advisee systems; students teaching other students; interdisciplinary, problem-solving curricula; etc.) were selected. Graduates of these two schools were found to be "strikingly more successful."
>
> One of the schools in the Eight-Year Study was the Ohio State University Lab School. The students who graduated in 1938 wrote a book called *Were We Guinea Pigs?*". . . Many years later a thorough follow-up study of the "guinea pigs" [was carried out]. The Lab School graduates were then between 35 and 40 years old. The study found that the "guinea pigs" had been strikingly successful in life. They were then compared with subjects in the Lewis Terman study of genius and with graduates of Princeton University, where a similar follow-up had been conducted. The experimental school graduates came out ahead. They more often expressed satisfaction with life, were judged leaders in their professions, had more stable family lives, possessed better, self-accepting attitudes, and were mentioned more frequently in *Who's Who*.
>
> Since 1970 a number of schools have been established which make use of curriculum and organization ideas developed either by the experimental schools in the thirties or even earlier by John Dewey and other progressive educators. . . . The Career Study Center in St. Paul, Minnesota, was created for students who are unsuccessful in the city's traditional high schools, where they are typically truant half the time. Some 70% of CSC candidates are in trouble with the law. . . . After some time in the . . . program at CSC, attendance rises to 80%. Youngsters then get out of trouble, and their parents can hardly believe the change in attitude toward schooling. Ninety percent graduate, although the original prognosis was that only 10% would do so (p. 569).

Jennings and Nathan told about numerous other studies showing similar results. Always, the schools showing the superior results were those with the fewer restraints on the actions of students and faculty—the schools with the greater degrees of freedom. If some students in some schools do so much better than students in traditional schools, the question arises: What are students in the traditional schools doing? The answer seems to be: Not much. That is, students get about the same scores on achievement tests even when their days in school are severely reduced.

In 1972, because of financial trouble, a school district in Maine reduced its schooling week from five days per week to four. After comparing scores on achievement tests from that year with scores from previous years, Robert Drummond (1972) wrote that with the four-day week, "gains clearly outweighed losses when considering the grade-equivalent scores of all students tested." Jennings and Nathan (p. 570) also told about the experience during a strike of teachers in Philadelphia. Some of the schools were closed for eight weeks. Lytle and Yanoff (1973) reported that when scores on tests of achievement were compared between students who had missed eight weeks of school and those who had not, no significant differences were found.

An experience I had with a sophomore college student fits into this picture. The student complained to my teaching assistant that she could not seem to get a grade better than "B" on the quizzes. She said she had been getting grades of "A" all through her schooling. My assistant asked her how she studied. She said that, as she read the textbook, she noted the important phrases in each topic. On a quiz, she looked for one of those important phrases in the stem of the item and then looked for an important phrase from the same topic in the answers offered. My assistant asked whether she read the stem carefully to ascertain its meaning. She said she did not.

In other words, that woman had been an "A" student for years, not necessarily because of understanding anything—or even trying to understand it—but because of becoming skillful at sorting phrases according to whether they had (so to speak) fallen on the same page of a textbook! And apparently she had never before encountered quiz items written in a way that required understanding! When you read the previous few paragraphs and this one, I hope it becomes easy to understand why some writers use the word "absurd" in writing about traditional schooling.

In the paragraphs above, I have been using the criterion of scholastic performance. Remember that coming out ahead, on the average, of students at Princeton or of Terman's "geniuses" is not a remarkable achievement. For one thing, those are average results; some did not come out ahead. More important however, is the fact that being a "genius" for Terman or graduating from Princeton (or from MIT) is largely a matter of marking the right answers on multiple-choice tests. Remember my student who got A's by watching and listening for words that occurred together. Remember the physics students at MIT who could not wire a light bulb to a battery.

From traditional schools, however, *we should not expect much more* than guessing some answers "correctly" on tests. That is what we have been told by Wolcott, Gatto, and Schmuck and Schmuck. I will tell you later of similar reports from Ewens, Sarason, Sedlak, and Tyler. Meanwhile, though I have said the influence of teachers on students is not marvelous, I should point out that other professions do no better. Physicians write prescriptions for medicines and explain to their patients how to take them. The pharmacist puts a label on the bottle and writes out thorough directions (a crib sheet) for taking the medicine. Yet the failure of patients to follow all those careful instructions is one of the major reasons that antibiotics rapidly lose their effectiveness. Do we scold physicians and pharmacists for failing to teach their lessons more effectively? Do we threaten to reduce their pay if their patients do not take their medicine more faithfully? (See Chapter 29 under "Giving Advice.") Do we scold wardens of prisons because they do not succeed in reforming their inmates? For what percentage of their constituents or of the country at large do politicians succeed in improving their lives? What percentage of churchgoers go and sin no more, compared to the unministered? Note that the clients in all these examples, in contrast to the usually fugitive urgencies of schooling, are often facing very serious difficulties, sometimes matters of life and death. Yet those clients seem to learn their lessons, so to speak, not much better than the students. If you have in mind the Requisites for a Particular Act, none of that is very surprising.

Parenthetically, I am not mentioning the writings of Ewens, Gatto, Sarason, Schmuck and Schmuck, Sedlak, Tyler, and Wolcott to argue the correctness of PCT. None of them mentioned PCT nor showed any awareness of circular causation. They were, however, well aware of the self-motivation and the uniqueness of humans and of the need for degrees of freedom among environmental opportunities if inner conflict is to be minimized. In citing the research and insights of those authors, I am saying that if you look around you, you can see, as they did, the pervasiveness of restricted degrees of freedom in schools. And I remind you that schools are in that regard representative of our organizations in general.

Now let me return to my listing of good schools.

In Rabun County, Georgia (USA), the students at the high school have been writing history and ethnography of that region of Georgia for thirty years. Their information comes from their extensive interviewing of residents and their examinations of artifacts. They have produced more than fifty semiannual issues of the *Foxfire Magazine*. You can find more information about the *Foxfire* in Wigginton (1985), Puckett (1989), Starnes (1999), and the magazine.

Montessori and Waldorf schools enroll thousands of students and have been long and widely known for their much-enriched environments (despite being often underfunded) and their encouragement of initiative and active learning. For the Montessori, see, for example, R. A. L. Wentworth (1998), and Harry Morgan (1999). For Waldorf, see M. Spock (1985), R. E. Kotzsch (1990), and Maher and Shepherd (1995).

The English boarding school Summerhill was founded in 1921 by A. S. Neill. Whether it still exists, I do not know. It is or was sufficiently radical that I cannot take the space to do it justice here. I can say that it is called a "free school," that classes are optional, and that everyone, staff or student, has equal rights. But if that is not enough, you can find a brief description in Ewens (1984, pp. 256–260) and more in the book by Hemmings (1973). Neill died in 1973, but Jonathan Croall edited a book of his writings in 1983 and issued a biography in the same year, and Albert Lamb edited a book in 1992.

Still another current of reform that seems to be sensitive to the need for sufficient degrees of freedom in social behavior is the strategy of the "brain-based" schools. This movement stems from Leslie A. Hart's two books *How the Brain Works* (1975) and *Human Brain and Human Learning* (1989). Here are some directions in which Hart's theory points, quoted from an article of his in the Phi Delta Kappan (1989a):

1 *Working from theory.* . . . Theory frees teachers and others from ritualistic ways of doing things and allows them to create and develop [their own] instructional directions. . . .
2 *Not teaching content in bits.* [The theory emphasizes] . . . the ability to perform tasks in realistic situations.
3 *Not teaching in isolated blocks of time.* Learning advances . . . as a flow of experience, exploration, and integration. Efforts to move a fixed group of students along in lock step (a frustrating impossibility in any case) give way to allowing students to progress without hindrance.
4 *Gradually dumping unproductive practices.* Giving grades burdens teachers but has little to do with learning; it has much more to do with control [of students] and punishment.
5 *Shifting to activities that promote learning.* . . . creating a non-threatening climate, letting students work together often (as humans do almost everywhere but in schools), and moving from correcting to coaching.
6 *Not isolating teachers.* . . . teachers can be allowed to move out of classroom cells in which isolation takes a depressing toll. . . . Working collegially, the "lonely teacher" vanishes (pp. 241–242).

Wayne Jennings of the University of St. Thomas in St. Paul, Minnesota, who has edited the *Brain-Based Education Networker* for a good many years, tells me that there were in 1997 several schools following Hart's theory in Wisconsin, a number in Kansas City, Missouri, and "dozens if not hundreds" in the nation. For more information, look into the *Brain-Based Education Networker*, CARE office, Rockhurst College, 1100 Rockhurst Road, Kansas City MO 64110–2561. There is also a book: Caine and Caine (1997).

In TIME magazine, Andrew Goldstein (2001) told about Mary Catherine Swanson, who, 21 years ago, undertook to supplement an English course with an additional hour with the students every day, during which she gave them tutoring and support and, in general, helped them discover how to do the academic things students do in the regular meeting of the course. Now, many thousands of teachers are conducting courses this way. Among students who do poorly in traditional courses—getting mostly Cs and Ds—almost all do better with this kind of attention and help. Since 1980, more than 93 percent of these students go to college, and two years after entering college, 85 percent (compared to 70 percent of all college entrants) are still enrolled.

Only students who volunteer are accepted for Swanson's kind of course. The auxiliary meeting is not graded. You can see that the students have the first Requisite: they *want* to perceive themselves doing well academically and do not yet perceive that. The opportunities to satisfy the second Requisite are greatly expanded in the tutoring period and also outside it. Not only do they get tutoring from the teacher, but I am supposing they get some tutoring from each other outside that period, also.

A practice you will find in all good schools is "cooperative learning," in which students cooperate in groups instead of competing or working alone. If students are actually enabled to help one another, a great deal of conflict vanishes. Books on the topic are plentiful; for example, D. W. Johnson and F. P. Johnson (1987), D. W. Johnson and R. T. Johnson (1987a, 1989, 1995), Slavin (1990), Sharan (1994), Johnson, Johnson, and Holubec (1994, 1994a), Bunker, Rubin, and others (1995), Sharan, Shachar, and Levine (1999), and Deutsch and Coleman (2000). As to outcomes, here is the abstract to an article by Qin, Johnson, and Johnson in a 1995 issue of the *Review of Educational Research*:

> The impacts of cooperative and competitive efforts on problem solving were compared. In order to resolve the controversy over whether cooperation promotes higher- or lower-quality individual problem solving than does competition, 46 studies, published between 1929 and 1993, were examined. The findings from these studies were classified in 4 categories according to the type of problem solving measured: linguistic (solved through written and oral language), nonlinguistic (solved through symbols, math, motor activities, actions), well-defined (having clearly defined operations and solutions), and ill-defined (lacking clear definitions, operations, and solutions).... Members of cooperative teams outperformed individuals competing with each other on all 4 types of problem solving.... These results held for individuals of all ages.... The superiority of cooperation [was greater] on nonlinguistic than on linguistic problems (p. 129).

As a final example of where you might find schooling with increased degrees of freedom, I will mention home schooling. I am supposing that on the average, children schooled at home will work with more degrees of freedom than children schooled in school buildings, because those at home usually have the greater bargaining power.

According to an article by Cloud and Morse in a 2001 issue of TIME, the number of students being schooled at home was estimated by "the government" at 345,000 in 1994 and at 850,000 in 1999. The latter number was four percent of all U.S. students in kindergarten through high school. That may seem a small number, but it is more than the number of students attending "all the public schools in Alaska, Delaware, Hawaii, Montana, New Hampshire, North Dakota, Rhode Island, South Dakota, Vermont, and Wyoming combined" (p. 47). For comparison with the figure of 850,000, about 500,000 students were in "charter" schools in 1999, and about 65,000 were receiving vouchers (p. 48).

The average SAT score in 2000 for home-schoolers was 1100, and for the general population 1019. The average home-schooler scores at the 75th percentile on the Iowa Test of Basic Skills (the 50th percentile is the national average) (p. 51). I tell about the scoring on those tests only because I suppose some of my readers will care about such a statistic. As you read Chapter 38, you will find that I myself have no admiration for academic tests.

In Orange County, California, two school districts have opened schools that offer programs for home-schoolers (p. 53). Nationally, one home-schooler in five enrolls in at least one class in a public or private school. About three-quarters of universities in the United States now have policies offering at least some welcome to home-schooled applicants. Harvard admissions officers attend home-schooling conferences to look for applicants (p. 51).

I said that home schooling would be my final example, but I meant the final example in this chapter. I have reserved Chapter 39 to tell you about a program that can radically improve the degrees of freedom for everybody within existing schools.

HOW IT FEELS

Now I will try to portray more concretely how it feels to be "schooled," on the one hand, and enabled to explore one's world, on the other. I'll begin with an ingenious experiment by Wild, Enzle, and Hawkins (1992). They arranged to give a piano lesson to 32 undergraduates—17 males and 15 females—who had no previous musical training.

An experimenter introduced each student, in turn, to the teacher and explained that the purpose of the experiment was to examine the performance of the *teacher*. Half of the students were then further instructed as follows:

> "I don't know if you happened to see our ad in the newspaper, but that's how we hired [teacher's name] to do your lesson today. Just bear with me while I get [his] money ready."

> The experimenter [then] took $25 from the petty cash envelope and signed a letter of acknowledgment. He then placed the money and the letter in a second envelope, on which he wrote the teacher's name.

The other half of the students underwent these instructions:

> "I don't know if you happened to see our ad in the newspaper, but that's how [teacher's name] came to volunteer to do your lesson today. Just bear with me while I sign the department's thank-you letter for [him]."
>
> [The experimenter] signed the letter and put it in an envelope, on which he wrote the teacher's name. To keep the [teacher] blind to the [instructions], $25 had been placed in the envelope . . . before the session (p. 247).

After that, regardless of instruction, experimenter and student then went to the piano and the lesson proceeded. I have omitted from my retelling many details of the care the experimenters took to be sure the teacher never knew the instruction given to the student and to keep the manner of giving the lesson as similar as possible from student to student.

After the student had learned to play a short version of "God Rest Ye Merry, Gentlemen," once through without error, the teacher summoned an interviewer, who

> explained that he or she would take the teacher and [student] in turn to his or her office for a 10-minute interview. The teacher was to be first, and the . . . interviewer asked the [student] to remain in the room until his or her turn. . . . The concealed audiorecorder recorded [the student's] piano-playing . . . during the 10-minute interval.
>
> After the free-play interval, the interviewer returned to the room and asked the [student] to complete a questionnaire. . . (p. 248).

Despite the care taken to keep the lessons uniform in every possible way for all the students and despite the random assignment of students to the two experimental conditions, the students believing they had a volunteer teacher rated the teacher, on the average, as more innovative and spontaneous than did the students believing they had a paid teacher. The students with the volunteer teacher also believed the teacher enjoyed giving the lesson more, on the average, than did the students with the paid teacher.

Students with the volunteer rated their enjoyment of the lesson higher, on the average, and rated their enjoyment of playing the song more, and rated their desire to learn more about playing the piano, than students with the paid teacher.

The recordings of the free-play time (when the students were left alone) were analyzed for whether the student was repeating "God Rest Ye Merry, Gentlemen" or was doing something new. Among the students with the volunteer, 63 percent initiated new kinds of playing, while only 13 percent stuck to the lesson. Among the students with the paid teacher, however, only 19 percent initiated new playing, while 50 percent stuck to the lesson. Those are remarkably different proportions.

Wild, Enzle, and Hawkins offered this reasoning concerning the expectations of the students:

> We propose that in extrinsically constrained educational contexts, students will expect their teachers to show little intrinsic interest in the activity and to be rigid and nonspontaneous in their teaching style. Students will also expect that they will find the learning experience to be boring and the new skill to be of little interest. . . . Students in relatively unconstrained educational contexts should therefore be more likely to enjoy the process of learning and to value newly acquired skills (p. 246).

This is a good example, by the way, of the fact that more than one explanation of a phenomenon can be available, some people preferring one, some the other. Wild, Enzle, and Hawkins seemed to claim that the students made a distinction between intrinsic and extrinsic motivation and believed the theory about those presumed motivational states set forth in traditional textbooks. Perceptual Control Theory, however, does not make that distinction; "motivation" is not even one of its technical terms. I prefer to take the study by Wild, Enzle, and Hawkins as an example of Requisite 2b, concerning the kind of aid the person expects to find in the environment.

Requisite 2b reminds us that if you perceive the teacher as taking the traditional role, it will be easy for you to fall into the complementary role, that of the traditional student. When the teacher is out of the room, there is "nothing to do." The traditional student must wait for the teacher to return before resuming "the lesson," each step of which is conducted at the teacher's initiative. If the student, alone, were to do something with the piano, it might not fit with the teacher's "lesson," and the time would be "wasted." Indeed, the teacher might even scold you for continuing without permission!

I think that the information that the teacher was being paid was enough, on the average, for the students to refer to their stereotypes about teachers and to look for the traditional dullness rather than delight in what the teacher was doing with them. One might say that one of the effects of certifying teachers as qualified to teach is simultaneously to certify them—at least in the belief of the students in this study—to bring dullness into the lives of their students. Merely the perception that the piano teacher might not act like a traditional teacher (was instead an unpaid volunteer) seemed enough to allow a larger percentage of these students to explore more pianistic possibilities—to think that the piano might offer a means of controlling more perceived variables than those brought to their attention in the experimental situation. The proportions I cited four paragraphs back seem to me a fair measure of the loss of potential we get when we bureaucratize schooling.

I will now describe a comparison of two kinds of teaching that differ in the degrees of freedom they give to children. This comes from Seymour Sarason's book *Schooling in America* (1983, p. 112). He quotes a science lesson described by Kamii and DeVries (1978). The first way is the standard way. It begins with the assumption that nothing good can happen before the teacher writes out "behavioral objectives":

Theme: Crystals.
Behavioral objective: At the end of the experience, the child will be able to

1   Pick out crystals when shown a variety of things.
2   Define what a crystal is.
3   Discuss the steps in making crystals at school.

The objective speaks of "the child." That sounds as if it means *every* child. But what the children do, in the end, will differ a little or a lot from what the teacher was imagining they might do, or what she hoped some would do, as she wrote those words. Dozens, maybe hundreds, of textbooks for aspiring teachers urge you to think that way about what you want to do and to write that way, too, despite the disappointment you store up for yourself when you do that.

When that teacher wrote out that behavioral objective, she described what *she* wanted to perceive eventually, not what the children wanted to perceive. Thereupon, in accordance with the traditional theory, it became her duty to bring about what she wanted to perceive, all quite without any thought of what the children wanted to perceive, then or later. Here is how she went about it:

The teacher will show the children different crystals and rocks. She will explain what a crystal is and what things are crystals (sand, sugar, salt, etc.). Then she will show some crystals she made previously.

There you see the traditional belief that the best way to "get" students to "learn" something—get them, that is, to say what the teacher says or do what the teacher does—is first to give them words, oral or written, about the topic and only then, if at all, let them use their hands to do something concerned with the topic. That idea of the correct or best way to enable students to acquire knowledge is accepted by most people the world over as the best way to organize a ten-minute lesson, a 50-minute lesson, a week, a semester, and a collegiate curriculum. That idea is so thoroughly and widely accepted that thousands of students (not all the students, but thousands of them) every year graduate from four years of college and thousands more from three or four years more of graduate school having talked and read *about* something they hoped to be doing eventually but never actually having done it.

The teacher writing out her behavioral objectives was trying to be a good teacher. I, too, spent many years trying to be a good teacher. It took me too many years to learn a few things. For example, the way to help students learn what they want to learn is *not* to try to be an interesting lecturer or even a dramatic demonstrator. I won't take space to elaborate on that statement here beyond saying that a good guide for the student, even if stated over-simply, is "Act first, then think." You might want to use your own imagination concerning the possibilities. Sarason (1983, pp. 126–129) gives a good example in the training of clinical psychologists. But let us return to our traditional teacher and her crystals:

The children are given materials . . . so they can make crystals to take home.
Method . . . Mix 1/2 cup each of salt, bluing, water, and 1 [tablespoon] ammonia. Pour over crumpled paper towels. In 1 hour crystals begin to form. They reach a peak in about 4 hours and last for a couple of days. . . .

Presumably, all the children did that, and all at the same time. There is nothing in the description giving any recognition to the facts that various children will see and hear different aspects of the lecture and the mixing experience, that they will have different reasons (criteria, internal standards) for paying attention or not, that they will feel differing urges to extend a curiosity to other aspects of their own lives in or out of school, and so on.

Kamii and DeVries say that one of their colleagues read the above lesson, modified the way of going about it, and wrote the following account of her own teaching about the topic:

> I decided to try it . . . like a cooking activity. I told them that we didn't know why it happened, but they got the idea that when some things mix together, sometimes something extraordinary happens. The activity was such a success that for days individual children were showing others how to make crystals, and some made their "own" to take home.

That teacher's account does not tell the length of her introductory lecture. Considering what happened afterward, however, I am guessing that she limited herself to a few sentences. Note that the emphasis in this description is precisely what the earlier description ignores; namely, what happens to individual children:

> This experiment inspired other experiments and a whole atmosphere of experimentation. One boy, during cleanup, decided to pour the grease from the popcorn pan into a cup of water and food coloring. He put it on the windowsill until the next day. He was sure "something" would happen and was surprised when nothing much did. Another child said she knew an experiment with salt, soap, and pepper (which she had seen on television). She demonstrated for those who were interested.

"For those who were interested." That child learned something very valuable about talking to others. By the time she underwent 12 or 16 years of schooling, however, she may have forgotten she ever had that experience.

> A third child was inspired by the soap experiment to fill a cup with water and put a bar of soap in. She was astonished by the change in water level and then tested other things in the water—a pair of scissors, chalk, crayon, and her hand to see the change in water level.

> The next day, one child brought a cup filled with beans, blue water, styrofoam packing materials, and a Q-Tip. "This is my experiment. Cook it," he said. So I asked what he thought would happen to each of the things in the cup. He made a few predictions, and I told him we could cook it the next day.

> (I wanted to experiment first to see if there might be anything dangerous involved.) At group time, he told everyone about his experiment, and the group made predictions which I wrote on the blackboard. Among these were: "The whole thing will get hot," "The water will change color," "The beans will get cooked, and you can eat them," and "The beans will grow." When I asked, "Will anything melt?" the children predicted that the styrofoam would not melt, but that the Q-Tip would. The next day, the child did his cooking experiment, and wrote down the results with my help. Many of his predictions were found to be true, but there were some surprises: It smelled terrible, the Q-Tip did not melt, and the whole thing bubbled.

I wish I had had a teacher like that when I was that age. I did have a teacher like that when I was a sophomore in college, but not as part of my authorized curriculum.

Looking over that narrative written by the colleague of Kamii and DeVries, many, perhaps most professional teachers and school administrators, would say something like, "But you must try to make sure that all the students learn the same lesson. How else can we maintain standards?" Look at all the assumptions in that remark: (1) that it is possible for all the students—or all except the hopelessly defective ones—to learn the same lesson, (2) that it is possible to ascertain whether a student has learned "the" lesson, (3) that the teacher, the test-maker, the administrators, the parents, and the legislators, or even a majority of them, will agree on what is meant by "the lesson," and (4) that putting scores on students in this way has some sort of usefulness in selecting people to work in a salt-packaging company or to be admitted to the freshman ranks at Harvard. No doubt you can see a few more assumptions in there.

As to whether all the students can learn the same lesson, I repeat the remark by the mathematics teacher who won the Presidential Award: "It doesn't work." Remember, too, the students who passed all the pre-college standards, presumably with flying colors, were admitted to the prestigious Massachusetts Institute of Technology, took physics courses there, graduated, and then failed abysmally the little quiz on hooking up a light bulb to a battery!

Students do not, never have, and never can all learn the same lesson. What, then, can one mean by "maintaining standards"? The answer to that seems to be obvious to almost everybody: Give a test. Allow only those who pass the test to be allowed to go to MIT or even to have the job at the salt-packaging company. That practice makes life easier for personnel directors and admissions officers, but the effects of that practice on society as a whole are not simple to think about—not for me, anyway. Some people think that testing and grading, even though they can never be relied on to give information of any reliable use about the tested person, do persuade the tested person, after some repeated failing scores, that he or she deserves employment only at boringly simple tasks. People who think that way believe, I assume, that society must be organized in a way that requires a great many jobs of that sort—a sort I described in Chapter 34 under "Assembly Line."

As a final item in this section, I will remind you that the vital part of growing up mentally is reorganization. The children in the example just above were certainly gathering surprises that could lead to reorganization for some of them, but reorganization is unpredictable in every individual instance, and it can be encouraged in only two ways: (1) offer lots of possibilities (degrees of freedom) and (2) be patient. Here I will copy an example by John E. Pfeiffer (1978, p. 372). I do not mean that the mother in this example is offering lots of possibilities during this quoted part of the example. The richness of the example came in the talking the child heard in the months following this conversation. This example, no matter how surprisingly simple it may seem to adults, shows elegantly how reorganization proceeds.

> . . . at about the age of 3, when they are still learning the rules of negation, they tend to produce double negatives. . .:
>
> *Child:* Nobody don't like me.
> *Mother:* No, say "nobody likes me."
> *Child:* Nobody don't like me.

This mother put up a game but losing fight. After the above dialogue was repeated word for word eight times in a row, she tried one last time:

> *Mother:* No, now listen carefully; say "nobody likes me."
> *Child:* Oh! Nobody don't likes me.

> . . . no matter how often its mother repeated the correct sentence starting with *nobody*, the child heard something else. It listened with its inner ear and heard not one but two words, *nobody don't*. Of course, within a month or two it had learned to hear differently and was producing perfect negatives.

That mother could have slapped the child for disobeying her. She could have put it in an institution for the feeble-minded. She could have implored a spouse or teacher or psychiatrist, "I can't make him say it; *you* make him!" I am glad that Pfeiffer told me that she simply waited for the child to learn in its own time.

To learn to pay attention to (to control) a new aspect of our experience, we must often give up controlling previously controlled aspects. In very early life, one can predict that most children will acquire certain kinds of new internal standards (mostly lower-level) by a certain age (see Chapter 9 under "Developmental Stages in Infancy"), but predicting later ages at which a particular child will acquire internal standards at the higher levels remains chancy. As we grow older, the internal standards at the higher levels become more resilient, more adaptable (encompassing) to varieties of perceptions. They become more difficult to predict and more difficult to replace.

The manner of functioning of reorganization, however, remains the same at all ages. Just as the child could not actually hear the words its mother was saying, so the eminent psychologist Carl Rogers, recipient of the Distinguished Scientific Contribution Award from the American Psychological Association, wrote for the original jacket of Powers's 1973 book: "It is delightful to have . . . a unique theory of the way in which behavior is controlled . . . by the individual" even while the title of the book, "Behavior: The Control of Perception," stood stark before his eyes.

UNCHAINING SCHOOLS

I struggled to think of a good heading to put on this section. I thought "deschooling" might be good, but then I thought that word might scare off teachers, principals, politicians, and too many parents. I am not going to propose here that we dismiss all the teachers tomorrow and set the torch to the buildings. (Neither did Ivan Illich, who published a book in 1971 describing the benefits of a "deschooled" society.) But I will say some things here about clocks, classrooms, and compulsion. I begin with what the eminent psychologist Leona Tyler wrote in 1978:

> . . . I began my career as an English teacher half a century ago in a small-town junior high school. I shall never forget the overgrown boys in the back row sullenly waiting out the excruciatingly long weeks until their 16th birthdays. "Just try to make me read *Julius Caesar*," they seemed to be saying. . . . Students at that time used to refer to school as a prison, and I suspect they still do. . .(p. 57).
>
> We tend to forget how new compulsory education is in the history of civilization. We have to go back only two or three centuries to arrive at a time when it was generally assumed that only a very small proportion of the population needed any schooling at all. The great European universities arose during times when the majority of Europeans were illiterate, and nobody worried about it. In this country we have to go back only a little more than a century to encounter a situation in which elementary schools were provided, but boys and girls were not required to attend them. . . . We have no guarantee anywhere in human history that secondary education for everybody is feasible or that it can be accomplished by requiring boys and girls to stay in school until they are 16 or 18 (p. 58).

Tyler went on to give some detailed evidence of the small effects that schooling in America has on most youngsters. I will cite a little of her evidence later. Here, I will quote some of her conclusion:

> My first objective would be to reduce the amount of compulsion in the system and create situations in which individuals would be more likely to choose, plan, and take responsibility for their own learning processes. Guidance and counseling programs were initially intended to serve this purpose, but I have come to the conclusion that they do not and perhaps cannot accomplish it. . . (p. 64).
>
> Our civilization has few niches for persons unable to read, write, and handle numbers. . . . it is reasonable to require that [those] be learned.

> What is not reasonable is to require that every person spend a specified number of *years* in a certain kind of school in order to do this. . . . I suspect that the strong dislikes many of the respondents [to Project TALENT] expressed . . . would not have developed if elementary schools had facilitated mastery of essentials without anxiety and boredom. . . . Only a thorough individualized program of elementary education with the opportunity for the student to move on to the wider options of secondary education as soon as the standards are attained, whether he or she is 10 or 16, will suit the purpose (p. 65).

Tyler goes on to offer numerous measures to bring about truly "wider options of secondary education." Throughout her chapter, she was calling for putting a rich array of opportunities within reach not only of school-age children, but of people of all ages, as did Illich. I disagree with Tyler only on the point of "requiring" people to learn to read and write. If we make the help easily available, almost all will learn in their own time, most in childhood.

In an older time, most parents and teachers believed it was a good thing to threaten students with being beaten. Now, George W. Bush and the other proponents of standardized testing seem to believe that it is a good thing to threaten the students—and their teachers, principals, and superintendents, too—with being tested. It is not difficult to find comment on the effects of threat. For example, Ewens (1984, p. 183) quotes Farson (1974, p. 97): "The enforced, threatening quality of education in America has taught people to hate school, to hate the subject matter, and tragically, to hate themselves." A good many of us remember the jingle we chanted as children at the end of the school term: "No more lessons, no more books; no more teachers' dirty looks." I will say more about the threat of testing in Chapter 38.

Here are some of Ewens's comments about compulsory education:

> . . . there is little that can be done to improve the effectiveness of compulsory schooling, and, indeed, many of our reform efforts will unfortunately make these involuntary institutions even worse p. 186).
>
> . . . the school establishment has not been ordained by nature; instead, schools were originally constructed during the [19th] century, as we have

seen, by people who believed it was necessary to create obedient children who would become productive factory workers and feel comfortable in the newly emerging industrial hierarchy (pp. 186–187).

> ... as long as schools remain compulsory, effective teaching methods must at some level remain either manipulative or directly coercive and the classroom atmosphere that of
> involuntary servitude.... Compulsory school attendance should therefore be abolished, along with the mechanisms and procedures that allow schools to rank, certify, and label students (p. 187).

Sedlak, Wheeler, Pullin, and Cusick (1986, p. 61) point to the large percentages of high-school students who turn to part-time employment as means for controlling variables to which schooling is irrelevant or hostile:

> Many adolescents are seeking out marketplace experience by working for pay. Many prefer the sense of responsibility and autonomy that even dead-end employment provides, a sharp contrast to their experience in school, where they must ask permission even to go to the restroom. Indeed, a sizeable number of students indicated in interviews that their jobs enabled them to escape the "dullness and boredom of school."

The attraction [and often economic necessity] of part-time employment was corroborated by McCaslin and Good (1992, p. 5):

> ... about one-half of high school sophomores, two-thirds of juniors, and three-quarters of seniors have jobs during the school year. The average senior works more than 20 hours a week in addition to spending hours in school [despite the fact that] students who work more than 10 hours per week are more likely to have problems with substance abuse and poorer school performance than are students who work less or not at all.

McCaslin and Good also note that less than five percent of students in Japan work during the school year.

Now I turn to the argument made by Seymour Sarason (1983)[2,3] about the inappropriateness of schools:

> ... everything being said and proposed [in 1983] was said, proposed, and acted upon earlier as a reaction ... when the Soviet Union orbited the first sputnik in 1957 (p. 4).

That is still true in 2002, though the proposal being bruited most often seems to be simply to demand that matters improve and to demand that tests be administered to see whether matters do improve. Sarason continued:

> 1   ... the ability of schools to engender and sustain student interest, especially in junior high and high schools, steadily decreased [even while] "Let's make schools interesting" became the basis for a new industry ... subsidized by federal, state, and local ... monies.
>
> 2   The efforts... were based on an axiom as potent as it was unverbalized: *education best takes place in classrooms in school buildings.* That axiom ... was both unfounded and self-defeating.
>
> 3   ... schools in the post-World War II era sought ways to give young people experience outside schools that spoke directly to student curiosity and needs. These programs ... were "add-ons" to the curriculum and were hardly related, if at all, to core subject matter.
>
> 4   What if it were illegal to teach math or science or biology in a classroom in a school? Where and how would you teach these subjects? If you start with such questions, unimprisoned by the imagery of a classroom and a school ... once you begin to be liberated from a world view in which is embedded a picture of where and how schooling "should" take place, you begin to understand why efforts to improve schools are doomed.
>
> 5   Isn't this notion [to put the greater part of schooling outside classrooms and school buildings] wildly impractical?

My response to Sarason's last question (a rhetorical one) is: Isn't present-day schooling wildly impractical? In summing up 100 follow-up interviews with former high-school students surveyed by Project TALENT, Leona Tyler (1978) wrote:

> Did education contribute to career success? In 62 percent of the cases in my sample, the answer is a clear "No" (p. 59).
>
> [As] to the influence of education on what people do with their leisure time.... Do they read good books, visit museums, go to concerts, and take part in the cultural life of their communities? The data indicate that only a minority [22 percent] of these 30-year-olds do any of these

things. Most of them watch television, perhaps read a newspaper or a popular magazine, watch sports events, but seldom engage in anything more demanding (p. 61).

To what extent do individuals coming through the U.S. school system participate in public affairs at any level—community, state, or national? . . . I found that even by [an undemanding] standard, my sample of 100 did not measure up very well. . . . Only 19 were participating in a more active way [than voting] or at least expressing an intention to do so (p. 62).

I think a system of education that gobbles up 12 years and more of a young person's life, gobbles up stupendous amounts of money, and produces the results reported in those paragraphs should properly be called "wildly impractical." Here is more from Sarason (1983):

As long as efforts to improve schools begin with the imagery of classrooms and school buildings; as long as that imagery bifurcates student experience into an inside and outside world; as long as that imagery pictures teachers as the major fount of intellectual knowledge and stimulation; as long as subject matters are presented as contents and skills to be learned independently of their experienced manifestation and importance in individual and social living . . .; as long as the structure and organization of subject matter is presented without reference to intellectual history—as long as we remain imprisoned in a world view in which education has these features, we set ourselves up for disappointment (p. 135).

"What would we *do* with the children if we [did not teach] them the basics in the early grades?" . . . The answer, of course, is that there is a staggeringly large universe of possibilities from which one could draw in order to capitalize on the interests and curiosity of children, possibilities that derive from their question asking about a world they are trying to understand. . . . What is basic is not subject matter but those matters in which the child is interested (pp. 136–137).

The evidence is overwhelming that middle and high school students can be engaged in structured and supervised experiences outside school that maintain their interest and have obviously beneficial intellectual and educational consequences (p. 140).

. . . one should not begin with our customary imagery of a school building and its physical and organizational structure, but rather with this question: in what activities can students engage that will capture their interest; expand and stimulate their knowledge of, and experience with, the social and physical world; serve to illuminate and develop skills and concepts that have broadened and transformed human experience; and sustain the sense of purpose, mastery, and continuity? (p. 148).

Here is a similar sentiment from Powers, taken from a posting to the CSGnet on 11 February 1997:

Why educate children? One of our big problems is that there is a social system in place that adults feel is very important, which says that there is a whole list of things that every child must learn by a certain age. The system says that these things have to be learned out of books, while sitting in rows in a school building and obeying the teachers. This may be OK with some children, but it is far from OK with most of them. There is something about "education" that is very unpleasant for lots of children, that seems unnatural and unsatisfying. Education is designed primarily to satisfy the adults who run the educational system.

What would happen if we just did away with education altogether? Would children stop learning? Is it possible that they might play for ten years, learning games, and then suddenly demand to know about things like reading and math just because adults are doing them? How long would it really take to learn math if you just waited until it looked interesting, and then tackled it the same way you learn other things you're interested in? Isn't most of school before high school a waste of time, a dreary repetition of the same subjects again and again, year after year? Isn't most of a school year devoted to baby-sitting?

If you really want to study education from the PCT standpoint, I think that the first thing has to be to haul out all the old cultural and academic assumptions and give them a good hard look. When it's very hard to get children to do something, maybe we need to reconsider what we're trying to make them do.

A couple of years ago my wife and I were discussing some topic of world affairs with another couple, and it turned out that one of them did not know the location of Taiwan. "I didn't have that in school," she said. She was quite serious. Her implication seemed to be that some kinds of knowledge are to be found only in a school, and if you don't "have" them there, nobody should think it odd if you don't have them later.

If you are required (by law or parents or both) to go to a building day after day where you will "have" the location of Taiwan on Thursday the 18th of October at 10:30, it can easily happen that your mind will be pointing elsewhere at that moment. According to Tyler's data, that can be the case with about 80 percent of the students sitting with you. Under that way of organizing schooling, it can only continue to be absurd. Some students, it is true, will be alert for whatever the teacher is saying at 10:30 on Thursday. Some will be fascinated by the information about Taiwan. In a "good" school, the portion listening and fascinated may even exceed the 20 percent that Leona Tyler found in her data.

As a final subtopic for this section, I will include a short note about going to college. Like every other feature of the environment, college can serve some of the purposes of some of the people some of the time, but it cannot serve even the high-priority purposes of all the people anytime. Even as long ago as 1975, in her book *The Case Against College,* Caroline Bird reviewed the many hopes and disappointments people experience with colleges. She showed, for example, that most people would be better off financially if, instead of spending money on college, they were to invest that same amount at, let's say, 7–1/2 percent interest. In an excerpt of the book published in *Psychology Today*, Bird wrote, "The most charitable conclusion we can reach is that college probably has very little, if any, effect on people and things at all." I do not go as far as to say "if any." There have been a few careful studies showing, for example, that college students on the average shift to more liberal attitudes during their college years. Those shifts, however, appear only among some of the students, not all; the shifts are not, on the average, remarkably large; and it is difficult to be sure, even in the careful studies, how much of the shift is due to what the professors did and how much to other concurrent events. The findings should not persuade Ms. Bird to retract her assessment of "very little effect." In 1993, Catherine Gysin reviewed two books published a year or so previous that took much the same view as Ms. Bird's.

REQUISITES

I have been spending a lot of space on degrees of freedom. One's success in controlling perceptions depends on one's knowledge and inventiveness, it is true, but those inner resources go for naught if the environment is bare of opportunities. Attention to degrees of freedom in a school (as in any social arrangement) is therefore vital. Whether the degrees of freedom in the environment are sufficient for your needs depends on your needs (internal standards) as well as the physical conditions in the environment, but I classify degrees of freedom under the second Requisite for a Particular Act. You might want to put a bookmark at the page in the Introduction to Part VI where the list of Requisites appears.

An environmental feature that is an opportunity to one person may be a nothingness to another. One person's meat is another's poison—or neither one, just a nothingness. If I am not hungry, a banana is just an uninteresting yellow blob. I think that is a good part of what Sarason meant when he wrote, "What is basic . . . is those matters in which the child is interested." This is where Requisite No. 1 comes in. If the student has no internal standard that has anything to do with the location of Taiwan, nothing the teacher can do will help the student to control a perception having anything to do with the location of Taiwan, because the student is not controlling such a perception.

That fact has long been understood, though in different language. It is common for a teacher or a parent to be controlling for a perception of the student saying something like, "Taiwan is a large island lying off the southeast coast of China!" So the teacher or parent does something to bring that perception nearer the internal standard. If persuasion doesn't work, maybe threats will—perhaps threats about sitting in the corner, staying after school, doing extra homework, getting a low grade, being sent to the principal, not being promoted, not graduating, not getting into college, or not getting a good job. And if threats do not work, maybe a strapping will, or maybe a beating with a cane. Indeed, that technique may in many cases, "persuade" the student to utter that sentence about Taiwan at least once. But that technique may also persuade the student to stay away, in future, from Taiwan, islands, China, teachers, the Pacific Ocean, books about any of those, books about anything at all, or all of the above.

Instead of Sarason's word "interests," maybe "desires" would serve better for this topic. To be helpful to students, the teacher will usually gain little from attracting the attention of students to what the teacher has in mind for the end result. It will be more helpful to provide opportunities for what the students want to *do* to become sufficiently visible to all so that teacher and students can act cooperatively. In the case of making crystals, the first demonstration may not have attracted the attention of all the pupils, perhaps not even most, but some of the pupils saw "something extraordinary"—that is, something that disturbed their expectation of what should happen when you pour a few common things together. Other children saw that those first children were having fun. Perhaps the onlookers wanted to find out what was fun so that they, too, could enjoy it. Or perhaps they did not want to be left out of an activity that was occupying the attention of an admired friend. It helps, of course, if students have found during earlier days that teacher often does things that some students find interesting.

One child was sure "something" would happen but wanted to find out. Another wanted to perceive herself demonstrating an experiment. Another felt an urge to put things in water and was astonished at the changes in water level, whereupon she wanted to see what changes other things, immersed, would produce. Another wanted to see a mixture cooked. And so on. I have no doubt there were some perceptions akin to camaraderie that were being controlled, too. But most of the children were not, I think, paying much attention to a grade they might get or to a punishment they might avoid. I don't think there was even as much striving to get approval from the teacher as one finds in a traditional classroom. Judging from the description, I think there was an unusual amount of attention to subject matter—to the physics of what those materials did and did not do. There was a lot of control of perception at fairly high levels going on and very little conflict among the actions the children took to maintain control.

The matter of conflict brings us to Requisite No. 3. The teacher arranged the environment so that children could test the physics of materials mostly at their own schedules, with materials of their own choice, and in satisfaction of their own immediate assortment of internal standards, with the result that the schoolroom provided always an adequate supply of degrees of freedom and therefore freedom from conflict. I'm not saying this kind of facilitation is easy; that teacher was very skillful. Loving, too.

CODA

A school is a social system in which, as in all social systems, the people are using one another in the pursuit of their own purposes. Some purposes are selfish, some altruistic, but they all serve to match perceptions with internal standards. In every continuing organization, schools included, patterns of social interaction (norms) arise. Concerning the norms for coexistence in school, Sedlak, Wheeler, Pullin, and Cusick (1986, p. 186) wrote:

> There are reasons why teachers are so dependent upon the goodwill of students, why they are allowed to bargain away standards and academic learning. Bargaining and disengagement are not just surface events. They are bound up with the differential return on academic learning and incentives for teaching in the United States as well as in the way we organize, finance, and administer our schools. They are embedded in our attitudes toward social class, toward what we expect of our children and the children of others, toward equal opportunity, and toward the place of schooling in these vital issues.

Students are at the bottom of the levels of influence. In bargaining for better working conditions, their best bargaining chip is that of refraining from making life difficult for their teachers. The bargains that result are those that increase the degrees of freedom for both teachers and students in controlling their perceptions—especially their perceptions at the higher levels of their control hierarchies. In those bargains, academic content rarely carries much value.

To illustrate the struggle for sufficient degrees of freedom in schools, I have mentioned and quoted literature from decades ago, even several centuries ago. You will have noticed that the contentions at the time of Horace Mann (1796–1859) and Henry VIII (reigned 1509–1547) sounded very familiar, as if you could have read them in this morning's newspaper. That is not surprising, since most people then, as now, believed that close supervision, "requirements," and threat can compel the actions of students—and of teachers and principals. Unless a fair portion of us come to realize that students, teachers, and principals all act to control their perceptions, we can expect newspaper headlines a century from now to cry, as if it is a bright new idea, "President Underbrush Urges Tough Tests for Schools."

If, instead of being dragooned into what we now think of as schools, children could have easy access to places or jobs or excursions or colllections of books where they could learn to do the things they want to do at the times they choose, I am sure our populace would be at least as well educated as it is now. Remember what Mr. Gatto told us about the effect of compulsory education on literacy in Massachusetts.

ENDNOTES

[1] For the citation, look in the list of references under "Teachers' . . .".

[2] This and later quotations from Sarason's 1983 book are reprinted with the permission of The Free Press, an imprint of Simon and Schuster Adult Publishing Group, from *Schooling in America: Scapegoat and salvation* by Seymour B. Sarason. Copyright © by The Free Press.

[3] Sarason has been an insightful and prolific writer on schooling. For more, see Sarason (1990, 1996, 1996a, 1997, 1998), Sarason and Lorentz (1998), and Sarason, Hill, and Zimbardo (1964).

Chapter 38

Mental testing

Most people seem to have an unquestioning faith that a paper-and-pencil test has about the same correspondence with what is named on the cover of the booklet as the correspondence of a tape measure with the girth of your waistline. Many people (not all) may doubt the validity of a test when they fail one, and perhaps even more when a child of theirs fails one, but most seem to go on thinking that those tests are valid for *other* people.

In this chapter, I am going to say some disparaging things about standardized mental tests, and you may wonder why, on earlier pages, I have sometimes mentioned average scores on standardized tests by students undergoing various sorts of schooling. I did so because some readers will put value on test scores. For reasons I gave in Chapter 26, test scores are of very little value in dealing with individuals, but averages over several dozen people can have respectable reliabilities (no matter what the scores actually stand for, if anything), even if only temporarily, and I was writing about averages. In Chapter 37, for example, I was saying that if you care about test scores, there are some groups or circumstances in which the averages will be higher than in others. I will point out a few more groups and circumstances in this chapter. I prefer the kind of criteria used in the Eight-Year Study and in the study by Leona Tyler, both of which I mentioned in Chapter 37. Because, however, so many reformers seem to want to rest their arguments on test scores, I will use this chapter to point out some dangers in doing that.

I said a good deal in Chapter 26 about personality questionnaires, and those questionnaires are a variety of psychological (or "mental") test. Much of what I said in Chapter 26 about personality tests applies as well to tests used to assess scholastic aptitude and progress. By "test," I will mean here such devices as school-achievement tests, scholastic aptitude tests, intelligence tests, and all such "mental" tests as are customarily assessed in the *Mental Measurements Yearbook* edited by James V. Mitchell (1998).

The technicalities of testing apply, too, to the test a horizontal bar can give to your high-jumping ability and the test a tape measure can give to your waistline, but those tests are not usually given the over-interpretation or the social consequences that are given the mental tests. When I write "test" here, think of the mental tests, especially tests used in schools.

HOW CAN YOU GET A HIGH SCORE ON A TEST?

Here I will again make use of the Requisites for a Particular Act.

Requisite 1

To get a high score on a test, you must first *want* to get a high score. That is, getting a high score must enable you to control a perception for which you have an internal standard. Reasons people have for wanting a high score are many: for the fun of seeing how high they can score, for the pleasure of seeing the pleasure of teacher or parent, for fear of seeing the censure of teacher or parent, to feel that they have taken another step toward (or out of) college, for the pleasure of being admired, and to maintain their pictures of themselves as knowledgeable, intelligent, or scholarly. No doubt you can think of other reasons.

Most tests are built so as to produce a wide array of scores, but not all tests are built that way. For example, almost all applicants pass the test for a driver's license, and few care how "high" the score was.

Teachers, principals, parents, college admission officers, and others want tests to yield a wide spread of scores, or at least five gradations. Grades and tests serve important purposes for those adults, but not many purposes for children. Grades and tests are not necessary for children to acquire knowledge and skill. Out of school, children acquire knowledge and skill greedily ever day. Grades and tests serve the purposes of students only to the extent that the students adopt the purposes of school and college personnel and participate purposely in the competitive movement from grade to grade. Some do, some don't.

Turning now to those students who so want to get a high score, Requisites 1b and 2a are given by the testing situation. As to 1b, you must perceive that you have not yet reached some goal such as passing the course, winning the scholarship, or getting admitted to the college. As to 2a, the suitable object (the test) awaits in the testing situation.

Requisite 2b

To get a high score, you must believe that getting a high score is a means by which you can reduce the distance to your goal.

Requisites 2c and 2d

You must believe you are capable of giving enough right answers. This belief comes easily to test-takers who have succeeded frequently in the past. For students who do not know enough answers and know that they do not, taking the test will not move them toward their goals. If the high score on the test is a necessary step, they must find some way to get a high score other than by reaching into their memory. It turns out that most of them help themselves in the ways we call "cheating."

I put "cheating" in quotation marks because it is not so much a name for what a student does as it is a name for a kind of disturbance to the perceptions of authorities. When you have forgotten how to make potatoes Patagonia and you look in a cookbook to find the answer to your perplexity, you don't call that cheating. Or when you wonder where Taiwan might be, and you look in an Atlas. Or when you want to win a scholarship, and you ask someone to help you answer the questions on the application form. What you do to answer a test question gets called cheating when somebody puts you into a competitive situation—promises you something you want if you find better answers or find answers sooner than some other people—*and* tells you that getting the answers from anyplace other than your memory at this brief moment is against the rules.

The school people tell you that you are not going to find a respectable place in society unless you come to school and get good scores on all those tests. Then they tell you that you have to memorize all those details the tests are going to ask you about, *and* you have to succeed in memorizing more details than, let's say, about half the other students taking the tests. (You could get a passing grade by memorizing more than about a tenth of what the others memorize, but that, they tell you, won't get you to a really respectable place.) You might think the young people in Patagonia are not placed in such a bind, but you don't want to go there, because you don't speak the language, and you might have an even harder time finding a respectable job if you went there. The obvious thing is to bring a few helps to the testing situation, maybe too a friend who has done a lot of memorizing. Living things do what they can to reach their goals.

In October of 1994, the Office of the Dean of Students at the University of Oregon sent to faculty members a memorandum containing this paragraph:

> In a recent questionnaire sent to 500 University of Oregon sophomores, juniors, and seniors as part of a survey study of 10 U.S. state universities, 91% of U of O students admitted to cheating on a written assignment or exam while attending the U of O. The mean for the total sample of U.S. state universities was 89%.

Every few years, it seems to me, a newspaper or magazine reports a cheating "scandal" at one of the nation's military academies. Turning to high schools, here is a report by Mathews and Argetsinger (2000) of *The Washington Post*, reprinted in my local newspaper:

> A nationwide poll of 20,000 students released in 1998 by the Josephson Institute of Ethics in Marina Del Rey, Calif., said seven out of 10 high-schoolers admitted to having cheated on a test (p. 14A).

Games such as football, basketball, and ice hockey draw millions of fans, and the players are much-admired celebrities. Several referees are posted at those games to award penalties to players who cheat—that is, commit fouls. The fans do not look upon those referees as symbols of shame. On the contrary, that kind of cheating seems to be an accepted and honorable part of the game. A player who succeeds in a foul without getting caught is considered by most fans not to be dishonorable, but simply skillful. If those much-admired and wealthy citizens get a high score for their teams by slipping one past the referee, I don't think we should be surprised that the same kind of thing happens regularly and frequently in classrooms.

Some test-takers will find that cheating puts them into conflict with other internal standards and will refrain. Others will try to cheat but fail from lack of skill. Most cheaters will not get caught, and many of the uncaught will get higher scores than they would have got without cheating. In sum, from knowledge, from skill at cheating, or a combination of both, you must, if you want to pursue a high score, estimate that your chance of a good score is good enough to justify taking the test seriously (2d).

Many students know from past experience that when they reach into their memories, they will find many fewer right answers than other students will find. Many of those students know, too, that their resources for cheating are also poor. Many, too, have no desire for the consequences of a high score that entice other students—perhaps admission to college or even approval of parents. Furthermore, if some of those students do come upon resources for cheating, such as a knowledgeable friend, they know that if they were to get a score considerably above their customary low level, the teachers would immediately suspect them of cheating. Many students with these disabilities have fallen into hopelessness. They are merely waiting for the legal age to leave school. They fail Requisite No. 1, not to speak of No. 2. On the answer sheet, many will mark all first or last answers, or mark the sheet in an artistic pattern, or simply hand in a blank sheet. This condition is similar to "learned helplessness," which I described under "Quantification" in Chapter 20.

Another thing that will interfere with your getting a high score is anxiety. You can be anxious because you worry that someone will scold you, or because of the threat of competition, or a dozen other reasons. In Chapter 21 under "Reorganization," I cited research by Hill and Sarason (1966) showing that test anxiety is as accurate a predictor of school grades as achievement tests.

Still another source of lower scores is the manner of administering the tests. For example, a principal can improve the standing of her school among others if she declares those students who get very low scores to be "special education" students and therefore exempt from taking the test. The ranking of her school among all schools in the district or state will go up, and the relative position of some other schools will go down. If you happen to be a student in one of those other schools, you will emerge from the test with a poorer standing than you would have had if the principal in that first school had not used that ploy. You would probably not even know that was happening to you.

There are other kinds of bad luck. Here are a few discovered by Bernard McKenna (1977) in schools he studied:

[Training for teachers in] the nature of the tests and... the most appropriate means for their administration ranged from minimal to none (p. 222).

Social studies teachers were required to administer reading tests with which they had little or no familiarity.

Some tests were mass administered over public-address equipment; instructions were difficult to understand and communication between the test administrator and the many room monitors was impossible, so monitors responded to student confusion and questions in varied and inconsistent ways.

... some teachers and counselors answered questions, gave clues, and repeated instructions, while others did not, making for potential unreliability of results.

... teachers were required to administer tests to special education students and to those for whom English was a second language, tests from which ... these students should have been exempt.

Some of the teachers who were most threatened by veiled warning of being evaluated on the basis of test results gave the tests rapidly and with little precision in the fall, hoping that, with more time and care, an appropriate amount of progress could be shown in the spring (p. 223).

In some schools, teachers were told that if scores were inordinately poor, they should not be turned in (p. 224).

McKenna went on to say,

Some students took the tests seriously and did their best. But teachers ... noted that many students reacted negatively... Some early elementary children cringed and cried when they entered the room and saw that it was arranged for testing again. Some developed nosebleeds. At upper elementary and high school levels, students shared answers, rushed through the tests or made no effort to complete them, marked the same answer columns throughout the whole test, and generally expressed the attitudes that the tests were stupid and did no good and that they were sick of tests (p. 224).

I have classified these characteristics of the testing situation and these matters of "bad luck" under Requisite 2, because they affect the suitability of the environment for controlling perceptions.

Requisite 3

To get a high score, with or without cheating, doing so must not arouse conflict with another internal standard. Do you, for example, perceive yourself as a person who would get a high score on a test like this? If you also want to avoid being perceived by your friends as an academic nerd, you may settle for a score below the level of academic nerdity. If you are female, you might be careful not to get too high a score on a mathematics test, thus avoiding conflict with a wish to be perceived as feminine. If you want to show disdain for your teachers, you might want to get a low score and thereby show that you do not care about what they care about. Or maybe you know your friend Josephine is going to fail this grade, and you don't want to leave her next year. Or you might want to avoid attracting attention for any reason at all and therefore might strive for a mediocre score on everything. You can think of other conflicts. Or you might want to look back at "The Living Environment" in Chapter 27 and review the hints about the internal standards of adolescents hanging out at the mall.

I will sum up by reversing the question and ask: What will give you a low score? First, not controlling any perception for which getting a high score on the test will serve as a means. For example, not caring whether you graduate. Second, not knowing enough, not believing that you know enough, not being willing to cheat, or being unlucky enough to be in circumstances that give other test-takers even higher scores than yours. Third, having one or more internal standards in conflict with getting a high score, such that getting a high score will create more total error of control than getting a lower score.

You can see that the answer to a question on a test indicates to some extent the answer the person could give if conditions were optimum, but it also indicates (is affected by) the states of the three Requisites. And the percentile ranking a test-taker is assigned depends not only on the score the test-taker achieves, but also the scores all other test-takers achieve. (Percentile ranking, whether of individuals, schools, or states, is the measure most used for practical purposes, such as getting a grant of money or boasting to parents. Percentile ranking, note, is competitive; a student or a school ranks higher only if others rank lower.)

WHAT CAN THE TEST INDICATE?

So far, I have been discussing the psychological and environmental conditions affecting the ability of the test-taker to get a high score. I have not said a word about the test-maker's claim concerning *content*—the "thing" the test is advertised to measure. In Chapter 26, I commented on what personality tests can indicate about personality, and in Chapters 29 and 31 I said some things about clinical diagnoses, with and without the *Diagnostic and Statistical Manual* of the American Psychiatric Association. I will not repeat here the general principles I set forth in Chapter 26. I will give only a brief report on one of the favorite tests used by school administrators to compare their students and schools to others; namely the Scholastic Aptitude Test (SAT), and another brief report on a test used to screen teachers; namely, the National Teachers Examination (NTE). My point is that most people are far too confident of the usefulness of mental tests—school administrators, members of state departments of education, college admissions officers, parents, legislators, and President George W. Bush.

The Scholastic Aptitude Test

Katz, Lautenschlager, Blackburn, and Harris (1990) made a study of the Reading Comprehension section of the SAT:

Does the Reading Comprehension (RC) task on the Scholastic Aptitude Test (SAT) measure factors having nothing to do with reading comprehension? There is a simple way to find out.

> [T]he task is intended to measure the ability to read and understand short English prose passages. Examinees must answer a group of multiple-choice items based on what is stated or implied in a passage... (p. 122).

In other words, the RC presented a prose passage and a few questions about it, then another prose passage and a few question about that one, and so on. Examinees were instructed to pick the right answer among those offered.

> Now suppose that information crucial to performance on the RC task were removed.... One has only to administer the RC test items *without the passages*. If the task requires the passages, performance on the passageless task should not exceed what would be expected by chance alone.... Better-than-chance performance without the passages would imply that test items can be answered correctly for reasons that have nothing to do with passage comprehension (p. 122).

The authors administered the RC to 61 members of college classes in introductory psychology. Some got RCs in the standard form, with prose passages preceding the questions; others got RCs with the prose passages removed.

As you might hope, the students given the RCs with the passages got considerably better scores than the students given RCs without the passages. Nevertheless, the students answering without passages *did far better than would be expected by chance*. The chance score for the RC used in this experiment was 20. The average score of the students without passages was 37.6 in one version of the experiment and about 46 in the other.

The students' scores were about double what chance could have given them. Maybe this reminds you of my student who had not needed to understand anything she read to get A's, but needed only to match words she thought the teacher would think "important." The authors of the study of the SAT concluded cautiously:

> The evidence presented here... should be enough to alert those who take the SAT, or make use of its results, that it may tell a story different from the one now generally accepted (p. 127).

My own conclusion is that although the SAT indicates a little of this and a little of that concerning the test-taker and the testing situation in an always unknown mixture, it indicates very little of any importance to anyone. It correlates chiefly with other tests that indicate a little of this and a little of that, including school grades. It does not tell us anything about the kind of criterion Leona Tyler cared about or most of the criteria used in the Eight-Year Study (for both of which, see Chapter 37).

When I say that the SAT and other such tests indicate little of importance to anyone, I do not mean that it has no important uses. Your shoes may indicate little about your musical ability, your preference in literature, or your future occupation, but they do have an important use. I am sure you would not want to be without them. Similarly, tests such as the SAT enable school administrators and college admissions officers to make quick choices among thousands upon thousands of prospective students, drawing few accusations of personal bias, and appearing to most people, I suppose, of being "scientific." There are other ways to do that task, of course, but a standardized test is very handy and less expensive than some. Still, Andrew Goldstein (2000, p. 52) reported recently in TIME that "A record 280 colleges and universities... will ignore the SAT in admissions this fall for some or all applicants...".

The National Teachers Examination

My information about this screening device comes from Sedlak, Wheeler, Pullin, and Cusick (1986):

> By 1985, nearly 70 percent of the states required prospective teachers to pass the National Teachers Examination [NTE] or other competency tests before receiving initial certification (p. 143).

Those authors go on to tell about a lot of other occasions for being tested that teachers must face, but the NTE alone will serve my purpose here. My chief reason for mentioning this test is the corroboration it gives to my remark about the unquestioning confidence so many people place in a published test. Take a deep breath; then read this (also from Sedlak and colleagues):

> Taking a pencil-and-paper test... requires different abilities than teaching in a classroom. [That] argument is supported by the Educational Testing Service, author and sponsor of the NTE,

which has stated that the test is not the best measure of competency for practicing teachers, for whom evaluation of on-the-job performance is a better measure of teaching competence (p. 144).

All those school districts in all those states (and probably more by now) were using a test for a purpose *the maker itself advised against*. That is what I call unquestioning confidence. Before I take medicine, I read the warnings on the bottle.

EFFECTS ON TEACHERS AND SCHOOLS

What do we mean by "raising standards"? I think most people mean demanding that schools in some way do better than they are doing now, and most people seem to be willing to settle for evidence given by standardized tests—that is, tests that can compare schools across the city, the state, and the nation.

But just how do you use tests to make those comparisons? You cannot compare how many "passed," because some districts or states or universities may want to declare a passing score different from what another wants to declare. Test-makers get around that difficulty by providing percentile rankings. They collect some thousands of scores and then tell the next test-taker, Alfred, what percentage of those thousands of scores his score exceeds. That is called "norm-referenced" scoring.

The college admissions officer does not care about Alfred's absolute score. The officer wants to know whether Alfred is among the cream of the crop. The percentile tells the thickness of the cream.

Accordingly, when you demand that Edison High School show a better average on a standardized test, you are demanding an increase in the average percentile rank. You are demanding that the students at Edison show higher scores *in relation to thousands of others elsewhere*. But do you want students at Edison to be compared with other students who took the test 20 or 30 years ago? Of course not. You can think of many reasons that would be an undesirable comparison. Fewer students attended high school back then, for example.

The test-makers revise the norms every few years. They calculate percentiles using the most recent scores. Every school's scores are pooled with thousands upon thousands of others in the new "norm" tables. If students at Edison get higher scores, they will move up in the norms *unless students in other schools also get higher scores*. If all schools get higher scores, they will all stay at the same percentile ranks at which they stood before they went to all that trouble.

Rate-busters

In an industrial company, employees who produce more units per hour than their fellows are called rate-busters. In settling down to a routine for manufacturing, a company will settle on a wage of so much per piece (a piece rate) so that the number of pieces most employees can turn out per hour will result in an acceptable weekly wage. Since pay is by the piece, however, some employees will try to make more money by turning out more pieces. (That corresponds in this comparison to a school turning out higher scores.) Then, however, a company often lowers the pay per piece so that the employees have to work faster to get the same pay they previously got. The other employees resent the rate-buster.

I am not saying that schools should not try to improve. I am saying that if they use percentile rankings on tests as a measure of improvement, they are going to run into the rate-busting phenomenon. For example, Sedlak and colleagues (1986) wrote:

> The minimum competency testing movement implies that it relies upon a set of standards defined to be so minimal as to be both essential . . . and attainable. . . . Results of the examinations indicate, however, that not all students are yet able to succeed in the new testing programs . . ., and that when 100 percent success rates are either approached or realized, the standards for passing them are consistently increased. . . (p. 30).

> [T]hose who expect an improvement in achievement test scores to translate into a competitive advantage or eventual economic success for their children will be disappointed. The value of the credential will simply be redefined to recreate the variance, and the variance will be associated with social class, just as it always has been (p. 188).

This rate-busting effect is not going to have equal effects on all classes of students. As usual, the stresses will be greater on students already disproportionately stressed:

> Educationally disadvantaged students . . . operate at a disadvantage when minimum competency

tests are used to make significant decisions about their lives. . . . In addition, the new test requirements have resulted in substantial numbers of students dropping out of school in expectation of unsuccessful test performance or of disappointment or embarrassment on graduation day. . . (p. 30).

Similarly, Jodie Morse (2000, p. 54) wrote this about the testing movement in George W. Bush's home state of Texas:

. . . more and more Texas minority students have dropped out of high school since the introduction of the state exam, from about 35% in the late 1980s to 40% through most of the 1990s.

More History

I said earlier that educational reform has a long and sobering history. Philip Sherwood (1977) has provided some cautionary information from 19th-century England. He tells us that in 1861, Queen Victoria ordered an inquiry "into the revenues and management of certain colleges and public schools and the studies pursued and the instruction given therein" (p. 230). There were nine public schools. The queen's Royal Commissioners proposed a written examination of the fifth forms (grades) of those schools. When the commissioners wrote to the nine headmasters, they received terse and hostile replies. Here are two of them.

> Your letter appears to me so seriously objectionable that I must beg to decline to entertain the proposal. The Dean of Westminster concurs with me.
> —Reverend Charles R. Scott, Westminster (p. 230).

> This interference with the authority of the headmaster is calculated to cause evil.
> —Dr. Balston, Eton (p. 231).

The elementary schools of England began to be supported by state funds in 1838. Her Majesty's Inspectors each applied his own standards, but rarely denied grants. Sherwood wrote:

> . . . in 1862 the era of "payment by results" (performance contracting) began. . . . [I]t was always explicit that the teachers' salaries depended on how their children performed. . . . It was undoubtedly a pernicious system, and it is easy to catalogue its baneful influence on education. [I]n self-defense the teacher will seek means to outwit the evaluator (p. 232).

> . . . at one school . . . bright pupils were borrowed from a neighboring school and presented for inspection so that the school's examination prospects might be improved.

> In some way the schoolmasters would get possession of the [answers] in advance of the exams. By exchange of information, they could work on the answers along with the scholars to be examined. . . .

> Teachers developed techniques of signaling information at examination time. For example: hands in pockets = multiplication, hands behind back = subtraction (p. 233).

> Education degenerated into a year of slaving to cover every possible permutation on the inspector's ragbag of questions (p. 234).

> By [1900] "performance contractings" and "evaluation" were so discredited that they passed from our [England's] educational history, leaving a teaching profession hostile to anything vaguely suggestive of testing to national standards (p. 235).

Notice that Sherwood does not call those practices "cheating." He calls them "outwitting the evaluator." When humans find themselves beset with inner conflict, they will cast about for some way to relieve it, and they will usually be more ingenious than the inspector.

Despite England's 60 years of agony, we here in the United States seem poised to spend untold hours, monies, and agonies to repeat England's history. One wonders whether it will take us 60 years also. In January of 2002, George W. Bush signed a law requiring schools to administer standardized tests to grades 3 through 8. The article in TIME reporting that distressing news, written by Jodie Morse (2002), gave still more examples of effects like those I have described in this chapter. It included a photograph of the dolorous faces of a roomful of students just presented with fat books on tricks for passing tests.

Cheating

Just as students cheat when teachers put them into competition to win a prized goal such as graduating, so teachers, principals, and superintendents cheat when legislators put them into competition to win a prized goal such as keeping their jobs or even winning a bonus. Here are some of the incidents described by Mathews and Argetsinger (2000, p. 14A):

> ... a Woodland, Calif., teacher ... allegedly broke the rules by using an old test to prepare his students....
>
> In the past two years alone, schools in New York, Texas, Florida, Ohio, Rhode Island, Kentucky, and Maryland have investigated reports of improper or illegal efforts by teachers, principals, and administrators to raise test scores. The rise in incidents of coaching has matched the surge of standardized testing....
>
> In Austin, a deputy superintendent was indicted on 16 counts of criminal tampering after central administrators and principals boosted scores by changing the ID numbers of students whose failing grades they didn't want counted.
>
> In New York City, cheating was found to be so rampant that it led to the resignation of the schools chief.
>
> ... a seventh-grade teacher allegedly left answers near the pencil sharpener, then urged her students to sharpen their pencils.

When schools in Texas get high averages on the Texas Assessment of Academic Skills (TAAS), Jodie Morse (2000) says, administrators can get bonuses of as much as $25,000. (This is like the situation in England 140 years ago.) On the other hand, low-scoring schools can suffer the ignominy of a "takeover" by the state. With those threats, it is not surprising that administrators put special attention on the TAAS. Morse (2000, p. 52) reported:

> Even as TAAS scores have skyrocketed, SAT marks have lagged. According to figures released last week by the College Board, the average math SAT score for Texas students increased just one point over last year, and the verbal score dropped to its lowest level in four years....

McGill-Franzen and Allington (1993) describe techniques for removing the test scores of low-achieving students from the averages reported to the public. One is retention in grade, or flunking. The standardized tests are usually not given in every grade. Suppose next year the tests will be given to the third grade but not the second. If the second-graders likely to get low scores on the test are not promoted to the third grade, but are retained in the second, they will not be present at the testing, and the average score will be considerably higher than if they had been promoted.

A second scheme is to classify those low-achieving students as handicapped or "special education" students. No state, as far as I know, requires those students to undergo standardized testing. McGill-Franzen and Allington(1993, p. 22) wrote:

> ... the contamination of accountability reports with unethical placement practices contributes to lost expectations for low-achieving children, lost opportunities for children's development, and lost resources for genuine inquiry and change.

For more on the harm done from retaining students in grade, see Roy P. Doyle (1989).

WHAT CAN BE DONE?

Unreliable tests are useful if you do not care about assessing an individual. If the inaccuracies and unreliabilities are randomly distributed among the individuals, you can still get an estimate of the relative frequencies of variable X in group A and in group B, along with an estimate of the degree of unreliability. As I have said elsewhere, that can be useful information.

When, however, you want to get reliable information about a single individual, an unreliable test is just that—unreliable. I said a good deal about using tests (such as personality inventories) in Chapter 26. Testing experts, by the way, would be more likely in this context to use the word "valid" than "reliable." They reserve "reliable" to mean giving repeatable results, regardless of what the score might indicate. But for my purpose here, the common meaning of "reliable" is good enough: that it can be counted on to do what you want it to do.

That is the first question one should ask about a test. What do you want it to do? When a chemistry teacher gives a paper-and pencil test to his chemistry students, what might he want that test to do? Well, he might want to declare that the six-sevenths of the class who got the higher scores have passed the course, which is a declaration the principal and the school district require of him. There are, of course, dozens of other things he could do to separate the students into those passing and failing. He could read roasted caribou bones, throw dice, give a French test, or give a high-jump test. Few people ever ask whether the chemistry teacher's test does better than those other possibilities.

But suppose he clings to the idea that it is a good idea to give a chemistry test to a chemistry class. Then, in interpreting the scores, he is faced with all those uncertainties I listed in the first part of this section on testing. Knowledge of something about chemistry certainly contributes some part to what some of the students, maybe most of them, do with the items on the test, but nobody will ever be able to find out how much of a particular student's "X" beside an answer is due to the student's knowledge about chemistry, or in what way. The teacher can be condoned for saying, though with some unknown degree of justice, "There are a lot of answers on those answer sheets that I just don't think those students could have given at the start of the term."

Well, very few teachers give a pretest at the start of the term. But let's give the teacher the benefit of the doubt and suppose it is true that if he had given the test at the start, fewer of those "correct" answers would have been there. For any given student, the teacher still does not know the proportions of the correct answers that got there when the student made guesses without any actual knowledge, when she marked answers hopelessly in some irrelevant pattern, when she copied them from her crib sheet, or when she actually selected the right answer from her fund of chemical knowledge. And the teacher does not know the proportion of correct answers *that* particular student could have given on a pretest. The teacher can say, "She got seven of the ten items right." The teacher should not be confident that she would get seven out of ten right tomorrow if he gave her the same test, nor that she would get the same items right, nor that she would get seven right out of ten new items. Nor that she would get seven out of ten right if some other teacher gave her the test. And so on.

So much for what you can know about where the *score* of a particular test-taker comes from—about how much of the score came from the test-taker's knowledge. But now that we have a score from that student, let's suppose, just for argument's sake, that some part of that score did come from knowledge about the items. What can we do with the score (besides pass the student)?

Let's start with a question that might be simpler: What does the score tell us about the student's knowledge of chemistry? Does it tell us that the student knows six-sevenths of what there is to know about chemistry? Six-sevenths of all there is to know about senior high-school chemistry? Of all there is in the textbook? Of all there is in this term's chemistry? You see my difficulty. You can't answer any of those questions with any confidence, either.

You may be saying, "Well, the poor teacher does the best he can, that's all." Of course he does. His job requires him to do it. But I am not writing here about how he can keep his job; I am writing about what use can be made of a test. You can see that I am working very hard to find a cheerful answer to that question. I'll say this: If the student wants to answer a test, by all means give her one. If it is voluntary and timed on her timetable, I am confident she will profit from it. But if you are required to give tests and she is required to answer them on a timetable convenient to the superintendent or the college admissions officer, then I can only say that when the superintendent is making use of his social environment (including the student) for his purposes and the student is making use of her social environment (including the superintendent) for her purposes, the superintendent is getting more than his share of the benefit.

The fact is that what the chemistry test, or the SAT, or any other mental test predicts best is more scores on tests. Grades predict grades, tests predict tests, tests predict grades, and not much else, and the correlations are low. In other words, the predictions have a poor likelihood of being accurate. You remember from Chapter 26 that a correlation must be very, very close to 1.00 if it is not going to be very, very risky to use it in giving advice to an individual test-taker.

What do you want a test to do? Well, beyond comparing the likely relative frequencies of something (perhaps scores on a chemistry test) between two groups of people, there is very little you *should* ask it to do. You should not use an academic or mental test to give advice to an individual or to require an individual to be put in a particular sort of class, or occupation, or mental hospital, or marriage. When I say, "should not," I mean that your predictions of benefit to the person will far too often be in error, and the predictions themselves will therefore be unethical.

That paragraph, I am sure, will horrify a lot of people who make their living giving mental tests and sorting other people into academic tracks, mental hospitals, and so on. You might want to look again at Chapter 31. I do not speak in such a blanket way about medical tests, but even there one finds too much ignorance about such statistical matters as false positives and false negatives, base rates, and so on.

How is it that so many people believe that guiding their acts by test scores is helping them to achieve their desired ends? (In that question, you will recognize Requisite 2b.) First, few people understand how the tests are made and validated (or invalidated). If the cover of the leaflet reads "Test of Academic Aptitude," more people than not seem ready to believe that there is such a "thing" as academic aptitude and that this test can measure it precisely on anybody at any time anywhere. Maybe the belief is helped along if the author's name is followed by "Ph.D.," or even better, "M.D." Second, many people believe that it must be useful to know somebody's score on academic aptitude—otherwise why would those Ph.D.'s make a test like that?

Third, you can actually see what seems to be evidence that it is profitable to have a high score on that stuff. George's daughter got herself a high score on one of those things, and the people at the university took a gander at it, opened the door, and said, "Step right in here, young lady!" Furthermore, Jim's daughter, who got a low score, was rejected even though she had a glowing recommendation from the mayor of the town. With evidence like that, it is easy to believe that the test has some magic in it.

Indeed, that is what it does have—magic. If, for example, you believe that job applicants who wear neckties are more likely to get the job, and the personnel director believes that applicants who wear neckties are more serious, ambitious, and eager to present a proper appearance, then your necktie will get you the job. It won't matter how the necktie was made or whether there is such a thing as necktieness. The magic will work, because the magic is in the heads of you and the personnel director, not in the necktie. Remember Abbas Khan in Chapter 24. I was glad to see Andrew Goldstein's (2000) report that some hundreds of colleges and universities were relinquishing their SAT requirements.

What can you do? Like the hundreds of colleges and universities that Goldstein wrote about, you can stop making decisions about the lives of children in school on the basis of magic—very unreliable magic. You can turn to helping instead of requiring. Requiring brings you the kind of helping the teachers in England gave in the 1860s and that many teachers today are giving in Texas and elsewhere.

I could go on saying what I have said here in several additional ways. I will content myself with mentioning two more places where you can read more horror stories about testing: Alfie Kohn (2000) and Paul Houts (1977).

Chapter 39

Ford's Responsible Thinking Process

In schools, prisons, assembly lines, and other settings where people are required to spend long hours under considerable restriction, the inmates often get "out of hand." They break rules and do things annoying or even dangerous to others, sometimes accidentally, sometimes deliberately. Typically, the authorities conclude that "discipline" is needed. Most people believe that discipline can be improved only by tighter restriction and more severe enforcement of the rules. PCT prescribes the opposite; the way to achieve greater discipline is to *increase* the degrees of freedom—to increase the opportunities of students or inmates to control their perceptions, to reach more of their goals.

I will begin by offering an observation from a janitor in an Australian school in Brisbane, Queensland. One day, he realized that he had been noticing some changes in the behavior of the students. He mentioned to the principal the changes he had been noticing. He then learned from her that the teachers had been learning to cope with disruptions in their classrooms by a new method. The principal asked the janitor to write down his observations, and she invited him to join the new program. Here, from Chapter 29 of a book by Edward E. Ford (1997, p. 218), is what the janitor wrote:

> I have found that since RTP began in our school a number of things have changed. . . . There are now no cheeky kids around. No basketballs on the walkways. No more smart aleck remarks. Children like to help with the rubbish. They like to help clean the school yard and the gardens. It's a pleasure to walk around the school yard and see how the atmosphere in this school has changed. The kids seem a lot happier now, not so much fighting and arguing. The children keep the toilets cleaner. They treat one another with more respect and play better together. I think this works on teachers as well, as I don't hear them yelling any more.

Ed Ford was long a social worker and counselor in Arizona when, in 1982, he was one of seven people attending the first meeting of the Control Systems Group. Ford found that PCT illuminated the thinking he had been doing about his work, and he attracted the interest of Tom Bourbon, whose work you have encountered earlier in this book. In a foreword to Ford's second book on discipline, Bourbon (1999) wrote this:

> In 1982, I organized the first meeting of people interested in perceptual control theory (PCT). . . . Ed Ford was one of the seven people who attended. . . . In his *Discipline for Home and School, Book One*, Ed described his Responsible Thinking Process (RTP). Ed tried to ground RTP in PCT science. When some of his colleagues from the public schools joined Ed at our meetings and described dramatic positive results using RTP, some of us who do laboratory research on PCT wondered if it could be as good as they said. Then I went to Phoenix to visit several schools that used RTP. . . .
>
> What I saw on that first visit convinced me that some of the dramatic accounts of RTP were correct, but I also saw schools where it did not work all that well. . . . [Now], after visiting dozens of schools, we have a clear picture of why RTP works extremely well in some schools, less well in others, and very poorly in a few. . . . RTP works best when the educators understand that children and educators alike behave to control their own experiences. When RTP works best, children understand that, too (pp. xi-xii).

In Chapter 19 of that same book, Bourbon (1999a, p. 149) wrote:

Ford's RTP is designed to help students and teachers control their own perceptions, in school, without unnecessarily disturbing other people. When one person does disturb someone else, perhaps unavoidably or unknowingly, RTP provides a way to deal with the disturbance in a way that minimizes conflict. In schools where RTP is used well, teachers and students [and janitors] are equally likely to say that their lives have changed for the better.

No doubt you can see from even those few words that the RTP is not a procedure for repairing minds that are presumably defective, nor is it a grab-bag of stimuli that will presumably cause children to learn the multiplication table or say "Yes, ma'am," when spoken to. It is a process of facilitation and enabling—a belief and attitude and process with which staff and students in a school can enlarge their degrees of freedom for making use of the social and curricular aspects of their environment in pursuing their purposes.

In any organization, people will inevitably stumble over one another. Timothy and Margaret Carey (2001) put it this way:

> . . . rules in an RTP school can be considered in terms of learning, safety, and social laws. The rules in an RTP school are designed to create environments where participation in learning opportunities can occur optimally. . . . Disruptions in RTP schools are [defined] to occur in the context of these three areas. . . . Other things, however, are *not* considered disruptions in RTP schools. For example, a daydreaming student can be frustrating for an educator who would prefer the daydreamer to be working at a faster rate. But [in RTP] daydreaming is not considered to be a disruption to safety, social laws, or the learning priorities of other individuals (p. 13).
>
> Within this environment, it is recognized that, from time to time, some people will disturb the experiences of other people. . . . Sometimes such disturbances are quickly resolved; at other times, the disturbances recur. For a small percentage of students, these recurrences become chronic.
>
> . . . when people share [an] environment, the way for them to interact harmoniously is to learn how they can experience what they intend while at the same time minimizing the extent to which they prevent other people from doing the same. . . . The point is that disturbances *always occur* in social groups. RTP has not been established to eliminate these disturbances. Educators using RTP . . . spend time exploring ways of minimizing the extent to which they occur and ways of resolving them respectfully when they do occur (p. 15).

HOW IT CAN LOOK

Now I turn to another writing by W. Thomas Bourbon (1998). This is what you see happening:

> When a student disrupts, the teacher asks a few simple questions, in a calm and respectful voice:
>
> "What are you doing?"
>
> "What is the rule?" or "Is that OK?"
>
> "What happens if you break the rule?"
>
> "Is that what you want to happen?"
>
> "What will happen the next time you disrupt?"
>
> The questions afford a choice to a student who disrupts: either he can stop disrupting and remain in the class, or he can continue to disrupt, and thereby choose to leave the classroom and go to the Responsible Thinking Classroom (RTC). For students who stop disrupting when they answer the questions the first time, nothing else happens. After teachers use the RTP for a while, the first question is often all they need. When they hear that question, most students who are disrupting immediately stop and indicate that they understand what they are doing and how it violates guidelines for the ways people should treat one another. On the other hand, if a student continues to disrupt after hearing the questions the first time, the teacher says, calmly, "I see you have chosen to go to the RTC" (p. 155).

That response was standard in 1998. In an e-mail to me of 16 August 2002, Ed Ford said that the response currently recommended is: "What are you doing?" What did you say would happen the next time you disrupted?" Then, "Where do you need to be now?" Bourbon (1998) continued:

> While they are in the RTC . . . students can sit quietly, or read, or do homework, or sleep. They can do anything, so long as they do not disrupt the RTC. Whenever a student decides she is ready, she works on a plan for how to return to class (p. 156).

That may sound familiar to you: A student gets out of hand, so the teacher sends him to the office or to a detention room. The student is required to apologize to the teacher, and after serving his sentence of so many hours or days, the student is sent back to the classroom to see whether he can "behave."

But that is *not* the way it is in the RTP. The teacher in the classroom is not allowed to send a student to the RTC whenever she feels annoyed by the student. She is permitted to send the student only if the student is interfering with what the rest of the class is doing, understands that he is interfering, and wants to go on interfering. The teacher in the classroom is *not* encouraged to scold or shame the student, to threaten the student with punishment, or to offer a reward to him to cease interfering. Any of that would in itself be an interference with instruction (as it is every hour in thousands upon thousands of schools today).

The teacher in the RTC is not a jailer. The student is not serving a sentence, is not being punished. There are no scoldings or recriminations. Whenever the student feels ready to make a plan for avoiding similar trouble, the RTC teacher will help him make a practicable plan, and he can take the plan to the classroom teacher for approval. Carey and Carey (2001) write:

> Students are not referred to the RTC to convince them, to motivate them, or to encourage them to act more appropriately in the environment they just left. Students attend the RTC because that is the only option available to them once they have violated the learning or safety rights of others (p. 101).
>
> Students go to the RTC because, within the constraints of the [school] context, that is the best that can be done in allowing the students, as well as the other people in the environment they left, to control their own experiences. Remember that, according to PCT, we [function so as to] control our own experiences. We are not designed to act in ways that are specified by other people. The RTC is provided so that teachers do not have to attempt to manage students' actions, and, as far as is practical within the school setting, so that they can treat students . . . according to [the assumptions of] PCT (p. 111).

You will have noticed there the phrases "the only option available to them," "within the constraints of the school," "the best that can be done," and "as far as is practical within the school setting." Seymour Sarason and others are well aware of the constraints of the school setting and urge us to carry on education in ways as little like present-day schools as possible. In Chapter 37, I took pains to paint the terrible wastage of time and the terrible burden of boredom that students and teachers now undergo every day.

I also took pains to tell you a little about some schools that had found ways to reduce the wastage and the burden. Of all those hopeful efforts, it seems to me that the RTP exemplifies the most powerful way to do it. The RTP is powerful not because it has the right recipe for actions to be taken, but because it has the right theory. Tom Bourbon is doing the RTP a great service when he goes from school to school and from conference to conference to help people learn how to think on their feet with PCT.

When the student goes back to the classroom, she does not necessarily go back to apologize; she goes back with a plan for avoiding the disturbance in the future. The interview with the classroom teacher is one of solving a problem, not one of contrition. Contrition and apology may come about through a natural feeling about a particular situation, but they are not a required part of the RTP.

To give you more flavor of what can happen, here is a conversation in an RTC told by Carey and Carey (2001, pp. 66–67). Tim Carey was visiting the school and got into this conversation with a boy from the seventh grade.

Tim: Hi, David, what brings you to the RTC?

David: My teacher sent me down here.

Tim: Oh. How did that happen?

David: Well, she hates me. She's always picking on me.

Tim: What does she do to pick on you?

David: She asks me these dumb questions all the time. "David, what are you doing?"

Tim: Oh, OK. So when does she pick on you by asking you those questions?

David: All the time.

Tim: But how does she know to pick on you? Does she start as soon as you walk in the door, or does she wait until you're doing some work that you're interested in? How does she know?

David: No. She asks me the questions just for stupid things. Like if I ask my friend if I can borrow his ruler, or if I walk over to the book stand to get a book. Then she says, "What are you doing?"

Tim: So, is it OK to do those things in class?

David: Well, no, we're not supposed to. But it doesn't hurt anyone.

Tim: But does it disrupt?

David: Yes.

Tim: So, when you do those things, do they interfere with other students who are trying to learn?

David: Well, yes. I suppose so.

Tim: So, does she ask you these questions when you're breaking the rules and interfering with other students' abilities to learn?

David: Yes.

Tim: Does she ask other students those same questions whenever they break the rules and disturb other students?

David: Sure she does. She asks everyone. They're really dumb questions.

Tim: Does she ever ask you the questions when you're not breaking the rules and not disturbing others?

David: What do you mean?

Tim: If you were just sitting there doing your work, would the teacher pick on you by asking you the questions?

David: No, 'course not. You have to break the rules and disturb others to get asked the questions.

Tim: So, are you telling me that even if the teacher *really* wanted to pick on you and *really* wanted to send you to the RTC, she couldn't do that if you were following the rules and not disrupting?

David sits quietly for a moment looking down. Then smiles: Hey! She couldn't. I'd never thought about it like that before.

Tim did not accuse David of being slow-witted or immoral. He did not tell David to do any particular thing. Tim's side of the conversation consisted entirely of questions. But Tim did not quiz David; that is, Tim was not hunting for "correct" answers to his questions. He wanted only to help David find a few more degrees of freedom among *his* perceptions of the situation—*whatever David's perceptions might be.* Tim's questions focused David's attention on the social situation in the classroom and on the perceptions David and the teacher could have of that situation. In the end, David discovered something he could *do* to reduce the conflict. Until Tim helped him see that the teacher also had rules to go by, David had been thinking himself at the mercy of the teacher's unfriendly whims. And incidentally, Tim gave David a nice demonstration of logical deduction.

I will not pretend to give you a 5-minute manual for how to establish RTP, keep it vigorous, and evaluate its success. I hope you will read more in the writings I refer to here. I will, however, give you some evidence of success and tell you about a few more features of RTP.

HOW IT CAN HELP

Bourbon and Ford (1999) have written a short piece on what they look for to judge whether a school staff is using the RTP effectively. I will abbreviate here the questions they ask:

1. Do teachers and other responsible people use the RTP questions at the appropriate times?
2. Do teachers send appropriately completed referral forms when students go to the RTC?
3. Do teachers negotiate effectively and consistently when students bring plans from the RTC?
4. Does the RTP administrator understand and drive the process. . .?
5. Does the RTC teacher help students create effective plans?
6. Does the RTC teacher consistently follow the process in the RTC?
7. Is the RTC a calm, quiet, and effective room?
8. Are intervention teams called promptly and in the appropriate cases?
9. Are the intervention teams effective?
10. What happens to students who make frequent visits to the RTC (the "frequent flyers")?
11. Is there evidence that people understand what RTP is about and what PCT is about?
12. What do students, of any age, say [about RTP]?. . .

There you have it. Asking questions is a big part of what we do when we evaluate a school. Of course, we also look at any statistics and written reports that are available. *But we learn the most by looking and listening and asking simple questions.*

I will give you a few statistics, and after that I will tell you a few things that Tom and Ed saw and heard. I draw first upon Bourbon (1998).

Ford's RTP was first tried in 1994 in a school in Phoenix, Arizona, containing grades 4, 5, and 6. In the first year of the program, compared to the year before, the declines in disruptions were:

Disruption	Percent decline
Possession of weapons	100
Fights	69
Physical assaults	62
Thefts	27

In the first year of the RTP at a school in Illinois, kindergarten through 5th grade, "serious acts of misbehavior" declined 65 percent from the previous year, and suspensions (sent home) declined 66 percent.

At a prison for juvenile males in Arizona, the high school there began to use the RTP in 1997. Comparing the same four-month periods in 1996 and 1997, disruptions decreased 52 percent in the school and 42 percent *in the remainder of the prison.* Among programs for organizational change, a spontaneous spread of effect from an original part of the organization to other parts is a rare case. You will remember, too, what the janitor in Brisbane said about the behavior of the children in his school *outside* their classrooms.

When you count the students who go to the RTC with various frequencies, you find the swooping curve familiar to social statisticians: many students go rarely if at all, and a few go frequently. In the typical RTP school, most students never go to the RTC, and two to five percent of the students make about a third of all the visits made.

At the prison in Arizona during the month of May, 1997, 56 percent of the 132 high-school students went to the RTC not at all, and 5 percent (7 of them) went to the RTC four or more times.

At the school in Illinois mentioned above, two percent of the students made 32 percent of all the visits to the RTC.

A school with kindergarten through the sixth grade in Arkansas began using the RTP in 1996–1997. In that year, a majority of the students never went to the RTC. Two percent made over one-third of the visits.

A middle school in Arizona with grades four through eight began the RTP in 1995. During 1996–97, 46 percent never went to the RTC. Three percent made a third of all the visits.

In most RTP schools, students who go to the RTC noticeably more often than most are called "frequent flyers." Bourbon (1998, pp. 164–165) writes:

> The way that faculty members deal with "frequent flyers" depends on how well they understand the basic principles of RTP and of Perceptual Control Theory. In some schools, the faculty literally destroys RTP in an attempt to "make all of those students stop going to the RTC so often." Using the logic from [linear] cause-effect theories of behavior, they believe RTP should "fix" the students, or the school, so that no one will ever disrupt again. Those adults do not understand that everyone acts to control perceptions, and sometimes they cannot avoid disturbing others. [Those adults] sacrifice the entire RTP program because they want to completely control the behavior of the most troubled two percent to five percent of the students. In the process, they ignore the large majority of students who do not disrupt at all, or who disrupt only once or twice a year.
>
> When people understand the basic concepts of RTP, they interpret frequent trips to the RTC as evidence that a student is trying to control perceptions of a serious problem in his or her life. The adults then devote special attention and resources to helping the "frequent flyer" make it through a difficult time.

The "difficult times" for some students are very difficult indeed, difficult in ways that cause most people of any age to reach out for special attention and help. Every day, some students suffer physical and sexual abuse, witnessing murders in the family, being diagnosed as mentally hopeless, and other horrors. Bourbon (1998) tells about some ways that students in "special education" make use of the RTP:

> Ed Ford's RTP has been used successfully with many special education students whose "diagnoses" are intended to imply that the child cannot tell right from wrong, or that they cannot learn to "control their own actions.". . . In schools that use RTP, very young students with special needs

demonstrate that they know when they took a toy from another child. They also select plans for them to share toys, or to take turns playing with them. Those students know when they have hit someone else, and they select plans that call for them to keep their hands to themselves, or to move away from people they might hit. What is more, the students are eager and proud to show the teacher, or visitors, that they are following their plans. Those students do not conform to what the experts say they can and cannot do, or to what teachers were trained to expect from them, or to what their parents came to believe were their limitations.

At a school in Texas, many emotionally-disturbed students had spent several *years* [emphasis PJR's] in special units, without ever returning to the regular classroom. A few months after RTP was introduced into their units, some of those students were in regular classrooms for three or more periods each day.

At a school in Mississippi, a young man diagnosed with autism and four other major disorders was referred to a special unit that had just started to use RTP. When he arrived in the unit, he disrupted his class so often that he made as many as six visits a day to the room [used as] the RTC. By the end of the year, the young man went to the special room no more than once every two or three weeks (pp. 161–162).

Similarly, [in another school] a disruptive autistic student was allowed to go to the RTC, [and] he remained there quietly for a while. He decided to return to the classroom, and he made a plan to do so. Had the staff tried to prevent that student from leaving the regular classroom, he would have behaved as though he wanted to leave and to be alone; he would have confirmed traditional ideas about what autistic children do and why they do it (p. 163).

Time and again, teachers who use RTP with special education students discover that the children can do much more than mental health professionals believe. Often, it becomes clear that traditional diagnoses create expectations that everyone helps students meet (p. 162).

Relevant here is a testimonial I found on the back cover of Ford's 1999 book. It came from Tammy Mason of Whittier Elementary School in Amarillo, Texas:

As a teacher in a Behavior Adjustment Classroom, I have seen RTP give all types of children the courage and power to become responsible, able, and self-confident students. The students in my classroom have multiple diagnoses such as ED, LD, ADD, ADHD, and/or ODD, and despite their behavior and learning problems, RTP has enabled them to learn, plan, and feel successful in daily activities. RTP allows students to be in control of their own lives and future.

Giving some added degrees of freedom to those "multiply diagnosed" students allowed them to find their own ways out of their difficulties and to avoid, many of them, being incarcerated in mental hospitals and jails for much of their lives. In regard to diagnoses of mental "disorders" by professionals and the treatments they recommend, you might want to review Chapters 29 and 31.

ASSUMPTIONS

Here I will comment on a few assumptions that could be in the minds of those people Bourbon told us about who did not want to perceive "all those students . . . going to the RTC so often." There is a tradition in schoolteaching that the essential skill, the sine qua non, is that of maintaining discipline without help from others. When a teacher finds himself losing a battle with a student (perceives the situation to be like that), the only recourse, usually, is to haul up the big gun—to send the student to an administrator. The administrator is usually no happier at having to cope with the student than the teacher was. Administrators, almost to a person, are firm in their opinion that a good teacher maintains discipline without sending students to an administrator.

Within that tradition, it is easy to think of sending students to the RTC as a mark of poor ability to maintain discipline, of repeated visits by any one student as a shameful failure to "discipline" a student properly, and of the RTC itself as a blot on the school's escutcheon. And just as killing the messenger is a quick way to do away with bad news, so abolishing the RTP is a quick way to do away with evidence of poor discipline.

I leave it as an exercise for the reader to look back over the previous two paragraphs and note the assumptions about human nature that support the traditional view of school discipline.

Faculty who understand the RTP do not look upon visits to the RTC as punishment. They look upon them as opportunities for students and faculty to build cooperative social relationships. Furthermore, students can practice skills of civilized society that many of them could not previously imagine. Remember the student to whom Tim Carey talked: "Hey! She couldn't. I'd never thought about it like that before."

The traditional view would be that a student going to the RTC is a sign that a bad thing has happened in the school, that continuing visits constitute a sign that bad things go on in this school all the time, that the teachers are lacking in an essential skill, and that they are trying to palm off their responsibility onto the RTC. The RTP view is that the RTC removes disturbances from the classroom immediately, it provides a means for student and staff to resolve difficulties, and for those students who need it (and want it), the RTC provides a means of learning important social skills.

The traditional view is that if a student shows insufficient social skill in the classroom, the student should be punished. The RTP view is that the student should be offered a means of improving that skill. In both the traditional and the RTP view, a student with insufficient skill in arithmetic should continue to visit the arithmetic class. Similarly, in the RTP view, a student with insufficient social skill should continue to visit the RTC.

COUNTERCONTROL

I wrote about countercontrol under that heading in Chapter 28. Students in schools often defend themselves by using countercontrol. Bourbon (1998, pp. 162–163) gives an example of a student who was skillful at countercontrol:

> ...a student in Michigan, diagnosed with "attention deficit hyperactivity disorder," was said to be "so out of control that he cannot function unless the teacher stands next to him." In fact, the young man was controlling the teacher's behavior by "making" the teacher stand where the student wanted him to stand. When the student was treated like all others in the Responsible Thinking Process, he quickly "gained control over his own actions."

There you see, in microcosm, an example of what Albee meant in the quotation you saw in Chapter 29: "... the problems of deeply distressed people are largely social in origin." The pattern of action was being maintained jointly by the teacher and the student. The psychiatric label was thoroughly misleading. Since the pattern was jointly maintained, it could be broken by *either* party. In the psychiatric view, it could be broken only by coaxing the student somehow, possibly by years of talking and drugging, to change *his* behavior. In this case, the pattern was broken by the student's teachers, requiring little talk and a few days. Here is an excerpt from what Carey and Carey (2001, pp. 44–45) have to say about countercontrol:

> Since all it takes to create the conditions under which countercontrol can occur is for one person to determine how another person must act, [school] settings can be considered target-rich environments for the phenomenon of countercontrol. While rules and standards are certainly important in RTP schools, the crucial difference in these schools is that the decision to adhere to the rules is left up to the student. It is not considered a teacher's responsibility to promote "rule-following" behavior of students. If students are not interested in adhering to the rules of an environment at any point ... and, in so doing, are compromising the learning and safety of others or societal laws, they go to another environment. They stay in that environment until they want to return to the environment from which they came and want to minimize the extent to which they disturb others in that environment. The ability to provide different environments for people goes a long way towards eliminating countercontrol in RTP schools.

The Embedding

I once saw a class of fifth-graders who had been taught skills for managing a meeting so well that they could manage themselves for more than an hour while the teacher sat quietly in the back of the room, doing absolutely nothing but sitting there. Those youngsters were so skillful, in fact, that when they were invited to take over a group of about 40 adults to demonstrate what they were doing, they took about five minutes to organize themselves, and then drew the adults into a demonstration of how they (the adults) could learn to do what they (the students) did.

I was enthralled. Those students had learned by the fifth grade what I had not learned until I was almost fifty years old. I thought it was marvelous that those youngsters were going to be able to participate with such skill in the meetings they would encounter throughout their lives. How lucky they were, in that small town on the McKenzie River in western Oregon! Unfortunately, one of the students proudly told his parents about what he and his classmates were learning in that class. That didn't sound like schooling to the father. The father told the principal he wanted his son to concentrate on reading, writing, and arithmetic, and not on that stuff about talking in meetings. The principal told the teacher to stop doing that. And that was that.

When I was an organizational consultant in schools, I saw that sort of tragedy every now and then. A faculty can work hard for a year or more to find a wonderful new way to work together, a way they can carry into their classrooms and their homes, and the whole lovely thing can be killed overnight by a new principal who comes in and says, "Stop that!"

The RTP exists under that same threat. It is wonderful that a process as radical as the RTP can be learned as quickly as some faculties do learn it, and it is wonderful that the RTP can make such strong and gratifying changes in a school, even though it is surrounded by people who believe the very opposite of the assumptions that the RTP and PCT make about how humans must function. Now and then, however, tragedy stalks the corridors. Too many of the faculty find incomprehensible the idea that people act so as to control perceptions; or a principal, frightened by a parent, a superintendent, or his own ghosts, says, "Stop that!" In a message to the CSGnet of 27 April 1998, Tom Bourbon carried on a dialog with William Powers:

Powers: The handicap under which RTP operates is that it's embedded in a coercive system that can't be changed without abolishing the whole concept. The student who is determined to disrupt is in fact punished upon being handed over to the juvenile "justice" system, which does not use RTP. The system is basically coercive once the student fails to take advantage of the special classroom as expected, and it's coercive in that the student is simply not allowed the choice of staying in the classroom AND disrupting it. The adults are physically in charge, and the cops are called if there is any effective resistance to this state of affairs.

Bourbon: All true. RTP must operate in exactly that kind of system.

Powers: In my view, what is most effective about RTP is that the teachers are taught that the students are autonomous control systems and basically can't be controlled by peaceable means—*so the teachers don't have to try* [my emphasis—PJR]. The teachers are taught not to lose their cool, and to treat the students (even offenders) with respect. What choices are actually possible for the student to make are treated as real choices; if the student doesn't want to make a plan for returning to the classroom, for example, that's OK..

Bourbon: Yes. The feeling of calm that pervades a school where teachers learn this "secret" is amazing. . . .

Powers: The main thing that changes as a result of RTP, I believe, is the way teachers treat the students. Classroom conflicts are de-escalated, and there is less for the children to fight back about. The level of tension seems to decrease markedly. Since the most dramatic initial change induced by RTP is in the teacher's behavior, I think we can conclude that by and large, there's not much wrong with the kids except ignorance of practical social behavior. RTP is really an adult therapy program, although you can't sell it by calling it that.

Bourbon: Excellent comments. As soon as most students sense the change in the way teachers treat them, they change the way they treat teachers. A few hold-outs, on either side, continue to treat people on "the other side" in the same old ways.

CODA

Paulo Freire (1973) says that we can perceive our troubles to arise from three sources. (1) We can believe that the world is a "vale of tears," and there is no escape from it. God made it that way. (2) We can believe that somebody did it to us. One way to get rid of trouble is to kill the person who is bringing it upon us. (With that reasoning, people sometimes kill themselves.) (3) We can believe that we live in a social web that constrains all of us—a system we all act to maintain. We can lessen our troubles by joining with others in the system to alter the way the system works. Like any classification of living reality, you can find awkwardnesses in Freire's triplet, but I find that it alerts me to possibilities. David, for example, was blaming his teacher for his troubles, and Tim helped him to perceive the system.

The RTP is Freire's third sort; it changes the social system. It is fragile, because it is beleaguered by the surrounding society. With only a little protection, however, it can maintain itself, because its members can teach it quickly enough to newcomers (both students and teachers).

For a finale, here are some reasons that Powers sees for the effectiveness of the RTP. He wrote this to the CSGnet on 31 January 1998:

> First, the teacher is acting as a good control system, and the social setup actually allows good control. The teacher has to tolerate a disruption only once, and then can turn the problem over to someone else. This avoids conflict, and the teacher doesn't have to resort to extreme efforts and emotions.
>
> Second, as a consequence of the above, disruptions are handled in a calm, matter-of-fact way. There is no need for anger or punishment; the child is not labeled as "bad." The setup makes it perfectly clear that disruptions have inevitable consequences, but the consequences are not harmful to the child, degrading, or discouraging. The child's self-esteem is not under attack; past misbehaviors are not used cumulatively against the child in the attempt to control the child. Forgiveness is taken for granted, because blame is never used. There is no "debt to society" to be paid.
>
> This means that a great deal of the effectiveness of the program comes from things that are *not* done to the child, things that are commonplace in traditional classrooms. And in order not to do these things, the teacher and the administrators must undergo a radical change of attitudes and methods. They must see the futility of trying to control the children by force, intimidation, punishment, and invalidation—and also by bribes, rewards, and false praise. They must stop feeling that they need to control the children, and thus stop *wanting* to control them. Ed's program gives them not only a reason to change their relationship to the children, but a simple and direct means of doing so. Each teacher who learns to carry out the program successfully will inevitably experience a personal transformation—as many of them have, in fact, said.
>
> Another very important factor is that when the classroom is calm and relaxed, children have far less reason to want to be elsewhere or to spoil the atmosphere.... The point of going to the RTC is not to punish or to earn back rewards that have been withdrawn, but to work out problems in a way that's useful and satisfying. There is a positive reason for going to the RTC, a net gain for the child. In the RTC, the child learns about controlling and being controlled, about working out problems with other people so they cease to be problems, about how people and societies work when they are working well. And then, upon going back to the classroom, the child sees a small example of such a society in action, so what is learned in the RTC is validated

If you want to read more about the RTP, look into Ford (1997, 1999), Bourbon (1998), and Carey and Carey (2001). To find further information, visit www.responsiblethinking.com. Wherever you choose to begin, you will find fascinating stories. Here is one of many more told by Carey and Carey (2001, p. 12):

> During October, 2000, it was reported on the ANANOVA web site . . . that [some] schoolchildren in Bulgaria were buying head lice to infest themselves. While this might seem like bizarre behavior, when students were discovered to have lice, they automatically received three days off school. Apparently, the students were using head lice to experience time at home. Even though their actions might seem inappropriate to some people, their behavior created the experience they [wanted].

Chapter 40

Society

Our beliefs about human nature shape the ways we try to use other people in controlling our perceptions. When our beliefs about human nature are erroneous, our social actions lead to reductions of control. Some widespread beliefs are erroneous.

It is not true that conditions or events (stimuli) in the environment can cause an individual, in a stimulus-and-response manner, to perform a particular act. It is not true that *you* (or others) can reliably cause an individual to perform a particular act. I wrote about this topic in Chapter 5 under "Controlling Others," in most of Chapter 29, in all of Chapter 33, in Chapter 34 under "Organizational Management," and in Chapter 35 under "Leadership." See also Powers (1973, p. 260 ff.).

You can, it is true, increase or decrease the *proportion* of people (rarely with much precision) who will buy tickets to a performance by the Ungrateful Livers or purchase a tube of ToothShine this month, but it is not true that you can reliably predict whether *this* person will buy a ticket or a tube in September.

It is not true that inner proclivities (personality traits, if you prefer that term) can alone cause an individual to perform a particular act. It is not true, either, that we all have the same needs, personality factors, or some other internal standards. We all have internal standards, but mine are unique and so are yours. (See Chapter 26 on personality and Chapters 20 and 21 on the reorganizing system.)

It is not true that people are necessarily intending to do what they seem to you to be doing. See Chapter 7 under "What Is the Person Doing?" Furthermore, all actions have unintended effects. Similarly, it is not true that two people can perceive the "same" event, and it is not true that anyone can perceive what something or event "really is."

It is not true that you can get rid of bad behavior by getting rid of bad people, because the people who have not yet done those bad things are still out there. If you think, oh, well, the fittest will survive, remember that in the evolutionary context, "fittest" merely means able to bring more offspring to maturity than somebody else does. There is no guarantee that the people who bring the most children to maturity will be good neighbors to you or me or our grandchildren.

It is not true that inner conflict is good for you. If a competitive situation brings you inner conflict, that competition is not good for you. I wrote about competition in Chapter 28.

It is true that much inner conflict is of small moment, occasioning little emotion and triggering little reorganization. Though a small conflict may bring us no benefit, it usually brings little or no harm, either. It is even possible that the prospect of the *risk* of inner conflict can be pleasurably exciting. That may be part of the attraction of adventurous fiction. I do not think, however, that inner conflict can be depended upon to strengthen the will, hone judgment, temper the soul, or do any other thing to improve our skill in avoiding further conflict. Sufficiently severe inner conflict sets off reorganization which *may or may not* enable us to avoid conflict more skillfully in the future.

What you can do is limited by what you can perceive to be possible. (Requisite 2.) If you hold wrong ideas, you will take actions that will increase the likelihood of consequences that you do not want. You will try, for example, to bring about particular actions on the part of other people by rewards and threats. Since rewards lose their effect unless they escalate, and since threats almost always escalate in the damage they promise, those techniques bring benefits to a few at the expense of the many. Since many people believe that the way things *do* work is the only way they *can* work, many of us, maybe most, go on to believe that the only way to be safe, comfortable, and happy is to exploit other people, and the more the merrier.

When we seek to achieve our goals by controlling the behavior of other people, we try, for example, to conduct therapy by the kind of diagnosis and prescription that I described in Chapter 31, and we try to manage organizations by using the kind of restrictions on behavior of which I gave examples on Chapters 34, 37, and 38. We try to reduce crime by putting more people in prisons. I quoted James Gilligan (1995, p. 95) in Chapter 27 that the United States imprisons a percentage of its population *five to twenty times* greater than any other country on earth except Russia. Somehow, most of us still sing unblushingly of "the land of the free."

I wrote about degrees of freedom in Chapters 27 and 34 and under "Powers" in Chapter 35. When the environment contains sufficient degrees of freedom for all of us, our actions need bring little conflict among ourselves or within ourselves. Threat and counterthreat can become much less likely than now. That is the key to improving society, to enabling it to become nurturing instead of punishing.

People with ample degrees of freedom are less likely, on average, to act so as to reduce *your* degrees of freedom; they are not the people, on average, whose internal conflicts and reorganizations cause them to flail in all directions (and possibly in yours) in their search for solutions to their conflicts. This is not to say that rich people, for example, are never dangerous to the rest of us. People who sometimes *seem to us* to have ample degrees of freedom may in their own perception have too little. In other words, people we think must surely be contented with their circumstances do turn out sometimes to be a danger to us. But I am speaking here in averages. Because people with ample freedom of action will not be using time and energy struggling against conflicts, they can turn some of that time and energy to helping to maintain a society that will provide ample freedom for their children and grandchildren—and for you and yours. Paul Wachtel (1983) in his book *The Poverty of Affluence* does a good job of explaining that in the United States and other developed nations, most people achieve sufficient affluence not to need to dream of wresting further wealth from other people.

That is the gist of what I want to say in this final chapter. In an earlier version of the manuscript for this book, I included four chapters telling about the planetary conditions we have produced that are inimical to our physical welfare, about what we do to worsen them, what we do to meliorate them, and how our human nature helps and hinders us. I came to see, however, that I might be distracting your attention from my chief purpose in this book, which is to explain the theory of perceptual control and some of its implications for what is possible and impossible in human behavior. I will briefly pursue here only two topics: (1) caring for the future of our species and (2) freedom.

THE FUTURE

Though we are very skillful, as a species, in controlling our perceptions now and in the very near future, we have little skill in controlling perceptions of movement toward a world in which we would be glad to live next year or next decade, not to speak of a world for our grandchildren to live in. This lack of skill has allowed our forebears to transmit to us a physical world which they and *their* forebears had steadily degraded and polluted, and a social world fraught with declining degrees of freedom and with increasing conflict.

Gary Gardner (2001) writes:

> Indeed, this threshold moment is virtually unprecedented in world history. Only the Agricultural Revolution that started 10,000 years ago and the Industrial Revolution of the past two centuries which brought unparalleled prosperity as well as environmental pathologies to a large share of humankind rival the current era as moments of wholesale change in human societies. But those global transformations unfolded much more slowly, and began in different regions at different times. The changes under way today are compressed to just a few decades and are global in scope. The question facing this generation is whether the human community will take charge of its own cultural evolution and implement a rational shift to sustainable economies, or will instead stand by watching nature impose change as environmental systems break down (p. 190).

According to PCT, humans deal with the far future in the imagination mode (for which see Chapter 19). *The imagination mode does not itself call for immediate action.* What we do with our imaginings depends on the internal standards (reference conditions) we form at the very highest levels of principle and system-concept. Whether we store up grain against the approach of winter depends on having a reference perception, for example, of our wives and children happily munching away while the snow gets deeper outside the cave. Whether we study the literature on air pollution, the voting record of our senator, and the operation of our local industries depends on having a reference perception, for example, of our children's children, free of emphysema, happily taking deep lungfuls of air as they walk down Main Street. The book by Ornstein and Ehrlich (1989) is chock full of illustrations of the difficulty we find in acting in the interests of our grandchildren.

Ten thousand years ago, John Pfeiffer (1977, p. 42) tells us, the entire world contained about ten million humans—not many more than *half* the population of the metropolitan area that includes New York City today. They had been making a living by hunting and gathering, and they had thrived:

> Hunting-gathering had been an effective adaptation to the wilderness, permitting man to spread from the African woodlands and savannas [to] all continents and all climates. It offered more leisure, more easy and spontaneous living, than any system devised since. . .
>
> Perhaps the world could have been made forever safe for hunting-gathering . . . if man had been able to change his breeding habits, if populations had stopped rising. He multiplied instead and, in the process, changed nature (p. 20).
>
> . . . man did what no other species had ever done. He created new environments which could support far greater populations than the wilderness, and succeeded too well. He found himself swept up in a round of changes as populations continued to climb, each change producing new conditions which demanded further change (pp. 32–33).
>
> The last ten millenniums have been a struggle to [win] a race between rising population and increasingly sophisticated use of the land, with population always running ahead [p. 21].

Those long-ago people could not have been expected to foresee a world in which millions of their descendants would routinely die of atmospheric pollution and in which daily warnings would be broadcast concerning the danger of skin cancer. Now, however, in a world of widespread scientific knowledge and almost instantaneous communication the world round, persons charged with our welfare still display sluggish skills at forecasting.

As a stark example, most members of our Congress still seem unconvinced that the ozone hole is a danger to us more than a decade after the situation was bad enough for the following news item to appear in the *Chicago Tribune* (quoted in *World Watch* for September–October 1990, page 6):

> For Australians, the Antarctic ozone hole is beginning to hit home. Shifting patches of ozone-depleted atmosphere above Australia have occasionally pushed ultraviolet radiation 20 percent above normal. Television stations now air daily ultraviolet readings and warnings for Australians—who already have the world's highest rate of skin cancer—to stay inside during bad spells.

The action (or inaction) of congresspersons is constrained by their high-level internal standards. The attention of most congresspersons is apparently attuned to the demands of constituents and lobbyists—so, at least, say Dye and Zeigler (1990, p. 135). The citizens of Australia are not constituents, and the lobbyists at the U.S. Congress for salubrious air in Australia are few and weak in comparison to the lobbyists defending polluters of air in the U.S. The idea that U.S. congresspersons and Australians breathe the same ocean of air and walk under the same ozonosphere may be quite beyond the comprehension of many congresspersons. As with everyone else, the actions of a congressperson follow his or her priorities—in this case, the relative levels of control for getting reelected versus the health of anybody's grandchildren.

Every year since 1984, the Worldwatch Institute has published a report called *State of the World*. In Worldwatch's 2001 report, Gary Gardner tells a story from which we learn that the U.S. Congress has not been as laggard as the British Navy:

> In 1601, an English sea captain named James Lancaster conducted an important experiment. Commanding four ships on a voyage from Eng-

land to India, he served lemon juice every day to the crew of one of the ships. Most remained healthy. But on the other three ships, 110 of the 278 sailors died of scurvy by the journey's midpoint. The experiment was of immense import to seventeenth-century seafarers, since scurvy claimed more lives than any other single cause, including warfare and accidents.

Surprisingly, however, this vital information had little impact on the British Navy. The Navy did not conduct its own experiments until 1747, nearly 150 years later, and did not stock citrus fruits on its ships until 1795. And the British merchant marine followed suit only in 1865, some two-and-a-half centuries after the first experiment with lemon juice was carried out. Despite the magnitude of the scurvy problem, and despite the availability of a simple solution, people were slow to respond (p. 189).

Today as in 1601, most people still dispose of statistical information according to their prejudices. In Chapter 25 under "Base Rates" and in Chapter 31 under "Diagnosis" and "Efficacy," I gave examples of many circumstances in which psychologists and psychiatrists have acted on prejudice rather than statistics. But psychologists, psychiatrists, or even long-dead British sea captains are not exceptional. Bankers, reputedly a "hard-headed" bunch, also have difficulty thinking about statistics and the future. David Roodman (2001), in his chapter in the Worldwatch report, writes:

> In late 2000, the World Bank released its latest *World Development Report*, which "urges a broader approach to reducing poverty" that embraces not only economic growth but "opportunity, empowerment, and security." Implicit in this message is a self-criticism, an admission of past mistakes, but the very implicitness indicates how reluctant the institution still is to examine those mistakes openly. The learning process has been welcome, but it is fair to ask why it took 50 years to reach this point (p. 156).
>
> ... the development effort of the last 50 years has spawned enough economic, environmental, and human rights disasters to fill many books and run up a lot of unpayable debt. To take one extreme example, the World Bank lent $72 million to Guatemala to help finance a dam on the Chixoy River in 1978. In 1980, local people whose land would be flooded by the reservoir, mostly indigenous Achi Indians, began protesting against the project. In response, soldiers massacred at least 294 villagers, after raping or torturing many of them. But the World Bank, "did not consider it to be appropriate to suspend disbursements." Five months after the reservoir filled, the dam stopped working because of poor design. The Bank lent another $47 million for repairs. Such problems, along with corruption, eventually raised the dam's total cost from $340 million to $1 billion, accounting for a substantial fraction of Guatemala's current $4.6–billion foreign debt (pp. 156–157).

Our evolution has bequeathed us our wonderful ability to think and imagine. With that ability, we can envision goals that we estimate to lie years, decades, lifetimes, even millennia in the future. We can imagine programs of subgoals to take us to new kinds of societies. Not only do we have the ability to think about these matters, but today enough dependable knowledge is available to enable us to take effective steps toward many of our distant social goals if enough of us desire to do so.

Despite the age-old belief in the profits to be made from overpowering and oppressing others, some people everywhere have always dreamed of an expanded liberty, and some people in every age have banded together in the hope of bringing it about. Gary Gardner (2001) gives these examples of movement toward freedom:

> Orchestrating change is not easy, but neither is it impossible. After millennia in which human servitude was commonplace, for example, freedom has been increasingly secured in the last 150 years. The abolition of slavery in the United States [and elsewhere], the nonviolent movement for India's independence, the end of apartheid in South Africa, the peaceful transition away from communist rule in Eastern Europe and the former Soviet Union—in each case leaders, organizations, and citizens demonstrated the flexibility and courage needed to respond to the moral imperatives of their day and to work for change. This generation will need to summon the same courage and commitment—and then some, given the daunting challenge facing the human family today (p. 190).

Social and political freedom for women lagged behind that for men, even in the places where freedom for men was increasing more rapidly. The first convention in the U.S. on women's rights was held at Seneca Falls, New York, in 1848; the 19th amendment to the Constitution, granting women suffrage, was ratified in 1920. Another example that required more than a generation to permeate American society was the movement (unfortunate, in my view) for compulsory education, which also began in the U.S. before the Civil War and blanketed the country in the early part of the twentieth century.

Some changes can occur rapidly. During the first half of the 1900s, most pediatricians believed that feeding babies from bottles of "formula" milk was better than feeding them from the breast. John McKnight (1984) tells the story of a woman in a Chicago suburb in the 1950s who

> . . . began a search throughout the area for someone who might still remember . . . the information necessary to begin the flow of milk. From that faint memory, breast-feeding began its long struggle toward restoration in our society [with the founding of] La Leche League. This incredible popular movement reversed the technological imperative in only one generation (p. 249).

The steps toward a salubrious society require not continual edicts from above, but intimate coordination in small collectivities—two or three hundred seems about the maximum number for good maintenance of perceptual control. I am not saying that the world should consist of small villages with no overarching government. I am saying that we need small organizations—neighborhoods, businesses, volunteer civic organizations, and all other sorts of associations—as units for governance at a human scale. The more we impose coordination only through governments of large cities, states, and the nation and through gigantic corporations, the more difficult it is to avoid the tyranny of bureaucracy.

An inspiring story of what can be done is that of Kerala, one of the poorest of the states of India. Chris Peacock (1991, p. 24) wrote that if Kerala were a separate country, "it would place ninth on a list of the world's poorest countries." Here are excerpts from an article about Kerala by Bill McKibben (1996):

> The life expectancy for a North American male . . . is 72 years, while the life expectancy for a Keralite male is 70.
>
> . . . the United Nations in 1991 certified Kerala as 100 percent literate. Your chances of having an informed conversation are at least as high in Kerala as in Kansas.
>
> Kerala's birth rate hovers near 18 per thousand, compared with 16 per thousand in the United States—and is falling faster.
>
> There is chronic unemployment, a stagnant economy . . ., and a budget deficit that is often described as out of control. But these are the kinds of problems you find in France [and recurrently in the United States]. Kerala utterly lacks [however] the squalid drama of the Third World—the beggars reaching through the car window, the children with distended bellies, the baby girls left to die.
>
> Development experts use an index . . . that runs on a scale from zero to a hundred and combines most of the basic indicators of a decent human life. In 1981, Kerala's score of 82 far exceeded all of Africa's, and in Asia only the incomparably richer South Korea (85), Taiwan (87), and Japan (98) ranked higher. . . By 1989, [Kerala's] score had risen to 88, compared with a total of 60 for the rest of India (p. 103).
>
> Kerala—and [some] other spots around the world . . . makes clear that coercion is not necessary [to reduce birth rates]. In Kerala the birth rate is 40 percent below that of India as a whole and almost 60 percent below the rate for poor countries in general. In fact, a 1992 survey found that the birth rate had fallen to replacement level. . . And, defying conventional wisdom, it has done so without rapid economic growth. . . (p. 108).
>
> Kerala's attitude toward female children is an anomaly as well. Of 8,000 abortions performed at one Bombay clinic [in contrast to Kerala] in the early 1990s, 7,999 were female fetuses. . . There are, in short, millions and millions of women missing around the world—women who would be there were it not for the dictates of custom and economy. So it is a remarkable achievement in Kerala to say simply this: There are more women than men. In India as a whole, the 1991 census found that there were about 927 women per 1,000 men; in Kerala, the number was 1,036 women. . . [T]he female life expectancy in Kerala exceeds that of the male, just as it does in the developed world (p. 109).
>
> Kerala does not tell us precisely how to remake the world. But it does shake up our sense of what's obvious, and it offers a pair of messages to the First World. One is that sharing works. Redistribution

> [of wealth] has made Kerala a decent place to live, even without much economic growth. The second ... lesson is that some of our fears about simpler living are unjustified (p. 112).

Here is an example of how a nongovernmental, volunteer organization not representing a business interest can save us from a terrible step in the wrong direction. John Isaacs, of the Council for a Livable World, sent e-mail on 6 September 2000 to contributors to the Council. Isaacs narrated the history of President Clinton's decision not to deploy the National Missile Defense (NMD). Isaacs mentioned numerous individuals and organizations who urged Clinton not to deploy the NMD. From the six pages of Isaacs's e-mail, I will quote only these paragraphs:

> Another part of the arms control organizations' plan was to encourage highly respected "validators" to advise that a deployment decision should be put off. The most significant hit was by a letter urging deferral that was put together by the Carnegie Corporation and the MacArthur Foundation. The letter stated: "We respectfully urge you to defer a decision to deploy, and not be forced by artificial deadlines, but to further the debate that has now begun in earnest." Signers of the letter who had special credibility with the Clinton Administration included former Defense Secretary William Perry, former Senator Sam Nunn, retired Generals John Shalikashvili and Andrew Goodpaster and retired Admiral William Owens. On June 9, 33 experts on Russia sent a letter to the [U.S.] President urging no deployment at this time, followed by a June 29 letter signed by 45 experts on China. Both letters had been organized by the Council for a Livable World Education Fund. A similar statement signed by 50 Nobel Laureates that had been organized by Federation of American Scientists was released at a July 6 press conference.
>
> All these letters garnered significant press attention, added to the credibility of the opposition and were cited many times during the subsequent debate.

In democracies, the tradition of free speech encourages the flow of information. Untrammeled news media, especially what we call in the United States the "alternative press," greatly enhance and protect this tradition. There have always been writers who protested the doings of the powerful; Cicero (106–43 B.C.), Rabelais (1494?–1553), and Voltaire (1694–1778) are examples. In the United States, the abolitionists were notable in the nineteenth century. In the early twentieth century, Upton Sinclair (1878–1968) wrote novels of social protest. In 1940, George Seldes founded the weekly newsletter In fact, devoted to exposing wrongdoing and knavery of all sorts. More recent examples of the alternative press are The Nation, Mother Jones, Utne Reader, and Ms. Most recently, even the mainstream periodicals have taken up investigative journalism—TIME, Newsweek, and others.

As an instance, two members of the alternative press, Mark Green and Gail MacColl, compiled a book of Ronald Reagan's "ignorance, amnesia, and dissembling": *There He Goes Again: Ronald Reagan's Reign of Error* (1983). Writing in 1994 in the *Utne Reader* (p. 154), Jay Walljasper saw the situation this way:

> Ronald Reagan was ... widely admired as a statesman in 1985.... The Reagan presidency was indelibly stained a few months later when a Beirut-based magazine revealed details of his administration's role in illegally brokering arms deals between the government of Iran and Nicaragua's contra rebels. The fact that an obscure Middle Eastern publication, not the *Washington Post* or *CBS News*, broke the story says a lot about the state of the mainstream press in the mid–1980s.

Writing in *Mother Jones* (p. 9) in 1987, Mark Green told about the reaction to his and MacColl's book:

> *Reagan's Reign of Error* ... provoked an indignant response from some loyalists and journalists.... Until Irangate. Now that President Reaan has had to repudiate nearly everything he originally said about the Iranian arms sale, everyone other than perhaps the First Lady knows that our 40th president is a chronic dissembler.... But given the Niagara of evidence of Reagan's previous deceptions, why did it take six years for an acquiescing public and press to catch on?

Considering the time it took the British navy to include citrus fruits in the diets of its sailors, perhaps six years is lightning rapidity. A free press is a priceless watchdog. It is easy for people in power to lose sight of the rest of us—and out of sight, out of mind. Amartya Sen (1993, p. 43) made the point that a free press can ward off famine:

Famine is entirely avoidable if the government has the incentive to act in time. It is significant that no democratic country with a relatively free press has ever experienced a major famine (although some have managed prevention more efficiently than others). This generalization applies to poor democracies as well as to rich ones. A famine may wipe out millions of people, but it rarely reaches the rulers. If leaders must seek reelection and the press is free to report starvation and to criticize policies, then the rulers have an incentive to take preemptive action.

Many people now want to protect the capability of our physical environment to nourish us. Many of them are aware of the controlling function of principles and system-concepts. Using the term "worldview," Alan Durning (1997) wrote:

> In the Pacific Northwest, as elsewhere in North America, the commonly held worldview is an old one from the frontier. It comes from the rear-view mirror, reflecting times when the world was big and the people were few... The emerging worldview, held as yet by a minority of citizens, is grounded in the reality of the present: a time when the world is small and the people are many. Through this lens, the world looks full and fragile.
>
> Worldviews are parts of culture; [they] change over time. They are influenced by what parents teach their children, by what young people learn in school, by what adults learn from peers, books, and social institutions such as churches. They are also influenced by mass media. The politics of sustainability, therefore, is about changing not only laws and habits, but also—even primarily—worldviews (p. 30).

Gary Gardner and Alan Durning are not the only ones to believe we are at a fateful turning point in the history of humankind. Some who concur, however, are more optimistic. Marvin Weisbord, the organizational consultant whom I mentioned in Chapters 11, 32, 36, and 40, thinks of the negotiation of degrees of freedom in terms of cooperation. In his 1987 book, he wrote:

> We are in the midst of an unstoppable historic shift from global competition to cooperation... The new paradigm, I think, will one day be understood as a revolutionary turning point in human history—from expert problem solving circa 1900 to everybody improving whole systems in 2000 A.D. and beyond. We have been slow to recognize how quickly this model is replacing the old one in the workplace... How people perceived autos in 1915 is, roughly speaking, the way we perceive work redesign, teamwork, and search processes today. Many people have heard of them, few have actually seen one, not everybody wants one, and those who do are still focussed on speed and cost more than a high-quality ride. As for those who own one, they find learning to operate it stimulating, irritating, threatening, miraculous, inspiring, frustrating, unpredictable, demanding, and hard to describe (pp. 368–369).

The eminent historian of warfare, John Keegan (1994), believes that war, too, is at a turning point. It is possible that humankind as a social order has a greater need of the skills of war now than ever before, if only because a few of us can do such cataclysmic damage to the rest of us. The kind of capability we need, however, is not the war of old. Keegan writes:

> Prospects of peace-making may be illusory... Yet the fact that the effort is being made betokens a profound change in civilization's attitude to war. [H]umanitarianism has not before been declared a chief principle of a great power's foreign policy, ... nor has it [until now] found an effective supra-national body to give it force, as it has recently in the United Nations, nor has it [until now] found tangible support from a wide body of disinterested states, willing to show their commitment to the principle by the despatch of peace-keeping ... forces to the seat of conflict (p. 58).
>
> I am impressed by the evidence. War, it seems to me, ... may well be ceasing to commend itself to human beings as a desirable or productive, let alone rational, means of reconciling their discontents... Throughout much of the time for which we have a record..., mankind can clearly be seen to have judged that war's benefits outweighed its costs... Now the computation works in the opposite direction. Costs clearly exceed benefits... Some of these costs are material... The human costs of actually going to war are even higher (p. 59).
>
> ... a world without armies—disciplined, obedient, and law-abiding armies—would be uninhabitable. Without their existence mankind

would have to reconcile itself... to a lawless chaos of masses warring... "all against all."... [O]ver the course of 4000 years of experiment and repetition, warmaking has become a habit... Unless we unlearn the habits we have taught ourselves, we shall not survive (pp. 384–385).

[The style in which peace-keeping warriors fight for] civilization—against ethnic bigots, regional warlords, ideological intransigents, common pillagers, and organized international criminals—cannot derive from the Western model of warmaking alone... There is a wisdom in the principles of intellectual restraint and even of symbolic ritual that needs to be rediscovered. There is even greater wisdom in the denial that politics and war belong within the same continuum (p. 392).

Still another vision of maintaining sufficient degrees of freedom for all is the justification Hugh Gibbons (1984) gives for the "deep structure" of American law. He begins with the axiom that—

The will of each person is entitled to respect. This axiom, I suggest, is the implicit core of American law (p. 175).

The process of will begins with a purpose, the formulation of an idea of a perception that is desired (p. 172).

The axiom says three very important things. First, that it is our will, not our feelings, or bodies, or activities, or well-being, that is at the heart of legal concern (pp. 175–176).

The second aspect elaborated on "respect":

It requires that we defer to the person, that we allow him that which we demand for ourselves—the right to pursue his prospects where he finds them (p. 177).

The third aspect emphasizes universality, that the law applies equally to everyone:

Relativistic comparisons of the will of one against another are irrelevant (p. 178).

Law, then, is about liberty... A person is at liberty to the extent that he is free from the sensible constraints of others.

It follows from the axiom that the interaction between people is to be characterized, to the greatest extent possible, by cooperation... We must accept [others'] autonomy, seeking out occasions on which we can cooperate (p. 179).

Those are some of our visions. In a communication to the CSGnet on 15 February 1996, Powers commented on our capabilities:

My view is that we, as a species, have developed some powerful new abilities that other species possess only in a limited way, and that we are still trying to learn how to ride this bucking bronco. One of our great mistakes was to assume that we could control other human beings in the same way we control animals, plants, and non-living objects. We can't truly comprehend social systems until we... realize that trying to control other control systems that have abilities equal to our own simply can't work the same way.

... people have a limited comprehension of the fact that their actions have consequences in addition to the ones they intend. So people tend to see their worlds through small windows where only immediate effects are considered, and less direct effects (in space and time) are ignored. This is one of the reasons we have environmental problems, and why we often progress from simple goal-seeking actions to world wars.

The human species is actually trying to correct its errors, trying to learn to use higher levels of control to its benefit instead of its detriment. If you get too pessimistic about the future of the human species, you will forget that there are many, many human beings who are trying to learn to deal with these problems, trying to teach and influence others to deal with them better. They don't necessarily go about this in the best possible way, but the intention is there, and learning always happens.

Education of almost any sort widens perceptions of what is possible. Actions become possible that were previously inconceivable. This was the ancient idea of the "liberal" education—to liberate, to free. Tyrants have seized upon schooling as a propaganda device, a device for indoctrination. But people who can read propaganda do sometimes also read the liberating books. And not even books are necessary for widening the possibilities for action. Preaching and even television can sow the seeds of revolt.

The care of the earth and human society requires cooperative work. Cooperation requires (1) the perception that the joint task will be of personal benefit and (2) trust that the cooperators will in this matter, for a sufficient period of time, put the cooperative venture higher in the control hierarchy than many other goals that might otherwise interfere with the cooperation. Since the care of the earth and its human societies requires long-term projects and commitments embracing many people, an individual must not only have a vision of the future, but must have conceptions of a good number of intermediate programs of social action to be used as subgoals. An individual usually requires, I think, some years to construct such an organization of principles and system-concepts; it does not appear in childhood. Furthermore, building the necessary trust among the cooperators almost always requires a lengthy period of joint work, successfully carried out.

The belief that one can control other people allows people to believe that they can put their welfare into the hands of someone who is especially skillful at controlling other people. Most people seem ready, even eager, to believe that a "great leader" on the throne or in the presidency can rescue them from their troubles. But no great leader, no matter how great, ever succeeded in improving the lives of the common people as well as the muddling democracies have done, not Alexander the Great, Augustus Caesar, Charlemagne, Genghis Khan, Peter the Great, Louis XIV, or Joseph Stalin. The most obvious activity of most of those great leaders was to control other people by killing them.

Those of us who seek a society with stronger norms of helping and being helped—regardless of race, creed, sex, nationality, and all the rest of the stereotypes—should not be surprised if the new pattern we seek fails to occur during our lifetimes. The campaign can be maintained only by adopting internal standards for the way-stations along the road. We must believe that the actions we are taking and the immediate norms in which we are participating are making more likely the kind of norms that will shape the society in which we would like to live and in which we hope our grandchildren will live. We must form a system-concept that connects today's intentions and next week's achievements with that goal imagined every day but never to be seen. We must maintain the perception of progress along the road despite difficulties and divagations.

In sum, we are capable of imagining the future, but few of us seem to be capable of acting as if we can imagine a future society different in kind from the present or as if a new kind is even now coming upon us. We should make opportunities for more people to think how we and our children can act in a salubrious way as our physical and social worlds keep changing. This is not a new idea, but it remains vital.

EXPERTS

Where there is division of labor, experts arise. In a complex society, everyone is an expert in something, to some degree. Perforce, we trust others to have knowledge and skill we ourselves do not. That trust is necessary, inevitable, but dangerous. Experts are tempted to pretend to knowledge of which they are ignorant. Indeed, when we beseech experts to bless us with their special knowledge, most experts are persuaded that they have that special knowledge, even when they have only the poorest evidence that they do indeed possess it. In Chapter 31, I gave evidence of that for the case of psychotherapists.

Being CEO of a large corporation or holding an academic degree from Oxford does not guarantee a proper knowledge, acumen, or sagacity. You can find witlessness and fecklessness everywhere, from the ward heeler to the president of the country, from the kindergarten child to the president of the university. That fact, however, does not mean that you should feel free to believe only what brings you pleasure to believe. You should find all the facts you can spare the time to find, and you should hold your beliefs as tentatively as you can, ready to alter them as you acquire new facts. You can protect yourself against wrong ideas, your own or the experts', only by putting that tentativeness higher in your control hierarchy than the beliefs you are entertaining.

I have described some ways that experts fall into wrong ideas in Chapters 4, 5, 15, 24, 25, 26, 29, 31, 36, and 39. In an early manuscript for this book, I used six pages to give still further examples of expert mistakes, mostly illustrating the proclivity of experts to assume that they are aware of all the relevant facts. Here I will give only two examples, and very briefly.

The gargantuan, billion-dollar Aswan dam on the Nile River in Egypt is generating electricity and providing irrigation. It also permits a great increase in bilharzia-carrying snails, and the disease has increased tenfold. Further, salts have accumulated in the soil, *reducing* fertility, and the yield of some crops is declining.

Ornstein and Ehrlich (1989, pp. 247-248) give an example from the science of chemistry:

Sir Robert Robinson, a British Nobel Laureate in chemistry [in 1971] wrote a letter to the London *Times*, claiming that there would be no problem for oceanic plankton from adding leaded compounds to the oceans. He wrote, "Neither our

'Prophets of Doom,' nor the legislators who are so easily frightened by them, are particularly fond of arithmetic" and then proceeded to do what he called some "simple arithmetic." He calculated what the dilution of lead would be in the vast volume of water in the oceans, showing that the lead would have no biological effect.

Unfortunately for Sir Robert, his arithmetic was just too "simple." Lead and other toxic substances are often not evenly diluted in the environment, because organisms have the power to concentrate them. . . result[ing] in the poisons being tens of thousands of times more concentrated in animals' bodies than they are in the animals' environment.

When I speak with dubious respect about experts, someone is sure to ask, "Do you want just anybody to take out your appendix?" No, I would prefer a practiced surgeon to do it, but not just anyone with an M.D., and not without a readiness to call the whole thing off if, at any point before the entry of the scalpel, I find myself feeling that this surgeon is treating me like a cadaver rather than as a live person with a mind of my own. I am happy to report that a recent newsletter from my health maintenance organization urges clients to confer actively with their physicians about treatments, not merely to sit passively and accept whatever the physician pronounces. I have mentioned before the organizational consultant Marvin Weisbord; concerning experts, he says (1987, p.184):

> If clear-air turbulence puts your plane in a dive, rely on the pilot to pull out of it. If peritonitis sets in, trust the surgeon to remove your appendix. And yet I find even these examples unsatisfying. The way a pilot handles emergencies influences whether people fly with that airline again. A surgical patient's mental attitude, influenced by what the surgeon says and does, may be critical to the patient's recovery. . . Over time, the process makes as much difference as technical skill.

Finally, I remind you that I am myself an expert. If you *believe* what I have been saying about experts, you will distrust what I have said. But if you distrust what I have said, you will *disbelieve* it. That is a paradox of the sort about which expert logicians ever since Zeno of Elea (495?-430? B.C.) have racked their brains. I leave you to dispose of it as you will.

FREEDOM

Humans have detailed and fairly long-term memories. They are able to remember places and arrangements of the environment in which they have in the past found ample freedom with which to control their perceptions. With those memories, they can imagine other places and environmental resources that might also bring ample freedom. Imagining those possible conditions of ample resources for controlling perceptions is what I take to be the essential component in the yearning for freedom.

As I have said earlier, we all perceive other people as rich sources of degrees of freedom—if they will only do what we want them to do. One way to increase your degrees of freedom is to wrest them from other people—to try to control the actions of others. Competition is a name for the situation in which two people or two groups are trying to increase their own degrees of freedom at the expense of another's. Cooperation is the name for an arrangement in which people can join their efforts to maintain or increase the degrees of freedom for everyone. I wrote about competition and cooperation earlier.[1]

We cannot doubt that hunters and gatherers, many thousands of years ago, were aware of the ease with which dominance of one person over another can come about in social life and aware, too, of the conflicts that usually follow. We can get a hint of how our forebears might have dealt with the threat of dominance and interpersonal conflict by looking at a modern people who live by hunting and gathering—the Bushmen of the Kalahari Desert. The Bushmen reduce the likelihood of conflict by living in small bands of about 25; nevertheless, John Pfeiffer (1977) wrote about them:

> The people know very well what they are doing: "We like to get together, but we fear fights." [There are] records of eighteen killings, and all but three of them occurred in larger-than-average groups of forty or more individuals (p. 61).

Some large corporations nowadays limit the sizes of their plants to about 200 employees. If they want to expand, they establish another plant elsewhere. Contrast that practice with the Chicago high school I discovered in 1957 containing 5000 students! For some further thoughts on governable sizes of communities, see Bryan and McClaughry (1991) and John Papworth (1991).

Against the ever-present possibility of violence, the Kalahari Bushmen also carefully follow social customs for maintaining peace:

> "When a young man kills much meat he comes to think of the rest of us as his servants or inferiors. We can't accept this. We refuse one who boasts, for some day his pride will make him kill somebody. So we always speak of his meat as worthless. This way we cool his heart and make him gentle."
>
> Hunter-gatherers live by the principle that the only way to get along together is in a society of equals. Accumulating wealth in any form is unheard of. Social pressure is so intense that an individual with a valuable object feels ill at ease and guilty for having something that others want. He gives it away with a sigh of relief, a passion that exceeds generosity. A Bushman who received a sweater from a visitor gave it to his son a few weeks later. Within a month it had passed to the brother of the son's wife, and when last seen it was being worn by a cousin in a band twenty miles away (Pfeiffer 1977, pp. 62–63).

With the advent of agriculture, towns and cities arose in which a resident would encounter many more than 25 other persons during an ordinary day. There were customs for courtesy and maintaining peace, but few of the new customs were fitted to relationships among equals. By then, I suppose, most customs regulated relationships of dominance and subservience between bosses and bossed. No longer could people hope to reduce arrogance by making disparaging remarks, as the Bushmen do, to someone who was gathering more riches than they.

If our society is to lessen conflict, it is vital that children learn that cooperation does not come about simply because some people are "by nature" cooperative and some not, but instead that cooperation is a special pattern for living and working with others that anyone can recognize, learn, and practice. Some people now know ways of introducing cooperative work into schools. William Kreidler (1989) wrote a book describing "200 activities for keeping peace in the classroom, K–6." Morton Deutsch (1993), Deutsch and Coleman (2000), Shlomo Sharan (1994), Sharan, Shachar, and Levine (1999), and Schmuck and Schmuck (2002) all wrote books about improving cooperation and reducing conflict in schools. Nancy and Theodore Graves edit the *Newsletter of the International Association for the Study of Cooperation in Education*.

SUMMING UP

It is possible, even here at the end of the book, that you might want to say something like this: "You tell me you don't want to tell me any particular thing to do. But sooner or later somebody has to *do* something." Of course. I have told you about a good many things that people have done or could do, both helpful and unhelpful. But I cannot tell you just when, under circumstances that *you* will encounter, any particular act will be helpful, unhelpful, or null. When you choose your own act for your own circumstances, you will act continuously to keep yourself going toward your goal. If, instead, you try to do what I tell you to do, your attention will be on doing the "right" thing in the "right" way; you will wait too long before correcting your course of action. You will realize too late that you are, as always, on your own. You should act on your own inventiveness. If you can use what I have said here to stimulate your own inventiveness, I will be proud.

Actions to increase the available degrees of freedom must occur in families, workplaces, schoolrooms, and neighborhoods. But those intimate actions must be permitted and encouraged by actions in boardrooms, city councils, and legislatures. And what is right for Peoria today is not necessarily right for Pasadena, Berlin, or Nairobi tomorrow. I cannot guess here what would be the right actions in any of those places. Persons to consult are the more forward-looking of the local and regional politicians and the more skilled among neighborhood organizers. Names of those people can be found easily in the alternative press.

Here I will offer three statements to comprise the policy, the habit of thought, the principles I think will be useful for the next hundred years or so. I phrase these in the imperative mood to suggest action, but I do not mean for you to rush out tomorrow morning and *do* something. Rather, I offer these statements as principles you may wish to adopt as criteria for any action you may contemplate. They are not prescriptions for particular actions; they are criteria for selecting actions to serve whatever purposes are pressing.

1 *Degrees of freedom:* Try to arrange environments, physical and social, so that everyone can achieve sufficient degrees of freedom. Do this everyplace: in your kitchen, on the street, at your place of work, in the halls of Congress.

2. *The future:* Whatever other internal standards are guiding your action, try to make use of some that concern the welfare of your grandchildren.
3. *Reality:* Try to keep in touch with scientific findings about the physical and social worlds. Remember modeling. Remember PCT. Remember the Method of Levels. Remember, too, that experts are helpful and dangerous at the same time.

I wish you well.

ENDNOTE

[1] I treated cooperation and competition in Chapter 28 under "Cooperation and Competition," and in Chapter 37 under "Good Schools." I treated the matter of controlling others in Chapter 5 under "Controlling Others," in Chapter 28 under "Interpersonal Conflict," in Chapter 29 under "Social Life, Conflict, and Deviant Behavior," in Chapter 33 under "Controlling Others," in Chapter 35 under "Leadership Styles," in Chapter 36 under "Consulting," and in Chapters 28 and 39 under "Countercontrol." I told how to *avoid* controlling others in Chapter 27 under "Opportunities" and in all of Chapter 30.

References

Abelson, Robert P., Elliot Aronson, William J. McGuire, Theodore M. Newcomb, Milton J. Rosenberg, and Percy H. Tannenbaum (Eds.) (1968). *Theories of cognitive consistency: A sourcebook*. Chicago: Rand McNally.

Aiken, Wilford (1942). *Story of the eight-year study*. New York: Harper.

Albee, George W. (1992). Genes don't hurt people: People hurt people. Review of *Social causes of psychological distress* by John Mirowsky and Catherine E. Ross. *Contemporary Psychology*, 1992, 37(1), 16–17.

Alderfer, Clayton P. (1972). *Existence, relatedness, and growth*. New York: Free Press.

Allport, Gordon W. and H. S. Odbert (1936). Trait-names: A psycho-lexical study. *Psychological Monographs*, 47)1, Whole No. 211).

American Heritage Dictionary, 2nd College Edition. Boston: Houghton Mifflin.

American Psychiatric Association (1994). *Diagnostic and statistical manual of mental disorders*. Washington, D.C.: Author.

American Psychological Foundation (1993). APF gold medal awards and Distinguished Teaching of Psychology award. *American Psychologist*, 48(7), 717–725.

Anfuso, Dawn (1999). Core values shape W.L. Gore's innovative culture. *Workforce*, March, 48–53.

Aoki, Ted T. (1984). Interests, knowledge, and evaluation: Alternative curriculum evaluation orientations. In T.T. Aoki (Ed.), *Curriculum evaluation in a new key*. Edmonton, Alberta, Canada: Department of Secondary Education, Faculty of Education, University of Alberta.

Appley, Mortimer H. (1991). Motivation, equilibration, and stress. Pages 1–66 in Richard A. Dienstbier (Ed.) *Perspectives on motivation. Nebraska symposium on motivation, 1990*. Lincoln: University of Nebraska Press.

Arnold, Matthew (1970). *Dover beach*. Columbus OH: Merrill.

Arrow, Holly, Joseph E. McGrath, and Jennifer L. Berdahl (2000). *Small groups as complex systems: Formation, coordination, development, and adaptation*. Thousand Oaks CA: Sage.

Asch, Solomon E. (1940). Studies in the principles of judgments and attitudes: II. Determination of judgments by group and ego-standards. *Journal of Social Psychology*, 12, 433–465.

Asch, Solomon E. (1951). Effects of group pressure upon the modification and distortion of judgment. In H. Guetzkow (Ed.), *Groups, leadership, and men*. Pittsburgh: Carnegie Press. Also in Dorwin Cartwright and Alvin Zander (Eds.), *Group dynamics: Research and theory*. Evanston, IL: Row, Peterson.

Asch, Solomon E., H. Block, and M. Mertzman (1938). Studies in the principles of judgments and attitudes: I. Two basic principles of judgment. *Journal of Psychology*, 5, 219–251.

Ashby, W.R. (1952). *Design for a brain*. New York: Wiley.

Atchison, Thomas J. (1991). The employment relationship: Un-tied or re-tied? *Academy of Management Executive*, 5(4),52–62.

Axelrod, Robert (1984). *The evolution of cooperation*. New York: Basic Books.

Banker, Rajiv D., Joy M. Field, Roger G. Schroeder, and Kingshuk K. Sinha (1996). Impact of work teams on manufacturing performance: A longitudinal field study. *Academy of Management Journal*, 39(4), 867–890.

Barker, Roger G., Paul V. Gump, and others (1964). *Big school, small school: High school size and student behavior*. Stanford CA: Stanford University Press.

Barrow, John D. (1992). *Pi in the sky*. Oxford: Clarendon.

Barry, David (1991). Managing the bossless team: Lessons in distributed leadership. *Organizational Dynamics*, 20(1), 31–47.

Bartlett, F. C. (1932). *Remembering*. Cambridge UK: Cambridge University Press.

Beckmann, Petr (1971). *A history of Pi*. New York: Dorset.

Bentzen, Mary M. and associates (1974). *Changing schools: The magic feather principle*. New York: McGraw-Hill.

Berkowitz, Leonard and P.G. Devine (1989). Research traditions, analysis, and synthesis in social psychological theories: The case of dissonance theory. *Personality and Social Psychology Bulletin*, 15(4), 493–507.

Berman, Jeffrey S. and Nicholas C. Norton (1985). Does professional training make a therapist more effective? *Psychological Bulletin*, 98, 401–407.

Berman, Paul, Milbrey McLaughlin, and others (Vols. 1–5 in 1975, Vols. 6–8 in 1977). *Federal programs supporting educational change*. Santa Monica CA: Rand Corporation.

Bernard, Claude (1859). *Lecons sur le proprietes physiologiques et le alterations pathologiques des liquides del'organisme* (2 vols.). Paris: Balliere.

Berscheid, Ellen (1983). Emotion. Chapter 4 (pp. 110–168) in Harold H. Kelley, Ellen Berscheid, Andrew Christensen, John H. Harvey, Ted L. Huston, George Levinger, Evie McClintock, Letitia Anne Peplau, and Donald R. Peterson. *Close relationships*. New York: W.H. Freeman.

Bird, Caroline (1975). College is a waste of time and money. *Psychology Today*, 8(12), 29–31, 33, 35, 78, 81.

Black, H. S. (1934). Stabilized feedback amplifiers. *Bell System Technical Journal*, 13, 1–18.

Blake, Robert R. and Jane S. Mouton (1957). The dynamics of influence and coercion. *International Journal of Social Psychiatry*, 2, 263–305.

Blake, Robert R. and Jane S. Mouton (1985). Effective crisis management. *New Management*, 3(1), 14–17.

Blase, Joseph J. (1986). A qualitative analysis of sources of teacher stress: Consequences for performance. *American Educational Research Journal*, 23(1), 13–40.

Bohl, Don L., Fred Luthans, John W. Slocum, Jr., and Richard M. Hodgetts (1996). Ideas that will shape the future of management practice. *Organizational Dynamics*, 25(1), 7–14.

Bolman, Lee G. and Terrence E. Deal (1984). *Modern approaches to understanding and managing organizations*. San Francisco: Jossey-Bass.

Bolman, Lee G. and Terrence E. Deal (1997). *Reframing organizations: Artistry, choice, and leadership* (2nd ed.). San Francisco: Jossey-Bass.

Boulding, Kenneth E. (1978). *Ecodynamics: A new theory of societal evolution*. Thousand Oaks CA: Sage.

Boulding, Kenneth E. (1990). *Three faces of power*. Newbury Park CA: Sage.

Bourbon, W. Thomas (1989). A control-theory analysis of interference during social tracking. Chapter 10 in W.A. Hershberger (Ed.) *Volitional action: Conation and control*. Amsterdam: North-Holland.

Bourbon, W. Thomas (1990). Invitation to the dance: Explaining the variance when control systems interact. *American Behavioral Scientist*, 34(1), 95–105.

Bourbon, W. Thomas (1994). Program-level selection of a sequence of perceived relationships: Perceptual control theory in the cognitive domain. Pages 1–11 in Marcos A. Rodrigues and Mark H. Lee (Eds.), *Perceptual control theory*. Aberystwyth UK: University of Wales.

Bourbon, W. Thomas (1995). Perceptual control theory. Pp. 151–172 in H. L. Roitblat and J-A. Meyer (Eds.), *Comparative approaches to cognitive science*. Cambridge: MIT Press.

Bourbon, W. Thomas (1996). On the accuracy and reliability of predictions by perceptual control theory: Five years later. *Psychological Record*, 46, 39–47.

Bourbon, W. Thomas (1998). An application of PCT: The responsible thinking process. Pages 153–166 in William T. Powers, *Making sense of behavior: The meaning of control*. New Canaan CT: Benchmark.

Bourbon, W. Thomas (1999). Foreword. Pages xi-xii in Edward E. Ford, *Discipline for home and school, Book two*, (rev. ed). Scottsdale AZ: Brandt.

Bourbon, W. Thomas (1999a). Perceptual control theory, reality therapy, and the responsible thinking process. Chapter 19 in Edward E. Ford, *Discipline for home and school, Book two*, (rev. ed). Scottsdale AZ: Brandt.

Bourbon, W. Thomas and Ed Ford (1999). What do we look for when we evaluate schools? Chapter 11 in Edward E. Ford, *Discipline for home and school, Book two*, (rev. ed). Scottsdale AZ: Brandt.

Bourbon, W. Thomas and William T. Powers (1993). Models and their worlds. *Closed Loop*, 3(1), 47–72.

Bourbon, W. Thomas and William T. Powers (1999). Models and their worlds. *International Journal of Human-Computer Studies*, 50, 445–461. (A later version of the 1993 paper.)

Bourbon, W. Thomas, Kimberly E. Copeland, Vick R. Dyer, Wade K. Harman, and Barbara L. Mosley (1990). On the accuracy and reliability of predictions by control-system theory. *Perceptual and Motor Skills*, 71, 1331–1338.

Bowen, David E., Gerald E. Ledford, Jr., and Barry R. Nathan (1991). Hiring for the organization, not the job. *Academy of Management Executive*, 5(4), 35–50.

Bower, Bruce (1985). Different strokes. *Science News*, 128(19), 301–302.

Bower, Bruce (1994). Growing up poor. *Science News*, 146(2), 24–25.

Bower, Bruce (1997). New schizophrenia therapy shows promise. *Science News*, 152(19), 293.

Bower, Bruce (1998). Incriminating developments. *Science News*, 154(10), 153–155.

Bower, Bruce (2001). Back from the brink: Psychological treatments for schizophrenia attract renewed interest. *Science News*, 159(17), 268–269.

Bowles, Samuel and Herbert Gintis (1976). *Schooling in capitalist America*. New York: Basic Books.

Brehm, Sharon S. and Jack W. Brehm (1981). *Psychological reactance: A theory of freedom and control*. New York: Academic Press.

Bronfenbrenner, Urie (1958). Socialization and social class through time and space. Pp. 400–424 of E. E. Maccoby, T. M. Newcomb, and E. L. Hartley (Eds.), *Readings in social psychology*. New York: Holt, Rinehart, and Winston. Condensed on pp. 349–365 of H. Proshansky and B. Seidenberg (Eds.), *Basic studies in social psychology*, New York: Holt, Rinehart, and Winston.

Brown, J. and Ellen Langer (1990). Mindfulness and intelligence: A comparison. *Educational Psychologist*, 25(3 & 4), 305–335.

Brown, Roger (1973). Schizophrenia, language and reality. *American Psychologist*, 28, 395–403.

Brown, Walter A. (1998). The placebo effect. *Scientific American*, 278(1), 90–95.

Bryan, Frank and John McClaughry (1991). From Vermont, a radical blueprint to reinvigorate American democracy. *Utne Reader*, January-February, 50–57.

Bunker, Barbara Benedict and Billie T. Alban (Eds.) (1992). Special issue ("Large group interventions") of *Journal of Applied Behavioral Science*, 28(4).

Bunker, Barbara Benedict, Jeffrey Z. Rubin, and associates (Eds.) (1995). *Conflict, cooperation, and justice: Essays inspired by the work of Morton Deutsch*. San Francisco CA: Jossey-Bass.

Caine, Renate Nummela and Geoffrey Caine (1997). *Unleashing the power of perceptual change: The potential of brain-based teaching*. Alexandria VA: Association for Supervision and Curriculum Development.

Cannon, Walter B. (1927). The James-Lange theory of emotions: A critical examination and an alternative theory. *American Journal of Psychology*, 39, 106–124.

Cannon, Walter B. (1929). *Bodily changes in pain, hunger, fear, and rage*. New York: Appleton.

Cannon, Walter B. (1932). *The wisdom of the body*. New York: W.W. Norton.

Carey, Thomas A. (1996). Review of *Business and management in Russia* by Sheila M. Puffer and associates. Brookfield VT: Edward Elgar. In *Academy of Management Executive*, volume and issue number missing, pages 111–112.

Carey, Timothy A. (1999). What makes a psychotherapist effective? *Psychotherapy in Australia*, 5(3), 52–59.

Carey Timothy A. (2000). *Psychotherapy: A story about what's wrong and how to make it better*. Unpublished manuscript.

Carey, Timothy A. and Margaret Carey (2001). *RTP intervention processes: A new approach for working with students whose successful participation in educational activities is chronically reduced*. Coorparoo, Queensland, Australia: Andrew Thomson.

Carroll, Lewis (1936). Introduction to "The Hunting of the Snark" in *Complete Works of Lewis Carroll*. New York: Modern Library.

Carroll, Lewis (1982). *Lewis Carroll's Alice's adventures in wonderland*. Berkeley: University of California Press. Preface and notes by James R. Kincaid, illustrated by Barry Moser. Originally published in 1865.

Chapanis A. (1961). Men, machines, and models. *American Psychologist*, 16, 113–131.

Chomsky, Noam (1969). *Deep structure, surface structure, and semantic interpretation.* Bloomington IN: Indiana University Linguistics Club.

Chomsky, Noam (1986). *Knowledge of language: Its nature, origin, and use.* New York: Praeger.

Chomsky, Noam (1995). *The minimalist program.* Cambridge MA: MIT Press.

Cloud, John and Jodie Morse (2001). Home sweet school. *TIME*, 158(8), 46–54.

Cofer, C.N. and M.H. Appley (1964). *Motivation: Theory and research.* New York: Wiley.

Cohen, David (1985). Japanese face up to stress on the job. *APA Monitor*, 16(12), 8.

Cole, Robert E. (1989). *Strategies for learning: Small-group activities in American, Japanese, and Swedish industry.* Berkeley CA: University of California Press.

Cole, Wendy (2000). The (Un)Therapists. *TIME, 156(16),* 95.

Collins, Linda N., J.W. Graham, W. B. Hansen, and C. A. Johnson (1985). Agreement between retrospective accounts of substance use and earlier reported substance use. *Applied Psychological Measurement*, 9, 301–309.

Combs, Arthur W. (1979). *Myths in education.* Boston: Allyn and Bacon.

Conant, James Bryant (1956). *The overthrow of the phlogiston theory.* Cambridge: Harvard University Press.

Conger, Jay A., Gretchen M. Spreitzer, and Edward E. Lawler III (Eds.) (1999). *The leader's change handbook: An essential guide to setting direction and taking action.* San Francisco CA: Jossey-Bass.

Cooper, J. and R.H. Fazio (1984). A new look at dissonance theory. In L. Berkowitz (Ed.), *Advances in experimental social psychology* (vol. 17). Orlando PL: Academic Press.

Corcoran, Elizabeth (1993a). Learning companies. *Scientific American*, 268(2), 105–108.

Coruzzi, Celeste A. (1987). Review of *Moments of truth* by Jan Carlzon. *Academy of Management Executive*, 1(3),256–257.

Costa, Paul T., Jr. and Robert R. McCrae (1988). Personality in adulthood: A six-year longitudinal study of self-reports and spouse ratings on the NEO personality inventory. *Journal of Personality and Social Psychology*, 54, 853–863.

Council for a Livable World Education Fund (1998). *Caution: Military-industrial complex at work.* Washington DC: Author.

Croall, Jonathan (1983). *All the best, Neill: Letters from Summerhill.* London: A. Deutsch.

Croall, Jonathan (ed.) (1983a). *Neill of Summerhill: The permanent rebel.* New York: Pantheon.

Cronbach, Lee J. (1975). Beyond the two disciplines of scientific psychology. *American Psychologist*, 30, 116–127.

Cronin, Mary (1992). Gilded cages. *TIME*, 139(21), 52–54.

Cross, Arthur Lyon (1914). *A history of England and Greater Britain.* New York: Macmillan.

Culbert, Samuel A. and John J. McDonough (1980). *The invisible war: Pursuing self-interests at work.* New York: Wiley.

Cziko, Gary (2000). *The things we do: Using the lessons of Bernard and Darwin to understand the what, how, and why of our behavior.* Cambridge MA: MIT Press.

Danziger, Kurt (1990). *Constructing the subject: Historical origins of psychological research.* Cambridge, UK: Cambridge University Press.

Davis, Stanley M. (1987). *Future perfect.* Reading MA: Addison-Wesley.

Dawes, Robyn M. (1988). *Rational choice in an uncertain world.* San Diego CA: Harcourt Brace Jovanovich.

Dawes, Robyn M. (1994). *House of cards: Psychology and psychotherapy built on myth.* New York: Free Press.

Dawkins, Richard (1982). *The extended phenotype.* Oxford: W.H. Freeman.

Dawkins, Richard (1987). *The blind watchmaker.* New York: W.W. Norton.

Dember, William N. (1974). Motivation and the cognitive revolution. *American Psychologist*, 29(3), 161–168.

Dember, William N. and Joel S. Warm (1979). *Psychology of perception* (2nd ed.). New York: Holt, Rinehart, and Winston.

Dement, William C., P. Henry, H. Cohen, and J. Ferguson (1969). Studies on the effect of REM deprivation in humans and animals. In K. H. Pribram (Ed.), *Mood, states, and mind.* Baltimore: Penguin.

Derber, Charles (1979). *Pursuit of attention: Power and individualism in everyday life.* Boston: G. K. Hall. Reprinted in paperback in 1983 by Oxford University Press, New York.

Deutsch, Morton (1993). Educating for a peaceful world. *American Psychologist*, 48(5), 510–517.

Deutsch, Morton and Peter T. Coleman (Eds.) (2000). *Handbook of conflict resolution: Theory and practice.* San Francisco: Jossey-Bass.

Dobzhansky, Theodosius, Francisco J. Ayala, G. Ledyard Stebbins, and James W. Valentine (1977). *Evolution.* San Francisco: W.H. Freeman.

Dowling, William F. (1978). Conversation: An interview with Fletcher Byrom. *Organizational Dynamics*, 7(1), 37–60.

Doyle, Roy P. (1989). The resistance of conventional wisdom to research evidence: The case of retention in grade. Phi *Delta Kappan*, 71(3), 215–220.

Drake, Stillman (1974). *Galileo Galilei: Two new sciences.* Madison WI: University of Wisconsin Press.

Drummond, Hugh (1987). And they make house calls. *Mother Jones*, 12(4), 16, 18.

Drummond, Robert (1972). *Preliminary report: Research and evaluation team.* University of Maine, Orono Achievement Testing Program for MSAD No. 3, mimeographed.

Dryden, Windy and Colin Feltham (1992). *Psychotherapy and its discontents.* Buckingham, England: Open University Press.

Durkheim, Emile (1897). *Suicide.* Published in English by Free Press of Glencoe, Illinois in 1951.

Durning, Alan Thein (1997). After the deluge: The changing world view. *World Watch*, 10(1), January-February, 25–31.

Dye, Thomas R. and Harmon Zeigler (1990). *The irony of democracy* (8th ed.). Monterey CA: Brooks-Cole.

Dyson, Freeman J. (1958). Innovation in physics. *Scientific American*, 199(3), 74–82.

Easterbrook, J.A. (1959). The effect of emotion on cue utilization and the organization of behavior. *Psychological Review*, 66, 183–201.

Editors of Time-Life Books (1989). *Powers of the Crown: Time Frame 1600–1700.* Alexandria VA: Time-Life Books.

Edney, J.J., C.A. Walker, and N.L. Jordan (1976). Is there reactance in personal space? *Journal of Social Psychology*, 100, 207–217.

Educated IQ. *Science News*, 1991, 140(12), 187.

Educational seduction. *Behavior Today*, 1973, 4(30), 2.

Edwards, Paul (Ed.-in-Ch.) (1967). *Encyclopedia of philosophy.* New York: Macmillan.

Ehrlich, Paul R. and Anne E. Ehrlich (1990). *The population explosion.* New York: Simon and Schuster.

Eisenberg, Daniel (2002). The coming job boom. *TIME*, 159(18), 40–44.

Eisner, Donald A. (2000). *The death of psychotherapy: From Freud to alien abductions.* Westport CT: Praeger.

Eldredge, Niles (1998). *Life in the balance: Humanity and the biodiversity crisis.* Princeton NJ: Princeton University Press.

Emrick, John A. and Susan M. Peterson (1977). *A synthesis of findings across five recent studies of educational dissemination and change.* Menlo Park CA: Stanford Research Institute.

Emrick, John A. with Susan M. Peterson and Rekha Agarwala-Rogers (1977). *Evaluation of the National Diffusion Network, Vol. I Findings and Recommendations.* Menlo Park CA: Stanford Research Institute.

Everett, Melissa (1989). *Breaking ranks.* Philadelphia: New Society Publishers.

Ewens, William L. (1984). *Becoming free: The struggle for human development.* Wilmington DE: Scholarly Resources.

Eysenck, Michael W. (1994). *Perspectives on psychology.* Hillsdale NJ: Lawrence Erlbaum.

Farson, Richard (1974). *Birthrights.* New York: Penguin.

Far West Laboratory for Educational Research and Development (1975). *Educational programs that work.* San Francisco CA: Author.

Faust, D., K. Hart, T.J. Guilmette, and H. R. Arkes (1988). Neuropsychologists' capacity to detect adolescent malingerers. *Professional Psychology: Research and Practice*, 19, 508–515.

Festinger, Leon (1957). *A theory of cognitive dissonance.* Evanston IL: Row, Peterson.

Festinger, Leon (1983). *The human legacy.* New York: Columbia University Press.

Festinger, Leon, Henry W. Riecken, and Stanley Schacter (1956). *When prophecy fails.* Minneapolis: University of Minnesota Press. Reprinted by Harper and Row, New York, in 1964.

Fischer, Kurt W. and Rose (1994). Chapter 1 (pp. 3–66) in G. Dawson and K. W. Fischer (Eds.), *Human behavior and the developing brain.* New York: Guilford.

Foa, Uriel G. and Edna B. Foa (1974). *Societal structures of the mind.* Springfield IL: Charles C. Thomas.

Ford, Edward E. (1997). *Discipline for home and school, Book one* (Rev. ed.). Scottsdale AZ: Brandt Publishing.

Ford, Edward E. (1999). *Discipline for home and school, Book two* (Rev. ed.). Scottsdale AZ: Brandt Publishing.

Ford Foundation (1987). *And gladly teach.* New York: Author.

Forssell, Dag C. (2000). *Management and Leadership: Insight for Effective Practice.* Hayward CA: Living Control Systems Publishing.

Forward, Gordon E., Dennis E. Beach, David A. Gray, and James Campbell Quick (1991). Mentofacturing: A vision for American industrial excellence. *Academy of Management Executive,* 5(3), 32–44.

Fosmire, Frederick and Carolin Keutzer (1968). Task-directed learning: A systems approach to marital therapy. Paper presented to the 1968 meeting of the Oregon Psychological Association and the Western States Psychological Association.

Francisco, Richard (1979). The Documentation and Technical Assistance Project in urban schools. *Theory Into Practice,* 18(4), 89–96.

Freed, A. M., P. J. Chandler, Jane S. Mouton, and R. R. Blake (1955). Stimulus and background factors in sign violation. *Journal of Personality,* 23, 499. (Abstract.)

Freire, Paulo (1973). *Education for critical consciousness.* New York: Seabury Press.

Freud, Sigmund (1915). *Instincts and their vicissitudes.* London: Hogarth.

Fricke, Benno G. (1956). *Prediction, selection, mortality, and quality control.* College and University, 32, 34–52.

Friedman, Miles I. and George H. Lackey, Jr. (1991). *Psychology of human control: A general theory of purposeful behavior.* New York: Praeger.

Furnham, Adrian (1996). *All in the mind: The essence of psychology.* London: Whurr.

Galambos, Robert (1961). Changing concepts of the learning process. In J.F. Delefresnaye (Ed.), *Brain mechanisms in learning.* Springfield IL: Charles C. Thomas.

Galbraith Jay R. and Edward E. Lawler (1993). Interview by Tom Brown. Advertisement from Jossey-Bass, Publishers.

Galton, Francis (1884). Measurement of character. *Fortnightly Review,* 36, 179–185.

Gardner, Gary (2001). Accelerating the shift to sustainability. Chapter 10 in Lester R. Brown and others. *State of the world, 2001.* New York: W.W. Norton.

Gardner, Martin (1981). Introduction. In James Tanis and John Dooley (Eds.), *Lewis Carroll's "The hunting of the snark".* Los Altos CA: William Kaufman.

Gatto, John Taylor (1990). Teacher of the year calls for "ferocious debate on education aims and methods." *News from New Society Publishers,* Santa Cruz CA, pp. a, b, c. For a somewhat abridged but more accessible version, see Gatto, John (1990). An award-winning teacher speaks out. *Utne Reader,* No. 41 (September-October), pp. 73–76. Reprinted from *The Sun,* (Chapel Hill NC), June 1990.

Gibbons, Hugh (1984). Justifying law: An explanation of the deep structure of American law. *Law and Philosophy,* 3,165–279.

Gibbs, Jack P. (1989). *Control: Sociology's central notion.* Urbana IL: University of Illinois Press.

Gilligan, James G. (1996). *Violence: Our deadly epidemic and its causes.* New York: Putnam

Glazer, Edward M. and Samuel H. Taylor (1973). Factors influencing the success of applied research. *American Psychologist,* 28(2), 139–146.

Glazer, Myron Peretz and Penina Migdal Glazer (1986). Whistleblowing. *Psychology Today,* 20(8), 37–39, 42–43.

Goldberg, Lewis R. (1993). The structure of phenotypic personality traits. *American Psychologist,* 48(1), 26–34.

Goldman, Emma (1931). *Living my life* (vol. 1). New York: Knopf. Reissued in 1970 by Dover.

Goldman, Paul, Diane M. Dunlap, and David T. Conley (1993). Facilitative power and nonstandardized solutions to school site restructuring. *Educational Administration Quarterly,* 29(1),69–92.

Goldstein, Andrew (2000). No SAT scores required. *TIME,* 156(11), 52–53.

Goldstein, Andrew (2001). The upgrader. *TIME,* 158(11), 84.

Gonzales, Susan (1991). Letter to the editor of *Utne Reader,* January-February, page 12.

Goodlad, John I. (1975). *The dynamics of educational change: Toward responsive schools.* New York: McGraw-Hill.

Gottfredson, Gary D. (1981). Schooling and delinquency. In S.E. Martin, L.B. Sechrest, and R. Redner (Eds.). *New directions in the rehabilitation of criminal offenders*. Washington DC: National Academy Press. Pp. 424–469.

Gray, Peter (1991). *Psychology*. New York: Worth.

Green, Mark (1987). Amiable dunce or chronic liar? *Mother Jones*, 12(5), 9–12, 14, 16–17.

Green, Mark and Gail MacColl with Robert Nelson and Christopher Power (1983). *There he goes again: Ronald Reagan's reign of error*. New York: Pantheon. (Revised edition in 1987.)

Greenhouse, Steven (2001). Americans spending more time on the job. *The Register-Guard* (Eugene, Oregon), 1 September 2001, pp. 1A, 9A.

Gregory, Bruce (1988). *Inventing reality: Physics as language*. New York: Wiley.

Gregory, Richard L. (1988). *Odd perceptions*. London: Routledge.

Gwynne, S. C. (1990). The right stuff. *TIME*, 136(18), 74–78, 81, 84.

Gysin, Catherine (1993). Big man (& woman) off campus. *Utne Reader*, No. 56, 18–19.

Haasen, Adolf (1996). Opel Eisenach GMBH – Creating a high-productivity workplace. *Organizational Dynamics*, 24(4), 80–85.

Hackman, J. Richard (1985). Doing research that makes a difference. In E. E. Lawler III and associates, *Doing research that is useful for theory and practice*. San Francisco: Jossey-Bass.

Hackman, J. Richard (1987). The design of work teams. In J.W. Lorsch (Ed.), H*andbook of organizational behavior*. Englewood Cliffs NJ: Prentice-Hall. Pp. 315–342.

Hall, A. D. and R. E. Fagen (1956). Definition of system. Pp. 18–28 in *General Systems I*. Reprinted on pp. 81–92 of W. Buckley (Ed.), *General systems research for the behavioral scientist*. Chicago: Aldine.

Hammond, George S. and W. Murray Todd (1975). Technical assistance and foreign policy. *Science*, 189(4208), 1507–1509.

Hamper, Ben (1986). I, rivethead. *Mother Jones*, 11(6), 27–32, 55.

Harrison, Roger (1978). How to design and conduct self-directed learning experiences. *Group and Organizational Studies*, 3(2), 149–167.

Harrison, Roger (1983). Strategies for a new age. *Human Resource Management*, 22(3), 209–235. Another version appears as "Leadership and strategy for a new age: Lessons from 'conscious evolution,'" chapter 7 in J. D. Adams (Ed.), *Transforming work*. Alexandria VA: Miles River Press.

Harrison, Roger and James Kouzes (1980). The power of organization development. *Training and Development Journal*, 34(4), 44–47.

Harrison, Roger and Herb Stokes (1992). *Diagnosing organizational culture*. San Diego: Pfeiffer and Company.

Hart, Leslie A. (1975). *How the brain works*. New York: Basic Books.

Hart, Leslie A. (1989). *Human brain and human learning*. Oak Creek AZ: Books for Educators.

Hart, Leslie A. (1989a). The horse is dead. *Phi Delta Kappan*, 71(3), 237–242.

Harvey, Jerry B. (1988). The Abilene paradox: The management of agreement. *Organizational Dynamics*, 17(1), 17–43.

Hasek, Glenn (2000). The right chemistry. *Industry Week*, March 6, 36–39.

Hazen, R.L. (1934). Theory of servomechanisms. *Journal of the Franklin Institute*, 218(Sept.), 279–331.

Heider, Fritz (1946). Attitudes and cognitive organization. *Journal of Psychology*, 21, 107–112.

Heider, Fritz (1958). *The psychology of interpersonal relations*. New York: Wiley.

Heller, Kenneth (1993). A festschrift for psychology's most persistent social critic and prevention advocate. Review of *The present and future of prevention* by Marc Kessler, Stephen E. Goldston, and Justin M. Joffe (Eds.) *Contemporary Psychology*, 38(8), 784–786.

Heller, Kenneth (1996). Coming of age of prevention science. *American Psychologist*, 51(11), 1123–1127.

Helson, R and P. Wink (1992). Personality change in women from early 40s to early 50s. *Psychology and Aging*, 7, 46–55.

Hemmings, Ray (1973). *Children's freedom: A. S. Neill and the evolution of the Summerhill idea*. New York: Schocken.

Hershberger, Wayne A. (1989). *Volitional action: Conation and control*. Amsterdam: North-Holland.

Herzberg, Frederick (1968). One more time: How do you motivate employees? *Harvard Business Review*, 46(1), 53–62.

Hilgard, Ernest R., Richard C. Atkinson, and Rita L. Atkinson (1975). *Introduction to psychology* (6th ed.) New York: Harcourt Brace Jovanovich.

Hill, K.T. and S.B. Sarason (1966). The relation of test anxiety and defensiveness to test and school performance over the elementary years. *Monographs of the Society for Research in Child Development*, 31 (Serial No. 104).

Hirschman, Albert O. (1970). *Exit, voice, and loyalty.* Cambridge MA: Harvard University Press.

Hoffman, Mark S. (Ed.) (1993). *World almanac and book of facts.* New York: World Almanac.

Hofstede, Geert (1993). Cultural constraints in management theories. *Academy of Management Executive*, 7(1), 81–94.

Hogan, Robert T., Joyce Hogan, and Brent W. Roberts (1996). Personality measurement and employment decisions. *American Psychologist*, 51(5), 469–477.

Hollin, Clive R. (ed.) (1995). *Contemporary psychology: An introduction.* London: Taylor and Francis.

Holmes, Michael (of the Associated Press) (1997). Tapes reveal Johnson's torment over Vietnam. *The Register Guard*, February 15, pp. 1A, 5A.

Horgan John (1995). Get smart, take a test. *Scientific American*, 273(5), 12, 14.

Houts, Paul L. (ed.). (1977). *The myth of measurability.* New York: Hart.

Huffman, Karen, Mark Vernoy, and Judith Vernoy (1994). *Psychology in action* (3rd ed.). New York: Wiley.

Human Interaction Research Institute (1976). *Putting knowledge to use: A distillation of the literature regarding knowledge transfer and change.* Los Angeles: Author.

Hummel, Ralph P. (1987). Behind quality management: What workers and a few philosophers have always known and how it adds up to excellence in production. *Organizational Dynamics*, 16(1), 71–78.

Illich, Ivan (1971). *Deschooling society.* New York: Harper and Row.

Illich, Ivan (1973). Outwitting the "developed" countries. Pages 123–131 in R. Theobald and S. Mills (Eds.), *The failure of success.* Indianapolis IN: Bobbs-Merrill.

Isaacs, William N. (1993). Taking flight. *Organizational Dynamics*, 22(2), 24–39.

James, William (1884). What is emotion? *Mind*, 9, 188–204.

James, William (1890). *Principles of psychology.* (2 vols.) New York: Henry Holt.

Jamieson, David and Julie O'Mara (1991). *Managing workforce 2000.* New York: McGraw-Hill.

Jennings, Wayne and Joe Nathan (1977). Startling, disturbing research on school program effectiveness. *Phi Delta Kappan*, 58(7), 568–572.

John, Oliver P. and Sanjay Srivastava (1999). The big five trait taxonomy: History, measurement, and theoretical perspectives. In *Handbook of personality: Theory and Research*, 2nd edition, edited by Lawrence A. Pervin and Oliver P. John. New York: Guilford.

Johnson, David W. and Frank P. Johnson (1987). *Joining together: Group theory and group skills* (3rd ed.). Englewood Cliffs NJ: Prentice-Hall.

Johnson, David W. and Roger T. Johnson (1987). *Learning together and alone: Cooperative, competitive, and individualistic learning.* Englewood Cliffs NJ: Prentice-Hall.

Johnson, David W. and Roger T. Johnson (1989). *Cooperation and competition: Theory and research.* Edina MN: Interaction Book Company.

Johnson, David W. and Roger T. Johnson (1995). *Reducing school violence through conflict resolution.* Alexandria VA: Association for Supervision and Curriculum Development.

Johnson, David W., Roger T. Johnson, and Edythe J. Holubec (1994). *The new circles of learning: Cooperation in the classroom and school.* Alexandria VA: Association for Supervision and Curriculum Development.

Johnson, David W., Roger T. Johnson, and Edythe J. Holubec (1994a). *Cooperative learning in the classroom.* Alexandria VA: Association for Supervision and Curriculum Development.

Judd, Joel B. (1992). *Second language acquisition as the control of non-primary linguistic perception: A critique of research and theory.* Doctoral dissertation, University of Illinois at Urbana-Champaign.

Kagan, Jerome and Julius Segal (1995). *Psychology: An introduction* (8th ed.). Forth Worth TX: Harcourt Press.

Kamii, C. and R. D. DeVries (1978). *Physical knowledge in preschool education: Implications of Piaget's theory.* Englewood Cliffs NJ: Prentice-Hall.

Kaplan, Michael (1997). You have no boss. *Fast Co.*, October-November, 226–227.

Karasek, Robert A. (1979). Job demands, job decision latitude, and mental strain: Implications for job redesign. *Administrative Science Quarterly*, 24(2), 285–308.

Katz, Stuart, Gary J. Lautenschlager, A. Boyd Blackburn, and Felicia H. Harris (1990). Answering reading comprehension items without passages on the SAT. *Psychological Science*, 1(2), 122–127. Published by Blackwell Publishing, Oxford, UK.

Keegan, John (1994). *A history of warfare.* New York: Knopf.

Kennaway, Richard (1997). Correlations cannot be used for reliable individual prediction. Unpublished paper, School of Information Systems, University of East Anglia. 19 September 1997.

Kennaway, Richard (1998). Population statistics cannot be used for reliably individual prediction. Unpublished paper, School of Information Systems, University of East Anglia. 17 April 1998.

Kershner, R.B. and L.R. Wilcox (1950). *The anatomy of mathematics.* New York: Ronald.

Kimble, Gregory A. (1996). *Psychology: The life of a science.* Cambridge MA: MIT Press.

Kirk, Stuart A. and Herb Kutchins (1992). *The selling of DSM: The rhetoric of science in psychiatry.* New York: Aldine de Gruyter.

Kofman, Fred and Peter M. Senge (1993). Communities of commitment. *Organizational Dynamics*, 22(2), 5–23.

Kohn, Alfie (1986). *No contest: The case against competition.* Boston: Houghton Mifflin.

Kohn, Alfie (1993). *Punished by rewards.* Boston: Houghton Mifflin.

Kohn, Alfie (1999). *The schools our children deserve: Moving beyond traditional classrooms and "tougher standards."* Boston: Houghton Mifflin.

Kohn, Alfie (2000). *The case against standardized testing: Raising the scores, ruining the schools.* Portsmouth NH: Heineman.

Kornhauser, Arthur (1965). *Mental health of the industrial worker.* New York: Wiley.

Koshland, D. E. (1980). *Behavioral chemotaxis as a model behavioral system.* New York: Raven.

Koslowsky, Meni, and Gardner Locke (1986). Decision rules for increasing the rate of successfully classified respondents. *Journal of Applied Behavioral Science*, 22(2), 187–193.

Kotzsch, Ronald E. (1990). Waldorf schools: Education for the head, hands, and heart. *Utne Reader*, September-October, No. 41, 84–90. Excerpted from *East West* (Box 57320, Boulder CO 80322–7320) of May 1989.

Krech, David and Richard S. Crutchfield (1948). *Theory and problems of social psychology.* New York: McGraw-Hill.

Kreidler, Willliam J. (1989). *Creative conflict resolution: More than 200 activities for keeping peace in the classroom.* Glenview Ill: Scott, Foresman.

Kuhn, Thomas S. (1969). Energy conservation as an example of simultaneous discovery. Pages 321–356 in M. Clagett (Ed.), *Critical problems in the history of science.* Madison WI: University of Wisconsin Press.

Laing, Ronald David and Aaron Esterson (1964). *Sanity, madness, and the family.* Baltimore: Penguin.

Lamb, Albert (ed.) (1992). *The new Summerhill / A. S. Neill.* London: Penguin.

Lamiell, James T. (1997). Individuals and the differences between them. Chapter 5 (pp. 117–141) in Robert T. Hogan, John Johnson, and Stephen Briggs (eds.), *Handbook of personality psychology.* San Diego CA: Academic Press.

Langer, Ellen J. (1983). *The psychology of control.* Thousand Oaks CA: Sage.

Lawler, Edward E. III (1986). *High-involvement management.* San Francisco: Jossey-Bass.

Lawler, Edward E. III (1991). The new plant approach: A second generation approach. *Organizational Dynamics*, 20(1), 5–14.

Lawler, Edward E. III (1992). *The ultimate advantage: Creating the high-involvement organization.* San Francisco CA: Jossey-Bass.

Lawler, Edward E. III (1994). *Motivation in work organizations.* San Francisco CA: Jossey-Bass.

Lawler, Edward E. III (1996). *From the ground up: Six principles for building the new logic corporation.* San Francisco CA: Jossey-Bass.

Lawler, Edward E. III (2000). *Rewarding excellence: Pay strategies for the new economy.* San Francisco CA: Jossey-Bass

Lawler, Edward E. III and Jay R. Galbraith (1994). Avoiding the corporate dinosaur syndrome. *Organizational Dynamics*, 23(2), 5–17.

Lawler, Edward E. III, Susan Albers Mohrman, and Gerald E. Ledford, Jr. (1995). *Creating high-performance organizations: Practices and results of employee involvement and Total Quality Management in Fortune 1000 companies.* San Francisco CA: Jossey-Bass.

Lawler, Edward E. III, Susan Albers Mohrman, and Gerald E. Ledford, Jr. (1998). *Strategies for high-performance organizations: The CEO report: Employee involvement, TQM, and reengineering programs in Fortune 1000 corporations.* San Francisco CA: Jossey-Bass.

Lazarus, Arnold A. (1994). Civilized but disgruntled, displeased, and distressed. Review of *Psychotherapy and its discontents* by Windy Dryden and Colin Feltham on pages 409–410 of *Contemporary Psychology*, 1994, 39(4).

LeBon, Gustave (1895). *The crowd.* Paris. English translation published by T. Fisher Unwin in London in 1896.

Lefkowitz, M., R.R. Blake, and Jane S. Mouton (1955). Statusfactors in pedestrian violation of traffic signals. *Journal of Abnormal and Social Psychology*, 51, 704–706.

Lindsay, Cindy P. and Bobby L. Dempsey (1983). Ten painfully learned lessons about working in China: The insights of two American behavioral scientists. *Journal of Applied Behavioral Science*, 19(3), 265–276.

Living longer: Conscientious kids. *Science News, 1995, 147(8), 124.*

Loftus, Elizabeth F. (1997). Creating false memories. *Scientific American*, 277(3), 71–75.

Lore, Richard K. and Lori A. Schultz (1993). Control of human aggression. *American Psychologist*, 48(1), 16–25.

Lynch, Lisa M. and Sandra E. Black (1996). *Beyond the incidence of training: Evidence from a national employer survey.* (Microform distributed to depository libraries in microfiche.) Philadelphia PA: National Center on the Educational Quality of the Workforce. Also U.S. Department of Education, Office of Educational Research and Improvement, Educational Resources Information Center. Available from the Superintendent of Documents, U.S. Government Printing Office.

Lytle, James H. and Jay M. Yanoff (1973). The effects (if any) of a teacher strike on student achievement. *Phi Delta Kappan*, December, p. 270.

Maher, Stanford and Ralph Shepherd (Eds.) (1995). *Standing on the brink: An education for the 21st century: An anthology of essays on Waldorf education.* Cape Town: Novalis.

Main, Jeremy (1994). *Quality wars: The triumphs and defeats of American business.* New York: Free Press.

Mansbridge, Jane J. (1991). *Beyond self-interest.* Chicago: University of Chicago Press.

Manz, Charles C. and Henry P. Sims, Jr. (1987). Leading workers to lead themselves: The external leadership of self-managing work teams. *Administrative Science Quarterly*, 32, 106–128.

Manz, Charles C., David E. Keating, and Anne Donnellon (1990). Preparing for an organizational change to employee self-management: The managerial transition. *Organizational Dynamics*, 19(2), 15–26.

Marken, Richard S. (1980). The cause of control movements in a tracking task. *Perceptual and Motor Skills*, 51, 755–758. Reprinted on pp. 61–66 of Richard S. Marken (1992), *Mind readings: Experimental studies of purpose.* Gravel Switch KY: Control Systems Group.

Marken, Richard S. (1982). Intentional and accidental behavior: A control theory analysis. *Psychological Reports*, 50, 647–650. Reprinted on pp. 35–39 of Richard S. Marken (1992), *Mind readings: Experimental studies of purpose.* Gravel Switch KY: Control Systems Group.

Marken, Richard S. (1983). "Mind reading": A look at changing intentions. Psychological Reports, 53, 267–270. Reprinted on pages 41–47 of Richard S. Marken (2002). *More mind readings.* St. Louis MO: newview.

Marken, Richard S. (1985). Selection of consequences: Adaptive behavior from random reinforcement. *Psychological Reports*, 56, 379–383. Reprinted on pp. 79–85 of Richard S. Marken (1992), *Mind readings: Experimental studies of purpose.* Gravel Switch KY: Control Systems Group.

Marken, Richard S. (1986). Perceptual organization of behavior: A hierarchical control model of coordinated action. *Journal of Experimental Psychology: Human Perception and Performance*, 12(3), 267–276. Reprinted on pp. 159–184 of Richard S. Marken (1992), *Mind readings: Experimental studies of purpose.* Gravel Switch KY: Control Systems Group.

Marken, Richard S. (1986a). Human factors and human nature: Is psychological theory really necessary? *Human Factors Society Bulletin*. Reprinted on pp. 207–212 of Richard S. Marken (1992), *Mind readings: Experimental studies of purpose*. Gravel Switch KY: Control Systems Group, 1992.

Marken, Richard S. (1988). The nature of behavior: Control as fact and theory. *Behavioral Science*, 33, 196–206. Reprinted on pp. 11–31 of Richard S. Marken (1992), *Mind readings: Experimental studies of purpose*. Gravel Switch KY: Control Systems Group.

Marken, Richard S. (1989). Behavior in the first degree. In Wayne A. Hershberger (Ed.), *Volitional action: Conation and control*, pp. 299–314. Amsterdam: North-Holland (copyright Elsevier Science Publishers B.V.). Reprinted on pp. 41–58 of Richard S. Marken (1992), *Mind readings: Experimental studies of purpose*. Gravel Switch KY: Control Systems Group.

Marken, Richard S. (1990). A science of purpose. *American Behavioral Scientist*, 34(1), 6–13. Reprinted on pages 11–21 of Richard S. Marken (2002), *More mind readings*. St. Louis MO: newview.

Marken, Richard S. (1990a). Spreadsheet analysis of a hierarchical control system model of behavior. *Behavior Research Methods, Instruments, and Computers*, 22(4), 349–359. Reprinted on pp. 133–156 of Richard S. Marken (1992), *Mind readings: Experimental studies of purpose*. Gravel Switch KY: Control Systems Group, 1992.

Marken, Richard S. (1991). Degrees of freedom in behavior. Psychological Science, 2(2), 92–100. Reprinted on pp. 185–204 of Richard S. Marken (1992), *Mind readings: Experimental studies of purpose*. Gravel Switch KY: Control Systems Group.

Marken, Richard S. (1992). *Mind readings: Experimental studies of purpose*. Gravel Switch KY: Control Systems Group.

Marken, Richard S. (1993). The hierarchical behavior of perception. *Closed Loop*, 3(4), 33–54. Reprinted on pages 85–112 of Richard S. Marken (2002), *More mind readings*. St. Louis MO: newview.

Marken, Richard S. (1994). My life as a control theorist. *Closed Loop*, 4(1), 49–56.

Marken, Richard S. (2002). *More mind readings: Method and models in the study of purpose*. St. Louis MO: newview.

Marken, Richard S. and Wm. T. Powers (1989). Random-walk chemotaxis: Trial and error as a control process. *Behavioral Neuroscience*, 103, 1348–1355. Reprinted on pp. 87–105 of Richard S. Marken. *Mind readings: Experimental studies of purpose*. Gravel Switch KY: Control Systems Group, 1992.

Marken, Richard S. and Wm. T. Powers (1989a). Levels of intention in behavior. Chapter 18 in Wayne A. Hershberger (Ed.), *Volitional action: Conation and control*, pp. 409–430. Amsterdam: North Holland (copyright Elsevier Science Publishers B.V.). Reprinted on pp. 109–132 of Richard S. Marken (1992), *Mind readings: Experimental studies of purpose*. Gravel Switch KY: Control Systems Group.

Martin, Shan (1983). *Managing without managers*. Beverly Hills: Sage.

Maslow, Abraham H. (1970). *Motivation and personality* (2nd ed.). New York: Harper and Row.

Mathews, Jay and Amy Argetsinger (2000). High test scores often come at high cost. *The Register-Guard* (Eugene, Oregon), June 3, pp. 1A and 14A. (Reprinted from *The Washington Post*)

Maxwell, James Clerk (1868). On governors. Pages 105–120 in *The scientific papers of James Clerk Maxwell*, Vol. 2. Reprinted by Dover, New York, 1965.

McCaslin, Mary and Thomas L. Good (1992). Compliant cognition: The misalliance of management and instructional goals in current school reform. *Educational Researcher*, 21(3), 4–17.

McClelland, Kent (1994). Perceptual control and social power. *Sociological Perspectives*, 37(4), 461–496.

McClelland, Kent (1994a). On cooperatively controlled perceptions and social order. Pages 29–40 in *Perceptual control theory: Proceedings of the 1st European workshop on perceptual control theory*, edited by Marcos A. Rodrigues and Mark H. Lee. Aberystwyth: University of Wales.

McClelland, Kent (1996). The collective control of perceptions: Toward a person-centered sociology. Working paper, Grinnell College.

McConnell, James V. and Ronald P. Philipchalk (1992). *Understanding human behavior*. (7th ed.) Fort Worth TX: Harcourt Brace Jovanovich.

McCrae, Robert R. and Paul T. Costa, Jr. (1987). Validation of the five-factor model of personality across instruments and observers. *Journal of Personality and Social Psychology,* 52(1), 81–90.

McCrae, Robert R. and Paul T. Costa, Jr. (1999). A five-factor theory of personality. In *Handbook of personality: Theory and research,* (2nd ed.), edited by Lawrence A. Pervin and Oliver P. John. New York: Guilford.

McDougall, William (1908). *An introduction to social psychology.* London: Methuen. 23rd edition in 1936. Reprinted by Barnes and Noble in 1960.

McDougall, William (1920). *The group mind.* New York: Putnam.

McDougall, William (1923). *Outline of psychology.* New York: Scribner's.

McGill, Michael E. and John W. Slocum, Jr. (1993). Unlearning the organization. *Organizational Dynamics,* 22(2), 67–79.

McGill-Franzen, Anne and Richard L. Allington (1993). Flunk 'em or get them classified. *Educational Researcher,* 22(1),19–22.

McGrath, Joseph E. (1984). *Groups: Interaction and performance.* Englewood Cliffs NJ: Prentice-Hall, 1984.

McKenna, Bernard M. (1977). A tale of testing in two cities. Pages 218–229 in Paul L. Houts (Ed.), *The myth of measurability.* New York: Hart.

McKibben, Bill (1996). The enigma of Kerala. *Utne Reader,* No. 74, 103–112.

McKnight, John L. (1984). John Deere and the bereavement counselor. Fourth Annual Schumacher Lecture, New Haven, printed as "The destructive side of technological development" on pp. 245–251 of Gareth Morgan (1989). *Creative organization theory.* Newbury Park CA: Sage.

McLaughlin, Margaret L., Michael J. Cody, and Stephen J. Read (Eds.) (1992). *Explaining one's self to others: Reason-giving in a social context.* Hillsdale NJ: Erlbaum.

McLaughlin, Milbrey Wallin (1975). *Education and reform.* Cambridge MA: Ballinger.

McPhail, Clark (1991). *The myth of the madding crowd.* New York: Aldine de Gruyter.

McPhail, Clark (2000). Collective action and perceptual control theory. Pp. 461–465 in David L. Miller, *Introduction to collective behavior and collective action.* (2nd ed.) Prospect Heights IL: Waveland Press.

McPhail, Clark and R. Wohlstein (1986). Collective locomotion as collective behavior. *American Sociological Review,* 51,447–463.

McPhail, Clark, William T. Powers, and Charles W. Tucker (1992). Simulating individual and collective action in temporary gatherings. *Social Science Computer Review,* 10(1), 1–28.

Messick, David M. and Wim B. G. Liebrand (1997). Levels of analysis and the explanation of the costs and benefits of cooperation. *Personality and Social Psychology Review,* 1(2), 129–139.

Miles, Matthew B. (1981). *Learning to work in groups.* (2nd ed.). New York: Teachers College Press.

Milford, Maureen (1996). A company philosophy in bricks and mortar. *New York Times,* September 1. R5.

Miller, James Grier (1978). *Living systems.* McGraw-Hill.

Millon, Theodore (1990). The disorders of personality. Chapter 13 in Lawrence A. Pervin (Ed.) *Handbook of personality: Theory and research.* New York: Guilford

Mills, Peter K. and James H. Morris (1986). Clients as "partial" employees of service organizations: Role development in client participation. *Academy of Management Review,* 11(4),726–735.

Minds of Our Own (1997). Produced by the Harvard-Smithsonian Center for Astrophysics, 1997. South Burlington VT: Annenberg/CPB Math and Science Collection. Video cassette, ISBN 1-57680-064-4.

Miner, Anne S. (1987). Idiosyncratic jobs in formalized organizations. *Administrative Science Quarterly,* 32(3),327–351.

Mintzberg, Henry (1994). *The rise and fall of strategic planning.* New York: Free Press.

Mirowsky, John and Catherine E. Ross (1989). *Social causes of psychological distress.* New York: Aldine de Gruyter.

Mitchell, James V. (Ed.) (1998). *Mental measurements yearbook,* vol. 13. Lincoln NE: Buros Institute of Mental Measurements, University of Nebraska. A series begun in 1940 by Oscar K. Buros. J.V. Mitchell became editor in 1985.

Mohandessi, K and Philip J. Runkel (1958). Some socio-economic correlates of academic aptitude. *Journal of Educational Psychology,* 49, 47–52.

Mohrman, Susan Albers, Jay A. Galbraith, Edward E. Lawler III, and associates (1998). *Tomorrow's organization: Crafting winning capabilities in a dynamic world.* San Francisco CA: Jossey-Bass.

Morey, Leslie C. (1997). Personality diagnosis and personality disorders. Chapter 35 (pp. 919–946) in Robert T. Hogan, John Johnson, and Stephen Briggs (Eds.), *Handbook of personality psychology*. San Diego CA: Academic Press.

Morgan, Harry (1999). *The imagination of early childhood education*. Westport CT: Bergin & Garvey.

Morse, Jodie (2000). Does Texas make the grade? *TIME*, 156(11), 50–54.

Morse, Jodie (2002). Test drive. *TIME*, 159(5), 53–54.

Mundy, John H. and Peter Riesenberg (1958). *The medieval town*. Princeton NJ: D. Van Nostrand.

Myers, David G. (1986). *Psychology*. New York: Worth.

Neisser, Ulric (1998). *The rising curve: Long-term gains in IQ and related measures*. Washington, DC: APA Books.

Newcomb, Theodore M. (1953). An approach to the study of communicative acts. *Psychological Review*, 60, 393–404.

Newcomb, Theodore M. (1959). Individual systems of orientation. Pages 384–422 of Sigmund Koch (Ed.), *Psychology: A study of a science*. (vol. 3). New York: McGraw-Hill.

Norman, W.T. (1967). *2800 personality trait descriptors: Normative operating characteristics for a university population*. Ann Arbor: University of Michigan, Department of Psychology.

Ornstein, Robert and Laura Carstensen (1991). *Psychology: The study of human experience*. San Diego: Harcourt Brace Jovanovich.

Ornstein, Robert and Paul Ehrlich (1989). *New world new mind*. New York: Simon and Schuster.

Palmore, E.B. (1969). Predicting longevity, a follow-up controlling for age. *Gerontologist*, 9, 247–250.

Papworth, John (1991). The best government comes in small packages. *Utne Reader*, January-February, 58–59.

Paris, Scott G., Theresa A. Lawton, Julianne C. Turner, and Jodie L. Roth (1991). A developmental perspective on standardized achievement testing. *Educational Researcher*, 20(5), 12–20.

Paulos, John Allen (1988). *Innumeracy: Mathematical illiteracy and its consequences*. New York: Hill and Wang.

Pavloski, Raymond P., Gerard T. Barron, and Mark A. Hogue (1990). Reorganization: Learning and attention in a hierarchy of control systems. *American Behavioral Scientist*, 34(1), 32–54.

Peacock, Chris (1991). Kerala's quiet revolution. *Utne Reader*, November-December, No. 48, 24.

Perez-Lopez, Jorge F. (1991). *The economics of Cuban sugar*. Pittsburgh: University of Pittsburgh Press.

Pervin, Lawrence A. and Oliver P. John (eds.) (1999). *Handbook of personality: Theory and research*. New York: Guilford.

Pfeiffer, John E. (1977). *Emergence of Society*. New York: McGraw-Hill.

Pfeiffer, John E. (1978). *Emergence of man*. New York: Harper and Row.

Pierce, Jon L. and Randall B. Dunham (1992). The 12-hour workday: A 48-hour, eight-day week. *Academy of Management Journal*, 335(5), 1086–1098.

Plooij, Frans X. (1984). *The behavioral development of free-living chimpanzee babies and infants*. Norwood NJ: Ablex.

Plooij, Frans X. (1990). Developmental psychology: Developmental stages as successive reorganizations of the hierarchy. Chapter 9 in R.J. Robertson and W.T. Powers. *Introduction to modern psychology*. Gravel Switch KY: Control Systems Group.

Plooij, Frans X. and Hedwig H. C. van de Rijt-Plooij (1989). Vulnerable periods during infancy, hierarchically reorganized systems control, stress, and disease. *Ethology and Sociobiology*, 10, 1–18.

Plooij, Frans X. and Hedwig H. C. van de Rijt-Plooij (1989). Evolution of human parenting: Canalization, new types of learning, and mother-infant conflict. *European Journal of Psychology of Education*, 4(2), 177–192.

Plooij, Frans X. and Hedwig H. C. van de - (1990). Developmental transitions as successive reorganizations of a control hierarchy. *American Behavioral Scientist*, 34(1), 67–80.

Plooij, Frans X. and Hetty H. C. van de Rijt-Plooij (1994). Learning by instincts, developmental transitions, and the roots of culture in infancy. Pages 357–373 in R. Allen Gardner, Beatrix T. Gardner, Brunetto Chiarelli, and Frans X. Plooij (Eds.), *The ethological roots of culture*. Dordrecht, the Netherlands: Kluwer Academic.

Powers, William T. (1972). Emotion. Pages 31–40 in W.T. Powers (1992), *Living control systems II*. Gravel Switch KY: Control Systems Group, distributed by Benchmark Publ., New Canaan CT.

Powers, William T. (1973). Behavior: *The control of perception*. Chicago: Aldine; now distributed by de Gruyter, New York.

Powers, William T. (1973a). Feedback: Beyond behaviorism. *Science*, 179(4071), 351–356. Reprinted on pp. 61–78 of Wm. T. Powers (1989). *Living control systems*. Gravel Switch KY: Control Systems Group, distributed by Benchmark Publ., New Canaan CT.

Powers, William T. (1978). Quantitative analysis of purposive systems: Some spadework at the foundations of scientific psychology. *Psychological Review*, 85(5), 417–435. Reprinted on pp. 129–165 of Wm. T. Powers (1989). *Living control systems*. Gravel Switch KY: Control Systems Group, distributed by Benchmark Publ., New Canaan CT.

Powers, William T. (1979). Degrees of freedom in social interactions. In Klaus Krippendorff (Ed.), *Communication and control in society*. New York: Gordon and Breach, pp. 267–278. Reprinted on pp. 221–235 of Wm. T. Powers (1989). *Living control systems*. Gravel Switch KY: Control Systems Group, distributed by Benchmark Publ., New Canaan CT.

Powers, William T. (1979a). The nature of robots, Part I: Defining behavior. *Byte*, 4(6), 132–134, 136, 138, 140–141, 144; Part II: Simulated control system. 4(7), 134–136, 138, 140, 142, 144, 146, 148–150; Part III: A closer look at human behavior. 4(8), 94–96, 98, 100, 102–104, 106–108, 110–112, 114, 116; Part IV: Looking for controlled variables. 4(9), 96, 98–102, 104, 106–110, 112.

Powers, William T. (1980). A systems approach to consciousness. In J. M. Davidson and R.J. Davidson (Eds.), *The psychobiology of consciousness*. New York: Plenum. Pp. 217–242.

Powers, William T. (1983). Deriving closed-loop transfer functions for a behavioral model, and vice versa. Pages 145–160 in W.T. Powers (1992), *Living control systems II*. Gravel Switch KY: Control Systems Group, distributed by Benchmark Publ., New Canaan CT.

Powers, William T. (1983a). Learning and evolution. Pages 161–170 in W.T. Powers (1992), *Living control systems II*. Gravel Switch KY: Control Systems Group, distributed by Benchmark Publ., New Canaan CT.

Powers, William T. (1988). An outline of control theory. Part II in Humberto R. Maturana, Wm. T. Powers, and Ernst von Glasersfeld. Conference workbook for "Texts in cybernetic theory". Printed by the American Society for Cybernetics for the conference in Felton CA in October 1988. Reprinted on pp. 253–293 of Wm. T. Powers (1989). *Living control systems*. Gravel Switch KY: Control Systems Group, distributed by Benchmark Publ., New Canaan CT.

Powers, William T. (1989). *Living control systems: Selected papers of William T. Powers*. Gravel Switch KY: Control Systems Group, distributed by Benchmark Publ., New Canaan CT.

Powers, William T. (1989a). Volition: A semi-scientific essay. Pages 21–38 in Wayne A. Hershberger (Ed.) (1989). *Volitional action: Conation and control*. Amsterdam: North-Holland (Copyright Elsevier Science Publishers B.V.).

Powers, William T. (1989b). Quantitative measurement of volition: A pilot study. Pages 315–332 in Wayne A. Hershberger (Ed.) (1989). *Volitional action: Conation and control*. Amsterdam: North-Holland (copyright Elsevier Science Publishers B.V.).

Powers, William T. (1990). A hierarchy of control. Chapter 5 in R.J. Robertson and W.T. Powers. *Introduction to modern psychology*. Gravel Switch KY: Control Systems Group, distributed by Benchmark Publ., New Canaan CT.

Powers, William T. (1992). CT psychology and social organizations. Pp. 91–127 in Wm. T. Powers (1992). *Living control systems II: Selected papers of William T. Powers*. Gravel Switch KY: Control Systems Group, distributed by Benchmark Publ., New Canaan CT. Originally a working paper distributed in 1980 by W.T. Powers and D. T. Campbell, then entitled "Control-theory psychology and social organizations: Background for a theory of the corruption of social indicators when used for social decision making. Working paper for a conference."

Powers, William T. (1992a). *Living control systems II: Selected papers of William T. Powers*. Gravel Switch KY: Control Systems Group, distributed by Benchmark Publ., New Canaan CT.

Powers, William T. (1994). An "articifial cerebellum" adaptive stabilization of a control system. Pp. 41–49 in Marcos A. Rodrigues and Mark H. Lee (Eds.), *Perceptual control theory: Proceedings of the 1st European workshop*. Aberystwyth, Wales: University of Wales.

Powers, William T. (1995). The origins of purpose: The first metasystem transitions. *World Futures*, vol. 45 (F. Heylighen, C. Joslyn, and V. Turchin, eds), pp. 125–138.

Powers, William T. (1998). *Making sense of behavior: The meaning of control*. New Canaan CT: Benchmark Publications.

Powers, William T. (1999). A model of kinesthetically and visually controlled arm movement. *International Journal of Human-Computer Studies*, 50, 463–479.

Powers, William T., Robert K. Clark, and Robert L. McFarland (1960). A general feedback theory of human behavior, Part I. *Perceptual and Motor Skills*, 11(1), 71–88; reprinted in Ludwig von Bertalanffy and Anatol Rapoport (Eds.), *General systems: Yearbook of the Society for General Systems Research 5*; Ann Arbor: Society for General Systems Research, 1960, 63–73. Reprinted in altered form, as "A general feedback theory of human behavior," in Alfred G. Smith (Ed.), *Communication and culture: Readings in the codes of human interaction*; New York: Holt, Rinehart, and Winston, 1966, 333–343. Part II, *Perceptual and Motor Skills*, 11(3), 309–323; reprinted in *General Systems: Yearbook of the Society for General Systems Research 5*, 1960, 75–83. Reprinted on pp. 1–45 of Wm. T. Powers (1989). *Living control systems*. Gravel Switch KY: Control Systems Group, distributed by Benchmark Publ., New Canaan CT.

Puckett, John L. (1989). *Foxfire reconsidered: A twenty-year experiment in progressive education*. Urbana IL: University of Illinois Press.

Qin, Zhining, David W. Johnson, and Roger T Johnson (1995). Cooperative versus competitive efforts and problem solving. *Review of Educational Research*, 65(2), 129–143.

Quick, Rebecca (2000). If you walk a mile in my shoes, you'll have sweaty feet. *Wall Street Journal*, July 18.

Rapoport, Anatol (1968). Foreword. Pp. xiii-xxii in W. Buckley (Ed.), *Modern systems research for the behavioral scientist*. Chicago: Aldine.

Renner, Michael G. (1996). Matters of scale: Who dominates the world? *World Watch*, November-December, 39.

Richardson, George P. (1991). *Feedback thought in social science and systems theory*. Philadelphia: University of Pennsylvania Press.

Rijt-Plooij, Hedwig H. C. van de and Frans X. Plooij (1986). The involvement of interactional processes and hierarchical systems control in the growing independence in chimpanzee infancy. In J. Wind and V. Reynolds (Eds.), *Essays in human sociobiology*, Vol. 2. Brussels: V.U.B. Press, Study Series No. 26. Pp. 155–165.

Rijt-Plooij, Hedwig H. C. van de and Frans X. Plooij (1987). Growing independence, conflict, and learning in mother-infant relations in free-ranging chimpanzees. *Behavior*, 101, 1–86.

Rijt-Plooij, Hedwig H. C. van de and Frans X. Plooij (1988). Mother-infant relations, conflict, stress, and illness among free-ranging chimpanzees. *Developmental Medicine and Child Neurology*, 30, 306–315.

Rijt-Plooij, Hedwig H. C. van de and Frans X. Plooij (1992). Infantile regressions: Disorganization and the onset of transition periods. *Journal of Reproductive and Infant Psychology*, 10, 129–149.

Rijt-Plooij, Hedwig H. C. van de and Frans X. Plooij (1993). Distinct periods of mother-infant conflict in normal development. *Journal of Child Psychology and Psychiatry*, 34(2), 229–245.

Robertson, Richard J. (1966). Factory reject personalities and the human salvage operations. *Journal of Applied Behavioral Science*, 2(3), 331–355.

Robertson, Richard J. (1990). Higher-order control systems: Personality and the self. Chapter 11 in R.J. Robertson and W.T. Powers (Eds.), *Introduction to modern psychology*. Control Systems Group, distributed by Benchmark Publ., New Canaan CT.

Robertson, Richard J. (1990a). Conflict between systems and reorganization of higher levels of the control hierarchy. Chapter 12 in R.J. Robertson and W.T. Powers (Eds.), *Introduction to modern psychology*. Gravel Switch KY: Control Systems Group, distributed by Benchmark Publ., New Canaan CT.

Robertson, Richard J. and Lawrence A. Glines (1985). The phantom plateau returns. *Perceptual and Motor Skills*, 61, 55–64.

Robertson, Richard J. and Wm. T. Powers (Eds) (1990). *Introduction to modern psychology.* Control Systems Group, distributed by Benchmark Publ., New Canaan CT.

Robertson, Richard J., David M. Goldstein, Michael Mermel, and Melanie Musgrave (1987). Testing the self as a control system: Theoretical and methodological issues. Unpublished paper, Department of Psychology, Northeastern University.

Robertson, Richard J., David M. Goldstein, Michael Mermel, and Melanie Musgrave (1999). Testing the self as a control system: Theoretical and methodological issues. *International Journal of Human-Computer Studies,* 50, 571–580.

Rodin, Judith, Carmi Schooler, and W. Warner Schaie (eds.) (1990). *Self-directedness: Cause and effects throughout the life course.* Hillsdale NJ: Lawrence Erlbaum.

Rogers, Carl R. (1951). *Client-centered therapy.* Boston: Houghton Mifflin.

Rosenbaum, M. and R.R. Blake (1955). Volunteering as a function of field structure. *Journal of Abnormal and Social Psychology,* 50, 193–196.

Rosenblueth, Arthur, Norbert Wiener, and Julian Bigelow (1943). Behavior, purpose, and teleology. *Philosophy of Science,* 10, 18–24. Reprinted on pages 221–225 of Walter Buckley (Ed.)(1968). *Modern systems research for the behavioral scientist: A sourcebook.* Chicago: Aldine.

Rosenhan, David L. (1973). On being sane in insane places. *Science,* 179, 250–258.

Royce, Josiah (1913). *The problem of Christianity.* Vol. 2. New York: Macmillan.

Runkel, Philip J. (1956). Equilibrium and "pleasantness" of interpersonal situations. *Human Relations,* 9(3), 375–382.

Runkel, Philip J. (1990). *Casting nets and testing specimens.* New York: Praeger.

Runkel, Philip J. David B. Peizer (1968). The two-valued orientation of current equilibrium theory. *Behavioral Science,* 13(1), 56–65.

Runkel, Philip J., Richard A. Schmuck, Jane H. Arends, and Richard P. Francisco (1979). *Transforming the school's capacity for problem solving.* Eugene, OR: Center for Educational Policy and Management, College of Education, University of Oregon.

Sahakian, William S. (1982). *History and systems of social psychology* (2nd ed.) Washington DC: Hemisphere.

Sarason, Seymour B. (1983). *Schooling in America: Scapegoat and salvation.* New York: Free Press.

Sarason, Seymour B. (1990). *The predictable failure of educational reform: Can we change course before it's too late?* San Francisco CA: Jossey-Bass.

Sarason, Seymour B. (1996). *Barometers of change: Individual, educational, and social transformation.* San Francisco CA: Jossey-Bass.

Sarason, Seymour B. (1996a). *Revisiting: The culture of the school and the problem of change.* New York: Teachers College Press.

Sarason, Seymour B, (1997). *How schools might be governed and why.* New York: Teachers College Press.

Sarason, Seymour B. (1998). *Charter schools: Another flawed educational reform?* New York: Teachers College Press.

Sarason, Seymour B. and Elizabeth M. Lorentz (1998). *Crossing boundaries: Collaboration, coordination, and the redefinition of resources.* San Francisco CA: Jossey-Bass.

Sarason, Seymour B., Kennedy T. Hill, and Philip G. Zimbardo (1964). *A longitudinal study of the relation of test anxiety to performance on intelligence and achievement tests.* Chicago: University of Chicago Press.

Sarason, Seymour B., Kenneth S. Davidson, Frederick F. Lighthall, Richard R. Waite, and Britton K. Ruebush (1960). *Anxiety in elementary school children.* New York: Wiley.

Sarbin, Theodore R. (1982). Ideological constraints on the science of deviant conduct. Chapter 10 in Vernon L. Allen and Karl E. Scheibe (Eds.), *The social context of conduct.* Westport CT: Praeger.

Sarbin, Theodore R. (1995). On the belief that one body may be host to two or more personalities. *International Journal of Clinical and Experimental Hypnosis,* 43(2), 163–183.

Schacter, Stanley (1959). *The psychology of affiliation.* Stanford CA: Stanford University Press.

Schacter, Stanley (1964). The interaction of cognitive and physiological determinants of emotional state. In Leonard Berkowitz (Ed.), *Advances in experimental social psychology* (vol. 1). New York: Academic Press.

Schein, Edgar H. (1993). On dialogue, culture, and organizational learning. *Organizational Dynamics,* 22(2), 40–51.

Schmuck, Patricia and Richard A. Schmuck (1990). Democratic participation in small-town schools.

Educational Researcher, 19(8), 14–19.

Schmuck, Richard A. (1997). *Practical action research for change*. Arlington Heights IL: Skylight Training and Publishing.

Schmuck, Richard A. (2000). *Practical action research: A collection of articles*. Arlington Heights IL: Skylight Training and Publishing.

Schmuck, Richard A. and Philip J. Runkel (1970). *Organizational training for a school faculty*. Eugene OR: Center for Educational Policy and Management, University of Oregon.

Schmuck, Richard A. and Philip J. Runkel (1994). *Handbook of organization development in schools and colleges* (4th ed.). Prospect Heights IL: Waveland Press.

Schmuck, Richard A. and Patricia A. Schmuck (2002). *Group processes in the classroom* (8th ed.) Boston MA: McGraw-Hill.

Schmuck, Richard A., Donald Murray, Mary Ann Smith, Mitchell Schwartz, and Margaret Runkel (1975). *Consultation for innovative schools*. Eugene, OR: Center for Educational Policy and Management.

Schoggen, Phil (1989). *Behavior settings: A revision and extension of Roger G. Barker's "Ecological psychology."* Stanford CA: Stanford University Press.

Schumacher, E.F. (1973). *Small is beautiful: Economics as if people mattered*. New York: Harper and Row (Perennial Library).

Schumacher, E.F. (1979). *Good work*. New York: Harper and Row.

Sedlak, Michael W., Christopher W. Wheeler, Diana C. Pullin, and Philip A. Cusick (1986). *Selling students short*. New York: Teachers College Press.

Semler, Ricardo (1989). Managing without managers. *Harvard Business Review*, 67(5), 76–84.

Sen, Amartya (1993). The economics of life and death. *Scientific American*, 268(5), 40–47.

Senchuk, Dennis M. (1991). *Against instinct: From biology to philosophical psychology*. Philadelphia: Temple University Press.

Sharan, Shlomo (1994). *Handbook of cooperative learning methods*. Westport CT: Greenwood Press.

Sharan Shlomo, Hanna Shachar, and Tamar Levine (1999). *The innovative school: Organization and instruction*. Westport CT: Bergin and Garvey.

Sherif, Muzafer and Carolyn Sherif (1969). *Social psychology*. New York: Harper and Row.

Sherif, Muzafer, O.J. Harvey, B. Jack White, William R. Hood, and Carolyn W. Sherif (1961). *Intergropup conflict and cooperation: The Robbers Cave experiment*. Norman OK: Institute of Group Relations, University of Oklahoma.

Sherman, Joe (1993). *In the rings of Saturn*. Cambridge MA: Oxford University Press.

Sherwood, Philip (1977). Evaluation, the English experience. Pages 230–238 in Paul L. Houts (Ed.), *The myth of measurability*. New York: Hart.

Shipper, Frank and Charles C. Manz (1992). An alternative road to empowerment. *Organizational Dynamics*, 20(3), 48–61.

Sieber, Sam D., Karen Seashore Louis, and Loya Metzger (1972). *The use of educational knowledge: Evaluation of the Pilot State Dissemination Program*. New York: Bureau of Applied Social Research, Columbia University.

Skinner, Burrhus Frederick (1953). *Science and human behavior*. New York: Free Press.

Slavin, Robert E. (1990). *Cooperative learning: Theory, research, and practice*. Englewood Cliffs NJ: Prentice-Hall.

Slocum, John W., Jr., Michael McGill, and David T. Lei (1994). The new learning strategy. *Organizational Dynamics*, 23(2), 33–47.

Smith, Laurence D. (1992). On prediction and control: B.F. Skinner and the technological ideal of science. *American Psychologist*, 47(2), 216–223.

Smith, M.L. and Glass, G V. (1977). Meta-analysis of psychotherapy outcome studies. *American Psychologist*, 32(9), 752–760.

Snyder, Mark (1993). Basic research and practical problems: The promise of a "functional" personality and social psychology. *Personality and Social Psychology Bulletin*, 19(3), 251–264.

Social ties and length of life. *Science News*, 1980, 118(25 and 26), 392.

Soldani, James (1989). Effective personnel management: An application of control theory. Pages 515–529 in Wayne A. Hershberger (Ed.) (1989). *Volitional action: Conation and control*. Amsterdam: North-Holland (copyright Elsevier Science Publishers B.V.).

Spitzer, Robert, J. Forman, and J. Nee (1979). DSM-III field trials: I. Initial interrater diagnostic reliability. *American Journal of Psychiatry*, 136, 815–817.

Spock, Marjorie (1985). *Teaching as a lively art.* Spring Valley NY: Anthroposophic Press.

Starnes, Bobby Ann (1999). *The Foxfire approach to teaching and learning: John Dewey, experiential learning, and the core practices* (microform). Charleston WV: Clearinghouse on Rural and Small Schools, Appalachia Educational Laboratory.

Stavrianos, L.S. (1976). *Promise of the coming dark age.* San Francisco: W.H. Freeman.

Stress-linked suicides claim more police lives than the "line of duty." *Behavior Today,* 1979, 10(42), 3–4.

Stromeyer, C. (1970). Eidetikers. *Psychology Today,* 4, 77–80.

Strupp, Hans H. and Suzanne W. Hadley (1979). Specific versus nonspecific factors in psychotherapy. *Archives of General Psychiatry,* 36, 1125–1136.

Suarez, Enrique M. and Roger C. Mills (1982). *Sanity, insanity, and common sense.* West Allis WI: Med-Psych Publications.

Sutton, R.I. (1991). Maintaining norms about expresssed emotions: The case of bill collectors. *Administrative Science Quarterly, 36(2), 245–268.*

Szamosi, Geza (1986). *The twin dimensions: Inventing time and space.* New York: McGraw-Hill.

Taylor, Martin M. (1995). Effects of Modafinil and Amphetamine on tracking performance during sleep deprivation. *Proceedings of the 37th annual conference of the International Military Testing Association, Toronto, Canada, October 1995,* pp. 97–102.

Teachers' work spotlighted amid confusion. *Cutting Edge,* 1987 (June), 16, pp. 1, 3, 4.

Teplin, Linda A., Karen M. Abram, and Gary M. McClelland (1994). Does psychiatric disorder predict violent crime among released jail detainees? A six-year longitudinal study. *American Psychologist,* 49(4), 335–342.

Thalhammer, Bryan (2000). *Self-image maintenance in an educational setting: A perceptual control theory study of technical training.* Unpublished dissertation, University of Illinois.

Thatcher, Robert W. (1994). Cyclic cortical reorganization: Origins of human cognitive development. Chapter 8 (pp. 232–266) in Geraldine Dawson and Kurt W. Fischer (Eds.), Human *behavior and the developing brain.* New York: Guilford.

Thompson, Dana (1994). Russia: Taking League skills to Moscow. *National Voter* (Journal of the League of Women Voters), 43(4), p. 8.

Thurstone, L.L. (1934). The vectors of mind. *Psychological Review,* 41, 1–32.

Trist, Eric (1981). *Evolution of socio-technical systems.* Toronto: Ontario Quality of Working Life Centre. Also published as "The evolution of sociotechnical systems as a conceptual framework and as an action research program." In Andy Van de Ven and Wm. Joyce (Eds.) (1982), *Perspectives on organization design and behavior.* New York: Wiley.

Trotter, Wilfred (1917). *Instincts of the herd in peace and war.* New York: Macmillan.

Tucker, Charles W., David Schweingruber, and Clark McPhail (1999). Simulating arcs and rings in gatherings. *International Journal of Human-Computer Studies,* 50, 581–588.

Tye, Kenneth A. and J. M. Novotney (1975). *Schools in transition: The practitioner as change agent.* New York: McGraw-Hill.

Tyler, Leona E. (1978). Do we have to have compulsory education for adolescents? In J. C. Flanagan (Ed.), *Perspectives on improving education.* New York: Praeger.

U.S. Department of Labor, Employment and Training Administration, U.S. Employment Service (1991). *Dictionary of occupational titles* (2 vols., 4th ed. rev.). Indianapolis IN: JIST Works.

Uttal, Wm. R. (1978). *Psychobiology of mind.* Hillsdale NJ: Erlbaum.

Vanderijt, Hetty and Frans Plooij (2003). *The wonder weeks: How to turn your baby's 8 great fussy phases into magical leaps forward.* Emmanus, PA: Rodale.

Vial, L.G. (1940). Stone axes of Mount Hagen. *Oceana,* 11,158–163.

Viscio, Randolph Louis (1994). Mall rats. *Utne Reader,* No. 66, 131–132. Reprinted from Viscio's article in *As We Are,* Winter 1993–94 (Box 380048, Cambridge MA 02238). Excerpted from the book *Under the Bridge.* Cow Pasture Productions, 1993.

Wachtel, Paul L. (1983). *The poverty of affluence.* New York: Free Press.

Walker, Elaine F. (1993). The nature of schizophrenia: Disease, hypothetical construct, or sociocultural myth? Review of *What is schizophrenia?* by William F. Flack, Jr., Daniel R. Miller, and Morton

Wiener (Eds.) *Contemporary Psychology*, 38(9), 951–952.

Wall, James A. (1986). *Bosses*. Lexington MA: Lexington Books.

Wall, Toby D., Nigel J. Kemp, Paul R. Jackson, and Chris W. Clegg (1986). Outcomes of autonomous workgroups: A long-term field experiment. *Academy of Management Journal*, 29(2), 280–304.

Walljasper, Jay (1994). What does alternative mean these days? *Utne Reader*, No. 66, 154–155.

Watson, John B. and William McDougall (1929). *The battle of behaviorism: An exposition and an exposure*. New York: W.W. Norton.

Weick, Karl E. (1977). Organization design: Organizations as self-designing systems. *Organizational Dynamics*, 6(2), 31–46.

Weick, Karl E. (1979). *Social psychology of organizing* (2nd ed.). Reading MA: Addison-Wesley.

Weick, Karl E. (1980). Blind spots in organizational theorizing. *Group and Organizational Studies*, 5(2), 178–188.

Weick, Karl E. (1995). *Sensemaking in organizations*. Thousand Oaks CA: Sage.

Weisbord, Marvin R. (1985). Participative work design: A personal odyssey. *Organizational Dynamics*, 13(4), 5–20.

Weisbord, Marvin R. (1987). *Productive workplaces*. San Francisco: Jossey-Bass.

Weisbord, Marvin R. (Ed.) (1992). *Discovering common ground*. San Francisco: Berrett-Koehler.

Wentworth, Roland A. Lubienski (1998). *Montessori for the new millenium: Practical guidance on the teaching and education of children of all ages, based on a rediscovery of the true principles and vision of Maria Montessori*. Mahwah NJ: Erlbaum.

Wiener, Norbert (1948). *Cybernetics: Control and communication in the animal and the machine*. New York: Wiley.

Wigginton, Eliot (1985). *Sometimes a shining moment: The Foxfire experience*. Garden City NY: Anchor-Doubleday.

Wild, T. Cameron, Michael E. Enzle, and Wendy L. Hawkins (1992). Effects of perceived extrinsic versus intrinsic teacher motivation on student reactions to skill acquisition. *Personality and Social Psychology Bulletin*, 18(2), 245–251.

Wilkins, L.T. (1965). *Social deviance: Social policy, action, and research*. Englewood Cliffs NJ: Prentice-Hall.

Will, Rosalyn, Jonathon Hickman, and Amy Muska (1998). International role models: The 1998 Corporate Conscience Awards. *Council on Economic Priorities Research Report*, May-June, 1–5.

Winters, Rebecca (2000). A job for a superhero? *TIME*, 155(5), February 7, pp. 70–71.

Wolcott, Harry F. (1974). The teacher as an enemy. Chapter 21 in G. Spindler (Ed.), *Education and cultural process: Toward an anthropology of education*. New York: Holt, Rinehart, Winston, pp. 411–425. Reprinted in the second edition, 1987, published by Waveland Press.

Wolfgang, M.E., R.M. Figlio, and T. Sellin (1972). *Delinquency in a birth cohort*. Chicago: University of Chicago Press.

World psychosis. *Science 80*, 1980, 1(6), 7.

Worrall, Alyson M. (1995). Suit-able for promotion: A game of snakes and ladders. Pages 165–179 in Diane M. Dunlap and Patricia A. Schmuck (eds.). *Women leading in education*. Albany: State University of New York Press.

Young, Rupert (2000). *Visual control in natural and artificial systems*. (Doctoral dissertation) Guildford, Surrey, U.K.: University of Surrey; School of Electronic Engineering, Information Technology, and Mathematics; Centre for Vision, Speech, and Signal Processing.

Zander, Alvin (1983). *Making groups effective*. San Francisco: Jossey-Bass.

Zeigarnik, B. (1927). Uber das Behalten von erledigten und unerledigten Hendlungen. *Psychologische Forschung*, 9, 1–85.

Name index

This index does not extend to the pages numbered with Roman numerals and not to the list of references.

Abbott, Bruce 22, 179, 303
Abelson, Robert P. 42
Abram, Karen M. 342
Agarwal, G.C. 158, 165
Agarwala-Rogers, Rehka 436
Aiken, Wilford 443
Alban, Billie T. 435
Albee, George W. 340, 363, 473
Alderfer, Clayton P. 386
Alexander the Great 421, 485
Allington, Richard L. 464
Allport, Gordon W. 292
Anfuso, Dawn 408
Aoki, Ted T. 434
Appley, Mortimer H. 40, 41, 43, 58, 181
Aquinas, Thomas 175
Arends, Jane H. 432, 436
Argetsinger, Amy 458, 464
Aristarchus of Samos 265
Aristotle 171, 276, 280
Arnold, Matthew 330
Arrow, Holly 411, 417
Asch, Solomon E. 259
Ashby, W.R. 41, 229
Atchison, Thomas J. 410
Atkinson, Richard C. and Rita L. 99, 199
Axelrod, Robert 325
Bach, J.S. 251
Balston, Headmaster 463
Banker, Rajiv D. 410
Barker, Roger G. 32, 441
Barron, G.T. 97, 143, 153
Barrow, John D. 442
Barry, David 423
Bartlett, F.C. 219
Baughman, Fred 365
Bayer, R. 357
Beach, Dennis E. 330, 408
Beckmann, Petr 124
Beethoven, Ludwig van 39, 169, 193, 209

Benne, Kenneth 382
Benny, Jack 373
Bentzen, Mary M. 436
Berdahl, Jennifer L. 411, 417
Berk, R. 100
Berkowitz, Leonard 42
Berman, Jeffrey S. 361
Berman, Paul 436
Bernard, Claude 40, 229, 459
Berscheid, Ellen 246
Bird, Caroline 454
Bizzi, E. 140, 153
Black, H.S. 58, 405
Blackburn, Boyd 460
Blake, Robert 11, 420, 421
Blase, Joseph J. 440
Bohl, Don L. 300
Bohr, Niels 176, 288, 289
Boisjoly, Roger 258
Bolman, Lee G. 421
Boulding, Kenneth E. 14, 386, 387
Bourbon, W. Thomas 81–87, 94–97, 111, 134, 136, 137, 139, 142, 150, 153, 167, 168, 192, 303, 319, 322, 323, 366, 367, 379, 392, 467, 468–475
Bourke, Vernon J. 175
Bowen, David E. 300, 408
Bower, Bruce 338, 339, 340, 362, 371
Bowles, Samuel 442
Braden, Spruille 277, 278
Bradford, Lee 382
Brahms, J. 253
Brehm, Jack W. 41
Brehm, Sharon S. 41
Bronfenbrenner, Urie 126
Brown, J. 301
Brown, Roger 284
Brown, Walter A. 363
Brownstein, A. 140, 153
Bryan, Frank 76, 486
Bundy, McGeorge 258

Bunker, Barbara Benedict 435, 446
Burns, Robert 225, 382
Bush, George W. 417, 451, 460, 463
Byrom, Fletcher 430
Cader, John 311
Caesar, Augustus 485
Caesar, Julius 451
Caine, Geoffrey 445
Caine, Geoffrey 445
Campbell, Charles 407
Cannon, Walter B. 229, 244
Carey, Margaret 323, 468, 469, 473, 475
Carey, Timothy A. 323, 350, 351, 354, 357, 364, 365, 379, 410, 468, 469, 473, 475
Carlzon, Jan 308
Carroll, Lewis 257, 375, 377
Carstensen, Laura 49
Chandler, P. J. 11
Chapanis, A. 58
Chaparral Steel 408
Charlemagne 485
Chomsky, Noam 373
Churchland, P.S. 141, 153
Churchland, P. M. 141, 153
Cicero 482
Clark, Robert K. 211, 213, 232
Clark, William 214
Clausewitz, Carl von 266
Cloud, John 446
Cody, Michael J. 286
Cofer, C. N. 40
Cohen, David 410
Cole, Robert E. 410
Cole, Wendy 353, 439
Coleman, Peter T. 446, 487
Collins, Linda N. 219
Combs, Arthur W. 325
Conant, James Bryant 98, 132
Confucius 403
Conger, Jay A. 410
Conley, David T. 422
Cooper, J 42
Copeland, K. C. 81–84, 142, 150, 153
Copernicus, Nicolaus 98, 177, 265
Corcoran, Elizabeth 410
Coruzzi, Celeste A. 308
Cosmides, Leda 339
Costa, Paul T. 293, 295, 299, 300
Couch, C. V. 100
Croall, Jonathan 445
Cronbach, Lee J. 126
Cronin, Mary 397

Cross, Arthur Lyon 442
Crutchfield, Richard S. 220
Culbert, Samuel A. 308, 423, 425, 426
Cusick, Philip A. 452, 455, 461
Cziko, Gary 24, 34, 64, 88, 232, 379
Danziger, Kurt 302
Dar, R. 151, 153
Darwin, Charles Robert 204
Davidson, J. M. 246
Davis, Stanley M. 334
Dawes, Robyn M. 218, 219, 221, 226, 281, 282, 283, 284, 337, 358, 359, 361, 362, 364, 367, 370
Dawkins, Richard 205, 302
Deal, Terrence E. 421
Dember, William N. 43, 199, 265
Dement, William C. 226
Dempsey, Bobby L. 410
Derber, Charles 375
Deutsch, Morton 446, 487
Devine, P. G. 42
DeVries, R. D. 448, 449
Dewey, John 443
Diener, E. 100
Disney, Walt 184
Dobzhansky, Theodosius 204, 205
Donahoe, J. W. 140, 153
Donnellon, Anne 423
Dowling, William F. 430
Doyle, Arthur Conan 412
Doyle, Roy P. 464
Drake, Stillman 132
Drummond, Hugh 338
Drummond, Robert 444
Dryden, Windy 369
Dunham, Randall B. 397
Dunlap, Diane M. 422
Durkheim, Emile 100
Durning, Alan Thein 483
Dye, Thomas R. 416, 479
Dyer, Vick R. 81, 82, 83, 84, 142, 150, 153
Dyson, Freeman 185
Easterbrook, J.A. 246
Edney, J. J. 42
Edwards, Paul 175
Ehrlich, Paul R. 362, 394, 479, 485
Einstein, Albert 98, 131, 178, 259
Eisenberg, Daniel 391
Eisner, Donald A. 362
Eldredge, Niles 394
Emrick, John A. 436
Enzle, Michael E. 446, 447
Espinas 100

Esterson, Aaron 367
Everett, Melissa 258
Ewens, Willam L. 310, 404, 444, 445, 451
Eysenck, Michael W. 49
Fagen, R. E. 111
Farragut, D. G. 385
Farson, Richard 451
Faust, D. 361
Fazio, R. H. 42
Feltham, Colin 369
Festinger, Leon 100, 235, 267, 403
Feynman, Richard 173
Field, Joy M. 338, 410
Figlio, R. M. 341
Fischer, Kurt W. 228
Flood, Daniel J. 277, 278
Flynn 301
Foa, Edna B. 386
Foa, Uriel G. 386
Ford, Edward E. 467, 468, 470, 475
Ford, Henry 404
Ford Foundation 441
Forssell, Dag C. 27, 66, 88, 255, 379, 425
Forward, Gordon E. 408
Fosmire, Frederick 399, 400
Francisco, Richard 327, 432, 436
Freed, A. M. 11
Freire, Paulo 475
Freud, Sigmund 99, 362
Frick, Henry Clay 404
Fricke, Benno G. 56
Friedman, Miles I. 49, 55
Frietchie, Barbara 39
Furnham, Adrian 49
Galambos, Robert 230, 233
Galbraith, Jay R. 410
Galen 99
Galilei, Galileo 65, 132, 133, 137, 177
Gall, F. J. 99
Galton, Francis 291, 292
Gardner, Gary 478, 479, 480, 483
Gardner, Martin 375
Gatto, John Taylor 439, 440, 444, 456
Gellner 369
General Motors 408
Genghis Khan 485
Gibbons, Hugh 484
Gibbs, Jack P. 55
Gilbreth, Frank 310, 311
Gilbreth, Lillian 310, 311
Gilligan, James G. 307
Gintis, Herbert 442

Giszter, S. 140, 153
Glazer, Myron P. 258
Glazer, Penina M. 258
Glines, Lawrence A. 214
Goldberg, Lewis R. 293
Goldman, Emma 404
Goldman, Paul 422
Goldstein, Andrew 445, 461, 466
Goldstein, David M. 74
Gonzales, Susan 438
Good, Thomas L. 452
Goodlad, John I. 436
Goodpaster, Andrew 482
Gore, Bob 406, 407
Gore, Vieve 406
Gore, William 406, 407
Gore (W. L. Gore and Associates)
  406, 407, 408, 418, 420, 429
Gottfredson, Gary D. 341
Gottlieb, G. L. 158, 165
Graham, J. W. 18
Graves, Nancy 487
Graves, Theodore 487
Gray, Peter 44, 49, 408
Gray, Thomas 100
Green, Mark 482
Gregory, Bruce 173, 176, 268, 285, 351
Gregory, Richard L. 219, 232, 233
Gump, Paul V. 441
Gwynne, S. C. 408
Gysin, Catherine 454
Haasen, Adolf 408
Hackman, J. Richard 416, 423, 434
Hadley, Suzanne W. 361
Hall, A. D. 111
Hammond, George S. 436
Hammurabi 403
Hamper, Ben 396
Harman, W. K. 81–84, 142, 150, 153
Harris, Felicia H. 460
Harrison, Roger 386, 421
Hart, Leslie A. 445
Hartmann, Lawrence 339
Harvey, Jerry B. 380
Hasek, Glenn 408
Hawkins, Wendy L. 446, 447
Hayes, S. C. 140, 153
Hazen, R. L. 58
Hegel, G. W. F. 100
Heider, Fritz 42
Heller, Kenneth 340, 342
Helson, R. 299

Hemmings, Ray 445
Henry V 197
Henry VIII 442, 455
Hershberger, Wayne A. 141, 153, 155, 319
Herzberg, Frederick 386
Hickman, Jonathon 408
Hilgard, Ernest 99, 199
Hill, K. T. 246, 456, 459
Hirschman, Albert O. 326
Hitler, Adolf 53
Hodgetts, Richard M. 300
Hoffman, Mark S. 333
Hofstede, Geert 410
Hogan, Robert T. 299, 300
Hogue, M. A. 97, 143, 153
Holerton, Arthur C. 192
Hollin, Clive R. 49
Holmes, Michael 258
Holubec, Edythe J. 446
Horgan, John 302
Houts, Paul 466
Howell, James 220
Hubis, Dan 407
Huffman, Karen 49
Hummel, Ralph 309, 310
Hutchinson, Neal 309, 311
Illich, Ivan 440, 451
Isaacs, William N. 410, 482
James, William 40, 141, 303
Jamieson, David 410
Jennings, Wayne 443, 444, 445
John, Oliver P. 293, 300
Johnson, David W. 18, 258, 446
Johnson, Lyndon 258
Jordan, N. L. 42
Judd, Joel B. 372, 375, 379
Kagan, Jerome 49
Kamii, C. 448, 449
Kaplan, Michael 408
Karasek, Robert A. 309
Katz, Smart 460
Keating, David E. 423
Keegan, John 266, 267, 324, 483
Kennaway, Richard 97, 234, 296–300, 303
Kepler, Johannes 177
Kershner, R. B. 276, 278
Keutzer, Frederick 399, 400
Khan, Abbas 264, 266, 466
Kihlstrom, J. F. 140, 153
Kimble, Gregory A. 49
Kipling, J. R. 180
Kirk, Stuart A. 353, 357, 358, 359

Klerman, Gerald 358
Koch, Edward 258
Kofman, Fred 410
Kohn, Alfie 326, 466
Kornhauser, Arthur 308
Koshland, D. E. 70, 232
Koslowsky, Meni 56
Kotzsch, Ronald E. 445
Kouzes, James 386, 421
Kraepelin, Emil 357
Krech, David 219
Kreidler, William J. 487
Kugler, P. N. 142, 153
Kuhn, Thomas S. 132
Kurtzer, Isaac 268, 269
Kutchins, Herb 353, 357, 358, 359
Lackey, George H. Jr. 49, 55
Laing, Ronald David 367
Lakatos 153
Lamb, Albert 445
Lamiell, James T. 301
Lancaster, James 479
Langer, Ellen J. 55, 301
Laubach, Vincent 258
Lautenschlager, Gary J. 460
Lavoisier, Antoine 98, 131, 132, 213
Lawler, Edward E. III 308, 309, 410, 423
Lawton, Theresa A. 246
Lazare, Mark 351
Lazarus, Arnold A. 369
LeBon, Gustave 100
Ledford, Gerald E. Jr. 300, 408, 410
Lefkowitz, M. 11
Levin, Irwin 258
Levine, Tamar 446, 487
Lewin, Kurt 382
Lewis, Meriwether 214
Liebrand, Wim B. G. 325
Lighthall, Frederick E. 246
Lindsay, Cindy P. 410
Lippitt, Ronald 382
Locke, Gardner 56
Lofland, J. 100
Loftus, Elizabeth F. 221
Lore, Richard K. 307
Lorentz, Elizabeth M. 456
Louis, Karen Seashore 436
Louis XIV 421, 485
Luthans, Fred 300
Lynch, Lisa M. 405
Lytle, James H. 444
MacColl, Gail 482

Mackay, Charles 100
Maher, Stanford 445
Main, Jeremy 433
Mann, Horace 455
Mansbridge, Jane 266
Manz, Charles C. 406, 407, 423
Marken, Richard S. 41, 65, 70–74, 78, 88–91, 93, 94, 97, 104, 106, 137, 138, 141, 153, 156, 165, 179, 180, 183, 192, 211, 214, 231, 232, 239, 257, 261, 268, 269, 303, 351, 379, 393
Martin, Shan 397, 419
Maslow, Abraham 386
Mathews, Jay 458, 464
Maturana, Humberto 173
Maxwell, James Clerk 58
McCaslin, Mary 452
McClaughry, John 486
McClelland, David C. 308
McClelland, Gary M. 342
McClelland, J. L. 153
McClelland, Kent 97, 100, 112–117, 118, 119, 120, 303, 379, 388, 389, 392
McConnell, James V. 49
McCrae, Robert R. 293, 295, 299, 300
McDonald, Allan 258
McDonough, John J. 308, 423, 425, 426
McDougall, William 31, 32, 40, 53, 99, 100
McFarland, Robert L. 211, 213, 232
McGill, Michael E. 408, 410
McGill-Franzen, Anne 464
McGrath, Joseph E. 179, 403, 411, 417
McKenna, Bernard 459
McKibben, Bill 481
McKnight, John L. 481
McLaughlin, Margaret L. 286
McLaughlin, Milbrey 432, 436
McPhail, Clark 100, 101, 109–112, 392, 413, 415
Meehl, P.E. 151, 153
Mermel, Michael 74
Messick, David M. 325
Metzger, Loya 436
Middleton, Harry 259
Milbrey, Wallin 432
Miles, Matthew 381
Milford, Maureen 408
Millard, Glenn 153
Miller, James Grier 43, 100, 111
Millon, Theodore 367
Mills, Peter K. 330, 359, 367, 368
Miner, Anne S. 408
Mintzberg, Henry 430

Mirowsky, John 340, 341
Misner, C. 203
Mitchell, James V. 457
Mohandessi, K. 56
Mohrman, Susan Albers 410
Morey, Leslie C. 368
Morgan, Harry 445
Morse, Jodie 446, 463–465
Moscovicci, S. 100
Mosely, B. L. 142
Mosley, B. L. 81–84, 150, 153
Mouton, Jane S. 11, 420, 421
Moynihan, Daniel 432
Mozart, Wolfgang Amadeus 284
Mundy, John H. 394
Musgrave, Melanie 74
Muska, Amy 408
Mussa-Ivaldi, F. A. 140, 153
Myers, David G. 24, 55
Nathan, Barry R. 300, 408
Nathan, Joe 443, 444
Neisser, Ulric 302
Newcomb, Theodore M. 42, 100
Newell, A. 141, 153
Newton, Isaac 13, 14, 17, 19, 25, 131, 133, 178, 208
Norman, W.T. 292
Norton, Nicholas C. 361
Novotney, J. M. 436
Nunn, Sam 482
O'Mara 410
O'Neill, W. D. 158, 165
Odbert, H. S. 292
Opel 408
Ornstein, Robert 49, 362, 479, 485
Osafa-Charles, E. 158, 165
Ostermann-Tolstoi 266
Oticon A/S 408
Owens, William 482
Palmer, Dan 140, 153, 303
Palmore, E. B. 338
Papanicolaou, Andrew C. 153
Papworth, John 486
Paris, Scott G. 246
Paulos, John Allen 280
Pavloski, Raymond. P. 97, 143, 153, 192, 230
Peacock, Chris 481
Peiser 42
Pepito, A. 100
Perez-Lopez, Jorge F. 278
Perry, William 482
Pervin, Lawrence A. 300

Peterson, Susan M. 436
Peter the Great 485
Petrie, Hugh 233
Pfeiffer, John E. 374, 408, 450, 479, 486, 487
Philipchalk 49
Pierce, Jon L. 397
Pilgrim 369
Pinkerton, Robert A. 404
Plooij, Frans X. 104–108, 192, 211, 214, 228, 338
Pol Pot 53
Pope, Alexander 1
Powers, Mary 195, 226, 244, 245, 351
Powers, William T. 22, 23, 25, 27, 36, 37, 39, 41, 45, 47, 52, 58, 64, 66, 67, 69, 70, 71, 77, 85, 87, 88, 97, 101, 105, 108–111, 116, 119, 127, 133, 136–139, 141, 143, 153, 156, 168, 171–173, 179, 180, 183, 188, 192, 193, 195, 197, 198–201, 204, 206–218, 222–238, 239, 243, 244, 246–248, 255, 260–262, 264, 267, 270, 271, 273, 275, 285, 286, 303, 311, 331, 337, 346, 349, 350, 351, 364, 374, 375, 379, 388, 389, 392–396, 415, 416, 425, 426, 428, 429, 453, 474, 475, 477, 478, 484
Priestley, Joseph 98, 131, 132, 188
Ptolemy 177
Puckett, John L. 445
Pullin, Diana C. 452, 455, 461
Qin, Zhining 446
Queen Victoria 463
Quick, James C. 408
Quick, Rebecca 408
Rabelais, Francois 482
Rapoport, Anatol 111
Read, Stephen J. 286
Reagan, Ronald 280, 482
Real, L. A. 140, 153
Richardson, George P. 41
Richters, John E. 339
Riesenberg, Peter 394
Rijt-Plooij, Hedwig H. 104–108, 192, 211, 214, 228
Roberts, Brent W. 299, 300
Robertson, Richard J. 64, 74, 75, 76, 134, 183, 214, 230, 259, 301, 340
Robinson, D. N. 151, 153
Robinson, Robert 485
Rockefeller, John D. 404
Rodin, Judith 49
Rogers, Carl R. 345, 374, 382, 450
Rosenbaum, M. 11
Rosenhan, David L. 359, 360, 362
Ross, Catherine E. 340, 341
Rossini, G. A. 15

Roth, Jodie L. 246
Royce, Josiah 56, 100
Rubin, Jeffrey Z. 446
Ruebush, Britton K. 246
Rumelhart, D. E. 153
Runkel, Margaret 436
Runkel, Philip J. 9, 20, 42, 52, 56, 64, 97, 143, 153, 211, 265, 323, 379, 381, 400, 401, 420, 432, 435, 436
Ruskin, John 240
Russell, Richard 258
Sahakian, William S. 100
Sahl, Mort 280
Sarason, Seymour B. 246, 444, 448, 452, 453, 454, 456, 459, 469
Sarbin, Theodore R. 221, 359, 367
Sasser, James 282
Schacter, Stanley 244
Schaffle, O. 100
Schein, Edgar H. 410
Schmuck, Patricia A. 441, 444, 487
Schmuck, Richard A. 379, 381, 400, 401, 420, 432, 435, 436, 441, 444, 487
Schoeppel, Andrew F. 277, 278
Schoggen, Phil 32
Schroeder, Roger G. 410
Schultz, Lori A. 307
Schumacher, E. F. 408, 434
Schweingruber, David 109, 392
Scott, Charles R. 463
Scott Bader 408
Sedlak, Michael W. 444, 452, 455, 461, 462
Segal, Julius 49
Seldes, George 482
Sellin, T 341
Semco 408
Semler, Ricardo 408
Sen, Amartya 482
Senchuk, Dennis M. 99
Senge, Peter M. 410
Shachar, Hanna 446, 487
Shakespeare, William 197
Shalikashvili, John 482
Sharan, Shlorno 446, 487
Shepherd, Ralph 445
Sherif, Muzafer 117
Sherman, Joe 408
Sherwood, Philip 463
Shimp, C. P. 140, 151, 153
Shipper, Frank 406, 407
Sieber, Sam D. 436
Sims, Henry B. 423

Sinclair, Upton 482
Skinner, Burrhus Frederick 53, 101, 140, 145, 152, 153, 322, 323, 385
Slavin, Robert E. 446
Slocum, John W. 300, 408, 410
Smith, Laurence D. 55
Smith, M. L. 361, 362
Smith, Mary Ann 436
Smullyan, Raymond 280
Snyder, Mark 44
Soldani, James 405, 406
Spitzer, Robert J. 357, 358
Spock, Marjorie 445
Srivastava, Sanjay 293
Stalin, Josef 53, 485
Starnes, Bobby Ann 445
Stavrianos, L. S. 403
Steinmetz, Charles Proteus 3
Stokes, Herb 421
Stromeyer, C. 217
Strupp, Hans H. 361
Suarez, Enrique M. 330, 359, 367, 368
Sutton, R. I. 248
Swanson, Mary Catherine 445
Szamosi, Geza 203
Takashima 266
Taylor, Frederick 311, 419
Taylor, Frederick W. 310
Taylor, Martin M. 87, 88, 135, 168, 303, 310, 379, 414, 415, 436
Teplin, Linda A. 342
Terman, Lewis 260, 443, 444
Thalhammer, Bryan 76
Thatcher, Robert 228
Thompson, Dana 410
Thoreau, Henry David 3, 4
Thorne, K 203
Thurstone, L L. 293
Tilly, C. 100
Todd, W. Murray 436
Tooby, John 339
Trist, Eric 310, 423
Trotter, Wilfred 99
Tucker, Charles W. 64, 109, 110, 111, 392
Turner, Julienne C. 246
Turvey, M. T. 142, 153
Tye, Kenneth A. 436
Tyler, Leona E. 444, 451, 452, 454, 457, 461
U.S. Department of Labor 333
Uttal, Wm. R. 36, 99
van de Rijt, Hetty
  See Rijt-Plooij

Vial, L. G. 267
Viscio, Randolph Louis 306, 307
Vives, Ludovico 442
Voltaire 482
Wachtel, Paul 478
Waite, Richard R. 246
Walker, Elanie F. 42, 367
Wall, James A. 410, 422
Wall, T. D. 309
Wallace, Alfred Russell 204
Walljasper, Jay 482
Warm, Joel S. 199
Washington, George 417
Watson, John B. 31, 32, 67, 71, 101
Watt, James 58
Weick, Karl E. 219, 262, 285, 309, 393, 394, 407, 429, 434, 436
Weisbord, Marvin R. 135, 382, 435, 483, 486
Wentworth, Roland A. 445
Wheeler, Christopher W. 452, 455, 461
Wheeler, V. 203
Whittier, John Greenleaf 39
Wiener, Norbert 40, 58
Wigginton, Eliot 445
Wilcox, L. R. 276, 278
Wild, T. Cameron 446, 447
Wilkins, L. T. 341
Will, Rosalyn 408
Williams, Gregory 153
Williams, William D. 153
Winters, Rebecca 441
Wohlstein, R. 392
Wolcott, Harry E. 438, 444
Wolfgang, M. E. 341
Wordsworth, William 238
Worrall, Alyson M. 395, 396, 397
Wundt, Wilhelm 57, 98
Yanoff, Jay M. 444
Young, Rupert 97, 199, 270, 379
Zander, Alvin 381
Zeigarnik, B. 202
Zeigler, Harmon 416, 479
Zeno of Elea 486
Zimbardo, Philip G. 100, 456
Zocher, Wolfgang 97, 303, 379

Subject index

This index does not extend to the pages numbered with Roman numerals and not to the list of references.

Abnormalities, biochemical 340
Abuse as child, memory of 221
Accuracy and reliability of PCT model 81
Action
    requisites for 5–7
    springs of 3–11
    See also Requisites for a particular act
Action output 46
Acts, intended and unintended effects 253
Adolescents at the mall 306
Advice giving 335–336
Affluence and poverty 387–388
American Psychiatric Association 339, 354, 357, 460
American Psychological Foundation 340
Animate things
    See Living things
Anxiety 248
Appearances, what a person is doing 67–69
Arms race 390
Assessing personality 291
Assumptions
    about living things 122
    beliefs, theory and action 52
    by teachers and administrators about
      human nature 472
    in experimentation, examples 69, 71, 74, 75,
      86, 93, 96–99
    of PCT 57
    of traditional theory and PCT 356
    pernicious 55
    that some people portray all people 185
    widespread but not true 477
    See also Postulates
Aswan dam 485
Asymmetry of interaction between person
  and environment 24–27
Automatic mode 224
Awareness and consciousness 238
Bargaining, selfish 380

Base rates 282
Behavior
    See also Action
Behavior, deviant 336
Behaviorism 31–34
Behavior not probabilistic 167–169
Beliefs, self-fulfilling 266
Big Five personality traits 295
British Navy 479
Bushmen of the Kalahari 486
Caribou bones 262
Category, perception of 206
Causation
    circular 20
    linear 20
Change
    accelerating in society 478
    in ourselves 235
    sluggishness in perceiving 479
    speed of large-scale 120
Characteristics of living things 122
Chemotaxis, experiment 70
Child abuse, memory of 221
Chimpanzees 106
Clients, captive 363
Coin game 64
Collective control of perceptions 112–120
Communication 375
    oral 313–320
Communicative task and process 381
Communities, governable sizes of 486
Comparator 39
    subtraction in 45, 224
Competition
    and cooperation 323–327
    and degrees of freedom 486
    and motivation 325
    and self-deception 326
    commercial avoidance of 326
    likelihood 324

Configuration, perception of 105, 106, 200
Confirmation and falsification 131–136
Conflict
   and longevity 260
   deviant behavior 336
     examples of 257–260
     inner 257–262
       and drugs 237
       and emotion 245
       and health 338
       and psychological health 338–342
       solutions for 261
       three levels 261
     interpersonal 321–323, 336
       resolving 116–117, 327
Consciousness and awareness 238
Consultation of boss with underling
   time required 420–421
Consulting, organizational, increasing
 degrees of freedom 435–436
Control
   action to achieve 255
   collective, of perceptions 112–120
   evidences of 105
   model-based 269–272
   See also Hierarchy of control
Control, speed of 105–106
Controlled quantity, full description of
 test for 76–78
Controlling
   future events 271–272
   gradually changing conditions 272
Controlling others 54, 322, 418–421
   as producing emotional disturbances 472
   See also Countercontrol
Control loop 31–49
   more about 253–256
Control mode 222
Control of controllers 473
Control system, superhuman 119
Control Systems Group 153, 319, 467
Control theory, perceptual
   See PCT
Cooperation
   and competition 323–327
   and degrees of freedom 486
   and trust 328, 484
   in a printing company 135–137
Cooperation and competition, simultaneous 115
Cooperative learning 446
Coordination
   non-oral 318
   social 411–426

Correlations
   and probability, reasoning with 280–284
   as commonly predicted in psych. research 56
   interpreting 295–298
Costs, sunk 282
Council for a Livable World 482
Counseling a disruptive student, examples 469
Countercontrol 322–323, 389–390, 473
Counterpunishment 389–390
Crime rates 307
Crowds, social interaction in 108–112
Crying in infancy 247
Cuban sugar 277
Dante robot 270
Defined and undefined terms 278
Degrees of freedom 203, 305–311, 393–410
   few in a county government 397
   few on an assembly line 396
   increasing 308
   increasing and decreasing simultaneously
     393–395
   increasing by a new work schedule for police 397
   increasing by getting promoted 395–396
   increasing in new style of prison design
     and management 397–398
   limited by limited expectations (perceptions)
     of others 399–404
Delinquency and schooling 341
Democracy 419
Demonstrations
   coin game 64
   of PCT on www 66, 88, 121
   rubber band 61
   with outstretched arm 210–212
Development in infancy 103–108
Diagnosis 357–360
   by extremes 368
   in PCT 365–367
Diagnostic and Statist. Manual of Mental Disorders
   See DSM
Diagnostic success rates 283–284
Dissonance 42
Disturbance 48
Dreaming 226
Drugs
   and emotion 248
   and inner conflict 237
   and surgery 363–364
DSM 339, 354, 368, 460
DSM-I 357
DSM-II 357
DSM-III 357, 358, 359, 360
DSM-IV 357, 360

Education, at MIT 240
Educational Testing Service 461
Effectiveness
    definition 434
    organizational 434–435
Effects, irrelevant 48
Eidetic recall 217
Emotion 243–249
    and communication 380
    and drugs 248
    and energy from viscera 244
    and esthetics 247
    and reorganization 246
    in infancy 247
    two components 244
Energy
    from viscera 244
Environment
    interaction with person 21–29
    living 305
    social 313–332
Environmental disturbances in research 134
Epiphany 239
Error signal 27, 28, 39, 45, 194, 195, 197, 204, 222, 231, 234, 236, 237, 388
    chronic 337
    intrinsic 247
Escherichia coli 70–75, 205, 232, 427
Esthetics and emotion 247
Ethics in using the test for the controlled quantity 78
Events, perception of 107, 201
Evolution 204–206
    of neural hierarchy 205–206
Exchange, integration and threat in social life 387
Experience and training do not improve psychotherapists 361–363
Experimens and studies in PCT, examples
    See Chapters 6, 7, 8, 9, 12 and 13
Experimental replications 81
Experimental use of the test for the controlled quantity 156–165
Experts 185, 219, 240, 281–284, 310, 311, 345, 351, 359, 433, 485, 486, 488
Explaining data 180
Explaining ourselves 285
Explaining things 284
Explaining traditional concepts 178–180
Factor analysis 293
Fallacy, pathetic 57
Falsification and confirmation 131–136
Famine and the free press 483
Far West Laboratory for Educational Research 436

Feedback, environmental 47
Feedback loop 22
Flicker fusion 201
Force as influence 388–389
    at Carnegie's Homestead mill 404
    at the Colorado Fuel and Iron Company 404
    defined 388
Foundations of PCT, some experiments 67–79, 137–153, 155–165
Fox, Dr. 377–378
Freedom 486
Freedom, degrees of
    See Degrees of freedom
Functionalism 44
Future 478
Genes 104, 205, 228, 231, 234, 253, 254, 289, 290, 295, 301, 302
Goal, organizational 416
Goal, resetting 268–269
Groups
    autonomous 310
    self-managed 405
Group expectation survey 399–404
Group mind 100
Hazards of research in PCT 133
Health 338
Helping 333–343, 360–365, 368
    by leaders 422
    definition 334
    disruptive students 467–475
    institutionalized 363
    in love 342–343
    service sector 334
    social roles for 334
    See chapters 30, 31, 37, 39
Helplessness 233, 459
Hierarchies of purpose 375–376
Hierarchy, neural experiment 88, 104
Hierarchy of control
    and reorganization 193–214, 227–241
    features of 192
    interconnections of levels 194–197
    of linguistic control systems 372
    perceiving lower levels 263
High-involvement companies 405
Historical and ahistorical methods 127
Homeostat 229
Home schooling 446
Human Interaction Research Institute 436
Hypothesis in PCT 55
Illogic in diagnosing schitzophrenia 284
Imagination and memory 215–226

Imagination mode 222
   and action 479
   and dreaming 226
   and thinking 225
Inanimate things
   See Nonliving things
Individual and organization 425–426
Influence 385–390, 418–421
   styles of 421
   undefined 385
Input quantity 36
Inside and outside 19–29
Instinct 99
Integration, threat and exchange in social life 387
Intelligence 301–302
Intensity, perception of 104, 198
Intention experiment 71
Interactions among people, experiments
  and studies 94–97, 103–120
Internal functions 34–48
Internal processing 34, 45
Internal standards 21, 27, 39, 46, 54, 57, 74, 78,
  104, 117, 193, 210, 234, 246, 258, 264–269, 289,
  319, 357, 369, 381, 392, 394, 414, 417, 455, 459,
  488
   adopting 330–332
   borrowing 267–268
   changeability 168
   conflict among 258
   highest 235
   See also Reference Signal
Intrinsic processes, health of 338
Intrinsic signals
   error 247
   reference 285
Irrelevant effects 48
Job descriptions 15
Just say no 235
Kerala (India) 481
Knowing something 123
Knowledge, and levels of perception 175
Language 371–383
   heritability of 373
Law and liberty 484
La Leche League 481
Leaders
   great 485
   influence of and on 53
Leadership 112, 418–421
   how to do it 424
   styles 421–424
Learned helplessness 233, 459

Learning 239–242
   about physics at MIT 240
   and reorganization 239–241
   some simple ideas about it 240–241
Learning organization 408
Lemonade 200
Levels, method of 345–390
Levels of control
   See Hierarchy of control
Levels of perception in reifying 174
Liberty, statue of 57
Living Systems (Miller) 43
Living things 13–18
   characteristics of 122
Logic 275–286
   application of 279
   fallacies and unrealities 277–278
   four cautions about 275
   its place in the control hierarchy 286
   syllogism 276
Longevity and conflict 260
Loop, control
   See Chapters 4 and 22
Love 329–330, 342–343
   Ah, love, let us be true 330
Magna Carta 403
Management
   at W.L. Gore 406
   industrial, best sort 431–432
   in Soldani's factory 406
   myth of 419–420
Meaning 375
Membership in an organization 417–418
Memory, warps of 217–220
Memory and imagination 215–226
Mental testing 457–466
Method of levels 345–351, 354, 360
   how it works 348
   on www 351
   role of the guide 350
Mind, group 100
Minds of Our Own 240
Model, minimum requirements 45
Model-based control 64, 269–272
Models
   as hypothesis 55
   as theory 141–142, 151
   examples of in PCT 85, 90, 95–98, 157–163
   in other writings 97
   experimental tests of 81–101, 137–153
   generative and descriptive 141
   using to test theory 189

Modes of connection in the feedback loop 222–226
Monopoly to avoid competition 326
Motives 54
    multiple 319
National Teachers Examination (NTE) 461–462
National Training Laboratories 382
Needs 54
Neural hierarchy
    See Hierarchy of control
Nonliving things, treating people like 15
NTE
    See Natinonal Teachers Examination
NTL Institute for Applied Behavioral Science 383
Organizational goal 416
Organization and individual 425–426
Orthogonality of levels of control 192
Output quantity 46
Ozone hole 479
Paraphrasing 379
Participative management, spread of 424
Passive observation mode 222
Pathetic fallacy 57
PCT
    and self-interest 266
    assumptions of 57
    diagnosis in 365, 365–367
    examples of studies in
        See Chapters 6, 7, 8, 9, 12, 13
    features of 103
    grandeur of as theory 213–214
    in Soldani's factory 406
    on www 66, 88, 121, 233
    principles at W.L. Gore 406
    uniqueness 426
Perceiving change, examples of sluggishness 479
Perception 36
    at lower and higher levels 199, 263
    defined 22
    levels of in reifying 174
    of category 206
    of configuration 105, 106, 200
    of events 107, 201
    of intensity 104, 198
    of principle 208
    of program 107, 207
    of relationships 105, 107, 202
    of sensation 104, 106, 199
    of sequence 207
    of social organization 210
    of system concept 209
    of transition 105, 107, 201

Perceptual Control Theory
    See PCT
Perceptual signal 36
Perceptual signals and reference signals 217
Personality 287–302
    assessing 291
Personality tests
    validity of 298–302
    what can be learned from them 293–295
Person and environment 21–29
    asymmetry of interaction 24–27
Planning 427–436
    as a predicted program 269–272
    long-range 430
Plans
    foolproof, requisites for 427–429
    how to make a good one 429–430
    Russian 5-year 430
Postulates of PCT 392
Postulates of traditional theory and PCT 355–357
Poverty and affluence 387–388
Predicting and controlling other people 8
Predicting behavior 17
Press, freedom of 482
Principles
    and system concepts, examples 265
    for coordination 323
    for the future 487–488
    perception of 208
Prisoner's dilemma 325
Probability and correlation, reasoning with 280–284
Probability of behavior 167–169
Program, perception of 107, 207
Psychotherapists identifying normal person
    as schizophrenic 359
Psychotherapy 353–370
    as help 368–370
    efficacy of 360–365
    more training and experience
        do not improve efficacy 361–363
    understanding the client 368
Purpose 13, 19, 39, 40–45
    hierarchies of 375–376
Reactance 41
Reading as part of science 51
Reality 37
    in PCT 171–176
    perception of 264–266
Reasoning
    bizarre 362
    fallacious, of U.S. Congress 277

Reasons, multiple 319
Reference level, virtual 119
Reference perceptions 168, 356, 366, 413, 414
   changeability 168
Reference signal 25, 27, 35, 39, 57, 99, 104–110, 156, 180, 194–197, 210–212, 222–224, 229, 230, 234, 235, 247, 261, 345, 349, 356, 392, 414
   See also Internal standards
Reification 56, 174–176, 367
Reinforcement 57
Relationships, perception of 105, 107, 202
Reliability and accuracy of the PCT model 81
Reorganization 107, 108, 134, 178, 212, 227–241, 260, 261, 337, 369, 379, 450
   and emotion 246
   and learning 239–241
   as a hazard in research 133
   as learning or understanding 379
   random 232
   what it can and cannot do 237
Reorganizing system 229–233
Replication of PCT experiments 81
Requisites 2a -2d for acts such as skipping school 339
Requisites for a foolproof plan 427–429
Requisites for a particular act 54, 301, 335, 339, 355, 381, 386, 421, 427, 429
   2b 399
   applied to getting a high test score 457–461
   defined 5–7
   examples of 4–10
   listed 304, 318
   to be helpful 318
Requisites for good schools 454–455
Response times 210–212
Responsible Thinking Classroom 468
Responsible Thinking Process 467–475
   effectiveness of 470
   on www 475
   reasons for effectiveness 475
Rewards 425
   and punishment 389
Robot Dante 270
Rubber bands 61
Russian five-year plans 430
Schitzophrenic illogic used by psychiatrists 284
Scholastic Aptitude Test 460–461
Schooling 437–456
   and delinquency 341
   at home 446
   cheating on tests 458–460, 463–464
   embedding of its culture in society's culture 473
   how it feels 438–442, 446–450
   restraint in 438
   testing in English schools in 1860s 463
   what tests can indicate 460–462
Schools
   as wildly impractical 453–454
   good 442
   unchaining 450–454
School improvement by teacher-led committees 422
School Reform 432–433
School testing
   norm-referenced 462
   rate-busters in 462
Science
   as a way to avoid fooling yourself 127
   reading and writing as part of 51, 72, 183–185, 189, 220
Self, changing 235
Self-concept, experiment 74
Self-fulfilling beliefs 266
Self-interest and PCT 266
Self-managing groups 405, 423
Self-preservation 254
Sensation 106
   perception of 104, 199
Sensitivity training 382
Sensor 36
Sequence, perception of 207
Service sector 391
Similarities and differences of behavior among persons 287–288
Sizes of governable communities 486
Social interaction, experiments and studies 94–97, 103–120
Social interaction in crowds, modeling 108–112
Social life, 8 features 392
Social ties and length of life
Speed of control and level of action in the neural hierarchy 105–106
Speed of large-scale change 120
Springs of action 3–11
Stability, social 416–417
Standard, internal
   See Internal standards
Standardized mental tests 457
Statistics
   See Correlations
Statue of Liberty 57
Stereotypes 16, 186
Stress-linked suicides 262
Subtraction in the comparator 45, 224
Success, keys to 430–433

Summerhill 445
Sunk costs 282, 429
Surgery and drugs 363–364
Switching connections in the feedback loop 222–226
Syllogisms in logic 276
System concept, perception of 209
T-groups 382
Talking about people 14–17
Tangibility and speculation 187
Task and process 381
Teaching
    cooperative learning 446
    failure of physicians, pharmacists, wardens, clergy to teach 444
Teamwork 423, 483
Terms, defined and undefined 278
Testing, mental 457–466
Testing in schools
    and bad luck 459–460
    effects on teachers and schools 462
    in English schools in 1860s 463
    what can be done 464–466
Tests
    calculating rates of diagnostic success 283–284
    getting a high score on 457–461
    what they can indicate 460–462
Test for the controlled quantity 72, 88, 103, 105, 121, 128, 265, 289, 318, 356
    ethics in 78
    experimental use 156–165
    full description 76–78
    using it in the natural setting 134
Texas Assessment of Academic Skills 464
Theories and models 97
Theory
    and reality 171–176
    in physical science 98
    in wholly conceptual systems 101
    revising 181
    synonyms 52
Theory to produce a working model 45
Therapy 353–370
    method of levels 345–351
Thinking 225, 373
Threats, exchange and integration in social life 387
Tiger 201, 202
Total Quality Management 405
Training
    aircrews 420
    fire fighters 421
    sensitivity 382

Training and experience do not improve psychotherapists 361–363
Traits 54, 288–291
    big Five 295
    disinterest of PCT 289
    examples 288
    forced 294
Transition, perception of 105, 107, 201
Trust 376
    and cooperation 117–118, 328
Try harder 235
TV set 33
Understanding 377
Understanding among physics students at MIT 378
Upper levels of control 263–273
Validity of personality tests 298–302
Variable, dependent and independent 53, 56
Vietnam war 267
Virtual reference level 119
Volition 156–165
Voting, appeal to, as validation 124
War, hope of its decline 483
What a person is doing 67–69
Will power 261, 268
Women's rights 481
World Bank 480
Writing
    and reading 189, 220
    as part of science 72, 183–185
    be alert and sceptical 51–58, 183–190
WWW
    Baughman 365
    method of levels 351
    PCT on 66, 88, 121, 233
    www.responsiblethinking.com 475